Constructed Wetlands for Water Quality Improvement

Constructed Wetlands for Water Quality Improvement

Edited by
Gerald A. Moshiri, Ph.D.

Professor of Biology and Director
Wetlands Research Laboratory
and
Institute for Coastal and Estuarine Research
University of West Florida
Pensacola, Florida

Library of Congress Cataloging-in-Publication Data

Moshiri, Gerald A.
Constructed wetlands for water quality improvement / Gerald A. Moshiri
p. cm.
Includes bibliographical references and index.
ISBN 0-87371-550-0
1. Constructed wetlands—Congresses. 2. Water quality management—Congresses. 3. Constructed wetlands—Case studies—Congresses. I. Title.
TD756.5.M67 1993
628.3′5—dc20 92-46759
CIP

Lewis Publishers is an imprint of CRC Press

International Standard Book Number 0-87371-550-0
Library of Congress Card Number 92-46759
Printed in the United States of America 1 2 3 4 5 6 7 8 9 0
Printed on acid-free paper

Preface

In recent years, heightened interest in the application of constructed wetlands for water quality improvement has been demonstrated by the organization of a number of major national and international symposia and conferences on this topic.

Currently, there are numerous instances of the application of this approach to water pollution abatement at the design, construction, or operational stages world-wide. This attests to the acceptance of wetland treatment across a wide range of municipal, industrial, and agricultural uses as an alternative to traditional and conventional engineering methods.

The Pensacola conference was planned to bring together internationally known researchers and practitioners in constructed wetlands technology in order to exchange past and current experiences across a broad spectrum of topics, methodologies, and applications. Represented were scientists from federal, state, and local governmental agencies; scientific research entities; private industry; and engineering and consulting organizations.

The excellent response to this conference is reflected by 85 oral presentations of which 68 appear in this volume. Topics discussed included theory, application, engineering, and processes. The information presented in this volume will be a valuable supplement to current knowledge in this field, and will, hopefully, serve as a basis for similar future endeavors.

Gerald A. Moshiri, Ph.D.
Professor of Biology and
Director, Wetlands Research Laboratory
Institute for Coastal and Estuarine Research
University of West Florida
Pensacola, Florida

Acknowledgments

A number of individuals and organizations are recognized for their contributions to the success of the Pensacola conference and the preparation of its proceedings for publication.

The staff of the Wetlands Research Laboratory and the Institute for Coastal and Estuarine Research (WRL & ICER), at the University of West Florida, contributed in a number of ways toward the success of this event. Particular appreciation is extended to Craig D. Martin for his collaboration throughout all aspects of this symposium. His unstinting and tireless efforts during the planning, organization, execution, and research phases of this event were paramount in its success. Special recognition is also extended to Mrs. Linda Matthews for her exceptional efforts in the organization and management of all correspondence and manuscript-related matters. Ron Withrow, and his staff from the Office of Extended Learning, performed commendably in all issues relating to finances and logistics.

The first and second rounds of reviews were undertaken by session moderators and WRL & ICER affiliates Drs. Sneed Collard, David Gibson, and Paul Hamilton, respectively. The efforts of all of these individuals are hereby gratefully acknowledged.

A number of organizations provided various forms of support. From this list, special recognition must be afforded Mr. Rod Kendig, Manager, the City of Pensacola; and Mr. Charles Miller, Director, Escambia County Department of Solid Waste Management, for their unfailing support. In addition, the following organizations provided various forms of assistance:

- Browning-Ferris Industries
- CH_2M Hill
- Escambia County Utilities Authority
- Florida Department of Environmental Regulation
- Tennessee Valley Authority
- The Monsanto Corporation
- The University of West Florida
- U.S. Environmental Protection Agency
- U.S. Forest Service
- West Florida Regional Planning Council
- Westinghouse Electric Company

The organizers also gratefully acknowledge the support of the following exhibitors:

- Bromwell and Carrier, Inc.
  Lakeland, Florida 33807
- Kester's Wild Game Food Nurseries, Inc.
  Omro, Wisconsin 54963
- Lemna Corporation
  St. Paul, Minnesota 55120
- Living Waters
  Warren, Vermont 05674
- Plants for Tomorrow, Inc.
  Loxahatchee, Florida 33470-9430

The Editor

Gerald A. Moshiri, Ph.D., is Professor of Biology and Director of the Wetlands Research Laboratory and the Institute for Coastal and Estuarine Research at the University of West Florida. Dr. Moshiri received his B.A. and M.A. in botany from Oberlin College, and his Ph.D. in aquatic ecology from the University of Pittsburgh.

He has held research appointments at the University of California, Davis, and at Michigan State University's Gull Lake research laboratories. He has served as a consultant to numerous private firms, as well as a number of local, state, national, and international environmental regulatory agencies, committees, and organizations. He holds memberships in 14 professional scientific associations, and has functioned as scientific proposal reviewer for the U.S. Department of the Interior, and the National Science Foundation.

Dr. Moshiri's research interests have focused on the use of natural processes in the solution of water resources and water pollution problems, particularly in economically disadvantaged countries. He has served as manuscript reviewer for various journals in aquatic sciences, and has published over 70 scholarly books, reports, and journal articles.

Contributors

Contributors are listed alphabetically by surname. The chapter(s) to which each contributed are indicated in brackets after the surname.

J. Allen [40] U.S. Fish and Wildlife Service, National Wetlands Research Center, Slidell, LA

D. Antonson [17] Minnesota Department of Natural Resources, Division of Minerals, St. Paul, MN

R. F. Ashworth [24] School of Civil Engineering, Portsmouth Polytechnic, Portsmouth, Hampshire, United Kingdom

M. M. M. Baghat [24] Faculty of Science, Suez Canal University, Ismailia, Egypt

W. Bastiaens [7] Department of Chemical Engineering, University of Michigan, Ann Arbor, MI

R. K. Bastian [5] Office of Water, U.S. Environmental Protection Agency, Washington, DC

H. J. Bavor [22] Head, Water Research Laboratory, University of Western Sydney, Hawkesbury, Richmond, New South Wales, Australia

R. A. Beggs [27] Senior Engineer, Nolte and Associates, Sacramento, CA

S. Bird [10] Civil Engineer, ASCI, Athens, GA

A. E. Birkbeck [52] British Columbia Research Corporation, Vancouver, British Columbia, Canada

R. Bouchard [37] Biologist, Maine Department of Environmental Protection, Augusta, ME

C. E. Boyd [67] Department of Fisheries and Allied Aquacultures, Auburn University, Auburn, AL

J. Boyd [49] Stone Container Corporation, Paperboard Division, Hodge, LA

C. Breed [38] Tennessee Valley Authority, Muscle Shoals, AL

K. C. Brinkmann [44] Department of Biology, University of Tennessee at Chattanooga, Chattanooga, TN

C. R. Britt [12] Environmental Scientist, Fossil Fuels, Tennessee Valley Authority, Knoxville, TN

H. Brix [2, 41] Department of Plant Ecology, Botanical Institute, Aarhus University, Nordlandsvej, Risskov, Denmark

G. A. Brodie [12, 13, 15] Environmental Engineer, Tennessee Valley Authority, Generating Group, Fossil & Hydro Power, Fossil Fuels, Fuels Property, Chattanooga, TN

L. Buddhavarapu [42] Lemna Corporation, St. Paul, MN

J. E. Butler [24] School of Civil Engineering, Portsmouth Polytechnic, Portsmouth, Hampshire, United Kingdom

K. D. Choate [13, 55, 56] Tennessee Valley Authority, Resource Group, River Basin Operations, Water Resources, Water Quality Department, Chattanooga, TN

D. W. Combs [54] Commonwealth of Kentucky, Department of Health Services, Environmental Sanitation Branch, Frankfort, Kentucky

R. B. Conway [49] Institute for Environmental Studies, Louisiana State University, Baton Rouge, LA

G. Coombs [52] British Columbia Research Corporation, Vancouver, British Columbia, Canada

P. F. Cooper [21] WRC plc, Swindon, Wiltshire, United Kingdom

P. D. Cottingham [32, 64] Water Studies Centre, Monash University, Victoria, Australia

C. B. Craft [28] Duke University Wetland Center, School of the Environment, Duke University, Durham, NC

R. W. Crites [27] Director — Water Resources, Nolte and Associates, Sacramento, CA

W. G. Crumpton [29] Associate Professor, Wetlands Research Laboratory, Iowa State University, Ames, IA

A. J. Cueto [9] Cameron County Engineering and R.O.W., Brownsville, TX

A. J. Danzig [31] Tennessee Valley Authority, Resource Group, River Basin Operations, Water Resources, Water Quality Department, Chattanooga, TN

T. H. Davies [32, 64] Water Studies Centre, Monash University, Victoria, Australia

J. Davison [16] President, Lambda Bioremediation Systems, Inc., Columbus, OH

A. Dewedar [24] Faculty of Science, Suez Canal University, Ismailia, Egypt

P. H. Dohms [8] Condor Earth Technologies, Inc., Pensacola, FL

J. N. Dornbush [63] Professor Emeritus, Civil Engineering Department, South Dakota State University, Brookings, SD

J. Eason [35] Animal and Dairy Science, Auburn University, Auburn, AL

B. Ebersole [10] Chief, Coastal Processes Branch, Coastal Engineering Research Center, U.S. Army Engineer Waterways Experiment Station, Vicksburg, MS

M. E. Edwards [44] Department of Biology, University of Tennessee at Chattanooga, Chattanooga, TN

P. Eger [17] Minnesota Department of Natural Resources, Division of Minerals, St. Paul, MN

M. El-Housseini [24] Soils and Water Research Institute, Cairo, Egypt

J. B. Ellis [43] Urban Pollution Research Centre, Middlesex University, London, United Kingdom

S. P. Ellner [67] Biomathematics Program, North Carolina State University, Raleigh, NC

D. L. Ferlow [39] Stearns & Wheeler, Darien, CT

E. S. Filho [34] Construtora Ambiental, Ltda. (Constructional Engineering), Rua Moraes Barros, Paracicaba, São Paulo, Brazil

S. W. Fisher [29] Research Assistant, Wetlands Research Laboratory, Iowa State University, Ames, IA

M. G. Ford [24] School of Biological Sciences, Portsmouth Polytechnic, Portsmouth, Hampshire, United Kingdom

S. J. Formica [18] FTN Associates, Ltd., Little Rock, AR

R. J. Freeman, Jr. [6] U.S. Environmental Protection Agency — Region IV, Atlanta, GA

T. M. Frostman [20] Principal Scientist, STS Consultants, Ltd., Minneapolis, MN

L. C. Gandy [18] FTN Associates, Ltd., Little Rock, AR

R. A. Gearheart [62] Professor, Environmental Resources Engineering, Humboldt State University, Arcata, CA

S. Geiger [33] Scientific Resources, Inc., Lake Oswego, OR

M. B. Green [57] Process Development Group, Severn Trent Water, Ltd., Birmingham, United Kingdom

M. A. Gross [18] University of Arkansas at Little Rock, Little Rock, AR

G. R. Guntenspergen [40] ASCI, EPA Environmental Research Laboratory, Duluth, MN

L. Hales [10] Research Hydraulic Engineer, Program Management Office, Coastal Engineering Research Center, U.S. Army Engineer Waterways Experiment Station, Vicksburg, MS

D. A. Hammer [5, 35, 48] Regional Waste Management Department, Tennessee Valley Authority, Knoxville, TN

S. J. Hancock [42] Lemna Corporation, St. Paul, MN

B. T. Hart [64] Water Studies Centre, Monash University, Victoria, Australia

K. Hatano [60] Department of Forestry, North Carolina State University, Raleigh, NC; Visiting Scholar, Takeda Chemical Industries, Osaka, Japan

D. Hatcher [49] Stone Container Corporation, Paperboard Division, Hodge, LA

R. S. Hedin [19] Biotechnology Section, Environmental Technology Group, U.S. Bureau of Mines, Pittsburgh Research Center, Pittsburgh, PA

J. Hestir [18] University of Arkansas at Little Rock, Little Rock, AR

M. J. Higgins [37] Research Assistant, Department of Civil Engineering, University of Maine, Orono, ME

M. Higley [62] Wildlife Biologist, Hoopa Indian Council

B. L. Hilton [11] Grenish & Olson, Inc., Madeira, CA

C. H. House [60] Research Assistant, Constructed Wetland Specialist, Department of Soil Science, North Carolina State University, Raleigh, NC

F. G. Howell [46] Department of Biological Sciences, University of Southern Mississippi, Hattiesburg, MS

R. Hunter [52] British Columbia Research Corporation, Vancouver, British Columbia, Canada

T. M. Isenhart [29] Research Assistant, Wetlands Research Laboratory, Iowa State University, Ames, IA

J. Jackson [33] Unified Sewerage Agency, Hillsboro, OR

R. B. Jacquez [68] Professor of Civil Engineering, Civil, Agricultural, and Geological Engineering Department, New Mexico State University, Las Cruces, NM

C. A. Johnston [30] Natural Resources Research Institute, University of Minnesota, Duluth, MN

R. H. Kadlec [4, 7, 23] Department of Chemical Engineering, University of Michigan, Ann Arbor, MI; Wetland Management Services, Chelsea, MI

J. R. Keough [40] EPA, Environmental Research Laboratory, Duluth, MN

R. L. Knight [4] CH_2M Hill, Gainesville, FL

M. D. Krom [67] Israel Oceanographic and Limnological Research, Institute of Oceanography, Tel Shikmona, Haifa, Israel

J. W. Lee [48] Bowater, Inc., Southern Division, Calhoun, TN

C. C. Lekven [27] Associate Engineer, Nolte and Associates, Sacramento, CA

D. K. Litchfield [53] Superintendent, Environmental Control & Safety, Amoco Oil Company, Mandan, ND

R. Lofgren [65] Wetlands Project Director, San Antonio, TX

J. Luzier [33] Luzier Hydrosciences, Lake Oswego, OR

E. S. Manfrinato [34] Hidric Resources and Applied Ecology Center, College of Engineering of São Carlos, University of São Paulo, São Carlos, Brazil

C. D. Martin [51] Wetlands Research Laboratory, University of West Florida, Pensacola, FL

E. May [24] School of Biological Sciences, Portsmouth Polytechnic, Portsmouth, Hampshire, United Kingdom

T. A. McCaskey [35] Animal and Dairy Science, Auburn University, Auburn, AL

O. McDonald [49] Stone Container Corporation, Paperboard Division, Hodge, LA

D. K. McMurry [48] Bowater, Inc., Southern Division, Calhoun, TN

G. Melchert [17] Minnesota Department of Natural Resources, Division of Minerals, St. Paul, MN

C. C. Miller [51] Solid Waste Management Department, Escambia County, Florida

G. A. Moshiri [51] Wetlands Research Laboratory, University of West Florida, Pensacola, FL

R. W. Nairn [19] Biotechnology Section, Environmental Technology Group, U.S. Bureau of Mines, Pittsburgh Research Center, Pittsburgh, PA

A. Neori [67] Israel Oceanographic and Limnological Research, National Center for Mariculture, Eilat, Israel

R. Netter [25] Institute of Water Quality Control and Waste Management, Technical University of Munich, Am Coulombwall, Garching, Germany

S. M. Nielsen [61] Danish Land Development Service, Slagelse, Denmark

M. H. Ogden [58] Southwest Wetlands Group, Inc., Santa Fe, NM

V. W. E. Payne [35] U.S. Department of Agriculture, Soil Conservation Service, Auburn, AL

J. H. Peverly [50] Soil, Crops, and Atmospheric Sciences Department, Cornell University, Ithaca, NY

R. J. Portier [49] Institute for Environmental Studies, Louisiana State University, Baton Rouge, LA

B. P. Pullin [35, 48] Regional Waste Management Department, Tennessee Valley Authority, Knoxville, TN

R. P. Reaves [46] Department of Biological Sciences, University of Southern Mississippi, Hattiesburg, MS

S. Reed [4] EEC, Norwich, VT

D. M. Revitt [43] Urban Pollution Research Centre, Middlesex University, London, United Kingdom

C. J. Richardson [28] Duke University Wetland Center, School of the Environment, Duke University, Durham, NC

C. A. Rock [37] Professor and Chair, Department of Civil Engineering, University of Maine, Orono, ME

R. W. Ruble [4] CH_2M Hill, Gainesville, FL

E. Salati [34] Department of Ecology, University of São Paulo, UNESP, Rio Claro, São Paulo, Brazil

W. E. Sanford [50] Environmental Sciences Division, Oak Ridge National Laboratory, Oak Ridge, TN

T. J. Schulz [22] Manager, Centre for Wastewater Treatment, University of New South Wales, Kensington, New South Wales, Australia

R. B. Shutes [43] Urban Pollution Research Centre, Middlesex University, London, United Kingdom

T. S. Steenhuis [50] Agricultural and Biological Engineering Department, Cornell University, Ithaca, NY

G. R. Steiner [54, 55, 56] Tennessee Valley Authority, Resource Group, River Basin Operations, Water Resources, Water Quality Department, Chattanooga, TN

E. Stengel [45] Institute for Biotechnology, Forschungszentrum Jülich GmbH, Jülich, Germany

J. M. Surface [50] U.S. Geological Survey, Ithaca, NY

D. Surrency [36] Plant Materials Specialist, U.S. Department of Agriculture Soil Conservation Service, Athens, GA

H. N. Taylor [12, 13] Environmental Engineer, Field Engineering, Tennessee Valley Authority, Resource Group, River Basin Operations, System Engineering, Data Systems, Field Operations, Chattanooga, TN

G. Tchobanoglous [3] Department of Civil and Environmental Engineering, University of California at Davis, Davis, CA

K. J. Tennessen [66] Tennessee Valley Authority, Vector Control Program, Muscle Shoals, AL

T. B. Terry, III [59] Registered Sanitarian, Lincoln Trail District Health Department, Hardin County Health Center, Elizabethtown, KY

R. P. Tettleton [46] Department of Biological Sciences, University of Southern Mississippi, Hattiesburg, MS

R. N. Thut [47] Weyerhaeuser Company, Federal Way, WA

T. M. Tomaszewski [12] Environmental Engineer, Water Quality Department, Tennessee Valley Authority, Chattanooga, TN

C. C. Trettin [60] Department of Forestry, North Carolina State University, Raleigh, NC; Environmental Sciences Division, Oak Ridge National Laboratory, Oak Ridge, TN

J. Upton [57] Process Development Group, Severn Trent Water, Ltd., Birmingham, United Kingdom

D. T. Urban [7] Wetland Research, Inc., Chicago, IL

J. Vymazal [26] Water Research Insitute, Prague, Czechoslovakia; Duke University Wetland Center, School of the Environment, Durham, NC

J. Wagner [17] Minnesota Department of Natural Resources, Division of Minerals, St. Paul, MN

R. Walton [10] Supervising Engineer, Ebasco Environmental, Bellevue, WA

J. T. Watson [23, 31, 44, 55, 56] Tennessee Valley Authority, Resource Group, River Basin Operations, Water Resources, Water Quality Department, Chattanooga, TN

B. Wengrezynek [37] Biologist, Soil Conservation Service, U.S. Department of Agriculture, Orono, ME

R. G. Wetzel [1] Bishop Professor of Biology, Department of Biological Sciences, University of Alabama, Tuscaloosa, AL

J. B. Williams [24] School of Civil Engineering, Portsmouth Polytechnic, Portsmouth, Hampshire, United Kingdom

S. R. Witthar [14] Project Manager, Black & Veatch, Kansas City, MO

A. G. Wollum, II [60] Professor, Soil Microbiology, Department of Soil Science, North Carolina State University, Raleigh, NC

W. H. Zachritz, II [68] Program Manager, Environmental Systems, Southwest Technology Development Institute, New Mexico State University, Las Cruces, NM

T. T. Zhang [43] Urban Pollution Research Centre, Middlesex University, London, United Kingdom

Contents

PART 9 — Small Systems

PART 10 — Case Studies

Constructed Wetlands for Water Quality Improvement

PART 1

GENERAL CONSIDERATIONS

CHAPTER 1

Constructed Wetlands: Scientific Foundations Are Critical

R. G. Wetzel

INTRODUCTION

Our understanding of the physiology, life history characteristics, and general biology of higher aquatic plants has increased enormously in recent decades. Several excellent monographs have summarized this information and provided syntheses with varying degrees of success (e.g., see References 1 and 2). Research among natural littoral zone and wetland ecosystems of lakes and streams has accelerated enormously in the past several decades.[3-6] From every quarter, aquatic ecologists are recognizing and quantifying the fundamental importance of the land-water interface region of aquatic ecosystems as a major, if not the dominant, source of organic matter and energy for the down-gradient water bodies. In addition, the land-water interface regions, in particular wetlands, are being recognized as metabolically active ecosystem components that influence the loadings of nutrients and other regulators of pelagic productivity.

The exponentially increasing demands of human expansion and resource exploitation have led to rapid exploitation of wetlands for the regulation of sediment, nutrient, or pollutant loadings to surface and ground waters. It has been recognized for some time that natural wetland ecosystems do not always function efficiently for purposes of permanent storage or controlled discharge of nutrients or pollutants. Various conservation efforts have further inhibited the use of natural wetlands for many applied purposes. These and many other factors have led to the rapid development of constructed wetlands in order to simulate and hopefully enhance the optimal properties of natural wetlands in performing the desired functions.[7] My ensuing remarks reflect upon my intensive study of natural wetland and littoral zone ecosystems over the past 2 decades. I suggest that we would be well served by incorporating some fundamental operational characteristics of natural systems into the design, operations, and expectations of wetlands.

WETLAND CHARACTERISTICS

As we examine the enormous variety of natural wetland gradients from uplands through down-slope aqueous communities, the great *diversity and heterogeneity* in the flora and their distribution emerges. Attempting to define wetlands in a traditional vein is fraught with problems. Clearly, traditional and perhaps practical definitions necessarily rely upon the plant species that are morphologically and physiologically adapted to tolerate saturated soils, at least periodically, for appreciable periods of time. The hydrology of wetlands must be considered from the standpoint of the aquatic plants. In view of the plant requirements, permanent water areas, including surface water or ground water in most of the root zone, and alternately wet and dry areas that are periodically sufficiently wet to create low oxygen conditions, have wetland hydrology.[8] Clearly, hydrology is the driving factor regulating the presence, characteristics, biology, and productivity of wetlands. From a legal viewpoint, however, because hydrology is so variable seasonally and spatially as well as difficult to evaluate quantitatively, it is much more prudent to use plants as indicators of the presence and functional efficacy of wetlands. The aquatic plants are relatively consistent in their physiological tolerances and requirements, and they integrate the environmental characteristics of a wetland over considerable time

0-87371-550-0/93/$0.00+$.50

periods. For many practical considerations, the presence of well-characterized indicator plants is likely among the most reliable means of rapidly evaluating the presence/absence of wetlands. Ideally, the legal characterization should extend much further, and combine soil and hydrological characteristics with the plant communities and biology. Although such characterization is often not practical for administrative purposes in order to analyze what is or is not a wetland, such detailed analyses should be routine for constructed wetland applications. Most constructed wetlands do not possess or simulate wetland hydrology.

In addressing what are the differences between natural and constructed wetlands, I avoid the question. Functionally, there should not be differences. Of course, there are certain characteristics of wetlands that could be enhanced under intensively managed conditions in order to maximize certain functions without compromising the composite functioning of the whole. Upon extensive reading in the constructed wetland literature, however, I sense that either the *integrated* operation of natural wetlands is not fully appreciated or that important characteristics are not being incorporated into the design and management of constructed systems.

Macrophyte Functions

First, let me address my appreciation of what some of the more important functional characteristics of the macrophytes in wetlands are. Why is this question important? To be candid, I am concerned with engineering-based statements that "all constructed wetlands are attached-growth biological reactors" and that "performance is based on first-order, plug-flow kinetics" (see Reference 9, among others). That is certainly *not* true for natural wetlands, and is likely not so for constructed wetlands that are not loaded so extremely that they cannot possibly function as wetland ecosystems. Additionally, *if it is true* for constructed systems, then the macrophytes have been usurped of their natural function, and one should not confound the system with macrophytes for any tangential reason, such as aesthetics. Trickling and other filtration systems that utilize microbial processing of organic matter loadings are much more efficient and manageable than are wetland ecosystems if one only wishes a reactor system. The microbial systems are likely more economical in the long term than are intensively disturbed wetland ecosystems. I will return to the latter point below.

What, then, are some of the functions of macrophytes in wetland ecosystems? First, in natural ecosystems, the macrophytes serve as *major storage sites for carbon and nutrients*. Many statements occur in the literature, particularly for constructed wetlands, that removal of the above-ground foliage of wetland plants, particularly emergent plants, is not practical for removal of nutrients from the system because the amounts are so low (e.g., ca. 5% of what is loaded artificially to the wetland). That statement is not true for natural wetlands. Statements that little is stored in the plants must be based on the entire plant, and for *most* emergent macrophytes such stages are in the below-ground tissue. Generally, the below-ground tissue is the recipient of two thirds or more (up to 90%) of the plant productivity. Only rarely is that below-ground productivity measured by isotopic methodology.[10] Much of the organic photosynthetic productivity is below-ground in a highly reducing anaerobic environment and much of it is permanently interred in this site with extremely slow decomposition rates. Many important nutrients such as N and P also remain bound with these organic constituents.

In addition, the ratio of above- to below-ground biomass and storage functions is quite responsive to environmental variables. For example, in nutrient-poor sediments (e.g., high marl content), root/rhizome development is much more extensive than in nutrient-rich sediments.[10] Foliage development can also be reduced by exposure to high and continual wind action.

Second, the *nutrient content of tissue fluctuates* widely. As the emergent plants enter growth stages of maturity and senescence, much of the critical nutrient content is translocated to the root/rhizome tissue for storage and use in subsequent cohort development because most aquatic plants are herbaceous perennials. If above-ground tissue is harvested at the time of maximum biomass for a given dominant cohort, one would anticipate that appreciable amounts of the critical nutrient content, particularly P and N, would have been translocated to and organically stored in the rooting tissues. In addition, senescence occurs gradually in phased sequences. During this time of maturation as well as during active growth, photosynthetic carbon fixation continues. As the tissue senesces, an appreciable amount of the organic content is released in dissolved form and includes considerable nutrients as well.

Third, *most natural wetlands do not consist of monocultures* of a single species. Although one or two species (e.g., *Typha*, *Scirpus*, and *Phragmites*) may totally dominate productivity in specific areas, even these apparent monocultures can contain many other species, particularly in the understory.[11] Most macrophytes — emergent, floating, and particularly submersed — experience more or less continuous senescence and sloughing of a portion of their foliar and rooting tissues. Within a population, even the same clone, cohorts can be phased over a growing season with continual and variable ascension and decline of phased cohorts.[12]

Hence, a primary function of macrophytes in natural wetlands *is to generate photosynthetically large amounts of organic carbon.* Organic carbon enters the wetland as both particulate material into the rooting and detrital pools, and a very appreciable amount of the organic matter is released as dissolved organic matter (DOM). The latter DOM can constitute as much as 30 to 40% of the total net productivity and be released within hours of senescence. If constructed wetlands are utilized primarily to extract organic matter from sewage (high biochemical oxygen demand [BOD] and chemical oxygen demand [COD]), that decomposition is being done microbially and not appreciably by the macrophytes. Therefore, the organic matter being produced by the macrophytes is competing with the loaded DOM and reducing the efficiency of the system to *process* the imported DOM.

A fifth important function of macrophytes is in regard to the *quality of organic matter that is loaded by the higher plants to the wetland ecosystem.* During the complex sequence of the degradation of polymers constituting structural and other plant tissues, many humic and fulvic compounds are released. These compounds have many functions within the wetland as well as within the lake or stream that receives them in effluents from the wetland. Despite the relative chemical recalcitrance of these organic compounds, they are metabolized by microbiota gradually. Due to the large quantity emanating from very high macrophytic productivity, the compounds serve as a major carbon and energy source for most aquatic ecosystems.[5,13,14]

Other functions occur that can suppress system metabolism. For example, humic and fulvic acids from macrophyte structural tissues can reversibly complex with and inactivate enzymes.[5,15,16] Important extracellular and membrane-bound enzymes that function in recycling of nutrients can be inactivated and stored for long periods, even permanently. Phosphatases are particularly affected by this process. As more information is gained on these interactions, conditions could be maximized in constructed wetlands to induce storage and burial of organically complexed phosphorus compounds and suppress associated enzyme activity that is required to access phosphorus in organic compounds.

The important point is that the wetland macrophytes are providing, via their very high productivity, large quantities of both labile, readily decomposed dissolved organic matter, and more recalcitrant organic matter from structural and storage tissues. The former labile DOM forms a major energy and carbon source for the microbial consortia of the detritus and sediments. The latter DOM is of a much more stable chemical composition that is slowly decomposed and utilized within the wetland and other ecosystem components down-gradient, such as lakes, rivers, and ground waters.

A sixth important function of macrophytes in wetlands focuses upon *conduction of gases to and from the sediments.* It is now well recognized that emergent and floating-leaved macrophytes rapidly (within a few hours) senesce and die when access to the atmosphere is eliminated to foliar and rooting tissues.[17,18] Rooting tissues enter into fermentative metabolism and rapidly form toxic end-products as oxygen is depleted. A small amount of oxygen is released into the microrhizosphere surrounding roots; this oxidized microzone counteracts toxic fermentative end-products such as hydrogen sulfide and volatile fatty acids, and can support microaerophilic bacterial communities. However, the oxygenation serves the purpose of oxygenating the rooting tissue, not the sediments. Wetlands are sites of high organic sedimentation with high rates of decomposition. Diffusion of gases within sediments is slow; oxygen entering the sediments is usually consumed microbially, essentially instantly. Hence, expectations that macrophytes can be sufficiently efficient to aerate saturated organic-rich sediments are not realistic.

The gas lacunal systems, often constituting as much as 70% of the plant volume, are extraordinary morphological adaptations to service the high biochemical respiratory demands of rooting tissues in anaerobic soils. Although the mechanisms driving gas exchange are physical, based on heat-pressure differences,[19,20] other gases generated by the tissue, as well as diffusing into the tissue from the sediments, are also transported to the atmosphere. In particular, CO_2, CH_4, and N_2 can be ventilated to the air via the plants.

If constructed wetlands are designed, for example, to maximize denitrification to encourage N_2 evasion, then other desirable properties may be sacrificed. The necessary oxygenation would encourage rapid decomposition, nutrient regeneration, and formation of humic organic acid compounds, which may or may not be desirable.

Attached Microbial Consortia

The most fundamental characteristic of wetlands is that their functions are almost solely *regulated by microbiota and their metabolism.* The retentive properties of soil and sediment particles are governed by the microenvironmental redox conditions, which in turn are controlled almost exclusively by bacterial metabolism. That metabolism is in turn regulated by the chemical composition of the organic matter loadings, the quantities of organic matter and essential nutrients, the subsurface interstitial hydrology, and other factors. The geochemical structure of the soil particles is only one factor, and often subordinate to other factors at regulating adsorptive and exchange properties.

Second, it must be recognized that *the microbial biomass is a major sink and repository* for organic carbon and many nutrients. The collective mass of microbiota, with high reproductive and growth rates, on the enormous surface areas of detritus and soil particles of wetlands is very large. Under anaerobic conditions, a significant portion of this mass is interred and decomposed at extremely slow rates.

A third point in regard to the attached microbiota of wetlands is the *propensity toward recycling of essential nutrients, including carbon.*[5] The photosynthetic productivity of attached algae and cyanobacteria can be very high. Assimilation by algae and bacteria of nutrients is the primary influx site. Once obtained, nutrients are intensively retained *by the community*, incorporated into the organic/inorganic detritus (i.e., the soil), and recycled. New growth is largely supported by incoming loadings from up-gradient sites.

Therefore, the microbiota function as efficient recyclers of nutrients. Critical nutrients such as phosphorus are effectively retained within the community. A fourth important characteristic of this community is its capacity for *very rapid growth responses* to changes in loadings, particularly to macrophyte senescence.

CONSTRUCTED WETLANDS AS INTEGRATED ECOSYSTEMS

The strong implications of the brief preceding remarks are that effective constructed wetlands must be regarded as, and designed as, natural integrated ecosystems. These dominant features can be summarized as follows.

1. The organisms are integrated and interdependent. The macrophytes in natural wetland ecosystems depend on the microbial activities, which they support with organic matter, for nutrient regeneration and their availability from organic detrital and soil sources.
2. The macrophytic photosynthetic process functions by providing recalcitrant organic matter. Substituting the functions of the aquatic plants by high BOD labile organic matter loadings decreases the stabilizing functions of the macrophytes and their roles in slow but massive decomposition.
3. Storage functions, particularly under anaerobic conditions, are not readily compatible with enhanced aerobic processes such as denitrification. Serially constructed wetlands, or wetlands in tandem with filtration microbial systems, would provide much more efficient multiple-functioning regulated ecosystems.
4. Natural wetlands export large quantities of relatively recalcitrant organic matter in dissolved form. If that export is not desirable in constructed systems, additional manipulations are needed; for example, chemical flocculation.

Any ecosystem, natural or artificial, has limits of capacity to accept disturbances (e.g., loadings, changes, perturbations, etc.). The responses can occur in many directions and at different rates. Certain changes can be rapid: for example, collapse of portions of the ecosystem from toxicity of metal or organic pesticide accumulations. Other changes can be much more gradual but equally pervasive: for example, a shift in macrophyte species composition with altered organic loadings.

Several features of natural wetlands could be utilized to maximize retentive or processing functions over the long term.

1. The macrophytes should be kept in r-growth stages by intentional, programmed disturbances.
2. Multiple species diversity would be generally more responsive to loading variations than would monocultures.
3. Detrital and sediment surface areas should be maximized to enhance microbial growth and sedimentation/storage functions.
4. Anaerobic conditions generally maximize overall retention of both organic matter and nutrients contained within it.
5. Hydrology should be used to maximize microbial access to dissolved organic matter, growth, and storage functions. This is one area in which constructed wetlands could greatly improve over the generally channelized hydrology of wetlands under natural conditions.
6. Alternative electron acceptors to oxygen could be added to constructed wetlands, along with other catalysts to manipulate pH and precipitation, in order to maximize retention.

Most evaluations of the use and efficacy of natural or constructed wetlands to improve water quality are little more than input-output analyses. Very little is understood about the control and regulatory capacities of the physical and biological properties within the "black box". The great heterogeneity among wetland ecosystems and the variety of hydrology and loadings entering these systems precludes effective generalizations. One finds a great deal of crude, even erroneous, statements based upon poor or problematic data from highly diverse ecosystems. I am adamant, however, that *commonality exists in wetland ecosystems but at the functional process, often physiological, levels*. Use of wetlands will continue and even intensify for management applications. Clearly, however, effective, highly efficient use will not emerge until a greater understanding of the functional regulatory processes is known. To argue that the systems are largely physical soil kinetics and adsorptive processes is myopic — these soil processes are driven and controlled by biotic metabolisms that alter the microenvironments and properties of particles. The systems are biologically mediated and we must have the scientific foundations of that biological control in an integrated ecosystem if we are to manage these ecosystems effectively and maximize their treatment capacities.

REFERENCES

1. Sculthorpe, C. D. *The Biology of Aquatic Vascular Plants*. St. Martins Press, New York, 1967.
2. Hutchinson, G. E. *A Treatise on Limnology. Vol. 3. Limnological Botany*. John Wiley & Sons, New York, 1975.
3. Wetzel, R. G. *Limnology*. W. B. Saunders, Philadelphia, 1975.
4. Wetzel, R. G. *Limnology*. 2nd Ed. Saunders College Publishing, Philadelphia, 1983.
5. Wetzel, R. G. Land-water interfaces: metabolic and limnological regulators. Baldi Memorial Lecture. *Verhand. Internat. Verein. Limnol.* 24:6, 1990.
6. Wetzel, R. G. and A. K Ward. Primary productivity, in *Rivers Handbook*. Vol. 1. P. Calow and R. E. Petts, Eds. Cambridge University Press, Cambridge (in press).
7. Hammer, D. A., Ed. *Constructed Wetlands for Wastewater Treatment. Municipal, Industrial and Agricultural*. Lewis Publishers, Chelsea, MI, 1989.
8. Tiner, R. W. The concept of a hydrophyte for wetland identification, *BioScience*. 41:236, 1991.
9. Watson, J. T. and J. A. Hobson. Hydraulic design considerations and control structures for constructed wetlands for wastewater treatment, in *Constructed Wetlands for Wastewater Treatment*. D. A. Hammer, Ed. Lewis Publishers, Chelsea, MI, 1989, 379.
10. Grace, J. B. and R. G. Wetzel. Phenotypic and genotypic components of growth and reproduction in *Typha latifolia*: experimental studies in marshes of differing successional maturity, *Ecology*. 62:789, 1981.
13. Wetzel, R. G. The role of the littoral zone and detritus in lake metabolism, in *Symposium on Lake Metabolism and Lake Management*. G. E. Likens, W. Rodhe, and C. Serruya, Eds. *Ergebnisse Limnol., Arch. Hydrobiol. Beih.* 13:145, 1979.
14. Wetzel, R. G. Detrital dissolved and particulate organic carbon functions in aquatic ecosystems, *Bull. Mar. Sci.* 35:503, 1984.
15. Wetzel, R. G. Extracellular enzymatic interactions in aquatic ecosystems: storage, redistribution, and interspecific communication, in *Microbial Enzymes in Aquatic Environments*. R. J. Chróst, Ed. Springer-Verlag, New York, 1991, 6.
16. Wetzel, R. G. Gradient-dominated ecosystems: sources and regulatory functions of dissolved organic matter in freshwater ecosystems, in *Dissolved Organic Matter in Lacustrine Ecosystems: Energy Source and System Regulator. Developments in Hydrobiology* (in press).
18. Sale, P. J. M. and R. G. Wetzel. Growth and metabolism of *Typha* species in relation to cutting treatments, *Aquatic Bot.* 15:321, 1983.
19. Dacey, personal communication, 1987.
20. Sebacher, D. I., R. C. Harriss, and K. B. Bartlett. Methane emissions to the atmosphere through aquatic plants, *J. Environ. Qual.* 14:40, 1985.

CHAPTER 2

Wastewater Treatment in Constructed Wetlands: System Design, Removal Processes, and Treatment Performance

H. Brix

INTRODUCTION

Ecosystems dominated by aquatic macrophytes are among the most productive in the world, largely as a result of ample light, water, nutrients, and the presence of plants that have developed morphological and biochemical adaptations enabling them to take advantage of these optimum conditions.[1] Free-floating tropical macrophytes, such as the water hyacinth (*Eichhornia crassipes*), have been reported[2,3] to have annual biomass productions of up to 106 to 165 t dw ha^{-1} $year^{-1}$, but also rooted emergent aquatic macrophytes (e.g., *Typha* sp. and *Phragmites* sp.) have high productivities (18 to 97 t dw ha^{-1} $year^{-1}$) both in temperate[4] and tropical[5] areas. Fully submerged aquatic plants have a much lower productivity due to the light attenuation in the water column and poor inorganic carbon availability.[1,6] However, some species of submerged plants have productivities of up to 20 t dw ha^{-1} $year^{-1}$ (Figure 1).[7] This high productivity of ecosystems dominated by aquatic macrophytes results in a high microbial activity in these systems and therefore a high capacity to decompose organic matter and other substances.

The permanently water-covered or water-saturated conditions in wetlands markedly reduces gas exchange rates between the sediments and the atmosphere.[8] As a result, the sediments become largely anoxic or anaerobic.[9] The rates of decomposition and mineralization of large quantities of organic matter produced by the primary producers within the wetlands are significantly reduced under anaerobic conditions, and organic matter tends to accumulate on the sediment surface.[10] The resultant organic sediments have very low bulk density, a high water holding capacity, and very high cation exchange capability. Furthermore, the layers of litter overlying the sediments and the emergent macrophytes themselves provide a huge surface area for attached microbial growth. Wetlands therefore have high potential to accumulate and transform organic material and nutrients.[11]

Both natural and constructed wetlands have been used as wastewater treatment systems; it is generally found that both systems may act as efficient water purification systems and nutrient sinks.[12-15] Long retention times and an extensive amount of sediment surface area in contact with the flowing water provide for effective removal of particulate matter. The sediment surfaces are also where most of the microbial activity affecting water quality occurs, including oxidation of organic matter and transformation of nutrients.

Natural wetlands are characterized by extreme variability in functional components, making it virtually impossible to predict responses to wastewater application and to translate results from one geographical area to another. Although significant improvement in the quality of the wastewater is generally observed as a result of flow through natural wetlands, the extent of their treatment capability is largely unknown. The performance may change over time as a consequence of changes in species composition and accumulation of pollutants in the wetland.[16] Therefore, the treatment capacity of natural wetlands is unpredictable, and design criteria for constructed wetlands cannot be extracted from results obtained in natural wetlands. There are still too few data from natural systems to allow confident predictions of the treatment performance of the systems and the effects of wastewater discharge on receiving ecosystems.[17] Exploitation of this untested technology could lead to disastrous results in ecosystems where recovery from long-term damage could take many decades. Thus,

0-87371-550-0/93/$0.00+$.50

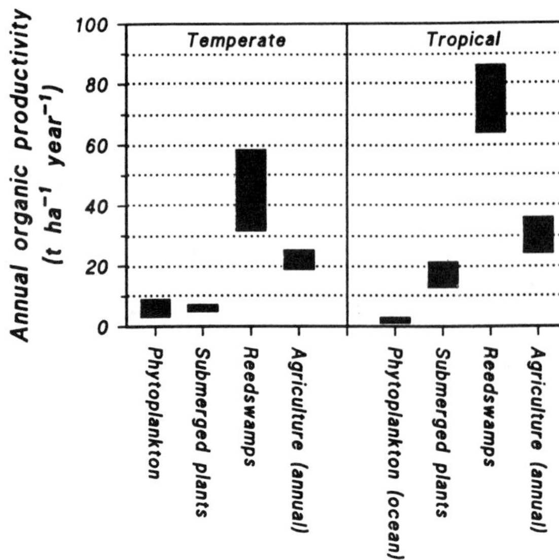

Figure 1. Probable annual average net primary productivity of fertile temperate and tropical sites. Selected data from Westlake (1963). (From Westlake, D. F. *Biol. Rev.* 38:385, 1963. With permission.)

natural wetlands should not be used deliberately as wastewater treatment systems, but should be preserved for environmental conservation.

Constructed wetlands can be built with a much greater degree of control, thus allowing the establishment of experimental treatment facilities with a well-defined composition of substrate, type of vegetation, and flow pattern. In addition, constructed wetlands offer several additional advantages compared to natural wetlands, including site selection, flexibility in sizing, and most importantly, control over the hydraulic pathways and retention time. The pollutants in such systems are removed through a combination of physical, chemical, and biological processes including sedimentation, precipitation, adsorption to soil particles, assimilation by the plant tissue, and microbial transformations.

Constructed wetlands for wastewater treatment may in some locations have several advantages compared to conventional secondary and advanced wastewater treatment systems. Some of these advantages are: (1) low cost of construction and, especially, maintenance; (2) low energy requirements; (3) being a "low-technology" system, they can be established and run by relatively untrained personnel; and (4) the systems are usually more flexible and less susceptible to variations in loading rate than conventional treatment systems.[15] The major disadvantages of constructed wetland treatment systems are the increased land area required, compared to conventional systems, and the possible decreased performance during winter in temperate regions. Therefore, the disposal of wastewater into constructed wetlands is an especially attractive alternative to conventional wastewater treatment technologies for small- to medium-sized communities, in sparsely populated areas, and in developing countries.

This article reviews different designs of constructed macrophyte-based wastewater treatment systems, the most important removal processes, and the treatment performance of the different systems. The possibility of achieving better treatment performance in improved designs of macrophyte-based systems is discussed.

REMOVAL MECHANISMS IN MACROPHYTE-BASED WASTEWATER TREATMENT SYSTEMS

A *macrophyte-based wastewater treatment system* may be defined as a wastewater treatment system in which aquatic macrophytes have a key function in relation to the cleaning of wastewater. Most macrophyte-based wastewater treatment systems consist of one or more shallow ponds in which one or more species of aquatic macrophyte are grown. The water flow within the system can occur as surface flow or as subsurface

Table 1. Removal Mechanisms in Macrophyte-Based Wastewater Treatment Systems

Wastewater constituent	Removal mechanisms
Suspended Solids	Sedimentation/filtration
BOD	Microbial degradation (aerobic and anaerobic)
	Sedimentation (accumulation of organic matter/sludge on the sediment surface)
Nitrogen	Ammonification followed by microbial nitrification and denitrification
	Plant uptake
	Ammonia volatilization
Phosphorus	Soil sorption (adsorption-precipitation reactions with aluminum, iron, calcium, and clay minerals in the soil)
	Plant uptake
	(Phosphine production)
Pathogens	Sedimentation/filtration
	Natural die-off
	UV radiation
	Excretion of antibiotics from roots of macrophytes

From Watson, J. T., S. C. Reed, R. H. Kadlec, R. L. Knight, and A. E. Whitehouse. in *Constructed Wetlands for Wastewater Treatment*. D. A. Hammer, Ed. 319, 1989. With modification.

flow. In all systems, the pollutants are removed by a complex variety of physical, chemical, and biological processes.[18] Biologically, the aquatic macrophyte-based wastewater treatment systems are far more diverse than present-day mechanical treatment systems, and many of the treatment processes are attributed to microorganisms living on and around the macrophyte. The macrophytes remove pollutants by (1) directly assimilating them into their tissue,[19] and (2) providing surfaces and a suitable environment for microorganisms to transform pollutants and reduce their concentrations. Oxygen transfer by aquatic plants into the rhizosphere is also a requisite for certain microbial pollutant-removing processes to function effectively.[20-21]

The most important removal mechanisms in constructed wetlands are listed in Table 1. Settleable and suspended solids are removed primarily in the mechanical pretreatment unit, which is usually installed in front of the actual wetland. The suspended solids that remain in the wastewater after mechanical pretreatment are removed in the wetland by sedimentation and filtration. These purely physical processes also remove a significant proportion of other wastewater constituents [BOD (biochemical oxygen demand), nutrients, pathogens].

Soluble organic compounds are, for the most part, degraded aerobically by bacteria attached to plant and sediment surfaces. However, anaerobic degradation may in some cases be significant.[8] The oxygen needed to support the aerobic processes is supplied directly from the atmosphere via diffusion through the sediment or water-atmosphere interface, by photosynthetic oxygen production within the water column, and by oxygen leakage from macrophyte roots.[20] Anaerobic degradation will occur during periods with oxygen depletion in the water column and in anaerobic sediments.

The major removal mechanism for nitrogen in constructed wetlands is nitrification-denitrification.[21,22] Ammonia is oxidized to nitrate by nitrifying bacteria in aerobic zones, and nitrates are converted to dinitrogen gas (N_2) by denitrifying bacteria in anoxic zones. The oxygen required for nitrification is delivered either directly from the atmosphere through the water or sediment surface, or by leakage from plant roots. Nitrogen is also taken up by the plants and incorporated into the biomass. Plant uptake of nitrogen is, however, generally of less importance than denitrification. Conversion of ammonium to ammonia gas and subsequent volatilization may be of importance in systems with open water, under conditions where algal photosynthesis increases the pH levels in the water column to values greater than the pK_a of ammonium (9.3).[23]

Phosphorus removal in constructed wetlands occurs mainly as a consequence of adsorption, complexation, and precipitation reactions with aluminum (Al), iron (Fe), calcium (Ca), and clay minerals in the sediment.[24] Alternate wet and dry periods enhance the fixation of phosphorus in the sediments.[25,26] Plant uptake may be significant in systems where the area specific loading rate is low.[19,27] Studies by Czinki[28] have indicated that reduction of phosphates to gaseous hydrogen phosphides (phosphine and diphosphine) under anaerobic conditions and subsequent release to the atmosphere might be of importance, but these results have been questioned by Nush et al.[29] for thermodynamical reasons.

Pathogens are removed in wetlands during the passage of wastewater through the system by sedimentation and filtration, and as a consequence of natural die-off in the — for pathogens — unfavorable environment.[18,30,31] Furthermore, root metabolites from the macrophytes have been reported to have an antibiotic effect on bacteria[32] although no real evidence for this has been reported. Ultraviolet radiation has a significant effect in systems with open water.[33]

Trace metals have a high affinity for adsorption and complexation with organic material and will be accumulated in the wetland sediment. Plant uptake and microbial transformations may also be of importance.[18]

DIFFERENT DESIGNS OF MACROPHYTE-BASED WASTEWATER TREATMENT SYSTEMS

Aquatic macrophyte-based wastewater treatment systems may be classified according to the life form of the dominating macrophyte into:

1. Free-floating macrophyte-based treatment systems
2. Rooted emergent macrophyte-based wastewater treatment systems
3. Submerged macrophyte-based wastewater treatment systems
4. Multistage systems, consisting of a combination of the above-mentioned concepts and other kinds of low-technology systems (e.g., oxidation ponds and sand filtration systems)

Aquatic macrophytes can be used in single-cell ponds, in series pond systems, and in parallel pond systems with alternating loading. Different types of macrophyte-based wastewater treatment systems may be combined with each other or with conventional treatment systems in order to exploit the specific advantages of the different systems. The quality of the final effluent from the systems improves with the complexity of the facility.

Free-Floating Macrophyte-Based Systems

Free-floating macrophytes are highly diverse in form and habit, ranging from large plants with rosettes of aerial and/or floating leaves and well-developed submerged roots (e.g., water hyacinth, *Eichhornia crassipes*) to minute surface-floating plants with few or no roots (e.g., duckweeds, *Lemna*, *Spirodella*, *Wolffia* sp.).[15]

Water Hyacinth-Based Systems

The water hyacinth (see Figure 2) is one of the most prolific and productive plants in the world. In tropical and subtropical regions, the water hyacinth is in general a severe weed, blocking irrigation canals, hindering boat traffic, increasing waterborne disease, and clogging rivers and canals, and thus making drainage impossible.[34] This high productivity is, however, exploited in wastewater treatment facilities. Two different concepts are applied in water hyacinth-based wastewater treatment systems. (1) Tertiary treatment systems (i.e., nutrient removal) are those in which nitrogen and phosphorus are removed by incorporation into the water hyacinth biomass. The biomass is harvested frequently to sustain maximum productivity and to remove incorporated nutrients.[3] Nitrogen may also be removed as a consequence of microbial denitrification. (2) Integrated secondary and tertiary treatment systems (i.e., BOD and nutrient removal) are those in which degradation of organic matter and microbial transformations of nitrogen (nitrification-denitrification) proceed simultaneously in the water hyacinth ecosystem.[35,36] Harvesting of water hyacinth biomass is only carried out for maintenance purposes. The latter system should include aerators, that is, areas with a free water surface where oxygen can be transferred to the water from the atmosphere by diffusion and where algal oxygen production can occur. The retention time in the systems varies according to wastewater characteristics and effluent requirements, but is generally on the order of 5 to 15 days.[37,38]

The role of water hyacinths in the process of suspended solids removal is well documented. Most suspended solids are removed by sedimentation and subsequent degradation within the basins, although some sludge might accumulate on the sediment surface.[37] The dense cover of water hyacinths effectively reduces the effects of wind mixing and also minimizes thermal mixing. The shading provided by the plant cover restricts algal growth, and hyacinth roots impede the horizontal movement of particulate matter.[39] Further-

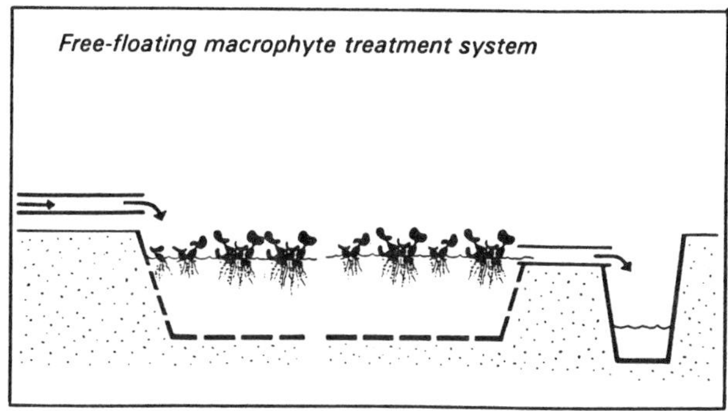

Figure 2. Schematic representation of a free-floating macrophyte-based wastewater treatment system. The species illustrated is water hyacinth (*Eichhornia crassipes*).

more, electrical charges associated with hyacinth roots are reported to react with opposite charges on colloidal particles such as suspended solids, causing them to adhere to plant roots, where they are removed from the wastewater stream and slowly digested and assimilated by the plant and microorganisms.[40] The efficiency of water hyacinths in removing BOD and in providing good conditions for microbial nitrification is related to their capability of transporting oxygen from the foliage to the rhizosphere.[41,42] The extensive root system of the water hyacinth provides a huge surface area for attached microorganisms, thus increasing the potential for decomposition of organic matter.

Water hyacinth-based wastewater treatment systems are sufficiently developed to be successfully applied in the tropics and the subtropics. Process design criteria have been published by Reed et al.[43] and Weber and Tchobanoglous,[44] among others. Water hyacinths are severely affected by frost, and the growth rate is greatly reduced at temperatures below 10°C. Consequently, in temperate regions, water hyacinth-based systems can only be used in greenhouses or outdoors during summer. Pennywort (*Hydrocotyle umbellata*), on the other hand, has a high growth rate and a high nutrient uptake capacity, even during relatively cold periods in subtropical areas. It has been suggested that water hyacinths and pennywort can be alternately cultured, winter and summer, in order to maintain the performance at a high level year-round.[45,46] No data exist on the performance or cost effectiveness of such systems.

Duckweed-Based Systems

Duckweeds (*Lemna*, *Spirodella*, and *Wolffia* sp.) have been investigated much less than water hyacinths for use in wastewater treatment.[47,48] Duckweeds, however, have a much wider geographic range than water hyacinths, as they are able to grow at temperatures as low as 1 to 3°C. Duckweeds, compared to water hyacinths, play a less direct role in the treatment process as they lack extensive root systems and therefore provide a smaller surface area for attached microbial growth. The main use of duckweeds is therefore in recovering nutrients from secondary treated wastewater. A dense cover of duckweed on the surface of water inhibits both oxygen entering the water by diffusion and the photosynthetic production of oxygen by phytoplankton because of the poor light penetration. The water consequently becomes largely anaerobic, which in turn favors denitrification. The light absorption of duckweed cover restricts growth of phytoplankton and therefore the production of suspended solids. Duckweed is easily harvested from the water surface. The nutritive value of the produced biomass is high compared to that of the water hyacinth, as it contains at least twice as much protein, fat, nitrogen, and phosphorus.[49,50] Duckweed-based systems may be plagued by problems, as high winds may pile the duckweed into thick mats and eventually completely sweep the plants from the water. Therefore, in large systems, it is necessary to construct some kind of barrier on the water surface to prevent this. The retention time in duckweed-based wastewater treatment systems depends on wastewater quality, effluent requirements, harvesting rate, and climate, but it varies typically from 30 days during summer to several months during winter.[47]

Emergent Aquatic Macrophyte-Based Systems

Rooted emergent aquatic macrophytes are the dominant life form in wetlands and marshes, growing within a water-table ranging from 50 cm below the soil surface to a water depth of 150 cm or more (see Figure 3).

In general, they produce aerial stems and leaves, and an extensive root and rhizome system. The depth penetration of the root system and thereby the exploitation of sediment volume is different for different species. Typical species of emergent aquatic macrophytes are the common reed (*Phragmites australis*), cattail (*Typha latifolia*), and bulrush (*Scirpus lacustris*). All species are morphologically adapted to growing in a water-logged sediment by virtue of large internal air spaces for transportation of oxygen to the roots and rhizomes. Most species of emergent aquatic macrophytes possess an extensive internal lacunal system that may occupy 50 to 70% of the total plant volume. Oxygen is transported through the gas spaces to the roots and rhizomes by diffusion and/or by convective flow of air.[51] Part of the oxygen may leak from the root system into the surrounding rhizosphere, creating oxidized conditions in the otherwise anoxic sediment and stimulating both decomposition of organic matter and growth of nitrifying bacteria.[21] Emergent macrophyte-based wastewater treatment systems can be constructed with different designs. Three of these will be described in some detail here.

Emergent Macrophyte-Based Systems with Surface Flow

One of the oldest concepts of constructed wetlands with surface water flow has been used in the Netherlands for nearly 30 years.[52] The design typically consists of 3 to 5-m wide and more than 100-m long ditches planted with bulrushes (*Scirpus lacustris*). Wastewater treatment is favored by the presence of submerged portions of stems and litter which serve as substrate for attached microbial growth (Figure 3a). A considerable proportion of the wastewater may drain out from the ditches through the unsealed bottom.

Emergent Macrophyte-Based Systems with Horizontal Subsurface Flow

The concept of treating wastewater in constructed wetlands with subsurface flow was developed in Germany in the 1970s.[53,54] The design typically consists of a bed planted with the common reed *Phragmites australis* and underlaid by an impermeable membrane to prevent seepage (Figure 3b). The medium in the bed may be soil or gravel. During the passage of the wastewater through the rhizosphere of the reeds, organic matter is decomposed microbiologically, nitrogen may be denitrified, and phosphorus and heavy metals fixed in the soil. The reeds have two important functions in the process: (1) to supply oxygen to the heterotrophic microorganisms in the rhizosphere, and (2) to increase and stabilize the hydraulic conductivity of the soil. The quantitative significance of the uptake of nutrients in the plant tissue is negligible, as the amount of nutrients taken up during a growing season constitutes only a few percent of the total content introduced with the wastewater. Moreover, the nutrients bound in the plant tissue are recycled in the system upon decay of the plant material. At present, there are several hundred systems of this design in operation in Denmark, Germany, and the U.K. The experience obtained thus far shows that suspended solids and BOD are generally removed effectively, effluent very nearly attaining advanced secondary treatment quality.[55,56] Removal efficiencies for nitrogen and phosphorus are variable and dependent on loading rate, type of substrate, and type and composition of wastewater. Surface run-off is a general problem in the soil-based treatment facilities — it prevents the wastewater from coming into contact with the rhizosphere. Furthermore, the oxygen transport capacity of the reeds seems to be insufficient to ensure aerobic decomposition in the rhizosphere and deliver the oxygen needed for quantitatively significant nitrification.[8]

Emergent Macrophyte-Based Systems with Vertical Subsurface Flow

In a vertical flow system (Figure 3c) the requirements for sufficient hydraulic conductivity in the bed medium and improved rhizosphere oxygenation can be established.[57] A design consisting of several beds laid out in parallel with percolation flow and intermittent loading will increase soil oxygenation several-fold compared to horizontal subsurface flow systems. During the loading period, air is forced out of the soil; during the drying period, atmospheric air is drawn into the porespaces of the soil, thus increasing soil oxygenation. Furthermore, diffusive oxygen transport to the soil is enhanced during the drying period, as the diffusion of oxygen is approximately 10,000 times faster in air than in water. This design and operational regime provides alternating oxidizing and reducing conditions in the substrate, thereby stimulating sequential nitrification-denitrification and phosphorus adsorption.[58] The sparse information available on the treatment performance of such systems indicates good performance with respect to suspended solids, BOD, ammonia, and phosphorus.

Submerged Macrophyte-Based Systems

Submerged aquatic macrophytes have their photosynthetic tissue entirely submerged (see Figure 4). The morphology and ecology of the species vary from small, rosette-type, low-productivity species growing only

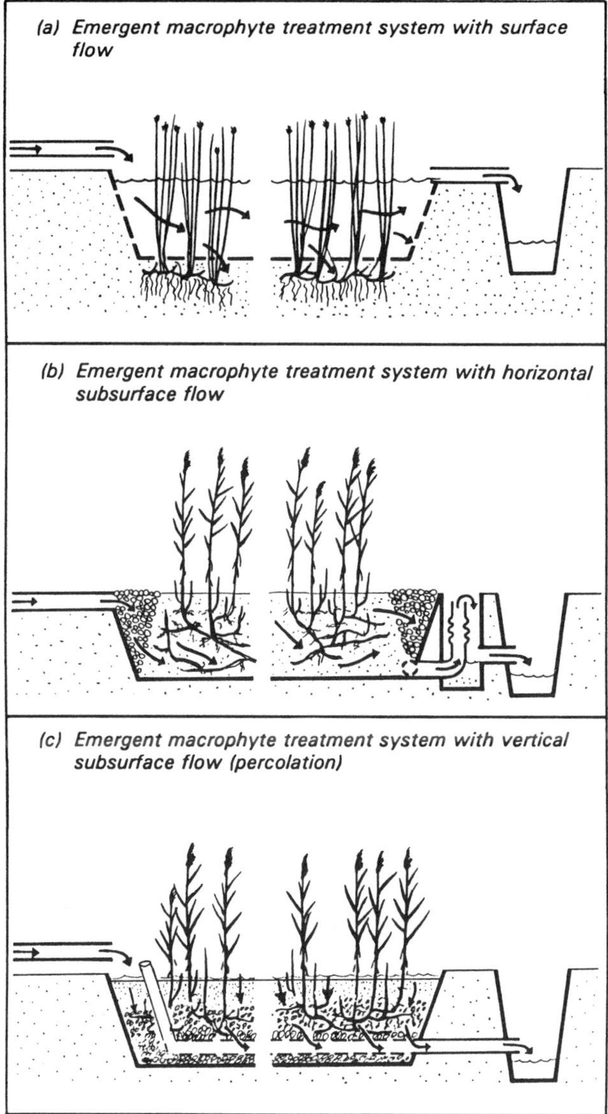

Figure 3. Schematic representation of emergent macrophyte-based wastewater treatment systems: A, system with surface flow. The species illustrated is bulrush (*Scirpus lacustris*); B, system with horizontal subsurface water flow. The species illustrated is the common reed (*Phragmites australis*); and C, system with vertical subsurface water flow (percolation). The species illustrated is the common reed (*Phragmites australis*).

in oligotrophic waters (e.g., *Isoetes lacustris* and *Lobelia dortmanna*) to larger eloedid-type, high-productivity species growing in eutrophic waters (e.g., *Elodea canadensis*).

Submerged aquatic plants are able to assimilate nutrients from polluted waters. Submerged plants, however, only grow well in oxygenated water and therefore cannot be used in wastewater with a high content of readily biodegradable organic matter because the microbial decomposition of the organic matter will create anoxic conditions. The prime potential use of submerged macrophyte-based wastewater treatment systems is therefore for "polishing" secondarily treated wastewaters, although good treatment of primary domestic effluent has been obtained in an *Elodea nuttallii*-based system.[59] The presence of submerged macrophytes depletes dissolved inorganic carbon in the water and increases the content of dissolved oxygen during the periods of high photosynthetic activity. This results in increased pH, creating optimal conditions for volatilization of ammonia and chemical precipitation of phosphorus. High oxygen concentrations also create

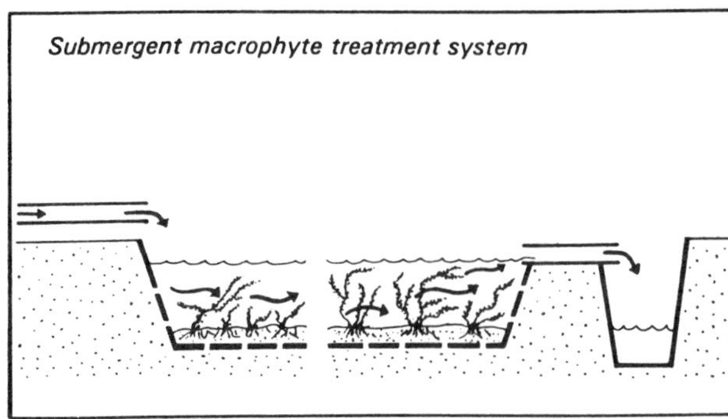

Figure 4. Schematic representation of a submerged macrophyte-based wastewater treatment system. The species illustrated is elodea (*Elodea canadensis*).

favorable conditions for the mineralization of organic matter in the water. The nutrients assimilated by the macrophytes are largely retained within the rooting tissues of the plants and by the attached microflora. Losses from the foliage of plant nutrients upon senescence of the macrophyte tissues are readily taken up by the periphytic community so that very little leaves the littoral detritus and macrophyte-epiphyte complexes. Much of the detrital matter produced in these systems will be accumulated and retained in the sediments.

The use of submerged macrophytes for wastewater treatment is still in the experimental stage, with species like egeria (*Egeria densa*), elodea (*Elodea canadensis* and *Elodea nuttallii*), hornwort (*Ceratophyllum demersum*), and hydrilla (*Hydrilla verticillata*) being the most promising.[61] Present knowledge suggests that their prime area of application will be as a final step in multistage systems.[43]

MultiStage Macrophyte-Based Treatment Systems

Different types of the above-mentioned macrophyte-based wastewater treatment systems may be combined with each other or with conventional treatment technologies. An example could be a multistage system consisting of: (1) a mechanical clarification step for primary treatment; (2) a floating or emergent macrophyte-based treatment system for secondary treatment; and (3) a floating, emergent, or submerged macrophyte-based step for tertiary treatment. The type of secondary and tertiary treatment step will, among other factors, depend on wastewater characteristics, the treatment requirements, the climate, and the amount of available land.

Some multistage systems have already been tested. One of them is the Marsh-Pond-Meadow system.[62] This concept consists of (1) a bar screen and an aeration cell using a floating surface aerator; (2) a lateral-flow marsh planted with cattails (*Typha latifolia*) in a sand medium; (3) a facultative pond with aquatic macrophytes (*Lemna* sp., *Sagittaria* sp., *Nuphar* sp., and *Anacharis* sp.) and herbivorous fish (*Cyprinus carpio*, *Ictalurus nebulosis*, and *Lepomis* sp.); (4) a meadow planted with reed canary grass (*Phalaris arundinacea*); and finally, (5) a chlorination chamber. The removal efficiency in this system is reported to be 77% for ammonia nitrogen and 82% for total phosphorus (yearly mean).

Another multistage system, realized in France, is the so-called Max-Planck-Institute-Process.[63,64] This design consists basically of four to five stages in cascade, each with several basins laid out in parallel and planted with emergent macrophytes in gravel (*Phragmites australis*, *Scirpus lacustris*, and *Iris pseudacorus*). The flow pattern in the first two stages is vertical, with the beds being loaded alternately. The later stages have horizontal water flow. The treatment performance in the system is good with regard to suspended solids and BOD, but the removal of nitrogen and phosphorus is generally poor, probably because of the rather high loading rates (2.5 m^2 per person equivalent). This design has, however, been the model for a system implemented in Oaklands Park in the U.K.[65] This system produces a fully nitrified effluent and 98% reduction of BOD and suspended solids, with an area use of only 1 m^2 per person equivalent.

DISCUSSION

The applicability of the different macrophyte-based wastewater treatment systems depends, among other things, on climatic conditions, effluent quality required, wastewater characteristics, availability and price of land, conservation regulations, etc. Water hyacinth-based systems can, for instance, only be used in warm climates, as the plant does not tolerate frost. The fact that the water hyacinth is a serious persistent weed in tropical and subtropical areas may also restrict the use of hyacinth systems. Effluent quality criteria may vary from location to location. In some situations, reduction of suspended solids and BOD may be regarded as sufficient; while in other situations, removal of nitrogen and phosphorus will be required as well. It is therefore very complicated and nearly impossible to present a general comparative evaluation of the applicability of the different systems. Therefore, I have chosen to suggest, based on information from the literature and theoretical considerations, how the treatment performance in a macrophyte-based system may be optimized and intensified; that is, how to get the most efficient removal of suspended solids, BOD, nitrogen, and phosphorous using the smallest area per person equivalent.

In many of the presently realized macrophyte-based system designs, the removal processes are supposed to occur simultaneously in a single unit of constructed wetland. This means that aerobic degradation of BOD, microbiological nitrification, denitrification, fixation of phosphorus, etc. are supposed to take place within the same "reactor". Although most of the key processes involved in the different types of macrophyte-based wastewater treatment are qualitatively well documented, quantitative information on the rates of these processes and the factors which affect them is sparse. Furthermore, the relative rates of the treatment processes for different wastewater constituents vary. This means that the design criteria have to be based on the constituent with the lowest removal rate in order to secure good performance with respect to all parameters. One major drawback of the many "one-unit" systems, where the cleaning processes are integrated in time and space, is the lack of control and regulation possibilities of the single biological and chemical processes. Once established, nothing can be done to improve performance. Therefore, in order to be able to intensify the treatment performance and to design the systems in relation to specific needs, the cleaning processes should be separated into different steps in multistage systems. For treatment of domestic sewage, a system could consist of the following units (Figure 5):

1. Mechanical pretreatment. A bar screen and a settlement tank or septic tank; settleable solids are removed by sedimentation.
2. Vertical-flow constructed wetlands with rooted emergent macrophytes (e.g., *Phragmites australis*); several (e.g., 4) beds laid out in parallel with alternating loading (2 days of loading and 6 days of drying). The substrate should be graded sand/gravel, with coarse gravel in the bottom for effective drainage, and a layer of fine sand at the surface in order to secure a good distribution of the sewage; the substrate will be kept largely aerobic by oxygen diffusion from the atmosphere through the substrate surface and from the bottom via the drains. Furthermore, air is drawn into the substrate during the drainage cycle.[57] Suspended solids and a great proportion of the readily degradable organic matter will be removed. During the drying period, the suspended solids on the substrate surface will dry out and thus counteract clogging. Further, the growth and motion of the macrophytes, as a consequence of wind, will help keep the substrate open for percolation.[66]
3. Two vertical-flow beds with rooted aquatic macrophytes and alternating loading (e.g., 2 days of loading and 2 days of drying). Construction as the previous unit. The substrate should contain a medium with a high capacity for fixation of phosphorus (e.g., sand with a high content of iron[67] or an artificial medium[68]); this unit will produce a fully nitrified effluent, and phosphorus will be fixed in the substrate.
4. A horizontal subsurface-flow bed with continuous water saturation. Substrate should be sand with high hydraulic conductivity and phosphorus fixing capacity. This bed will be anoxic and have a high capacity for denitrification. It may be necessary to supplement this unit with some primary effluent in order to secure a sufficient organic carbon source for denitrification.
5. A vertical-flow unit with emergent aquatic macrophytes and continuous loading. The substrate should be relatively coarse gravel and the drainage system should produce continuous good aeration of the substrate (as in a conventional trickling filter). The macrophytes will take up some of the remaining nutrients in the effluent and the water will be aerated during percolation through the filter. This unit may be replaced with some other kind of plant-based system in which equivalent processes will occur (e.g., a pond stocked with submerged macrophytes).

The multistage system sketched in Figure 5 will be able to operate under temperate conditions. The initial experiences with systems of this type have shown that the treatment processes are significantly intensified as compared to the processes in one-unit systems. The surface area required to reach secondary standards may be less than 2 m^2 per person equivalent.[65] Furthermore, it is possible to regulate the loading pattern in the different units in accordance with sewage production and composition, and it is possible to take one or several of the units out of operation for maintenance purposes without having to disconnect the sewage from the

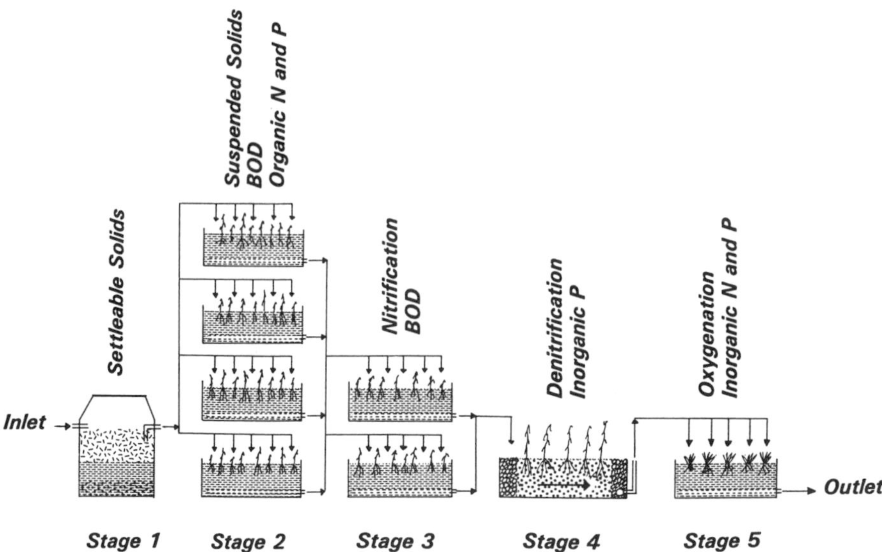

Figure 5. Hypothetical optimal design of a multistage macrophyte-based system. The system consists of (a) a mechanical pretreatment step (stage 1); (b) four beds with vertical water flow (percolation) and alternating loading (stage 2); (c) two beds with vertical flow and alternating loading (stage 3); (d) a horizontal subsurface-flow bed (stage 4); and (e) a vertical flow bed (stage 5). (See text for further explanation.)

system. Further studies are needed to evaluate the optimal loading regimes and the specific area needed in the separate units for getting the most efficient treatment performance.

The fundamental difference between conventional and macrophyte-based wastewater treatment systems is that in conventional systems, wastewater is treated rapidly in highly managed, energy intensive environments; in macrophyte-based systems, treatment occurs at a comparatively slow rate in essentially unmanaged "natural" environments. The consequences of the above differences are: (1) conventional systems require more construction and equipment but less land than macrophyte-based systems; and (2) conventional processes are subject to greater operational control and less environmental influence than macrophyte-based processes. In multistage macrophyte-based systems, an attempt is made to mimic the possibility of operational control of the single removal processes in conventional systems. This is done by separating the cleaning processes into different steps in order to be able to optimize the treatment performance in relation to the specific need. Future construction of macrophyte-based wastewater treatment systems should combine the specific qualities of the single macrophyte-based process types. The systems should include several small units containing different kinds of macrophytes, different substrates, different flow patterns, and perhaps some degree of water recirculation. Further studies are required to improve the understanding and the capacity of the key cleaning processes before the best-suited macrophyte-based systems or combination of systems can be selected and applied extensively, at least in temperate regions.

SUMMARY

The high productivity and nutrient removal capability of wetlands has created substantial interest in their potential use for improving wastewater quality. Both natural and constructed wetlands may act as efficient water purification systems and nutrient sinks. Natural wetlands, if used for wastewater treatment, will change in an often irreversible manner because of the increased nutrient loading and the change in water regime. As a consequence, former natural wetlands will be destroyed and turned into a man-made type of managed eutrophicated wetlands. Natural wetlands should therefore not be used deliberately as wastewater treatment systems, but should be preserved for environmental conservation.

Constructed wetlands may be classified according to the dominant type of macrophyte within the system: (1) free-floating macrophyte-based treatment systems, (2) emergent macrophyte-based treatment systems, (3) submerged macrophyte-based treatment systems, and (4) multistage macrophyte-based treatment systems. Different constructed wetlands may be combined with each other or with conventional treatment systems in order to exploit the specific advantages of the different systems. The pollutants are removed by a complex variety of biological, chemical, and physical processes. The actual role of the macrophytes in the removal processes varies according to the system design. The macrophytes remove pollutants by (1) directly assimilating them into their tissue, and (2) providing a suitable environment for microbial activity. Aquatic macrophytes transport oxygen into their rhizosphere, thereby stimulating aerobic degradation of organic matter and growth of nitrifying bacteria.

One of the main differences between conventional advanced wastewater treatment plants and constructed wetlands is the possibility of specific control and regulation of single biological and chemical processes in the former. This is only partly possible in constructed wetlands, where the cleaning processes are usually integrated in time and space. In the future, efforts should be made to separate the cleaning processes into different steps in multistage systems in order to be able to optimize the treatment performance in relation to specific need. Initial experience has shown that treatment processes can be significantly intensified in multistage systems consisting of a combination of several emergent macrophyte-based systems laid out in parallel, with alternating loading and horizontal subsurface flow systems. Further studies are needed to evaluate optimal loading regimes and the specific area needed in separate units for getting the most efficient treatment performance.

REFERENCES

1. Westlake, D. F. Comparisons of plant productivity, *Biol. Rev.* 38:385, 1963.
2. Reddy, K. R. and D. L. Sutton. Water hyacinths for water quality improvement and biomass production, *J. Environ. Qual.,* 13:1, 1984.
3. Trivedy, R. K. and V. R. Gudekar. Water hyacinth for wastewater treatment: a review of the progress, in *Current Pollution Researches in India.* R. K. Trivedy and P. K. Goel, Eds. 1985, 109.
4. Schierup, H.-H. Biomass and primary production in a *Phragmites communis* TRIN. swamp in North Jutland, Denmark. *Verh. Int. Verein. Limnol.* 20:94, 1978.
5. Gopal, B., K. P. Sharma, and R. K. Trivedy. Studies on ecology and production in Indian freshwater ecosystems at primary producer level with emphasis on macrophytes, in *Glimpses of Ecology.* J. S. Singh and B. Gopal, Eds. International Scientific Publications C-70, Jaipur, 1978, 349.
6. Madsen, T. V. and K. Sand-Jensen. Photosynthetic carbon assimilation in aquatic macrophytes. *Aquat. Bot.* 41:5, 1991.
7. Westlake, D. F. The primary productivity of water plants, in *Studies on Aquatic Vascular Plants* J. J. Symoens, S. S. Hooper, and P. Compère, Eds. Royal Botanical Society of Belgium, Brussels, 1982, 165.
8. Brix, H. Gas exchange through the soil-atmosphere interphase and through dead culms of *Phragmites australis* in a constructed reed bed receiving domestic sewage, *Water Res.* 24:259, 1990.
9. Ponnamperuma, F. N. The chemistry of submerged soils, *Adv. Agron.* 24:29, 1972.
10. Rogers, K. H., Wetlands as accreting systems: organic carbon. *J. Limnol. Soc. S. Afr.* 9:96, 1983.
11. Kelly, J. R. and M. A. Harwell. Comparisons of the processing of elements by ecosystems. I. Nutrients, in *Ecological Considerations in Wetlands Treatment of Municipal Wastewaters.* P. J. Godfrey, E. R. Kaynor, S. Pelczarski, and J. Benforado, Eds. Van Nostrand Reinhold, New York, 1985, 137.
12. Nichols, D. S. Capacity of natural wetlands to remove nutrients from wastewater, *J. Water Pollut. Control Fed.* 55:495, 1983.
13. Gersberg, R. M., B. V. Elkins, and C. R. Goldman. Wastewater treatment by artificial wetlands, *Water Sci. Technol.* 17:443, 1984.
14. Brodrick, S. J., P. Cullen, and W. Maher. Denitrification in a natural wetland receiving secondary treated effluent, *Water Res.* 22:431, 1988.
15. Brix, H. and H.-H. Schierup. The use of aquatic macrophytes in water pollution control, *Ambio.* 18:100, 1989.
16. Richardson, C. J. and D. S. Nichols. Ecological analysis of wastewater management criteria in wetland ecosystems, in *Ecological Considerations in Wetlands Treatment of Municipal Wastewaters.* P. J. Godfrey, E. R. Kaynor, S. Pelczarski, and J. Benforado, Eds. Van Nostrand Reinhold, New York, 1985, 351.

17. Richardson, C. J. and J. A. Davis. Natural and artificial wetland ecosystems: ecological opportunities and limitations, in *Aquatic Plants for Water Treatment and Resource Recovery*. K. R. Reddy and W. H. Smith, Eds. Magnolia Publishing, Orlando, FL, 1987, 819.
18. Watson, J. T., S. C. Reed, R. H. Kadlec, R. L. Knight, and A. E. Whitehouse. Performance expectations and loading rates for constructed wetlands, in *Constructed Wetlands for Wastewater Treatment, Municipal, Industrial and Agricultural*. D. A. Hammer, Ed. Lewis Publishers, Chelsea, MI, 1989, 319.
19. Breen, P. F. A mass balance method for assessing the potential of artificial wetlands for wastewater treatment, *Water Res.* 24:689, 1990.
20. Moorhead, K. K. and K. R. Reddy. Carbon and nitrogen transformations in wastewater during treatment with *Hydrocotyle umbellata* L, *Aquat. Bot.* 37:153, 1990.
21. Reddy, K. R., W. H. Patrick, Jr., and C. W. Lindau. Nitrification-denitrification at the plant root-sediment interface in wetlands, *Limnol. Oceanogr.* 34:1004, 1989
22. Gersberg, R. M., B. V. Elkins, and C. R. Goldman, Nitrogen removal in artificial wetlands, *Water Res.* 17:1009, 1983.
23. Eighmy, T. T. and P. L. Bishop. Distribution and rate of bacterial nitrifying populations in nitrogen removal in aquatic treatment systems, *Water Res.* 23:947, 1989.
24. Richardson, C. J. Mechanisms controlling phosphorus retention capacity in freshwater wetlands, *Science* 228:1424, 1985.
25. Bayley, S. E. The effect of natural hydroperiod fluctuations on freshwater wetlands receiving added nutrients, in *Ecological Considerations in Wetlands Treatment of Municipal Wastewaters*. P. J. Godfrey, E. R. Kaynor, S. Pelczarski, and J. Benforado, Eds. Van Nostrand Reinhold, New York, 1985, 180.
26. Sah, R. N. and D. S. Mikkelsen. Transformations of inorganic phosphorus during flooding and draining cycles of soil, *Soil Sci. Am. J.* 50:62, 1986.
27. Reddy, K. R. and W. F. DeBusk. Nutrient removal potential of selected aquatic macrophytes, *J. Environ. Qual.* 14:459, 1985.
28. Czinki, L. Nitrat- und phosphateliminierung mit "Eutrostop", *Korrespondenz Abwasser*. 33:440, 1986.
29. Nusch, E. A., J. Poltz, and K. Bucksteeg. "Eutrostop" — eine neue alternative zur gewässersanierung? *Korrespondenz Abwasser*. 34:1083, 1987.
30. Gersberg, R. M., S. R. Lyon, R. Brenner, and B. V. Elkins. Fate of viruses in artificial wetlands, *Appl. Environ. Microbiol.* 53:731, 1987.
31. Lance, J. C., C. P. Gerba, and J. L. Melnick. Virus movement in soil columns flooded with secondary sewage effluent, *Appl. Environ. Microbiol.* 32:520, 1976.
32. Seidel, K. The cleansing of bodies of water by higher plants, *Garten und Landschaft*. 1:9, 1978.
33. Moeller, J. R. and C. Calkins. Bactericidal agents in wastewater lagoons and lagoon design, *J. Water Pollut. Control Fed.* 52:2442, 1980.
34. National Academy of Sciences. *Making Aquatic Weeds Useful: Some Perspectives for Developing Countries*. Washington, D.C., 1976.
35. Dinges, R. Upgrading stabilization pond effluent by water hyacinth culture, *J. Water Pollut. Control Fed.* 5:833, 1978.
36. Dinges, R. and J. Doersam. The Hornsby Bend hyacinth facility in Austin, Texas, *Water Sci. Technol.* 19(10):41, 1987.
37. DeBusk, T. A., K. R. Reddy, T. D. Hayes, and B. R. Schwegler, Jr. Performance of a pilot-scale water hyacinth-based secondary treatment system, *J. Water Pollut. Control Fed.* 61:1217, 1989.
38. John, C. K. Treatment of agro-industrial wastes using water hyacinth, *Water Sci. Technol.* 17:781, 1984.
39. Dinges, R. *Natural Systems for Water Pollution Control*. Van Nostrand Reinhold, New York, 1982.
40. Wolverton, B. C. Aquatic plant/microbial filters for treating septic tank effluent, in *Constructed Wetlands for Wastewater Treatment. Municipal, Industrial and Agricultural*. D. A. Hammer, Ed. Lewis Publishers, Chelsea, MI, 1989, 173.
41. Reddy, K. R., E. M. D'Angelo, and T. A. DeBusk. Oxygen transport through aquatic macrophytes: the role in wastewater treatment, *J. Environ. Qual.* 19:261, 1989.
42. Jedicke, A., B. Furch, U. Saint-Paul, and U.-B. Schlüter. Increase in the oxygen concentration in Amazon waters resulting from the root exudation of two notorious water plants, *Eichhornia crassipes* (Pontederiaceae) and *Pistia stratiotes* (Araceae), *Amazonia*. 11:53, 1989.
43. Reed, S. C., E. J. Middlebrooks, and R. W. Crites. *Natural Systems for Waste Management and Treatment*. McGraw-Hill, New York, 1988.

44. Weber, A. S. and G. Tchobanoglous. Prediction of nitrification in water hyacinth treatment systems, *J. Water Pollut. Control Fed.* 58:376, 1986.
45. Reddy, K. R. and W. F. DeBusk. Growth characteristics of aquatic macrophytes cultured in nutrient-enriched water. I. Water hyacinth, water lettuce and pennywort, *Econ. Bot.* 38:229, 1984.
46. DeBusk, W. F. and K. R. Reddy. BOD_5 removal in floating aquatic macrophyte-based wastewater treatment systems, *Proc. Int. Conf. Waste Stabilization Ponds.* Lisbon, Portugal, 1987.
47. Ngo, V. Boosting pond performance with aquaculture, *Operations Forum.* 4:20, 1987.
48. Sutton, D. L. and W. H. Ornes. Phosphorus removal from static sewage effluents using duckweed, *J. Environ. Qual.* 4:367, 1975.
49. Cully, D. D. and E. A. Epps. Use of duckweed for waste treatment and animal feed, *J. Water Pollut. Control Fed.* 45:337, 1973.
50. Hillman, W. S. and D. D. Culley, Jr. The uses of duckweed, *Am. Sci.* 66:442, 1978.
51. Armstrong, W., J. Armstrong, P. M. Beckett, and S. H. F. W. Justin. Convective gas-flows in wetland plant aeration, in *Plant Life Under Oxygen Deprivation.* M. B. Jackson, D. D. Davies, and H. Lambers, Eds. SPB Academic Publishing bv, The Hague, The Netherlands, 1991, 283.
52. Greiner, R. W. and J. de Jong. *The Use of Marsh Plants for the Treatment of Wastewater in Areas Designated for Recreation and Tourism,* RIJP Report No. 225. Lelystad, The Netherlands, 1984.
53. Brix, H. Treatment of wastewater in the rhizosphere of wetland plants — the root-zone method, *Water Sci. Tech.* 19:107, 1987
54. Kickuth, R. Degradation and incorporation of nutrient from rural wastewaters by plant rhizosphere under limnic conditions, in *Utilization of Manure by Land Spreading*, Comm. of the Europ. Communities, EUR 5672e, London, 1977, 335.
55. Findlater, B. C., J. A. Hobson, and P. F. Cooper. Reed bed treatment systems: performance evaluation, in *Constructed Wetlands in Water Pollution Control.* P. F. Cooper and B. C. Findlater, Eds. Advances in Water Pollution Control, Pergamon Press, Oxford, 1990, 193.
56. Schierup, H.-H., H. Brix, and B. Lorenzen. Wastewater treatment in constructed reed beds in Denmark — state of the art, in *Constructed Wetlands in Water Pollution Control.* P. F. Cooper and B. C. Findlater, Eds. Advances in Water Pollution Control, Pergamon Press, Oxford, 1990, 495.
57. Brix, H. and H.-H. Schierup. Soil oxygenation in constructed reed beds: the role of macrophyte and soil-atmosphere interface oxygen transport, in *Constructed Wetlands in Water Pollution Control.* P. F. Cooper and B. C. Findlater, Eds. Advances in Water Pollution Control, Pergamon Press, Oxford, 1990, 53.
58. Hill, D. E. and B. L. Sawhney. Removal of phosphorus from wastewater by soil under aerobic and anaerobic conditions, *J. Environ. Qual.* 10:401, 1981.
59. Bishop, P. L. and T. T. Eighmy. Aquatic wastewater treatment using *Elodea nuttallii*, *J. Water Pollut. Control Fed.* 61:641, 1989.
60. Wetzel, R. G. Land-water interfaces: metabolic and limnological regulators, *Verh. Int. Verein. Limnol.* 24:6, 1990.
61. McNabb, C. D. The potential of submerged vascular plants for reclamation of wastewater, in *Biological Control of Water Pollution.* J. Tourbier and R. W. Pearson, Eds. The University Press, Philadelphia, 1976, 120.
62. Conway, T. E. and J. M. Murtha. The Iselin Marsh Pond Meadow, in *Constructed Wetlands for Wastewater Treatment. Municipal, Industrial and Agricultural.* D. A. Hammer, Ed. Lewis Publishers, Chelsea, MI, 1989, 139.
63. Lienard, A., C. Boutin, and D. Esser. Domestic wastewater treatment with emergent hydrophyte beds in France, in *Constructed Wetlands in Water Pollution Control.* P. F. Cooper and B. C. Findlater, Eds. Advances in Water Pollution Control, Pergamon Press, Oxford, 1990, 183.
64. Boutin, C. Domestic wastewater treatment in tanks planted with rooted macrophytes: case study; description of the system; design criteria; and efficiency, *Water Sci. Technol.* 19:29, 1987.
65. Burka, U. and P. C. Lawrence. A new community approach to waste treatment with higher water plants, in *Constructed Wetlands in Water Pollution Control.* P. F. Cooper and B. C. Findlater, Eds. Advances in Water Pollution Control, Pergamon Press, Oxford, 1990, 359.
66. Nielsen, S. M. Sludge dewatering and mineralisation in reed bed systems, in *Constructed Wetlands in Water Pollution Control.* P. F. Cooper and B. C. Findlater, Eds. Advances in Water Pollution Control, Pergamon Press, Oxford, 1990, 245.

67. Netter, R. and Bischofsberger, W. Sewage treatment by planted soil filters, in *Constructed Wetlands in Water Pollution Control*. P. F. Cooper and B. C. Findlater, Eds. Advances in Water Pollution Control, Pergamon Press, Oxford, 1990, 525.
68. Roques, H., L. Nugroho-Jeydy, and A. Lebugle. Phosphorus removal from wastewater by half-burned dolomite, *Water Res.* 25:959, 1991.

CHAPTER 3

Constructed Wetlands and Aquatic Plant Systems: Research, Design, Operational, and Monitoring Issues

G. Tchobanoglous

INTRODUCTION

Over the past 15 years, great strides have been made in understanding how constructed wetlands bring about the attenuation of pollutants.[2,4,9,11,19,20] However, if significant new advances are to be made in understanding how these systems work and how to design and implement them effectively, a more systematic approach to their study will be required. The purpose of this paper is to highlight some of the important design, research, operational, and monitoring issues that must be addressed by the scientific and engineering communities.

DESIGN ISSUES

Of the many design issues that must be resolved, the use of design equations based on nonspecific parameters and the need for hydraulic process control, new physical designs, and combined systems for nitrogen removal are considered in the following discussion.

Use of Design Equations Based on Nonspecific Parameters

Perhaps one of the most serious limitations in the design of constructed wetlands is the use of design equations based on nonspecific parameters, specifically, the 5-day biochemical oxygen demand (BOD) and suspended solids (SS). The use of these parameters is limiting because no fundamental information is obtained on the operative removal mechanisms.

Biochemical Oxygen Demand

The results of BOD tests are now used to determine (1) the approximate quantity of oxygen that will be required to stabilize biologically the organic matter present, (2) the size of wastewater treatment facilities, (3) the efficiency of some treatment processes, and (4) compliance with wastewater discharge permits.[16] With respect to the design of constructed wetlands, the principal limitation with the use of BOD as a design parameter is that the nature of the organic substance that is causing the oxygen demand is usually unknown. As shown in Figure 1A, the nature of the residual organic matter that causes a BOD is continually changing with time. The only point of real interest is the stoichiometric point where all of the biodegradable organic matter is converted to cell tissue, a phenomenon which seldom occurs exactly at 5 days. Since the nature of the organic matter is unknown in the BOD test, the use of BOD values in first-order removal equations (see Figure 1B) does not shed any light on the nature of the attenuation mechanisms that might be operative. For example, if the substances that comprise the BOD are highly adsorbable and/or filtrable (e.g., cell tissue), they may be removed by sorption or filtration, as opposed to bacterial conversion. The lack of stoichiometric validity, at all times, reduces the usefulness of the BOD as a design parameter.

0-87371-550-0/93/$0.00+$.50

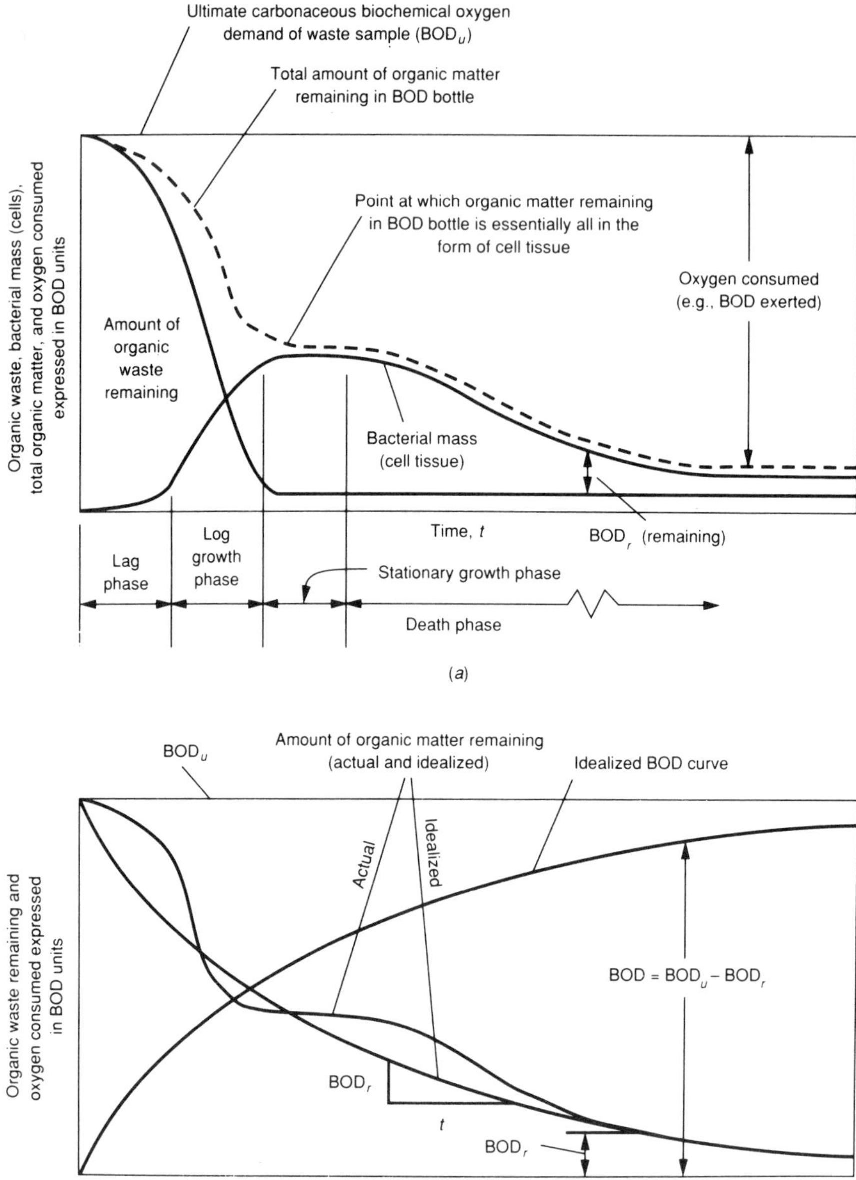

Figure 1. Functional analysis of the BOD test: A, interrelationship of organic waste, bacterial mass cell tissue, total organic waste, and oxygen consumed in BOD test; and B, idealized representation of the BOD test results. (See Reference 13.)

Suspended Solids

Suspended solids (SS) is also a lumped parameter because no information is available on the nature of the particles and the distribution of particle sizes that comprise the measured SS values. Since the distribution of particle sizes is unknown, it is difficult to say anything about the particle removal mechanisms that may be operative. The distribution of particle sizes in domestic wastewaters is typically bimodal (see Figure 2A), with one peak occurring at about 1 μm and a second peak occurring in the range 20 to 100 μm.[5-7] The specific location of the second peak will depend on the operating characteristics of the pretreatment facilities (see Figure 2B).

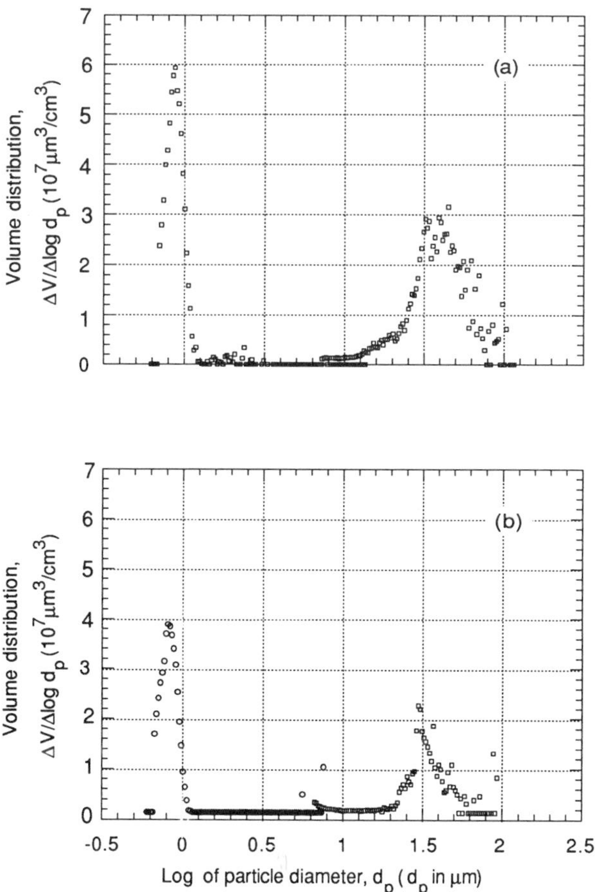

Figure 2. Typical particle size distributions in secondary effluent from the UC-Davis wastewater treatment plant: A, unfiltered and B, filtered through a granular-medium filter.

Hydraulic Process Control

In most constructed wetlands, there is essentially no control over the hydraulic process. If these systems are to be managed, some means of controlling the process must be developed. One of the best methods is to use internal flow recycle and effluent recycle. The beneficial effects of using recycle flows — (1) to reduce the concentration of the BOD, (2) to control the transport of contaminants to achieve effective treatment, and (3) to control the production of odors — are just beginning to be appreciated more fully. To take advantage of the beneficial effects of flow recycle, new pond configurations (see Figures 3C and 3D) are needed to minimize the cost of pumping.[15] By selectively reducing the concentration of the BOD, the treatment potential of constructed wetlands can be enhanced. For example, if the concentration of BOD and suspended solids is too great, the oxygen transported to the plant roots may be wasted in treating sludge that accumulates around the roots, as opposed to treating the organic matter in the fluid bulk.

Design of Physical Facilities

Unfortunately, to date, the design of constructed wetlands has been derivative rather than innovative. The configuration of the early constructed wetlands was typically plug flow.[4,9,11,12] Many of the early systems were often nothing more than oxidation ponds in which either emergent or floating aquatic plants were placed. The subsurface flow systems that have become popular in recent years are essentially copied from European practice, with little or no attention to system hydraulics.[10] As a consequence, a number of the constructed wetland treatment systems have failed to meet expectations.

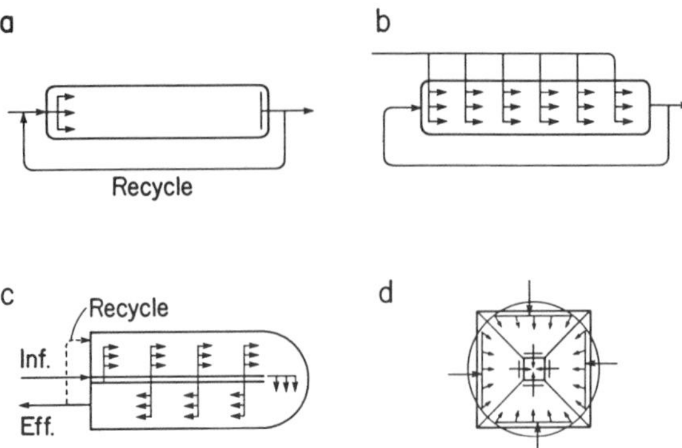

Figure 3. Alternative flow diagrams for constructed wetlands: A, plug-flow with influent distribution and recycle; B, step-feed with recycle; C, step-feed with recycle in wrap-around pond; and D, peripheral feed center drawoff. (See References 14 and 15.)

Aspect Ratio

The *aspect ratio*, defined as the length-to-width ratio, used for constructed wetlands, typically varies from 4:1 to 10:1. However, based on mounting evidence,[4,11,15] it appears that aspect ratios approaching 1:1 may be more appropriate (see Figure 3D). Two reasons that larger aspect ratios were used in the early designs included familiarity with plug flow reactors and lack of usable kinetic design information. Unfortunately, because of the continued use of lumped design parameters such as BOD and SS, little usable information has been developed on the operative attenuation mechanisms.

Surfacing Flow

A major problem with subsurface flow systems (SFS) has been the surfacing of the applied wastewater before reaching the effluent end. Here again, the design process has been derivative rather than innovative. Most of the systems appear to have been designed and constructed without proper computation of the subsurface flow hydraulics. The problem is further compounded, because little or no information is available on the characteristics of the wastewater being applied, and the role of the plants has not been considered adequately. For example, the fact that the porosity of the medium used in SFS systems is changing continually due to the growth of the plant tubers is seldom, if ever, considered in their design.

Input/Output Facilities

Another area of design that has not received sufficient attention in the implementation of constructed wetlands is the proper configuration of the inlet and outlet facilities. Proper design of these facilities is of critical importance if the systems are to function without serious short-circuiting. Ideally, the inlet facilities should be designed to distribute the flow over as wide an area as possible (Figure 3D), consistent with the design objectives. Influent distribution is especially critical as aspect ratios begin to approach 1:1.

Combined Designs For Nitrogen Removal

Based on mounting experimental and field evidence, it can be concluded that achieving both nitrification and denitrification in a single constructed wetland treatment system will not be cost effective. What is required is to combine different types of systems to achieve the desired effect. For example, a free water surface (FWS) constructed wetland could be used for BOD removal. The FWS would be followed by a slow sand filter for nitrification, followed by a subsurface flow system for denitrification. Many other combinations are possible, depending on the nature and characteristics of the wastewater to be treated.[10] If the wrap-around configuration is used for the constructed wetland, the nitrified effluent could be denitrified by recycling a portion of the incoming flow (Figure 3C).

Table 1. Analytical Techniques Applicable to Particle Size Analysis of Wastewater Contaminants

Observation and measurement techniques	Separation and analysis techniques
Microscopy	Sedimentation
Light	Centrifugation
Transmission electron	Membrane filtration
Scanning electron	Field flow fractionation
Image analysis	Ultrafiltration
Particle counters	Gel filtration chromatography
Conductivity difference	
Equivalent light scattering	
Light blockage	

From Reference 6. With permission.

RESEARCH ISSUES

Two of the principal research issues with constructed wetlands are related to the need for improved characterization of the wastewaters that are applied to wetlands and a more thorough understanding of the transport and fate processes that are operative in constructed wetlands.

Improved Waste Characterization

Improved waste characterization is needed to define the nature of the waste contaminants applied to constructed wetlands and the operative removal and/or attenuation mechanisms. Once the wastewater characteristics and the operative removal mechanisms are defined more clearly, it will be possible to design systems to optimize the removal of specific constituents. As a first step, it is suggested that the particle size distribution of the waste be characterized and that the organic content of the various size fractions be defined.

Particle Size Distribution

Particle size analysis can be used as an effective first step in characterizing the wastewater to be treated. The analytical techniques available for particle size measurement vary in complexity, utility, and expense. Some techniques appropriate for analysis of the contaminants in wastewater are listed in Table 1. In general terms, particle size distribution analysis is based on either observation and measurement techniques or separation and analysis techniques. Observation and measurement techniques are useful for rapid assessment of particle size, while techniques for separation and analysis can be used to correlate size with other physical and chemical characteristics.

Owing to the varied nature of wastewater particles, not all techniques are equally effective for characterizing size distributions of wastewater particulate contaminants. Each size measurement method is based on classifying particles due to physical or chemical differences and is usually only effective for measuring particles in a specific size range. Since particle size measurements are based on different particle properties, it is difficult to compare particle size data obtained from different measurement techniques. As reported in the literature,[5-7] significant variation in particle size distributions in untreated wastewater exists due to the source of the wastewater, time of year, degree of infiltration and inflow, the quantity of grit in the collection system, and a number of other factors. Particle size characterization techniques are described in greater detail in References 5 and 6.

Biodegradability Based On Particle Size Distribution

If a separation and analysis technique such as membrane filtration is used to quantify the size distribution of the solids in the influent wastewater, the results can be correlated to observed oxygen (BOD) uptake rates, determined using a respirometer (Table 1). As reported in Table 2, the observed BOD uptake rates are significantly affected by the size of the particles in wastewater. As shown in Table 2, wastewaters with significantly different particle size distributions will behave differently when applied to constructed wetlands.

Table 2. Effect of Particle Size on Observed BOD Uptake Rates for the Particles found in Wastewater

Fraction	Size range (mm)	k (base 10) (d^{-1})
Settleable	>100	0.08
Supracolloidal	1–100	0.09
Colloidal	0.1–1.0	0.22
Soluble	<0.1	0.39

Adapted from Reference 1.

Determination of the oxygen (BOD) uptake rate and the stoichiometric point, when all of the biodegradable organic matter has been converted to cell tissue, is best accomplished using an instrumented large-volume (1.0 l) electrolysis cell or a Gilson respirometer (Figure 1A).[13,16]

Improved Characterization of Waste Transport and Fate Processes

Under normal conditions, pollutants discharged to constructed wetlands are attenuated by a wide variety of mechanisms, including bacterial conversion, sorption, sedimentation, natural decay, volatilization, and chemical reactions (Table 3). The following general equation is used to relate these factors in a one-dimensional (horizontal) direction.

$$\alpha(1+\beta)\frac{\partial C_A}{\partial t} = -V_x\frac{\partial C_A}{\partial x} + D_x\frac{\partial^2 C_A}{\partial x^2} + G \tag{3–1}$$

where α = total porosity, m^3/m^3
β = retardation factor accounting for sorption and phase change
C_A = concentration of compound A, g/m^3
V_x = average fluid velocity in the X-direction, m/s
D_x = effective longitudinal diffusion coefficient, m^2/s
G = lumped parameter used to account for all generation terms, $g/m^3/s$
x = distance, m

The lumped parameter term G in Equation 1 is comprised of a series of rate expressions, one for each of the mechanisms that either attenuates or causes a release of a given constituent (see Table 3). Appropriate rate expressions must be defined for each type of constructed wetland. First-order, saturation, and Monod-type rate expressions are commonly used to define the processes occurring in constructed wetlands.[4,9,11,12,19,29]

For some slowly biodegradable EPA organic priority pollutants, adsorption is the most common way in which these materials are removed as they move through constructed wetlands. If hydrodynamic dispersion is neglected, the general materials balance for a contaminant subject only to adsorption is given by the following modified form of Equation 1.

$$\frac{\partial S}{\partial t}\frac{\rho_b}{\alpha} + \frac{\partial C}{\partial t} = -v_x\frac{\partial C}{\partial x} \tag{3–2}$$

where S = mass of solute sorbed per unit mass of dry material (composite function may be required, depending on type of wetland), g/g
ρ_b = bulk density of the material(s) comprising the solid surfaces in the wetlands, g/m^3
α = porosity
C = concentration of contaminant in the liquid phase, g/m^3
v_x = average fluid velocity in the X-direction, m/s

Table 3. Fate Processes in the Constructed Wetlands

Processes	Comments
Bacterial conversion	Bacterial conversion (both aerobic and anaerobic) is the most important process in the transformation of contaminants discharged to constructed wetlands. The exertion of CBOD and NBOD are the most common examples of bacterial conversion encountered in water quality management. The depletion of oxygen in the aerobic conversion of organic wastes is also known as *deoxygenation*. Solids discharged with treated wastewater are partly organic. Upon settling to the bottom, they decompose bacterially, either anaerobically or aerobically, depending on local conditions. The bacterial transformation of toxic organic compounds is also of great significance.
Gas absorption/desorption	The process whereby a gas is taken up by a liquid is known as *absorption*. For example, when the dissolved oxygen concentration in a body of water with a free surface is below the saturation concentration in the water, a net transfer of oxygen occurs from the atmosphere to the water. This transfer (mass per unit time per unit surface area) is proportional to the amount by which the dissolved oxygen is below saturation. The addition of oxygen to water is also known as *reaeration*. Desorption occurs when the concentration of the gas in the liquid exceeds the saturation value and there is a transfer from the liquid to the atmosphere.
Sedimentation	The suspended solids discharged with treated wastewater ultimately settle to the bottom. Settling is enhanced by flocculation and hindered by ambient turbulence. In some wetlands, turbulence is often sufficient to distribute the suspended solids over the entire water depth.
Natural decay	In nature, contaminants will decay for a variety of reasons, including mortality in the case of bacteria and photooxidation for certain organic constituents. Natural decay generally follows first-order kinetics.
Adsorption	Many chemical constituents tend to attach or sorb onto solids. The implication for wastewater discharges is that a substantial fraction of some toxic chemicals is associated with the suspended solids in the effluent. Adsorption combined with solids settling results in the removal from the water column of constituents which might not otherwise decay.
Volatilization	Volatilization is the process whereby liquids and solids vaporize and escape to the atmosphere. Organic compounds that readily volatilize are known as VOCs (volatile organic compounds). The physics of this phenomenon are very similar to gas absorption, except that the net flux is out of the water surface.
Chemical reactions	Important chemical reactions which occur in wetlands include hydrolysis, photochemical, and oxidation-reduction reactions. Hydrolysis reactions occur between contaminants and water. Solar radiation is known to trigger a number of chemical reactions. Radiation in the near-UV and visible range is known to cause the breakdown of a variety of organic compounds.

Adapted from References 8 and 17.

Where the partitioning of the contaminant between the solid surfaces in the wetlands and the wastewater can be described adequately by the distribution coefficient $K_{SD,}$ Equation 2 can be written as follows.

$$-v_x \frac{\partial C}{\partial x} = \left(1 + \frac{\rho_b}{\alpha} K_{SD}\right) \frac{\partial C}{\partial t} \tag{3–3}$$

The retardation of the contaminant front, relative to the fluid, can be described with the following relationship.

$$R = \frac{v_x}{v_{xc}} = \left(1 + \frac{\rho_b}{\alpha} K_{SD}\right) \tag{3–4}$$

where R = retardation factor, unitless
v_x = average fluid (wastewater) velocity in the X-direction, m/s
v_{xc} = average velocity of the $C/C_o = 0.5$ point of the retarded contaminant concentration profile, m/s

Clearly, the research agenda for constructed wetlands must include gathering, through both laboratory and field studies, the data and information that will make it possible to quantify the retardation and subsequent fate of specific contaminants.

MANAGEMENT ISSUES

While improved designs for constructed wetlands and greater understanding of the processes occurring in them are important, of equal or even greater importance is the long-term operation and maintenance of the wetland treatment systems once they are constructed. If constructed wetlands are to be accepted by regulatory agencies for the treatment of wastewater, their operation and maintenance must be definable. Improved management techniques must be developed that will allow for operational changes to be made in response to changes in the wastewater characteristics, effluent quality, climatic conditions, and effluent discharge requirements. Depending on the design objectives, the major operational and management issues associated with constructed wetlands include:[18]

- Vector (pest) control
- Vegetation control (manipulation)
- Wildlife habitat management
- Water quality management
- Regulatory requirements
- Hydraulic controls
- Structural integrity
- Water quality structures
- Education
- Recreation

For each of the above operational and management issues, a clear set of operating instructions must be developed for each constructed wetland treatment system so that corrective actions can be taken as the need arises. In dealing with each of the above issues, it will be important to define (1) the operating goals, (2) the basis for problem identification, (3) the causative factors, (4) the appropriate management strategies, (5) the lead time, and (6) the method(s) that will be used to evaluate the effectiveness of the control. Examples of the type of information that must be developed for vector and vegetation control for constructed wetlands are identified in Tables 4 and 5, respectively. Having a set of instructions for the operation of a constructed wetland will improve the operation and performance of these systems.[2,3,18]

MONITORING ISSUES

If improved techniques for the operation and maintenance of constructed wetlands are to be formulated, improved monitoring programs must be developed. Monitoring issues are related to regulatory requirements, process performance and control, the fate of specific constituents, and the development of a database. These issues are discussed briefly below.

Table 4. Operational Issue — Vector Control

Item	Objective/action
Operating goal	Control of mosquitoes
Problem identification	Increased counts in resting box, emergence traps, dip samples
Causative factors	Excessive plant growth, lack of predators
Management strategies	Draw water surface down, use biological controls, use conventional controls (e.g., Bear oil 2000)
Lead time	2 to 3 weeks, depending on sampling frequency
Evaluation of control	Reduced larval count

Adapted from Reference 18.

Table 5. Operational Issue — Vegetation Control

Item	Objective/action
Operating goal	Process performance
Problem identification	Clogging of flow paths, odors from decomposition, short-circuiting, low density, poor plant health
Causative factors	Aggressive growth, lack of vegetative management, excessive water depth, poor water flow patterns, seasonal variation, grazing
Management strategies	Reduce water depth, soil enhancement, supplemental planting, controlled burns, periodic harvesting
Lead time	Growing season
Evaluation of control	Vegetation surveys, vegetation maps, photographic records

Adapted from Reference 18.

Regulatory Requirements

At present, the monitoring requirements for most constructed wetlands are based on conventional parameters that typically include: BOD, SS, bacteria, nutrients, and vectors. Unfortunately, as noted earlier, information on these parameters does little to enhance current understanding of the operative removal mechanisms. In the future, as constructed wetlands are used to meet inland surface water standards for metals, EPA priority pollutants, other organics, and biotoxicity, it is anticipated that the data collected will be more useful in defining the operative fate processes.

Process Performance and Control

Another important reason for monitoring the performance of constructed wetlands is to collect data that can be used to develop process performance and control strategies. A suggested list of monitoring parameters for constructed wetlands is presented in Table 6. The specific parameters that need to be monitored will depend on the design objectives, local conditions, and regulatory requirements. In addition to meeting regulatory reporting requirements, monitoring data should be used to assess process stability and performance. For example, process stability can be assessed by plotting selected parameter values on either arithmetic- or logarithmic-probability paper and noting the slope of the line and the extreme values. If the process is found to be unstable during certain times of the year, corrective measures (such as effluent recycle) can be undertaken to improve the stability of the process. The most effective use of process control charts is in spotting trends before they become problems. While both probability plots and process control charts have been used for years in other fields, their use for monitoring constructed wetlands is relatively new.

Fate of Specific Constituents

Long-term monitoring can be used to define the the fate of specific constituents, which, in turn, can be used to define the dynamics of constructed wetlands. Long-term data are needed to define the removal, retardation,

Table 6. Summary of Suggested Monitoring Parameters for Constructed Wetlands

Parameter	Project phase (pre, or during, or construction or ongoing)	Location	Frequency of collection
Water quality[a]			
Dissolved oxygen	Ongoing	In, out, along profile	Weekly
Hourly dissolved oxygen	Ongoing	Selected locations	Quarterly
Temperature	Pre, during, ongoing	In, out, along profile	Daily/weekly
Conductivity	Pre, during, ongoing	In, out,	Weekly
pH	Pre, during, ongoing	In, out	Weekly
BOD	Pre, during, ongoing	In, out, along profile	Weekly
SS	Pre, during, ongoing	In, out, along profile	Weekly
Particle size distribution	Pre, during, ongoing	In, out, along profile	Weekly
Nutrients	Pre, during, ongoing	In, out, along profile	Weekly
Chlorophyll A	Ongoing	Within wetland, along profile	Quarterly
Metals (Cd, Cr, Cu, Pb, Zn)	Pre, during, ongoing	In, out, along profile	Quarterly
Bacteria (total and fecal coliform)	Pre, during, ongoing	In, out	Monthly
EPA priority pollutants	Pre, during, ongoing	In, out, along profile	Semiannually
Other organics	Pre, during, ongoing	In, out, along profile	Annually
Biotoxicity	Pre, during, ongoing	In, out	Semiannually
Sediments			
Redox potential	Pre, during, ongoing	In, out, along transects	Quarterly
Salinity	Pre, ongoing	In, out, along transects	Quarterly
pH	Pre, during, ongoing	In, out, along transects	Quarterly
Organic matter	Pre, post	In, out, along transects	Quarterly
Biota			
Plankton (zooplankton tow)	Ongoing	Within wetland, along transects	Quarterly
Invertebrates	Ongoing	Within wetland, along transects	Quarterly
Fish	Ongoing	Within wetland, along transects	Quarterly
Birds	Pre, during, ongoing	Within wetland, along transects	Quarterly
Endangered species	Pre, during, ongoing	Within wetland, along transects	Quarterly
Mosquitoes	Pre, during, ongoing	Within wetland, selected locations	Weekly during critical months
Wetland development			
Flowrate	Ongoing	In, out	Continuous
Flowrate distribution	Ongoing	Within wetland	Annually
Water surface elevations	Ongoing	Within wetland	Semiannually
Marsh surface elevations	Ongoing	Within wetland	Quarterly

[a] Water quality for pre- and during construction refers to the wastewater that is to be applied to the wetland.

Adapted from Reference 18.

transformation, and movement of specific constituents within the wetland. This type of information will be needed as constructed wetlands are used to meet the restrictive inland water quality standards that are based on specific constituents. For most of the compounds such as heavy metals and refractory organic compounds, it will be mandatory to know their fate in the wetland.

Development of Database

Although it is difficult to get operating agencies to collect anything but the bare minimum operating data, every effort should be made to collect data that can also be used to develop a database on the operation of the system.

Ultimately, the database can be used to develop (1) improved process designs for new systems, (2) improved process control measures, (3) improved long-term management strategies, and (4) information that will be of use to the profession.

SUMMARY

Although the beneficial use of constructed wetlands for the treatment of wastewater is well established, the current state of knowledge concerning the dynamics and long-term operation of these systems is in its infancy. As the uncertainties now associated with the use of constructed wetlands are resolved, this technology will assume its rightful place alongside more conventional technologies for the treatment of wastewater. The challenge for the attendees of this *International Symposium on Constructed Wetlands For Water Quality Improvement* is to help develop the data and information that will allow for the development of rational design and operational parameters, based on defendable scientific principles. Specifically, the influent wastewaters must be characterized more fully, more rational design parameters must be developed and tested, and a more fundamental understanding of the transport and fate processes must be developed. In addition, improved management techniques and more effective monitoring programs must also be developed to insure the long-term use of these systems.

REFERENCES

1. Balmat, J. L. Biochemical oxidation of various particulate fractions of sewage, *Sewage and Industrial Wastes*: 29(7): 1957.
2. DeBusk, W. F. and K. R. Reedy. Wastewater treatment using floating aquatic macrophytes: contaminant removal processes and management strategies, in *Aquatic Plants for Water Treatment and Resource Recovery*. K. R. Reedy and W. H. Smith, Eds. Magnolia Publishing, Orlando, FL, 1987, 643–656.
3. Gearhardt, R. Constructed free surface wetlands to treat and receive wastewater — pilot project to full scale, presented at International Conference on Constructed Wetlands for Wastewater Treatment, Chattanooga, TN, 1988.
4. Hammer, D. A. Ed. *Constructed Wetlands for Wastewater Treatment: Municipal, Industrial and Agricultural*. Lewis Publishers, Chelsea, MI, 1989.
5. Levine, A. D., G. Tchobanoglous, and T. Asano. Characterization of the size distribution of contaminants in wastewater: treatment and reuse implications, *J. Water Pollut. Control Fed.* 57(7):205, 1985.
6. Levine, A. D., T. Asano, and G. Tchobanoglous. Particulate contaminants in wastewater: a comparison of measurement techniques and reported particle size distributions, Fluid Particle Separation *J. Am. Filtration Soc.* 4(2):89, 1991.
7. Levine, A. D., T. Asano, and G. Tchobanoglous. Size distributions of particulate contaminants in wastewater and their impact on treatability, *Water Res.* 25(8):911, 1991.
8. Linsley, R. K., J. B. Franzini, D. Fryberg, and G. Tchobanoglous. *Water Resources Engineering*. 4th ed. McGraw-Hill, New York, 1991.
9. Reed, S. C., E. J. Middlebrooks, and R. W. Crites: *Natural Systems for Waste Management and Treatment*. McGraw-Hill, New York, 1988.
10. Reed, S. C. Subsurface flow constructed wetlands systems, presented at the The Natural/Constructed Wetlands Treatment Systems Workshop, Sponsored by the U.S. Environmental Protection Agency, Region VIII and the Colorado Department of Health, Denver, CO, September 4–6, 1991.
11. Reedy, R. and W. H. Smith, Eds. *Aquatic Plants for Water Treatment and Resource Recovery*. Magnolia Publishing, Orlando, FL, 1987, 27–48.
12. Stowell, R., R. Ludwig, J. Colt, and G. Tchobanoglous. Concepts in aquatic treatment system design, *Journal EED*, Proceedings ASCE, 107(EE5):October, 1981.

13. Tchobanoglous, G. and E. D. Schroeder. *Water Quality: Characteristics, Modeling, Modifications.* Addison-Wesley, Reading, MA, 1985.
14. Tchobanoglous, G. Aquatic plant systems for wastewater treatment: engineering considerations, in *Aquatic Plants for Water Treatment and Resource Recovery.* K. R. Reedy and W. H. Smith, Eds. Magnolia Publishing, Orlando, FL, 1987, 27–48.
15. Tchobanoglous, G., F. Maitski, K. Thomson, and T. H. Chadwick. Evolution and performance of city of San Diego pilot scale aquatic wastewater treatment system using water hyacinths, *J. Water Pollut. Control Fed.* 61(11/12):1989.
16. Tchobanoglous, G. and F. L. Burton. *Wastewater Engineering: Treatment, Disposal, Reuse.* 3rd ed. McGraw-Hill, New York, 1991, 1–1350.
17. Tchobanoglous, G. Land based, constructed wetlands, and aquatic plant systems in the United States, in *Proc. Int. Conf. Ecological Engineering for Wastewater Treatment.* C. Etnier and B. Guterstam, Eds. Stensund Folk College, Sweden, March 27, 1991, Bokskogen, Gothenburg, Sweden, 1991.
18. Tchobanoglous, G., R. Gearhardt, and R. Crites. System operation and monitoring, presented at the The Natural/Constructed Wetlands Treatment Systems Workshop, Sponsored by the U.S. Environmental Protection Agency, Region VIII and the Colorado Department of Health, Denver, CO, September 4-6, 1991.
19. U.S. Environmental Protection Agency. *Design Manual for Constructed Wetlands and Floating Aquatic Plant Systems for Municipal Wastewater Treatment.* EPA 625/1-88-022, U.S. EPA, Cincinnati, OH, September 1988.
20. Water Pollution Control Federation. *Manual of Practice, Natural Systems for Wastewater Treatment.* February 1990.

CHAPTER 4

Wetlands for Wastewater Treatment: Performance Database

R. L. Knight, R. W. Ruble, R. H. Kadlec, and S. Reed

INTRODUCTION

Background

The use of wetlands for treatment of wastewaters is an emerging technology in North America and worldwide. There are at least 200 municipal, industrial, and stormwater wetland treatment systems in North America at the present time. These wetland systems have a wide variety of engineering designs, wetted areas, flow rates, inflow water qualities, plant communities, hydrologic regimes, effluent limitations, and monitoring requirements. Until recently, an engineer or regulator considering the use of wetland technology for a specific treatment need has had to search through a great deal of information to determine basic engineering guidelines for wetland areas and pretreatment levels necessary to achieve effluent criteria. Five recent handbooks[1-5] provide useful syntheses of existing knowledge concerning the design of new wetlands; however, the existing database from operational wetland treatment systems is growing so fast that these handbooks may be quickly outdated by new empirical results.

Information on the effects of wetlands on water quality and the effects of wastewaters on wetland biota have been collected from many operational wetland treatment systems. This information base is widely scattered in scientific journal articles, monitoring reports to agencies, consultant reports, and private databases. A framework to record and update this expanding knowledge base is necessary to make information available to engineers and scientists nationwide to eliminate duplication of effort and to continue to refine empirical wetland design approaches now in use.

Questions remain concerning wetland treatment system longevity, effects on biota, design innovations to enhance pollutant assimilation, and other aspects of this emerging technology. The answers to some of these questions will come from pilot and monitoring studies being conducted nationwide. A number of these questions may also be answered more cost effectively by organizing and examining existing data from operational wetland treatment systems. The basis for these answers and the identification of the gaps in our knowledge require cataloging existing information; summarizing data into a consistent, unified database; and distributing this database to the engineering, scientific, and regulatory communities.

This paper describes the first phase of the development of a database cataloging information available from wetland treatment sites throughout North America. Approximately half of the existing wetland treatment systems in North America are included in this preliminary effort. These systems include wetlands receiving municipal and industrial wastewaters and storm waters. Systems that are included treat wastewater volumes that are generally greater than 100,000 gallons per day (378 m^3/d), except for the inclusion of some smaller-scale pilot systems. Natural and constructed wetlands are included in this database. Under the constructed category, the database includes systems that are designed for free-water surface (FWS) flow as well as vegetated submerged bed (VSB) flow.

When complete, it is expected that the wetland treatment database will provide the most comprehensive summary of wetland treatment performance available. Continuing expansion of database contents and analysis of the data collected in this format will add to its widespread usefulness for the engineering, scientific, and regulatory communities. At this time, this database is incomplete and is being developed under contract with the Environmental Protection Agency's (EPA) Office of Research and Development and Office of Water. This article describes the organization of the database and presents some preliminary data summaries.

0-87371-550-0/93/$0.00+$.50

Purposes and Goals

The purpose of the wetland treatment database effort is to develop a database package that can be used to summarize existing information and that can be expanded to accommodate additional information. The primary goal of the wetlands treatment database is to provide information that will allow effective designs, cost-efficient construction, and reliable operation of new wetland treatment systems. Literature reviews are quickly out of date or are incomplete, and engineers frequently have access to a small fraction of the information available concerning the actual performance of wetlands receiving wastewaters under differing climatic, design, and operational conditions. A single, comprehensive database that is periodically updated will provide users with information concerning wetland performance and will help to reduce design errors and potential system failures.

A second goal of the wetland treatment database is to provide a research tool for scientific investigations into the subject of wetlands ecology. The completed database will provide a detailed data repository for some of the physical, chemical, and biological processes of treatment wetlands. This knowledge may help to direct new research efforts. The database is expected to be useful for calibration or verification of theoretical models (water quality and biological) related to wetland functions influenced by pollutants.

A third goal of the wetlands treatment database is to help establish standardization of the monitoring and reporting efforts in wetland treatment systems nationwide. Currently, permits require widely variable reporting requirements for wetlands receiving wastewaters, and researchers may omit key water quality parameters from monitoring or pilot programs. Examination of the operational data in the database will provide permit writers and researchers with an understanding of data gaps and an appreciation of the difficulty associated with data interpretation concerning wetlands with insufficient information. Such gaps may be acknowledged to encourage others to collect new data in those areas.

DESCRIPTION OF THE DATABASE FILES

Seven individual database files on the use of North American wetlands for water quality treatment have been developed. These files have been named according to their contents (within the 8-letter DOS limitation): sites.dbf, systems.dbf, permits.dbf, cells.dbf, operate.dbf, literat.dbf, and people.dbf. Separate database files have been developed due to the hierarchal nature of the data; for example, one site may have several systems or one system may have multiple cells. Separate database files obviate the need for repetitive (overlapping) data. In addition, data may be deposited in associated memo files. These memo files are variable length, full-screen editing files that have the same name as their associated database files, but with the filename extension "DBT". These seven memo files are updated by entering data into the comments field supplied with each database file. Table 1 presents a detailed list of the contents of each of the seven individual database files. A brief description of each database file follows.

Sites Database File

The sites database file (Sites.DBF) assigns basic information about each site, such as site name and number, state, community, and EPA region, to a site number. There are 63 fields and 546 characters in the site database file. The smallest fields are the "checkoff" fields, which contain an X to indicate that information for a particular parameter exists either in the seven database files or elsewhere, and have a width of one character. The largest fields are the manager and environmental engineer address fields (120 characters each). There are two indexes; three code fields defined by 5, 5, and 8 codes, respectively; and four reports. Indices are .NDX files, which sort a database by one or more fields. Code fields are three-character fields that contain predefined letters. Reports are .FRM files, which can be invoked from the main menu and provide listings of database contents to attached printers. Reports list data in the fields; database files with more fields or wider fields require more reports.

Systems Database Files

The systems database file (Systems.DBF) describes each system at a site. Systems are defined as wetland treatment areas that contain one or more cells and that have separate outflow monitoring stations. A single system may consist of multiple cells arranged in series.

Table 1. Field Designations in Seven Wetland Treatment Database Files

SITES.DBF	SYSTEMS.DBF	PERMITS.DBF	CELLS.DBF	OPERATE.DBF	OPERATE.BDF (Cond)	OPERATE (Cond)	LITERATE.DBF	PEOPLE.DBF
SITE_NO	SITE_NO	SITE_NO	SITE_NO	SITE_NO	MB_OGN	MA_BOD	SITE_NO	SITE_NO
SITE_NAME	SITE_NAME	SITE_NAME	SYSTEM_NO	SYSTEM_NO	MB_OGN_IN	MA_BOD_IN	SYSTEM_NO	SITE_NAME
COMMENTS	SYSTEM_NO	SYSTEM_NO	CELL_NO	CELL_NO	MB_OGN_OUT	MA_BOD_OUT	SYS_NAME	COMMENTS
COUNTRY	SYSTEM_NAM	DESGN_FLOW	COMMENTS	COMMENTS	MB_ORG_REM	MA_BOD_REM	LAST1	LAST_NAME
EPA_REGION	COMMENTS	LIMIT	HYDRL_TYPE	TIMEPERIOD	MB_ORG_EFF	MA_BOD_EFF	FIRST1	FIRST_NAME
STATE	TOT_NO_CEL	UNITS	AREA	NO_OF_DAYS	MB_TP	MA_TSS	LAST2	ORGANIZATION
COMMUNITY	ORIGIN	DURATION	VEG_TYPE	AF	MB_TP_IN	MA_TSS_IN	FIRST2	ADDRESS
AN_TEMP	HYDRL_TYPE	COMMENTS	VEG_SPC_1	AV_FLOW	MB_TP_OUT	MA_TSS_OUT	LAST3	PHONE
AN_RAINFA	AREA	CCODE	VEG_SPC_2	IF	MB_TP_REM	MA_TSS_REM	FIRST3	ROLE
SRCE_OF_WW	FLOW	TCODE	VEG_SPC_3	INFLOW	MB_TP_EFF	MA_TSS_EFF	TITLE	
NON_WWUSES	VEG_TYPE		LENGTH	OF	MB_DP	MA_TKN	CITATION	
POPULATION	START_DATE		WIDTH	OUTFLOW	MB_DP_IN	MA_TKN_IN	CIT_CONTIN	
OPER_SEASN	DOWN_DATE		DEPTH	RF	MB_DP_OUT	MA_TKN_OUT		
FORM_PRTRT	CAP_COST		SLOPE	OTHER_FLOW	MB_DP_REM	MA_TKN_REM		
NO_SYSTEMS	OM_COST		SUBSTRATE	SV	MB_DP_EFF	MA_TKN_EFF		
DESNG_FLOW	CURRENCY		VSB_GSIZE	SUPER_VELO	CN_BOD	MA_NH4		
MANAG_NAME			VSB_GTHICK	DP	CN_DOB_IN	MA_NH4_IN		
MANAG_ADDR			VSG_BSLOPE	DEPTH	CN_DOB_OUT	MA_NH4_OUT		
MANAG_PHON			VSB_TSLOPE	AR	CN_DOB_EFF	MA_NH4_REM		
ENVRG_NAME				AREA_WET	CN_TSS	MA_NH4_EFF		
ENVRG_ADDR				DT	CN_TSS_IN	MA_NO3		
ENVRG_PHON				DETEN_TIME	CN_TSS_OUT	MA_NO3_IN		
CAP_COST				HL	CN_TSS_EFF	MA_NO3_OUT		
OM_COST				HD_LD_RATE	CN_TKN	MA_NO3_REM		
CURRENCY				PR	CN_TKN_IN	MA_NO3_EFF		
BOD				PRECIPITAT	CN_TKN_OUT	MA_TN		
TSS				EP	CN_TKN_EFF	MA_TN_IN		
TKN				EVAP	CN_NH4	MA_TN_OUT		
N_NH4				GD	CN_NH4_IN	MA_TN_REM		
N_NO3				GRND_WATER	CN_NH4_OUT	MA_TN_EFF		
N_ORG				MB_BOD	CN_NH4_EFF	MA_OGN		
N_TOT				MB_BOD_IN	CN_NO3	MA_OGN_IN		
P_DSV				MB_BOD_OUT	CN_NO3_IN	MA_OGN_OUT		

Table 1. Field Designations in Seven Wetland Treatment Database Files (continued)

SITES.DBF	SYSTEMS.DBF	PERMITS.DBF	CELLS.DBF	OPERATE.DBF	OPERATE.BDF (Cond)	OPERATE (Cond)	LITERATE.DBF	PEOPLE.DBF
P_TOT				MB_BOD_REM	CN_NO3_OUT	MA_OGN_REM		
DO				MB_BOD_EFF	CN_NO3_EFF	MA_OGN_EFF		
F_COLIF				MB_TSS	CN_TN	MA_TP		
HYDROLOGY				MB_TSS_IN	CN_TN_IN	MA_TP_IN		
VEG				MB_TSS_OUT	CN_TN_OUT	MA_TP_OUT		
NOT_USED				MB_TSS_REM	CN_TN_EFF	MA_TP_REM		
SULFAT_IDE				MB_TSS_EFF	CN_OGN	MA_TP_EFF		
REDOX				MB_TKN	CN_OGN_IN	MA_DP		
TDS				MB_TKN_IN	CN_OGN_OUT	MA_DP_IN		
COND				MB_TKN_OUT	CN_OGN_EFF	MA_DP_OUT		
TURB				MB_TKN_REM	CN_TP	MA_DP_REM		
ALK				MB_TKN_EFF	CN_TP_IN	MA_DP_EFF		
PH				MB_NH4	CN_TP_OUT			
CHLRD				MB_NH4_IN	CN_TP_EFF			
METAL_COMM				MB_NH4_OUT	CN_DP			
METAL_TOXC				MB_NH4_REM	CN_DP_IN			
SYNTH_ORGA				MB_NH4_EFF	CN_DP_OUT			
PESTICIDE				MB_NO3	CN_DP_EFF			
HERBICIDE				MB_NO3_IN	CN_DP_CODE			
NON_FEC_BA				MB_NO3_OUT	CN_DO			
VIRUS				MB_NO3_REM	CN_DO_IN			
TEMP				MB_NO3_EFF	CN_DO_OUT			
VEG_BIOMAS				MB_TN	CN_DO_EFF			
VEG_CHEM				MB_TN_IN	CN_FC			
LIT_BIOMAS				MB_TN_OUT	CN_FC_IN			
LIT_CHEM				MB_TN_REM	CN_FC_OUT			
SED_BIOMAS				MB_TN_EFF	CN_FC_EFF			
SED_CHEM								
INVERTEBRT								
VERTEBRT								

Information entered into this database includes site name and number, system name and number, total number of cells, origin, hydrologic type, design area in hectares (ha), and flow in m^3/d. There are 16 fields and 156 characters in the systems database file. The smallest fields are the coded fields (width of 3 characters). The largest field is the system name field (35 characters). There are two indexes; three code fields defined by 4, 5, and 8 codes, respectively; and two reports.

Permits Database File

The permits database file (Permits.DBF) contains one record for each parameter permitted at a system. Multiple permit parameters, permit durations, or permit limits result in multiple records for the system. An extensive code system indicates the parameter being subjected to permit limitation. There are 10 fields and 106 characters in this database file. The smallest fields are the coded fields (width of three characters). The largest field is the site name field (25 characters). There are two indexes; two code fields defined by 33 and 3 codes, respectively; and one report.

Cells Database File

The cells database file (Cells.DBF) contains data for each cell in a system, including site number, system number, cell number, hydrologic type, plant species names for resident vegetation, and cell length and width. Cells are wetland areas that are clearly delineated from other treatment areas by dikes or uplands and have recognizable inlet and outlet points.

There are 19 fields and 191 characters in the cells database file. The smallest fields are the coded fields (width of three characters). The largest fields are the three vegetation species names fields (30 characters each). There are two indexes; two code fields defined by 4 and 8 codes, respectively; and two reports.

Operate Database File

The operate database file (Operate.DBF) is the largest of the seven database files, both in terms of number of records and column widths. It contains all the operational data for the cells within each system for a specified time period. Efforts were made to provide data on a seasonal basis, though data for monthly, annual, or other time periods have also been included.

Concentrations for 5-day biological oxygen demand (BOD_5), total suspended solids (TSS), total Kjeldahl nitrogen (TKN), total ammonia nitrogen (NH_3-N), nitrite+ nitrate nitrogen (NO_2+NO_3-N), total nitrogen (TN), organic nitrogen, total phosphorus (TP), dissolved phosphorus, dissolved oxygen (DO), and fecal coliforms are entered in this database file. Hydraulic loading rates (cm/d) and mass balances (kg/ha/d) are automatically computed. All concentrations except fecal coliforms are provided in units of milligrams per liter (mg/L). Fecal coliform data are entered as number of colonies per 100 milliliters.

There are 165 fields and 975 characters in the operate database file. The smallest fields are one-character code fields. The largest fields are flow, area, retention time, and hydraulic loading rate fields (12 characters each). There are three indexes; two code fields defined by 6 and 5 codes, respectively; and nine reports.

Literature Database File

The literature database file (Literat.DBF) contains selected references to printed documents for systems included in the seven database files. Up to three authors may be entered, allowing for their selective retrieval. There are 12 fields and 908 characters in the literature database file. The smallest field is the site number field (width of seven characters). The largest fields are the title and two citation fields (250 characters each). The database contains four indexes and one report.

People Database File

The people database file (People.DBF) is the smallest of the seven database files in terms of numbers of fields available. One name per record is entered. A coded field associates the recorded individual with his or her involvement with the wetland treatment system (e.g., researcher, engineer, designer, or operator). There are 9 fields and 295 characters in the people database file. The smallest field is the role field (width of one character). The largest field is the address field (120 characters). There are three indexes, one code field defined by seven codes, and one report.

DESCRIPTION OF THE SOFTWARE

Software Selection

Software selection was based on the ability to run in a DOS environment on an IBM compatible microcomputer, in order to provide as much access to the data as possible. It was also decided that spreadsheet programs are insufficient for large amounts of data, and that software for large database development should be employed. The software must be able to accept, edit, sort, search, display, and print large volumes of diverse information with relative ease.

Additional considerations for the selection of software include the cost of the license, quality of documentation, ease of product use, reporting capabilities, Structured Query Language support, third-party acceptance (i.e., "add-on" software availability), memory requirements, search and sorting performance, stability of native code, run-time distribution potential, and "network-ability".

The dBASE IV software was selected because it meets the requirements above, and because it is the most widely used microcomputer database package (3 million installations worldwide). In addition, the database file format (DBF) is the industry standard; thus, direct conversion and use of the data by many word processors, spreadsheet packages, and other database packages is possible. Therefore, it was decided that more users could access the database with dBase IV than with any other microcomputer software package.

dBASE IV v1.1 Features

dBASE IV is essentially a relational database compiler. This means that the software allows the user to define database structures, store vast amounts of information in a computer's memory in user-defined sequences, link (relate) database files, and process sequential commands provided by the user.

A database file may contain up to 1 billion records, with a maximum of 255 fields per record. The user may have up to 10 database files open at the same time (important for linkage considerations), with 47 indexes per file. An index is a separate file (.NDX) that serves as a pointer to allow sorting data without having to create a separate database. When the database is updated, all open indexes are updated simultaneously.

The sequential commands used by the compiler to interact with the database files can be provided in one of three increasingly complex and powerful ways: (1) the control center (known as the Assistant in dBASE III+), (2) the "dot prompt," and (3) variable length programs written by the user or commercially available elsewhere.

The control center is a pull-down menu environment that allows nonprogrammers to begin accessing database files right away. Novices may choose to explore lessons in the supplied electronic tutorial. Help facilities are on-line and may be accessed by a single key stroke.

To more fully interact with the database files, the user can leave the control center and work directly with the data from the dot prompt. This is essentially a one-line programming interface with which the user can execute many of the 440 database commands and functions presently supported.

There are additional programming possibilities using dBase IV commands and functions in simple sequential line statements. These statements are in text format and can be edited by the supplied dBASE word processor (called from the dot prompt) or by other word processors. The dBASE IV Developer's Edition contains a built-in applications generator for faster program construction and a runtime module for royalty-free distribution of completed applications. Programs may be compiled for faster execution. The runtime modules needed to run the programs are considerably cheaper than a full dBASE package; thus, they are potentially usable by more people. However, programs cannot be modified later or recompiled without the full dBASE package.

Future versions of dBASE are expected to continue the past trend of developing an increasingly sophisticated control center so that novices can access data more easily and more powerfully. Complete instructions for dBASE installation and operation are provided in the manufacturer's supplied manuals and are not discussed further here.

SUMMARY OF THE DATABASE CONTENTS

Quality Control

Data quality is an important consideration when preparing a database from numerous sources of information. Quality control efforts between operators and researchers at wetland treatment systems are highly variable. Many smaller utilities have no quality control other than the integrity and care of the operators collecting samples and running analyses with relatively unsophisticated analytical procedures. Other utilities or industries are well funded and utilize all aspects of credible quality control procedures for instrument calibration, duplicates, blanks, and data validation. Data were obtained by the authors through personal collection, telephone contacts, returned survey forms, electronic files, and printed reports. The data included in the wetland treatment database are from sources with highly variable quality assurance procedures; therefore, data quality is variable within the wetland databases.

Sites and Systems

The existing wetland treatment database includes 96 sites comprised of 127 separate systems. The higher number of systems reflects the presence of multiple systems (separate inflow and outflow points) at a number of sites. Table 2 presents a list of sites and systems in the existing database. This list includes 89 municipal systems, 22 industrial systems, 5 storm water systems, and 11 systems that are not easily categorized; 33% of these systems are natural wetlands and 66% are constructed, while 69% are of the FWS type and the rest are VSB or hybrid (HYB) systems.

Treatment wetlands within the database occur in all 10 EPA regions, with Region 4 having the largest number of sites and systems. VSB systems are most common in the southern U.S., in EPA Regions 4 and 6. Natural wetlands are most frequently found in EPA Regions 4 and 5.

Design Considerations

The existing treatment wetland database includes design information for 127 systems, with a total of 215 individual treatment cells. Multiple cells in a system may function in series or in parallel; 71% of the cells in the existing database are FWS and the remainder are VSB or hybrid systems. Cell areas range from 0.02 hectares (ha) to 1093 ha. Length-to-width ratios for FWS and VSB wetlands vary from about 0.2 to 83.

The most common vegetation species planted in treatment wetlands in North America are cattail (*Typha* spp.) and bulrush (*Scirpus* spp.). Other frequently planted plant species are pickerelweed (*Pontederia* spp.), duck potato (*Sagittaria* spp.), duckweed (*Lemna* spp.), and miscellaneous sedges and grasses. The most commonly utilized natural wetland treatment systems are dominated by cypress (*Taxodium* spp.), red maple (*Acer rubrum*), willow (*Salix* spp.), black gum (*Nyssa biflora*), and spruce (*Abies* spp.).

Design depth varies from 15 to 610 cm in FWS wetlands, with an average depth of 61 cm. Bed depth in VSB wetlands ranges from 25.4 to 100 cm, with an average value of 52 cm.

Bottom slopes in FWS wetlands vary from 0 to 15% with an average of about 0.23%. Bottom slopes in VSB wetlands vary from 0 to 1%, with an average of about 0.10%.

Operational Data

The Operate.DBF file is the largest individual file in the database, with 727 records. Most records include two or more water quality values for each of the water chemistry fields in the database. Table 3 presents a summary of overall average flow rates and inflow/outflow constituent concentrations from 84 systems. These summary data can be used as a convenient reference for the typical ranges for these systems, but should not be used for a quantitative analysis of wetland operational data. The averages in Table 3 are based on a variety of reporting and averaging periods, and the detailed records in the database present data at the level of seasonal averages whenever available. Seasonal averages consist of daily or monthly values averaged over the seasonal time periods of winter (December through February), spring (March through May), summer (June through August), and fall (September through November).

Operational data are reported for 69 FWS wetland systems and 15 VSB or HYB systems. The following average outflow concentrations were observed for these systems: BOD_5, 10.5 mg/L; TSS, 15.3 mg/L; NH_3-

Table 2. Listing of Wetland Sites and Systems in the Database

Site name	System name	Country	EPA Region	State	Community	WW Source[a]	Origin[b]	Hydrol type[c]	System area (ha)	Design flow (m^3/d)
Albany, LA	Albany, LA	USA	4	LA	Albany, LA	MUN	CON	HYB	0.11	132
Andrews	Black River Swamp	USA	4	SC	Andrews	MUN	NAT	FWS	185	7193
Apalachicola	Huckleberry (Whortleberry) Swamp	USA	4	FL	Apalachicola	MUN	NAT	FWS	63.7	3785
Armstrong Slough	Armstrong Slough	USA	4	FL	South Florida	STO	NAT	FWS	12.1	41880
Bellaire	Bellaire 1	USA	5	MI	Bellaire	MUN	NAT	FWS	18	600
Bellaire	Ballaire 2	USA	5	MI	Bellaire	MUN	NAT	FWS	20	600
Bellaire	Bellaire 3	USA	5	MI	Bellaire	MUN	NAT	FWS	38	600
Benton	Benton — Cattails	USA	4	KY	Benton	MUN	CON	FWS	1.5	1400
Benton	Benton — Woolgrass	USA	4	KY	Benton	MUN	CON	FWS	1.5	1400
Benton, KY	Benton, KY	USA	4	KY	Benton, KY	MUN	CON	VSB	1.46	341
Benton, LA	Benton, LA	USA	6	LA	Benton, LA	MUN	CON	VSB	0.48	1173
Biwabik, MN	Biwabik	USA	5	MN	Biwabik	MUN	NAT	FWS	40.5	1060
Black Diamond, WA	Black Diamond	USA	10	WA	Black Diamond	MUN	NAT	FWS		
Bradley, AR	Bradley, AR	USA	6	AK	Bradley, AR	MUN	CON	VSB	0.29	567
Brillion	Brillion Marsh	USA	5	WI	Brillion	MUN	NAT	FWS	156	5400
Cannon Beach, OR	Cannon Beach	USA	10	OR	Cannon Beach, OR	MUN	NAT	FWS	7	1174
Cargill (Frank Lake)	Frank Lake	Canada			High River, Alberta	IND	NAT	FWS	1093	
Carville, LA	Carville	USA	6	LA	Carville, LA	MUN	CON	VSB	0.260	568
Central	Central Slough	USA	4	SC	Central	MUN	NAT	FWS	31.6	4543
Clermont	Clermont Plot H	USA	4	FL	Clermont	MUN	NAT	FWS	0.2	27.4
Clermont	Clermont Plot L	USA	4	FL	Clermont	MUN	NAT	FWS	0.2	4.3
Clermont	Clermont Plot M	USA	4	FL	Clermont	MUN	NAT	FWS	0.2	10.6
Cobalt, Ontario	Cobalt Plot 1	Canada			Cobalt, Ontario	MUN	CON	FWS	0.1	17
Cottonwood, AL	Cottonwood, AL	USA	4	AL	Cottonwood, AL	MUN	CON	HYB	0.40	587
Crowley, LA	Crowley, LA	USA	6	LA	Crowley, LA	MUN	CON	HYB	17.00	13248
Cypress Domes	Cypress Domes	USA	4	FL	Gainesville	MUN	NAT	FWS	1.05	114
Deer Park	Deer Park	USA	4	FL	New Port Richey	MUN	NAT	FWS	50.6	3218
Degussa, Corp.	Degussa Corp., Theodore, AL	USA	4	AL	Theodore, AL	IND	CON	HYB	0.89	2040
Denham Springs, LA	Denham Springs	USA	6	LA	Denham Springs, LA	MUN	CON	VSB	6.15	11355
Des Plaines	Des Plaines 3	USA	5	IL	Wadsworth	OTH	CON	FWS	1.87	914
Des Plaines	Des Plaines 4	USA	5	IL	Wadsworth	OTH	CON	FWS	2.33	490

Des Plaines	Des Plaines 5	USA	5	IL	Wadsworth	OTH	CON	FWS	2.34	350
Des Plaines	Des Plaines 6	USA	5	IL	Wadsworth	OTH	CON	FWS	3.45	766
Dessau Mobile Home Park	Pflugerville, TX	USA	6	TX	Pflugerville, TX	MUN	CON	VSB	0.18	567
Doyline, LA	Doyline, LA	USA	6	LA	Doyline, LA	MUN	CON	VSB	0.28	416
Drummond	Drummond Bog	USA	5	WI	Drummond	MUN	NAT	FWS	6	9000
Fontanges, Quebec	Fontanges	Canada			Fontanges, Quebec	OTH	NAT	FWS		
Foothills Village, TN	Foothills Village, TN	USA	4	TN	Loudon, Co., TN	MUN	CON	VSB	0.10	67
Fort Deposit	Fort Deposit	USA	4	AL	Fort Deposit	MUN	CON	FWS	6	900
Great Meadows	Great Meadows National Wildlife Ref	USA	1	MA	Concord	MUN	NAT	FWS	56	2000
Gustine	Constructed Wetland at Gustine	USA	9	CA	Gustine	MUN	CON	FWS	9.6	3785
Hamilton Marshes	Tidal wetland	USA	2	NJ	Hamilton Township	MUN	NAT	FWS	500	
Hammond, LA	Hammond	USA	6	LA	Hammond, LA	OTH	CON	VSB	0.126	329
Hardin, KY	Hardin, KY	USA		KY	Hardin, KY	MUN	CON	VSB	0.64	378
Haughton, LA	Haughton, LA	USA	4	LA	Haughton,LA	MUN	CON	VSB	0.62	1324
Hay River	Hay River	Canada			Hay River, NW Territories	MUN	NAT	FWS	47	1000
Hidden Lake	Hidden Lake S.W.	USA	4	FL	Orlando	STO	NAT	FWS	3	
Hillsboro	Jackson Bottoms	USA	10	OR	Hillsboro	IND	CON	FWS	35.70	
Hilton Head Plantation	Whooping Crane	USA	4	SC	Hilton Head Plantation	MUN	NAT	FWS	36.5	1893
Hornbeck, LA	Hornbeck, LA	USA	6	LA	Hornbeck, LA	MUN	CON	VSB	0.09	231
Houghton Lake	full scale	USA	5	MI	Houghton Lake	MUN	NAT	FWS	75	10000
Houghton Lake	pilot	USA	5	MI	Houghton Lake	MUN	NAT	FWS	4	360
Hurtsboro	Hurtsboro	USA	4	AL	Hurtsboro	MUN	NAT	FWS	0.16	56
Incline Village	Incline Village	USA	9	NV	Incline Village	MUN	CON	FWS	150	5000
Ironbridge	Ironbridge	USA	4	FL	Orlando	MUN	NAT	FWS	494	75720
Island Lake	Island Lake	USA	4	FL	Longwood, FL	OTH	NAT	FWS	42	
Jasper	Basin Swamp	USA	4	FL	Jasper	MUN	NAT	FWS	24	
Johnson City, TX	Johnson City, TX	USA	6	TX	Johnson City, TX	MUN	CON	FWS	0.50	114
Johnson City, TX	Johnson City, TX	USA	6	TX	Johnson City, TX	MUN	CON	VSB	0.11	114
Kingston Power Plant, TN	Kingston Power Plant, Kingston, TN	USA	4	TN	Kingston, TN	MUN	CON	VSB	0.26	76
Kinross (Kincheloe)	Kinross	USA	5	MI	Kinross, MI	OTH	NAT	FWS	110	450
Lake Apopka	Wetlands Flowway	USA	4	FL	Apopka	OTH	NAT	FWS	750	733536
Lake Jackson	Lake Jackson	USA	4	FL	Tallahassee	STO	CON	FWS	2.31	0–86400
Lakeland	Lakeland	USA	4	FL	Lakeland	MUN	CON	FWS	498	52704

Table 2. Listing of Wetland Sites and Systems in the Database (continued)

Site name	System name	Country	EPA Region	State	Community	WW Source[a]	Origin[b]	Hydrol type[c]	System area (ha)	Design flow (m^3/d)
Leaf River	Leaf River Pond 1	USA	4	MS	New Augusta	IND	CON	FWS	0.13	224.83
Leaf River	Leaf River Pond 2	USA	4	MS	New Augusta	IND	CON	FWS	0.13	253.60
Leaf River	Leaf River Pond 3	USA	4	MS	New Augusta	IND	CON	FWS	0.13	220.29
Listowel	Listowel 1	Canada			Listowel, Ontario	IND	CON	FWS	0.0909	17
Listowel	Listowel 2	Canada			Listowel, Ontario	IND	CON	FWS	0.0941	17
Listowel	Listowel 3	Canada			Listowel, Ontario	IND	CON	FWS	0.1324	17
Listowel	Listowel 4	Canada			Listowel, Ontario	IND	CON	FWS	0.1324	17
Listowel	Listowel 5	Canada			Listowel, Ontario	IND	CON	FWS	0.344	17
Mandan (Amoco)	Mandan	USA	8	ND	Mandan, ND	IND	CON	FWS	16.6	
Mandeville, LA	Mandeville	USA	6	LA	Mandeville, LA	MUN	CON	VSB	2.61	5678
Mayo Peninsula, MD	Ann Arundal Co., MD	USA	3	MD	Ann Arundel Co., MD	MUN	CON	VSB	1.53	2990
Mays Chapel	Mays Chapel	USA	3	MD	Cockeysville, MD	STO	CON	FWS	0.24	160.4
Monterey, VA	Monterey	USA	3	VA	Monterey, VA	MUN	CON	VSB	0.023	76
Moodna Basin (Harriman)	Moodna Basin 1	USA	2	NY	Harriman, NY	MUN	CON	FWS	0.152	57
Moodna Basin (Harriman)	Moodna Basin 2	USA	2	NY	Harriman, NY	MUN	CON	FWS	0.152	57
Mt. View Sanitary	Mount View Marsh	USA	9	CA	Martinez	MUN	CON	HYB	8.5	
Mt. View Sanitary	Mount View Test Marsh/ Forest	USA	9	CA	Martinez	MUN	CON	HYB	0.05	
Norwalk	City of Norwalk	USA	5	IA	Norwalk	MUN	CON	FWS	6.1	3600
Orange County	Eastern Service Area	USA	4	FL	Orlando	MUN	HYB	FWS	89	13251
Paris Landing, TN	Paris Landing State Park, TN	USA	4	TN	Paris Landing State Park	MUN	CON	VSB	0.15	284
Pelahatchie, MS	Pelahatchie, MS	USA	4	MS	Pelahatchie, MS	OTH	CON	HYB	2.63	2157
Pembroke, KY	Pembroke A	USA	4	KY	Pembroke	MUN	CON	HYB	0.93	
Pembroke, KY	Pembroke B	USA	4	KY	Pembroke	MUN	CON	HYB	0.93	340
Pembroke, KY	Pembroke, KY	USA	4	KY	Pembroke, KY	MUN	CON	HYB	0.54	340
Phillips High School, AL	Phillips High School	USA	4	AL	Bear Creek, AL	MUN	CON	VSB	0.203	76
Poinciana	Boot	USA	4	FL	Poinciana	MUN	NAT	FWS	46.6	1325
Pottsburg	Pottsburg Creek Swamp	USA	4	FL	Jacksonville	MUN	NAT	FWS	100	14040
Provencal, LA	Provencal, LA	USA	6	LA	Provencal, LA	MUN	CON	VSB	0.14	344
Reedy Creek	Reedy Creek OFWTS	USA	4	FL	Lake Buena Vista	MUN	NAT	FWS	5.9	3786
Reedy Creek	Reedy Creek WTS1	USA	4	FL	Lake Buena Vista	MUN	NAT	FWS	35	16280

Reedy Creek	Reedy Creek WTS2	USA	4	FL	Lake Buena Vista	MUN	NAT	FWS	41.3	
Richmond, CA	Richmond	USA	9	CA	Richmond, CA	IND	CON	FWS	36	16000
Roswell, NM	Roswell Correctional Center	USA	6	NM	Roswell Corr. Ctr.	MUN	CON	VSB	0.004	15
Santa Rosa	Kelly Farm	USA	9	CA	Santa Rosa	MUN	CON	FWS	4.05	7570
Sea Pines	Boggy Gut	USA	4	SC	Sea Pines	MUN	NAT	FWS	20	3786
Seneca Army Depot	Seneca	USA	2	NY	Seneca Army Depot	MUN	OTH	FWS	2.5	950
Shelbyville, MO	Shelbyville	USA	7	MO	Shelbyville, MO	MUN	CON	HYB	0.162	280
Shelbyville, MO	Shelbyville, MO	USA	7	MO	Shelbyville, MO	MUN	CON	VSB	0.041	280
Sibley, LA	Sibley, LA	USA	6	LA	Sibley, LA	MUN	CON	VSB	0.21	492
Silver Springs Shores	Lake Coral	USA	4	FL	Silver Springs Shores	MUN	CON	FWS	21	3786
Smackover, AR	Smackover, AR	USA	6	AR	Smackover, AR	MUN	CON	VSB	3.51	1892
St. Joseph (MN)	St. Joseph	USA	5	MN	St. Joseph, MN	STO	NAT	FWS	18.6	900
Terry, MS	Terry, MS	USA	4	MS	Terry, MS	MUN	CON	HYB	0.52	378
University of Florida	Lake Alice	USA	4	FL	Gainesville	MUN	NAT	FWS	33	7500
USDA/NSCS	Nutrient Sediment Control System	USA	1	ME	Orono, ME	OTH	CON	FWS		
Utica, North, MS	Utica, MS	USA	4	MS	Utica, MS	MUN	CON	VSB	0.73	341
Utica, South, MS	Utica, MS	USA	4	MS	Utica, MS	MUN	CON	VSB	0.92	442
Vereen	Bear Bay	USA	4	SC	Little River	MUN	NAT	FWS	69	3786
Vermontville, MI	Vermontville	USA	5	MI	Vermontville, MI	MUN	CON	FWS	4.6	630
Waldo	Waldo	USA	4	FL	Waldo	MUN	NAT	FWS	2.6	226
West Jackson County	WJC System 1	USA	4	MS	Ocean Springs	MUN	CON	FWS	8.9	2272
West Jackson County	WJC System 2	USA	4	MS	Ocean Springs	MUN	CON	FWS	13.8	3786
West Jackson County	WJC System 3	USA	4	MS	Ocean Springs	MUN	CON	FWS	5.06	
Weyerhauser	Weyerhauser	USA	4	MS	Columbus	IND	CON	FWS	101.26	75720
Weyerhauser	Weyerhauser No 1	USA	4	MS	Columbus	IND	CON	FWS	12.08	
Weyerhauser	Weyerhauser No 2	USA	4	MS	Columbus	IND	CON	FWS	14.18	
Weyerhauser	Weyerhauser No 3	USA	4	MS	Columbus	IND	CON	FWS	11.05	
Weyerhauser	Weyerhauser No 4	USA	4	MS	Columbus	IND	CON	FWS	12.28	
Weyerhauser	Weyerhauser No 5	USA	4	MS	Columbus	IND	CON	FWS	12.28	
Weyerhauser	Weyerhauser No 6	USA	4	MS	Columbus	IND	CON	FWS	13.77	
Weyerhauser	Weyerhauser No 7	USA	4	MS	Columbus	IND	CON	FWS	12.72	
Weyerhauser	Weyerhauser No 8	USA	4	MS	Columbus	IND	CON	FWS	12.90	
Wildwood	Wildwood	USA	4	FL	Wildwood	MUN	NAT	FWS	204	3786

[a] WW Source = Source of Waste Water; MUN — Municipal; STO — Stormwater; IND — Industrial; OTH — Other; UNK — Unknown

[b] Orgin = Origin of Wetland; CON — Constructed; NAT — Natural; OTH — Other; UNK — Unknown

[c] Hyrol Type = Hydrologic Type; VSB — Veg. Subm. Bed; FWS — Free Water System; HYB — Hybrid; OTH — Other; UNK — Unknown

Table 3. Summary of Wetland Operational Data

System name	Type	Record (years)	Area (ha)	Flow (m^3/d)	BOD_5 (mg/l)		TSS (mg/l)		NH_3-N (mg/l)		TN (mg/l)		TP (mg/l)	
					IN	OUT	IN	OUT	IN	OUT	IN	OUT	IN	OUT
Lakeland	FWS	2	498	26978	3	2.5	4	3.5	0.9	0.36	9.04	1.83	9.46	4.07
Boggy Gut	FWS	2.6	20	2508	6.3	3	10.7	3	4.05	1.96	11.2	3.53	4.26	3.35
Clermont Plot L	FWS	1	0.2	4.44					1.67	0.15	7.54	2.09		
Clermont Plot M	FWS	1	0.2	11.05					1.67	0.25	7.54	2.29		
Clermont Plot H	FWS	1	0.2	27.4					1.67	0.26	7.54	2.16		
Pottsburg Creek	FWS	2	100	14040					1.03	0.29	5.81	2.43	3.03	1.7
Eastern Service Area	FWS	0.8	88.7	3407	1.5	1.26	0.5	6.2			3.41	1.68	0.5	0.05
Armstrong Slough	FWS	3	12.1								1.88	1.74	0.17	0.13
Cypress Domes	FWS	2.8	1.05	37.5					3.7	0.28	9.6	1.04	8.5	0.49
Basin Swamp	FWS	2	24.3	1810	21.1		19.9							
Reedy Creek WTS1	FWS	11.2	35.24	12058	5.3	1.9	8.9	2.4	2.98	0.72	8.55	1.87	1.4	1.78
Reedy Creek OFWTS	FWS	11.2	5.87	2423	5.8	1.6	10.9	2.4	3.29	0.12	9.77	1.17	1.8	0.79
Lake Coral	FWS	1	21	1363	11.6	2.6	6	1.5			20	1.6	6.2	5.2
Central Slough	FWS	4	32	5372	16.3	6.5	27.7	14.8	7.49	1.38	16.4	3.7	4.09	1.46
Ironbridge	FWS	1	486	34254	4.8	2.1	10.5	3.2			3.99	0.94	0.53	0.11
Hidden Lake SW	FWS	0.75	3		2.7	3	6.4	13	0.053	0.05	0.58	0.66	0.07	0.16
Hurtsboro	FWS	0.75	0.8	41	37.1	14.6			0.64	2.15	34.4	21.5		
Whooping Crane	FWS	1	36.5	1862	4.5	3	12		3.02	0.12				
Fort Deposit	FWS	0.67	6	628	29.9	5.4	78.7	10.4	13.59	1				
West Jackson Co.	FWS	0.5	8.91	1953	21.6	10.5	65.9	24.5	2.69	0.19	8.91	2.48	5.13	3.56
Boot	FWS	3.7	46.5	891	2.4	4.9					7.67	3.77	4.2	0.66
Bear Bay	FWS	3.25	68.9	733	13.4	1.6	14.9	3.8	1.44	0.28	18.5	2.2	4.2	0.2
Black River	FWS	0.15	185		26	7			7.7	0.4				
Jackson Bottoms	FWS	0.21	6.31	1719	5.1	3.1	6.1	6.6	9.9	3.1			7.31	4.3
Leaf River Pond 1	FWS	1.2	0.13	225	15.8	14	54.8	30.1	9.91	7.21	19.3	12.9	8.91	8.16
Leaf River Pond 2	FWS	1.2	0.13	254	15.8	15.7	54.8	34.9	9.91	6.33	19.3	12.4	8.91	8.16
Leaf River Pond 3	FWS	1.2	0.13	220	15.8	13.9	54.8	25.5	9.91	6.79	19.3	12	8.91	5.9
Kelly Farm	FWS	0.25	0.81	1363					8.4	0.1				
Waldo	FWS	1	2.6										1.4	1.3
Wildwood	FWS	1.75	204	946									0.4	0.1
Deer Park	FWS	1	50.6	388	13		1		2.82		3.4			
Bellaire	FWS	11	18	572					7.94	0.92			2.03	0.18
Cannon Beach	FWS	0.75	7	872	21.7	10	30	10						

Cobalt	FWS	1	0.1	49	20.7	4.6	36.2	28	2.95	1.04			1.68	0.77
Des Plaines 3	FWS	1	2.12	1185			19.4	4.6						
Des Plaines 4	FWS	1	2.17	338			19.4	6						
Des Plaines 5	FWS	1	1.76	886			19.4	4						
Des Plainers 6	FWS	1	2.9	484			19.4	5.5						
Fontanges	FWS	0.5		280	86.2	2.5			13.6	0.3	20.8	1.7	4.2	0.38
Houghton Lake	FWS	15	100	3374	14.1		17.2		4.42	0.3	3.32	0.09		
Kinross	FWS	15	110	1350	28	10	36	2	10	0.24				
Mandan	FWS		16.6	2596										
"Pembroke ,KY"	FWS	0.75	1.48	188	67.4	9.4	91.9	8.2	13.8	3.35			6.03	3.16
Shelbyville	FWS	0.5	0.16	250	73.6	31.8			6.84	2.87				
Apalachicola	FWS	3	63.7	3936	15.2	1.1	107	8	3.62	0.09	14.4	0.17	3	0.21
Benton - cattail	FWS	1.7	1.5	374	25.6	9.6	57.4	10.6	7.68	7.88			4.54	4.22
Benton - woolgrass	FWS	1.7	1.5	374	25.6	12.2	57.4	15.5	7.68	6.43			4.54	3.98
Brillion Marsh	FWS	1.2	156	16305	27.9	6.4	153	50.7					3.28	2.01
Drummond	FWS	6	6	265					0.37	0.32			2.88	0.43
Great Meadows	FWS	0.8	22	46.5	42.8	6			8.79	2.04			2.14	0.56
Gustine 1A	FWS	1	0.39	163	130	49.8	73.5	39.6	17.04	16.1				
Gustine 1B	FWS	1	0.39	82	130	26.8	81.2	23	16.31	17.9				
Gustine 1C	FWS	1	0.39	41	145	24.2	88.1	56.6	18.45	20.4				
Gustine 1D	FWS	1	0.39	164	141	30.5	98.2	20.2	19.65	22.9				
Gustine 2A	FWS	1	0.39	174	151	44.8	99.8	33.8	18	23.2				
Hamilton	FWS	2	100	22.5					1.08	0.82	27.5	20	4.32	4.13
Modna Basin	FWS	2	0.15	57	52.8	17.7	33.9	12.1	20.4	11.4				
Hay River	FWS	0.5	47	1000	118	2.7	180	5.8	10.3	0.39			11	0.26
Incline Village	FWS	3.7		2157	15.7				14.4				22	
Listowel #1	FWS	4	0.34	17	19.6	8	22.8	8.6	7.15	4.85			1.04	0.67
Listowel #2	FWS	4	0.09	17	19.6	11.3	22.8	9	7.15	5.1			1.04	0.76
Listowel #3	FWS	4	0.13	17	19.6	7.6	22.8	9.2	7.15	3.78			1.04	0.5
Listowel #4	FWS	4	0.13	17	56.4	9.6	111	8	8.58	6.12			3.18	0.62
Listowel #5	FWS	4	0.094	17	56.4	14.6	111	11.6	8.58	7.92			3.18	0.99
Mt. View Marsh	FWS	1	8.5		34	30.2	27	82.5	12.12	5.25				
Norwalk	FWS	0.1	6.1	1	229	9.5	232	33						
Seneca Army Depot	FWS	3	2.5	239	15.1	5	10.1	14.3	3.19	1.3				
Mays Chapel	FWS	1	0.24	160			85.4	33.9	0.23	0.07			0.33	0.19
Island Lake	FWS	1	42						0.23	0.01			0.23	0.03
Phillips High School	VSB	2	0.2	58.7	15.3	1	63.7	2	11	1.7	51	6.7	6	0.3
Denham Springs	VSB	1.5	2.05	2548	28.2	10.5	53	17		4.33				

Table 3. Summary of Wetland Operational Data (continued)

System name	Type	Record (years)	Area (ha)	Flow (m^3/d)	BOD_5 (mg/l)		TSS (mg/l)		NH_3-N (mg/l)		TN (mg/l)		TP (mg/l)	
					IN	OUT	IN	OUT	IN	OUT	IN	OUT	IN	OUT
Mandeville	VSB	1	2.61	2839		10		9		0.95				
Carville	VSB	3.2	0.26	318		8.4		3.7		2.51				
Hammond	VSB	0.7	0.13	378		1		1		0.03				
Monterey	VSB	1.1	0.023	83	38	15	32	7	9.33	8.67				
Degussa Corp.	HYB	3.1	0.89	1249	5.1	4.2	25.5	4.5	5.19	3.02				
Kingston	VSB	0.7	0.93	76	56	9	83	3	22	16			3.4	2.1
Ann Arundal Co.	HYB	0.8	0.385	464	13.3	9	9.6	3.2	9.69	6.48	34.2	16.2	6.01	4.31
Albany	HYB	1.1	0.06	21.9		21.4		29.2						
"Benton,La"	VSB	2.8	0.48		16.8	6.3	38.1	11.3	5.28	2.74				
Haughton	VSB	1.7	0.62		14.3	4.9	34.6	6.1	1.1	3				
Cottonwood	HYB	1.1	0.11			13.1		51.8		6.3				
Hornbeck	VSB	0.7	0.1	138		7.9		5						
Paris Landing	VSB	1.3	0.15	170		3.6		6.2						
Average		2.3	32.6	2159.1	38.8	10.5	49.1	15.3	7.5	4.2	14.0	5.0	4.2	1.9
Minimum		0.1	0.023	1	1.5	1	0.5	1	0.053	0.01	0.58	0.09	0.07	0.03
Maximum		15	498	34254	229	49.8	232	82.5	22	23.2	51	21.5	22	8.16
N		83	83	75	58	61	56	59	60	63	29	28	44	43

N, 4.2 mg/L; TN, 5.0 mg/L; and TP, 1.9 mg/L. The average concentration removal efficiencies (in percent) were as follows: BOD_5, 73; TSS, 69; NH_3-N, 44; TN, 64; and TP, 55.

Permits

Permit information is available for 47 systems in the existing treatment wetland database. Table 4 summarizes the permitted flow and water quality criteria at the wetland outflow for these systems. The permit limits summarized in Table 4 are the lowest values for each constituent; additional higher values may be in effect at some wetland systems on a shorter averaging basis or for a different season.

The most commonly permitted parameters for these wetland treatment systems are BOD_5 and TSS. Permit limits for BOD_5 are typically between 5 and 30 mg/L; and for TSS, between 10 and 30 mg/L. Permit limits for NH_3-N are included for 55% of the systems listed in Table 4. These limits range from 1 to 8 mg/L for municipal systems and up to 50 mg/L for an industrial facility. Limits for DO, pH, and fecal coliforms are also frequently included. Permit limits for TN and TP are relatively uncommon, except in wetland systems located in Florida.

People and Literature

The existing treatment wetland database includes 273 citations to scientific journal articles, system design and data reports, and other documents related to the wetland systems included in the database. These literature citations, listed and sorted by site number, should be consulted for more detailed information concerning each system in the database.

In many cases, no published information is available for operating wetland systems. In these cases, the best source of additional information is available from the operator or system manager, a researcher who may be working with the system, or the system engineer who knows the design considerations and may be involved in performance assessment. The existing database contains names and addresses for these key contacts.

EXAMPLE APPLICATIONS DERIVED FROM THE DATABASE

The wetland treatment database includes information concerning wetland design and performance. This information can be downloaded from the dBASE files to other software for individual database manipulations and numerical analyses. Also, information in the seven individual database files can be linked to provide an examination of relationships between files. A few examples of the types of analyses that can be conducted by an investigator are presented below. Due to the potential variability of data quality and validation during the preliminary effort, results of these example applications must be cautiously interpreted.

Assimilation of Biochemical Oxygen Demand

BOD_5 is one of the most frequently measured parameters in wastewater treatment plants and their discharges. This parameter integrates the processes of organic and chemical oxidation, which occur in a water sample containing solid and dissolved pollutants. BOD_5 is typically about 150 to 300 mg/L in raw municipal wastewaters and about 20 to 30 mg/L following conventional secondary treatment. Wetland treatment systems commonly receive inflow BOD_5 concentrations of 10 to 100 mg/L, depending on the degree of pretreatment.

Operational wetland data in the database indicate that the average wetland outflow BOD_5 concentration is about 10.5 mg/L, with values ranging from 1 to 50 mg/L. Since wetlands are frequently designed for BOD_5 removal, an important design question is often the amount of wetland area required to consistently achieve a specified BOD_5 discharge criterion. The permit database indicates that the most common BOD_5 outflow criterion is 10 mg/L. Process design consists of predicting the necessary wetland BOD_5 inflow concentration and wetland size required to meet a 10-mg/L BOD_5 outflow limit.

Figure 1 illustrates the relationship between the mass loading of BOD_5 in a wetland and the mass of BOD_5 removed. Typical BOD_5 mass removal efficiencies are near 70% or more at mass loading rates of up to 280 kg/ha/day. Lower removal efficiencies can occur, especially when mass loadings are less than 50 kg/ha/day. A linear regression was conducted to examine the predictability of BOD_5 outflow concentration as a function of BOD_5 inflow concentration and hydraulic loading rate. Based on 324 complete data records for these three

Table 4. Summary of Wetland System Outflow Permit Limits

System name	Type[a]	Flow (m^3/d)	DO (mg/l)	pH (s.u.)	BOD_5 (mg/l)	TSS (mg/l)	NH_3-N (mg/l)	TN (mg/l)	TP (mg/l)	FEC COLIF (col/100 ml)
Lakeland	FWS	53004			5(A)	10(A)	1(A)	4(A)		
Eastern Service Area	FWS	13251						2.2(A)	0.2(A)	
Reedy Creek WTS1	FWS	16280	6.0(I)	6-7.5(I)	5(M)	20(A)	1(M)	2.0(A)	0.5(A)	
Reedy Creek OFWTS	FWS	3786	6.0(I)	6-7.5(I)	5(M)	20(A)	1(M)	2.0(A)	0.5(A)	
Central Slough	FWS	4542								
Ironbridge	FWS	45432						2.3(A)	0.2(A)	
Fort Deposit	FWS	909	6.0(D)	6-9(D)	10(M)	30(M)	2(M)			
West Jackson Co.	FWS	6058	6.0(D)	6-8.5(D)	10(M)	30(M)	2(M)			2200(M)
Bear Bay	FWS	9462	1.0(D)		12(M)	30(M)	1.2(M)			
Deer Park	FWS	3029		3.3-7.5					0.2(A)	200(M)
Houghton Lake	FWS	6000							0.5	
Mandan	FWS	2650			75	52	50			
Pembroke	FWS	340	7(D)	6-9	10	30	4			200
Apalachicola	FWS	3785		6.5-8.5	20	20				200
Benton	FWS	1400	7(D)	6-9	25	30	4			200
Drummond	FWS	70			30	30				
Gustine	FWS	3785			30	30				
Mt. View	FWS	5300			30	30				
Seneca Army Depot	FWS	950	7	6-9	5(D)	10(M)	2(D)			
Phillips High School	VSB	76		6-9(M)	20(M)	30(M)	8(M)			
Denham Springs	VSB	11355	5(M)	6-9(M)	10(M)	15(M)	5(M)			200(M)
Mandeville	VSB	5678		6-9(M)	10(M)	15(M)	5(M)			200(M)
Carville	VSB	568			10(M)	15(M)	5(M)			200(M)
Hammond	VSB	329			10(M)	15(M)				
Monterey	VSB	76			30(M)	30(M)				
Degussa Corp.	VSB	2040			10(M)	10(M)	10(M)			
Albany	VSB	132			20(M)	20(M)				200(M)
Benton, La	VSB	1173			10(M)	15(M)				200(M)
Haughton	VSB	1324			10(M)	15(M)				200(M)
Hardin	VSB	378		6-9(M)	10(M)	30(M)	4(M)			200(M)
Foothills Village	VSB	67			30(M)	30(M)				
Bradley	VSB	568	5(M)		15(M)	15(M)	5(M)			

Cottonwood	VSB	587			10(M)	30(M)	2(M)			
Hornbeck	VSB	231			20(M)	20(M)				
Provencal	VSB	344			10(M)	15(M)				
Crowley	VSB	113248	5(M)		10(M)	15(M)	5(M)	7.5(M)		200(M)
Doyline	VSB	416	5(M)	6-9(M)	10(M)	15(M)	5(M)			200(M)
Paris Landing	VSB	284	1(M)	6-9(M)	30(M)	30(M)				200(M)
Terry	VSB	378			10(M)	30(M)	2(M)			
Utica North	VSB	331			10(M)	30(M)	2(M)			
Utica South	VSB	442			10(M)	30(M)	2(M)			
Smackover	VSB	1893	3(M)		15(M)	15(M)	5(M)			
Shelbyville	VSB	280			30(M)	30(M)				
Pelahatchie	VSB	2157			10(M)	30(M)	2(M)			
Sibley	VSB	492			10(M)	15(M)				200(M)
Dessau	VSB	568			10(M)	15(M)				
Johnson City	VSB	114			5(M)	5(M)	2(M)		1(M)	

Note: Permit limits are based on interval in parentheses ( ), where: A = annual, M = monthly, D = daily, and I = instantaneous. The most limiting interval or seasonal limit is reported.

[a] Type = Hydrologic Type; VSB — Veg. Subm. Bed; FWS — Free Water System; HYB — Hybrid; OTH — Other; UNK — Unknown

parameters in the database, this regression can be expressed as follows:

$$BODOUT = 0.097*HLR + 0.192*BODIN \quad R^2 = 0.72 \tag{4–1}$$

where BODOUT = BOD_5 outflow concentration, mg/L
BODIN = BOD_5 inflow concentration, mg/L
HLR = hydraulic loading rate, cm/d

From this equation, it can be surmised that HLR or effective land area does not have much of an effect on BOD_5 removal efficiency except at very high HLR values.

Total Nitrogen Assimilation

Total nitrogen is removed in wetlands primarily through the processes of nitrification and denitrification by aquatic microbes.[5] The process of nitrification is typically limited by DO availability, temperature, and adequate residence time. Denitrification is typically very rapid in most wetlands. Inasmuch as the achievement of TN removal in wetland treatment systems is frequently a goal, design equations that adequately predict these processes are critical.

Figure 2 illustrates the relationship between TN mass loading and removal for the wetlands in the existing wetland treatment database. Unlike BOD_5, TN mass removal efficiencies decline at mass loading rates above 20 kg/ha/day. Also, mass removal efficiencies are more consistent at lower mass loading rates than they are for BOD_5. A regression predicting TN outflow concentration based on HLR and TN inflow concentration was developed from 213 records in the database. This multiple linear regression can be expressed as follows:

$$TNOUT = 0.28*HLR + 0.33*TNIN \quad R^2 = 0.54 \tag{4–2}$$

where TNOUT = TN outflow concentration, mg/L
TNIN = TN inflow concentration, mg/L.

Based on this equation, TN removal is more dependent upon the effect of high HLRs in wetland treatment systems than is BOD_5 mass removal.

Size and Flow Distributions

Wetland treatment systems generally require large tracts of land and such natural systems are not generally used for large flows. The database can be used to examine the typical size and flows for current operational facilities.

Figure 3 gives the size distribution of the four principal types of wetlands: natural free-water surface (Nat FWS), constructed free-water surface (Con FWS), constructed hybrid (Con HYB), and constructed vegetated submerged bed (Con VSB). Sizes range from 0.004 ha at Roswell, New Mexico, to 1093 ha for the Frank Lake, Alberta wetlands. The peaks of these distributions are widely spread; Nat FWS wetlands are typically ten times larger than Con FWS wetlands, which are ten times larger than either Con VSB or Con HYB wetlands. Most natural systems are 10 to 100 ha, while most VSB systems are 0.1 to 1.0 ha.

The disparity among flows for the different wetland types is not as great. Figure 4 shows coincident peaks in the distributions for the three types of constructed wetland systems, all of which are in the range 100 to 1000 m^3 per day (m^3/day). Natural FWS wetlands peak at about 1000 to 10,000 m^3/day.

Therefore, a sharp distinction also occurs in hydraulic loading rate. Figure 5 shows that most natural systems are loaded at less than 2 m^3/day. Most Con FWS systems have hydraulic loading rates less than 20 m^3/d, while most Con HYB systems range from 5 to 20 m^3/day, and most Con VSB wetlands range from 2 to 30 m^3/day.

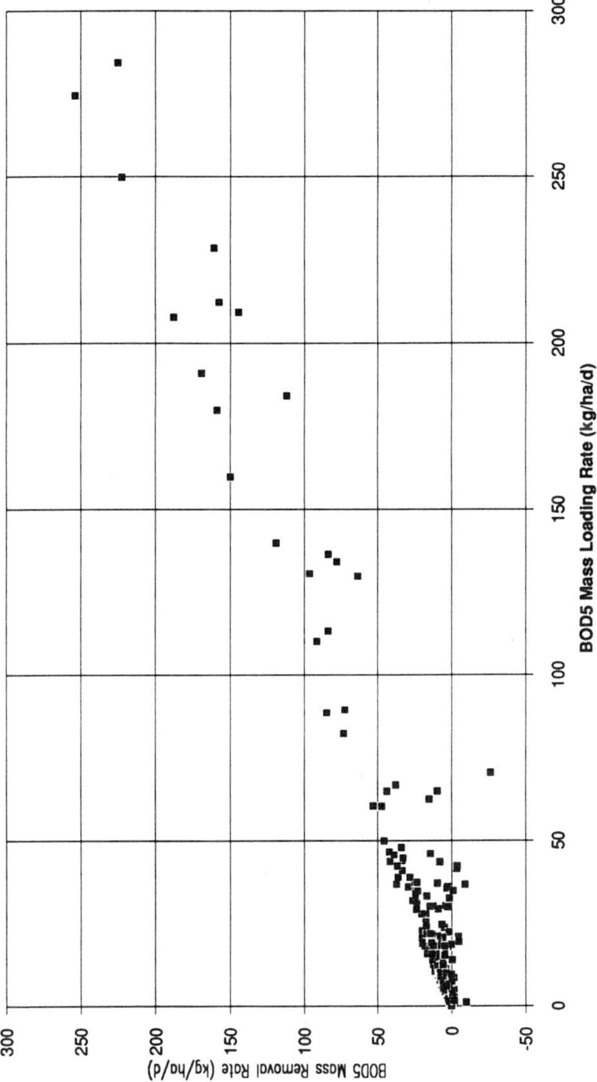

Figure 1. BOD_5 mass loading and removal rates in wetland treatment systems.

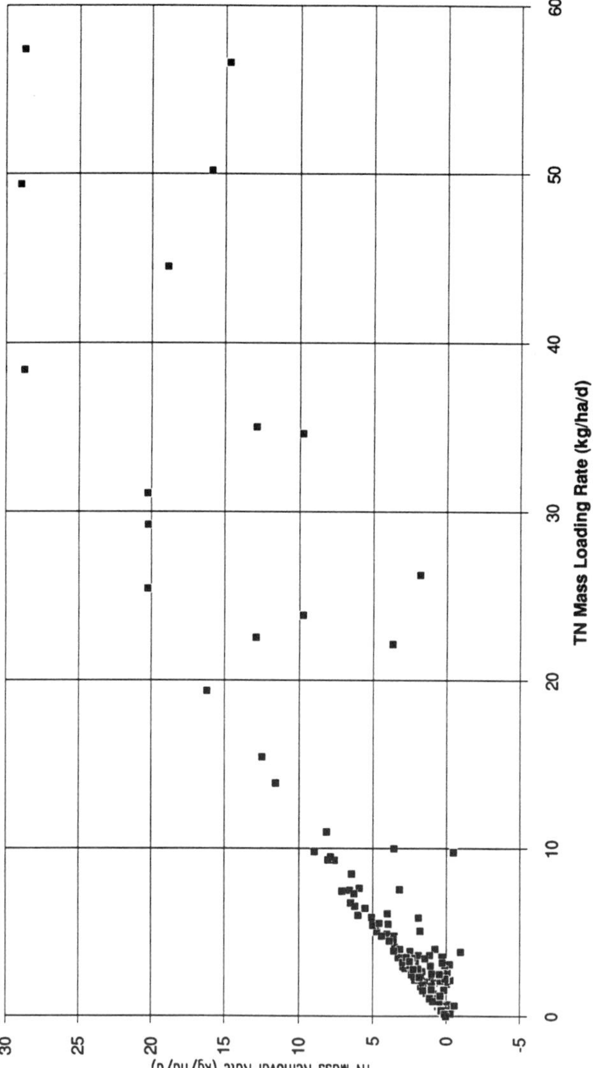

Figure 2. Total nitrogen mass loading and removal rates in wetland treatment system.

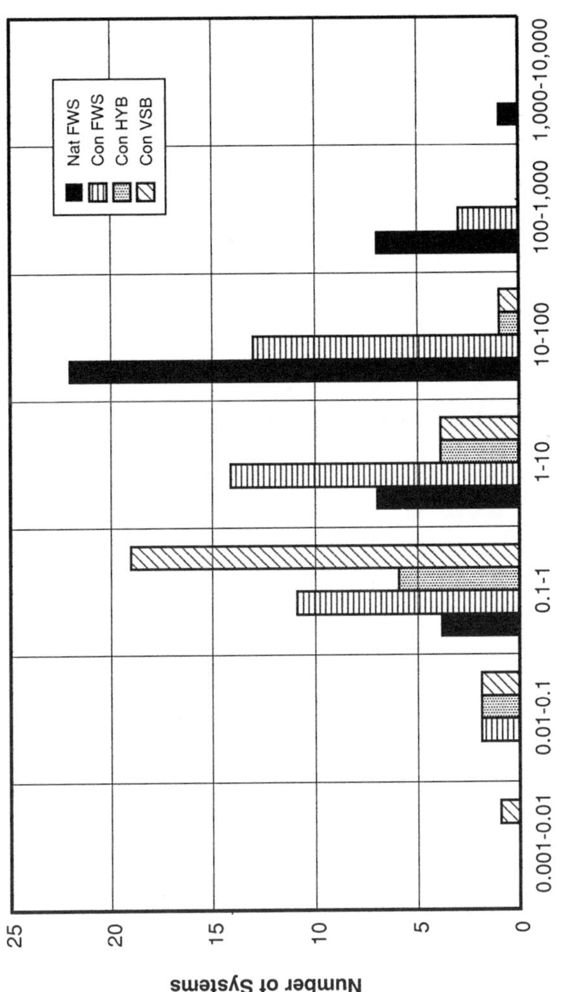

Figure 3. Size distributions of wetland treatment systems.

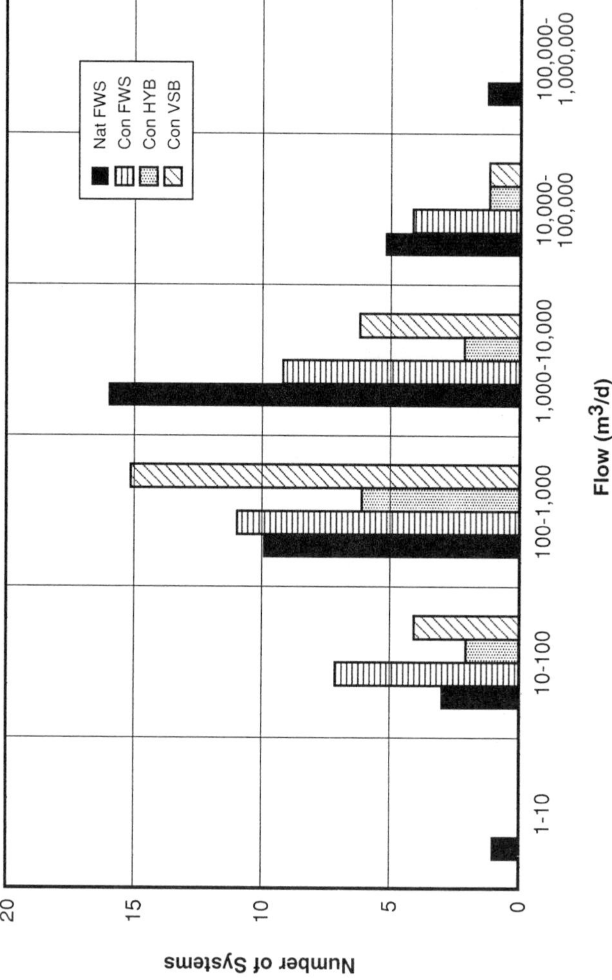

Figure 4. Flow distributions of wetland treatment systems.

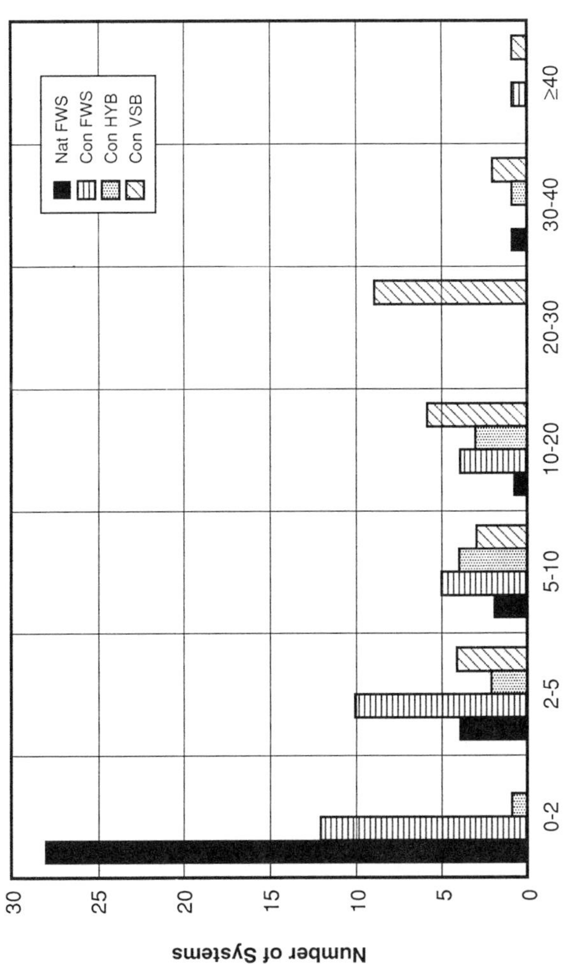

Figure 5. Hydraulic loading distributions of wetland treatment systems.

REFERENCES

1. Cooper, P. F. and B. C. Findlater, Eds. Constructed wetlands in water pollution control. *Proc. Int. Conf. on the Use of Constructed Wetlands in Water Pollution Control.* Pergamon Press, Oxford, U.K., 1990, 605.
2. Hammer, D. A., Ed. *Constructed Wetlands for Wastewater Treatment.* Lewis Publishers, Chelsea, MI, 1989.
3. Reed, S. C., E. J. Middlebrooks, and R. W. Crites. *Natural Systems for Waste Management and Treatment.* McGraw-Hill, New York, 1988.
4. U.S. Environmental Protection Agency. *Design Manual. Constructed Wetlands and Aquatic Plant Systems for Municipal Wastewater Treatment.* EPA-625/1-88/ 022, 1988, 83 pp. U.S. EPA, Cincinnati, OH, September 1988.
5. Water Pollution Control Federation. *Natural Systems for Wastewater Treatment. Manual of Practice.* FD-16, 1990, 270.

CHAPTER 5

The Use of Constructed Wetlands for Wastewater Treatment and Recycling

R. K. Bastian and D. A. Hammer

INTRODUCTION

Constructed wetlands are one of a series of engineered and managed "natural systems" that are receiving increased worldwide attention for wastewater treatment and recycling.[1-8] This appears to be, at least in part, due to a growing "green" environmental movement that supports more resource conservation and environmental protection, and greater reliance upon natural ecological processes and systems in preference to the more energy and chemical intensive "mechanical" waste management systems. Examples of the "green" technologies more extensively considered and used today include composting and direct land application of solid waste (especially yard wastes) and sludges (much of which had been disposed of by landfilling until recent years) and a wide variety of land treatment systems (i.e., slow-rate irrigation, rapid infiltration, and overland flow — or grass filtration), ponds, aquaculture and aquatic plant systems, as well as both natural and constructed wetlands systems, to treat effectively and recycle polluted waters from point and nonpoint sources.

Though not as thoroughly studied and understood as some of the other natural treatment systems, recent projects involving constructed wetlands have been established throughout the U.S. and in many other countries. However, by building upon the more extensive experience with land treatment systems, it is anticipated that many of these wetland systems will provide effective and reliable long-term, low-cost wastewater treatment, while in at least some cases providing wildlife and recreational benefits commonly associated with natural wetlands.

HISTORICAL DEVELOPMENT

Biogeochemical Cycles

Early studies of the process dynamics, movements, cycling and reservoirs for nutrients, metals, micronutrients, and trace elements established the theoretical basis for later studies of wetlands transformations of various types of wastewater. Studies by ecologists of biogeochemical cycles, the flux of materials between the living and nonliving components of the biosphere, in a variety of environments, quickly identified the important role of wetlands systems in material flux between important short-term and long-term sinks (reservoirs). Both freshwater and saltwater wetlands, due to their transitional location and reducing conditions, were found to have very significant roles in the natural cycling of organic and inorganic materials. Hutchinson,[9] E. P. Odum,[10] and H. T. Odum conducted extensive investigations of element cycles and also summarized the studies of many others in landmark publications from 1944 through 1971 that became the foundation for understanding material transformations in wetlands ecosystems and the important role of wetlands in the landscape. Productivity evaluations in Wisconsin marshes by John Kadlec,[11] Danish *Phragmites* marshes by Hans-Henrik Schierup,[12] and symposia at the Savannah River Ecology Laboratory in 1976, led by Adriano and Brisbin,[13] and in 1965 at Pallanza, Italy, edited by Goldman,[14] also contributed to the information base on material processing and transformations in wetlands ecosystems.

0-87371-550-0/93/$0.00+$.50

Experimental Evaluations

Early Studies in Germany

The first work deliberately investigating wastewater treatment by wetlands plants was conducted by Seidel at the Max Planck Institute in Plon, Germany. In 1952, she explored the removal of phenols from wastewater by *Scirpus lacustris* and in 1956 began testing dairy wastewater treatment with *S. lacustris*. From 1955 through the late 1970s, Seidel published numerous studies on water and wastewater treatment with wetlands plants.[15-17] A student, Kickuth, continued this experimental work but in 1975 also initiated monitoring at the Othfresen system where municipal wastewater was discharged into a natural *Phragmites* marsh.[18] Although the Othfresen results have been questioned, Kickuth and co-workers popularized this concept in Europe, resulting in nearly 200 municipal and industrial waste treatment systems.

U.S. Involvement

Throughout the 1970s land treatment alternatives as we know them today were developed with the support of a significant research and development effort funded by the U.S. Environmental Protection Agency, the U.S. Army Corps of Engineers, the U.S. Department of Agriculture, and other agencies. During that same period, the National Science Foundation supported a series of studies by John and Bob Kadlec[19] at the University of Michigan, and the Odum and Ewel[12] studies' use of natural wetlands for wastewater treatment at the University of Florida.[20] Their efforts, along with those of Sterns, Spangler, Fetter, Sloey, and Whigham in Wisconsin,[21] Valiela and Teal in New England,[22] and others in addressing the impacts of wastewater effluents on natural wetlands led to deliberate efforts to construct wetlands for wastewater treatment by Small[23] at Brookhaven National Laboratory in New York, and to develop various aquatic plant systems by Dinges[24] in Texas, and Wolverton[25] at the NASA facilities in Mississippi. Subsequently, intensive pilot studies were conducted by Gersberg[26] at Santee, California, and Gearheart[27] at Arcata, California, along with early operational systems at Listowel, Ontario,[28] Iselin, Pennsylvania,[29] and Arcata, California.

Acid Drainage Modification

Similarly, wetlands modification of acid drainage was first documented during Huntsman[30] studies to evaluate detrimental impacts to wetlands caused by mine drainage in Ohio and in West Virginia by Wieder and Lang.[31] Subsequently, the teams of Wieder, Lang et al., and Huntsman et al. initiated natural bog monitoring studies along with pilot experiments and small-scale constructed bogs designed to treat acid drainage waters. In the early 1980s, Pesavento constructed a number of operating systems with a "let's learn as we go" approach that significantly contributed to widespread implementation of wetlands for acid drainage control in Pennsylvania.[32]

Since natural wetlands treatment of mine drainage was found in *Sphagnum*-dominated bogs, most experimental work and early operating systems were attempts to construct such systems, but with highly variable results. Basic information on creation and management of bogs was severely limited. In1984, the initial TVA team of Hammer, Pearse, and Tomljanovich[33] designed *Typha* marshes, for acid drainage treatment because considerable information was available for marshes, and marshes were likely precursors to bogs in the Ohio and Pennsylvania natural wetlands. Subsequently, most acid drainage treatment wetlands have been *Typha* or *Scirpus* surface-flow marshes. During the mid to late 1980s numerous reports on experiments and operating systems appeared, and over 200 systems are operating in Appalachia alone.

Other Major Events

Other notable firsts include the use of constructed wetlands for treating:

1956 — livestock wastewaters — experimental; Seidel[16]
1975 — petroleum refinery wastewaters — operational; Litchfield[34]
1978 — textile mill wastewaters — operational; Kickuth[35]
1978 — acid mine drainage — experimental; Huntsman[30]
1979 — fish rearing pond discharge — operational; Hammer and Rogers[36]
1982 — acid mine drainage — operational; Pesavento[32]
1982 — reduction of lake eutrophication — experimental; Reddy[37]

1982 — urban stormwater runoff — operational; Silverman[38]
1983 — pulp/paper mill wastewaters — experimental; Thut[39]
1985 — photochemical laboratory wastewaters — experimental; Wolverton[25]
1985 — seafood processing wastewater — experimental; Guida and Kugelman[40]
1988 — compost leachate — operational; Pauly[41]
1988 — landfill leachate — experimental; Trautman and Porter[42]
1988 — livestock wastewaters — operational; Hammer and Pullin[43]
1989 — sugar beet processing plant wastewaters; Anderson[44]
1989 — reduction of lake eutrophication — operational; Szilagyi[45]
1990 — harbor dredged materials — experimental; Pauly[41]
1991 — pulp/paper mill wastewaters — operational; Thut[46]

Relevant conferences and symposia (Table 1) featuring wetlands and other aquatic plant systems have been held every few years starting in 1972, with Tourbier and Pierson's conference[47] on "Biological Wastewater Treatment" in Philadelphia, Pennsylvania, followed by the University of California-Davis wastewater aquaculture conference in 1979. The AWWA Water Reuse Symposium held in 1984 in San Diego, California, included several sessions featuring wetlands treatment projects. Beginning in 1976, a series of conferences focusing on wetlands systems have been held, including the University of Michigan, Ann Arbor conference in 1976, the workshop on ecological impacts at the University of Massachusetts in 1984, the University of Florida conference in Orlando, FL, in 1986, the TVA international conference in Chattanooga, Tennessee, and the Humbolt State University conference in Arcata, California, in 1988, the Cambridge, UK, international conference in 1990, and this meeting in Pensacola in 1991.

SYSTEM TYPES

The use of constructed wetlands for wastewater treatment and recycling seeks to take advantage of many of the same processes that occur in natural wetlands, but does so within a more controlled environment. However, rather than a single concept for constructed wetlands in wastewater treatment, wastewater and constructed wetlands have been combined in a number of approaches to accomplish wastewater treatment and recycling. These include: (1) systems receiving high loading rates built to maximize the amount of wastewater treated on the smallest area possible. For example, the test systems at Santee, California, and Listowell, Ontario, as well as the Gustine, California, and Harriman, New York, surface flow systems and the more recent subsurface flow systems in Mississippi, Louisiana, Germany, Denmark, and U.K. (2) Systems receiving relatively low loadings of pretreated effluents for further polishing while enhancing wildlife habitat. For example, Arcata, California; Incline Village, Nevada; Orlando and Lakeland, Florida; and 3) systems utilizing treated effluents as a water supply to enhance, restore, or create and maintain wetland habitat areas, for example, in Jackson County, Mississippi; Show Low, Arizona; and Martinez, California.

The two principal types of constructed wetlands managed for treating wastewater are surface flow wetlands and subsurface flow systems, also known as vegetated rock-reed filters and root-zone (RZM) systems.[1,2,4,5,57] Surface flow wetlands consist of shallow basins or channels with planted wetland vegetation through which water flows over the surface at relatively shallow depths. Such systems are similar to natural marshes and often manage to provide wildlife habitat as well as wastewater treatment. They have been established and maintained on substrates ranging from gravel or mine spoils to clay or peat. Some systems have been designed to recycle at least a portion of the treated wastewater by recharge of the underlying groundwater, while others have been designed as nondischarging systems.

Subsurface flow wetlands involve shallow basins or channels with planted wetland vegetation growing in soil, sand, rock, or other media designed so that the wastewater flows horizontally through the media with no above media or open surface flow. There is considerable debate over the "most effective" media type and size, length-to-width ratios, vegetation type, and percent cover. Many of the systems designed to operate as subsurface flow wetlands have been overloaded, but others have suffered media plugging and surface flow with lower application rates. A few recent European designs have incorporated vertical flow and batch loading in an attempt to allow for effective wetting, drying cycles, and entraining more oxygen to enhance nitrification.

Table 1. Important Conferences Featuring Constructed Wetlands for Wastewater Treatment

1. Freshwater Wetlands and Sewage Effluent Disposal — University of Michigan, Ann Arbor, May 1976[19]
2. Aquaculture Systems for Wastewater Treatment — University of California–Davis, Sept 1979[48]
3. Ecological Considerations in Wetlands Treatment of Municipal Wastewaters — University of Massachusetts, Amherst, June 1982[49]
4. Treatment of Mine Drainage by Wetlands — Penn State, August 1984[50]
5. Wetlands and Water Management on Mined Lands — Penn State, October 1985[51]
6. Aquatic Plants for Water Treatment and Resource Recovery — Orlando, FL, July 1986[1]
7. The Use of Macrophytes in Water Pollution Control — Piracicaba, Brazil, August 1986[52]
8. Mine Drainage and Surface Mine Reclamation — Pittsburgh, April 1988[53]
9. Constructed Wetlands for Wastewater Treatment — Chattanooga, TN, June 1988[5]
10. Wetlands for Wastewater Treatment and Resource Enhancement — Humboldt State University, Arcata, CA[54]
11. The 1990 Mining and Reclamation Conference and Exhibition — Charlestown, WV, April 1990[55]
12. Use of Constructed Wetlands in Water Pollution Control — Cambridge, U.K., September 1990[6]
13. Use of Created and Natural Wetlands in Controlling Nonpoint Source Pollution — Arlington, VA, June 1991[56]
14. Symposium on Constructed Wetlands for Water Quality Improvement — Pensacola, FL, October 1991
15. Wetlands Systems in Water Pollution Control — Sydney, Australia, December 1992

SYSTEM PERFORMANCE

The operation and pollutant removal mechanisms in wetlands are similar to those in overland flow systems. The latter bring thin-sheets of wastewater flowing down a gentle, uniformly sloping, grass-covered field in contact with both aerobic and anaerobic conditions in the soil/surface litter zone.[2,4,58] A major difference between these systems, however, is that the retention time for the wastewater is usually only hours in overland flow systems, in contrast to days for wetland systems.[59-61]

Under appropriate conditions, both natural and constructed wetlands treatment systems have achieved high removal efficiencies for BOD_5, SS, and nutrients from wastewater, as well as natural die-off of pathogens (Tables 2 and 3). While high removal rates for BOD_5, SS, and bacteria are commonly achieved by constructed wetlands, ammonia conversion/removal efficiency by nitrification and denitrification is variable depending upon system design, retention time, and oxygen supplies. Phosphorus removal rates also vary between projects and may be effective only for limited periods unless large areas or special media are involved. Excellent trace heavy metal removals have been demonstrated in tests at the pilot wetlands at Santee, California, with municipal wastewater,[26] and at numerous constructed wetlands treating acid drainage.[4,5] Few studies have investigated the fate of trace organic contaminants in wastewater effluents under wetland treatments, although such systems are employed as a part of treatment systems for oil refinery and pulp mill effluents.[5]

WETLANDS EXPERIMENTAL DESIGN

Numerous studies are planned or underway to explore applications of this new technology to different types of wastewaters and to begin to describe quantitatively processes, rates, components, and designs that should facilitate optimized design and operation of future systems. The authors have been involved in monitoring and evaluating laboratory-scale, pilot-scale, demonstration, and operation of constructed wetlands systems treating a variety of wastewaters for over 15 years. Throughout these efforts, it has become clear that certain aspects of wetlands ecosystems must be considered in developing the design for any investigations on these systems. One concern is that in many cases, operating systems have not been able to achieve and/or maintain the performance that was projected from the results of studies on small-scale and/or short-term projects.

Consequently, it is urged that constructed wetlands researchers more carefully consider the nature of wetlands wastewater treatment systems prior to designing future investigations. Some issues to consider should be that wetlands are:

Table 2. Range of Pollutant Removal Efficiencies Reported for Constructed Wetland Systems

	Removal (%)
BOD_5	50–90
SS	40–94
Nitrogen	30–98
Phosphorus	20–90

From References 2, 3, 4, and 60. With permission.

Table 3. Pathogen Removal in Constructed Wetland Systems

	System performance	
Location	Influent	Effluent[a]
Santee, CA; bulrush wetland[b]		
Winter season (Oct–Mar)		
Total coli, no./100 mL	5×10^7	1×10^5
Bacteriophage, PFU/mL	1900	15
Summer season (Apr–Sept)		
Total coli, no./100 mL	6.5×10^7	3×10^5
Bacteriophage, PFU/mL	2300	26
Iselin, PA; cattails and grasses[c]		
Winter season (Nov–Apr)		
Fecal coli, no./100 mL	1.7×10^6	6200
Summer season (May–Oct)		
Fecal coli, no./100 mL	1.0×10^6	723
Arcata, CA; bullrush wetland[d]		
Winter season		
Fecal coli, no./100 mL	4300	900
Summer season		
Fecal coli, no./100 mL	1800	80
Listowell, ONT; cattails[4]		
Winter season		
Fecal coli, no./100 mL	556,000	1400
Summer season		
Fecal coli, no./100 mL	198,000	400

[a] Undisinfected.
[b] Gravel bed, subsurface flow.
[c] Sand bed, subsurface flow.
[d] Free water surface.

1. Land intensive, relatively large, outdoor systems normally subject to the vagaries of weather and other important forcing functions. Small tanks isolated from sun, wind, precipitation, and atmospheric effects do not represent the environmental regime which operating-scale systems experience.
2. Dependent on proper flow patterns and accurate application/loading rates. The small-scale micro- or mesocosms often suffer significant influences from scale effects. Small-diameter pumps, piping, etc. are easily clogged, with important ramifications to the planned loading rates. In addition, small tanks or pilot systems have a considerably higher boundary zone or marginal area that entails influential edge effects.
3. Significantly influenced by the size of important components; for example, some species of wetlands plants grow to heights of 3 to 4 meters, especially in nutrient-rich wastewaters. Performance testing with a 3-m tall plant in a tank that is 2 m wide and 4 m long is unrealistic since the root mass of a single plant

may occupy a significant proportion of the entire substrate. In addition, wind throw and other factors causing plants to fall over could easily block much of the surface of the cell diverting most flow to the tank perimeter.

4. Likely to be affected by gradual development and continually changing components and processes. Few researchers would suggest that a constructed wetlands system include all of the important components in the appropriate quantities immediately after planting vegetation. Peak performance should not be anticipated until after the system has experienced two to three growing seasons to achieve some measure of maturity. However, like natural wetlands which are dynamic, ever-changing systems, constructed wetlands only appear to remain in an early phase because of external disturbances or forcing functions that retard major successional changes. Constructed wetlands also undergo significant changes; attempting to project results to long-term project performance from short-term (2 to 5 year) studies may prove unreliable. Since wetlands are transitional by nature, perhaps we need also to consider providing disturbance factors such as burning, periodic drying, etc. for some of the older operating systems.
5. Generally, competitive with other methods of wastewater treatment, as long as they are designed to be inexpensive to construct and simple to operate. Elaborate use of pumps and piping, chemicals, recirculating, specialized media, or other complex measures may result in a system that is expensive to construct and complex to operate. In such cases, costs and operational complexity could exceed the capabilities or commitments of potential users.

Therefore, we recommend that experimental cells designed for investigating the performance of constructed wetlands treatment of wastewaters consist of plots wherein the minimum dimension is not less than 10 times the size of the largest component of the system. For example, with 3-m high wetlands plants, the minimum cell dimension would be 30 m. Even in this case, a single wind-thrown plant could obstruct the flow to a significant proportion of the cell surface. In any case, the root mass of one plant should not influence a large percentage of the cell substrate. But both above- and below-surface effects could be significant when comparing results from a number of small cells.

The use of micro- or mesocosm studies should be limited to those processes not greatly impacted by space restrictions. Such an approach for investigations of complex wetlands systems would be appropriate as long as:

1. The results are further tested in projects of gradually increasing size and longevity and decreasing environmental control before they are applied to operating systems.
2. Physical, chemical, and biological processes that are not significantly influenced by macrophytes and larger animals are taken into consideration.

There are now hundreds of operating systems treating many types of contaminated waters distributed throughout North America and Europe from which considerable information is being obtained. Many different designs have been used and, more importantly, many different operating regimes have been employed even within the same system. Because of regulatory considerations, however most of these systems have been subjected to repeated attempts to improve performance by operational modifications without regard to the basic need for establishing a long-term record under one set of stabilized operating conditions. In addition, most of this database consists of simple in and out records on permit discharge parameters, with little or no information from within the system or on the universe of other measurements that may be important in understanding the significance of certain components, processes, or fluxes. These operating systems provide an important opportunity that should not be overlooked.

In reviewing the existing information from operating systems and the problems associated with small-scale, short-term studies on wastewater treatment in constructed wetlands, it is readily apparent that advances in this technology are unlikely to proceed more rapidly unless a new approach is undertaken. Therefore, it is suggested that intensive long-term evaluations be conducted at a series of operating systems representing the major categories of design, geographically distributed as widely as feasible and treating different types of contaminated waters (municipal, acid drainage, industrial, and agricultural). It is recommended that each system be operated under controlled conditions to develop a database under specific operating conditions. These systems should be made available for studies by qualified invesitgators in ways that would not conflict with the established operating conditions or ongoing research by others.

The value of simultaneous, long-term investigations of complex systems has been thoroughly documented by results from studies on the various biome and biosphere reserve programs, as well as studies at national laboratories. Since constructed wetlands students are confronted with a multitude of long-term processes in a complex system, it would appear that this approach may be as productive in all such future endeavors.

SYSTEM CONSTRAINTS, LIMITATIONS, AND OPPORTUNITIES

Potential users of constructed wetlands systems for wastewater treatment and recycling frequently face many technological as well as practical problems. The lack of long-term operational data creates uncertainty when establishing project designs and operation guidance. Seasonal variability in treatment performance and concerns over the potential impact on wildlife may create serious operational constraints and difficult permitting conditions. It is clear, however, that moderately loaded systems designed to duplicate natural marshes will provide low-cost, low-maintenance, efficient wastewater treatment to meet stringent discharge limits.[63] Constructed wetlands may also be the principal choice for controlling major non-point sources such as from agriculture and urban stormwater, since conventional methods are too expensive and unable to tolerate fluctuating loading conditions.[64] The problems associated with media plugging and surface flooding at many subsurface flow systems raise serious questions as to the viability of these systems as currently designed and operated.

However, in spite of such problems and uncertainties, the potential for realizing low-cost, reliable wastewater treatment along with the possibilities for linking wastewater management with wetlands creation and wildlife habitat enhancement seem, in many cases, worth the additional effort required to gain project approval. In fact, wildlife use may be a key to low cost, efficient wastewater treatment since wetlands systems capable of supporting appropriate wildlife also have the complexity and stability to support other important functional values, including water purification.[62] Efforts are currently underway to generate the data needed to overcome the constraints and limitations associated with the systems in operation today and to help develop better technical guidance for designing and operating constructed wetlands wastewater treatment and recycling systems.

REFERENCES

1. Reddy, K. R. and W. H. Smith, Eds. *Aquatic Plants for Water Treatment and Resource Recovery.* Magnolia Publishing, Orlando, FL, 1987.
2. Reed, S. R., E. J. Middlebrooks, and R. W. Crites. *Natural Systems for Waste Management and Treatment.* McGraw-Hill, New York, 1988.
3. Bastian, R. K. and J. Benforado. Waste treatment: doing what comes naturally, *Technol. Rev.* 86(2):58, 1983.
4. Water Pollution Control Federation. Manual of Practice, Natural Systems for Wastewater Treatment. Water Pollution Control Board, Alexandria, VA, 1990.
5. Hammer, D. A., Ed. *Constructed Wetlands for Wastewater Treatment: Municipal, Industrial and Agricultural.* Lewis Publishers, Chelsea, MI, 1989.
6. Cooper, P. F. and B. C. Findlater, Eds. *Constructed Wetlands in Water Pollution Control.* Pergamon Press, New York, 1990.
7. Etnier, C. and B. Guterstam, Eds. *Ecology Engineering for Wastewater Treatment.* Bokskogen/Stensund Folk College, Sweden, 1991.
8. Tchobanoglous, G. T. and F. L. Burton. *Wastewater Engineering: Treatment, Disposal, Reuse*, 3rd ed. McGraw-Hill, New York, 1991.
9. Hutchinson, G. E. *A Treatise on Limnology. Vol I, II.* John Wiley & Sons. New York, 1957.
10. Odum, E. P. *Fundamentals of Ecology.* W.B. Saunders, Philadelphia, 1971.
11. Kadlec, J. A. Effect of a drawdown on a waterfowl impoundment, *Ecol.* 43:267, 1962.
12. Schierup, H. H. Biomass and primary production in a *Phragmites communis* swamp in North Jutland, Denmark, *Verh. Intern. Vereinig. Limnol.* 20:94, 1978.
13. Adriano, D. C. and I. L. Brisbin, Jr., Eds. *Environmental Chemistry and Cycling Processes.* CONF-760429. NTIS, Springfield, VA, 1978.
14. Goldman, C. R., Ed. *Primary Productivity in Aquatic Environments.* University of California Press. Berkeley, 1978.
15. Seidel, K. Die Flechtbinse *Scirpus lacustris,* in *Okologie, Morphologie und Entwicklung, ihre Stellung bei den Volkern und ihre wirtschaftliche Bedeutung.* Schweizerbartsche Verlagsbuchhdlg, Stuttgart, 1955.

16. Seidel, K. Zur Problematik der Keim- und Pflanzgewasser, *Verh. Intern. Limnol.* 14:1035, 1961.
17. Seidel, K. Macrophytes and water purification, in *Biological Control of Water Pollution.* J. Tourbier and R. W. Pierson, Jr., Eds. Pennsylvania University Press, Philadelphia, PA, 1976.
18. Kickuth, R., Ed. Grundlagen und Praxis Naturnaher Klarverfahren — 10 Jahre Othfresen. Sammelband des Symposiums 31 January–1 February 1985, Liebenburg, Germany, 1985.
19. Tilton, D. L., R. H. Kadlec, and C. J. Richardson, Eds. Freshwater Wetlands and Sewage Effluent Disposal. *Symp. Proc. University of Michigan,* Ann Arbor, MI, 1976.
20. Ewel, K. C. and H. T. Odum. *Cypress Swamps.* University of Florida Press, Gainesville, 1984.
21. Spangler, F., W. Sloey, and C. W. Fetter. Experimental use of emergent vegetation for biological treatment of municipal wastewater in Wisconsin, in *Biological Control of Water Pollution.* J. Tourbier and R. W. Pierson, Jr., Eds. Pennsylvania University Press, Philadelphia, PA, 1976.
22. Valiela, I., J. M. Teal, C. Cogswell, J. Hartman, S. Allen, R. Van Etten, and D. Goehringer. Some long-term consequences of sewage contamination in salt marsh ecosystems, in *Ecological Considerations in Wetlands Treatment of Municipal Wastewaters.* P. J. Godfrey, E. R. Kaynor, S. Pelczarski, and J. Benforado, Eds. van Nostrand Reinhold, New York, 1985.
23. Small, M. Natural Sewage Recycling Systems. BNL 50630. Brookhaven Nat'l. Lab., Upton, NY, 1977.
24. Dinges, R. *Natural Systems for Water Pollution Control.* Van Nostrand Reinhold, New York, 1982.
25. Wolverton, B. C. Artificial marshes for wastewater treatment, in *Aquatic Plants for Water Treatment and Resource Recovery.* K. R. Reddy and W. H. Smith, Eds. Magnolia Publishing, Orlando, FL, 1987.
26. Gersberg, R. M., S. R. Lyon, B. V. Elkins, and C. R. Goldman. The removal of heavy metals by artificial wetlands, in *Future of Water Reuse, Proc. Water Reuse Symp. III.* AWWA Research Foundation. Denver, CO, 1985.
27. Gearheart, R. A., F. Klopp, and G. Allen. Constructed free surface wetlands to treat and receive wastewater: pilot project to full scale, in *Constructed Wetlands for Wastewater Treatment; Municipal, Industrial and Agricultural.* D. A. Hammer, Ed. Lewis Publishing, Chelsea, MI, 1989.
28. Herskowitz, J., S. Black, and W. Lewandowski. Listowel artificial marsh treatment project. in *Aquatic Plants for Water Treatment and Resource Recovery.* K. R. Reddy and W. H. Smith, Eds. Magnolia Publishing, Orlando, FL, 1987.
29. Davido, R. L. and T. E. Conway. Nitrification and denitrification at the Iselin marsh/pond/meadow facility, in *Constructed Wetlands for Wastewater Treatment; Municipal, Industrial and Agricultural.* D. A. Hammer, Ed. Lewis Publishing, Chelsea, MI, 1989.
30. Huntsman, B. E., J. G. Solch, and M. D. Porter. Utilization of a *Sphagnum* sp. dominated bog for coal acid mine drainage abatement. Geological Soc. of America (91st Annu. Meeting) Abstracts. Toronto, Ontario, 1978.
31. Wieder, R. K. and G. E. Lang. Modification of acid mine drainage in a freshwater wetlands, in *Symp. Wetlands of the Unglaciated Appalachian Region.* West Virginia Univ., Morgantown, WV, 1982.
32. Stone, R. W. and B. G. Pesavento. Micro- and macro-biological characteristics of wetlands removing iron and manganese, in *Wetlands and Water Management on Mined Lands.* R. P. Brooks, D. E. Samuel, and J. B. Hill, Eds. Pennsylvania State University, University Park, PA, 1985.
33. Hammer, D. A. Constructed wetlands for acid water treatment — an overview of an emerging technology, *Proc. Mining Drainage Symp.; Comb. Annu. Meeting — Geol. Assoc. of Canada and Mineralogical Assoc. of Canada.* Vancouver, BC, 1990.
34. Litchfield, D. K. and D. D. Schatz. Constructed wetlands for wastewater treatment at Amoco Oil Company's Mandan, North Dakota refinery, in *Constructed Wetlands for Wastewater Treatment; Municipal, Industrial and Agricultural.* D. A. Hammer, Ed. Lewis Publishing, Chelsea, MI, 1989.
35. Kickuth, R. Das Wurzelraumverfahren — ein kostengunstiges Klarverfahren fur den dezentralen Einsatz in Kommunen und Gewerbe. Der Tropenlandwirt, Zeitschrift fur die Landwirtschaft in den Tropen und Subtropen 83, 1982.
36. Hammer, D. A. and P. Rogers. Treating fish rearing pond discharge waters with artificial wetlands. 27 p. Mimeo. TVA. Knoxville, TN, 1980.
37. Reddy, K. R., K. L. Campbell, D. A. Graetz, and K. M. Portier. Use of biological filters for agricultural drainage water treatment. *J. Environ. Qual.* 11:591, 1982.
38. Silverman, G. S. Development of an urban runoff treatment wetlands in Fremont, California, in *Constructed Wetlands for Wastewater Treatment; Municipal, Industrial and Agricultural.* D. A. Hammer, Ed. Lewis Publishing, Chelsea, MI, 1989.

39. Thut, R. N. Utilization of artificial marshes for treatment of pulp mill effluents, in *Constructed Wetlands for Wastewater Treatment; Municipal, Industrial and Agricultural.* D. A. Hammer, Ed. Lewis Publishing, Chelsea, MI, 1989.
40. Guida, V. G. and I. J. Kugelman. Experiments in wastewater polishing in constructed tidal marshes: Does it work? Are the results predictable? in *Constructed Wetlands for Wastewater Treatment; Municipal, Industrial and Agricultural.* D. A. Hammer, Ed. Lewis Publishing, Chelsea, MI, 1989.
41. Pauly, U. personal communication, 1990.
42. Staubitz, W. W., J. M. Surface, T. S. Steenhuis, J. H. Peverly, M. J. Lavine, N. C. Weeks, W. E. Sanford, and R. J. Kopka. Potential use of constructed wetlands to treat landfill leachate, in *Constructed Wetlands for Wastewater Treatment; Municipal, Industrial and Agricultural.* D. A. Hammer, Ed. Lewis Publishing, Chelsea, MI, 1989.
43. Hammer, D. A. and B. P. Pullin. Constructed wetlands for livestock waste treatment, in *Proc. Natl. Non-point Conf. — The Water Quality Act — Making Non-point Programs Work.* St. Louis, 1989.
44. Anderson, P. personal communication, 1989.
45. Szilagyi, F., L. Somlyody, and L. Koncsos. Operation of the Kis-Balaton reservoir: evaluation of nutrient removal rates, *Hydrobiologia,* 191:297, 1990.
46. Thut, R. N. personal communication, 1991.
47. Tourbier, J. and R. W. Pierson, Jr., Eds. *Biological Control of Water Pollution.* Pennsylvania University Press, Philadelphia, 1976.
48. Bastian, R. K. and S. C. Reed, Eds. Aquaculture systems for wastewater treatment: Seminar Proc. and Engineering Assess. 430/9-80-006. U.S. Environmental Protection Agency, Washington, D.C., 1979.
49. Godfrey, P. J., E. R. Kaynor, S. Pelczarski, and J. Benforado, Eds. *Ecological Considerations in Wetlands Treatment of Municipal Wastewaters.* van Nostrand Reinhold, New York, 1985.
50. Burris, J. E. Treatment of mine drainage by wetlands. Conf. Proc. Contrib. 264. Dept. of Biol., Pennslyvania State University, University Park, PA, 1984.
51. Brooks, R. P., D. E. Samuel, and J. B. Hill, Eds. Wetlands and water management on mined lands. Conf. Proc. Pennsylvania State University, University Park, PA, 1985.
52. Athie, D. and C. C. Cerri, Eds. The use of macrophytes in water pollution control. *Water Sci. Technol.* Pergamon Press, New York, 19(10):1987.
53. U.S. Bureau of Mines. Mine Drainage and Surface Mine Reclamation. Vol. II: Mine Reclamation, Abandoned Mine Lands and Policy Issues. Conf. Proc. Info. Circ. 9183. U. S. Bureau of Mines, Pittsburgh, PA, 1988.
54. Allen, G. H. and R. A. Gearheart, Eds. *Proc. Conf. Wetlands for Wastewater Treatment and Resource Enhancement.* Humboldt State University. Arcata, CA, 1988.
55. Skousen, J., J. Sencindiver, and D. Samuel, Eds. *Proc. 1990 Mining and Reclamation Conf. Exhibit.* Vol. II. West Virginia University, Morgantown, WV, 1990.
56. Olson, R., Ed. Workshop Proceedings: The Role of Created and Natural Wetlands in Controlling Nonpoint Source Pollution. EPA/600/ Sept. 1991. Corvallis, OR, 1991.
57. U.S. Environmental Protection Agency. Process design manual for constructed wetlands and floating aquatic plant systems for municipal wastewater treatment. EPA 625/1-88-022. Center for Environmental Research Information. Cincinnati, OH, 1988.
58. U. S. Environmental Protection Agency. Process design manual for land treatment of municipal wastewater, supplement on rapid infiltration and overland flow. EPA 625/1-81-013a. Center for Environmental Research Information. Cincinnati, OH, 1984.
59. Reed, S. R., R. K. Bastian, S. Black, and R. Khettry. Wetlands for wastewater treatment in cold climates, in *Future of Water Reuse, Proc. Water Reuse Symp.* III. AWWA Research Foundation. Denver, CO, 962, 1985.
60. Reed, S. R. and R. K. Bastian. Wetlands for wastewater treatment: an engineering perspective, in *Ecological Considerations in Wetlands Treatment of Municipal Wastewaters.* P. J. Godfrey, E. R. Kaynor, S. Pelczarski, and J. Benforado, Eds. Van Nostrand Reinhold, New York, 1985.
61. Benforado, J. and R. K. Bastian. Natural waste treatment, in *McGraw-Hill Yearbook of Science and Technology.* McGraw-Hill, New York, 1985.
62. Hammer, D. A. *Creating Freshwater Wetlands.* Lewis Publishers, Chelsea, MI, 1991.

63. Hammer, D. A. Water improvement functions of natural and constructed wetlands, in *Proc. Newman Teleconference Semin. Ser. Protection and Management Issues for South Carolina Wetlands*. R. White, Coord. Columbia, SC, 1990.
64. Hammer, D. A. Designing constructed wetlands to treat agricultural nonpoint source pollution, in *Proc. The Role of Created and Natural Wetlands in Controlling Nonpoint Source Pollution*. R. Olson, Ed. EPA/600/Sept 1991. Corvallis, OR, 1991.

CHAPTER 6

Constructed Wetlands Experience in the Southeast

R. J. Freeman, Jr.

Constructed wetlands (CW) as a wastewater treatment technology have seen a dramatic increase in use throughout the U.S. in the last several years. In part due to the warmer climate and availability of land for siting, CW systems have been utilized in the "sun-belt" states more frequently than elsewhere. Of the over 140 municipal constructed wetlands systems (operational, under construction, and planned) inventoried in the U.S. in March, 1990, by Sherwood Reed[1] under contract to EPA, 89 were in EPA Regions IV and VI, the southeast and southwest. In spite of the number of these systems in use, the shortage of meaningful data and the lack of understanding regarding the basic physical and biochemical processes taking place has resulted in little progress toward a sound design approach. This problem has led to unexpected difficulties with a number of the CW systems in operation. Both gravel substrate (sub-surface flow) and soil substrate (surface flow) systems have encountered serious problems, in some cases jeopardizing the continued use of the CW technology at those locations. This paper will discuss some of these issues and the measures being considered to attempt to avoid such problems in the future.

Constructed wetlands have been generally grouped into two basic categories, the simplest being the systems in which rooted aquatic plants are planted in a soil substrate within a constructed earthen basin. The basin may be lined or not, depending on natural soil permeability and groundwater protection requirements. The systems are generally designed to allow wastewater effluent, following preliminary treatment to flow at a depth of 1 to 2 in. up to 12 to 18 in. through the basin in a plug flow pattern. The Water Pollution Control Federation (WPCF) design manual titled *Natural Systems for Wastewater Treatment*[2] designates these systems as free-water surface (FWS) systems referring to the nature of the surface flow. The second type of CW system is similar to the FWS systems except the basin is filled with aggregate such as gravel or crushed stone to a depth of 12 to 24 in., in which the aquatic vegetation is planted and through which the wastewater flows with no visible surface flow. These systems have likewise been designated as vegetated submerged bed (VSB) systems by the WPCF design manual.

A brief comparison of those of the two types of systems is shown below.

FWS Systems	VSB Systems
Lower installed cost/gal	Greater assimilation rate, less land required
Simpler hydraulics	No Visible flow, less nuisance, vector problems, odors
More natural wetland values can be incorporated into the system wildlife habitat, etc.	More cold tolerant

The 89 CW systems in Regions IV and VI are divided into 17 of the FWS type and 72 of the VSB type. The prevalence of the VSB type system is due, at least in part, to the success that Dr. Bill Wolverton, formerly with NASA at their southwest Mississippi test facility, had in researching and promoting the use of these systems. Of those 72 VSB systems identified, 49 are located in Mississippi and Louisiana.

The original and on-going popularity of the CW technology derives primarily from its two-fold promise of lower costs and little requirement for operation and maintenance compared to conventional technology. The EPA construction grants program has encouraged the use of the technology due to the Innovative/Alternative (I/A) grant bonus received by grantees selecting CW systems of either type. The Tennessee Valley Authority (TVA) has also played a significant role in encouraging these systems, especially in the Tennessee Valley area.

0-87371-550-0/93/$0.00+$.50

Benton, Kentucky, was assisted by TVA in the construction of a CW system to treat the wastewater from the approximately 4600 residents. Their previous system was a 10.5-ha (26-acre) facultative two-cell lagoon with a flow as high as 4500 m^3/day (1.2 MGD) due to infiltration/inflow. The original 4.0-ha (10-acre) second cell was converted to three equal-sized parallel cells, one filled to a depth of 0.6 m (2 ft) with crushed limestone, and the other two left as native soil, giving one VSB cell and two FWS cells (Figure 1).[3] The VSB cell was planted with softstem bulrush and the FWS cells were planted with woolgrass in one and arrowhead and cattail in the other. The system was designed for an average flow of 4160 m^3/day (1.1 MGD) with 50% of the flow intended to go through the VSB and the remaining 50% to be equally divided between the two FWS cells.

The city's NPDES permit required the following limits to be met:

	BODq	TSS	NH_3	DO (all mg/L)
Summer	25	30	4	7
Winter	25	30	10	7

During the construction of the wetland system and the FWS cells, all of the lagoon effluent was routed through the VSB cell. During the first year of full operation, 1988, the influent flows to the wetland cells averaged a BOD_5 of 25 to 30 mg/L, TSS of 40 to 70 mg/L, NH_3 of 6 to 9 mg/L, and flows averaging 1800 to 2600 m^3/d (0.5 to 0.7 MGD). The flows were within the design expectations, with the exception of rainfall-related high flows and the construction period diversion of all the flow through the VSB as mentioned.

The BOD_5 removals experienced were quite good, averaging 50 to 60% in the FWS cells and 80% in the VSB cell.[4] The NH_3 trend, however, was very disappointing. All three cells experienced a significant *increase* (30 to 50%) in NH_3 throughout the cells, yielding effluent consistently noncompliant with the summer NH_3 limits, and occasionally exceeding the winter NH_3 limit as well.[4] The effluent DO limits were also consistently violated, both summer and winter. This pattern continued in 1989, and in 1990 two major changes were made to the Benton system to attempt to remedy the lack of nitrification occurring. The first attempted solution was the implementation of a recycle system in one of the FWS cells in an attempt to increase the availability of DO to provide an environment in which the nitrification reactions could occur. This effort proved fruitless even at a 3-4Q recycle and using sprinklers to introduce DO. The second change was to dramatically reduce the loading in the VSB and one of the FWS cells and route the remaining flow through the remaining FWS cell. The original design of the FWS cells called for a loading of 1.5 ha/1000 m^3/d (14 ac/MGD). The reduced loading was set at 4.5 ha/1000 m^3/d (42 ac/MGD).[5] This reduced loading corresponds to the area required for ammonia conversion in an FWS cell as calculated using the formulations in the WPCF Natural Systems Manual. Unfortunately, the nitrification in that cell still did not appear to increase since the NH_3 in the effluent still exceeded that in the influent by 25% for the 2 months for which data are available.

It is not clear why the nitrification reactions are not more effective in reducing ammonia. Some H_3 is being oxidized, while the decrease in TN and TKN suggests that organic N is being converted to NH_3 at the same time that some nitrification is occurring. Possible deposition of organic material from the lagoon during the higher loading period may have created a "background" DO load that will take some time (months or years) to satisfy. Such oxygen demand does not leave enough DO for more complete nitrification. The often-cited ability of plant roots to "pump" oxygen down into the VSB bed also appears to be questionable. Several observation pits dug in the VSB cell showed little root penetration below the 0.15 to 0.25-m (6 to 10-in.) depth, contrary to the expected full-depth root penetration.[6] It may also be that short-circuiting could be occurring in the FWS cell, thus effectively reducing the area and increasing the net loading, such that nitrification is not promoted. Remedying any of these possible problems would be difficult.

The VSB cell also experienced surfacing of wastewater at the original design flow of 2080 m^3/d (0.55 MGD). A detailed investigation of the cause of the surface flow showed that the crushed limestone bed was heavily plugged, with an inorganic gel-like substance formed from an interaction of the limestone, silicon, and algae in the wastewater.[7] A reduction in loading to approximately 10% of design brought the water level back below the surface. The resulting loading was 11.7 ha/1000 m^3/d (105 ac/MGD) compared to the original design of 0.7 ha/1000 m^3/d (6.5 ac/MGD). The VSB cell is now finally achieving satisfactory nitrification, with effluent NH_3 values of 3 mg/L, well within the permit limit of 4 mg/L. At this loading rate, however, another 90 acres or so of VSB cells would be necessary to bring NH_3 at Benton into compliance. Clearly, this is economically prohibitive.

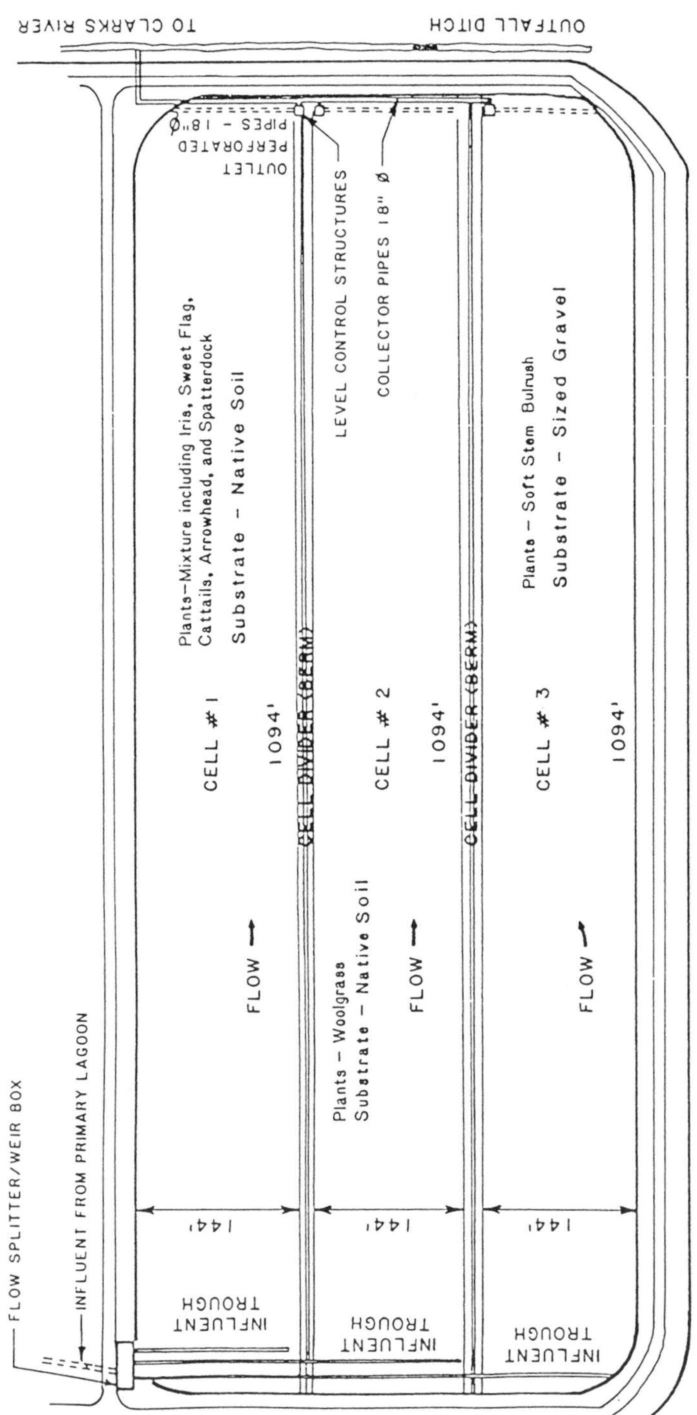

Figure 1. Gravel and surface flow marsh system, Benton, Kentucky.

These results at Benton are in line with the recommendations in the WPCF Manual of 1.2 ha/1000 m^3/d (11.2 ac/MGD) of the VSB cell, and 3 ha/1000 m^3/d (28 ac/MGD) of FWS cell for satisfactory BOD reduction. For significant ammonia conversion, however, the size of those cells needs to be increased to at least 4 ha/1000 m^3/d (40 ac/MGD). These results bring into serious question the commonly used assumptions regarding oxygen transport into the root zone in VSB systems, the loading rates used for those systems, and the effectiveness of CW systems for NH_3 removal, except perhaps at very low loadings.

Denham Springs, Louisiana, is an example of a VSB system designed and constructed under commonly used design criteria that this author believes come under serious question. The system was designed to treat 11,300 m^3/d (3.0 MGD) average flow following a 32 ha (80 ac) two-cell facultative lagoon to meet effluent limits of 10 mg/L BOD and 15 mg/L TSS monthly.[8] The CW system consists of three VSB cells of 2 ha (5 ac) each, 320 m (1050 ft) long and 66 m (215 ft) wide. The aggregate consists of 0.45 m (18 in) of 0.5 to 1.8-cm (1 to 4 in) limestone topped with 0.15 m (6 in) of 2.5-cm (1 in) gravel. Bulrush and canna lilies are planted in the cells. In 1989, the flow to the system averaged 9000 m^3/d (2.4 MGD) with an influent BOD of 27 mg/L. During that period, the 10-mg/L BOD permit limit was exceeded about half of the time, and a significant portion of the length of the cells experienced surfacing of wastewater. The design loading of this system is approximately 0.5 ha/1000 m^3/d (5.0 ac/MGD) at average flow, about 20% heavier than Benton, Kentucky, and about twice as heavy as the WPCF manual recommendation. It will be interesting to see how the performance of the Denham Springs system fares as it reaches design flow.

Two other troubling concerns with VSB systems are the large length/width ratios and the use of the larger size rock in the apparent hope that plugging will not be a problem. According to the design methodology in the WPCF manual, as the size of the aggregate goes up and the corresponding porosity goes down, larger area cells are required to give adequate opportunity for the biological processes to occur. In addition, the configuration of VSB cells should generally be much wider and not as long, a result based loosely on Darcy's Law. A system like Denham Springs, if designed along the lines of the WPCF manual for BOD removal only, would have the configuration shown below.

Denham Springs, LA Q average = 11,355 m^3/d (3 MGD)

Existing Design	New Design
Loading rate 0.5 ha/1000 m^3/d	1.6 ha/1000 m^3/d
(5 ac/MGD)	(15 ac/MGD)
Area Needed 6 ha (15 ac)	18 ha (45 ac)
Length × width 320 × 197 m	80 × 2270 m
(1050 × 645 ft)	(262 × 7440 ft)

It is obvious that a dramatically different design would result if the WPCF procedure were used. The important question is — which one is correct? Since the WPCF approach would require a 300% increase in land and in rock (the most expensive item in the system), its use would render the economics of the VSB, as described above, questionable as a viable alternative. The use of smaller aggregate would reduce the land area significantly and might bring the costs back to an acceptable level, but would still result in much wider and shorter cells. However, if the design were to provide for NH_3 removal, the size required would increase significantly. A FWS system, for comparison, would require approximately 25 ha (62 ac) for 11,355 m^3/d (3 MGD) using the WPCF approach for BOD removal but would increase to 70 ha (172 ac) for NH_3 removal.

The sobering implication of these large differences in the design of the Denham Springs system and the WPCF approach is that Denham Springs is typical of the design approach at a number of other operational or under-construction systems in Louisiana, as well as several in nearby states. An EPA-funded research effort presently underway is intended to evaluate more thoroughly some of these VSB systems and shed some new light on these systems.

A promising VSB system designed and constructed by TVA has been in operation since late 1988 at Phillips High School in Bear Creek, Alabama.[9] The system is designed to treat 76 m^3/d (20,000 gpd) of package treatment plant effluent to limits of BOD = 20 mg/L, TSS = 30 mg/L, and NH_3 = 8 mg/L. The VSB is sized at 0.2 ha (0.5 ac), resulting in a loading of 2.6 ha/1000 m^3/d (25 ac/MGD) with 0.3 m (12 in.) of pea gravel substrate. This loading rate is one fourth the original design of the Benton, Kentucky, VSB cell and less than one fifth the loading of the Denham Springs VSB, in spite of the fact that the influent wastewater is more highly treated (influent BOD averaged 13 mg/L, NH_3 10.7 mg/L). As a result, the performance of the Phillips VSB is excellent, as shown below:

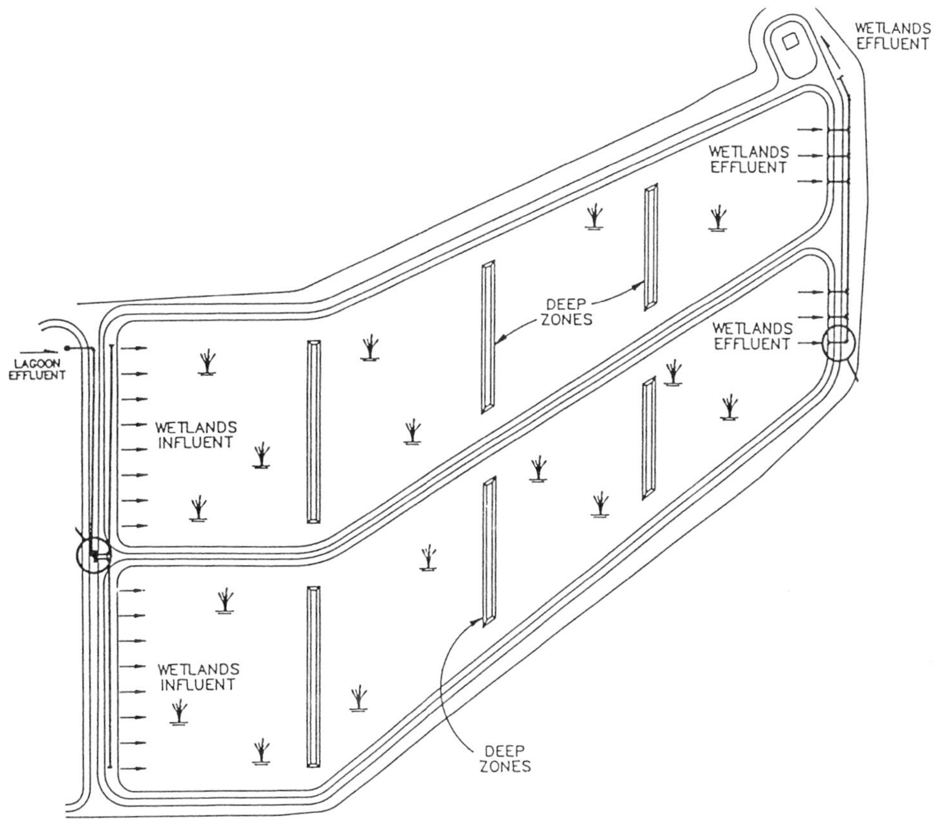

Figure 2. Fort Deposit, AL FWS Constructed Wetland Cells.

Phillips High School VSB

	BOD, mg/L	TSS, mg/L	NH_3, mg/L
Influent	13	60	10.7
Effluent	<1	<3	1.8

The high degree of nitrification is due, at least in part, to the low loading of the system. The nature of operation of the Phillips VSB may also contribute to its high-level treatment since the high school is not in session for most of the summer and therefore little or no loading is observed then. Low loading, combined with correspondingly low nutrient application and a relatively shallow gravel depth, could be responsible for the full bed root penetration that has been observed there, and unlike that observed at several other systems. This intermittent type of operation may assist the system in staying aerobic so the nitrification reactions can take place.

In FWS systems, two beneficial modifications that may address the problems of short-circuiting and inadequate oxygen transfer have been utilized in the design and construction of the cells. The Fort Deposit, Alabama, FWS system has two parallel cells that are subdivided by three deep zones that are approximately 1 m (3 ft) deeper than the rest of the cell to prohibit plant growth and redistribute the flow across the cell, (Figure 2). The zones are 5 to 8 m (15 to 24 ft) wide and may also serve as a reareation area to help keep the DO elevated. Fort Deposit is designed to treat 900 m^3/d (0.24 MGD) following a partially aerated lagoon to meet monthly limits of BOD = 10 mg/L, NH_3 = 2 mg/L on a year-round basis. The system has a total FWS area of 6.1 ha (15 ac) for a loading of 6.8 ha/1000 m^3/d (60 ac/MGD). This rate is in line with the WPCF approach, due to the fact Dr. Robert Knight of CH_2M-Hill designed the Fort Deposit FWS system and is the author of the Wetlands Systems chapter of the WPCF manual. The Fort Deposit system will begin a detailed performance evaluation this summer (1991) and results of the modifications utilized there will be forthcoming. Another FWS system designed by Dr. Knight for West Jackson County, Mississippi, incorporates the deep zones and also uses a shallow inlet area to maximize oxygen transfer. The FWS system is a 2300-m^3/d (0.6 MGD) two-cell (series) basin using 8.9 ha (22 ac) for a loading of 3.9 ha/1000 m^3/d (27 ac/MGD). The inlet zone in each cell is a very shallow 5 cm (2 in), gradually deepening to a normal depth of 0.3 to 0.5 m (12 to 18 in). This inlet area should increase the oxygen transfer capability to enhance nitrification and to meet the effluent limits of BOD = 10 mg/L, NH_3 = 2 mg/L monthly on a year-round basis. The loading rate in this system is relatively high for a nitrification system with limits this tight. The data collection effort, which will begin in earnest the summer of 1991, should provide an indication of the ability of systems of this type to meet strict discharge limits.

The story of CW systems in the southeastern part of the U.S. is obviously not over yet. It is also too soon to tell whether or not it will have a happy ending for the cities and towns that have built and are operating these systems. The evidence thus far suggests that the early optimism which produced the relatively heavily loaded VSB and FWS systems was not warranted and a serious reevaluation of those systems may be required. Hopefully, the newer developments in CW system design and construction will prove to be successful in remedying some of the problems described.

Until more definitive information is available regarding design protocols and expected performance, the people involved in the review, approval, design, and construction of these systems must exercise caution in their use. While under the right circumstances a CW system can be a very desirable treatment technology, care must be taken to assure that an appropriate system for the situation is selected.

REFERENCES

1. Reed, S. An Inventory of Constructed Wetlands Used for Wastewater Treatment in the United States, draft report prepared for U.S. Environmental Protection Agency Risk Reduction Engineering Laboratory, Cincinnati, OH, March, 1990.
2. Wetland Systems, *Natural Systems for Wastewater Treatment.* Manual of Practice FD-16, Water Pollution Control Federation, Alexandria, VA, 1990.
3. Steiner, G. R., J. T. Watson, and D. A. Hammer. Constructed Wetlands for Municipal Wastewater Treatment, presented at *Mississippi Water Resources Conference.* Jackson, MS, March 29–30, 1988.
4. Watson, J. T. unpublished data presented at *TVA-EPA Task Force Meeting.* Benton, KY, January 17–18, 1991.
5. Calculations performed by author based on unpublished data presented at *TVA-EPA Task Force Meeting.* Benton, KY, January 17–18, 1991.
6. Hammer, D. A. personal communication, 1990.
7. Kadlec, R. H. Analysis of gravel cell number three-Benton, KY Wetlands, report prepared under contract to TVA, December 18, 1990.
8. Rao, L. Unpublished presentation at *Constructed Wetlands Seminar.* EPA Region VI, Baton Rouge, LA, 1990.
9. Watson, J. T. *Design and Performance of the Constructed Wetland Wastewater System at Phillips High School, Bear Creek, AL.* TVA Technical Report Series, TVA/WR/WQ—90/5, Chattanooga, May, 1990.

PART 2

ENGINEERING

CHAPTER 7

Hydrological Design of Free Water Surface Treatment Wetlands

R. H. Kadlec, W. Bastiaens, and D. T. Urban

INTRODUCTION

Free-water surface wetlands are densely vegetated overland flow channels which function much like trickling filters. The main purpose of the vegetation in this system is to create sites for bacteria which degrade pollutants. Open water areas are often part of such systems, either by design or by virtue of natural events such as muskrat activity. A wide variety of shapes and depth distributions have been employed in more than 100 such systems. Attempts to correlate wetland performance for pollutant reduction with very simple design variables, such as hydraulic loading rate, retention time, and pollutant loading rate, have all failed to produce satisfying results.

It is widely presumed that simple first-order chemical rate laws apply for pollutant reduction and that constructed wetlands are approximately plug flow in their internal hydrology. These presumptions appear in all design manuals to date: the USEPA Manual,[1] the text of Reed et al.,[2] and the WPCF Manual of Practice.[3]

It is the purpose of this paper to show that the second of these presumptions is incorrect and that the design equations based upon it are therefore also incorrect.

Several recent studies have shown that constructed free-water surface wetlands are not plug flow devices (for example, see References 4–6). The mixing characteristics are intermediate between plug flow and well-mixed, even for long narrow wetlands.[7] Further, contact times are not always as great as the theoretical retention time calculated from the wetland empty volume and the volumetric flow rate. The hydraulic efficiency of a constructed treatment wetland is related to the basin morphology and the vegetation density patterns, as well as to the balance between evapotranspiration and precipitation.

Open water bodies display this same non-ideal flow character. Thackson et al.[8] studied residence time distributions in shallow basins of dredged material containment areas. They investigated the effect of the length-to-width ratio, width-to-depth ratio, and wind effects, and determined that the length-to-width ratio had the greatest effect on hydraulic efficiency, and was defined to be actual retention time divided by theoretical retention time. Seo[9] measured dispersion in long slender channels containing obstructions which displayed a high degree of internal mixing.

THEORY: NON-IDEAL CHEMICAL REACTORS

The theory of non-ideal flow patterns in chemical reactors is well understood: the ingredients were published several decades ago, and the subject has been in textbooks for 20 years.[10,11] Levenspiel introduces the subject as follows: "Thus far, we have restricted ourselves to two idealized flow patterns, plug flow and mixed flow. ... (But) deviation from ideality can be considerable. This deviation can be caused by channeling of fluid, by recycling of fluid, or by creation of stagnant regions in the vessel. ... Ignoring this factor can lead to gross errors in design. In this chapter we hope ... to deal with it in a rational manner consistent with present-day knowledge."

The basis for description is comparison of the real reactor to either or both of the ideal flow patterns: plug flow reactors (PFR) and continuous stirred tank reactors (CSTR). The conversion in the steady-flow PFR for a first-order, irreversible reaction is given by:

0-87371-550-0/93/$0.00+$.50

$$\frac{C}{C_i} = \exp(-k\tau) \tag{7–1}$$

where C = concentration, gm/m^3;
i = inlet;
k = first-order rate constant, L/hr or L/d;
and t = nominal retention time, d.

Retention time is the volume of liquid in the reactor divided by the volumetric flow rate. The formula for the CSTR is:

$$\frac{C}{C_i} = \frac{1}{1 + k\tau} \tag{7–2}$$

Long, narrow reactors are generally presumed to be PFRs; tanks with stirrers are presumed to be well mixed. If the device is close to being a PFR, the model for mixing and reaction is modified to contain a dispersion effect. The dispersion model was solved by Wehner and Wilhelm;[12] a simplified version for small amounts of dispersion is:

$$\frac{C}{C_i} = \exp(-k\tau) \cdot \exp\left[\frac{D}{uL}(k\tau)^2\right] \tag{7–3}$$

where D = dispersion coefficient, m^2/d;
u = velocity, m/d;
and L = length, m.

Because of the wide application of this model in chemical production plants, correlations for the dispersion constant have been developed, for both homogeneous (no packing) and packed bed pipe flow reactors.[10] For the wetland situation, there are both open water areas and zones of high stem and litter density which resemble packed beds. For almost all wetlands, packed bed correlations predict the correction term in Equation 7–3 to be close to unity, and therefore plug flow is predicted to prevail. In contrast, homogeneous flow correlations predict very large dispersion in the open water areas. Under these conditions, dispersion is too great for Equation 7–3 to apply. Experimentation has shown that, in fact, dispersion effects in wetlands are large.

The vegetated free-water surface flow conditions therefore require field experimentation to ascertain the relative importance of mixing. The principal experimental tool is the tracer study, in which an impulse of an inert substance is introduced into the reactor inflow and subsequently measured in the outflow. The actual retention time is the mean of the response pulse, and the degree of mixing is characterized by the variance.[10] The PFR response is a spike of the tracer exiting the device after one residence time (Figure 1), but dispersion blurs this spike. The CSTF response is a decreasing exponential. Intermediate degrees of dispersion lead to pulse responses of varying width. A second approach for describing nonideal flow patterns is the tanks in series model. The reactor is divided into N portions, each of which is considered to be a CSTR. The tracer response of this series approaches plug flow as the number of tanks increases (Figure 2).

More complicated reactors in general, and wetlands in particular, may be characterized by series and parallel combinations of PFR and CSTR portions. Open water areas are likely to be mixed, and densely vegetated areas are likely to be plug flow. The open water zones may exist on a variety of distance scales, ranging from large ponds in a macrophyte bed to the spaces between the plants in a vegetated zone. Any of these scales may be modeled, and have been in the non-wetland chemical reactor context.

The concentration reduction for a series or parallel combination of ideal units is readily calculated from Equations 1 and 2, with appropriate retention times for the component parts. For the simple case of the first-order irreversible reaction at steady state, two equal sized CSTRs in series produce

$$\frac{C}{C_i} = \left[\frac{1}{1 + k\tau / 2}\right]^2 \tag{7–4}$$

A CSTR and a PFR of equal size in series give:

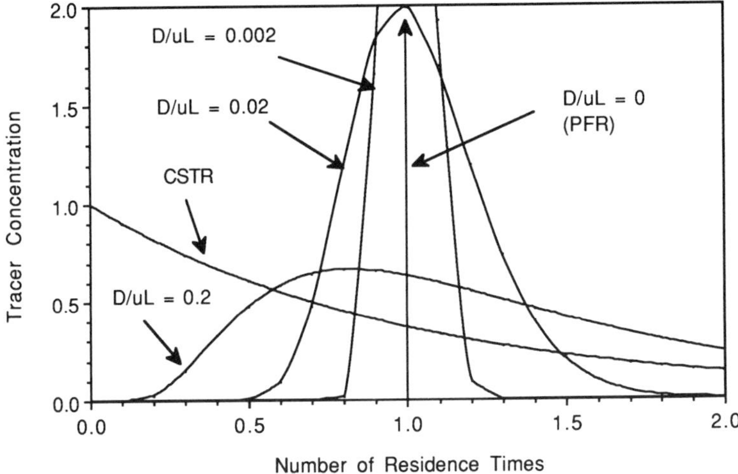

Figure 1. Tracer responses to an impulse input for ideal and partially back-mixed reactors.

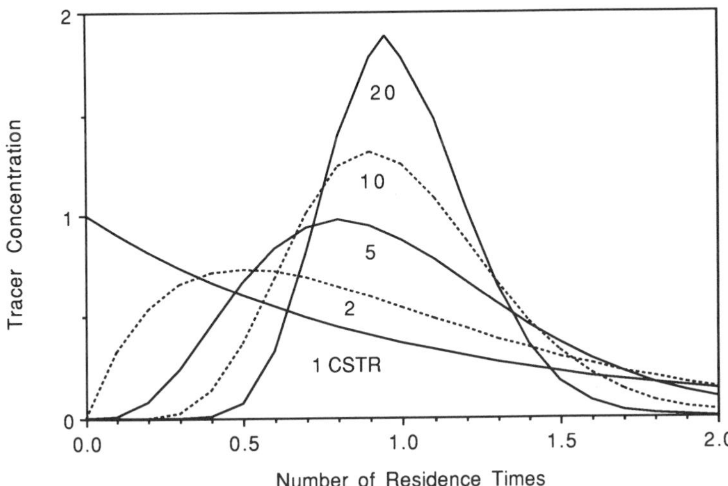

Figure 2. Tracer responses to an impulse input for tanks-in-series models.

$$\frac{C}{C_i} = \left[\frac{1}{1 + k\tau/2}\right] \cdot \exp(-k\tau/2) \qquad (7\text{–}5)$$

An alternative network model may be advanced, based on the concept of small deadwater zones which exchange pollutants slowly with the passing water in the flow path. This zonation of a densely vegetated wetland may occur on a scale approximating the plant clump size. Levenspiel[10] advances a discretized version of this model which contains three parameters: the fraction of deadwater, the exchange rate, and the number of stages. This modification also employs CSTR and PFR subunits.

THEORY: BOD REDUCTION IN FWS WETLANDS

The design literature[1-3] presumes BOD reduction to be a first-order, irreversible reaction occurring in a PFR. However, a certain fraction of the BOD is presumed to "settle out" in the "headworks" of the wetland. This results in a slight modification of Equation 1:

$$\frac{C}{C_i} = F \cdot \exp(-k\tau) \tag{7–6}$$

where F is the fraction of the BOD which does not settle out. The retention time in this equation is corrected for the fraction of wetland volume occupied by plants and litter. Kadlec[13] has pointed out that this equation is not valid for conditions of significant evapotranspiration or rainfall; nor is it valid for transient conditions, such as those occasioned by pump shutoffs or startups, rain events, and the diurnal cycle in evapotranspiration. The product $k\tau$ is further presumed to depend on specific plant surface area (A_v = area per unit volume, m^2/m^3), nominal hydraulic retention time (HRT = total volume/volumetric flow rate, d), and void fraction (ϕ):

$$k\tau = 0.7 \cdot K_T \cdot A_v^{1.75} \cdot HRT \cdot \phi \tag{7–7}$$

The constant K_T is expected to be temperature dependent, but there are wetland systems for which this is not true. It should be noted that the parameter A_v has never been measured in a wetland, although a value of 15.7 m^2/m^3 is recommended in design manuals.[1-3] The value given for ϕ is 0.75, but this parameter has not often been measured.

The uncertainties listed above lead to a large "site-specific" variability in expected performance. Some of this variability can be removed by properly dealing with the flow and mixing patterns. Data from several wetland sites are reviewed here to illustrate principles and results.

RESULTS

Benton, Kentucky Wetlands

The Benton, Kentucky wetland treatment system has been described in detail elsewhere.[14] Cell 2 is an FWS wetland, 44 × 331 m, and tapering from an inlet depth of 10 cm to a maximum depth at the outlet of 50 cm, for an average depth of 25 cm. During the period discussed here, early September 1990, the flow was 1500 m^3/d; thus the HRT was 3464 min. Relatively stable flow conditions prevailed throughout the test: diurnal variations were ±8%, and outflow and inflow differed only within ±5%. Rhodamine WT was used as a tracer in a pulse experiment begun on September 6, 1990. The response curve at the wetland outflow (Figure 3) is intermediate between well-mixed and plug flow. The mean residence time calculated from the response curve is 3344 min, indicating a flow porosity of 97%.

A patterns-of-flow model consisting of two stirred sections, each being 30% of the wetland volume, and a plug flow section comprising the remaining 40%, adequately describes the steady flow tracer response (Figure 4). No attempt to optimize the fit has been made. Given the slope of the cell bottom, approximately the first 210 m would be in plug flow, the remaining 120 m are described as two mixed cells in series. The last section of cell 2 does not contain dense emergent macrophytes. The mixing was observable: the duckweed on the surface of the last section of the cell was sometimes observed moving upstream under the influence of light breezes (ca. 0.4 m/s = 5 mph).

The spatial details of internal water movement are not contained in the overall system tracer response. Responses were also recorded at Benton at five locations, equidistant 250 m from the inlet and equally spaced across the wetland (Figure 5). There are considerable differences among these responses: the retention time to this distance is 1620 ± 240 min (±15%). But more importantly, the amount of dye passing these sample points varies by ±50%. This indicates large transverse variability in flow rates.

The average travel time to a specific point in the wetland is given by the mean of the tracer response curve; not by the peak in the response. The peak in Figure 3 is at about 2200 min; the mean is at 3344 min. Therefore, use of the peak travel time in any of the model equations will result in serious errors in the determination of the rate constant: on the order of 50% for this cell.

The wrong flow model also leads to errors in the determination of the rate constant. At the 90% conversion level, the difference between the PFR and (PFR + 2 CSTR) combination is 16%.

Listowel, Ontario Wetlands

This wetland facility was studied from 1980–1984.[14] A dye tracer study on cell 2 produced the same sort of results as at Benton. Cell 2 was 334 m long × 80 m wide × 20 cm deep, and received 17 m^3/d during the

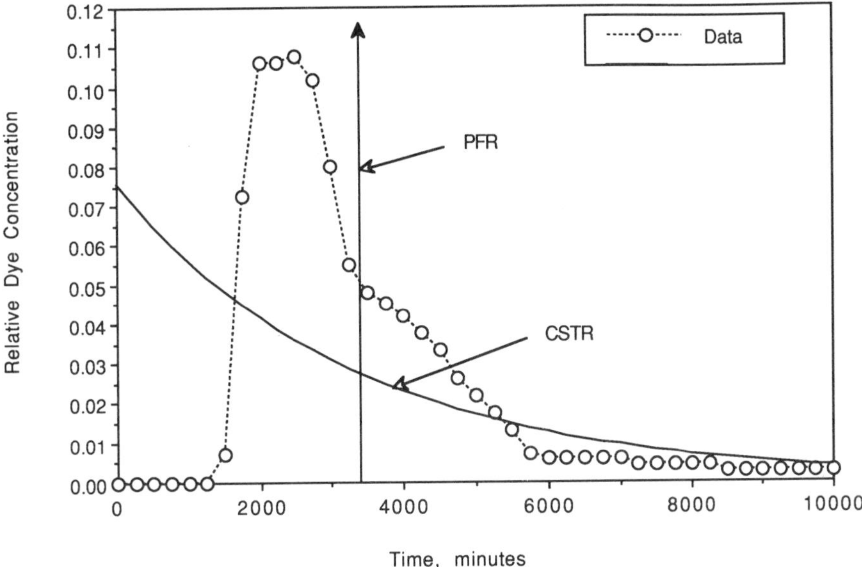

Figure 3. Response to a rhodamine WT pulse at the outlet of Benton cell 2 in September 1990.

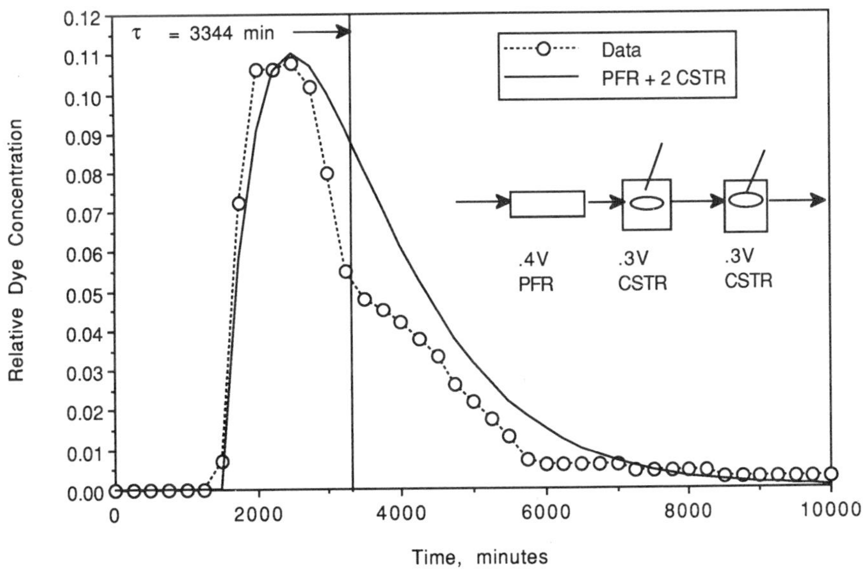

Figure 4. Patterns of flow model for Benton cell 2 compared to data.

test. The HRT was thus 10 days. The shape of the dye pulse for a test run in May 1982 shows considerable dispersion (Figure 6). From the width of the pulse, it may be determined that a reasonable flow model is four stirred tanks in series. The error in rate constant due to the wrong flow model would be about 25% at 80% conversion.

Data were acquired on actual contact time for five wetlands at Listowel, over a 4-season period of operation. The ratio of actual measured contact time to nominal contact time was 1.26 ± 0.8 (N = 27). The range was from 0.39 to 3.78.

Data were acquired for BOD reduction over the four seasons, using weekly and biweekly sampling at five points along the length of the wetland. There is little evidence of any temperature effect. The data trend downward as expected (Figure 7), and it is easy to presume first-order behavior. That model appears to fit

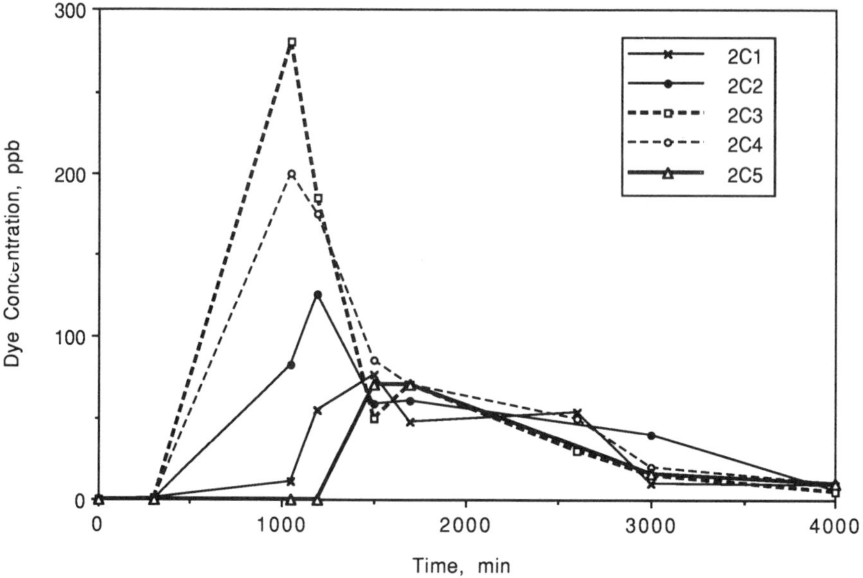

Figure 5. Transverse variability in tracer response at the 3/4 distance for Benton cell 2. Rows 1 to 5 are parallel to flow.

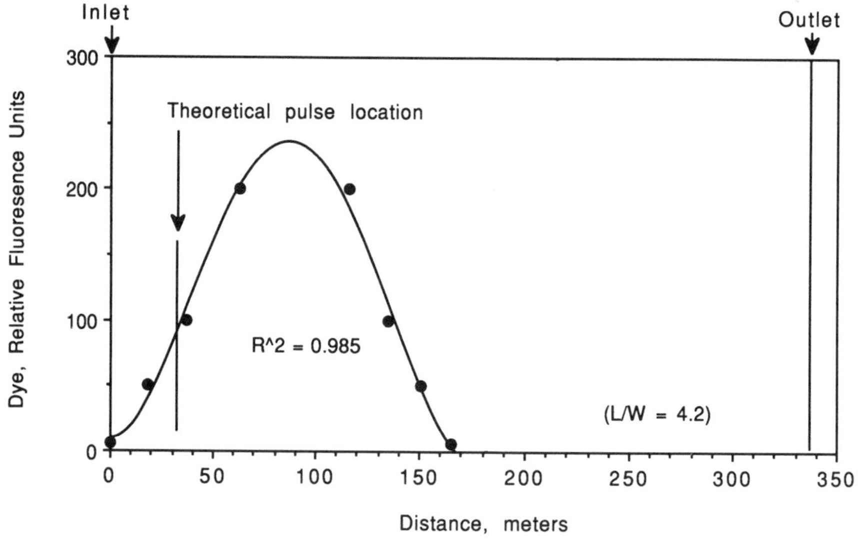

Figure 6. Internal measurement of a dye pulse along one row parallel to flow at Listowel, Ontario, cell 2.

reasonably well, except it under-predicts the feed concentration. A premultiplier of 0.6 appears on the exponential correlation. In fact, it may be easily shown theoretically that a PFR regression of CSTRs in series necessarily produces this apparent "headworks settling." The F-factor in the design manuals thus appears to be an artifact of an incorrect flow model.

Richmond, New South Wales Wetlands

Fisher[4] presents data on a wetland containing the submergent, *Myriophyllum aquaticum*. The wetland was 100 m long × 4 m wide × 50 cm deep and run at an HRT of 6.3 days for the tracer test. The tracer test produced a mean retention time of 5.3 days (Figure 8). Fisher[4] concluded that the pattern resembled "a completely mixed

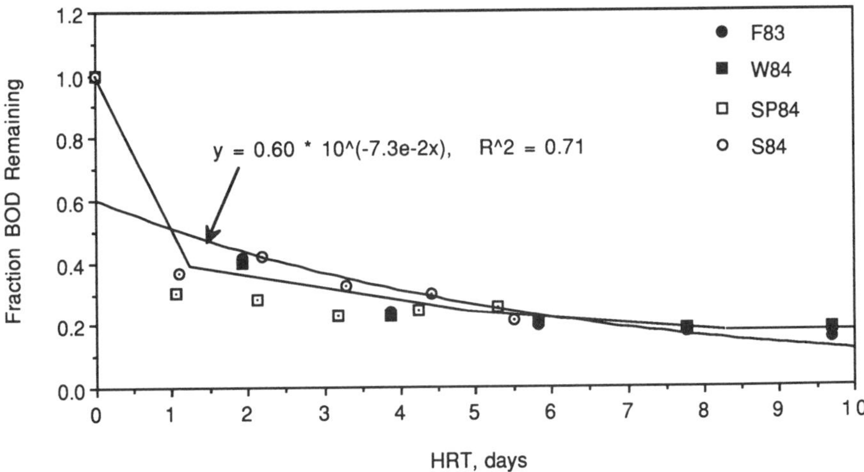

Figure 7. BOD reduction in cell 3 at Listowel. The regressed line is for a first-order reaction and the PFR assumption.

hydraulic regime after an initial lag of about 3.5 days." However, the pattern appears to more closely resemble a 2-day PFR in combination with two CSTRs. It is not possible to distinguish the internal sequence of these units from the data given; one possibility is shown in Figure 8. Unexplainably, after determining the flow patterns in his wetlands, Fisher proceeds to assume plug flow for chemical conversion analyses.

Des Plaines, Illinois Wetlands

Wetlands have been reconstructed at this northern Illinois site for purposes of river water quality improvement. Water is pumped from the Des Plaines River to the wetlands for 1 to 8 hours each weekday, and is allowed to drain back to the river from the wetland outfall. All water budget components are measured: inflow, outflow, infiltration, precipitation, and those meteorological variables that allow calculation of evapotranspiration. Complete dynamic water budgets are thus available for each of the four wetland cells. The dynamic water mass balance model calculates the water level in the wetland as a function of time. The closure of these overall water balance calculations has been excellent.[15]

Lithium chloride was used as a tracer in pulse test experiments. Approximately 15 kg of lithium chloride were dissolved in 100 L of water and dumped quickly into the inlet. Samples were withdrawn every 2 hours at the outlet and analyzed for lithium via atomic absorption spectrometry. Mass recovery of lithium was within experimental error. The tracer response of this wetland reflects the pumping activities as well as the overall effects of basin morphology. It is clear that this wetland is not operating in plug flow.

Cell EW3 (Figure 9) is kidney shaped and comprises about 2.33 ha (5.75 acres). It receives an average of 1000 m^3/d, and averages about 26 cm in depth. The central area is deep, with only sparse submergent vegetation, while the fringes are shallow and densely covered with cattail (*Typha latifolia*). A series network of three CSTR units is envisioned to describe the main deep water zone, followed by a PFR unit which describes the outlet region. The shallow side zones are not in the direct flow path and are envisioned to exchange water with the central units via wind-driven and gradient-driven flows. The water mass balance model for this network is required to yield identical, but time-varying water levels in all units because the water surface has been measured to be essentially flat. That time-varying water level is the result of the overall mass balance model calculations. This constrains the net interchange flows. Side zones were assumed to be of equal size and to experience equal interchange flows.

Due to intermittent pumping, rainfall, and evapotranspiration, the water and tracer mass balance models must be solved numerically. A fourth-order Runge-Kutta routine on a desktop computer was adequate for this task. Four model parameters were calibrated from data. The void volume was adjusted to produce the measured retention time; that void fraction was as low as 70%. A plug flow zone was added with retention time matching the delay in first appearance of tracer (Figure 10, about 1 day). The other adjustable parameters were the fraction of the wetland in the side zones and the interchange flow rate. These last two parameters were varied until the sum of the squares of errors was minimized.

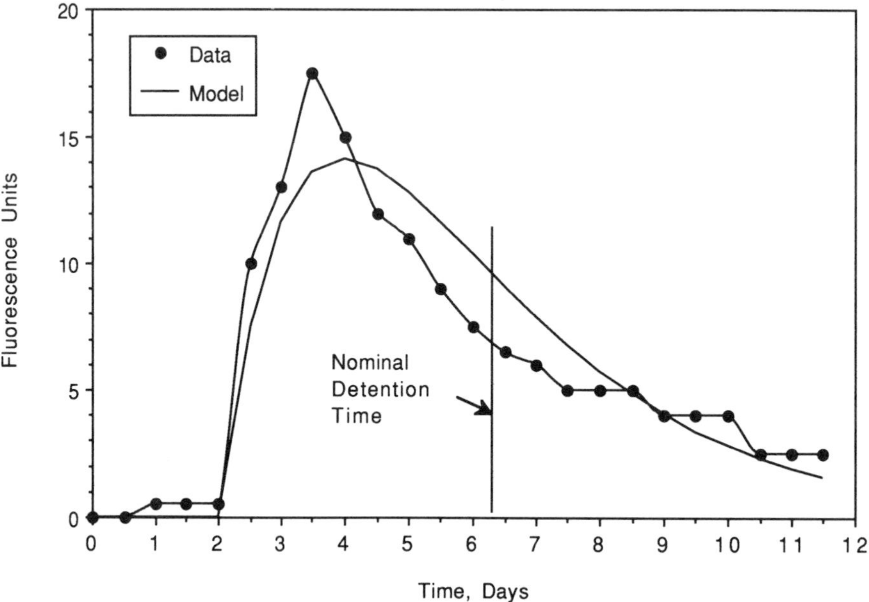

Figure 8. Response of the Richmond, NSW wetland, containing submerged aquatics, to an impulse tracer test.

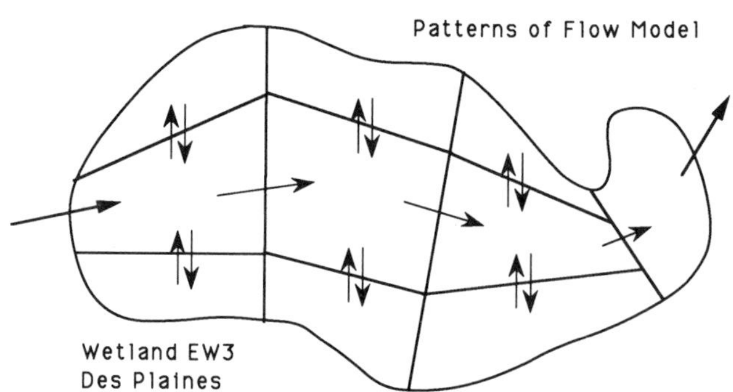

Figure 9. Patterns of flow model for the Des Plaines wetland cell EW3.

This network adequately describes the overall flow patterns in the wetland, as evidenced by the tracer response (Figure 10). It should be noted that there is a noticeable amount of "noise" in the tracer response curve. This is probably due to pumping events, wind events, and bioturbation, the latter being attributable to carp and researcher travel at this extensively studied site. There are interesting and significant implications of this micromixing for chemical reaction rates, but this is beyond the scope of the present study.

Predictions may be made of the response of this model to the actual transient conditions for first-order irreversible reaction kinetics (Figure 11). It is then possible to back-calculate the rate constant under the PFR assumption. For $k = 0.100\ h^{-1}$, the PFR assumption returns a value of $k = 0.023\ h^{-1}$. For $k = 0.0100\ h^{-1}$, the PFR assumption returns a value of $k = 0.0061\ h^{-1}$. The errors are factors of 4 and 2, respectively.

CONCLUSIONS

Flow and mixing in FWS wetlands follows a pattern describable by a small number of series and parallel CSTRs and PFRs. The flow distribution is not uniform from side to side in wetlands. Benton and Listowel both show large differences. A single PFR does not adequately describe many systems.

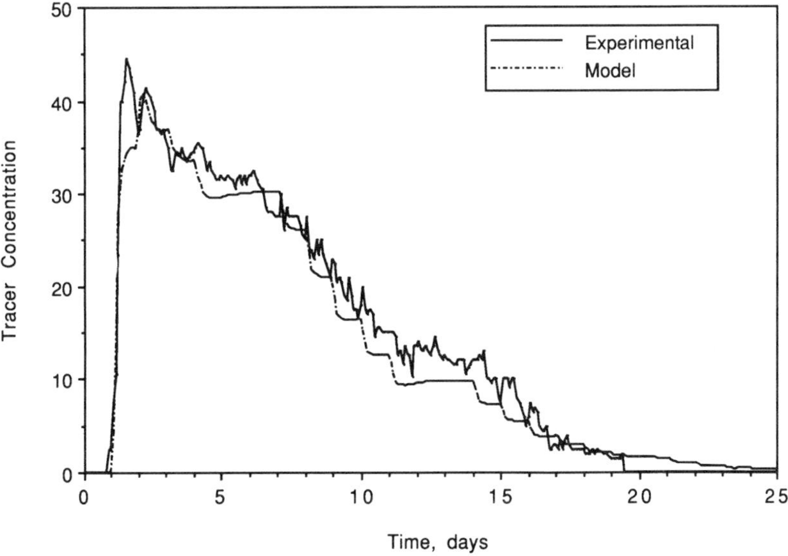

Figure 10. Tracer response model for Des Plaines cell EW3. Comparison of data and calibrated model.

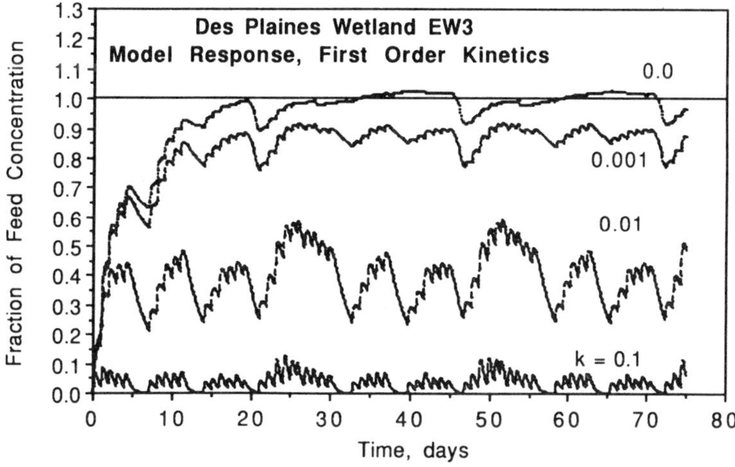

Figure 11. Des Plaines cell EW3: model calculations using first-order kinetics.

The above results contradict many of the popular design equations in the literature. Those equations were based on assumptions that appeared either reasonable or expeditious at the time of their presentation. There are now several dozen constructed wetland systems that are returning data from which the assumptions may be checked. The above results are representative of current FWS wetlands. The range of length to width ratios was from 2:1 (Des Plaines) to 25:1 (Richmond). Percentage cover by dense emergent macrophytes ranged from 0% (Richmond) to 100% (Listowel). Contact times varied from 1 to 10 days. Flows were nearly steady (Benton) in some cases, and extremely variable (Des Plaines) in other cases. Here are the implications of this recent detailed research.

1. *Assumption*: Plug flow properly approximates the flow pattern in FWS wetlands. *Conclusion*: It doesn't. Pulse tracer tests should produce a sharp concentration peak at the wetland outlet after one nominal retention time. A small amount of dispersion would symmetrically broaden the peak by a few percent of one residence time. The systems described here displayed very broad, asymmetrical response peaks.

2. *Assumption.* Even if the wetland is not plug flow, we can still approximate performance that way. It won't make much difference in the determination of rate constants, or conversely, in design. *Conclusion*: It does make a significant difference. Rate constants can be in error by as much as a factor of four.
3. *Assumption.* If we don't assume plug flow, calculations will be impossibly difficult. *Conclusion*: Calculations may still be made with pocket calculators or spreadsheets. The techniques are in textbooks.
4. *Assumption.* The void fraction of a FWS wetland is about 75%, and therefore the actual contact time is 75% of the nominal retention time. *Conclusion*: This is not correct. There is wide variability, depending on meteorology, vegetation, and short-circuiting.
5. *Assumption.* The rate constant should be proportional to some function of the submerged substrate density, which is 15.7 m^2/m^3. *Conclusion*: The first part of this may be correct, but the parameter is nearly impossible to measure. These parameters should be combined into an overall rate constant.
6. *Assumption.* The equation describing conversion in a PFR is the exponential form (Equation 1). *Conclusion*: This is incorrect if there is significant rainfall or evapotranspiration.[13]
7. *Assumption.* There is a fraction of the BOD which settles out in the headworks of a wetland. *Conclusion*: This is probably incorrect, and is more likely the result of presuming the wrong flow pattern in data reduction.

The above are all peculiar to the case of a first-order reaction. Nitrate reduction has been shown to follow first-order kinetics in wetlands, but other pollutant reductions may not. BOD is produced by wetland decomposition processes; the appropriate mechanism should therefore contain a production term as well as a disappearance term. Other substances, notably nitrogen compounds, follow a sequential set of kinetics which may be reactant limited. Both oxygen and carbon are involved in the nitrogen pathways. For these more complicated kinetic mechanisms, it is also necessary to know the non-ideal flow pattern.

REFERENCES

1. USEPA (U.S. Environmental Protection Agency). Design Manual: Constructed Wetlands and Aquatic Plant Systems for Municipal Wastewater Treatment. USEPA 625/1-88/022, 83 PF, 1988.
2. Reed, S. C., E. J. Middlebrooks, and R. W. Crites. Natural Systems for Waste Management and Treatment. McGraw-Hill, New York, 1988.
3. WPCF (Water Pollution Control Federation). Wetland systems, in Manual of Practice: Natural Systems. MOP FD-16 WPCF, 1990.
4. Fisher, P. J. Hydraulic characteristics of constructed wetlands at Richmond, NSW, Australia, in Constructed Wetlands in Water Pollution Control. Pergamon Press, New York, 1990, 21.
5. Urban, D. T. Methods of Determining Residence Time Distribution in a Reconstructed Wetland. MSE Thesis, Illinois Institute of Technology, 1990, 76.
6. TVA (Tennessee Valley Authority). Unpublished dye study results from September, 1990 at Benton, KY.
7. Hershkowitz, J. Listowel Artificial Marsh Project Report. Ontario Ministry of the Environment, Water Resources Branch, Toronto, 1986, 251.
8. Thackson, E. L., F. D. Shields, Jr., and P. R. Schroeder. Residence time distributions of shallow basins, J. Environ. Eng., ASCE. 113:1319, 1987.
9. Seo, I. W. Laboratory and numerical investigation of longitudinal dispersion in open channels, Water Resources Bull. 26:811, 1990.
10. Levenspiel, O. Chemical Reaction Engineering. 2nd ed. John Wiley & Sons, New York, 1972.
11. Fogler, H. S. Chemical Reactor Engineering. 2nd ed. Prentice-Hall, New York, 1991.
12. Wehner, J. F. and R. H. Wilhelm. Chem. Eng. Sci. 6(89):1956.
13. Kadlec, R. H. Hydrologic factors in wetland water treatment, in Constructed Wetlands for Wastewater Treatment. D. A. Hammer, Ed. Lewis, Chelsea, MI, 1989, 21.
14. Steiner, G. R., J. T. Watson, D. A. Hammer, and D. F. Harker. Municipal wastewater treatment with artificial wetlands — a TVA/Kentucky demonstration, in Aquatic Plants for Water Treatment and Resource Recovery. K. R. Reddy and W. H. Smith, Eds. Magnolia Publishers, Orlando, FL, 1987, 923.
15. WRI (Wetland Research, Inc.). Hydrologic monitoring, 1991, in The Des Plaines River Wetlands Demonstration Project. Progress Report to USEPA, Grant CR-817120-01-0. 1192, 1991.

CHAPTER 8

Hydrogeology and Ground Water Monitoring, Constructed Wetlands System, Perdido Landfill, Escambia County, Florida

P. H. Dohms

INTRODUCTION

Landfill leachate at the Perdido Landfill is managed through a unique system that incorporates a composting process and a constructed wetlands. This system reduces the volume of both solid waste and leachate and also provides leachate (wastewater) treatment. Discharge from the wetlands is returned to the primary leachate collection pond, thereby insuring a closed system. Regulatory concern, however, has been focused on the potential for the escape of leachate from the system, principally via infiltration through the bottom of the wetlands. The Florida Department of Environmental Regulation (FDER) specified that, notwithstanding the presence of a constructed clay liner, groundwater monitoring would be required to detect any evidence of degradation of groundwater quality.

Prior to designing the water quality monitoring system, it was necessary to undertake an exploration program of the constructed wetlands project site and its environs. A soil boring program was completed in 1988–1990 to provide lithologic and stratigraphic information. This was supplemented in 1990 by groundwater exploration wells and piezometers at the perimeter of the project site.

The constructed wetlands are underlain by a south to southwesterly dipping sequence of sands, silts, and clays. A shallow regime of perched groundwater was found above one of the clay horizons and was later removed during project construction.

The first aquifer of any importance underlies the project site at depths of 21 to 40 ft below the bottom of the constructed wetlands liner. The groundwater in this aquifer was found to be flowing west to southwest at a relatively gentle gradient; the direction of groundwater flow appears to be controlled by the lithology and attitude of the strata underlying the site. Both the quantity and rate of flow of water beneath the project area in this aquifer proved to be modest, and it is apparent that the formation is capable of yielding only small amounts of water for beneficial use.

The groundwater monitoring system was designed to meet applicable FDER rules as well as the need to fulfill certain objectives that would allow the early detection of significant leakage through the liner. Those monitoring system objectives include the ability to detect increases in the rate of flow of groundwater down-gradient from the site, any changes in groundwater chemistry, and the appearance of contaminants. The system that was installed in 1991 includes background, intermediate, and down-gradient monitor wells. These wells were designed to conform to rigorous construction specifications.

PERDIDO LANDFILL CONSTRUCTED WETLANDS

Perdido Landfill, located on the western boundary of Escambia County, Florida, is a municipal landfill that accepts some 1000 tons per day of industrial and municipal refuse (Figure 1).[1] Leachate generated within the landfill cell system is routed to a lined 30-million-gallon capacity primary treatment pond where it is joined by rainfall and approximately 30,000 gallons per day of domestic septage while undergoing constant aeration. From the primary treatment pond, leachate is first pumped into a small holding pond from where it is sprayed

0-87371-550-0/93/$0.00+$.50

over composting shredded municipal waste in a 10-acre area.[1] Leachate draining from the composting area is captured in a retention pond and is then pumped into an in-series constructed wetlands through which it flows by gravity back to the primary treatment pond.[1,2] While constructed wetlands treatment of municipal landfill leachate is becoming relatively common,[3,4] it is believed that the Perdido Landfill is one of the few where landfill leachate is also used in the reduction of solid waste volume by composting and the entire system is designed to be a closed loop. Figure 2 shows the general facility layout.

The wetlands area itself occupies some 15 acres and consists of 13 individual cells, plus one surge pond and two filtration ponds. A typical cell is 330 ft E-W by 50 ft N-S and is planted in a single species.[2] Inlet elevation to the wetland area is about 97 ft above sea level, while the invert at the outlet is at an elevation of 71 ft. The fall of 26 ft occurs through a leachate travel path of some 4200 ft (0.6% slope). This entire complex was built overlying a through-going constructed clay liner; the berms dividing the cells were built on top of the liner and are not intended to be water-tight.[1]

The clay liner is constructed of a natural clay that was excavated at Perdido Landfill. Its composition was determined through electron microscopy to comprise approximately equal parts of kaolinite, illite, and quartz flour.[5] Tests of the hydraulic conductivity of the clay have been underway for over a year, using Perdido Landfill leachate as the test fluid. No breakthrough has been observed in the permeameter, and latest calculations suggest that the natural hydraulic conductivity of this clay material is somewhat lower than 1×10^{-10} cm/s.[6] Strict quality control measures were employed during the placement of the liner.[7]

Regulatory Concern

When the Escambia County Solid Waste Department first proposed the use of a constructed wetlands system to treat landfill leachate at the Perdido Landfill, the FDER responded with caution. Based on review of the proposed construction specifications of the entire system, the concerns of FDER were centered on two specific subjects:

1. The potential for the escape of leachate through the floor of the constructed wetlands cells. This concern was based on the necessity to maintain a 2-ft depth of leachate in the wetlands — a depth which would place a hydraulic head on the clay liner. Imperfections could allow the discharge of potentially significant quantities of leachate through the liner.
2. The potential for the escape of leachate through the floor and walls of the two holding ponds adjacent to the constructed wetlands. This concern was based on the fact that significant depths of leachate would be maintained in each pond for operational reasons. Large hydraulic heads on the composite liners could allow the discharge of significant quantities of leachate into the ground via liner imperfections.

Based on these concerns, the FDER mandated that the Escambia County Solid Waste Department complete an investigation of the geology and hydrogeology of the constructed wetlands site and that a ground-water monitoring system be designed and installed.

SITE GEOLOGY

The Perdido Landfill is located on the western edge of Escambia County, within the Gulf of Mexico Coastal Plain. The site is underlain by sands, silts, clays, and (at depth) limestones of Mesozoic to Recent age. On a regional basis, this area is on the north flank of the Gulf Coast Sedimentary Basin and the east flank of the Mississippi Embayment. These features account for a regional southwestward dip and gulfward thickening of most formations from surface down to the basal Cretaceous deposits.[8,9] The regional stratigraphy is shown in Figure 3,[9] although the emphasis of the remainder of this presentation is on the uppermost of the units shown in that figure.

From a physiographic standpoint, the Perdido Landfill is near the western edge of the Western Highlands province of the Florida Panhandle. The site is located near the southerly edge of the uplands, a short distance north of the Coastal Lowlands.[10]

Pleistocene terrace deposits of the Citronelle Formation extend from land surface to a depth of 300 to 400 ft beneath the site. These deposits, of fluvial and deltaic origin, are predominantly white to light brown, poorly sorted sands, with numerous lenses of silty sands, silts, and clays.[8] The Citronelle contains hardpan layers, up to 4 ft thick, and consisting of sand cemented by iron oxide (known locally as "iron rock"). Occasional fragments of petrified wood are found associated with these hardpan layers.[11]

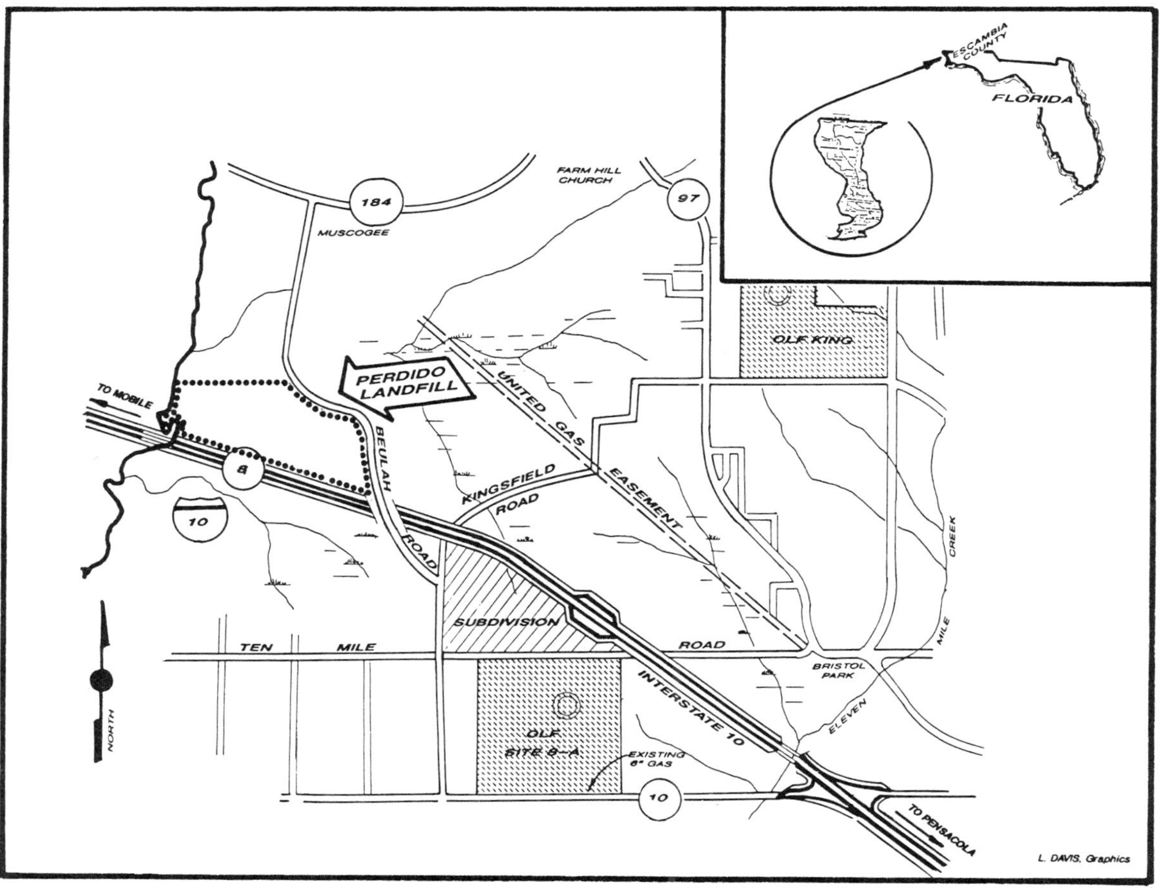

Figure 1. Location map for Perdido Landfill, Escambia County, Florida.

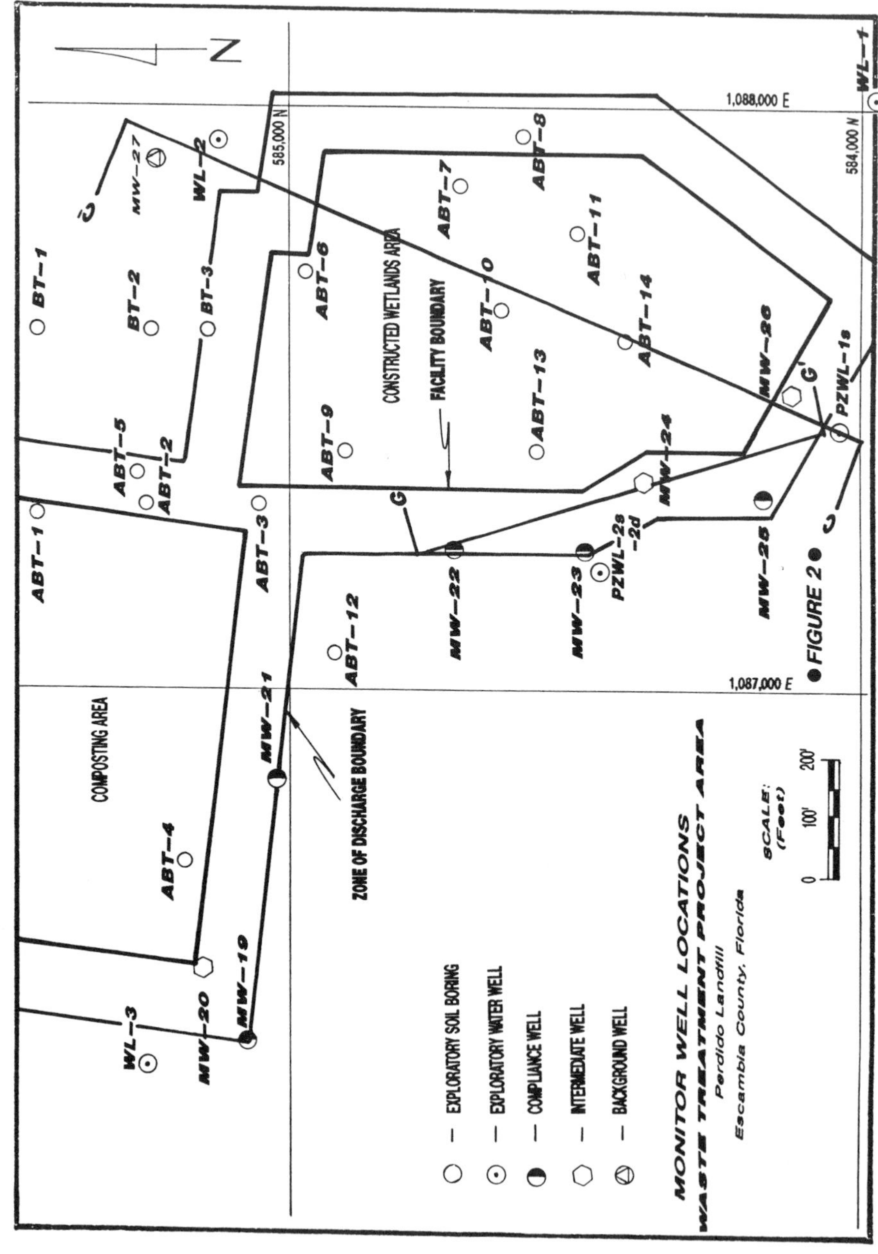

Figure 2. Generalized site map.

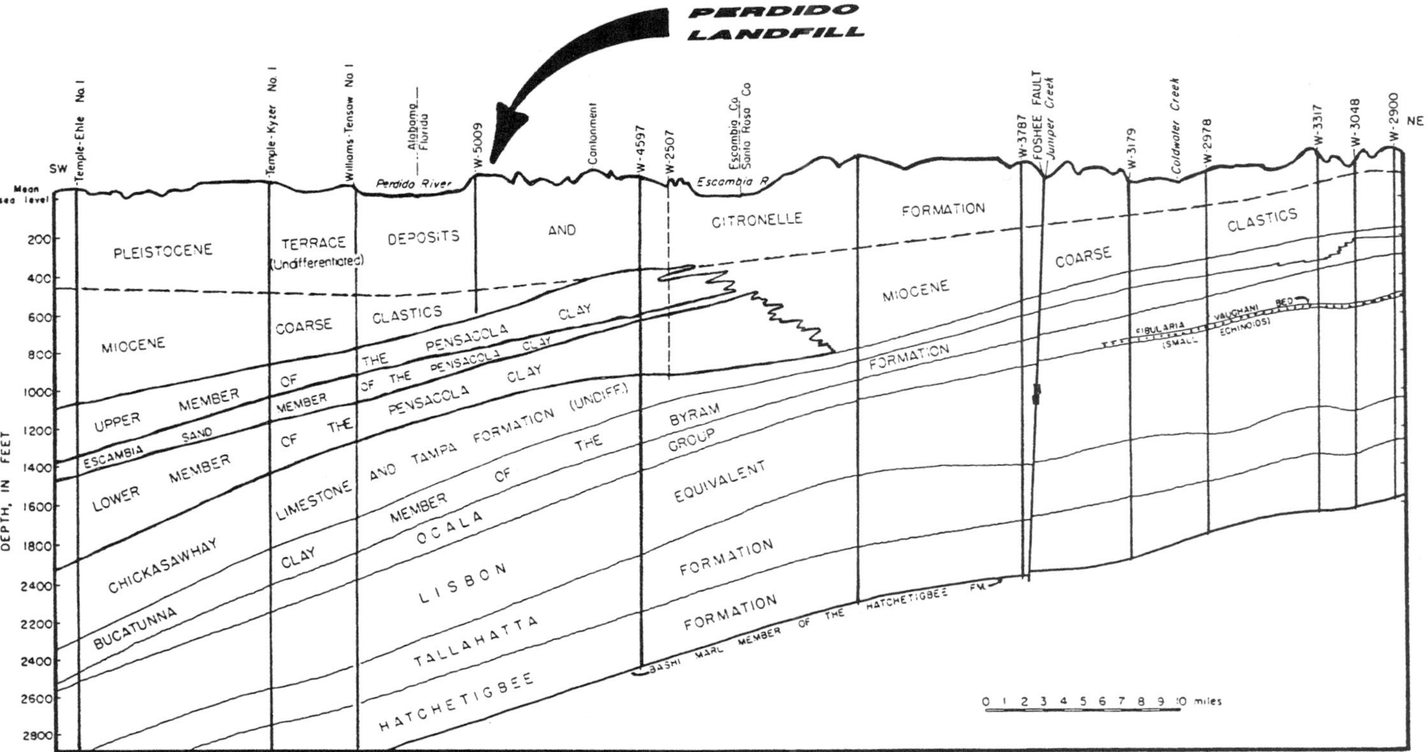

Figure 3. Geologic section across Escambia and Santa Rosa Counties. (See Reference 9.)

Geologic crosssections through the constructed wetlands are shown in Figures 4 and 5. Figure 4 is aligned from SW to NE, roughly parallel with the groundwater gradient beneath the wetlands area. Figure 5, aligned along the western margin of the wetlands, is roughly parallel with the geologic dip beneath the site. Strata immediately beneath the constructed wetlands have been the focus of this study; they are divided into four general groups of beds.

The uppermost horizon, absent over much of the constructed wetlands area, is comprised of a silty sand. Color was seen to be mostly grey, with varying proportions of tans, yellows, and browns. The thickness of this horizon varies from zero (absent) to 10 feet. Prior to construction of the wetlands, this horizon was seen to be occupied by a perched aquifer of limited areal extent.[12]

The next horizon beneath the wetlands is comprised of mixed lithologies ranging from stiff clays of an important upper confining layer to sands and poorly sorted gravels. The boundaries and topography of the top of the clay within this horizon are shown on Figure 6; this clay was found to be contiguous with the clay excavated for construction of the wetlands liner. Semi-confining silts and sandy clays are also present, and beds of the mixed horizon were found to be acting as confining layers in the monitor wells at the southwest edge of the wetlands. A general southerly inclination of about 3% is recognized (see Figures 5 and 6), with the thickness of the mixed zone varying from about 16 to 20 ft.[13] During construction of the clay liner which underlies the wetlands, clays within this horizon were exposed and graded to act as portions of the foundation layer beneath the constructed clay liner.

The second aquifer underlies the mixed horizon. This horizon, as described below, is the first through-going aquifer beneath the wetlands and was the principal object of regulatory concern. It is generally comprised of poorly sorted sands with some silty sands and occasional iron rock horizons. As can be seen in Figure 5, this horizon is wedge-shaped and thins significantly from NW to SE.

The fourth and deepest horizon that was explored during this program is the lower confining layer that underlies the second aquifer. Since no boring was allowed to penetrate this unit, its thickness is unknown. The clay within this formation was tested and was found to have geotechnical properties virtually identical to the clay used to line the wetlands. A short-term test of the undisturbed hydraulic conductivity, for instance, returned a value of somewhat less than 6.4×10^{-10} cm/s.[14]

SITE HYDROGEOLOGY

The hydrogeology of the constructed wetlands site at Perdido Landfill is relatively straightforward. The horizons of interest were found to comprise two aquifers and two confining layer systems. The upper, perched aquifer, as was mentioned, was largely removed and almost completely drained during wetlands construction; it is not discussed further. The rest of this section is a discussion of the second aquifer, the focus of regulatory concern.

As can be seen in Figure 5, the second aquifer is a wedge-shaped unit which diminishes in thickness towards the south. It is both overlaid and underlain by confining to semi-confining units. These have created an interesting feature. As can be seen from the water elevation plots in Figures 4 and 5, the second aquifer is under unconfined, water table conditions beneath the northern end of the constructed wetlands. This means that the top of the saturated zone of the aquifer is at the same elevation as the water table. Water in wells MW-27 and MW-22, which penetrate the unconfined portion of this aquifer, stands at the elevation at which it was intersected in the borehole (Figures 4 and 5).

Beneath the southern end of the wetlands, however, the second aquifer is confined. The confining layers of the mixed horizon, sloping southward, intersect and dip beneath the water table so that the water in the second aquifer is under pressure. Water elevations in wells MW-24, MW-25, and MW-26, therefore, rise significantly higher in each well than the elevation at which saturated strata were intersected in the respective well borings (Figure 5).

The other significant observation about the second aquifer is that it appears to be thinning towards the south. Indeed, the saturated thickness of the second aquifer is seen to decrease from a maximum of 40 ft in MW-22 to a minimum of 17 ft in MW-25 (Figure 5).[13]

The direction and gradient of groundwater flow were examined on a number of occasions. Groundwater flow directions of S60°W to S75°W have been determined and groundwater gradients of 0.003 to 0.009 feet per foot have been measured. Long-term studies are planned to track these parameters, as variations from a baseline might be indicative of discharge from the wetlands.

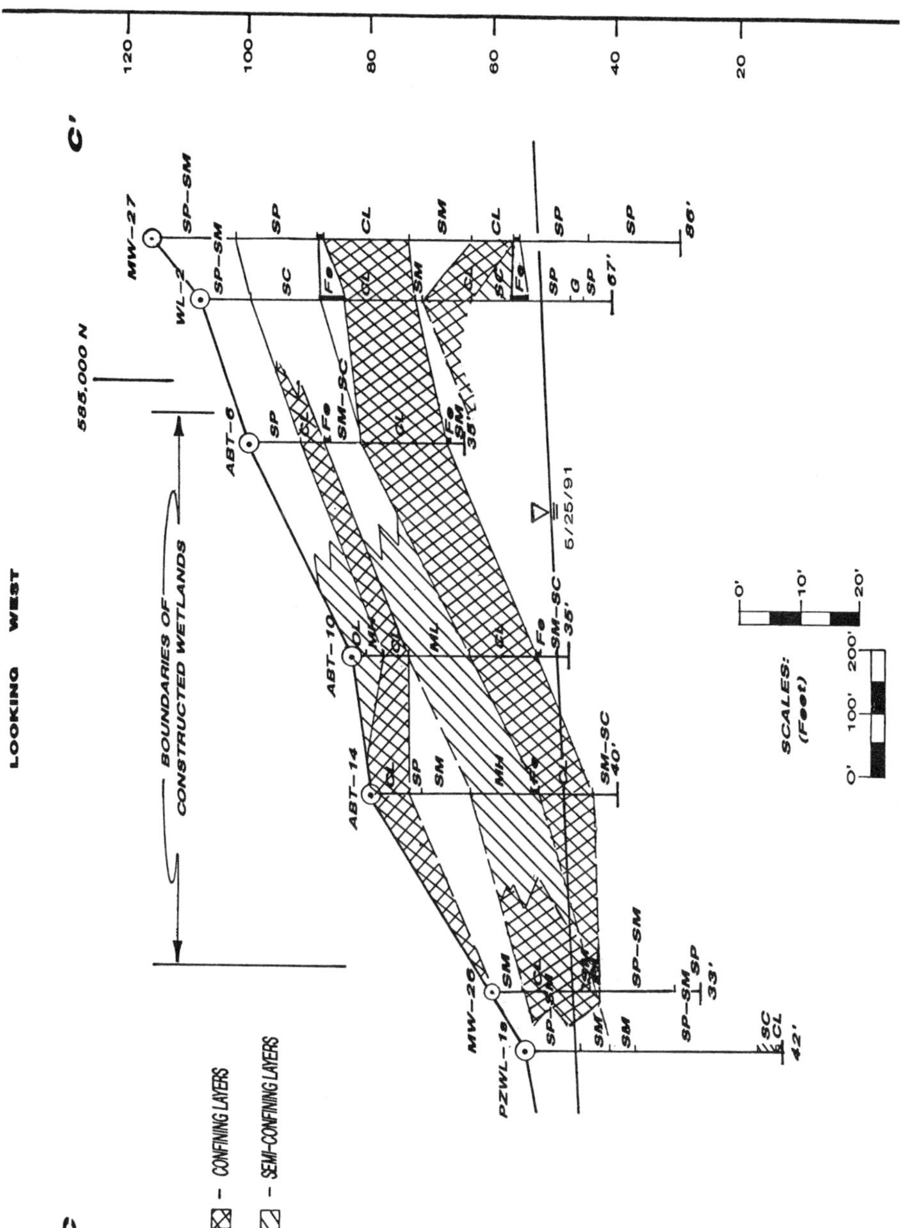

Figure 4. Geologic cross-section C—C′ through the constructed wetlands.

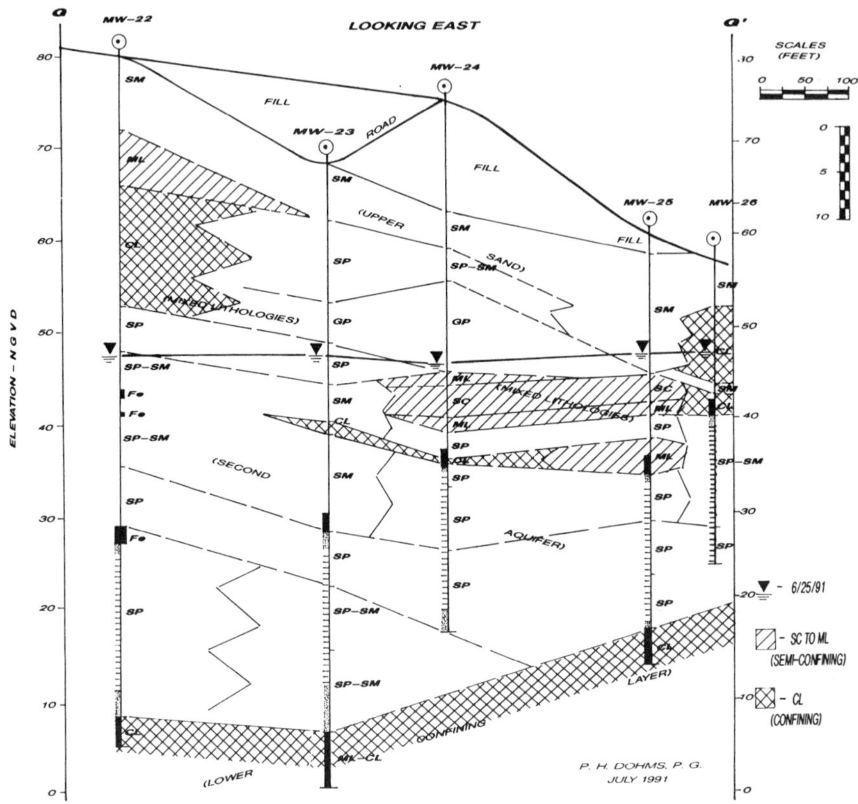

Figure 5. Geologic cross-section G—G′ through the constructed wetlands.

The hydraulic conductivity of the sands of the second aquifer was measured at three locations by slug testing.[15] Values of horizontal hydraulic conductivity varied considerably, with reported values ranging from 1.8×10^{-4} to 2.3×10^{-2} cm/s. Subsequent estimates of ground water flow velocity in the second aquifer were found to vary from 2 to 245 ft per year.[16]

Groundwater discharge beneath the constructed wetlands area, calculated by the Darcy Equation, is seen to lie within a range of 500 to 22,000 gallons per day. This is a modest quantity and demonstrates that the potential for beneficial use of this aquifer is small.

GROUNDWATER MONITORING

The principal regulatory agency with jurisdiction over this project is the FDER. This section briefly describes the process by which the monitoring system was defined through discussions with FDER, and then installed.

Florida Requirements

During early discussions between Escambia County and the FDER, it was concluded that the upper perched aquifer, being removed, would not require monitoring. All discussions were therefore focused on the second aquifer.

From the standpoint of the FDER, the facility they were being asked to regulate comprises a wastewater treatment system. Landfill leachate, domestic septage, and any constituents that might be donated through

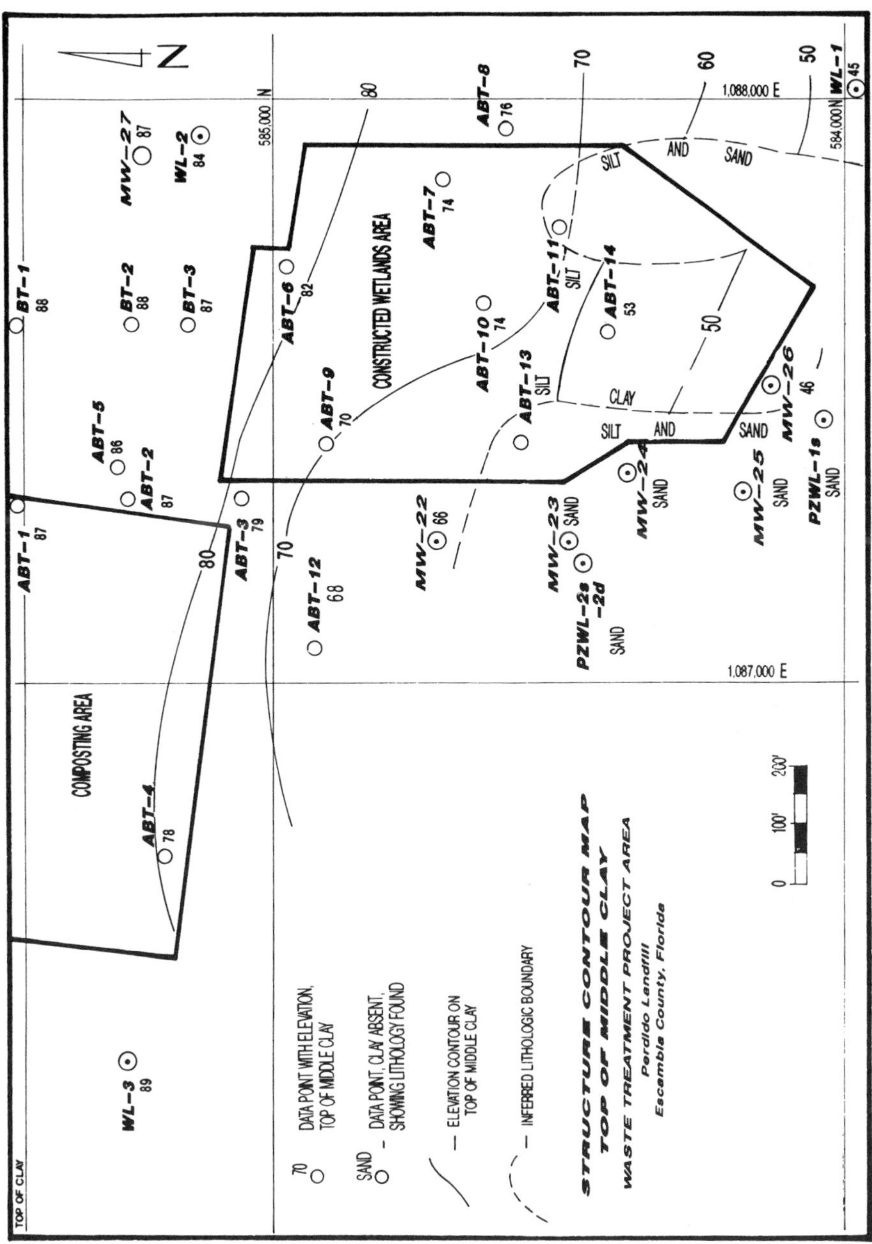

Figure 6. Structure contour map, top of principal clay.

leaching and/or biological activity in the composting beds would be drained into the constructed wetlands. Given the consequent potential for adverse impacts on the groundwater quality in the event of liner failure, the concerns that were expressed by the agency are unfounded.

Monitoring takes place under the provisions of Chapter 17-28.700 ("Installations Discharging to Groundwater; Permitting and Monitoring Requirements"), Florida Administrative Code (FAC). These regulations presume that discharge to groundwater will occur through any liner. On that basis, a "Zone of Discharge" (ZOD) is defined for each facility, extending as much as 100 ft from the facility boundary (Figure 2). Within this zone, groundwater quality standards can be exceeded, but all water quality standards have to be met at the boundary of the ZOD.

The requirements for monitor wells in Chapter 17-28.700, FAC, specify that "background" wells, "intermediate" wells, and "compliance" wells are to be installed for each monitored site. One background well is required for every facility in an area where no effects from the facility will be seen in groundwater. Intermediate wells are to be placed within the ZOD in locations that will allow the early detection of discharge. Compliance wells are to be placed at the down-gradient perimeter of the ZOD; groundwater quality standards are to be met at these locations.

Monitoring System

The monitoring system that was installed at the Perdido Landfill constructed wetlands facility fulfills all of the above-cited requirements. Figure 2 illustrates the locations of the respective monitoring wells. Well MW-27 is the "background" well; its location is up-gradient of the wetlands facility, as shown in Figure 4. Wells MW-24 and MW-26 are the "intermediate" wells; they were placed within the ZOD, relatively close to the facility boundary. Finally, wells numbered MW-22, MW-23, and MW-25 are the "compliance" wells; they are placed at the down-gradient boundary of the ZOD.

All wells were installed in 8 in. diameter boreholes that were drilled by mud rotary techniques. Construction details are shown in Figure 7. Well casings are 4 in. in diameter and the annulus between the well casing and the borehole wall is occupied by silica-sand filter pack, or a bentonite seal, or by cement grout, depending on location in the well. Well screen slot size was designed to the characteristics of the formation being tested and vary from 0.010 to 0.030 in. in width. Following their installation, all wells were developed by a combination of surge pumping, overpumping, and pumping until clear formation water was obtained. Each well was fitted with a heavy-duty steel protective vault set within a concrete pad in order to provide long-term protection from damage. After the new wells were developed, they were turned over to the Escambia County Solid Waste Department for incorporation into the overall Perdido Landfill monitoring program.

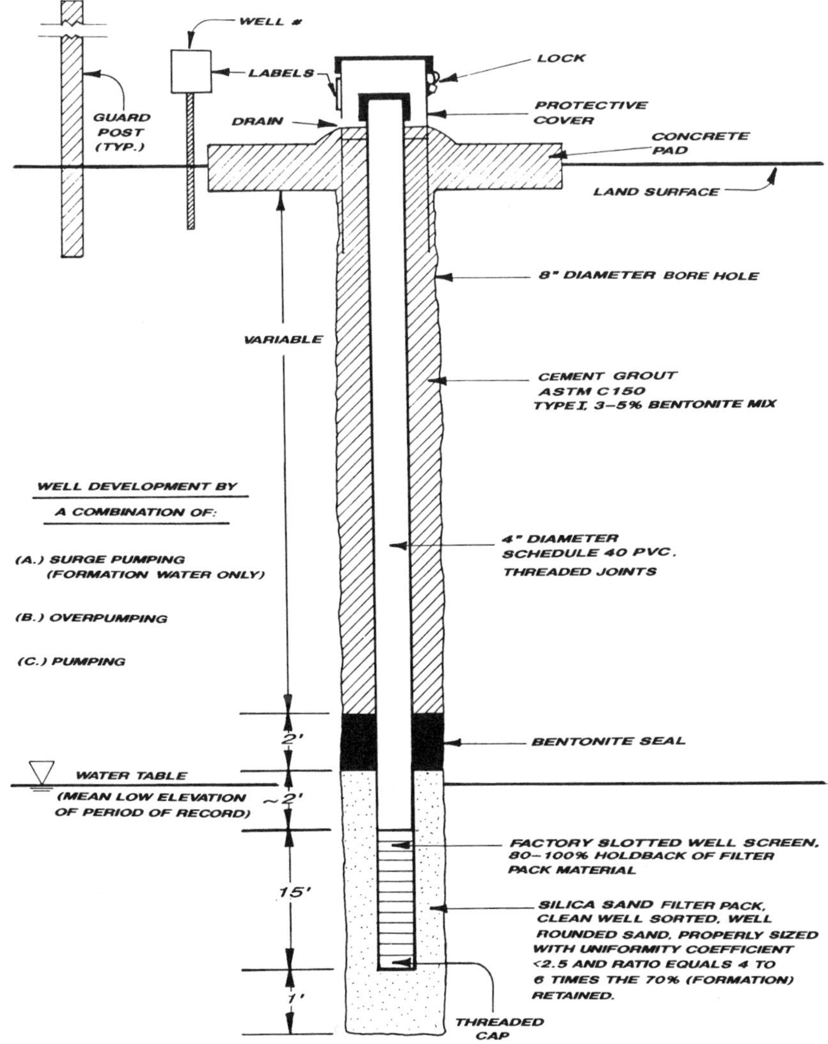

CONSTRUCTION DETAILS - TYPICAL 4" MONITOR WELL - WATER TABLE AQUIFER
WASTE TREATMENT PROJECT AREA
PERDIDO LANDFILL ESCAMBIA COUNTY, FLORIDA
(N. T. S.)

Figure 7. Construction details of a typical monitor well.

REFERENCES

1. Miller, C. C. Personal communication, 1991.
2. Martin, C. D., G. A. Moshiri, and C. Miller. Personal communication, 1991.
3. Bouldin, D. R. Personal communication, 1991.
4. Sanders, F. Personal communication, 1991.
5. Rebstock, J. M. Clay testing and analysis, Perdido Landfill, Escambia County, Florida. Private report to Escambia County Solid Waste Department by McCrone Associates, Inc., Atlanta, July 5, 1990.
6. Baily, J. R. Personal communication, 1991.
7. Baily, J. R. Wetlands Liner Quality Assurance Testing. Private reports to Escambia County Solid Waste Department, Pensacola, 1990–1991.
8. Miller, J. A. Hydrogeologic framework of the Floridan Aquifer System in Florida and in parts of Georgia, Alabama, and South Carolina, United States Geological Survey professional paper 1403-B, U.S. Government Printing Office, Washington, D.C., 1986.
9. Marsh, O. T. *Geology of Escambia and Santa Rosa Counties, Western Florida Panhandle.* Florida Geological Survey Bulletin No. 46, Tallahassee, FL, 1966.
10. Fernald, E. A. and D. J. Patton. *Water Resources Atlas of Florida.* Florida State University Institute of Science and Public Affairs, Tallahassee, FL, 1984.
11. VanLandingham, D. Personal communication, 1990.
12. Dohms, P. H. Report of investigation, hydrogeology of the proposed solid waste volume reduction area and constructed wetlands area, Perdido Landfill, Escambia County, Florida. Private report for the Escambia County Solid Waste Department, Pensacola, FL, 1990.
13. Dohms, P. H. Constructed wetlands monitor well installation, Perdido Landfill, Escambia County, Florida. Private report for the Escambia County Solid Waste Department, Pensacola, FL, 1991.
14. Baily, J. R. Reports of geotechnical testing, selected monitor well borehole samples, Perdido Landfill. Private report for the Escambia County Solid Waste Department, Pensacola, FL, 1991.
15. Jacobs, L. M. and S. Sanchez. Perdido Landfill Slug Tests. Private report for the Escambia County Solid Waste Department, Pensacola, FL, 1990.
16. Dohms, P. H. Proposed ground water monitoring plan, Perdido Landfill, Escambia County, Florida. Private report for the Escambia County Solid Waste Department, Pensacola, FL 1990.

CHAPTER 9

Development of Criteria for the Design and Construction of Engineered Aquatic Treatment Units in Texas

A. J. Cueto

The use of engineered wetland systems to treat municipal wastewater has been continuously increasing world-wide. Given the right application, these systems can be a cost-effective alternative to traditional wastewater treatment technology. The State of Texas has developed a set of guidelines for the design and construction of these systems, more commonly known as constructed wetlands. The guidelines are entitled *General Guidelines for the Design of Constructed Wetlands Filters for Use in Municipal Wastewater Treatment*. These guidelines set minimal acceptable standards for the design and construction of constructed wetlands and provide direction and recommendations to the consulting engineering community. This article discusses the objectives for the development of these guidelines. An outline of the formation, design, and contents of the guidelines is also presented.

As with any government action, the development of these rules was motivated by multiple objectives. However, the primary driving forces that led to the need for a low cost, low maintenance, reliable wastewater treatment alternative may be divided into the following three categories:

- Stricter water quality stream standards mandated by the Clean Water Acts
- Discontinuation of the EPA's Construction Grants program and the Innovative and Alternative technology funding program
- A general downturn in the regional economy during the last decade

Like the reasons for the guidelines, the objectives of these guidelines were also numerous. First and foremost, the guidelines hope to establish a minimum acceptable standard for the design and construction of constructed wetlands so that the waters of the state would be protected. Secondly, the guidelines sought to provide direction in the design of these units and to state recommended sizing to members of the engineering community not familiar with the design of these units. Also, the guidelines aspired to show acceptance for the use of constructed wetlands as a valid alternate treatment technology.

These criteria, titled *General Guidelines for the Design of Constructed Wetlands Filters for Use in Municipal Wastewater Treatment*, are found in *Tile 31 Texas Administrative Code, §317* (31 TAC §317). The section establishes minimum design standards for facilities which use constructed wetlands filters to treat and dispose of primarily domestic wastewater. The section defines minimum acceptable standards and recommended guidelines which must be met in order to obtain construction approval and, where applicable, rate the treatment capacity of the facility. The authority vested in this chapter was granted to the Texas Water Commission by 70th Legislature under the Texas Water Code §26.034 (Vernon's TCA Water Code, §26.034).[1]

The proposed rules will be included in Section §317 as §317.15 and reflect guidelines for the design and construction of constructed wetlands filter units. The guidelines are composed of four sections: (1) definitions, (2) general considerations, (3) submerged flow filter design, and, (4) free-water surface design. The design and construction of two types of wetland filters, submerged flow and free-water surface, are considered. Critical requirements and recommendations are outlined in an easy-to-follow didactic manner. The guidelines include recommended organic loading rates for both types of filters. The loadings are presented in the form of hydraulic retention times in days for the desired 5-day biochemical oxygen demand (BOD_5) reduction efficiency.

0-87371-550-0/93/$0.00+$.50

In order to understand these guidelines better, a brief overview of constructed wetland filters follows. The objective of wastewater treatment is to reduce the concentration of pollutants and nutrients in the waste stream to the level where the discharge will not adversely affect the environment.[2] For any given waste stream and specific location, the degree and type of treatment, therefore, are variables that require competent engineering decisions. Treatment processes vary greatly and can be classified by a vast number of methodologies. Generally, the degree of wastewater treatment can be divided into the three levels: primary, secondary, and tertiary.[2]

Each of the various treatment levels can be achieved by what can be termed "mechanical" and "natural" treatment methods. Although all wastewater treatment methods, to some degree, depend on natural responses (such as gravity) for sedimentation, or biological organisms for pollutant and nutrient removal, in a typical mechanical treatment facility these natural components are supported by a complex array of energy-intensive mechanical equipment. Therefore, the term "natural treatment" processes can be described as those processes that depend primarily on their natural components to achieve treatment.[3] There are many different methods of natural treatment, ranging from facultative ponds to water hyacinth basins. Included in these processes are constructed wetlands. These systems have been termed "constructed wetlands" because they are designed to mimic the ecological, physical, and hydrological conditions of natural wetlands. A natural wetland can be defined as land where the water surface is near the ground surface long enough each year to maintain saturated conditions, and grow vegetation associated with a wetland ecosystem. These wetlands are typically characterized by emergent aquatic vegetation such as cattails (*Typha*), rushes (*Scirpus*), and reeds (*Phragmites*). They can also contain floating plant species such as duckweed (*Lemna*) and water hyacinths (*Eichhornia*). Like natural wetlands, constructed wetlands are composed of similar species.

Constructed wetlands can be classified into two functional categories by the location of the wastewater surface. Like mechanical systems, they are also capable of achieving advanced secondary treatment levels. Critical system design objectives are:

- Maintenance of healthy plants and an environment conducive to growth
- Design of a plug flow waste stream
- Sufficient hydraulic conductivity for peak flows
- Maintenance of sufficient dissolved oxygen levels in unit and effluent

The first functional type of wetland is termed a "submerged flow system". The submerged flow constructed wetland is a lined basin that is filled with large gravel rock media in which aquatic plants are grown (Figure 1). The wastewater stream flows in one end at approximately 6 in. below the surface of the media, and then out the opposite end. As the wastewater flows through the system, the plants and associated microorganisms synergistically remove or transform a portion of the pollutants.

Although all designs should be based on sound engineering principles, field data, and extensive pilot studies, the guidelines recommend design parameters for constructed wetland loadings. The BOD_5 loading rates are presented in terms of hydraulic retention times in days and are outlined below. For the submerged flow filters, the times are given by the general equation:

$$\left(\frac{C_e}{C_o}\right) = \frac{(\ln t)}{3.684} + 0.524 \qquad (9\text{–}1)$$

where C_e = effluent BOD_5, mg/L
C_o = influent BOD_5, mg/L
t = average hydraulic residence time, days

The average hydraulic residence time which takes into account the filter's evapotranspiration and rainfall inflow is represented in Equation 2:

$$t = \frac{7.481(LWdn)}{\left(\frac{Q_1 + Q_2}{2}\right)} \qquad (9\text{–}2)$$

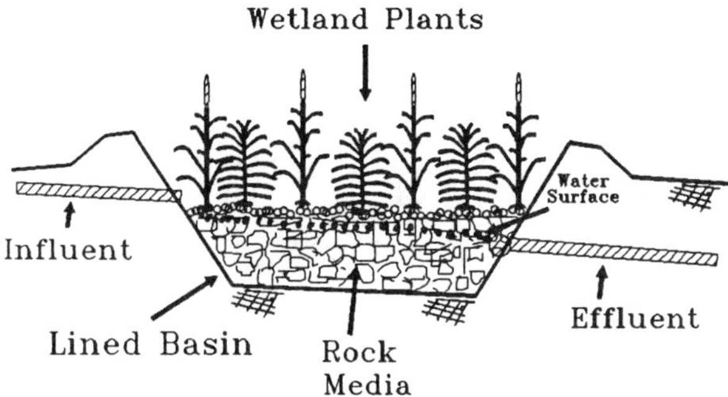

Figure 1. Submerged flow system.

where L = length, ft
W = width, ft
d = water depth, ft
n = porosity of the filter; i.e., volume of basin not occupied by plant material
Q_1 = influent flow, gpd
Q_2 = effluent flow, gpd

The removal efficiency of these units are a function of many factors including: wastewater characteristics, dissolved oxygen in the unit, and ambient temperature of the wastewater.[3]

The second functional category is labeled a free-water surface system (Figure 2). The free-water surface system functions very much like the submerged flow system with the exception that in the latter the basin is not filled with rock. The free-water surface system is composed of a lined basin with approximately 1 ft of sandy clay soil at the bottom. Wastewater flows through the basin at a depth of about 15 in. The water surface is exposed to the atmosphere, thus yielding the name free-water surface. Like the submerged flow system, as wastewater flows through the system, the plants and other microorganisms attached to the plant stems remove a portion of the pollutants.

The loading rates for the free-water surface filters are modeled by a first-order kinetic equation first presented in the EPA Design Manual *Constructed Wetlands and Aquatic Plant Systems for Municipal Wastewater Treatment*.[4] The general equation can be seen below in Equation 3:

$$\left(\frac{C_e}{C_o}\right) = \exp^{-C'(K_T t)} \tag{9–3}$$

where C_e = effluent BOD_5, mg/L
C_o = influent BOD_5, mg/L
C' = media constant dependent surface area
K_T = temperature-dependent first-order reaction rate constant, $days^{-1}$
t = hydraulic residence time, days

The first-order rate constant can be described by Equation 4:

$$K_T = K_{20}(1.1)^{(T-20)} \tag{9–4}$$

where K_{20} = rate constant at 20°C, $days^{-1}$
T = ambient temperature in the unit, °C

The media constant can be expressed as follows:

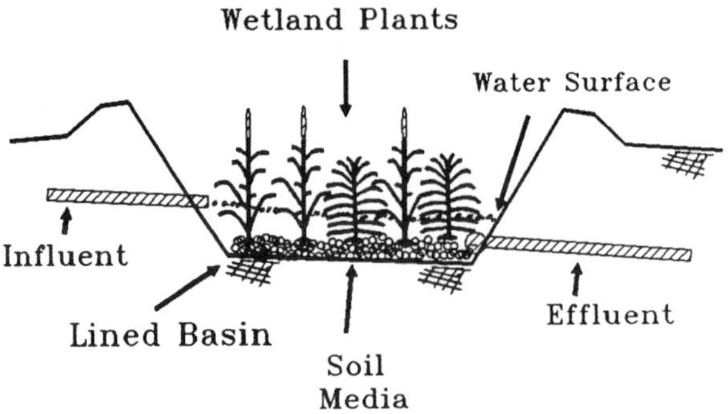

Figure 2. Free-water surface system.

$$C' = -0.7(A_v)^{1.75} \tag{9–5}$$

where A_v = specific surface area for microbial activity, assuming to equal 15.7 (m^2/m^3).

Assuming a winter low temperature of 7.5°C, Equations 3–5 can be combined and solved for t.[4]

$$t = -6.7 \ln\left(\frac{C_e}{C_o}\right) \tag{9–6}$$

The capability for reliable treatment and ease of operation of these systems has been demonstrated in several states. Table 1 shows a list of various constructed wetlands and their removal efficiencies for a variety of pollutants. The values offered in Table 1 represent average removal efficiencies as presented by Watson et. al.[5] Similar values for submerged flow systems in Louisiana have been reported by Jones.[6]

A preliminary study comparing the cost effectiveness of these systems to conventional treatment technology indicated their economic feasibility for communities with populations under 50,000.[7] Table 2 shows the cost comparisons between mechanical treatment systems and a constructed wetland system. The analysis is based on the following assumptions: a secondary effluent treatment level, discount rate of 10%, and a net project life of 20 years. The average construction costs for constructed wetlands used in this analysis were $1.60 per gallon of plant capacity. The average operations and maintenance costs were $0.40 per gallon of plant capacity per year of operation. These costs were higher than the actual calculated mean by one half of a standard deviation to compensate for errors in the reported data. Construction, operation, and maintenance costs for the conventional treatment plant design were taken from *Means Building Construction Costs Index: 1991 Edition*,[8] *U.S. Army Corp of Engineers CAPDET Program Users Guide*,[9] and *EPA Construction Grants Cost Summaries 1985*.[10] Each of the costs was corrected for inflation using the *Engineering News Record*[11] Index Numbers for the appropriate year and then the averages were determined.

Figure 3 is a graphical representation of Table 2. The curves can be seen crossing at approximately 5 million gallons per day (5 MGD) which would be the equivalent of a community of about 50,000. For flows less than 5 MGD, the net present worth of the costs for constructed wetland systems are less than the cost for the equivalent conventional systems, which would indicate the cost effectiveness of the wetland systems. It should be noted that a detailed cost-effective analysis of a constructed wetlands system has not been undertaken. Any conclusions derived from this preliminary study can, at best, only be a foundation for a more detailed engineering analysis.

Note that if the use of these systems lends itself to communities of under 50,000, other advantages of the systems become apparent. These systems are best suited to small, agriculturally-based rural communities; they do not require high levels of expertise to run, as do some conventional treatment plants. Rather, traditional agricultural know-how, so common in rural America, is sufficient for their operation. However, this by no means eliminates the requirement for operator certification. Also, these systems can function on limited power input. For the most part, the only energy inputs required are gravity, sunshine, and manpower.

Table 1. Removal Efficiencies for Pollutants at Various Constructed Wetlands Sites

Location	System type	BOD_5 removal efficiency	TSS removal efficiency
Listowel, Ontario	Free-water	0.72	0.76
Arcata, California	Free-water	0.53	0.85
Brookhaven, New York	Free-water	0.89	0.88
Santee, California	Subsurface	0.80	0.90
Iselin, Pennsylvania	Subsurface	0.82	0.92
Benton, Kentucky	Combination	0.58	0.77
Neshaminy, Pennsylvania	Combination	0.96	0.94

Table 2. Present Costs of Wastewater Treatment Facilities Conventional Treatment Plan

Plant flow (MGD)	Life (yr)	Disc. rate	Unit O&M costs ($/gal)	Total yearly O&M costs ($)	Unit capital costs ($/gal)	Total capital costs ($)	Net present worth of costs ($)
0.1	20	10.0%	$0.69	$69,000	$5.59	$559,000	$1,146,436
0.5	20	10.0%	$0.52	$260,000	$3.42	$1,710,000	$3,923,527
1.0	20	10.0%	$0.37	$370,000	$3.99	$3,990,000	$7,140,019
5.0	20	10.0%	$0.26	$1,300,000	$3.19	$15,950,000	$27,017,633
10.0	20	10.0%	$0.19	$1,900,000	$2.29	$22,900,000	$39,075,771
50.0	20	10.0%	$0.13	$6,500,000	$2.00	$100,000,000	$155,338,164
100.0	20	10.0%	$0.10	$10,000,000	$1.75	$175,000,000	$260,135,637
Present Costs of Wastewater Treatment Facilities Wetland Treatment Plan							
0.1	20	10.0%	$0.40	$40,000	$1.60	$160,000	$500,543
0.5	20	10.0%	$0.40	$200,000	$1.60	$800,000	$2,502,713
1.0	20	10.0%	$0.40	$400,000	$1.60	$1,600,000	$5,005,425
5.0	20	10.0%	$0.40	$2,000,000	$1.60	$8,000,000	$25,027,127
10.0	20	10.0%	$0.40	$4,000,000	$1.60	$16,000,000	$50,054,255
50.0	20	10.0%	$0.40	$20,000,000	$1.60	$80,000,000	$250,271,274
100.0	20	10.0%	$0.40	$40,000,000	$1.60	$160,000,000	$500,542,549

Constructed wetlands are not a panacea for the wastewater treatment problems that confront small towns. They have their limitations. The most familiar limitation of this technology is their extensive land needs. Therefore, an area with high land values may not be the best choice for a constructed wetland. Another limitation is the availability of suitable wetland or aquatic plants. Finally, these systems will use more water than an equivalent conventional system due to the increased rates of evapotranspiration.

These guidelines are among the first published state design criteria for constructed wetlands. However, they are by no means the only definitive criteria for the design of such systems. Therefore, it is hoped that they provide steps toward the formation of definitive criteria, and that through their implementation more scientific and research-oriented investigations will be undertaken.

ACKNOWLEDGMENTS

The author acknowledges the assistance of Ansle Jones, Ray Dinges, Sherwood C. Reed, Steve W. Manning, and Marvin Neal in the formulation of the stated guidelines.

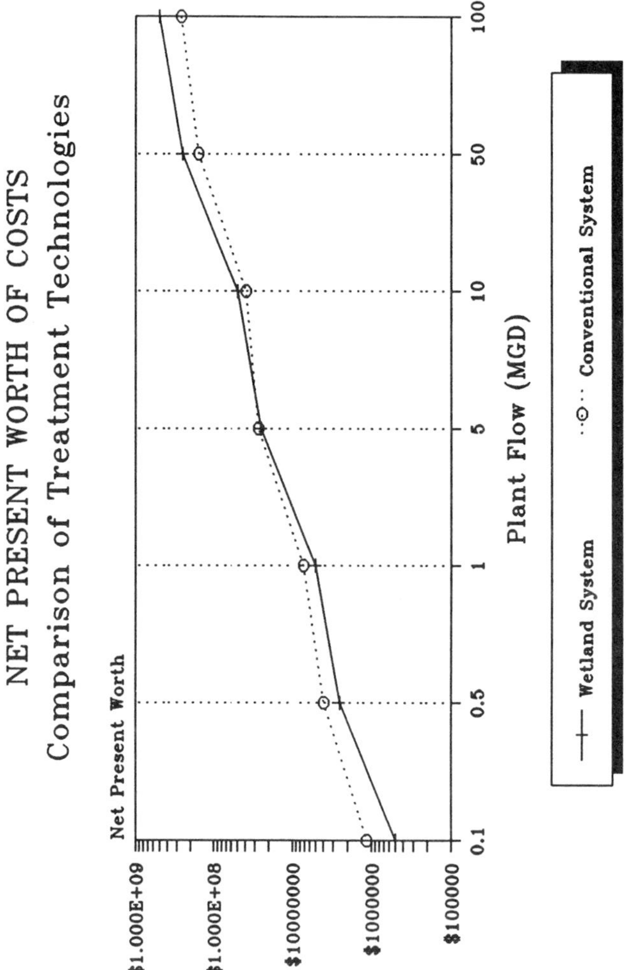

Figure 3. Net present worth of costs; comparison of treatment technologies.

REFERENCES

1. *Vernon, Texas Codes Annotated: Water*. Vol. 1A, West Publishing, St. Paul, MN, 1988.
2. Vesilind, P. and J. Peirce. *Environmental Engineering*, Butterworth Publishers, Inc., Boston, 1982.
3. Reed, S., J. Middlebroks, and R. Crites. *Natural Systems For Waste Management and Treatment.* McGraw-Hill, New York, 1988.
4. United States Environmental Protection Agency. *Constructed Wetlands and Aquatic Plant Systems for Municipal Wastewater Treatment: A Design Manual.* Washington, D.C, EPA Publication 625/1-88/022, September 1988.
5. Hammer, D. A., B. P. Pullin, and J. T. Watson. Constructed Wetlands For Livestock Waste Treatment, *Natl. Nonpoint Conf.*, St. Louis, MO, April, 1989.
6. Jones, A. *Constructed Wetlands for Municipal Wastewater Treatment.* Draft paper: EPA Publication, Dallas, TX, 1990.
7. Cueto, A. J. A Quantitative Evaluation Of Constructed Wetlands, Draft thesis, Southwest Texas State University, San Marcos, TX, 1990.
8. *Means Building Construction Cost Data: 1991 ed.* R. S. Means Company, Kingston, MA., 1990.
9. United States Army Corp Of Engineers. *Design of Wastewater Treatment Facilities, CAPDET, Program Users Guide.* Washington, D.C., U.S. Army Engineering Manual No. 1110-2-501, July 1980.
10. United States Environmental Protection Agency. *Construction Grants: 1985.* Washington, D.C, EPA Publication 430/9-84-004, July 1984.
11. ENR Market Trends, *Engineering News Record.* September 15, 1988, 148; September 24, 1981, 66; December 17; 1981, 96.

CHAPTER 10

Engineering and Environmental Assessment of Enhanced Bolsa Chica Wetland

R. Walton, S. Bird, B. Ebersole, and L. Hales

INTRODUCTION

Bolsa Chica is an unincorporated area of Orange County, California, located along the coastline approximately 9 miles (15 km) south of Long Beach, and lying within the City of Huntington Beach. The study area comprises approximately 1645 acres (666 ha) (Figure 1). At present, tidal flow enters Outer Bolsa Bay, Inner Bolsa Bay, and the California Department of Fish and Game (DFG) muted tidal cell only through Anaheim Bay and Huntington Harbour. Local rainfall is the only source of freshwater inflow. The proposed wetland enhancement area is currently a producing oil field. Dirt roads and dikes criss-cross the lowland, connecting oil pumping rigs, drill pads, related structures, and pipe networks. A unique feature of the site is that tidal culverts separate the three existing wetland regions. In Outer Bolsa Bay, the mean tidal range is 6 ft. However, this is reduced to 1.5 ft in Inner Bolsa Bay through three 28-ft long culverts, and to 1 ft in the DFG muted tidal cell.

Several alternatives have been proposed to develop the site by replacing the oil-producing region with tidal wetlands. Additional modifications include adding a second tidal entrance, connecting waterways, tidal culverts, and a marina. As this is a rather complex system, it was decided to develop a computer model of the site to (1) assess the performance of the various plan alternatives, (2) assess the impacts of the various alternatives on the adjacent system, and (3) examine the effects of external influences, such as floods and sedimentation, on the entire system.

MODEL DEVELOPMENT

A version of the Dynamic Estuary Model (DEM), called DYNTRAN,[6,7] was selected for application to the Bolsa Chica site because its pseudo-two-dimensional structure was well suited to represent the complex geometry. The model simulates the momentum equation along links, using the surface elevation gradients between adjacent nodes. The mass continuity and conservation of constituent mass equations are then solved at the nodes for tidal elevations and constituent concentrations, using flows in the links. As DYNTRAN is a link-node model, it is relatively easy to modify to incorporate hydraulic structures.

The Bolsa Chica site currently has two sets of tidal culverts, some with one-way flaps, that control the tidal exchanges between Outer Bolsa Bay, Inner Bolsa Bay, and the DFG muted tidal cell. The plan alternatives call for adding additional culverts to control water levels in newly created tidal wetland areas. DYNTRAN was modified to replace the momentum equation in culvert links with an equation that estimated the flow using heads at each end of the culvert[1,2]. The details are presented in Hales et al.[3,4] and Walton et al.[8]

DYNTRAN can simulate the flooding and drying of shoal areas. Generally, this is done by specifying links in the shoal areas and defining when they are wet and dry. However, as some tidal flats lie adjacent to some of the channels and the channel simply increases in width during the tidal cycle, we also modified the channel cross-section type and replaced it with a double trapezoid.[3,4] This permitted the channel to increase significantly in width, without the necessity for small times steps associated with placing links in relatively narrow overbanks areas.

0-87371-550-0/93/$0.00+$.50

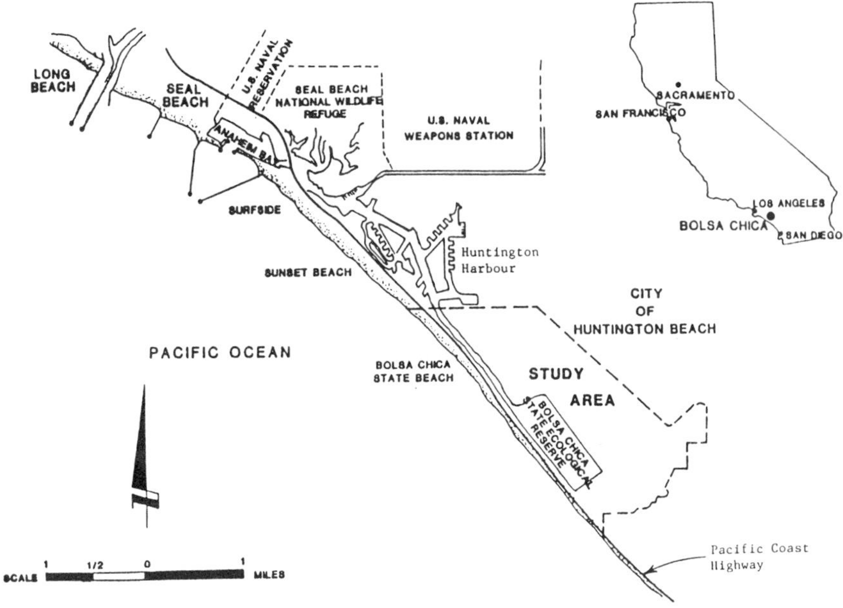

Figure 1. Bolsa Bay, California study location.

The modified program was then applied to the Bolsa Chica site. Figure 2 shows the link-node network. The model was calibrated and verified to tidal elevations and velocities measured at the site during August 1986.[5] Figure 3 shows that the model not only reproduces observations well, but that the culvert routine accurately simulates the more than 4 ft in tidal range reduction seen across the 28-ft culverts separating Outer and Inner Bolsa Bay.

DESCRIPTION OF ALTERNATIVES

Orange County, California, adopted a Land Use Plan (LUP) as part of the Local Coastal Program for the Bolsa Chica area in accordance with the California Coastal Act of 1976. This LUP includes a navigable ocean entrance system (Preferred Alternative), and a non-navigable ocean entrance system (Secondary Alternative). The principal landowner of the region, Signal Landmark, desired to implement the Preferred Alternative. Later, a third alternative, The Lake Plan, was submitted by Signal Landmark for additional analysis.

The Preferred Alternative contained, comprised (1) 915 acres (370 ha) of restored, high-quality, fully functioning full tidal, muted tidal, fresh, and brackish water wetlands, (2) 86 acres (35 ha) of existing or newly created environmentally sensitive habitat, and (3) at least 75 acres (30 ha) of mixed-use, marina and commercial area providing in-water berthing and dry storage for at least 1700 boats. The Secondary Alternative also contained 915 acres (370 ha) of wetlands, a nonnavigable tidal entrance to provide a second source of flushing water only, and a different internal use configuration that included a reduced marina in a different location. The third alternative, the Lake Plan, contained the same areas of wetlands, but with a smaller marina and more open water. The internal hydraulics were modified and a nonnavigable tidal entrance was included.

In addition to the three major alternatives, there were a number of variations, including whether the connector channel to Huntington Harbour would be navigable or nonnavigable.

ANALYSIS OF ALTERNATIVES

The purpose of this article is to examine some aspects of the analysis of the plan alternatives, rather than examining each design in detail. In particular, we wish to emphasize the need to examine both the effect of

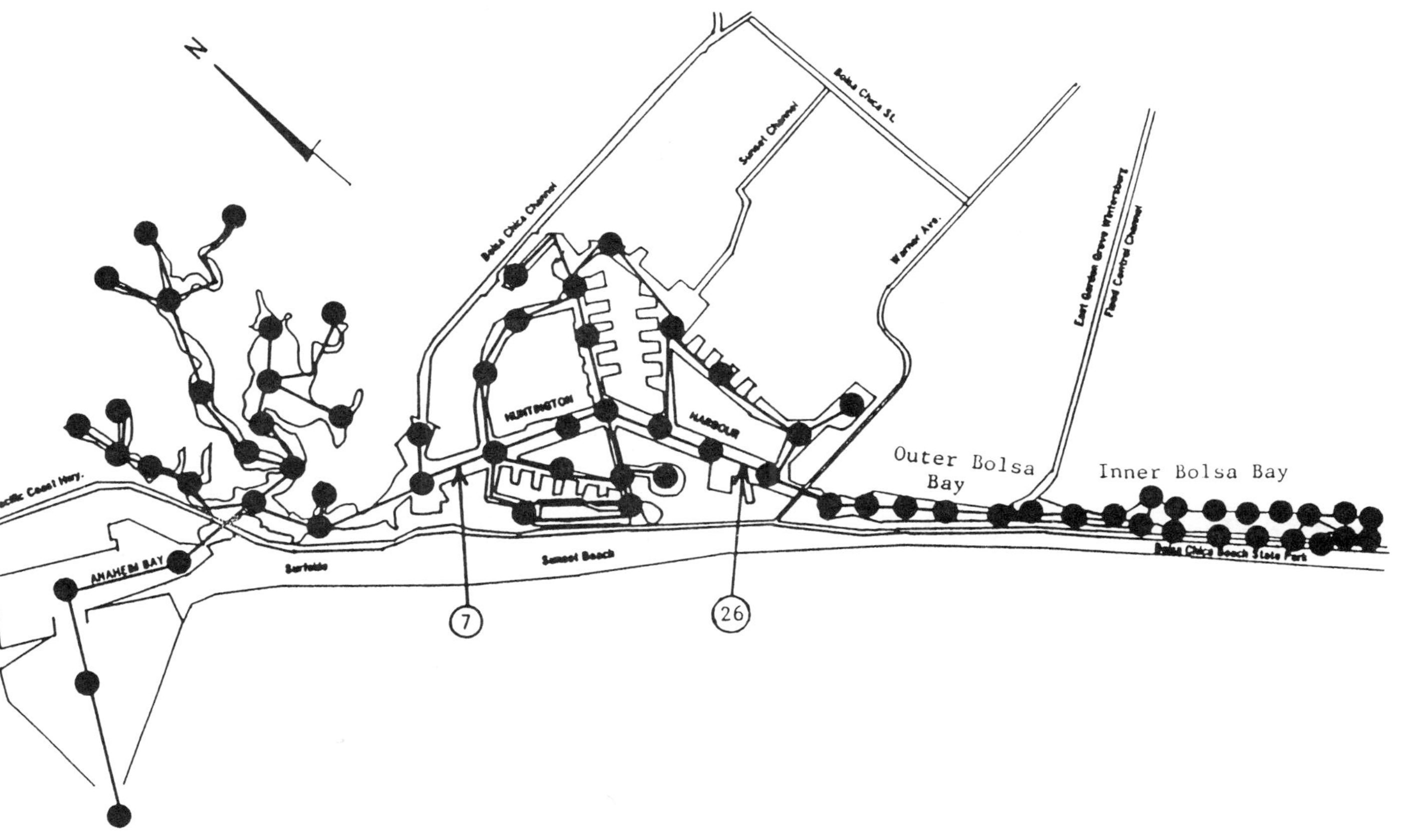

Figure 2. Location of links and nodes for numerical model DYNTRAN at Bolsa Bay, California. Links 7 and 26 velocities used for numerical model calibration.

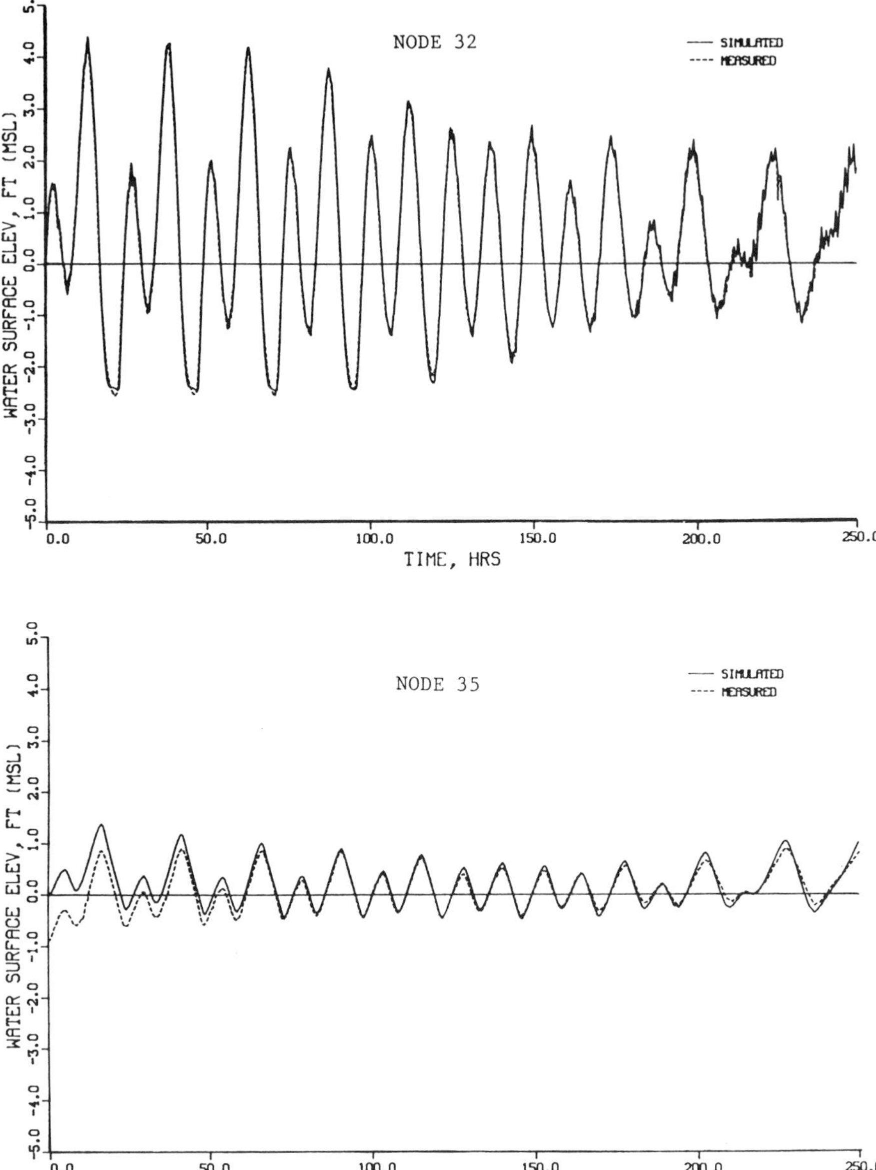

Figure 3. Simulated vs. measured water surface: node 32, Outer Bolsa Bay and node 35, Inner Bolsa Bay.

the design and the influence it may have on surrounding systems. To do this, we will examine three aspects of the analysis: (1) the performance of the system designed, (2) the effects on the surrounding system, and (3) the impacts of infrequent, but significant events.

Generally, with a system such as the plan alternatives for the Bolsa Chica site, there are sufficient controls (tidal culverts, for example) that the internal system can usually operate in the designed manner. That is not to say that a particular choice is correct — that is left for others to debate and negotiate. However, if we wish the interior wetland to experience a 1-ft tidal range for vegetative reasons, then we can design a system of controls to achieve this.

The second, more important, question is how this design affects the surrounding systems. For the Bolsa Chica system, the three plan alternatives and their variations were simulated using DYNTRAN, and the results analyzed. Figure 4 shows the effect on tidal circulation. When a second tidal entrance is introduced, whether navigable or not, the internal tidal hydraulics in Huntington Harbour are now influenced by two tidal entrances, not one. This means that there is the potential to create a "null" point where tidal velocities are significantly reduced (Figure 4), and thus tidal flushing potentially takes longer from this point, with inherent water quality impacts. Before the second entrance was introduced, all tidal exchange with Inner Bolsa Bay and the interior areas passed through Huntington Harbour. With the second tidal entrance, the interior areas are flushed from the new entrance.

Another way of looking at this, and other aspects of the system's performance, is to simulate mass transport in the study area. The companion mass transport model, TRAN, was used to simulate the "age" of water in the system. After each time step of the simulation, the water remaining in the system was aged by the time step. Newly entering water was flagged at time zero. For each of the plan alternatives, the mass transport model was run to a cyclic steady state, and the "age" of the water reported (Table 1). The results show that the introduction of a second, navigable tidal entrance and the improvement of the channel to Huntington Harbour can increase the flushing times in Huntington Harbour (nodes 9-17) by 35 to 50%.

Finally, the third question is how the new system responds to infrequent events, such as river flooding and closure of the tidal entrances. Figure 5 shows the effect for several of the plan alternatives of a flood event in the East Garden Grove-Wintersburg Flood Control Channel. From this figure, for a node in the DFG muted wetland, it can be seen that the impacts are mixed. For the existing system, this interior area would be flooded to a depth of about 7 ft (if the flood persisted for 50 h or more). However, note that the plan alternatives tend to reduce the level of inundation. In fact, for the alternative that includes a navigable second tidal entrance, little impact is seen because the entrance is deep enough that water can pass through without a significant backwater effect.

Another significant event considered was the possible closure of the second tidal entrance due to sedimentation from littoral processes. Figure 6 shows tidal velocities in Outer Bolsa Bay near the tidal culverts leading to Inner Bolsa Bay. The results show that entrance closure causes a return to existing velocities for cases in which the connector channel to Huntington Harbour has not been modified. When the connector channel has been deepened, the velocities are still small, but the phase is modified due to the deepening. However, when the interior hydraulics have been extensively modified, as for the Lake Alternative, flushing water to these new areas from the second tidal entrance are now replaced with water from the main tidal entrance, with a significant increase in tidal velocities.

DISCUSSION AND CONCLUSIONS

The analysis of the plan alternatives serves to illustrate some of the aspects of a study that should be examined when proposing extensive modifications to a hydraulic system. Using hydraulic structures, we have the ability to construct a system to operate generally as designed. However, it is important to consider the effects that these changes can have on nearby systems and whether the occurrence of infrequent, but important events increases or reduces the level of impact.

For the Bolsa Chica system, this is illustrated by examining the changes in flushing and tidal velocities in the adjacent Huntington Harbour system. If a navigable second tidal entrance is introduced, then a "null" may be found in Huntington Harbour, which could represent reduced water quality conditions. In addition, channels need to be correctly designed when significant new tidal areas are opened up so that souring does not occur.

The results also illustrated the effect of plan alternatives on the impact of flooding and entrance closure. In this case, a navigable second entrance may be beneficial in passing flood events without significant impounding in the interior wetland areas. Sedimentation and closure of the second tidal entrance may create an area of high tidal currents as the flushing patterns are altered.

The final design must consider all these consequences and determine what represents an appropriate design. That is, "appropriate" in the sense that the trade-offs in the performance of the new and existing systems are acceptable, and "manageable" in the sense that we understand some of the potential consequences and are prepared to respond to them as needed. A numerical model of such a complex system is a useful tool to evaluate a wide range of plan alternatives.

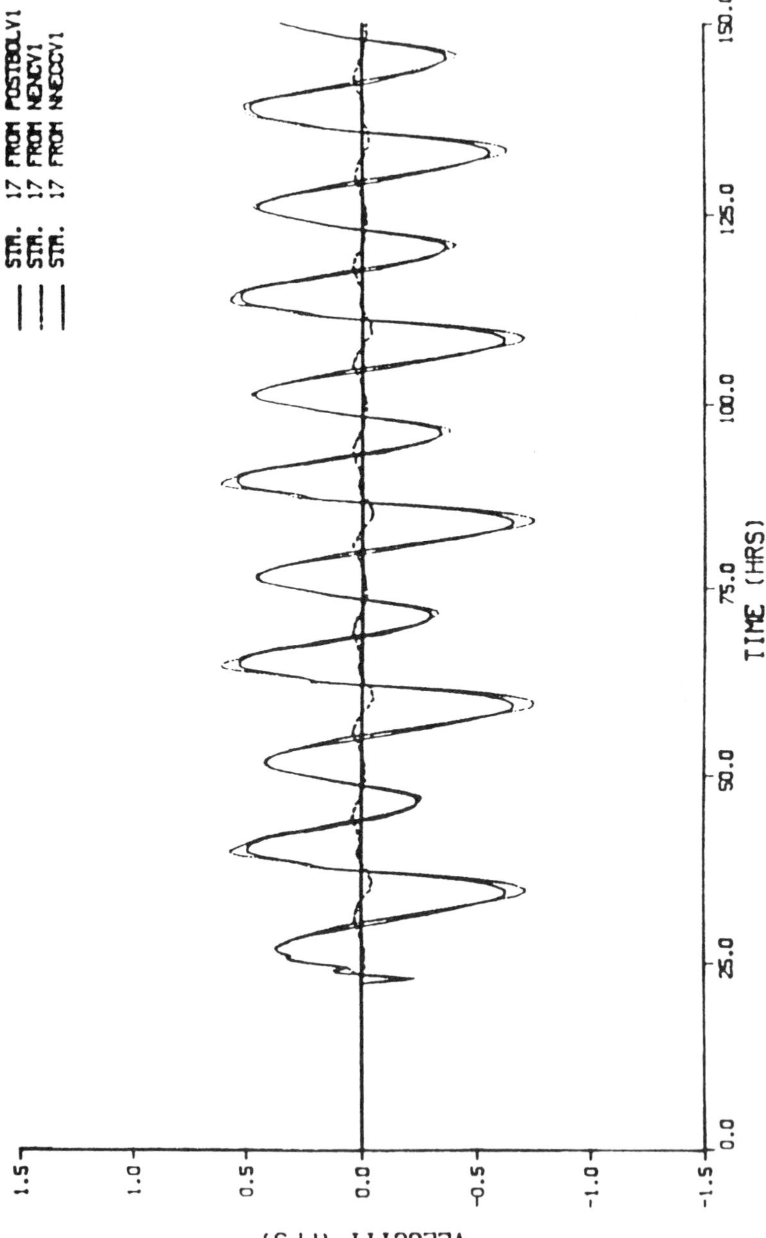

Figure 4. Average velocities in Huntington Harbour. POSTBOLV1 = existing conditions; NENCV1 = navigable (entrance and channel to Huntington Harbour); and NNECCV1 = nonnavigable (entrance and channel to Huntington Harbour).

Table 1. Water Age in Huntington Harbour and Bolsa Bay, California: Existing Conditions vs. Navigable Entrance Concepts, Average Age (h) for Final 25 h of Simulations

		Navigable entrance channel				
		Navigable connector channel to Huntington Harbour		Existing Outer Bolsa Bay channel to Huntington Harbour		
Node #	Existing condition	Wetlands connected	Wetlands not connected	Wetlands connected	Wetlands not connected	Supplemental channel to inner wetlands connected
9	281	343	343	173	173	173
15	425	576	576	289	289	289
17	435	637	637	302	302	303
24	434	366	366	289	289	290
29	487	274	275	339	339	338
37	684	256	220	321	284	160
40	751	357	293	423	360	155
54	855	466	397	531	465	230
111	—	383	395	451	463	357
122	—	430	447	495	512	442
129	—	394	417	460	482	415
134	—	390	421	455	486	420

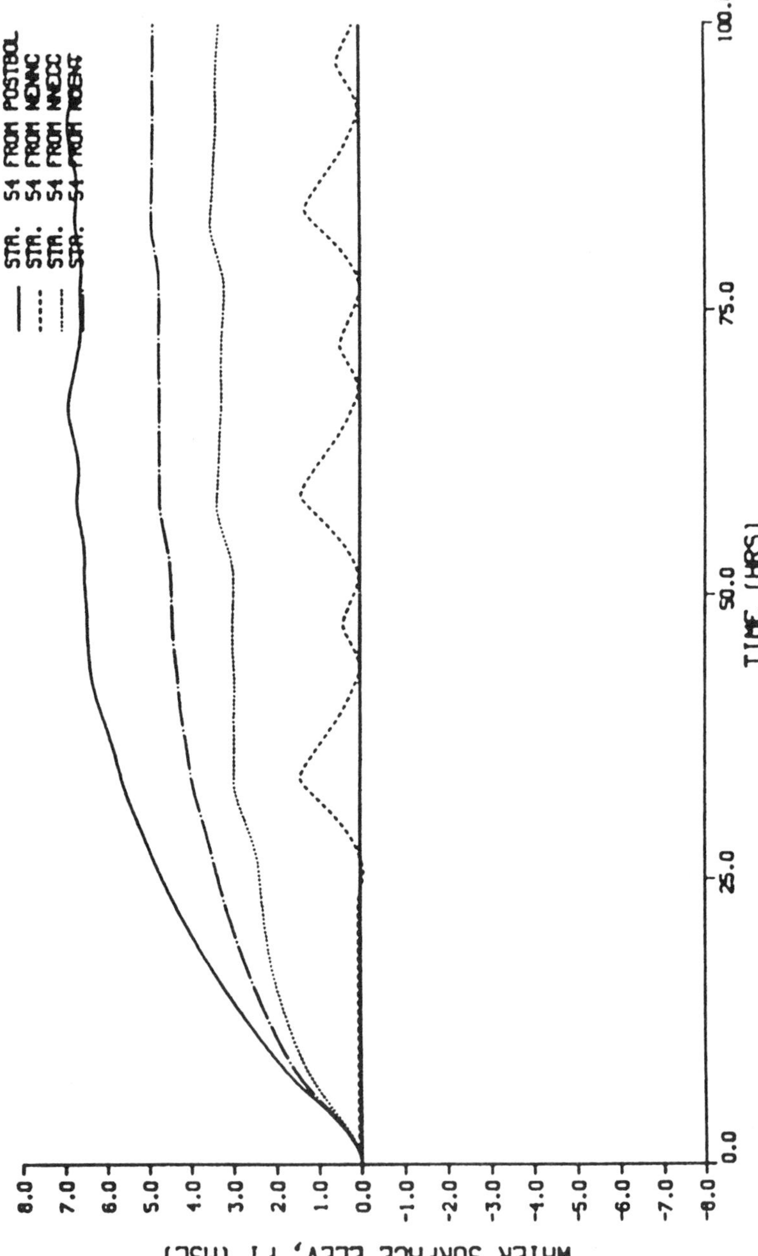

Figure 5. Water surface elevations in the DFG muted tidal cell at East Garden Grove-Wintersburg flood flow = 9710 cfs. POSTBOL = existing conditions; NENNC = navigable entrance; NNECC = nonnavigable entrance; and NOENT = no entrance.

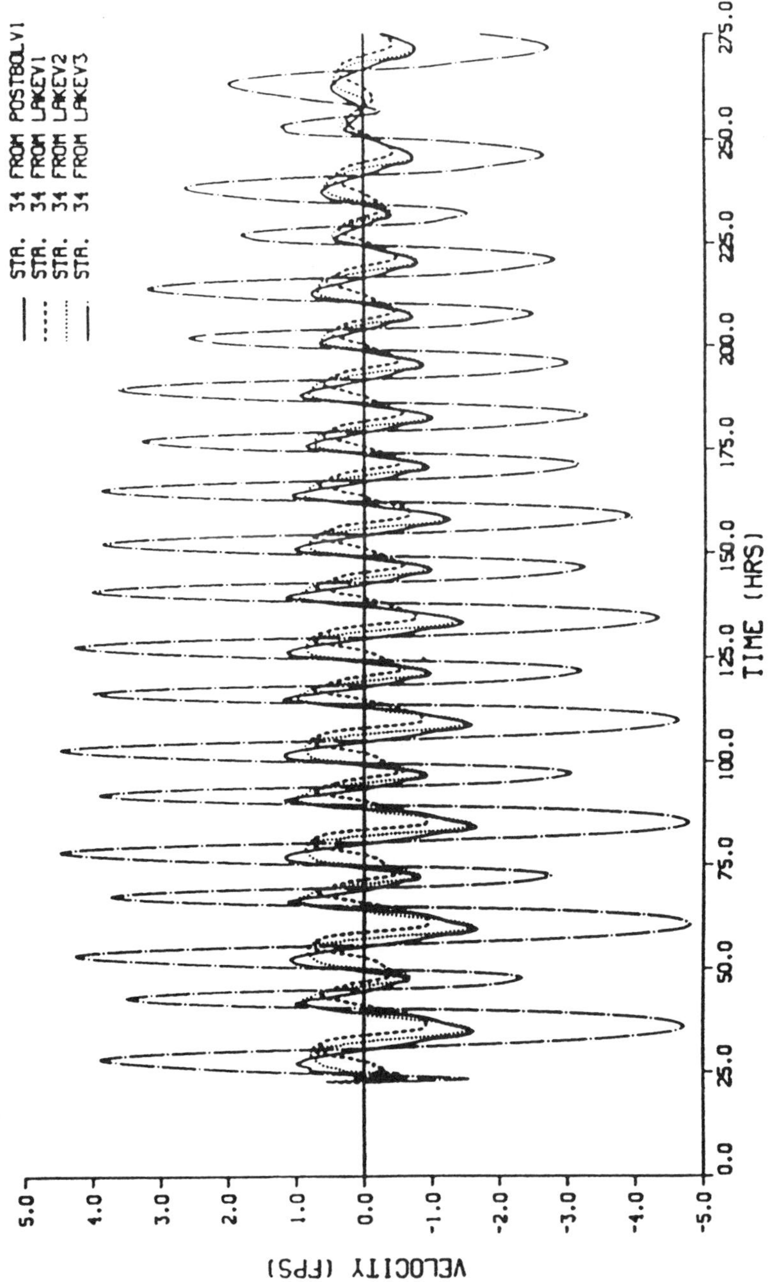

Figure 6. Average velocities under Warner Avenue Bridge. POSTBOLV1 = existing conditions; LAKEV1 = 350-ft wide entrance; LAKEV2 = 200-ft wide entrance; and LAKEV3 = no entrance.

REFERENCES

1. Bodhaine, G. L. Measurement of Peak Discharge at Culverts by Indirect Methods, in *Techniques of Water Resources Investigations at the USGS*, USGS, Washington, D.C., Book 3, chap. 3, 1968.
2. Chow, Ven Te. *Open Channel Hydraulics*. McGraw-Hill, New York, 1959.
3. Hales, L. Z., S. L. Bird, B. A. Ebersole, and R. Walton. Bolsa Bay, California, Proposed Ocean Entrance System Study: Report 3, Tidal Circulation and Transport Computer Simulation, and Water Quality Assessment, Misc. paper CERC-89-17, U.S. Army Engineer Waterways Experiment Station, Vicksburg, MS, March 1990.
4. Hales, L. Z., S. L. Bird, and R. Walton. Hydrodynamic modeling of Bolsa Chica, California, proposed wetland enhancements, *Proc. Natl. Conf. Hydraulic Engineering*. ASCE, Nashville, TN, July/August, 1991.
5. Moffatt and Nichols, Engineers, Data Collection Program for Bolsa Bay, 1986.
6. Moore, C. I. and R. Walton. DYNTRAN/TRAN User Manual, Camp, Dresser, & McKee, Inc., Annandale, VA, prepared for SRA Technologies, Inc., Arlington, VA, under contract to U.S. Navy, Annapolis, MD, October, 1984.
7. Walton, R., C. M. Adema, and P. F. Seligman. Mathematical modeling of the transport and fate of organotin in harbors, *Proc. Oceans '86*. Washington, D.C., September, 1986.
8. Walton, R., S. Bird, B. Ebersole, and L. Hales. Numerical modeling studies of wetland hydraulics for Bolsa Bay, California, *Proc. 3rd Natl. Conf. Hydraulic Engineering*. ASCE, New Orleans, LA, August, 1989.

CHAPTER 11

Performance Evaluation of a Closed Ecological Life Support System (CELSS) Employing Constructed Wetlands

B. L. Hilton

INTRODUCTION

Constructed wetlands, also called microbial rock reed filters, have been employed for treatment of sewage from systems as small as single-family residences to communities with flows of 3 MGD. Constructed wetlands are divided into two types: (1) free-water surface (FWS) systems and (2) subsurface flow systems (SFS), also called "root zone" or "rock-reed filters". The latter is the type of system used in the Biohome and is the subject of this study.

BACKGROUND

Constructed wetlands offer a promising alternative to conventional treatment plants because they are: (1) relatively inexpensive to construct and operate; (2) easy to maintain; (3) effective and reliable wastewater treatment systems; (4) tolerant of fluctuating hydrologic and contaminant loading rates; and (5) providers of green space, wildlife habitats, and recreational and educational areas. The SFS consists of channels or trenches with relatively impermeable bottoms filled with sand or rock media to support vegetation.[1] These low-cost, low-technology, low-maintenance systems provide a high-quality, polished effluent which meets U.S. EPA discharge standards for advanced secondary treatment. However, little is known about the actual microbial action, the kinetics, or the more definitive engineering design parameters.

A major portion of the development of this technology has been conducted at NASA's[1] Stennis Space Center in Mississippi. These units are used to treat the waste at the center from locations such as the laboratory at the Rouchon House. A small unit was installed in the Biohome, a one-person "closed ecological life support system" (CELSS) habitat. The CELSS concept is inherently important for future space travel due to the logistical impossibility of transporting sufficient resources (food and water) to permit single use or limited reuse of water, food, and other materials. Physical-chemical or reuse systems require an input of energy to convert materials, degrade wastes, or purify water. Current work with respect to the currently planned space station has resulted in a significant downsizing of energy related systems due to limitations in on-site power generation. Furthermore, food and fiber production require the production of biomass. Consequently, the most rational, energy-efficient method for the degradation of waste products, the synthesis of food and fiber, and the purification of water for on-board consumption is the incorporation of a CELSS system.

The low-technology, low-energy, low-maintenance process, as exemplified by the facility in the Biohome, can be used as a basis for future design and further development of a practical ecological life support system. It is understood that process adaptations would be required to compensate for the growth of microbial and macrobiotic systems under conditions of microgravity.

In constructed wetland and aquatic plant systems, sedimentation, filtration, gas transfer, adsorption, ion exchange, chemical precipitation, chemical oxidation and reduction, and biological conversion and degradation occur at lower rates than in conventionally engineered treatment systems.[1]

Disadvantages include (1) large land area requirements, (2) imprecise design and operating criteria, (3) biological and hydrological complexity, (4) possible problems with pests, and (5) plugging potential.[2]

0-87371-550-0/93/$0.00+$.50

Fundamental Considerations

Suspended Solids

In FWS systems, suspended solids are removed in part by sedimentation and in part by filtration through the living vegetation. Additional removal also occurs at the soil interface. In the SFS, suspended solids are removed primarily by filtration through the soil or subsurface media.[3]

Organic Matter

The organic matter in wastewater is removed by microbial degradation. In most cases, natural systems are designed to operate under aerobic conditions because aerobic decomposition tends to be more rapid and complete than anaerobic decomposition. Furthermore, odors caused by anaerobic degradation are avoided.[3]

Nitrogen

The removal of nitrogen depends on the form of nitrogen present (nitrate, ammonia, or organic nitrogen). Plants remove ammonia, nitrate, or nitrite, which are then incorporated into the cell mass. Under certain circumstances, ammonia can be absorbed temporarily through ion exchange reactions on soil particles and charged organic particles.[4]

Ammonia is converted to nitrites and nitrates by aerobic autotrophic microorganisms such as the genera *Nitrosomonas* (NH_3 to NO_2^-) and *Nitrobacter* (NO_2^- to NO_3^-) in the process of nitrification. On the other hand, the reduction of nitrates and nitrites to N_2 is called "denitrification". Microorganisms prefer an aerobic environment, but when oxygen is not available they can use bound oxygen in nitrites and nitrates.[5]

Microorganisms

Microorganisms are removed by die-off, filtration, sedimentation, entrapment, predation, radiation, desiccation, chlorination (when disinfection is applied), and adsorption.

Under anaerobic conditions, bacteria produce an abundance of organic acids and alcohols. These intermediates are further decomposed to carbon dioxide and methane.[6] In general, equations 1 and 2 apply:

$$\text{Organics} = \text{intermediates} + CO_2 + H_2O + \text{energy} \qquad (11\text{–}1)$$

$$\text{Intermediates} = CH_4 + CO_2 + \text{energy} \qquad (11\text{–}2)$$

Trace Elements

Removal of these elements, principally heavy metals, occurs mainly through sorption. Some metals are also removed by plant uptake.[7]

Physical and Chemical Characteristics

Excess water causes many physical and chemical changes in soils, and wetland hydrologic regimes can change from saturation to short-duration flooding. In permanently flooded wetland ecosystems (which include constructed wetlands), a wide range of oxidation-reduction reactions allow the wetland to function as an effective transformer of nutrients and metals.[4]

Wetland Vegetation

Potentially useful emergent forms include cattail, reed, rush, sedge, and grass species. Emergent plants are often grown in gravel beds to create suitable conditions for substrate oxidation.[8] Traditionally, plants have been used in the full-scale engineered systems.

Wetland Microbiology

Biodegradable organics are the energy-yielding substrates. If there is a nutrient deficiency, the microorganisms must adapt or perish. Many microorganisms become dormant until external conditions improve. Bacteria are very "patient" and can maintain this state for many years.[9]

Fixed-Film Reactors

Fixed-film reactors have the ability to immobilize the bacteria to stabilize the organics in wastewater efficiently. These processes accumulate large quantities of microbial biomass which are attached to the media (rock or plastic). Constructed wetlands are, in effect, large, fixed-film bioreactors with rocks, soil, and plant roots providing the surface upon which microorganisms attach or where they are entrapped.

There have been several studies on the successful operation and treatability of a variety of wastewaters with these reactors under anaerobic conditions.[10,11] However, no information exists on the relationship between waste concentration and biomass and their influences on process efficiency. This is due to the difficulty in measuring the biofilm concentration. The relationship between waste stabilization and reactor biomass is very important in order to characterize the reactor with regard to active biofilm thickness, specific substrate utilization, and biofilm sloughing.[10]

Fixed-film reactors could handle severe hydraulic overloading without serious problems. Kennedy and Van Den Berg[12] investigated the operation of fixed-film reactors at 10 to 35°C. These reactors were durable, regained normal performance 12 to 48 h after an upset, and were not prone to some of the operating instabilities associated with conventional anaerobic reactors.

OBJECTIVES

The purpose of this study was to evaluate the performance of the treatment system as constructed in the Biohome and to determine the kinetic parameters associated with this type of treatment system. To date, engineering practice with respect to design and construction of constructed wetlands has been based upon sizing units using the "black box" approach, with little understanding of the kinetics of substrate utilization, fate of compounds or components, or a significant grasp of the quantity and concentration of waste capable of being treated by one of these systems. The use of plants has been emphasized as being necessary, yet plants *per se* utilize inorganic rather than organic compounds. During the earlier part of this century in the U.S., and currently in many other countries such as Mexico, sewage farms take advantage of the filtering provided by plants. This filtration capability has permitted health authorities to approve irrigation of plants with poorly treated sewage, with little fear of disease transmission, so long as raw sewage does not come in contact with the fruit.

In this study, it was anticipated that measurements of carbon and nitrogen utilization and conversion would provide insight into the biological and biochemical aspects of waste utilization in these systems. Once these parameters were known, it would then be possible to formulate a set of design parameters using substrate utilization rates upon which further design could be based.

PROCEDURES

Two experiments were conducted. One was performed upon the existing system in the Biohome. The system (Figure 1) was constructed of six sections of 254-mm PVC pipe, 3.6 m long, with 6 tees per section for the planting of various plants. In this system, three organic loading rates were utilized. During the first 4 weeks of the study, the existing flow and nutrient input were used with the system. The average flow was 33 $l\ d^{-1}$ and the organic input was the daily (Monday to Friday) waste contribution from the technician in care of the facility. This waste contribution included a daily sanitary (toilet) input and showers, for a mass loading of 0.007 kg $TOC/m^3/d$.

During the second (3-week) period of the study, 38 L/d of Rouchon House septic tank supernatant were added to the waste stream in addition to the technician's contribution. The mass loading was 0.023 kg $TOC/m^3/d$.

During the last (3-week) period of the study, 125 L/d of Rouchon House septic tank supernatant were added to the waste stream, resulting in a mass loading of 0.082 kg $TOC/m^3/d$. Subsequent to the author's direct participation in the study, a greater quantity of Rouchon House septic tank influent was added to the waste stream. For all three periods of the study, the volumetric loading was less than 20% of the customary loading for low-rate anaerobic reactors (0.3 to 0.4 kg $TOC/m^3/d$) and less than 10% of the lowest loading rate of high-rate anaerobic reactors (1 to 4 kg $TOC/m^3/d$).

In the second experiment, four parallel reactors were constructed. One reactor (control) contained no rocks or plants, one contained rocks, one contained plants (Elephant Ears, *Colocasia* spp.), and the fourth contained plants and rocks. All were fed effluent from the septic tank for Rouchon House. The flow was 6 to 8 L/d for a mass loading of 0.08 to 0.3 kg $TOC/m^3/d$.

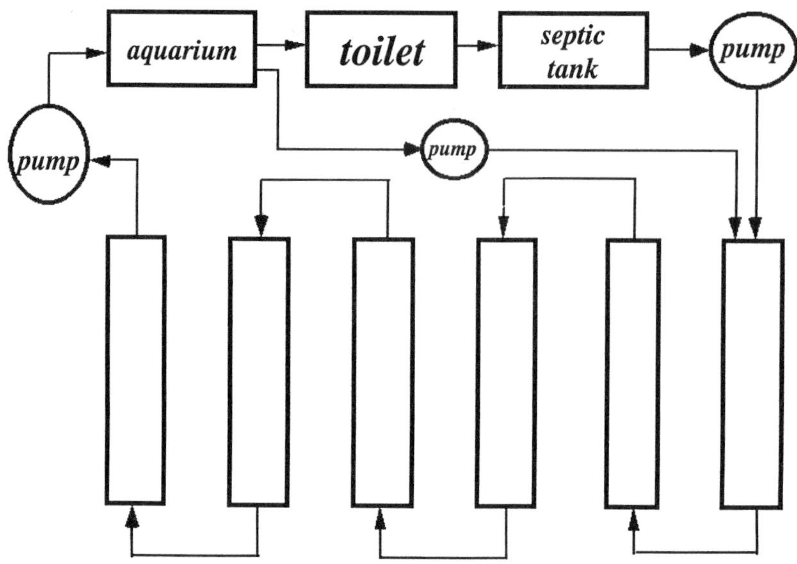

Figure 1. Biohome flow schematic diagram.

ANALYTICAL

Dissolved oxygen was measured using a dissolved oxygen meter (Yellow Springs Instrumentation Co.). Periodically, a Winkler titration was performed using a HACH dissolved oxygen test kit. Oxidation-reduction potential (ORP) was measured using an Orion platinum redox electrode using a Hach portable meter. Total organic carbon (TOC) was measured by a TOC analyzer using the persulfate reduction method (OI Analytical). Ammonia was measured on a Technicon Autoanalyzer in accordance with standard methods.[12] Carbonate, chloride, nitrate, nitrite, phosphate, and sulfate were measured using a Waters ion chromatography system. BOD_5 and fecal coliform were tested in accordance with the procedures set forth in Reference 12.

RESULTS AND DISCUSSION

Biohome

Carbon Removal

Two rates of carbon removal were noted: one for the sections with lava rock and another with a higher rate for the sections with granular activated carbon or calcium zeolite. For lava rock, the regression equation for removal was:

$$S_e = 0.673\ S_o + 0.43\ V/Q \tag{11-3}$$

and for granular activated carbon, the regression equation for removal was:

$$S_e = 0.512\ S_o + 0.825\ V/Q \tag{11-4}$$

where Q = flow, L/d
V = media volume, L
S_o = influent carbon, TOC, mg/L
S_e = effluent carbon, TOC, mg/L

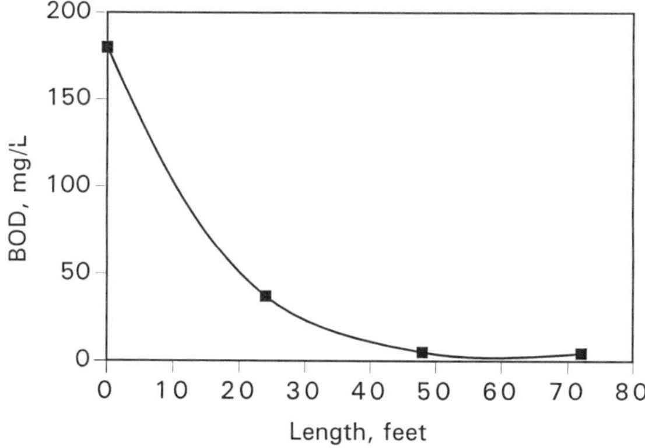

Figure 2. Effluent carbon concentrations.

This means that the carbon removal rate of the fine media (carbon or calcium zeolite) was greater than that of the course media (lava rock). Since removal rates for calcium zeolite and activated carbon were similar, and since both of these exhibited a greater surface area than the lava rock, one would conclude that the higher removal rate was a function of increased surface area and not something intrinsic to the nature of the media.

Acclimation to Changing Load

Daily testing for the TOC from each segment of the treatment unit in the Biohome was performed at each increase in influent carbon. Carbon removal, as exhibited by the effluent carbon concentration (segment 6), returned to the same value as noted during the previous testing period (Figure 2). The curve is similar to an exponential decay curve. However, if the curve were modified to show carbon removal rather than effluent carbon concentration, then the curve would approximate an exponential growth curve.

The exponential change in removal simulates growth in microorganism populations. It is hypothesized that if the quantity of biomass were measured, there would be a direct correlation between the removal of carbon and the quantity of biomass. From this, it is further hypothesized that if comparisons were made between various segments of the microbial populations, one would find an inverse relationship between microbial population densities and soluble organic carbon concentrations. Furthermore, it would also be hypothesized that a population shift from carbon-utilizing microorganisms to nitrifying organisms will occur when the concentrations of carbon decrease in the system. This latter phenomenon is best illustrated by the classic BOD curve showing oxygen demand first as a function of carbonaceous BOD, followed by nitrogenous BOD. Microorganisms utilizing ammonia for energy have a lower growth rate than those utilizing carbon; therefore, those utilizing ammonia would be expected to be out-competed under conditions of high carbon concentrations.

Ammonia Removal

Ammonia concentrations followed carbon concentrations. The variation in ammonia may have been a function of the carbon concentration or it may have paralleled it. However, the data neither verify nor disprove this. Plants were present in sections 2 to 6, and the total nitrogen decreased; therefore, the decrease in nitrogen may well have been a function of plant utilization (Figure 3).

Oxidation-Reduction Potential

Large negative potentials were measured throughout the system and were directly correlated with the concentration of ammonia present in the system (Figure 4). Such negative potentials are indicative of reducing (e.g., anaerobic) environments. These measurements, however, do not preclude the possibility of a microaerophilic environment at the interface of the root hair and the fluid, but do indicate that the predominant environment is anaerobic and not aerobic. This finding corresponds with that of other researchers.[13]

Rouchon House Performance

The effluent BOD_5 in the Rouchon House rock filter (1988–1990) (Figure 5) was relatively stable, with wide fluctuations in influent BOD_5. The basal level of effluent BOD_5 suggests that effluent quality was not

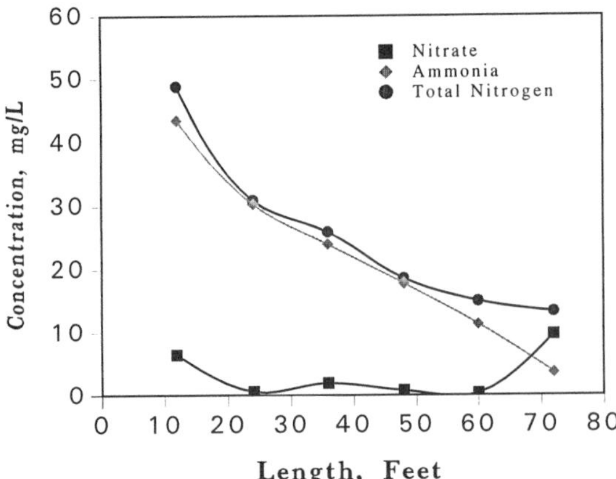

Figure 3. Ammonia concentrations in biohome system.

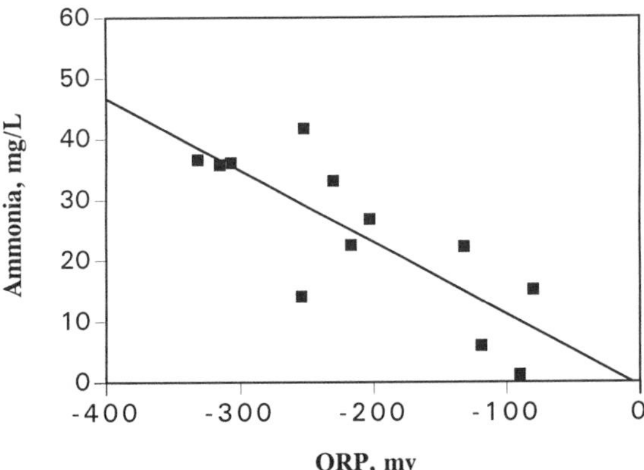

Figure 4. Oxidation Reduction Potential (ORP) relating to ammonia concentration found in the biohome unit.

significantly improved by decreasing the load to the unit. Conversely, the efficiency of carbon removal increased with increased carbon input, a finding which is contrary to conventional wisdom with regard to the performance of treatment units.

Parallel Reactors

The four parallel reactors were fed 0.09 to 0.4 kg TOC/m^3/d with an HRT of 3 to 7 h, which was a lower HRT and higher mass loading than these applied to the Biohome system. Longer HRTs would have been used; however, the pumping system used for the experiment was already operating at the lowest possible setting.

Current design practices for these systems dictate that plants be an integral part of construction and operation. Carbon removal in the reactors with plants, rocks, or plants and rocks was twice that of the control (Figure 6) which was merely an open pipe, yet there was no significant difference between removals in those three reactors. The only possible conclusion is that, under these conditions, plants did not enhance carbon removal. If this were not so, then the reactor with plants and rocks should have exhibited a carbon removal rate approximating the sum of the removal with plants alone and that with rocks alone. From these results, microbial degradation appears to be the most logical explanation for carbon removal in these systems.

On the other hand, the reactor with plants and rocks (Figure 6) exhibited a marked improvement in the removal of nitrogen. This suggests that if low concentrations of nitrogen are required in the effluent, then a careful selection of plants can enhance the treatment process. The engineer must be judicious in the selection of vegetation, as evidenced in a parallel experiment. In that experiment, which used Chinese bean plants, the concentration of nitrogen in the effluent was greater than in the influent. This reflects the traditional nitrogen fixation associated with legumes.

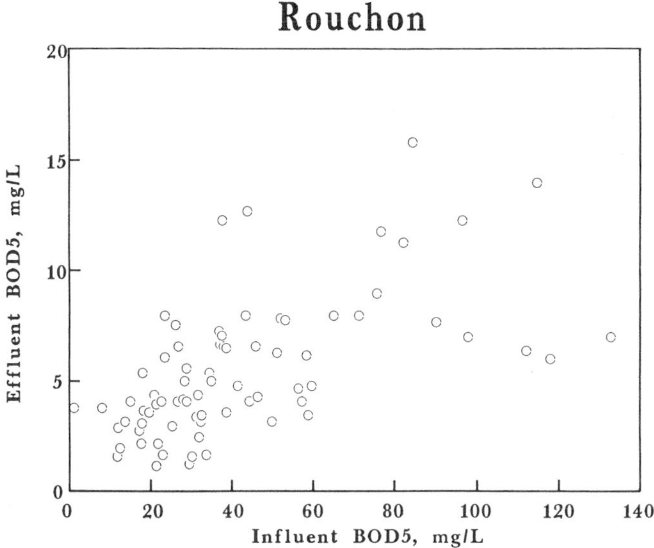

Figure 5. Effluent BODs in the Rouchon House rock filter.

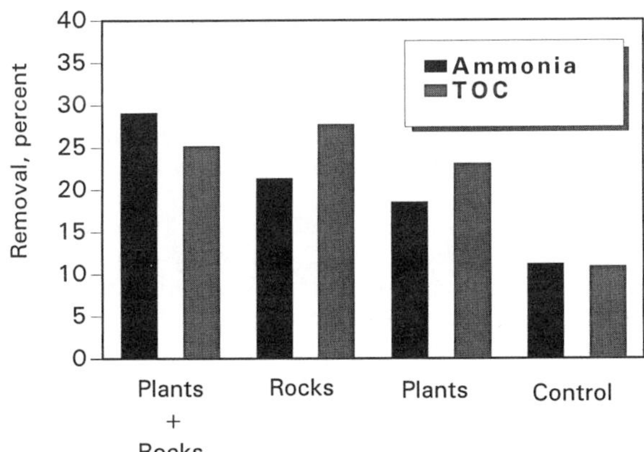

Figure 6. Carbon removal in the reactors with plants, rocks, or plants and rocks.

The low BOD_5 associated with the operation of these subsurface flow wetlands could be due, in part, to the removal of ammonia. The traditional test for BOD_5 does not distinguish between carbonaceous and nitrogenous oxygen demand unless the nitrifiers are intentionally suppressed. Consequently, it is plausible that the beneficial effects of plants in achieving a low BOD_5 effluent are due to their nitrogen.

Plugging

Recurring problems with reduced hydraulic conductivity in the system were experienced due to the growth of large masses of roots from the plants. Subsequent conversations with personnel operating full-scale systems have revealed this to be common. The plugging with roots affects the flow through the systems and results in surface ponding and short-circuiting. In a subsequent project,[14] plugging due to the growth of biomass has been encountered. The loading rates of these systems may be more limited by hydraulics than by carbon.

The minimal removal achieved in the end of the system raises the necessity of the extreme lengths of presently constructed systems (ca. 300 to 400 m). Due to the reduced flow effected by plugging with plant roots and microbial growth, consideration might be given to changing the dimensions, thereby increasing the width of the systems to compensate for the low hydraulic conductivity caused by the increased quantity of biomass.

DESIGN CRITERIA

The following list of design criteria is a compilation from this project and from current engineering practice. The recommendations for carbon removal, hydraulic retention time, and recirculation are based on the findings of this project. In general, recirculation has not been practiced in full-scale systems. Based on the results of this research, recirculation may significantly enhance performance of field units just as it does in conventional aerobic biotowers and trickling filter systems.

Design Criteria

Carbon removal (TOC)	10–20 g/m^3
Void ratio	40%
Hydraulic retention time	24–48 h
Media size	4–6 in.
Media depth	24–30 in.
Liquid depth	24–30 in.
Trench depth	30–36 in.
Pretreatment	Primary sedimentation
Recirculation	0–200%

CONCLUSIONS

1. Ammonia decreased throughout the system. In the first segments, this may have been associated with plant uptake; in the last segment, ammonia was converted to nitrate (e.g., oxidized).
2. Ammonia concentration was directly correlated with TOC concentrations.
3. Carbon (TOC) removal was 10 mg/L in the lava rock and 20 mg/L in the carbon and zeolite (mg TOC/L media volume).
4. Low oxygen levels were observed in all but the last segment. These were correlated with low redox potentials. This suggests that much carbon removal was by anaerobic metabolism.
5. Experiments with separate, parallel units showed minimal carbon removal in units with plants as compared with units containing only gravel. Carbon removal in units with plants and no gravel was less than units containing only gravel. Therefore, it may be concluded that the addition of plants to these systems will not significantly improve carbon removal but does enhance nitrogen removal.
6. Within the limits of the carbon loading of this experiment, the effluent carbon concentration varied minimally with significant variations in influent carbon concentrations. This suggests that the common engineering practice of preceding these units with aerobic lagoons may not significantly improve effluent quality, and may also suggest that effluent quality of those systems would be relatively unaffected by elimination of the preceding lagoons.
7. Effluent BOD_5 in the Rouchon House facility varied little with a wide range of variation in influent BOD_5. This followed the familiar Monod hyperbolic curve, and again suggests that significant pretreatment other than solids removal by primary sedimentation will not significantly improve effluent quality.

ACKNOWLEDGMENTS

This research was conducted during a 10-week NASA/ASEE summer faculty fellowship at the NASA Stennis Space Center in Mississippi. The assistance of NASA and Sverdrup Technologies personnel is gratefully acknowledged. The conclusions are solely those of the author.

REFERENCES

1. Reed, S. C., E. J. Middlebrooks, and R. W. Crites. *Natural Systems for Wastewater Treatment*. McGraw-Hill, New York, 1988.
2. Hammer, D. A. and R. K. Bastian. Wetlands ecosystems: natural water, in *Constructed Wetlands for Wastewater Treatment*. D. A. Hammer, Ed. Lewis Publishers, Chelsea, MI, 1989.

3. Metcalf and Eddy, Inc. *Wastewater Engineering: Treatment, Disposal, and Reuse.* McGraw-Hill, New York, 1991.
4. Faulkner, S. P. and C. J. Richardson. Physical and chemical characteristics of freshwater wetland soils, in *Constructed Wetlands for Wastewater Treatment.* D. A. Hammer, Ed. Lewis Publishers, Chelsea, MI, 1989.
5. Gaudy, A. F., Jr. and E. T. Gaudy. *Elements of Bioenvironmental Engineering.* Engineering Press, San Jose, CA, 1988.
6. Viessman, W., Jr. and M. J. Hammer. *Water Supply and Pollution Control.* Harper and Row, New York, 1985.
7. Wolverton, B. C. Water Hyacinths for Removal of Cadmium and Nickel, NASA: TM-X-72721, Bay St. Louis, MS, 1975.
8. Guntenspergen, G. R., F. Stearns, and J. A. Kadlec. Wetland vegetation, in *Constructed Wetlands for Wastewater Treatment.* D. A. Hammer, Ed., Lewis Publishers, Chelsea, MI, 1989.
9. Portier, R. J. and S. J. Palmer. Wetland microbiology: form, function, processes, in *Constructed Wetlands for Wastewater Treatment.* D. A. Hammer, Ed., Lewis Publishers, Chelsea, MI, 1989.
10. Kennedy, K. J. and R. L. Droste. Anaerobic fixed-film reactors treating carbohydrate wastewater. *Water Resources.* 20(6):685, 1986.
11. Kennedy, K. J., M. F. Hamoda, and S. G. Guiot. Anaerobic treatment of leachate using fixed film and sludge bed systems, *J. Water Pollut. Control Fed.* 60(9):1675, 1988.
12. APHA. *Standard Methods for the Examination of Water and Wastewater.* 17th ed. American Public Health Association, Washington, D.C., 1989.
12. Kennedy, K. J. and L. Van Den Berg. Stability and performance of fixed film reactors during hydraulic overloading at 10–35°C, *Water Resources.* 16:1391, 1982.
13. Skipper, D. G. A Rock-Plant Filter Bench-Scale Study and Computer Model, unpublished Ph.D. Dissertation, Louisiana State University, Baton Rouge, LA, 1990.
14. Nicholaou, S. A. Performance Evaluation of a Laboratory Scale Horizontal Flow Biological Filter. Unpublished M.S. Thesis, University of New Orleans, New Orleans, LA, 1992.

PART 3

ACID MINE

CHAPTER 12

Anoxic Limestone Drains to Enhance Performance of Aerobic Acid Drainage Treatment Wetlands: Experiences of the Tennessee Valley Authority

G. A. Brodie, C. R. Britt, T. M. Tomaszewski, and H. N. Taylor

INTRODUCTION

Staged aerobic constructed wetlands are an effective means of treating certain qualities and quantities of acid drainage from various mine spoils and refuse, and from coal ash disposal areas.[1] Wetlands efficiencies are related primarily to flow rates, pH, and concentrations of dissolved oxygen (DO), Fe, Mn, acidity, and alkalinity. Foremost are the relationships among influent DO, Fe speciation, alkalinity, and influent/effluent pH.

Metals removal mechanisms in wetlands are variable and numerous, but it is probable that microbially catalyzed Fe oxidation and hydrolysis are major reactions responsible for Fe removal in aerobic wetlands systems.[2] These reactions may dictate the ultimate success or failure of a wetlands system because of their acidity-producing natures, as summarized below.

$$2FeS_2 + 2H_2O + 7O_2 = 2Fe^{2+} + 4H^+ + 4SO_4^{2-} \qquad (12\text{–}1)$$

(Pyrite oxidation)

$$4Fe^{2+} + O_2 + 4H^+ = 4Fe^{3+} + 2H_2O \qquad (12\text{–}2)$$

(Ferrous oxidation)

$$Fe^{3+} + 3H_2O = Fe(OH)_3 + 3H^+ \qquad (12\text{–}3)$$

(Ferric hydrolysis)

Reactions 1 and 3 all occur in the backfill, while reactions 2 and 3 occur in the wetlands. Only reaction 3 acts to increase acidity in a wetlands. If sufficient buffering capacity is not present in the acid drainage, pH will decrease in a wetlands as ferric hydrolysis proceeds. Several TVA wetlands systems, although removing substantial amounts of Fe and Mn, have had influent pH near 6.0 and effluent pHs of less than 3.0.[1] TVA's 3-cell, 2.3 acre constructed wetland at the Kingston Fossil Plant in Roane County, Tennessee, receives, on average, 379 GPM of acid seepage from an ash disposal area with an influent pH of 5.5 and total Fe of 170 mg/L. Effluent from the wetlands system has a pH of 2.9 and total Fe of 83 mg/L; thus, the hydrolysis of about 87 mg/L of Fe is responsible for the decreased pH.

0-87371-550-0/93/$0.00+$.50

DISCUSSION

Numerous methods, primarily chemical treatment, have been used to meet effluent standards for treating acid drainage; however, these methods are expensive and require constant attention to maintain compliance with effluent limitations. Also, many chemical additives have an adverse impact on aquatic biota in receiving streams. Limestone treatment has many advantages over conventional chemical treatment, including lower sludge production, lower cost, and less potential for overdosing. However, the use of limestone rock as a buffering agent was abandoned because, in an oxidizing environment, Fe hydroxides coat the limestone surface and inhibit or eliminate the dissolution of the limestone, thus preventing effective buffering. This article discusses a passive methodology of introducing alkalinity to acid drainages in order to buffer pH decreases in constructed wetlands due to Fe hydrolysis.

In 1988, the Tennessee Division of Water Pollution Control (TWPC) constructed and evaluated prototype passive anoxic limestone drains (ALDs) at two mine sites in Tennessee.[3] At the same time, the TVA recognized a correlation between alkalinity and DO of wetlands influent and the relative success of TVA constructed wetlands, and conducted several relevant bench-scale studies on alkalinity enhancement to achieve National Pollution Discharge Elimination System (NPDES) compliance of acid drainage using passive aerobic wetlands systems.[4]

Results of a TVA study identified a correlation between wetlands influent alkalinity and effluent pH, Fe, and Mn. An "accidental" roadbed originated anoxic drain (AROAD) was identified at the TVA Impoundment 1 constructed wetlands, which treats high-Fe (>100 mg/L) acid drainage emanating from a fine-coal refuse disposal area at the Fabius Coal Preparation Plant site in northeastern Alabama (Figure 1). Air photo investigations showed that a 1400-ft-long, 25-foot-high earth dam was constructed in 1974 over an existing coal haul road built of crushed rock that was presumably obtained from the local Monteagle formation, a high $CaCO_3$, oolitic limestone. Although this limestone roadbed was not intended for water quality enhancement, it was hypothesized that it was pretreating acid drainage seeping through the dike by adding alkalinity and buffering capacity so that the wetlands was producing compliance-quality discharges. TVA drilled and sampled the dike and the AROAD, and confirmed the presence of a 4 to 11 in. roadbed consisting of gravel-sized limestone (peloid grainstones, peloid packstones, pellet packstones/grainstones, and coarse-grained bivalved grainstone) containing 46 to 58% Ca.[5] Split spoon samples of the limestone showed no evidence of mineralization or coatings on the stone. Dissolution was evidenced by the rounded edges of the recovered crushed stone. Monitoring wells were drilled and are being sampled to determine the changes in water quality as the flow passes through the earth dam. Figure 2 depicts preliminary water quality data from six monitor wells completed in the dam at various locations relative to the AROAD, the slurry lake, and the seepage.

Over 50 ALDs have now been constructed in the eastern U.S.[6] The ALD, which consists of a shallow, limestone-filled trench excavated into the spoil and sealed from the atmosphere, passively introduces buffering capacity, as alkalinity, into the acid drainage. Changes in pH due to acid production from Fe hydrolysis in the wetlands are buffered due to the high alkalinity in the influent. Detailed design and construction guidelines have been previously published.[7]

The basic design of an ALD is shown in Figure 3. The ALD consists of an open, unlined trench or excavation back-filled with gravel-sized, crushed, high-$CaCO_3$ limestone. The limestone is covered with plastic to preclude oxygen infiltration and CO_2 exsolution. It is desirable to place geotechnical fabric over the plastic for protection of the plastic from puncture by equipment or manpower. A clay soil is then placed over the fabric.

It is assumed that the following reactions are applicable within the ALD:

$$CaCO_3 + 2H^+ = Ca^{2+} + H_2CO_3 \quad (12\text{–}4)$$

$$CaCO_3 + H_2CO_3 = Ca^{2+} + 2HCO_3^- \quad (12\text{–}5)$$

$$CaCO_3 + H^+ = Ca^{2+} + HCO_3^- \quad (12\text{–}6)$$

Equation 4 reacts limestone with acidity (at pH < 6.4) present in the mine drainage to form free calcium and dissolved carbon dioxide (carbonic acid). The carbonic acid further reacts with limestone in Equation 5

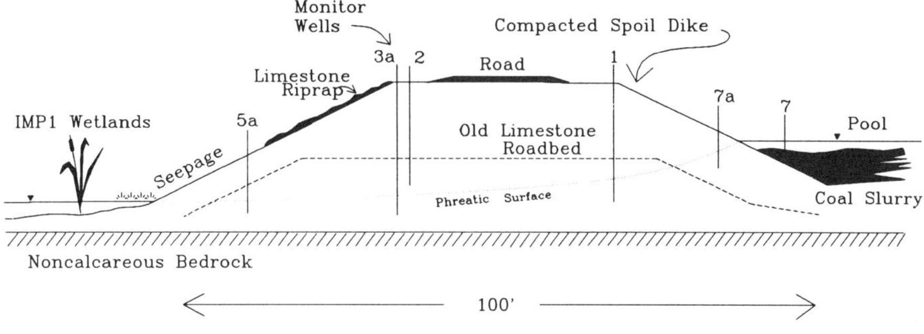

Figure 1. Schematic of slurry lake 2 earth dam "AROAD".

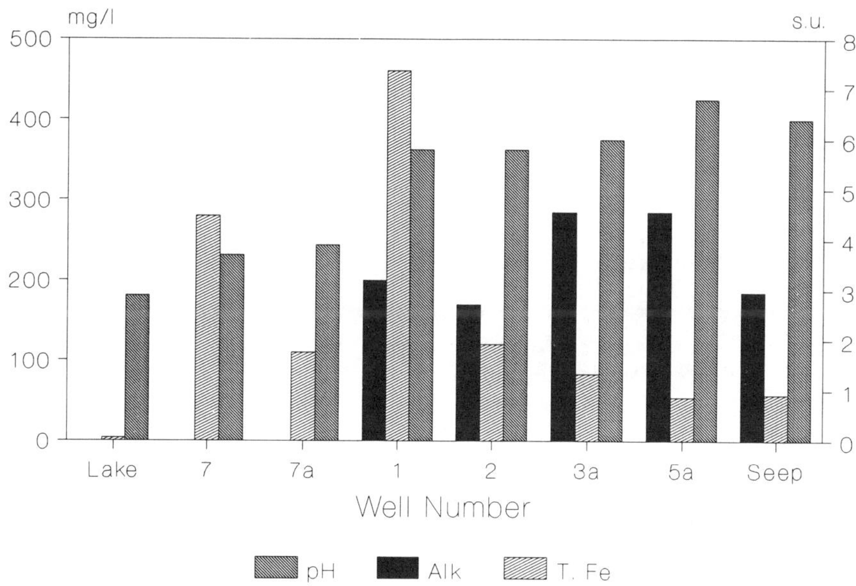

Figure 2. Preliminary data for monitoring wells located in slurry lake 2 dam.

to produce alkalinity. As reactions 4 and 5 act to increase pH above 6.4, Equation 6 becomes the major reaction, where bicarbonate is the dominant dissolved CO_2 species.[8] Systems observed to date indicate that if the ALD is properly constructed, limestone in the ALD will not armor with Fe or Al precipitates and alkalinity can be significantly increased in the ALD effluent.

DESIGN GUIDELINES

Guidelines to determine the utility for an ALD are suggested as follows.

Case 1. Alkalinity >80 mg/L, Fe > 20 mg/L; an ALD may be beneficial, but a wetlands system based on previously reported chemical loading rates is probably adequate[1]

Case 2. Alkalinity >80 mg/L, Fe < 20 mg/L; an ALD is not necessary, only an adequately designed wetlands

Case 3. Alkalinity <80 mg/L, Fe > 20 mg/L; an ALD is recommended and probably necessary as the initial stage in a constructed wetlands system

Case 4. Alkalinity <80 mg/L, Fe < 20 mg/L; an ALD is recommended, but not necessary

Case 5. Alkalinity ≤0 mg/L, Fe < 20 mg/L; an ALD will likely be necessary as Fe approaches 20 mg/L

Case 6. Dissolved oxygen > 2.0 mg/L *or* Fe oxidizing conditions exist (e.g., pH > 6.0, eH > 100 mV); an ALD should not be installed because of probable limestone armoring

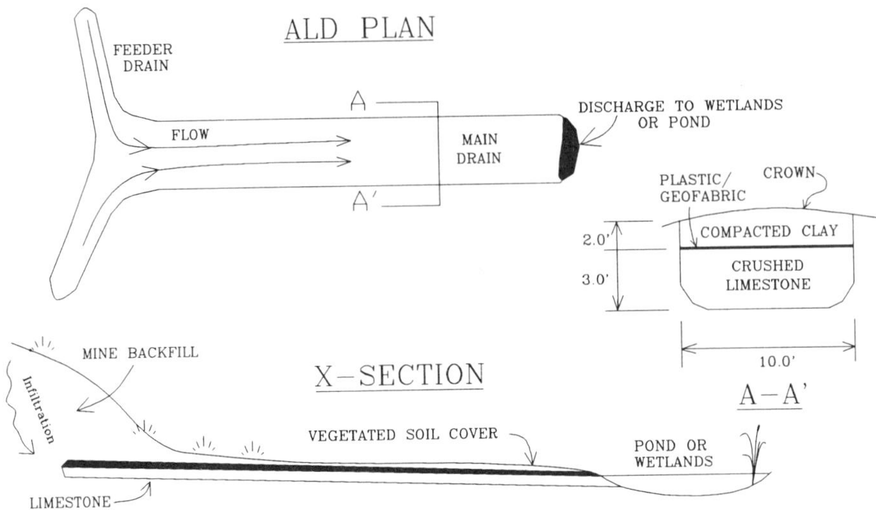

Figure 3. Generalized anoxic limestone drain schematic.

Case 7. Ferric iron present in appreciable concentrations in the unaerated seep sample; an ALD may not be appropriate due to potential limestone armoring

In order for an ALD to perform successfully, unaerated mine drainage must come into contact with the buried limestone. Discrete seeps or boils from embankments are generally good starting points for excavation of the ALD into the backfill. Non-point seep areas may require more innovative means of collection, such as specialized rock drains or the construction of an embankment to contain the ALD. Underground mine adits could be sealed and flooded, routing the drainage via a pipe to an ALD, or a mine adit could be backfilled with limestone and thereby delineate the boundaries of the ALD.

Maximum expected flow through the ALD should be determined to prevent hydraulic failures and disruption of the sealed nature of the ALD. Owing to the relative low cost of an ALD compared to conventional treatment systems, an over-designed ALD is recommended to ensure hydraulic stability and to maximize longevity. In fact, desired longevity of the ALD will likely dictate a design size far in excess of the hydraulic design requirements. A suggested crude, but conservative method for hydraulically designing the ALD is to calculate infiltration from the contributing drainage basin at the point source of the seepage and assume an attenuation factor based on the mine fill characteristics.

Another important preliminary consideration for ALD construction is water quality. The success of an aerobic wetlands system to meet effluent limitations is directly and primarily related to influent pH, acidity, alkalinity, Fe, and Mn concentrations. Alternatively, DO and oxidation-reduction potential (ORP) are critical for the proper operation of an ALD. All waters that would be routed through an ALD should be sampled and analyzed for the above-mentioned parameters. Samples should be collected directly from the seep via methodologies precluding sample aeration (e.g., peristaltic pumps, large syringes, etc.). DO and ORP must be field measurements. Fe^{2+} and Fe^{3+} are good indicators of ORP (and possible armoring). If samples are immediately acidified with HNO_3 (or another nonoxidizing acid), Fe^{2+} can be determined accurately in the lab. Al precipitation within the ALD has also been proposed as a potential plugging problem,[9] but has not been observed at TVA systems including the IMP1 AROAD where total Al is greater than 10 mg/L.

Side slopes of the excavation are not critical to the operation of the ALD and are best made near vertical to facilitate construction. High-calcium limestone (i.e., ≥90% available $CaCO_3$) should be used. TVA and TWPC have used $^3/_4$ to $1^1/_2$ in crushed rock in existing ALDs because of its high hydraulic conductivity, large surface area, and ready availability; however, a different gradation, a well-graded mixture, or a layering of several grades of rock may be suitable. Larger rock size or more dolomitic (i.e., $CaMgCO_3$) stone is not recommended due to the loss of surface area in the ALD with the large rock and the lower reactivity of dolomite. The depth of limestone backfill should be determined by (1) the need to accommodate the maximum probable flow, (2) the desired longevity of the ALD, and (3) a comfortable safety factor for both (1) and (2). Desired longevity is discussed below. TVA's and TWPC's ALDs have contained limestone depths of 2 to 5 ft. Since the very lowest portion of the backfill may be rendered ineffective due to its tendency to embed in the bottom of the excavation, about 6 in. additional depth of limestone should be included to account for this loss.

A minimum of 2 ft of soil cover should be placed and compacted over the plastic and fabric-covered limestone backfill. Soil should be sufficiently impermeable to oxygen migration. If available and practicable, clay soil is recommended. Alternatively, a clay, clayey-silt, or silty-clay loam should be adequate. The final cover should be slightly crowned and protected with erosion control fabric or adequately water-barred to prevent erosion. Crowning will allow for subsidence of the ALD over the years due to limestone dissolution. Ideally, the crowned ALD should be rip-rapped or revegetated with species such as sericea (*Lespedeza cuneata*) or crownvetch (*Coronilla varia*), which will discourage the establishment of trees whose roots could penetrate the ALD and render it ineffective.

Prototype installations of ALDs have used 5-mil plastic and/or filter fabric covered with clay soil to seal the limestone from the atmosphere.[3] More recent ALDs have used 20-mil plastic or double layers of 10-mil plastic, which is significantly less expensive than 20-mil and is more readily available. Geotechnical fabric should be of quality sufficient to protect the integrity of the plastic from puncture under loads from equipment and workers.

An additional design recommendation is the use of an oxidation basin immediately after the seep discharges from the ALD and prior to routing that flow into a constructed wetlands. The purpose of an oxidation basin is aimed at high Fe influents (e.g., >50 mg/L). The basin allows the newly aerated and highly alkaline ALD discharge water to react and precipitate the majority of its Fe load, which can be achieved physiochemically and without the use of a constructed wetlands. The U.S. Bureau of Mines has suggested a general empirical guideline which states that about 50 mg/L Fe can be removed in an aerobic wetlands cell before reaeration is required.[10] An oxidation basin will greatly enhance the efficiency and lifespan of a downstream constructed wetlands as the basin can be dredged, thus preventing excessive precipitate build-up in the wetlands. Such a pond needs simply to be designed according to existing guidelines with considerations of regulatory requirements, desired capacity, and maintenance. Alternatively, a modified marsh-pond cell with a major portion devoted to deep (3 to 6 ft) water might be more applicable to acid drainage with low to moderate Fe concentrations.

Longevity of an ALD has been estimated by TVA and TWPC laboratory and field investigations based on the dissolution rate of limestone.[3,4] These studies include small-scale pilot studies and evaluations of existing ALDs and indicate that ALDs can be expected to last 20 to 80 years or more if properly designed and constructed.

CASE HISTORIES

Impoundment 4

TVA's Impoundment 4 constructed wetlands (IMP 4) was built in October 1985 to treat acid drainage emanating from a fine coal refuse area at the Fabius Coal Preparation Plant in northeast Alabama.[1] The inflow water quality was characterized by pH = 5.5, total Fe = 65 mg/L, total Mn = 17 mg/L, acidity > 200 mg/L, and alkalinity < 30 mg/L. Flow averaged 34 gal/min. Although significant amounts of Fe and Mn were removed from the acid drainage by the 4-cell, 0.5-acre wetlands (Fe was reduced to less than 10 mg/L), the effluent pH and Mn levels required chemical treatment before final discharge to achieve permit limitations.

In April 1990, an ALD was installed upstream of IMP 4 (Figure 4). The ALD consisted of a trench 10 to 15 ft wide, 5 ft deep, and 260 ft long excavated into the refuse and seepage area with a swamp-tracked hydraulic excavator. Some 400 tons of $^3/_4$ to $1^1/_2$-in. crushed limestone were backfilled into the open trench. The limestone was covered with two layers of 10-mil plastic over which was placed geofabric and 2 ft of local soil (sandy-silty loam). The surface was slightly crowned, seeded with a mixture including sericea lespedeza, mulched, and fertilized.

Figure 5 shows typical pre-ALD and post-ALD water quality for IMP 4 discharge. The ALD has successfully increased alkalinity and enabled the IMP 4 wetlands to meet pH, Fe, Mn, and suspended solids concentrations to compliance levels from the initial operation of the ALD to date; wetlands effluent pH has risen from 3.1 to 6.3, acidity has been reduced from 350 to 40 mg/L, alkalinity has increased from 0 to 100 mg/L, Fe has decreased from 6.0 to 1.0 mg/L, and Mn from 1.6 to 0.2 mg/L.

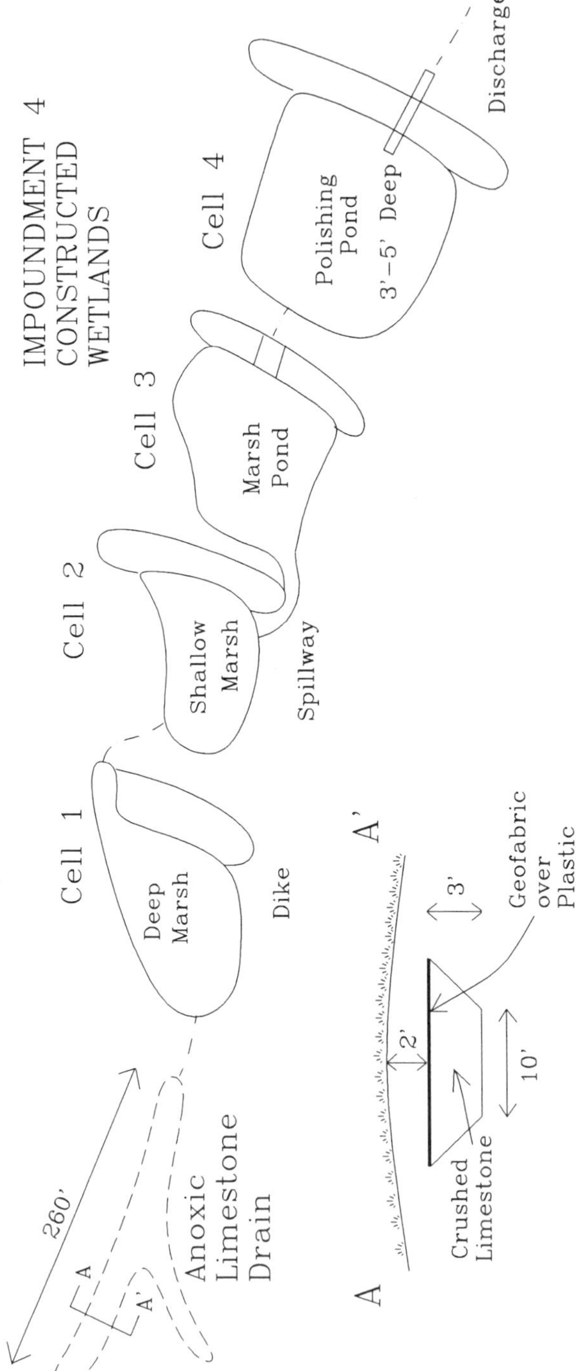

Figure 4. IMP4 schematic showing constructed wetlands and anoxic limestone drain.

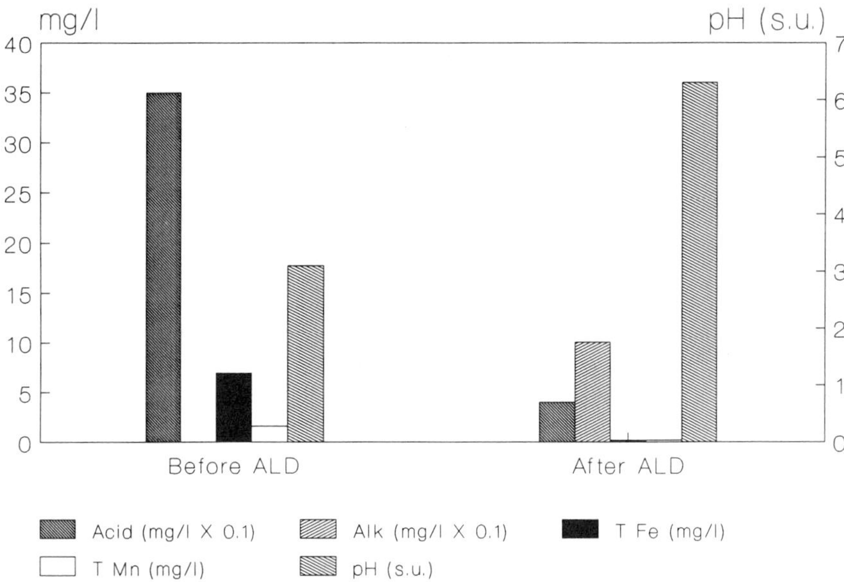

Figure 5. Average IMP4 effluent data before and after the ALD installation.

IMP 4 wetlands influent (i.e., the ALD effluent) had an average pH of 6.3 and concentrations of 85 mg/L Fe and 17 mg/L Mn before the ALD. After the ALD was installed, ALD effluent Fe and Mn concentrations decreased to very low concentrations (Figure 6). This phenomenon probably indicated oxidation and precipitation of Fe and Mn within the ALD. This was verified by a series of excavations into the ALD in May 1991. It was found that the upper portion of the limestone in the unsaturated zone of the ALD was moderately armored with Fe coatings. The stone in the saturated zone showed only traces of Fe precipitates. Two possible reasons for the armoring are hypothesized. First, the DO concentration in the ALD effluent was 1.0 to 2.0 mg/L, with an ORP of less than zero. Although these conditions, in water, may be conducive to keeping iron in the reduced, soluble ferrous form (Fe^{+2}), the atmosphere in the unsaturated zone probably had elevated oxygen levels due to several unsealed monitoring wells in the ALD. Periodic fluctuation of the water levels in this zone may have resulted in increased oxygen availability at the limestone surface for Fe oxidation/hydrolysis, resulting in armoring above the zone of saturation in the ALD.

A second hypothesis assumes that ferric hydrolysis (Equation 3) is occurring in the ALD because this reaction is not affected by DO levels. The Fe^{3+} concentration in the acid drainage is probably high enough for ferric hydrolysis to occur (i.e., $Fe^{3+}/Fe^{2+} > 0.25$). Whether this phenomenon is occurring is under evaluation by TVA[11] and the U.S. Bureau of Mines.[9]

The discharge from IMP 4 remained well within compliance levels (i.e., pH = 6–9, Fe < 3.0 mg/L, and Mn < 2.0 mg/L) during this period. It was assumed that if ferric hydrolysis was occurring in the ALD, the system would plug and eventually fail. However, if the ALD was simply being adversely impacted by oxygen invasion, then a solution to the problem might be to induce saturated conditions throughout the majority of the ALD. In June 1991, the monitoring wells were sealed and an earth dike (Figure 7) was constructed at the discharge point of the ALD to raise the water level in the ALD approximately 3 to 4 ft. Measurements in the monitoring wells confirmed saturation of the ALD in its upper portion, and Fe levels have increased significantly in the ALD discharge, as shown in Figure 6. Saturated conditions in the limestone appear to have ameliorated the precipitation of Fe within the ALD. Further evaluations are ongoing. Based on the above problems, it may be prudent to construct ALDs with high width-to-length ratios and shallower depths of limestone to ensure saturation.

The total installation cost of the IMP 4 ALD was about $19,000. This compares to annual costs of about $20,000 for NaOH and $10,000 for operation and maintenance associated with conventional treatment of the IMP 4 flow from October 1985 to May 1990.

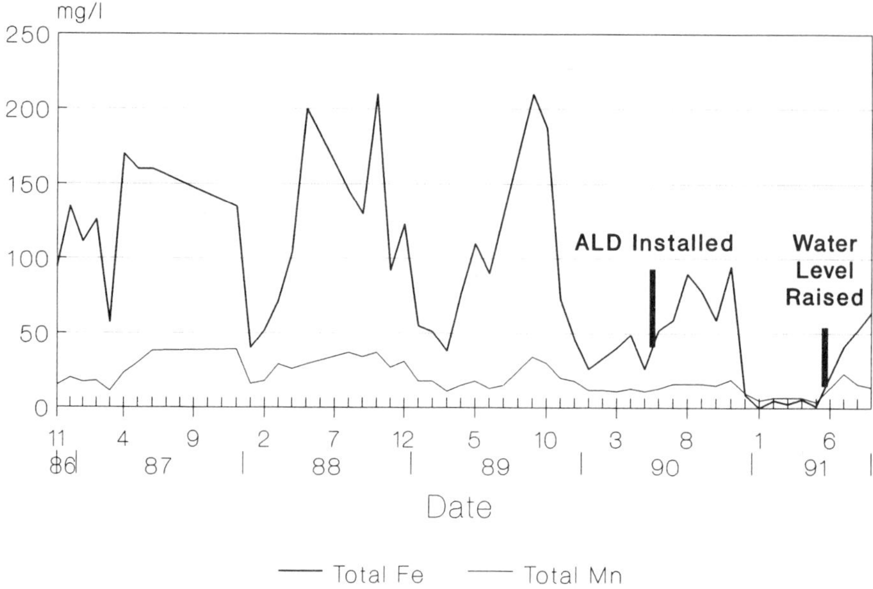

Figure 6. Fe and Mn data for IMP4.

KIF6 Case History

TVAs Kingston 006 constructed wetlands (KIF 006) was built in October 1987 to treat acid drainage emanating from an ash disposal embankment at the Kingston Fossil Plant in Roane County, Tennessee.[1] Inflow water quality was characterized by pH = 5.5, total Fe = 170 mg/L, total Mn = 4.4 mg/L, alkalinity <40 mg/L, and acidity >400 mg/L with flow averaging 408 gal/min. The 2.3-acre, 3-cell wetlands system consistently removed about 50% of the Fe loading and the pH at the final cell was 2.9. The final cell was pumped to the plant's active ash pond for treatment prior to discharge.

In April 1991, a 10-acre area within the ash disposal area up-gradient from KIF 006 was treated with the bactericide "ProMac", and reclaimed with lime-amended soil and a grass mixture. This area had been a site where water from the plants bottom ash sluice canal had pooled and presumably contributed significantly to the KIF 006 drainage. In September 1991, an ALD was installed upstream of KIF 006 in an existing seepage collection channel along the toe of the ash embankment. The ALD consisted of a trench approximately 5 ft wide, 5 to 10 ft deep, and 1400 ft long. The trench was backfilled with 3600 tons of $^3/_4$ to $1^1/_2$-in. crushed limestone which was placed on filter fabric. The stone was covered with geotechnical fabric and 2 to 6 ft of compacted local clay loam. The ash embankment and the ALD cover was sloped approximately 3:1 and seeded with a mixture including sericea lespedeza.

The first cell of the KIF 006 wetlands was converted to an oxidation precipitation pond by dredging to a depth of about 6 to 8 ft. A shallow subsurface rise transversing the midpoint of the cell allows for reareation of the water within the oxidation pond. It is estimated that this cell alone contains approximately 1 million gallons of acid drainage.

Since the recent completion of the ALD at KIF 006 (10/1/91), results are not available on any improvement in the water quality. Preliminary results show the pH of the ALD discharge is 6.5, compared to 5.5 for pre-ALD seepage. The total cost of the ALD, the ash pond reclamation with Pro Mac application, and wetlands modifications was approximately $116,000.

CONCLUSIONS

Anoxic limestone drains have been installed at sites in Tennessee, Alabama, West Virginia, and Pennsylvania to increase buffering capacity and alkalinity in acid mine drainage and prevent pH decreases in effluent due to Fe hydrolysis. Results to date are optimistic and in most instances alkalinity has been significantly

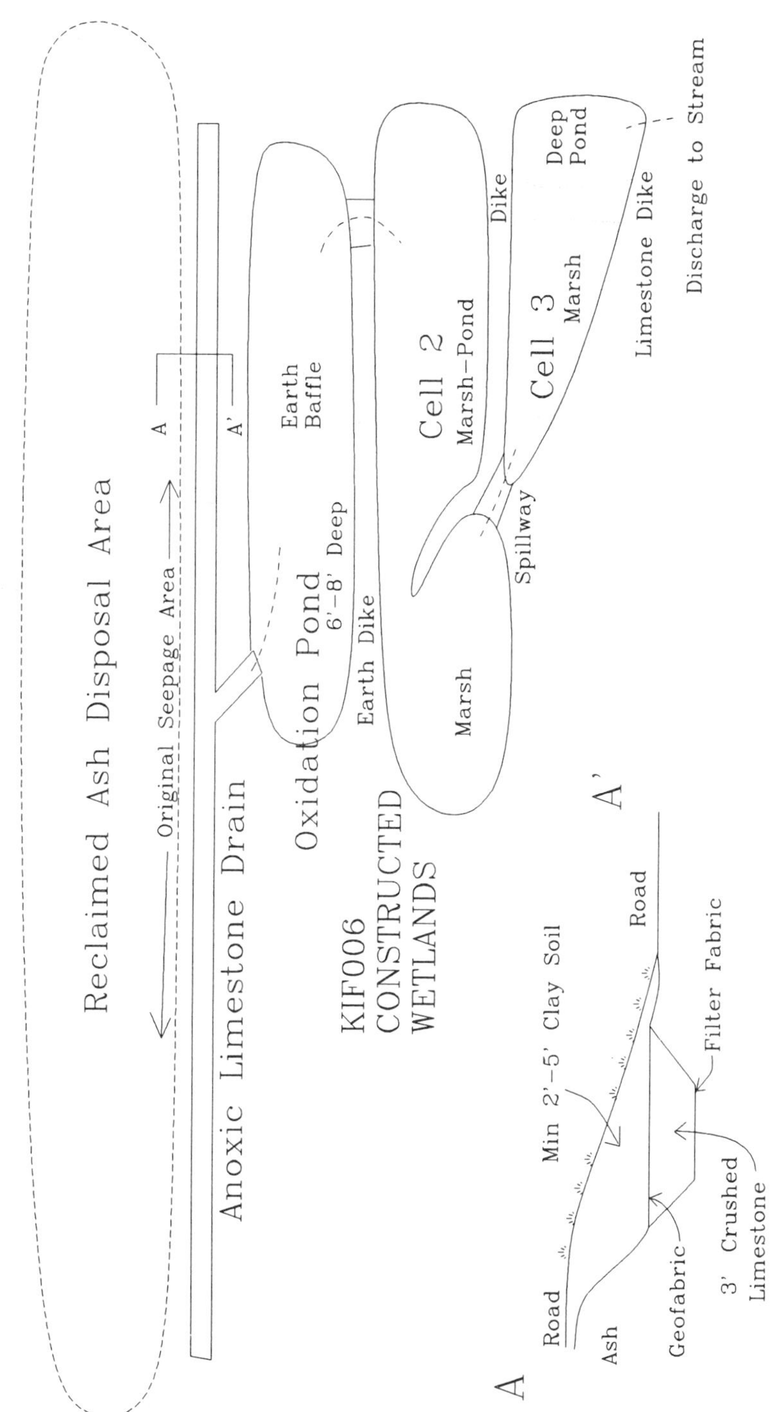

Figure 7. Kingston 006 constructed wetlands and anoxic drain schematic.

increased to levels sufficient to buffer the flow from pH decreases. Long-term results must be evaluated to establish the stability of an ALD over the duration of acid drainage at a particular site.

Design, utility, and longevity of an ALD require further examination to develop guidelines and expectations. Particularly, stone gradation and composition, dissolution rates, reaction mechanisms and products, and the nature of ALD-enhanced Mn removal need to be researched. However, the relative simplicity and low cost of the ALD and preliminary studies indicating lifespans of several decades suggest that the system may be an effective means of renovating inefficient, existing aerobic wetlands and an important component of staged wetlands system designs, providing the capability of treating poorer quality waters with less area. To meet effluent compliance limits, TVA advocates the use of ALDs only as a staged portion of aerobic acid drainage wetlands systems, and does not recommend their use as stand-alone systems or as a stage of an anaerobic wetlands system.

REFERENCES

1. Brodie, G. A. Achieving compliance with staged, aerobic constructed wetlands to treat acid drainage, in *Proc. 1991 Am. Soc. Surface Mining Reclamationists.* W. Oaks and J. Bowden, Eds. Durango, CO, 1991, 151.
2. Faulkner, S. P. and C. J. Richardson. Biogeochemistry of iron and manganese in selected TVA constructed wetlands receiving acid mine drainage, *Duke Wetland Center Publication*, Durham, NC, 90.
3. Turner, D. and D. McCoy. Anoxic alkaline drain treatment system; a low cost acid mine drainage treatment alternative, in *Proc. 1990 Natl. Symp. Mining*. D. H. Graves, Ed. University of Kentucky BU 153, Lexington, KY, 1990, 73.
4. Brodie, G. A., C. R. Britt, H. N. Taylor, T. M. Tomaszewski, and D. Turner. Passive anoxic limestone drains to increase effectiveness of wetlands acid drainage treatment systems, in *Proc. 12th Annu. Natl. Assoc. Abandoned Mined Lands Programs, Returning Mined Land to Beneficial Use*, 1990, 89.
5. Chung, Y. C. Fabius Mine-Slurry Lake Dam #2 investigation — Impoundment 1 anoxic drain, *Singleton Laboratories Report 30-02-0018A*, Louisville, KY, 1991.
6. Skousen, J. personal communication. University of West Virginia, Morgantown, WV, 1991.
7. Brodie, G. A., C. R. Britt, H. N. Taylor, and T. M. Tomaszewski. Use of passive anoxic limestone drains to enhance performance of acid drainage treatment wetlands, in *Proc. 1991 Am. Soc. Surface Mining Reclamationists*, Durango, CO, 1991, 211.
8. Cravotta, C. A., K. B. C. Brady, M. W. Smith, and R. L. Beam. Effectiveness of the addition of alkaline materials at surface coal mines in preventing or abating acid mine drainage. Part 1: Geochemical considerations, in *Proc. 1990 Mining and Reclamation Conf. Exhibition. Vol. 1.* J. Skousen et al., Eds. West Virginia University Publication Services, Morgantown, WV, 1990, 221.
9. Nairn, R. W. personal communication. U.S. Bureau of Mines, Pittsburgh, PA, 1991.
10. Kleinmann, R. L. P. personal communication. U.S. Bureau of Mines, Pittsburgh, PA, 1990.
11. Tomaszewski, T. M. personal communication. Tennessee Valley Authority, Chattanooga, TN, 1991.

CHAPTER 13

Storm Event Effects on Constructed Wetlands Discharges

H. N. Taylor, K. D. Choate, and G. A. Brodie

INTRODUCTION

Acid mine drainage is successfully treated by constructed wetlands at a mining site currently being reclaimed near Fabius, Alabama. However, the State of Alabama is interested in the fate of pollution parameters being removed from acid drainage. This study examines effects of storm water drainage through the constructed wetlands by evaluating effluent water quality. In particular, the study reviews total iron (TFe), total manganese (TMn), and total suspended solids (TSS) concentrations and pH levels vs. flow, rain intensity, and total rain. Two constructed wetland sites, Impoundment 1 (Imp 1) and Rocky Top 2 (RT 2), are investigated.

SITE DESCRIPTIONS

The two wetlands systems studied have generally similar characteristics for the wetland impoundments, drainage basins, and water characteristics, as summarized in Tables 1 and 2.[1,2] Imp 1, a four-cell system, collects seepage below an old slurry collection pond. This wetland, constructed in 1985, produces compliance-grade water. RT 2 collects seepage from a reclaimed strip mine. This three-cell system consists of a deep pond followed by two wetland cells. RT 2, constructed in 1987, was released from National Pollution Discharge Elimination System (NPDES) monitoring in 1990 because of its compliance record. Apparent system differences relevant to storm event study are discussed in the following sections.

DISCUSSION

Initially, storm samples were proposed to be collected quarterly because of observed seasonal fluctuations in NPDES effluent parameter concentrations. The storm events were collected Fall 1988 through Fall 1989. Summer storms were not collected, mainly due to mechanical problems. However, after comparing historical NPDES effluent and precipitation data with storm data collected during the study, study results appeared to be unaffected.

Parameter values associated with storm conditions had to be identified to develop true baseline values from historical NPDES data. Since historical NPDES effluent data were collected during all flow conditions at Imp 1 and RT 2, frequency analyses were conducted on the NPDES flows of both sites to help indicate storm flows. Additional investigations of historical precipitation verified that flows greater than 30 gpm (gpm = gallons per minute) for Imp 1 and 90 gpm for RT 2 were associated directly with storms. Greater flows and associated concentrations were removed from NPDES databases used to establish baselines for this study.

These flow and precipitation reviews were divided seasonally as initially prescribed. At this point, the historical data suggested summer and fall shared similar effluent and precipitation characteristics, as did winter and spring.

Effluent and hyetologic data collected during the study storms agreed with baseline trends. These hyetologic/hydrologic characteristics and effluent trends indicated that dry (summer/fall) and wet (winter/spring) season

0-87371-550-0/93/$0.00+$.50

Table 1. Impoundment and Drainage Basin Characteristics

	Impoundment		Drainage basin		
Wetland system	Area (acres)	Vegetation coverage (%)	Area (acres)	Slope (%)	Vegetation (%)
Imp 1	1.4	80	15	10	100
RT 2	1.8	30	38	8	100

[a] Vegetation coverage is 100% in trees, wood shrubs, and tall grasses.

Table 2. Wetlands Water Summary

Wetland system	Influent water parameters (mg/L)				Daily maximum NPDES values (mg/L)					Baseline effluent water parameters (mg/L or gpm)[b]				
	pH	TFe	TMn	TSS	pH	TFe	TMn	TSS	Season[a]	pH	TF	TMn	TSS	Flow
Imp 1	6.1	69.0	9.3	9.5	6–9	7.0	4.0	70.0	Dry	6.7	1	1.4	2.6	10
									Wet	6.7	0.8	2.1	2.1	16
RT 2	5.7	45.2	13.4		6–9	6.0	4.0	70.0	Dry	6.8	0.5	0.8	3.1	52
									Wet	6.8	0.8	2.9	3.6	85

[a] Dry = 6/21 through 12/20, Wet = 12/21 through 6/20.
[b] Storm flows and associated concentrations excluded from NPDES database; Imp 1 = 30 gpm, RT 2 = 90 gpm.

analyses would be more appropriate. The storm characteristics recorded during this study are summarized in Table 3.

A total of nine seasonal events were collected, four events at Imp 1 and five events at RT 2. The events collected at Imp 1 represented two wet-season storms (January and June) and two dry-season storms (October and November). The events collected at RT 2 represented three dry-season storms in November and two wet-season storms (January and June).

Precipitation was collected concurrently with the hydrographs and samples, however, not inside either drainage basin. These data were collected at a permanent weather station approximately 3 miles from either site. The weather station collected cumulative rainfall, total rainfall, humidity, and temperature. Cumulative rainfall and humidity were chart-recorded automatically for a 7-day period, while total rainfall and temperature (high and low) were recorded manually every 24 hours. Since cumulative rainfall was recorded with time, rainfall intensity could be determined.

Both study sites were equipped with automated samplers and hydrograph recorders with event-pen markers (i.e., marked the hydrograph when a sample was taken), enclosed in aluminum railroad-type shelters. A small impoundment with a 1-ft 90° V-notch weir spillway was constructed immediately downstream of the constructed wetland. The enclosure and equipment were elevated over this small impoundment so that a float for the water elevation (hydrograph) recorder could be suspended in the impoundment. The sampler was activated at a desired elevation above the base flow elevation. The recorded elevation was later converted to flow using the head-to-flow formula for a 90° V-notch weir:

$$Q \text{ (gpm)} = [2.50(H)**2.50]448.83$$

Up to 24 hourly samples were taken for each storm event. About half of the 24 samples were selected for laboratory analysis, the decision being based on field inspection of the recorded hydrograph. Generally, the sample selections included the first three samples taken, the samples taken at the peak of the hydrograph, and

Table 3. Constructed Wetlands Storm Summary

				Initial peak deviations				Peak deviations concurrent with maximum flow			
Storm	Intensity (in/hr)	Total rain (inches)	Max flow deviation (gpm)	TSS (mg/L)	TFe (mg/L)	TMn (mg/L)	pH	TSS (mg/L)	TFe (mg/L)	TMn (mg/L)	pH
Imp 1											
Oct 19, 1988	0.63	0.09	151.0	28.4	2.70	–0.5	0.7	20.4	0.60	–0.9	0.8
Jan 11, 1989	0.13	1.91	351.7	7.9	0.09	–0.9	0.2	7.9	0.09	–0.9	0.2
Jun 14, 1989	0.10	0.86	77.4	3.9	0.70	–0.8	0.3	3.9	0.50	–0.8	0.3
Nov 15, 1989	0.29	0.35	583.4	39.4	2.30	0.0	0.6	39.4	2.30	0.0	0.6
RT2											
Nov 4, 1988	0.21	0.00	441.6	87.9	7.3	0.9	0.1	50.9	3.0	0.7	0.1
Nov 20, 1988	0.22	0.77	1372.0	106.9	5.4	1.4	0.1	75.9	3.7	1.2	0.0
Jan 8, 1989		0.82	777.2	29.4	2.5	–0.3	0.1	29.4	2.5	0.1	0.2
Jun 14, 1989	0.10	0.86	662.4	19.4	1.4	0.8	0.4	67.4	4.0	1.4	0.4
Nov 15, 1989	0.29	0.35	1081.0	136.9	8.4	3.9	0.1	90.9	5.2	3.2	0.1

evenly spaced samples taken during the tailing off of the hydrograph (Figures 1 and 2). Field personnel collected the samples as soon as possible after a storm event and measured the pH values in the field. The discrete samples were divided into a total metal sample and a total mineral sample and transported to the TVA Environmental Chemistry Laboratory.

RESULTS

This discussion of results covers general responses of wetlands to storm event hydrology, simple regressions for measured parameters vs. factors affecting parameters, and comparisons between wetland regressions. Discussion of general responses covers certain aspects of flushing of parameters, indicated by concentrations above baseline values. The flushing aspects include how high any concentration increases went, when increases occurred during the storm, and what differences occurred between dry-season and wet-season storms. Simple regressions show possible correlations for factors that upset wetland performance. Factors selected for examination were maximum flow increases, rainfall intensities, and total precipitation 3 days prior to the storm event. Comparing regressions of Imp 1 and RT 2 demonstrates performance variations possibly due to physical differences between the systems.

Figures 1 and 2 show examples of wetland response to storm event hydrology for Imp 1 and RT 2, respectively. These figures showing TFe and flow vs. time generalize the parameter (TFe, TSS, TMn, and pH) responses of both systems for dry- and wet-season storms. All parameter concentrations exceeded seasonal baseline values during all storms at both sites, with the exception of TMn. Imp 1 TMn storm concentrations were less than seasonal baseline concentrations. The first pH values collected for each storm at Imp 1 were above seasonal baseline values. Subsequent pH values were slightly higher than this initial value. Initial storm pH values for RT 2 were less than seasonal baseline values, but increased above seasonal baseline values during the storm. Initial flushes in Imp 1 appeared to be alkaline while RT 2 appeared to be more acidic.

All parameter peaks occurred at similar times during the storms. As illustrated in Figure 1, dry storms generated the highest peak concentration before the maximum flow occurred (first flush), with another peak occurring close to maximum flow. Wet storms (Figure 2) often generated a peak concentration before the maximum flow, but the highest peak occurred with maximum flow. Dry-season storms produced higher concentrations than wet-season storms.

From a regulatory standpoint, the most important storm-related responses of constructed wetlands are the maximum concentrations reached during the storm (NPDES limits are defined by concentration). This maximum concentration response is important from the design standpoint; but to be useful, the time the peak occurred within the storm needs to be included, as well as type of storm (intensity and duration), to help identify factors that cause increases.

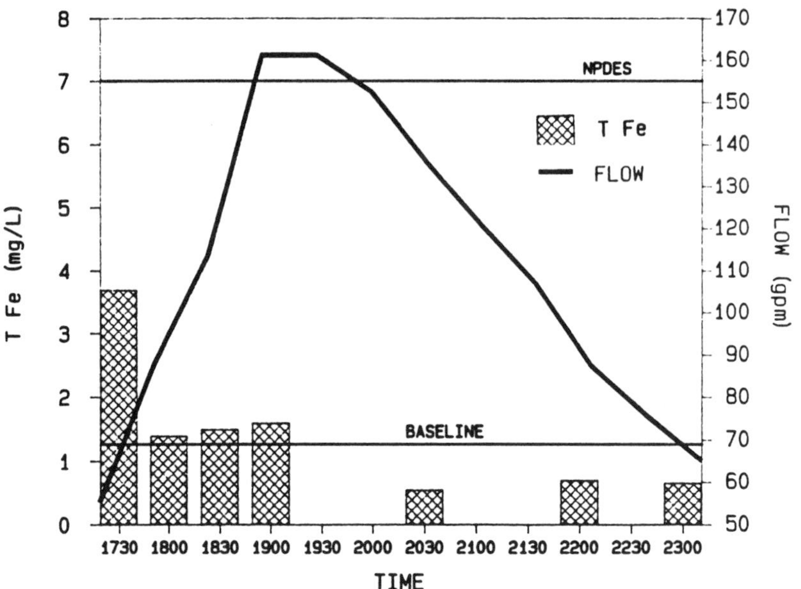

Figure 1. Dry-season storm response, Impoundment 1, October 18, 1988.

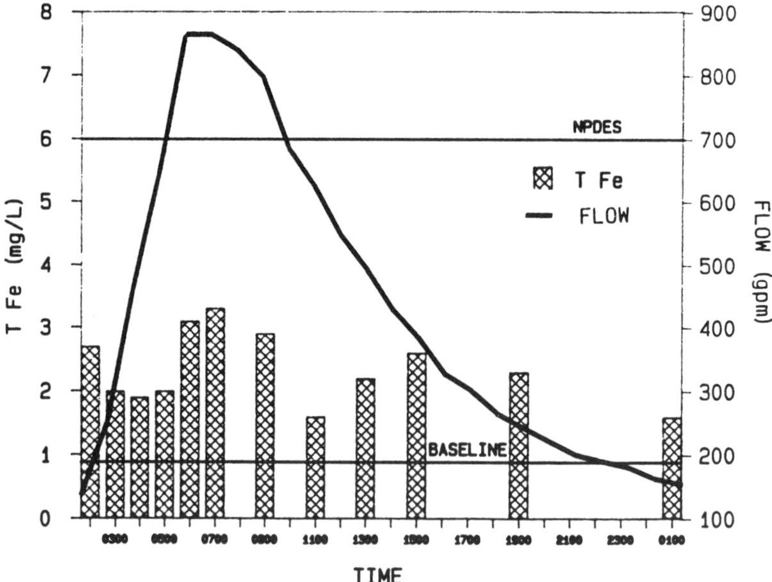

Figure 2. Wet-season storm response, Rocky Top 2, January 8, 1989.

Studying the precipitation and flows associated with the sampled storm event offered insights into some of the phenomena of the previous discussion. Simple regressions were conducted for the following: the first peak concentration deviation (concentration minus seasonal baseline concentration) with intensity, the peak concentration deviation that occurred during maximum flow deviation, and the maximum peak concentration deviation for the entire storm with total rain (3 days prior to storm event).

For both sites, general relationships existed between intensity, flow, and total rain vs. peak concentration deviations of TSS, TFe, and TMn. These observations were made, recognizing the fact that this study was conducted during seasons with greater-than-average precipitation.

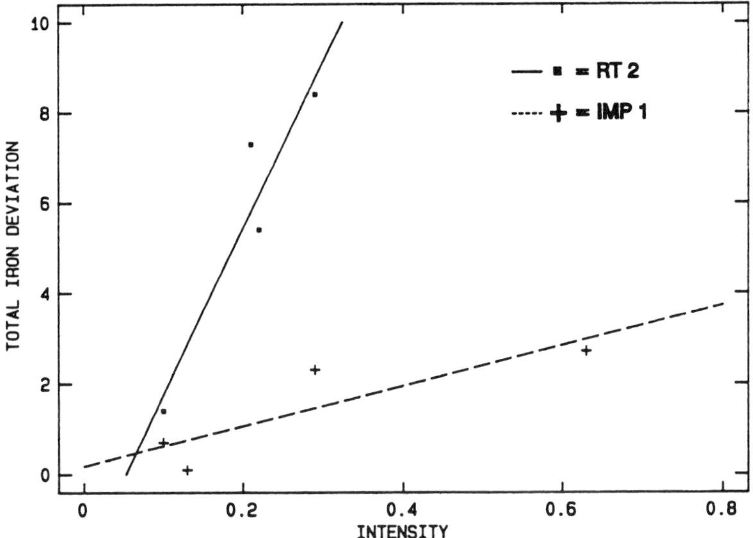

Figure 3. Linear regression analyses for total iron deviations vs. intensity.

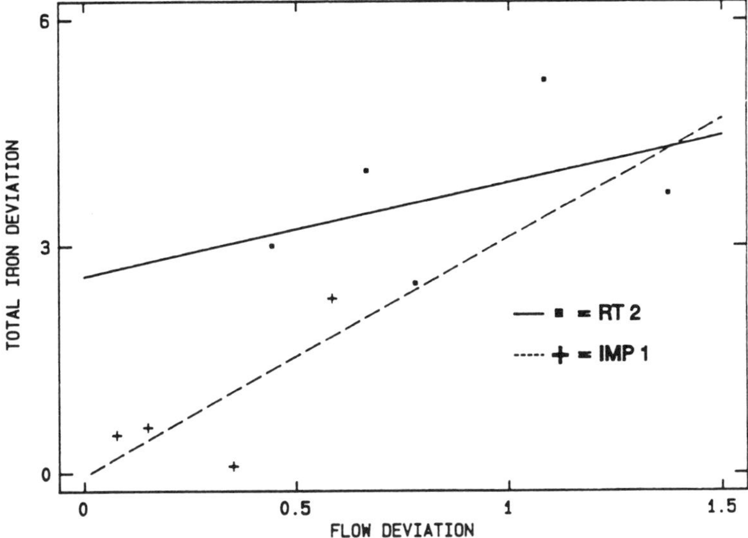

Figure 4. Linear regression analyses for total iron deviations vs. maximum flow deviations.

Peak concentration deviations increased with increase in intensity (Figure 3), and with increase in flow deviations (Figure 4); whereas deviations decreased with increase in total rain for both sites (Figure 5). However, limited data prevented conclusions of high confidence.

For both sites, storm events produced higher concentrations for TSS, TFe, and generally TMn. Values for pH were also higher than seasonal base lines. Deviations were much higher for RT 2 than Imp 1. Also, regression lines of RT 2 were steeper for deviations vs. intensity and total rain. These observations indicated that storm events impacted the normal performance of RT 2 more than Imp 1.

The relationships between pH deviations and intensity were different for each site (Figure 6). Deviations decreased with increases in intensity for Imp 1. The inverse relationship of Imp 1 was produced because a large portion of runoff travels over a limestone source. Also, a major seep into Imp 1 travels through an old limestone roadbed. Linear relationships were not identified for flow deviations and total rain.

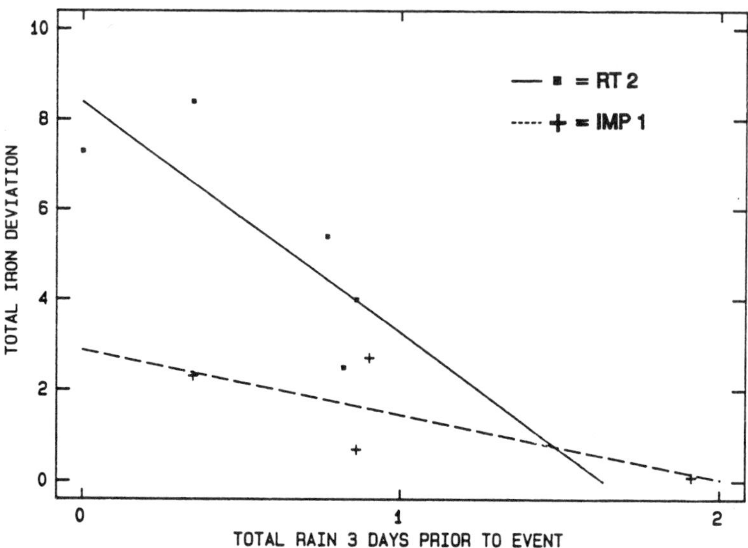

Figure 5. Linear regression analyses for total iron deviations vs. total rain.

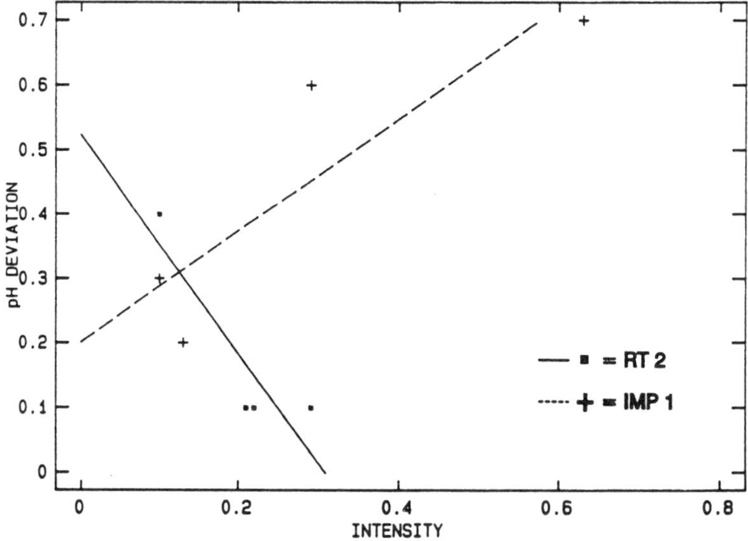

Figure 6. Linear regression analyses for pH deviations vs. intensity.

CONCLUSIONS

Storm events do affect performance of acid drainage constructed wetlands by elevating TFe, TMn, TSS, and pH values above baseline values. From the regulatory standpoint, dry-season storms appear to have more significance than wet-season storms because the former generate higher deviations. From a design point of view, the important storm characteristics appear to be the rainfall intensity and the first flush mechanism because (1) dry-season storms characteristically have more intense rainfall in this area, (2) dry-season storms always exhibited initial deviations that preceded maximum flow deviations, (3) these deviations were always higher than subsequent deviations, and (4) lower total rainfall during a season (i.e., dry conditions) tend to increase deviations.

Design of wetlands for acid drainage treatment must consider water chemistry, parameter loading rates, hydrology considerations, and settling basin technology. When considering baseline parameter removal, these two systems operate identically. However, storm effluent data indicate that Imp 1 is more conservatively designed than RT 2. Between these two systems, factors affecting storm event performance appear to be drainage basin size, baseline loading rates, and impoundment vegetation coverage. By equalizing hydrology effects (comparing effluents of storms of nearly equal overflow velocities) and assuming baseline loading design considerations are adequate (because baseline effluent data indicates adequate performance), impoundment vegetation coverage appears to have caused differences between storm performances of the two systems during this study.

RECOMMENDATIONS

When considering regulation, design, and supporting research, more storm event studies are needed to support the correlations drawn from this study. In addition to repeating these types of studies, influent data should be collected concurrently to allow calculation of removal efficiencies and to verify resuspension of material in the wetland instead of overwhelming by surface constituents.

REFERENCES

1. Brodie, G. A. Achieving Compliance with Staged, Aerobic, Constructed Wetlands to Treat Acid Drainage, in 1991 Natl. Meeting of the American Society for Surface Mining and Reclamation. Durango, CO, May 14–17, 1991.
2. Taylor, H. N. Enhancing Effluent Water of Sedimentation Basins Using Constructed Wetlands Technology, in *1991 ASCE Hydraulic Engineering: Proc. 1991 Natl. Conf.* R. M. Shane, Ed. Nashville, TN, August 1991.

CHAPTER 14

Wetland Water Treatment Systems

S. R. Witthar

INTRODUCTION

Constructed wetlands are effective, ecologically sound, and aesthetically pleasing wastewater treatment systems that require low maintenance. Wetlands decrease acidity, metals, pathogens, trace organics, and, to a lesser extent, nitrogen and phosphorus concentrations in wastewater, mine drainage, and other waters. Although design criteria for sizing the components of an effective wetland are affected by the types and levels of pollutants present, many of the components of a successful wetland treatment system are the same. Black & Veatch has provided conceptual, design, and construction administration for several mine drainage, storm, and municipal wastewater wetland treatment systems. Black & Veatch design considerations for two of these natural wastewater treatment systems currently operating are discussed. These two sites were selected for comparison due to their preconstruction similarity of acidic water with high metal contents.

The two wetland systems reviewed provide the only treatment for the acid mine drainage and range in size from a small wetland (<1 acre) to what is believed to be the largest constructed wetland designed to treat acid mine drainage in the U.S. The Roberts wetland treatment system has been treating water for over 4 years; the Indian Creek wetland treatment system has been in operation for over 2 years. Both have been very successful in decreasing acidity and metals concentrations in the water, to the extent that the Indian Creek wetland is now utilized as wildlife habitat and has ceased to be a source of off-site water pollution. Where prior to construction of the wetlands, the Indian Creek site was barren of fish and its drainage killed fish for miles downstream, all but one of the wetland cells now support fish populations. The fish support waterfowl, including herons, which nest onsite.

Drainage leaving both of these existing project sites typically had a pH of 3 before the wetlands were constructed. The pH of the water leaving the wetlands is now above 6. Over the same period of time, iron content of the drainage from each of these sites has decreased from over 10 mg/L to less than 1 mg/L.

WETLAND DESIGN CONSIDERATIONS

Constructed wetlands provide ecological improvement through multipurpose wastewater treatment systems that serve as wildlife habitats. Additionally, public and regulatory perceptions of wetland wastewater treatment systems are favorable when designed and presented correctly. Researchers who have assessed wetlands have recommended several criteria for designing effective wastewater treatment systems. Black & Veatch was one of the first organizations to develop full sized wetland wastewater treatment system designs. Recommended criteria have changed substantially since we designed our first system. Our operating systems have been successful and have proven more so each year of operation.

Wetland treatment systems fall into the following three major types.

- Free-water surface (FWS) systems. In a FWS, water flows at a low velocity over the top of the soil media and through wetland vegetation in shallow basins or channels.[1]
- Subsurface flow systems (SFS). An SFS is similar to the FWS system, but the water flows through the substrate. SFSs require less land area but are more difficult to maintain.[1]
- Aquatic plant systems (APS). An APS is similar to an FWS, but the water is in deeper ponds and the vegetation consists of floating or submerged plant systems. This type of system requires greater maintenance than an FWS.[1]

0-87371-550-0/93/$0.00+$.50

Wetland wastewater treatment systems commonly consist of two or three of these types. This article only discusses design criteria for FWS systems. There has been greater success with these due to the lesser maintenance required to be functional.

Although design criteria for sizing the components of an effective wetland are affected by the types and levels of pollutants present, many of the components of a successful wetland treatment system are the same. Wetland components developed by Black & Veatch and proven to be effective are described in the following four major areas: physical layout, flora, weather, and maintenance.

Physical Layout

The physical layout of a wetland treatment system is based upon the water chemistry, discharge expectations, and site physical features. The following seven considerations are important to designing an operational wetland: (1) area requirements, (2) water depth, (3) number of cells, (4) cell shape, (5) flow velocity, (6) wastewater retention time, and (7) substrate.

Area Requirements

The area required for the design of a wetland to treat wastewater is typically expressed in terms of square feet per gallon per minute (sf/gpm) and is based upon the following: (1) design flow rate, (2) type of waste to be removed, (3) concentration of waste in the wastewater, and (4) pollutant discharge requirements.

Preliminary results of a survey of existing wetlands treating mine drainage showed treatment areas ranging from 2.1 to 380 gallons per day per square meter (GDM) (41 to 7300 sf/gpm) and average 24.6 GDM (630 sf/gpm).[2] Wetlands using 77.5 GDM or more (200 sf/gpm or less) did not always produce acceptable water quality improvements, so area requirements were increased.[3] Further research has been performed to define better impact of concentration levels and removal efficiencies. These removal efficiencies are dependent upon the type and level of pollutant and expected removal efficiency for the pollutant. Black & Veatch used 22 to 38 GDM (400 to 700 sf/gpm) to lower 6 to 80 mg/L iron from acid mine drainage to 0.6 to 1.3 mg/L in the Roberts wetlands[4] and 9 to 26 GDM (600 to 1600 sf/gpm) to lower 20 to 585 mg/L iron to 0.3 to 0.6 mg/L in the Indian Creek wetlands. In general, regardless of the waste constituent, the wetland design that produces the best water quality is the one with the largest ratio of treatment area to base flow.

Water Depth

It is generally agreed that FWS systems function best when the depth of water is less than 18 in. The Roberts wetland utilizes surface flows that are 0 to 4 in. deep. The Indian Creek wetland varies from 0 to 9 ft in depth. Black & Veatch tries to create a more natural wetland system where feasible, which has varying water depths of 6 to 18 in. for treatment areas, and 2 to 10 ft for wildlife habitat and sediment trapment. Water should be of sufficient depth to allow fish to survive severe winters. Our Indian Creek wetland cells are providing habitat for amphibians, fish, and waterfowl by providing deep pools in each cell. The fish provide control of mosquito populations, which is important to adjacent land owners. Sediment is not a problem in the Roberts wetlands; but as sediment accumulates in the Indian Creek wetlands, it will cause the shallow areas over 1.5 ft deep to become wetland areas.

Number of Cells

Black & Veatch prefers to use multiple cells where topography and economics allow. The outlet of each cell can be designed to aerate the wastewater. This raises low levels of dissolved oxygen caused by high biochemical and chemical oxygen demands. The Indian Creek wetlands are a series of three or four cells with gabion outflow structures which allow the water to be aerated several times between cells.

Cell Shape

The shape of wetland cells is dependent upon flow, topography, and ownership constraints. A large length-to-width ratio is important to wastewater treatment to help assure that short-circuiting does not occur in large wetland systems. It increases the distance wastewater travels, resulting in greater contact with treatment processes. The Roberts wetland is small, with scattered seeps that restrict short-circuiting; thus, it is effective at 4:1. The Indian Creek cells have a 12:1 to 96:1 length-to-width ratios.[4]

Topography and land ownership boundaries impact the shape of a wetland. Since wetlands require flat terrain, severe topography makes it necessary for wetlands to be placed in valleys or follow contours. In large flat areas, baffles can simulate high length-to-width ratios.[3] The Tebo Creek wetlands, which Black & Veatch designed, and are currently under construction, have baffles that force the wastewater to weave through a triangular wetland with a resulting width-to-length ratio of about 20:1.

Flow Velocity

A low flow velocity is necessary to allow time for water treatment, with recommendations ranging from 0.1 to 1.0 ft/s.[5,6] Black & Veatch designed flow velocities for treating base flow range from 0.06 ft/s in the Roberts wetlands, to 0.1 to 0.9 ft/s in the Indian Creek wetland cells.[4] Treatment was designed for base flows only, since runoff during high flooding is not as acidic. Erosion control was designed for a 25-year, 24-hour storm event for the Roberts wetlands and an 80% possible maximum precipitation (PMP) event for the Indian Creek wetlands.

Wastewater Retention Time

The time that the wastewater is detained in the wetlands for treatment is dependent upon the treatment area and flow velocity. Research has shown that typical retention times range from 0.25 to 75 days and average about 5 days.[2] Retention times for the Roberts wetlands range from 1.1 to 3.8 days and for the Indian Creek wetlands range from 21 to over 356 days if stormwater discharge is not considered.

Substrate

Researchers have shown that spent mushroom compost or peat and crushed limestone are important to subsurface flow systems and have an impact on FWS systems.[7-10] However, Brodie et al. reported that TVA compared topsoil, natural wetland soil, acid wetland soil, clay, mine spoil, and pea gravel as substrate in lined cells.[11] The study compared water quality improvements and found similar, substantial reductions in iron, and inconclusive reductions in manganese concentrations and changes in pH between the different substrates.[11] Black & Veatch has utilized whatever soil is readily available for wetland treatment systems. Fertilizer and lime were added to provide nutrients for plant growth; successful wetland plant growth occurred on even coal slurry in the Indian Creek wetland cells.

Flora

Wetland flora remove or create an environment in which bacteria cause water pollutants to precipitate. Algae, bacteria, sphagnum moss (*Sphagnum*), cattails (*Typha*), bulrushes (*Scirpus*), and rushes (*Juncus*) have been shown to remove or support bacteria which remove nutrients, sulfate, and iron and other metals from polluted water.

Algae and Bacteria

Algae and bacteria have been shown to increase precipitation of iron, manganese, and possibly other metals. Bacterial activity from different types of bacteria promotes oxidation and reduction of pollutants. Bacteria also attach themselves to the root systems of flora and act in a manner similar to attached growth (trickling filter) systems.[4]

Aeration of surface waters promotes aerobic bacterial activity. Water entering a wetland at the water's surface often contains sufficient oxygen to promote aerobic activity. Small waterfalls or other aeration methods can be constructed at the inlet to a constructed wetland to aerate the water. As wastewater flows through the wetland, aerobic activity often removes sufficient oxygen to the extent that a reducing environment is created that supports anaerobic bacterial activity. Additionally, liquid depth can dictate aerobic/anaerobic activity. Anaerobic activity is usually greater at depths over 3 ft while aerobic activity is usually greater at shallower depths. Reducing environments exist in SFS where wastewater moves through the substrate. Anaerobic bacteria use sulfate and ferric iron to oxidize organic matter in an anoxic aquatic environment.[3]

Sphagnum Moss

Sphagnum moss (*Sphagnum*) was a common wetland flora in wetlands constructed prior to 1986.[2] Sphagnum moss in a wetland removes metals and increases the pH of wastewater streams. However, it is susceptible to toxic accumulations of iron, and changes in acidity, sediment loads, climate, water chemistry, depths, and flows. Therefore, sphagnum moss requires more attention and maintenance than other wetland vegetation.[3]

Cattails

Cattails (*Typha* sp.) are typically 4 to 6 ft tall, have 6 to 8-in. brown flower spikes, are easy to propagate, and produce a large biomass. They thrive under a wide range of water quality and other environmental conditions, including acidic and brackish waters and waters with iron concentrations of up to 100 mg/L. Cattails produce high radial oxygen loss, inducing oxidation of iron and causing it to precipitate, encrust its roots, and increase the binding of insoluble metal precipitates into the substrate.[12,13]

Cattails exist in most FWS systems with planting densities of 0.5 to 1.0 plant per square foot in most wetlands. The above-water portion of cattail plants typically dies back when transplanted, but rhizomes quickly provide new sucker growth.[3] Cattails have been seeded in the Roberts wetlands by shaking ripe spikes over shallow flowing water and along the shoreline of the Indian Creek wetland cells. Growth in areas seeded was very dense at both sites. Water even only a few inches deep prevented seed germination. Our experiences indicate that cattail seedings are successful only in shallow streams (<2 in.) and along shorelines. However, during the 1991 drought, many cattail seedlings started in areas previously flooded by 2 ft of water. Growth is such that even upon inundation, they may survive. Cattail sprigs and seeding of the shoreline are planned in our new Tebo Creek wetlands.

Bulrushes and Rushes

Bulrushes (*Scirpus*) and rushes (*Juncus*) are easy to propagate, produce a large biomass, and thrive under a wide range of water quality and environmental conditions.[14] Rushes are reported to be more susceptible to transplanting stress than cattails and bulrushes.[13] Some bulrush sprigs were planted near the shore of the Indian Creek wetlands. Their spread has been slower than that of the seeded cattails. A 50% planting of bulrushes is planned in the Tebo Creek wetlands to provide some diversity in the wetland flora.

Planting

Planting practices typically recommended include planting sprigs or transplanting root stock to establish wetland flora more quickly. After transplanting, many wetland plants die back but send out new sucker growth, thereby establishing wetland plants in deeper water faster than seed germination. This enables the wetland to become more effective sooner than by waiting for seed germination and spreading from the shore. Additional planting mechanisms include seeding and voluntary seeding of the wetland. The majority of the flora in a 2-year-old wetland will be voluntary seeded by runoff, air, and animals.[3] Many wetlands are vegetated by these voluntary mechanisms. Less than 0.1% of the areal extent of the Indian Creek wetland cells was seeded or planted with cattails or bulrushes, but a substantially larger area has wetland flora that has voluntarily germinated on-site.

Weather

Stormwater runoff can affect wetland wastewater treatment efficiencies. Wetland vegetation and structures such as dams and spillways must be designed to withstand erosion caused by runoff from expected storm events. The efficiency of the wetland will decrease during flooding caused by storms. This is due to increased flow velocities and decreased retention times. Typically, pollutants are diluted, but in some instances may increase if off-site pollutants are flushed through the wetland.

Stormwater can also transport pollutants from off-site sources such as farming, manufacturing, and waste disposal industries. These sources can produce pollutants which could impact wetland vegetation and waste-water treatment. Wastewater characterization programs performed to provide design input should identify pollutants originating off-site during storms, as well as those from the known waste stream. Therefore, treatment may have to reflect storm events if regulatory discharge requirements are to be met.

Maintenance

Wetlands are usually a low-maintenance wastewater treatment system. FWS wetland systems are usually self-maintaining if biomass is created quickly enough to fixate metal concentrates in sediments. Maintenance of SFS will include periodic removal and addition of new limestone and biomass. Over time, the accumulation of solids will reduce permeability of the crushed limestone and the effectiveness of the biomass will decrease through chemical reactions. Maintenance of all wetlands can include dredging, sediment build-up, mowing of impoundment structures to control trees, and pest control of insects, muskrats, and beavers. Black & Veatch clients are always recommended to keep impoundment structures mowed on at least an annual basis. Designs include techniques which minimize potential damage from muskrats and beavers.

BLACK & VEATCH WETLANDS

Black & Veatch was one of the first design firms to utilize wetland treatment systems in the treatment of acid mine drainage on abandoned mine lands. It is difficult to show that dilution of the wastewater by flow of relatively clean water from off-site is not responsible for improved water quality. It should be noted that in each case, the off-site water was polluted by the time it left the site prior to creation of the wetlands. The water at both sites is exempt from meeting regulatory discharge requirements since the sites are at abandoned mines. However, the water leaving both sites is at or above a pH of 7 and at or below 1 mg/L iron. The two sites discussed in this article are briefly described below.

Roberts Wetlands

The first wetland treatment system that Black & Veatch designed was the Roberts wetlands. The wetlands were part of reclamation of four abandoned mines on three sites. Figure 1 shows the post-construction hydrology of the site where the wetlands were planted. The site had a creek flowing through the middle of it and two acid seeps. Reclamation grading increased the pH of the seeps from about 3.5 to 4.5. During the reclamation process, the landowner agreed to the addition of two small wetlands in the bottomland areas. The small wetland areas were seeded with cattail spikes harvested near the site. The seeding consisted of simply shaking the ripe cattails above the flowing water. Hundreds of new cattails sprouted in the early spring before the adjacent grasses emerged.[3]

Before development of the two small wetland communities, iron and aluminum in the acid mine drainage precipitated approximately 100 ft downstream of the site where mixture with other waters occurred. The iron and aluminum were visible as red and white streaks for more than 1000 ft downstream. Since development of the wetland communities, iron and aluminum have been precipitating and fixating in on-site sediment.[3]

Table 1 summarizes pre- and post-wetland water quality on the site. During the first year, water from the seeps initially entered the wetlands with a pH of about 4 and increased to about 5 to 5.5 before exiting. Results of analyses performed subsequently have shown not only a steady increase in the pH of the water leaving the seeps, but also a steady increase in iron content. The pH of the water leaving the site has also steadily increased, with the iron content staying at or below 1 mg/L.

Indian Creek Wetlands

The Indian Creek wetlands were developed as the main effort in reclaiming an abandoned mine site. This site contained a slurry pond and coal refuse disposal area, as well as other abandoned mine land hazards. Additionally, acidic water flows continuously onto the site from an off-site abandoned mine into Wetland A, and from an acid seep into Wetland B. Table 2 summarizes pre- and post-wetland water quality. The intent of the project was to improve downstream water quality, create a wildlife refuge, and reduce on-site safety hazards.

Before construction, the pH of the water leaving the site typically averaged 3.3 and ranged from 2.4 to 5.6. Iron content of the water leaving the site was about 22 mg/L, with over 100 mg/L at most testing stations on-site. Indian Creek runoff flowed through the slurry pond where it became more acidic and carried coal refuse downstream. The acid and coal refuse killed fish for miles downstream and eroded portions of a new culvert downstream of the site.[3]

The coal refuse pile and adjacent spoil areas were sealed and vegetated to decrease infiltration into acid-generating waste. The slurry pond was developed into a series of five large wetland cells to improve water quality (see Figure 2). The wetland cells are separated by gabion spillways, which impound shallow water. The five new wetland cells added 70 acres of wetlands and ponds to the existing 60 acres on-site. Acid mine drainage flows onto the site through Wetland A. A baffle was installed in this cell to assure that there would be no short-circuiting portions of the treatment area.

Analyses of on-site water show that the only cell that has low pH water with a high iron content is Wetland B, into which the acid seep flows. The water in the other cells has a pH of over 6 and iron contents of less

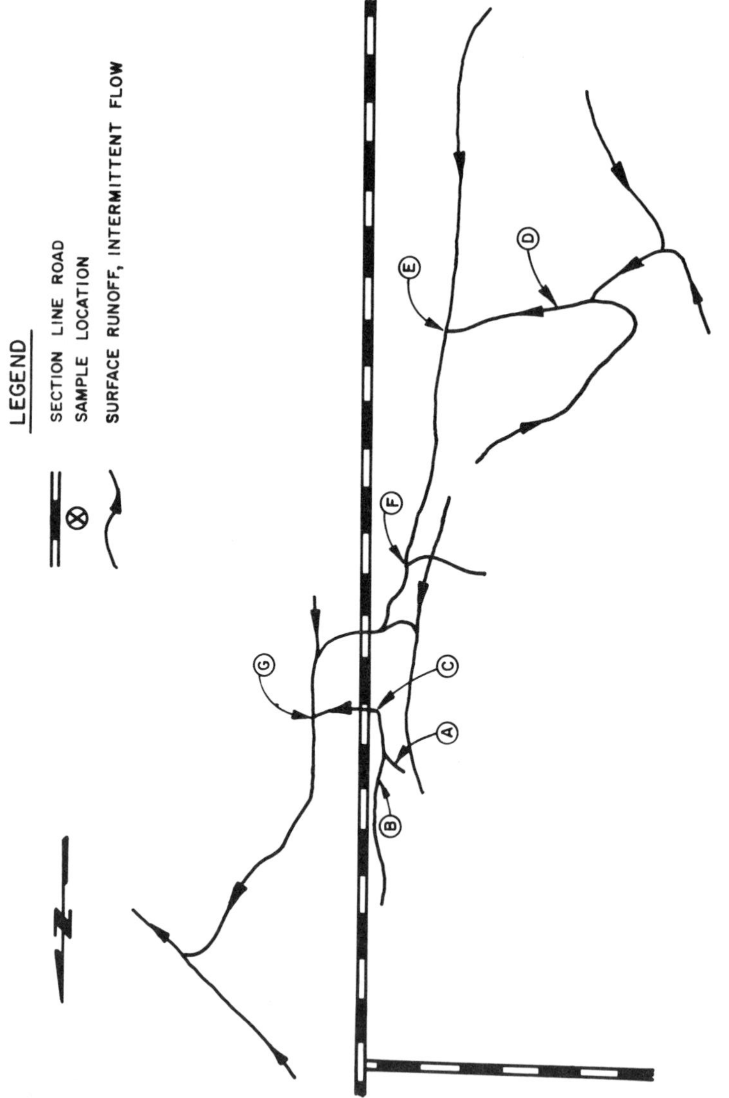

Figure 1. Post construction hydrology of the Roberts treatment wetlands.

Table 1. Robert Wetlands

	Wetland A			Wetland B			Water exiting site
	A	B	C	D	E	F	G
Pre-wetland data							
Base flow (GPM)	5	5	5	5	10	15	50
pH	4.0	4.0	4.0	3.3	6.2–7.8	3.3	6.4
Soluble iron (mg/L)	6–10+	6–10+	6–10+	18	0.0–0.6	46	2.4
Sulfate (mg/L)	NA	NA	NA	NA	387	1711	NA
Constructed-wetland							
Base flow (GPM)	5	5	5	5	10	15	50
Wetland area (acres)[a]	NA	0.2	NA	0.6	7.5	4.1	NA
pH	5.8–6.2	NA	6.8–7.4	3.3–5.3	6.7–7.4	6.5–7.2	NA
Soluble iron (mg/L)	23–78	NA	0.6–0.8	3–17	0.5–1.2	0.8–1.3	NA

[a] Area inundated by 0 to 2 ft of water.

Table 2. Indian Creek Wetlands

	Water entering site	Company pond	Wetland					Water exiting site
			A	B	C	D	E	
Pre-wetland data								
Base flow (GPM)	5	70	5	10	100	100	300	300
Pond area (acres)	NA	42.4	3.9	0.0	0.4	12.8	4.4	NA
pH	2.0–4.4	5.2–6.9	2.8–3.6	2.8–3.6	2.0–4.6	2.7–5.4	2.4–5.6	2.4–5.6
Soluble iron (mg/L)	585	0.8	134	173	122	29	20	22
Sulfate (mg/L)	1416	348	2093	2728	2763	2821	2877	264
Constructed-wetland data								
Base flow (GPM)	5	0	40	40	0	300	300	300
Wetland area (acres)[a]	NA	NA	2.0	1.4	1.8	7.5	4.1	NA
Base flow gdm of wetland area	NA	NA	7.1	10.2	NA	14.2	26.0	NA
Shallow area (acres)[b]	NA	NA	5.8	4.1	3.0	10.0	10.0	NA
Deep area (acres)	NA	51.2	0.0	0.0	1.4	16.4	13.6	NA
pH	3.0	7.6–8.0	6.6–7.8	5.3–6.4	7.0–7.4	6.9–7.4	7.2–7.5	7.3–7.5
Soluble iron (mg/L)	40	0.5–0.6	0.5–3.4	1.8–4.4	0.3–0.5	0.7–1.5	0.8	0.4–0.6

[a] Area inundated by 0 to 2 ft of water.

[b] Area inundated by 2 to 5 ft of water and subject to sediment loading from adjacent farm lands. Possible future wetlands areas.

than 4 mg/L. Recent tests in October 1991 showed a continuing trend of the pH increasing and iron content decreasing in all cells. Only water in Wetland B had a pH less than 7 and an iron content greater than 1.5 mg/L.

SUMMARY

Black & Veatch develops site-specific wetland designs. Our existing wetland wastewater treatment systems have been very successful using our present system. We are currently constructing one other large wetland treatment system. In addition, we have also designed wetlands for treatment of storm water and municipal wastewater. With the success we have had at our existing sites, we will continue to utilize wetlands for wastewater treatment in the future.

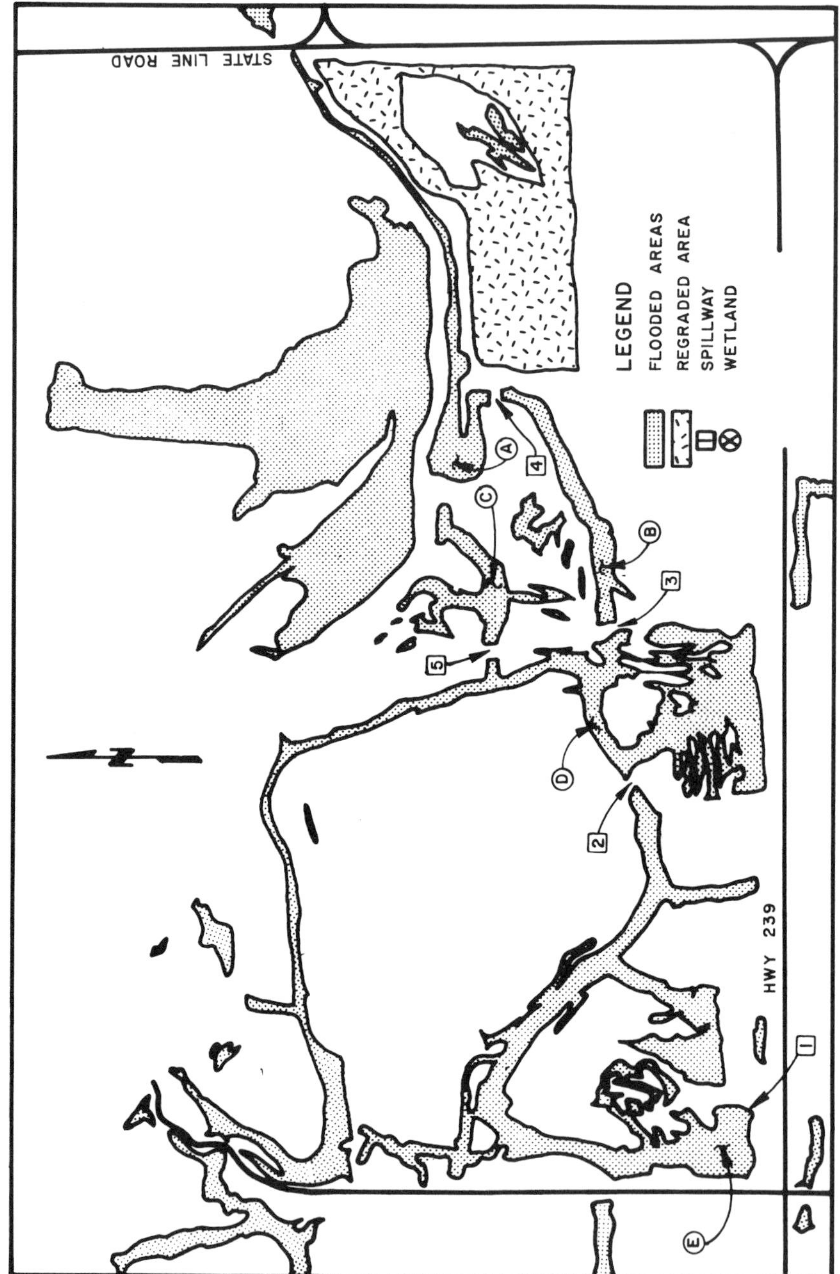

Figure 2. Indian Creek wetlands and associated hydrology.

REFERENCES

1. Crites, R. W., D. C. Gunther, A. P. Kruzic, J. D. Pelz, and G. Tchobanoglous. *Design Manual Constructed Wetlands and Aquatic Plant Systems for Municipal Wastewater Treatment.* EPA/625/1-88/022, U.S. Environmental Protection Agency, Cincinnati, OH, 1988, 4–6.
2. Girts, M. A. and R. L. P. Kleinmann. Constructed Wetlands for Treatment of Acid Mine Drainage: Preliminary Review, in National Symposium on Mining, Hydrology, Sedimentology, and Reclamation, University of Kentucky Press, 1986, 165–171.
3. Witthar, S. R. Wetland design for treatment of acid mine drainage, in *Proc. 11th Annu. Assoc. Abandoned Mine Land Programs Conf.* Williamsburg, VA, October 16–19, 1989.
4. Hunter, G. L., H. O. Andrews, and S. R. Witthar. The use of wetlands for the treatment of abandoned mine drainage, presented at Annual Kansas Water Control Meeting, Great Bend, Kansas, March 4–6, 1990.
5. Chironis, N. B. Mine-built ponds economically clear acid mine waters, *Coal Age*. January: 58, 1987.
6. Kleinmann, R. L. P. Treatment of Acid Mine Water by Wetlands, *IC 9027 Control of Acid Mine Drainage*. Bureau of Mines, United States Department of Interior, 1985, 48–52.
7. Hedin, R. S., D. M. Hyman, and R. W. Hammack. Implications of sulfate-reduction and pyrite formation process for water quality in a constructed wetland: preliminary observations. *IC 9183 Mine Drainage and Surface Mine Reclamation. Vol. I: Mine Water and Mine Waste*. Bureau of Mines, United States Department of Interior, 1988, 382.
8. Vogel, W. G. and F. M. Rothwell. Mushroom compost and paper mill sludge influence development of vegetation and endomycorrhizae on acid coal-mine spoils, *IC 9184 Mine Drainage and Surface Mine Reclamation. Vol. II: Mine Reclamation, Abandoned Mine Lands and Policy Issues*. Bureau of Mines, United States Department of Interior, 1988, 206.
9. Lappakko, K. and P. Eger. Nickel and copper removal from acid mine drainage by a natural wetland, *IC 9183 Mine Drainage and Surface Mine Reclamation. Vol. I: Mine Water and Mine Waste*. Bureau of Mines, United States Department of Interior, 1988, 301.
10. Lappakko, K. and P. Eger. Trace Metal Removal from Stockpile Drainage by Peat, *IC 9183 Mine Drainage and Surface Mine Reclamation. Vol. I: Mine Water and Mine Waste*. Bureau of Mines, United States Department of Interior, 1988, 291.
11. Brodie, G. A., D. A. Hammer, and D. A. Tomljanovich. An evaluation of substrate types in constructed wetlands acid drainage treatment systems, *IC 9183 Mine Drainage and Surface Mine Reclamation. Vol. I: Mine Water and Mine Waste*, Bureau of Mines, United States Department of Interior, 1988, 325.
12. Sencindiver, J. C. and D. K. Bhumbla. Effects of cattails (*Typha*) on metal removal from mine drainage, *IC 9183 Mine Drainage and Surface Mine Reclamation. Vol. I: Mine Water and Mine Waste*. Bureau of Mines, United States Department of Interior, 1988, 359.
13. Brodie, G. A., D. A. Hammer, and D. A. Tomljanovich. Constructed wetlands for acid drainage control in the Tennessee Valley. *IC 9183 Mine Drainage and Surface Mine Reclamation. Vol. I: Mine Water and Mine Waste*. Bureau of Mines, United States Department of Interior, 1988, 389.
14. Hunter, G. L. and J. L. Carter. Wetland Treatment, Description of Process & Effluent Limits. Internal Black & Veatch Memorandum, Kansas City, MI, May 3, 1989.

CHAPTER 15

Staged, Aerobic Constructed Wetlands to Treat Acid Drainage: Case History of Fabius Impoundment 1 and Overview of the Tennessee Valley Authority's Program

G. A. Brodie

INTRODUCTION

Staged, aerobic constructed wetlands offer an inexpensive, natural, low maintenance, and potentially long-term solution to treating acid drainage without chemical additives.[1] Since 1985, the Tennessee Valley Authority (TVA), the nation's largest electric utility, has constructed 14 wetlands systems for treating acid drainage at coal mining and processing facilities and coal-fired power plants. Twelve of these sites are now operational (one is under development and one was abandoned), and have been evaluated and manipulated in attempts to understand and refine the processes occurring in constructed wetlands; 9 of the 12 operational systems are meeting water discharge limitations without any form of additional treatment.

METHODOLOGIES

Guidelines for design, construction, and operation of aerobic acid drainage treatment wetlands have been developed.[2] These should be considered comprehensive due to the continued up-grading of the constructed wetlands technology and the need for site-specific designs which often restrict the use of standard engineering and construction methods. Detailed descriptions of the design, preliminary considerations, and construction of TVA's wetlands have been published elsewhere.[1-3]

Design and construction

Figure 1 shows a typical plan for a constructed wetlands. TVA's aerobic wetlands generally consist of a pretreatment stage (anoxic limestone drain and/or oxidation basin) followed by several cells of shallow to deep (0.1 to 2.0 m) cattail (*Typha* sp.) marsh ponds. Most of the systems have been constructed in groundwater gaining streams created by the acid drainage, although a few sites required diversions to rout the drainage to the wetlands system. Some systems are followed by a final polishing pond which may improve long-term capacity and minimize storm event flushing of Fe and Mn precipitates from a constructed wetlands.[4]

Based on TVA's results only, aerobic wetlands systems should be designed for 4.0 to 11.0 GDM of Fe removal depending on pH, alkalinity, and Fe concentrations of the inflow.[2] TVA's early wetlands systems (i.e., those before 1988) were sized hydraulically and then increased if the site allowed. Cell areas were arbitrarily increased in size if very poor quality water was to be treated. The most recent wetlands have been sized based on state-of-the-art guidelines,[2] but in most cases have been built larger than design size to increase the safety factor and lifespan of the systems.

To ensure long-term stability, dikes were sloped no steeper than 2:1 and rip-rapped or protected with erosion-control fabric on the slopes. Spillways were designed for handling the maximum probable flow with an ample safety factor and protected with either large (>30 cm) rip-rap or nonbiodegradable erosion-control fabric planted with species such as wool grass (*Scirpus cyperinus*), sedge (*Carex* sp.), or threesquare (*Scirpus americanus*).

0-87371-550-0/93/$0.00+$.50

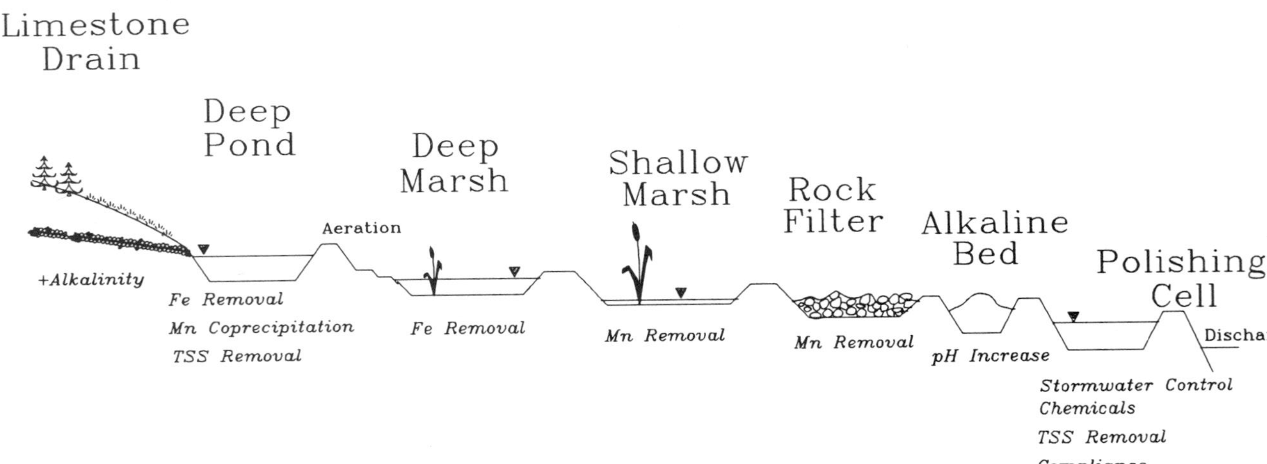

Figure 1. General schematic of staged aerobic constructed wetlands.

Wetlands shapes varied and were dictated by existing topography, geology, and land availability. Irregular shapes for wetlands cells enhanced their natural appearance and provided hydraulic discontinuity. The number of cells was determined by site topography, hydrology, and water quality. Level sites were amenable to large cells hydraulically chambered with rock or earthen finger dikes, large logs, or vegetated hummocks. Steeper slopes required more grading, or a system of several cells terraced downgrade. Staged treatment using several cells probably provides a more efficient system since chemical and hydraulic overloading of all cells is minimized. One wetlands cell should be constructed for each 50 mg/L Fe in the inflow due to the need for reaeration after oxidation of this amount of Fe.

Average water depth in TVA's wetlands ranged from 15 to 30 cm, with some deeper and shallower areas to provide for species diversification. Isolated deep pockets up to 2.0 m were included in many cells to provide for aquatic fauna refuge in drought events.

Vegetation was hand dug to obtain complete root balls/rhizomes and planted on the same day as digging. Cattail (*Typha* sp.) was set into the substrate at 0.3 m centers in early systems and 1.0 m centers in later ones. Stems were broken over at the water level to prevent windfall and to stimulate new growth from the rhizomes. Wool grass (*Scirpus cyperinus*), sedge (*Carex* sp.), and rush (*Juncus effusus*) clumps were simply placed in the desired location. Squarestem spikerush (*Eleocharis quadrangulata*) and scouring rush (*Equisetum hyemale*) were carefully set into the substrate. Complete wetlands installations were operating in 6 to 10 weeks, depending on their complexity. Most wetlands were completed in early summer, although successful installations were completed as late as October. Wetlands were fertilized generally only once with a phosphorous-potassium fertilizer such as 0-12-12 at 400 kg/ha.

Mosquitoes have not been a problem at TVA's wetlands; however, as a precaution, mosquitofish (*Gambusia affinis*) were stocked in all wetlands. Several bat houses have been installed at TVA to investigate the use of bats for mosquito control. Additionally, various bird houses (martin, screech owl, wood duck, or blue bird) were erected at the sites.

Post-construction activities included water quality monitoring, maintenance of dikes and spillways, and the addition of new ponds to further treat the wetlands discharge. Additional water, substrate, and vegetation sampling and biological monitoring was performed to quantify the wetlands development and treatment efficiencies.

Sampling and Analytical Techniques

NPDES monitoring requirements included pH, total Fe and Mn, and total suspended solids (TSS). Effluent samples from the wetlands were obtained during daylight hours, generally within the 2nd and 4th weeks of the month. Sampling was always initiated within 2 weeks of system start-up. All samples were collected and analyzed according to standard methods.

Total metals samples were collected in 500-ml acid-rinsed polyethylene bottles, preserved with HNO_3 to a pH of greater than 2.0, usually by tilting the bottle gently into the seep. The sample was placed on ice and transported to the laboratory for analyses. Samples were digested with concentrated, redistilled HNO_3 and HCl, reduced to 20 ml, diluted back to volume, centrifuged or filtered depending on solids, and then analyzed by atomic emission or atomic adsorption.

Total suspended solids (TSS) samples were collected in 1-l cubitainers. All air was expelled from the cubitainer and the sample was stored on ice for transport to the lab. Samples were filtered through glass fiber filters, dried at 102 to 105°C, with the difference of weight retained on filter and reported as TSS.

RESULTS

A summary of characteristics and water quality parameters for TVA's 14 constructed wetlands is presented in Table 1. Significant water quality improvement occurred at all of the 12 operating wetlands. Nine systems produced discharges that consistently met NPDES monthly average discharge limitations (pH = 6 to 9 s.u.; Fe < 3.0 mg/L; Mn < 2.0 mg/L; TSS < 35.0 mg/L) with no chemical treatment. Where regulatory limits were not entirely achieved, cost savings were realized as a reduction in chemicals needed for further metals precipitation or pH adjustment at IMP2, WCF6, and COF13. TVA's wetlands were, on average, hydraulically loaded between 0.02 and 0.24 L/day/m^2 of wetlands (LDM). Maximum hydraulic loading ranges were 0.06 to 1.47 LDM and averaged 0.42 LDM.

Ten of the 12 operating wetlands produce discharges in compliance with total Fe limitations, i.e., less than 3.0 mg/L (Table 1). Figure 2 shows Fe loading in the systems ranges from 0.03 to 41.4 GDM. Fe removed

Table 1. TVA Acid Drainage Wetlands Treatment Summary

Wetlands system	Date initiated operation	Area m²	Number cells	Influent water Parameters (mg/L)				Effluent water Parameters (mg/L or L/min)						Loading g/day/m²		L/day/m²
				pH	Fe	Mn	TSS	pH	Fe	Mn	NFR	Flow Ave	Flow Max	Fe	Mn	Ave flow
950	1-76	3400	3	5.7	12.0	8.0	20.0	6.5	1.1	1.6	5.4	83	341	0.42	0.28	0.04
IMP1	7-85	5700	4	3.1	69.0	9.3	9.5	6.7	0.9	1.8	3.0	73	693	1.27	0.17	0.02
IMP4	11-85	2000	3	5.7	65.0	16.8	21.0	6.3	0.4	0.6	6.0	131	693	6.13	1.58	0.09
WCF5	7-90	6600	4	—	—	—	—	8.4	2.2	0.7	—	973	2057	—	—	0.21
WCF6	6-86	4800	3	5.6	150.0	6.8	—	3.9	6.4	6.2	—	289	1495	13.0	0.59	0.09
WCF19	6-86	25,000	3	5.6	17.9	6.9	—	4.3	3.3	5.9	—	492	6360	0.50	0.20	0.03
IMP2	6-86	11,000	5	3.5	40.0	13.0	9.0	3.1	3.4	13.0	0.8	1016[a]	1540[a]	5.32	1.72	0.13
IMP3	10-86	1200	3	6.3	15.8	4.9	21.4	7.0	0.5	0.7	9.0	58	250	1.10	0.34	0.07
RT2	9-87	7300	3	5.7	45.2	13.4	—	6.8	0.6	1.8	3.2	277	1155	2.47	0.73	0.05
950NE	9-87	2500	4	6.0	11.0	9.0	19.0	6.9	0.6	0.8	5.0	385	1386	2.44	2.00	0.22
KIF6	10-87	9300	3	5.5	170.0	4.4	40.0	2.9	82.5	4.6	—	1574	2271	41.43	1.07	0.24
COF	10-87	9200	5	5.7	0.7	5.3	—	6.7	0.7	3.5	—	288	408	0.03	0.24	0.05
DLL	5-90	7550	4	6.2	10.0	5.5	23.0	6.4	2.1	2.2	10.0	385	7700	0.73	0.40	0.07
HR000	6-91	40,000	5	4.5	40.0	17.0	—	6.3	1.5	1.5	—	4000	—	—	—	—

[a] Also receives pumpage from slurry lake up to 4800 L/min.
[b] Under construction.

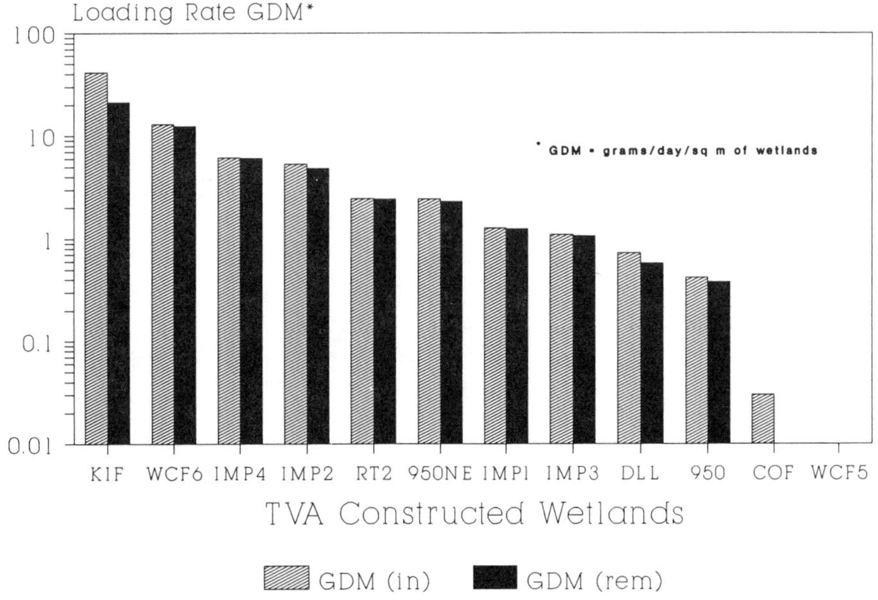

Figure 2. Fe loading and removal in TVA constructed wetlands.

ranges from 0.0 to 21.3 GDM, corresponding to 0 to 99% Fe removal. Note that the 0% removal is associated with very low Fe inflow (0.7 mg/L) at the COF wetlands; this data may be a result of sampling error or a lower limit of Fe removal possible at the COF wetlands. Fe removal in the wetlands is very efficient for loadings up to 13 GDM, which occurs at the WCF6 wetlands. Fe removal is less efficient (51%) at KIF where Fe loading exceeds 41 GDM. More data for Fe loadings between 13 and 41 GDM is needed to better assess the Fe loading limit for a constructed wetlands. There is no correlation in TVA's wetlands between Fe removal and influent alkalinity, Fe removal and wetlands size, or Fe removal and hydraulic loading.[1]

Figure 3 shows total Mn loading and removal in the TVA wetlands. Data for WCF5 is not available. Nine of the twelve operating wetlands produce discharges in compliance with total Mn limitations, i.e., less than 2.0 mg/L (Table 1). Mn loading ranges from 0.17 to 2.00 GDM. Mn removed ranges from 0.15 to 1.87 GDM, corresponding to 0 to 96% Mn removal. The low removal rates are all associated with low pH (2.9 to 3.9 s.u.) systems. Mn removal in the wetlands is very efficient for loadings as high as 2.0 GDM, which occurs at the 950NE wetlands. There is no correlation in TVA's wetlands between Mn removal and wetlands size, and Mn removal and hydraulic loading. There is a good correlation between Mn removal and influent alkalinity and acidity concentrations (Figure 4).[1,3] Systems with zero alkalinity have removed 0 to 16.5% Mn, while systems with alkalinity greater than 62 mg/L and with excess acidity as high as 248 mg/L have removed 85 to 97% of the Mn load. Inflows with zero alkalinity have always resulted in low pH in the wetlands. Low Mn removal is associated with zero alkalinity and thus, low pH in the wetlands. These data suggest that Mn coprecipitation on Fe-oxides at circumneutral pH is a likely mechanism of Mn removal.[5]

Nine of the 12 systems increase or maintain inflow pH to produce discharges in compliance for pH, i.e., 6.0 to 9.0 (Table 1). Three systems cause pH reductions due to Fe oxidation and hydrolysis; these systems are being modified with anoxic limestone drains.[3] All of the wetlands produce discharges in compliance with total suspended solids limitations, i.e., less than 35 mg/L.

CASE HISTORY OF IMPOUNDMENT 1 CONSTRUCTED WETLANDS

Impoundment 1 (IMP1), constructed in June 1985, was TVA's first acid mine drainage treatment wetlands. The system treats acid seepage emanating from an earth dike impounding 16 ha of coal slurry at TVA's reclaimed Fabius Coal Preparation Plant in Jackson County, Alabama.

Since construction, IMP1 has generally produced compliance-quality effluent. Figure 5 shows average water quality data during the period July 1985 to October 1991 for the wetlands inflow and the discharges from

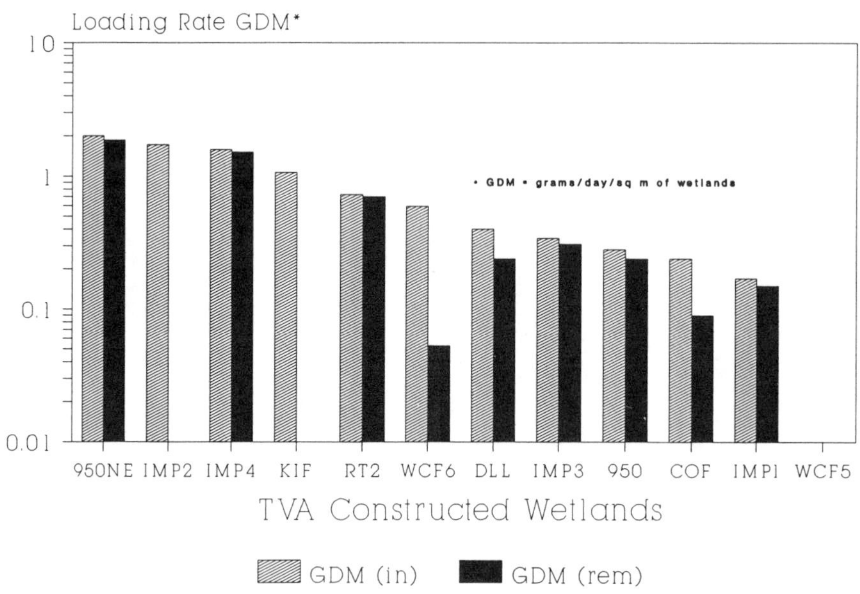

Figure 3. Mn loading and removal in TVA constructed wetlands.

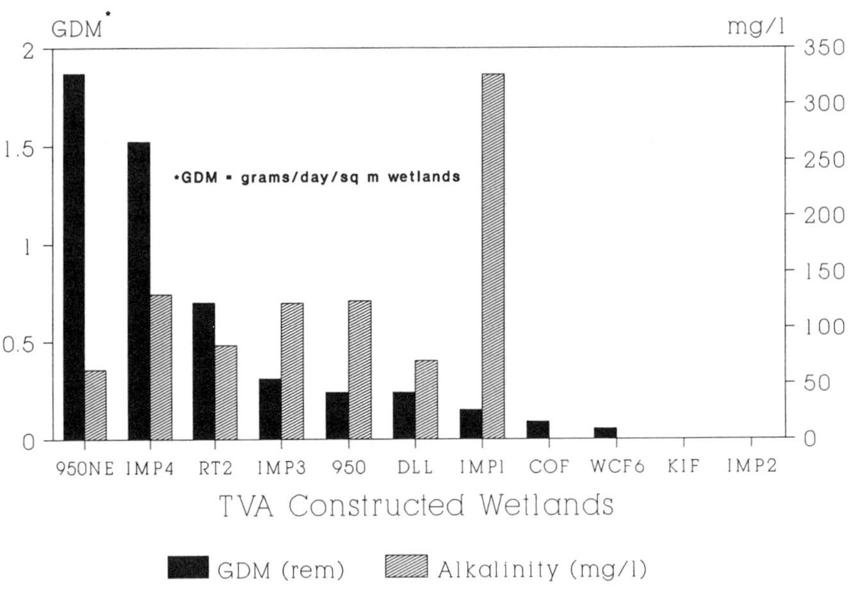

Figure 4. Mn removal vs. influent alkalinity in TVA constructed wetlands.

each of the four wetlands cells. Variations in flow from each cell were due, in part, to acid seeps encountered along a sandstone shelf underlying the site and in the leaky nature of the wetlands system. Effluent from the first cell alone has met discharge limitations 56% of the time. Figure 6 shows total Fe and total Mn concentrations in the effluent for the same period. From June to August 1988, and again from June to September 1989, total Mn concentrations increased to several times the IMP1 average discharge concentration of Mn. Similar but less drastic increases in Mn concentrations were noted in the summers of 1986 and 1987. When these anomalies were compared to rainfall records and wetlands flow, no correlations were apparent. Most of the other wetlands have exhibited similar patterns of Mn concentration variability. These increases

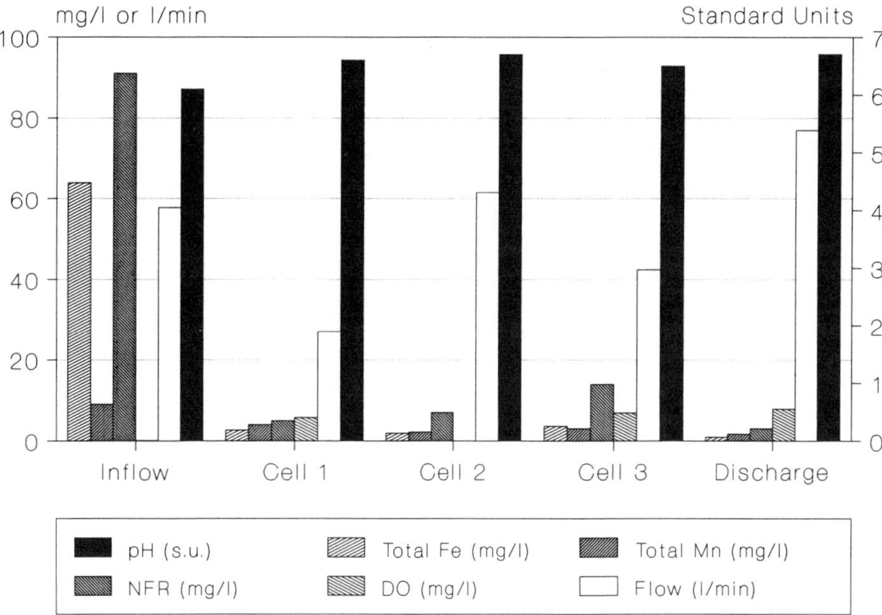

Figure 5. TVA Fabius IMP1 constructed wetlands average water quality data.

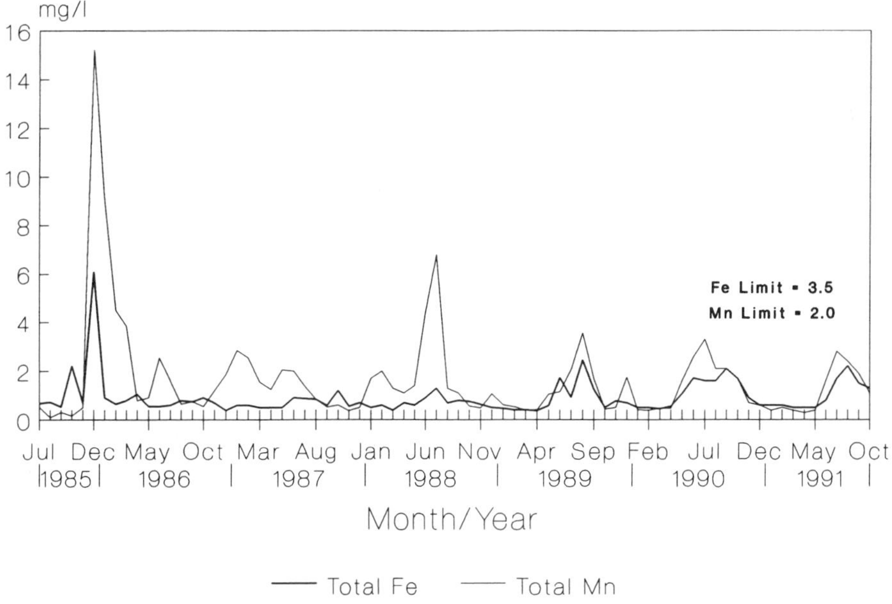

Figure 6. TVA Fabius IMP1 constructed wetlands effluent Fe and Mn data.

are probably seasonally related and could be due to numerous factors, including temperature, degree of mixing, redox conditions, nutrient and/or carbon availability, or photosensitivity of Mn-oxidizing bacteria.

IMP1 is one of two constructed wetlands receiving inflow total Fe concentrations exceeding 50 mg/L that has successfully produced compliance-quality discharges without chemical treatment. Other wetlands receiving greater than 50 mg/L total Fe (Table 1) have been impacted by low pH, high acidity, and the resultant inability to remove Fe and/or Mn to meet discharge limitations. Investigations into differences among the wetlands revealed that IMP1 influent had an alkalinity often exceeding 250 mg/L. The three other high-Fe wetlands had influent alkalinity ranging from 0 to 26 mg/L with high acidity. Further investigations into the IMP1 characteristics disclosed that the leaking coal slurry impoundment dike was constructed in 1974 over an existing limestone coal mine haul road, which may represent the oldest, working anoxic limestone drain (ALD).[3] Historically, local limestone has been quarried from the Monteagle Formation, an oolitic, high-calcium limestone. Apparently, this road is the source of the IMP1 influent alkalinity.

Stability problems at IMP1 resulted from inadequate spillway and dike designs. Each dike was repaired in late 1989 to increase the freeboard to over 30 cm. The spillways were reconstructed to provide long-term, erosion-resistant stability. Six species were originally planted in IMP1, including broadleaf cattail (*Typha latifolia*), wool grass (*Scirpus cyperinus*), rush (*Juncus effusus*), scouring rush (*Equisetum hyemale*), and squarestem spikerush (*Eleocharis quadrangulata*). Over 70 vegetative species have now been identified in IMP1, dominated by *Typha latifolia*, *Scirpus cyperinus*, *Juncus effusus*, *Eleocharis quadrangulata*, and rice cutgrass (*Leersia oryzoides*). The remaining stream of IMP1 originally was almost biologically dead (less than 5 invertebrate species). The stream now contains over 30 species of invertebrates, as well as mosquitofish (*Gambuzia affinis*) and other minnow species.

Total cost of the IMP1 wetlands was $43,000 (1985 dollars). Annual costs from 1985 to 1990 were about $13,000 due to repairs on the prototype design and to extensive monitoring. Operations and maintenance costs today are less than $1000 annually. Costs to chemically treat this acid drainage site instead of with wetlands treatment would have been approximately $250,000 from 1985 to 1991.

SUMMARY AND CONCLUSIONS

TVA has constructed 14 and operates 12 aerobic, staged wetlands systems for treating acid drainage from coal mine spoil, coal slurry and gob, and coal ash. These systems offer a preferred alternative to conventional methods of treating acid drainage from various coal-related sources.

Nine wetlands systems now produce effluents meeting all discharge limitations without chemical treatment, four of which have been released from NPDES monitoring requirements. These nine systems are associated with moderate inflow water quality (i.e., total Fe = 0.7 to 69 mg/L, total Mn = 5 to 17 mg/L), relatively high total Mn to total Fe ratios in the influent (average Mn/Fe = 0.44), significant inflow alkalinity (35 to 300 mg/L), and variable Fe loading (0.03 to 6.13 GDM).

Five systems have experienced high acidity production and low pH within the wetlands due to Fe oxidation/hydrolysis. Four of these systems are associated with high influent Fe concentrations (40 to 170 mg/L), high Fe loads (5 to 41 GDM), and zero to very low influent alkalinity. Two of these wetlands discharges require NaOH treatment to achieve compliance quality. One system, Imp4, was modified with an anoxic limestone drain, which has allowed cessation of chemical treatment and enabled the wetlands to produce compliance-quality discharges. Another system, KIF6, has been modified with an anoxic limestone drain, but due to its recent completion (10/1/91), only preliminary data have been collected.

One system (COF) which has experienced very little metals reduction is associated with low Fe (0.7 mg/L) and higher Mn (5.3 mg/L). The performance of this wetlands may be related to absence or inhibition of Mn-oxidizing bacteria or lack of Fe-Mn coprecipitation. TVA is currently investigating the use of rock filters to lower concentrations of Mn in low Fe acid drainage.

Fe and Mn removal efficiencies and pH improvement in the TVA wetlands do not correlate well with wetlands treatment areas for total Mn or flow. This, in part, is probably due to the effect of Fe oxidation/hydrolysis overwhelming other wetlands system mechanisms.

Many factors affect the ability of wetlands to ameliorate acid drainage, including hydrology, Fe and alkalinity concentrations, and various wetlands characteristics such as depth, area, hydraulics, vegetative and microbial species and extent, and substrate. Because of the interrelationships among these many factors and their effects on wetlands treatment efficiencies, it is difficult to develop treatment area design guidelines. Additionally, with the relatively new concept of the anoxic limestone drain, required areas and designs of wetlands to achieve compliance may be greatly affected. However, TVA's data show that even in the absence of alkalinity, wetlands are removing up to 21.3 GDM of Fe. Mn is being removed up to 1.9 GDM in the presence of alkalinity. TVA's experience suggests that these numbers may represent an upper-limit sizing criteria for aerobic wetlands as they are currently designed and constructed.

TVA's encouraging results suggest that staged treatment wetlands systems are preferred designs potentially capable of treating poor-quality acid drainage. Such staged treatment may consist of: (1) an initial anaerobic limestone trench at the source of the seepage to passively add alkalinity; (2) a large, deep settling basin to accumulate oxidized and precipitated Fe sludges; and (3) a two- or three-cell constructed wetlands for Mn and further Fe removal. TVA is currently investigating the use of passive alkaline beds to increase the pH of constructed wetlands discharges in cases where pH remains below 6.0. TVA plans to continue its research and evaluation of operational and experimental wetlands treatment systems, especially regarding methods to passively increase buffering capacity and pH in wetlands influents and effluents. As more information is made available by TVA and other operating systems and research activities, design guidelines for the components of staged-treatment wetlands systems should be improved.

REFERENCES

1. Brodie, G. A. Achieving compliance with staged, aerobic constructed wetlands, in *Proc. 1991 Annu. Meeting Am. Soc. Surface Mining Reclamationists*. Durango, CO, 1991, 151.
2. Brodie, G. A. Staged, Aerobic constructed wetlands for acid drainage and stormwater control. Manual for a short course, presented at the *34th Annu. Meeting Assoc. Engineering Geologists*. Chicago.
3. Brodie, G. A., C. R. Britt, H. N. Taylor, and T. M. Tomaszewski. Personal communication, 1991.
4. Taylor, H. N. and K. D. Choate. Personal communication, 1991.
5. Faulkner, S. P. and C. J. Richardson. Biogeochemistry of iron and manganese in selected TVA constructed wetlands receiving acid mine drainage. *Duke Wetland Center Publication*, Durham, NC, 1990.

CHAPTER 16

Successful Acid Mine Drainage and Heavy Metal Site Bioremediation

J. Davison

INTRODUCTION

Bioremediation is based on natural biological and physical principles operative for millions of years. These are the same principles which have been responsible for the deposition of ore bodies. In these geologically historic processes, biologically active microbes have fostered the accumulation and deposition of minerals and metals by removing them from their dissolved states in water and fixing them in the underlying soils in an insoluble state, thereby returning them to the ore cycle.

The Lambda Bio-Carb Process (patent pending) is an *in situ* bioremediation system utilizing site-indigenous, mixatrophic cultures hybridized for maximum effectiveness. Each site requires hundreds of microbes in a perfect microecological balance. This enables the system to be self-sustaining and thus provide permanent, passive treatment which is self-adjusting in response to influent changes. Further, the process can and has been utilized in conjunction with constructed wetlands in order to reestablish a basic macroecosystem. This speeds up natural succession, leading to even more rapid remediation. Over 6000 microorganisms have been catalogued by Lambda following successful field application.

The Lambda Process has been utilized successfully in the treatment of acid mine drainage (AMD) since 1984. The process has also proven effective in the treatment of ponds, estuaries, and contaminated soils. Results obtained at all sites are confirmed by independent laboratories and/or government agencies. Furthermore, both lab and field testing have confirmed the viability of the process in treating sites contaminated by heavy metals, hydrocarbons, organics, agricultural wastes, and other hazardous compounds.

PBS MINE SITE ASSESSMENT

The PBS Coal site in Somerset County, Pennsylvania was mined from the mid-1930s until 1985, first as a deep mine, then as a surface mine. The seams mined were the Lower Freeport, Lower Kittanning, and Mercer, with a mining area covering 500 ac. Prior to ceasing operation, the water treatment system covered less than 1 ac. Raw pit water from the sump was aerated and then pumped into a collecting pond where the water received constant chemical treatment. The residue (sludge) was removed and placed in a pit, west of the treatment area. After mining ceased, three additional ponds were added to the system. One pond was installed as a collection and aeration site preceding the existing pond. A caustic tank was installed between these two ponds which injected 10,000 gal of liquid sodium hydroxide per month. Two ponds, were added *beyond* the existing pond in order to increase retention time within the system.

A site assessment was performed in order to evaluate the sources and levels of problematic metals (Fe and Mn), pH, micro- and macroecosystem components, and geological/hydrogeological constraints. Soil and water samples were collected for laboratory testing from each pond and areas surrounding each pond. Laboratory testing revealed that the microbial population had been compromised by the contaminants of the AMD, as well as by the caustic chemical used in treatment.

A laboratory pilot tank was established to mirror site conditions and was monitored for relevant chemical parameters as well as for microbial activity. It was determined that bioremediation could be utilized to correct effectively the existing imbalances and eliminate the need for chemical treatment.

0-87371-550-0/93/$0.00+$.50

PBS DESIGN, PREPARATION, AND INOCULATION

Following the site assessment, the fourth pond was redesigned as a three-cell wetlands, and approximately three fourths of an acre was reexcavated. A total of 335 tons of two-crop mushroom compost was installed in the third pond and the wetland cells comprising the fourth pond. In addition, cattails were introduced to establish a wetland macroecosystem. The multi-pond system was designed to function as follows: all water is collected from an underground sump and run through a large aeration system and blown into pond 1 to oxygenate the water. Pond 1 is approximately 300 × 400 ft and 2 to 6 ft deep. From pond 1, water runs over a limestone spillway for reoxygenation before entering pond 2. Pond 2 is 50 × 250 ft and also 2 to 6 ft deep. It is piped from pond 2 and flows into pond 3, which is 75 × 75 ft and 6 ft deep. A pipe carries the effluent from pond 3 to the top of the entrance to the first wetlands cell (pond 4) and it cascades down a small rip-rap area into the wetlands. In order to establish and sustain a sulfate reduction sink necessary for a 6-9 pH and manganese reduction, the water transfer through the three cells is done by pipes that permit no further aeration. The entire wetlands (pond 4) system is approximately 200 × 200 ft and 2 to 3 ft deep. A fifth 190 × 210 ft and 2 to 4 ft deep pond was available to be used if needed. The system discharged directly from wetlands cell 3 of pond 4 into Coxes Creek.

The maximum influent flow rate of 400 gal/min, levels of heavy metals, population densities and viability of indigenous bacteria, algae, fungi and protozoa found in the system were each considered in calculating both the composition and volume of Bio-Carb necessary for site remediation. The Bio-Carb material consists of hybridized, site-indigenous microbes scaled up in high-nutrient, enzyme-enhanced media and embedded in activated charcoal. The charcoal is then bagged in burlap and boxed for shipping. The bags weigh approximately 100 lbs each and act as a "microincubators" that constantly replenish the microbes until balance is achieved in the ponds. The weight of the bags is sufficient to prevent washout at any flow rates found at mine sites to date. For shallow wetlands, 30–50 lb bags are used and laid in the compost. The larger bags in the deep ponds are dropped from a boat, or placed along shallow flow paths.

A total of 5 tons of Bio-Carb was shipped to PBS via common carrier and planted along with the wetlands plants on July 27, 1990. Chemical treatment ceased two weeks prior to site bioremediation and the caustic system was dismantled and moved to another mine site in September 1990, when it was determined that further chemical treatment was unnecessary to maintain compliance.

RESULTS AND DISCUSSION

Weekly chemical analyses were performed by Lambda and an independent laboratory for confirmation of results. Samples were obtained at the exit of each pond and at the discharge point. First week results indicated that compliance levels were achieved.

By the end of September 1990, extreme precipitation caused the sump to reach dangerously high levels that demanded 24-hour pumping. This forced a system designed to treat 400 gallons per minute of influent to respond to 800–1,000 gallons per minute. Under these conditions, pH and iron stayed in compliance. However, manganese fluctuated to 1–2 ppm above compliance levels beyond 800 gallons per minute. Increased retention time became a key factor. A system designed to provide 24 hours of retention was allowing less than 12. The influent could not make sufficient contact with the microbes or the wetlands plants. The fifth pond was brought into the system and water was pumped from cell 2 of pond 4 and reaerated as it was blown into the air to fall on pond 5, rather than run into it. Pond 5 has developed into a modified wetlands system, as did the discharge ditch into Coxes Creek.

Compliance levels have consistently been achieved since retention time was corrected, and no additional chemical treatment has been used at the PBS site since bioremediation was initiated. In addition to the chemical monitoring, monthly biomonitoring has been performed by Lambda Bioremediation Systems, Inc. to ensure that the proper microbial balance is sustained. The biological systems did respond admirably to the increased flow and heavy metal loading and continues to perform to and beyond our expectations for this system.

The graphs (Figures 1, 2, and 3) indicate the chemical parameters achieved using bioremediation to treat the PBS site. The discharge permit required effluent parameters as follows: pH = 6–9, Mn = 4, and Fe = 3.5–7.0 (monthly average to high).

IRON

P.B.S. COAL

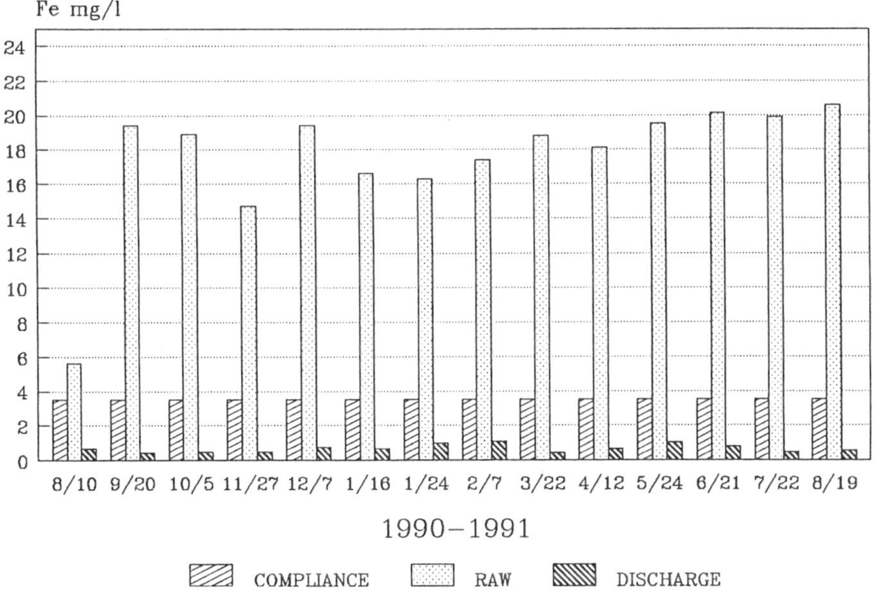

Figure 1. Iron concentrations at PBS bioremediation site.

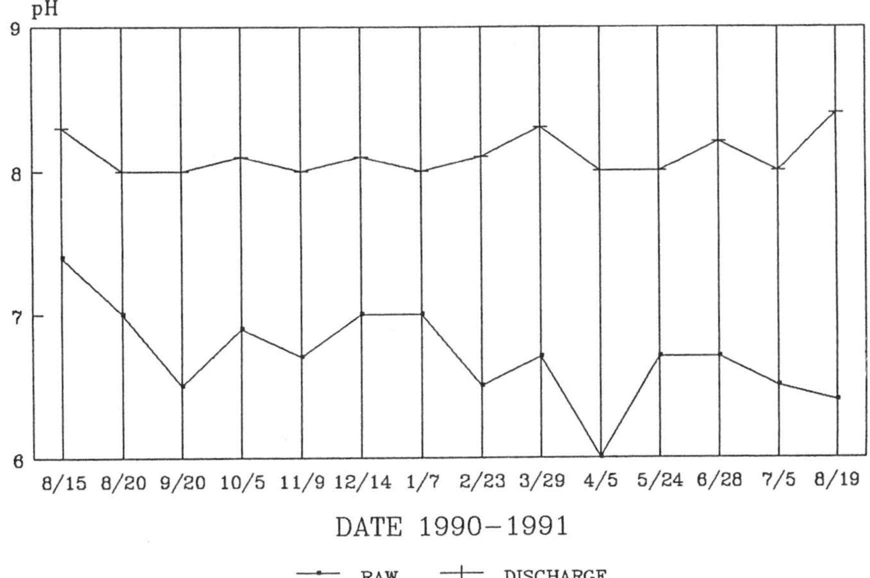

Figure 2. pH values at PBS bioremediation site.

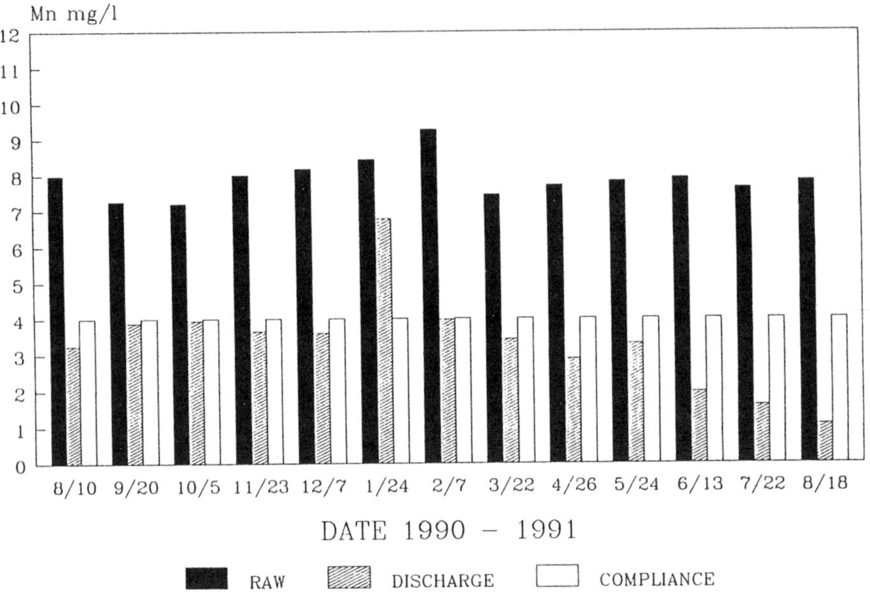

Figure 3. Manganese concentrations at PBS bioremediation site.

The use of pond 5 and an extension of pond 3 were necessary to provide a 24-hour retention period when flow rates increased dramatically two months after the system was installed. Also, one bag of NaOH briquets was placed in pond 5 the day the system changed over to include this pond, but nothing more was deemed necessary. The increased retention time was adequate to meet the discharge requirements.

The elimination of caustics expense *alone* exceeded the cost of installation in less than 2.5 years. As of the writing of this article, the system discharges well within N.P.D.E.S. standards, despite fluctuations in pollution loads and flow rates.

ACKNOWLEDGMENTS

A great deal of credit is due to Robert Deason, Senior Geologist at PBS, for his invaluable assistance and excellent suggestions. He is also to be commended for his imaginative applications of aeration systems. Further information about this site is available from Lambda or from Mr. Deason at PBS Mines in Friedens, Pennsylvania.

The Lambda process was able to achieve and maintain compliance standards. Additionally, it was determined that by combining our Bio-Carb technology with wetlands treatment, the following conclusions could be drawn. Less treatment space is required to treat more water and higher pollution loads than would be true using either technology alone, and microbial, charcoal, plant, compost, and construction costs are lower than they would be with either system alone. At the same time, higher volumes of flow rates can be accommodated with a faster reaction time to sudden fluctuations. The combined technologies function more effectively in the winter than does either system alone, but no chemical treatment is required when the systems are combined. Some chemical treatment is more likely to be required with either system alone. In addition, natural succession escalates and the entire system matures more quickly in tandem than alone. Other benefits include the following: sludge build-up is eliminated, reducing a significant removal cost to the mine operator; valuable wetlands are restored; wildlife is reestablished; and the site is aesthetically pleasing and may be utilized for other purposes.

CHAPTER 17

The Use of Wetland Treatment to Remove Trace Metals from Mine Drainage

P. Eger, G. Melchert, D. Antonson, and J. Wagner

INTRODUCTION

Drainage from mineralized Duluth Complex stockpiles located at LTV Steel Mining Company's Dunka Mine in northeastern Minnesota contains elevated concentrations of nickel, copper, cobalt, and zinc. This drainage has increased metal concentrations in nearby receiving waters to levels which are as much as 500 times natural background concentrations. A feasibility analysis, conducted in 1985, concluded that although an active treatment plant could generally achieve water quality guidelines, a more cost-effective, passive approach (low cost, low maintenance) might also be successful.[1]

Wetland treatment is a crucial aspect of this passive approach; and although previous work[2-4] had demonstrated the effectiveness of peat to remove trace metals from mine drainage, no field data from an actual treatment system existed.

In 1986, LTV Steel Mining Company and the Minnesota Department of Natural Resources began a cooperative program to develop data on optimal wetland treatment design and system life. The goal of the program was to collect data for the design of full-scale treatment systems for the stockpile drainages at the Dunka Mine by constructing and operating wetland treatment test plots.

SITE DESCRIPTION

The Dunka Mine is a large open pit taconite operation covering approximately 160 hectares (ha). At this location, the Duluth Complex, a metalliferous gabbroic intrusion, overlies the taconite ore and is removed and stockpiled along the east side of the open pit. The Duluth Complex material contains copper, nickel, and iron sulfides, and the stockpiles contain over 32 million tons and cover about 120 ha. Discrete seepages appear at the bases of the stockpiles and generally flow continuously from early April to late November. Average flows from the various seepages range from 0.5 to 14 L/s (8 to 220 gpm), but maximum flows exceeding 100 L/s (1600 gpm) were observed after periods of heavy precipitation.

Nickel is the major trace metal in the drainages, with annual median concentrations on the order of 3 to 30 mg/L. Copper, cobalt, and zinc are also present, but are generally less than 5% of the nickel values. Median pH ranges from 5.0 to 7.5, but most of the stockpile drainages have pH greater than 6.5.

Wetlands, primarily wet meadows dominated by sedges and grasses (Type 2, USFWS Circular 39), are located near every stockpile and appear to offer potential treatment areas for each seepage.[5] These wetlands are typical of the many small lowland areas in northern Minnesota and would generally be associated with any future mining area.

METHODS

Prior to beginning the study, each wetland was surveyed and its capacity to treat the associated mine drainage was determined. Estimated lifetimes based on input metal load, wetland area, and a removal zone

0-87371-550-0/93/$0.00+$.50

of 20 cm ranged from 20 to over 700 years.[5] Based on the survey work, a test area was selected and plots constructed.

Four cells were designed so that a variety of water levels, vegetation, and flow regimes could be tested (Figure 1). These cells were constructed in a natural wetland. Each cell was 6 m wide × 30.5 m long (20 ft wide × 100 ft long) and was surrounded by a compacted peat berm.

Stockpile drainage was collected at the toe of the stockpile (Site W3D), piped to the plots, and dispersed across each cell with a perforated PVC pipe, then collected with an open half pipe at the outflow. Based on the results of previous laboratory studies,[4] a design residence time of 40 to 48 h was selected for the cells.

The initial residence time in each cell was estimated from the volume of water above the peat surface. In 1990, rhodamine dye was used to measure the actual residence time in each cell.

Specific cell designs were as follows:

Cell 1. Unmodified natural wetland (wet meadow, type 2); water dispersed across natural wetland; vegetation was primarily sedges (*Carex* sp.) and grasses (*Calamograstis* sp.); water depth was 5 cm (2 in.).

Cell 2. Modified wetland; shallow trenches were constructed with a backhoe; these trenches were spaced about 4.5 m apart, were about 60 cm deep (15 ft and 2 ft), and were dug perpendicular to the flow path; the spoil material from the trench was cast downstream; sedges and grasses from the surrounding area were transplanted into the cell; water depth, 5 cm (2 in.).

Cell 3. Modified wetland; hay bales placed to create serpentine flow, 5 cm (2 in.) of straw placed on the bottom of the entire cell to encourage sulfate reduction; cell planted with cattails (*Typha latifolia*) (1 per square meter); water depth, 15 cm (6 in.).

Cell 4. Modified wetland; peat berms constructed across the cell, perpendicular to flow; cattails planted (1 per square meter); water depth, 15 cm (6 in.).

Initial water level measurements and dye studies indicated that up to 8 L/min (2 gpm) leakage was occurring through the berms and that the cells were interconnected. A cut-off wall was installed by mixing bentonite with sand and placing the mixture into a trench which was cut in the center of the berms with a Ditch Witch power trencher. The ditch was cut into the mineral soil underlying the cells so that any lateral flow into or out of the cells would be minimized. Subsequent dye and water level measurements indicated that seepage in all cells had been reduced to less than 10% of the input flow.[6]

The input stockpile drainage can be characterized as a high hardness, neutral drainage whose primary contaminant is nickel. Average hardness is around 2300 mg/L as $CaCO_3$, with a pH range of 6.5 to 7.9. Copper and zinc concentrations generally meet the water quality criteria established by the Minnesota Pollution Control Agency, while those of cobalt and nickel routinely exceed the criteria, sometimes by almost an order of magnitude (Table 1).

Data collection began in August 1989. Water quality samples of the inflow and outflow were collected about twice per week and analyzed for pH, specific conductance, alkalinity, acidity, calcium, magnesium, sodium, potassium, sulfate, copper, nickel, cobalt, zinc, iron, and manganese.

Metals were analyzed with a Perkin Elmer atomic absorption unit (Model No. 603) either in flame or flameless mode. Sulfate was analyzed with barium chloride turbidimetric method (Standard Methods 426C, HF scientific model DRT-100 turbidimeter), and alkalinity and acidity were determined by titration. Input and output flows from each cell were measured with Data Industrial electronic flow meters and recorded with a Campbell Scientific micrologger, which also recorded precipitation, temperature, and relative humidity. A settling barrel and screen was installed ahead of the flowmeters to prevent debris and aquatic organisms from plugging the flowmeters.

With the exception of Cell 4, all cells functioned according to design. The peat berms in Cell 4 did not transmit water at a rate equal to the inflow rate of 11.4 L/min (3 gpm), and the water level rose above the design level. To keep the water level in the cell below the top of the cut-off wall, slots were cut into the peat berms to provide a serpentine flow path. (Figure 1 depicts the final configuration of Cell 4.)

Drainage from Site W3D was treated in all cells from August to November 1989 and from May to September 17, 1990. Cells 2, 3, and 4 continued to treat W3D drainage through November 1990, while Cell 1 treated drainage with higher metal concentrations from Seep 1 and Aquifer X. Seep 1 water was mixed with W3D water on a 1:1 volume basis, while Aquifer X water was fed directly to Cell 1. The target residence time for the Seep 1 study was 48 h, while the residence time in the Aquifer X test was varied between 24 and 48 h.

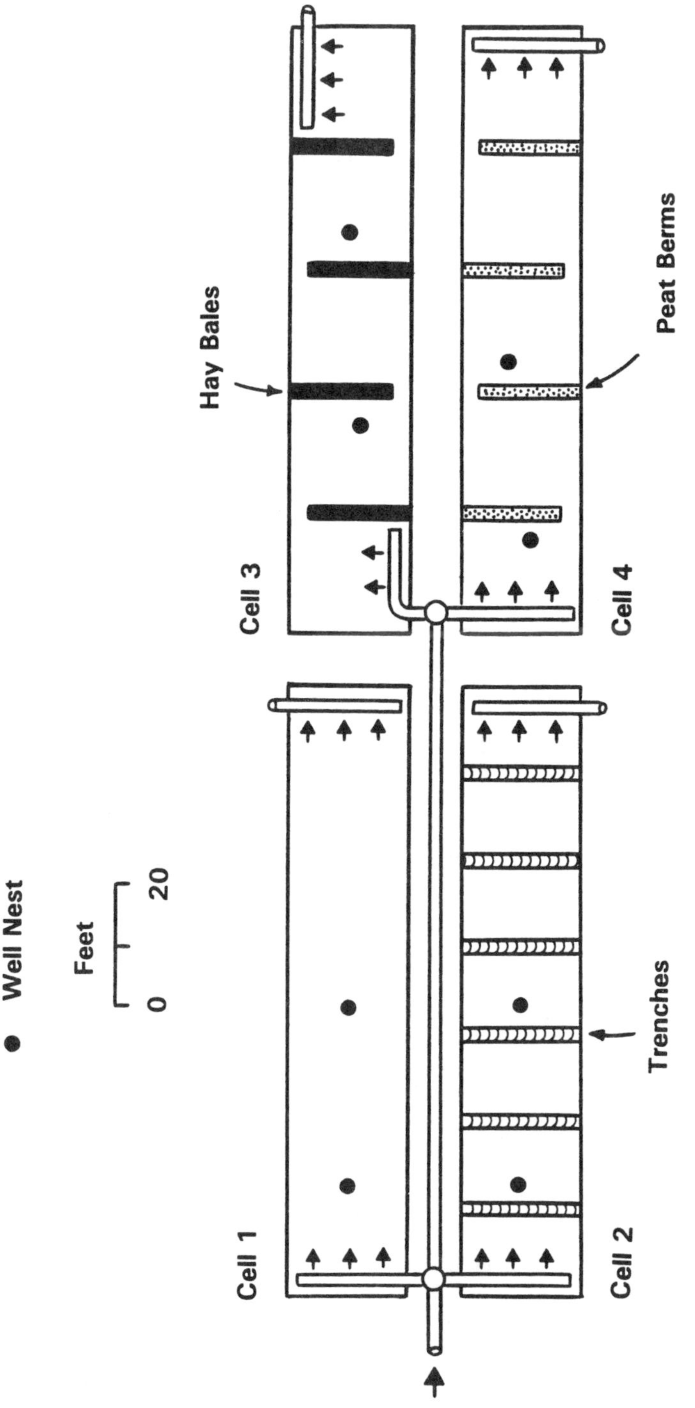

Figure 1. Test cell design.

RESULTS

Residence Time

The residence times in Cells 1 and 2 were 43 and 49 h, respectively, and were in the original design range. The residence times in Cells 3 and 4 were only 18 to 22 h, or about a factor of 2 below the design residence time. The input flow rates into Cells 3 and 4 were decreased from 11.4 to 5.3 L/min (3 to 1.4 gpm) so as to increase the residence time to around 48 h.

Although residence times were constant for most of the summer, problems with plugging of the feed lines and the flowmeters caused residence time to vary widely, particularly during the spring of 1990. As a result, the residence time of all the cells varied from less than 1 day to around 12 days.

Water Quality

Water quality data presented in this article are generally for filtered metal values. In general, the differences among filtered, unfiltered, and total digested values were less than 10%.[6] Concentrations in the outflow of the cells are compared to proposed water quality standards which must be met at the outflow of the wetland treatment system.

In general, for both years, there was little difference between the input and the output concentrations for the major cations and anions (calcium, magnesium, sodium, potassium, and sulfate), and for the trace metals copper and cobalt.

Output concentrations for iron, alkalinity, and acidity were slightly higher than the input values, while significant decreases in output concentrations were observed for pH, nickel, and zinc ($p < 0.001$, independent t test). Although the pH decreased by about 0.2 to 0.3 pH units as the water passed through the cell, all the outflow pH values met or exceeded the minimum pH criteria of 6.5. Average zinc concentrations decreased from 0.03 mg/L in the input water to 0.01–0.02 mg/L in the outflows of all the cells.

Nickel is the trace metal in the stockpile drainage with the highest concentration, and the wetland treatment system was primarily designed to remove nickel from solution. Every cell was successful in removing nickel from solution.

In general, the cells which contained 5 cm of water were more effective in reducing nickel concentrations than the cells with 15 cm of water (Figure 2). Cells 1 and 2, the unmodified wetland and the cell containing the trenches, were successful in removing over 80% of the nickel from the input W3D stockpile drainage, and over 90% of the effluent samples contained less than 0.2 mg/L nickel. Over the course of this study, these cells successfully treated input nickel concentrations which ranged from 0.11 to 2.1 mg/L.

Cells 3 and 4, the cells with serpentine flow, cattails, and 15 cm of water, although not so successful as cells 1 and 2 at reducing nickel concentrations, still removed 69 and 39% of the input nickel, respectively.

Removal of Metals as a Function of Input Drainage Quality

Water from the Seep 1 and Aquifer X sites contained higher concentrations of trace metals and was tested in the unmodified natural wetland cell (cell 1). Each test was run for 2 to 3 weeks.

During the first test (Seep 1), metal removal exceeded 90% for copper, cobalt, nickel, and zinc. Although nickel removal exceeded 90% for the entire test, outflow concentration increased steadily during the test. Midway through the test, nickel in the outflow from this cell exceeded the water quality guideline of 0.213 mg/L and continued to increase to 0.5 mg/L by the end of the test (Table 1).

During the second test (Aquifer X), copper, cobalt, and zinc removal remained high (82 to 95%), but nickel removal decreased to about 40% by the end of the test. None of the nickel concentrations in the outflow met water quality guidelines, and concentrations increased by a factor of 2 when the residence time was decreased from 34 to 22 h (Table 1).

Nickel Mass Removal

The total mass of nickel into and out of each treatment cell was determined by combining the total daily volume with the measured or an estimated daily nickel concentration. The mass removed in the cells is the difference between the input and output values.

Table 1. Summary of Wetland Treatment Results, Cell 1

Input source		Input concentration range (mg/L)	Output concentration range (mg/L)	Water quality standards	Overall mass removal (%)	Median residence times (hours)
W3D	pH	6.5–7.9	6.5–7.7	6.5–8.5	NA	18–48[a]
8/1/89–9/16/90	Cu	0.002–0.02	0.001–0.010	0.013	NS	
	Ni	0.11–2.1	0.020–0.39	0.213	87	
	Co	0.003–0.047	0.002–0.043	0.05	NS	
	Zn	0.02–0.07	0.006–0.08	0.149	33	
Seep 1	pH	6.6–7.4	6.6–7.45	6.5–8.5	NA	34
+	Cu	0.11–0.18	0.006–0.023	0.013	90	
W3D	Ni	4.27–6.12	0.04–0.63	0.213	94	
9/17/90–10/5/90	Co	0.27–0.5	0.006–0.050	0.05	90	
	Zn	0.65–0.94	0.019–0.037	0.149	96	
Aquifer X	pH	6.65–7.15	6.45–7.00	6.5–8.5	NA	34
10/11/90–10/19/90	Cu	0.08–0.24	0.005–0.007	0.013	95	
	Ni	1.75–1.98	0.56–0.64	0.213	59	
	Co	0.18–0.26	0.003–0.008	0.05	97	
	Zn	1.42–1.76	0.024–0.058	0.149	96	
Aquifer X	pH	6.73–7.30	6.73–7.35	6.5–8.5	NA	22
10/20/90–10/26/90	Cu	0.21–0.35	0.004–0.022	0.013	73	
	Ni	1.91–1.96	0.95–1.20	0.213	41	
	Co	0.21–.24	0.029–0.040	0.05	81	
	Zn	1.87–1.96	0.055–0.320	0.149	90	

[a] 30% of values exceeded 48 hours.
NA = not applicable; NS = no significant removal.

During the course of the study, total nickel input to the cells ranged from 1780 to 3370 g, and removal ranged from 1000 to 1700 g. The daily surface area loading and daily removal rates were 0.037 to 0.066 g nickel input/m^2/day and 0.020 to 0.035 g of nickel removal/m^2/day, respectively.

The greatest mass removal occurred in cell 3, a modified wetland cell with serpentine flow, 15 cm of water, and cattails. Cell 4, the other modified wetland cell with 15 cm of water, removed the least nickel. The cells containing 15 cm of water had larger input flows and therefore had a larger input mass than the cells with 5 cm of water. The cells with 5 cm of water, although they removed less mass, had a higher treatment efficiency (around 86%).

DISCUSSION

pH

As hydrogen ions are exchanged for metals during wetland treatment, solution pH tends to decrease. The total change in pH is a function of the character of the peat, the amount of ion exchange that occurs, and the alkalinity of the solution. In this study, the pH in the outflow from all cells was significantly lower than the input by about 0.1 to 0.3 pH units ($p < 0.001$, independent t test).

Since the ability of peat to remove metals from solution decreases as the pH of the solution decreases[4] and since effluent pH must remain above 6.5 (Table 1) to meet water quality standards, the input pH must be greater than 6.5. Most of the stockpile seepages at the Dunka Mine generally have pH values greater than 6.5, alkalinity greater than 80 mg/L as $CaCO_3$, and from 1 to 10 mg/L nickel. The outflow pH will likely remain above the standard if influent pH is 6.7 or greater. If the pH of the stockpile drainage is less than 6.7, the pH of the input could be raised using a limestone bed.[7] Effluent pH could also be adjusted, but there may be a greater possibility of plugging the limestone bed with organic material from the wetland.

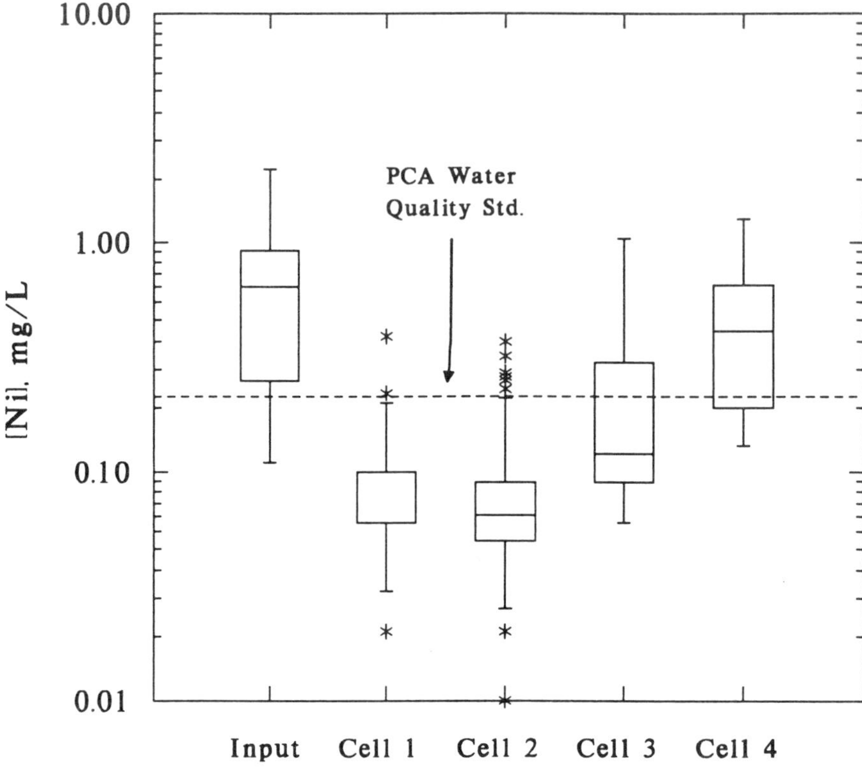

Figure 2. Nickel concentrations, wetland treatment cells 1989–1990.

Nickel

Although all cells consistently removed nickel from solution, the nickel concentrations in the outflow from the cells containing 5 cm of water (cells 1 and 2) were almost always less than those containing 15 cm of water (cells 3 and 4). Lower water levels and longer residence times in cells 1 and 2 probably account for the increased removal.

The outflow elevations in the cells were set to maintain an average water level of about 5 cm (2 in.) in cells 1 and 2 and 15 cm (6 in.) in cells 3 and 4. Since most of the flow in the treatment cell occurs across the surface, the deeper the water, the less contact there will be with the metal removal sites on the peat. The lack of contact between the input water and the peat in cells 3 and 4 was demonstrated in the residence time dye test.[6] Rhodamine dye, which was used as the tracer, can be adsorbed onto an organic substrate, and the low concentrations and overall dye recovery observed in cell 1 and 2 indicated that significant removal of dye had occurred. In contrast, large dye concentrations in the outflow of cells 3 and 4 indicated that significant dye removal had not occurred and that, therefore little contact occurred between the water and the organic matter.

Previous laboratory kinetic data indicated that nickel removal for a drainage with water quality similar to the W3D input was 68% complete after 24 h and 80% complete after 48 h. Initially, all cells were designed to have a residence time on the order of 42 h. The dye studies conducted in July 1990 demonstrated that the residence time in cells 1 and 2 were at the correct value, while the residence time in cells 3 and 4 were only about one half of the desired value.

Based on spot observations and measurements, the water level in the cells appeared to be similar to the design depth, so the "effective" or "exchangeable" volume of standing water must be less than the total volume. The difference in volume could be the result of either incomplete mixing or short-circuiting within the cell.

In cell 3, the original serpentine path was disrupted when the hay bales subsided. Water could flow across the top of the bales throughout most of the celi, thereby decreasing the effective volume in the cell. The layer

Table 2. Projected Nickel Removal in a Wetland Treatment Cell as a Function of Adsorption Capacity and Depth

Nickel removal capacity (mg nickel/kg peat)		1000	10,000	1000	10,000
Effective depth of removal (cm)	**Mass of dry peat (kg)**	**Total nickel removal (g)**		**Duration of treatment for entire cell (years)**	
1	190	190	1900	0.2	2
5	930	930	9300	0.8	8
20	3720	3720	37,200	3	30
Cell 2 data		1130 g removed		Greater than 1.3 yrs	

Note: Assumptions: nickel input = 1 mg/L, 3.8 L/min, 210 days of flow per year.

of straw in cell 3 is about 4 to 6 in. below the surface of the water and could restrict mixing of the water moving across the surface with the water in the straw layer.

In cell 4, the movement of dye was rapid, and portions of the cell did not contain dye, which indicated the effective volume would be less than the theoretical volume calculated from cell dimensions.

Treatment Capacity

The total amount of metal removal occurring in the treatment cells is a function of the input water quality, the residence time, the water temperature, the distribution of flow within the cell, the capacity of the peat to remove metals from solution, and the depth to which that removal occurs.

When the study began, the initial estimate of treatment system lifetime was based on the assumption that 1 kg dry peat can remove 10,000 mg of nickel, and metal removal occurs to a depth of 20 cm. The removal capacity was estimated from laboratory calculations and field measurements of the nickel content of peat, which ranged from 10,000 to 26,000 mg/kg.[6] The nickel capacity used to estimate treatment life represents the mass of nickel removed from solution and represents a maximum expected uptake value, not necessarily a value below which water quality guidelines will be consistently met.

The removal depth of 20 cm was based on metal concentrations measured in peat at a white cedar wetland at the Dunka mine.[2] Although metal concentrations were greater between 0 and 20 cm, elevated metal concentrations were also found in 20 to 50-cm samples.

Table 2 provides an estimate of total nickel removal and treatment life in a treatment cell as a function of removal capacity and removal depth and compares the observed removal in cell 2 to the projections.

Over 1100 g of nickel have been removed in cell 2 during the 1.3 years of the study. Outflow nickel values from cell 2 met the proposed water quality standard over 90% of the time, and there is no evidence that treatment has decreased over the course of the study.

Additional water quality and peat data are being collected to more accurately determine field removal capacity and depth of removal.

CONCLUSIONS

Wetland treatment reduces trace metal concentrations in stockpile drainage. Overall reductions for 1989–1990 in nickel mass ranged from 39 to 86%. In short-term tests, concentrations of copper, cobalt, and zinc were also reduced by 83 to 94%.

Over the course of the study, nickel removal ranged from about 6 to 10 g/m^2 or 0.020 to 0.035 $g/m^2/day$.

Additional work is underway to define overall treatment capacity and treatment lifetime better. Information is being collected on the effect of the following on metal removal: residence time, the cumulative mass of metal removed, biological and microbiological processes, and the addition of peat with low metal content to the wetland.

Adding peat or other organic material to the wetland can extend the treatment life by providing new adsorption sites or by increasing the rate of sulfate reduction. The material that has been added can be periodically removed, and new material can be added to increase the life of the system. With this type of a program, the wetland can be used indefinitely, and then the original wetland could be re-established when the treatment has been completed or when the input metal load has been reduced to a point that any residual metal can be handled by the re-established wetland.

ACKNOWLEDGMENTS

LTV Steel Mining Company, with the assistance of STS Consultants, constructed the cells and has provided funding to aid the data collection and analysis efforts. DNR staff including Cal Jokela, Al Klaysmat, Jean Matthew, Ann Jagunich, Kate Willis, Diane Melchert, and Linda Stajkowski have also been instrumental in generating and reporting the results presented in this article.

REFERENCES

1. Barr Engineering Company, Feasibility assessment of mitigation measures for gabbro and waste rock stockpiles, Dunka Pit Area, 1986.
2. Eger, P. and K. Lapakko. Nickel and copper removal from mine drainage by a natural wetland, in *Mine Drainage and Surface Mine Reclamation, Vol. I, Mine Water and Mine Waste.* Bureau of Mines IC, 1988, 301.
3. Lapakko, K. and P. Eger. Trace metal removal from mine drainage, in *Mine Drainage and Surface Mine Reclamation, Vol. I, Mine Water and Mine Waste.* Bureau of Mines IC 9183, 1988, 291.
4. Lapakko, K., P. Eger, and J. Strudell. Low cost removal of trace metals from copper-nickel mine stockpile drainage, Vol. 1, Laboratory and Field Investigations, USBM Contract Report JO205047, 1986.
5. Eger, P. and K. Lapakko. Use of wetlands to remove nickel and copper from mine drainage, in *Constructed Wetlands for Wastewater Treatment.* D. A. Hammer, Ed., 1989, 780.
6. Eger, P., G. Melchert, D. Antonson, and J. Wagner. The use of wetland treatment to remove trace metals from mine drainage at LTV's Dunka Mine, Minnesota Department of Natural Resources, Division of Minerals, 1991.
7. Lapakko, K. and D. Antonson. Pilot scale limestone bed treatment of the Seep 1 waste rock drainage, Minnesota Department of Natural Resources, Division of Minerals, 1990.

CHAPTER 18

A Comparison of Local Waste Materials for Sulfate-Reducing Wetlands Substrate

M. A. Gross, S. J. Formica, L. C. Gandy, and J. Hestir

INTRODUCTION

In bacterial sulfate-reducing constructed wetlands, the organic substrate is used to provide a carbon source for the sulfate-reducing bacteria that are responsible for catalyzing the desired sulfate reduction process in the wetland.[1,2] As the sulfate-reducing bacteria metabolize the organic carbon available in the substrate, sulfate is converted to hydrogen sulfide passing through the substrate. Inasmuch as the substrate plays an important role in sustaining bacterial sulfate reduction and in the treatment of acid mine drainage in constructed wetlands, the properties of the substrate must be considered in the selection of substrates for constructed wetlands. The most important property is that the substrate must contain available carbon as an energy source for the sulfate-reducing bacteria. Another important feature that should be evaluated includes the substrate's hydraulic conductivity. The hydraulic conductivity of the substrate must be sufficiently high to allow liquid to pass through the substrate at the design flow rate with an achievable liquid head over the substrate.

Mushroom compost is used as a substrate in anaerobic constructed wetlands[1,3-5] because it has performed well in anaerobic bacterial sulfate-reducing wetlands, neutralizing the acidity and raising the pH of the acid mine discharge (AMD). It also provides adequate available carbon. As early as 1969, researchers suggested that other substrates may perform well in providing carbon for bacterial sulfate reduction.[6] Alternative substrates studied to date for sulfate-reducing wetlands have included sawdust, aged manures, and decomposed wood products.[4]

In 1989, a sulfate-reducing constructed wetland was determined to be a feasible alternative to the chemical treatment of AMD at a clay mining facility in Perla, Arkansas. Specifically it was determined to be a feasible treatment method to raise the pH, reduce sulfate concentrations, and neutralize the acidity of the AMD. Mushroom compost is not readily available, however, in central Arkansas. The nearest source of mushroom compost is in Miami, Oklahoma, approximately 360 miles from the Perla site. Transportation costs to deliver the required volume of substrate to the site would have precluded the use of a sulfate-reducing constructed wetland at the site. Therefore, alternative substrates that were locally available were desirable. Organic materials other than mushroom compost are widely available in Arkansas. Arkansas is a major producer of rice and chickens. The timber and wood products industry is also a major industry in the state. By-products of these industries include rice hulls, chicken litter, and sawdust — all organic compounds with potential application as substrates for sulfate-reducing constructed wetlands.

The purpose of this study was to investigate the suitability of local organic materials for their use in sulfate-reducing constructed wetlands at the Perla site. Specifically, their suitability as an organic carbon source, to maintain hydraulic conductivity, and their effectiveness in increasing the pH and decreasing sulfate concentrations and acidity of the Perla site AMD were investigated in this study. This project is undergoing two stages of study: a bench-scale study in the laboratory and a pilot-scale study in the field. This article presents the results of the laboratory bench-scale study.

0-87371-550-0/93/$0.00+$.50

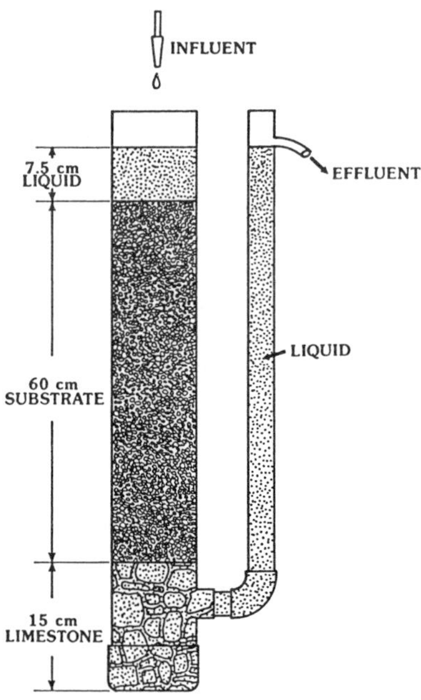

Figure 1. Typical column cross-section.

Table 1. Substrate Mixtures and Column Designation

Column	Substrate mixture (volume/volume)
A	100% mushroom compost
C	75% mushroom compost/25% rice hulls
D	50% chicken litter/50% rice hulls
E	50% chicken litter/50% sawdust
F	75% composted chicken litter/25% rice hulls
G	75% mushroom compost and composted chicken litter/25% rice hulls

MATERIALS AND METHODS

Six columns were constructed in the laboratory, and acid runoff from the clay mining site was applied to each column. Columns were constructed as shown in Figure 1. Column construction and operation have been previously discussed in detail by Gandy et al.[7]

Six substrate mixtures were used in the columns as noted in Table 1. A seventh column was constructed of 100% composted chicken litter, but was not used in this study because the hydraulic conductivity was too low to allow passage water through the column at the experimental rate. The components used to make the mixtures were analyzed at the University of Arkansas, Fayetteville.

Prior to loading the columns with acid runoff, tap water was passed through all columns at 2 to 4 mL/min for 5 days to wash labile organics and nutrients from the substrates. Sediment from a local wetland was used to seed each column with naturally occurring wetland organisms, and acid runoff was applied to the columns at 0.8 mL/min. The rate of application was calculated to give 5.1 days' residence time in the column based upon voids in the substrate and limestone and including the free water above the substrate. Limestone and substrate porosities were estimated at 60% based upon rough measurements made in the laboratory.

TABLE 2. Organic Substrate Analyses

Sample	Chicken litter	Composted chicken litter	Mushroom compost	Rice hulls
Analysis:				
pH	7.0	6.5	7.0	6.4
EC micromhos	3.100	7.800	6.500	200
% H_2O	44.4	8.9	54.4	0.27
% N[a]	1.50	3.03	2.50	0.36
% P	2.04	2.63	0.51	0.24
% K	0.67	1.88	1.38	0.73
% Ca	3.18	4.20	11.30	0.35
% Mg	0.70	0.87	0.41	0.18
% Na	0.08	0.33	0.15	0.00
Fe, μg/g	8105	5745	1033	75
Mn, μg/g	148	658	185	1743
Cu, μg/g	25.6	807.3	161.1	15.5
Zn, μg/g	278.3	597.0	692.6	83.2
B, μg/g	10	53	18	23
% S	0.29	0.71	1.86	0.18
NH_4-N, μg/g	1363	614	279	6
NO_3-N, μg/g	1037	1021	171	28
Total carbon %	14.2	20.4	22.9	31.0

[a] To convert to μg/g: %/10,000 = μg/g

For the first 47 days, the column influent and effluent were sampled weekly and analyzed for pH, acid neutralizing capacity (ANC), 5-day biochemical oxygen demand (BOD_5), dissolved oxygen (DO), total iron (Fe), nitrate-nitrogen (NO_3-N), chloride (Cl), calcium (Ca), temperature, conductivity, and sulfate (SO_4). Ammonium-nitrogen (NH_4-N) was occasionally analyzed. From day 48 through 109, effluent samples were not collected; however, DO, pH, temperature, and conductivity were analyzed in the column influent and effluent.

Analyses, collection, and preservation were performed according to Standard Methods for the Examination of Water and Wastewater[8] or EPA-approved methodologies.[9]

RESULTS AND DISCUSSION

Alternative substrates evaluated in this study are mixtures of several organic and inorganic materials. Chicken litter, for example, is often a mixture of chicken manure and sawdust, wood shavings, and/or rice hulls. Composted chicken litter is chicken manure composted with wood products and/or other organic materials. The principal components of spent mushroom compost are hay and a bedded animal manure. Limestone is added to mushroom compost during the mushroom growing process. The bedded animal manure in the mushroom compost used in this study was chicken litter.

The results of the chemical analysis of the substrates used in this study are shown in Table 2 (sawdust was not analyzed). The percentage of total carbon for the substrates analyzed ranged from 14.2 to 31.0%, with rice hulls having the highest carbon content. Mushroom compost had the highest Ca concentration, as expected, since limestone is a component of spent mushroom compost. These materials also had significant amounts of NH_4-N and NO_3-N, with chicken litter having the highest concentrations of 1363 and 1037 μg/g, respectively.

The results of the column test showed similar performance between columns in raising pH and neutralizing acidity of the AMD. Table 3 summarizes the results of analyses performed during the course of the laboratory study. Column influent pH values ranged from 3.0 to 3.7 and the pH of the AMD was raised to greater than 6 in all columns regardless of substrate mixture (Figure 2). Average influent acidity was 3.2 meq/L and the average effluent ANC ranged from 4.1 to 10.6 meq/L (Figure 3). The 75% composted chicken litter/25% rice hull column (F) had the highest average ANC value of 10.6 meq/L.

Differences in SO_4 removal were observed between the columns. Average SO_4 removal ranged from 34 to 79%. The highest average SO_4 removal was observed in column F, composed of 75% composted chicken litter/25% rice hulls.

The columns containing mushroom compost produced effluent with relatively high Ca concentrations compared to the columns without mushroom compost. The initial Ca concentration during start-up in the 100% mushroom compost column was 188 mg/L, and the lowest concentration observed in that column's effluent was 113 mg/L. The mushroom compost analysis showed 11.3% Ca by weight (Table 2). As a comparison, the highest Ca concentration in the effluent from columns containing no mushroom compost was 88 mg/L; it was found in the 50% chicken litter/50% sawdust column's effluent. Chicken litter analysis showed 3.18% Ca by weight. Although the Ca concentrations were high in the mushroom compost and in the effluent from the mushroom compost column, the ANC in the column's effluent quickly reached levels comparable to all other columns' effluent (Figure 3). The relative effect of limestone on Ca concentration, pH, and alkalinity for each column was not determined.

BOD_5, NH_4-N, and NO_3-N were monitored in the columns' effluent to determine the substrates' potential impact on water quality from these constituents. Average influent BOD_5 was 7 mg/L. During the first month the columns were operated, the BOD_5 concentrations ranged from 40 to 430 mg/L. By the 5th month, the effluent concentrations were less than 20 mg/L. Column effluent NH_4-N concentrations ranged from 0.6 to 12.5 mg/L. By the end of the study, effluent NH_4-N concentrations for the columns were all less than 3.2 mg/L except for the column E which contained 50% chicken litter/50% sawdust. The NH_4-N concentration in this column was still 12.5 mg/L. The average influent NO_3-N concentration was 0.2 mg/L. The column average effluent NO_3-N concentrations were less than 0.2 mg/L except for column F which contained 75% composted chicken litter/25% rice hulls. Column F's average NO_3-N concentration was 0.9 mg/L. Results of these analyses indicate that labile BOD and NH_4-N leaching from these substrates in a constructed wetland could initially cause water quality degradation; however, these impacts are temporal and disappear within a short period of time in most cases.

Hydraulic conductivity appeared to be adequate for all substrate mixtures during the course of this study. None of the columns used during the study clogged or showed any indication of not being able to maintain flow through the systems.

CONCLUSIONS

Locally available organic waste products have shown promise in the laboratory for use as substrates in wetlands to treat acid runoff. The chicken litter, rice hulls, sawdust, and composted chicken litter are available in Arkansas, and all performed similarly in the simulated wetlands columns. Local materials performed as well as mushroom compost to raise pH, reduce sulfate concentration, and reduce iron concentration.

Concerns regarding effluent BOD_5 and NH_4-N concentrations are valid; however, after approximately 4 to 5 months of operation, the BOD_5 and NH_4-N concentrations in the effluent are below 10 mg/L and 4 mg/L, respectively.

Table 3. Summary of Data Collected During Column Tests of AMD Influent and Treated Effluents

	Parameters						
	PH (su)	Acidity (meq/L)	ANC (meq/L)	Sulfate (mg/L)	Calcium (mg/L)	BOD_5 (mg/L)	NH_4-N (mg/L)
	3.0–3.7[a]	2.0–4.2 (3.2)[c]	—	141–122 (184)	1.6–2.6 (1.9)	<3–28 (7)	0.12–0.38 (0.24)
Acid runoff	16[b]	15	—	15	15	14	7
Column							
A Mushroom	6.5–7.8	—	4.4–9.7 (6.5)	49–191 (112)	113–188 (143)	<3–143 (43)	1.1–7.5 (3.2)
compost	16	—	14	14	15	15	7
C Mush. comp.	6.6–8.0	—	4.9–10.2 (6.6)	49–133 (97)	105–160 (125)	3.0–236 (60)	0.35–9.5 (2.3)
rice hulls	16		14	14	14	15	7
D Chicken litter/	6.7–7.7	—	1.2–9.1 (4.1)	7.5–194 (92)	10–78 (55.9)	<3–248 (56)	0.55–5.0 (1.6)
sawdust	17		14	14	14	15	7
F Composted	7.0–7.4	—	2.6–7.6 (4.7)	3.6–154 (81)	3.3–88 (47.6)	22.8–427 (72)	8.5–20 (12.2)
chicken litter/	16		14	14	14	14	7
Rice hulls	7.2–7.9	—	5–38 (11)	15–95 (53)	11–44 (22.8)	<3–430 (65)	0.54–10 (2.9)
G Mush. comp./	16		14	13	14	15	7
comp. chicken	6.7–7.8	—	3.9–11 (6.8)	43–162 (91)	28–95 (62.2)	<3–204 (42)	1.8–3.5 (2.7)
litter/rice hulls	16		14	14	14	15	7

[a] Range.
[b] Number of samples.
[c] Value in parentheses is mean.

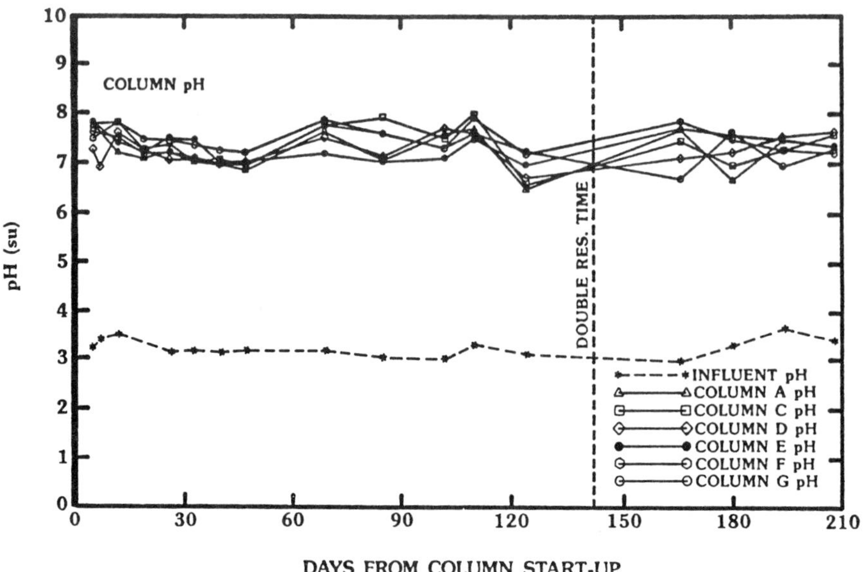

Figure 2. Column influent and effluent pH.

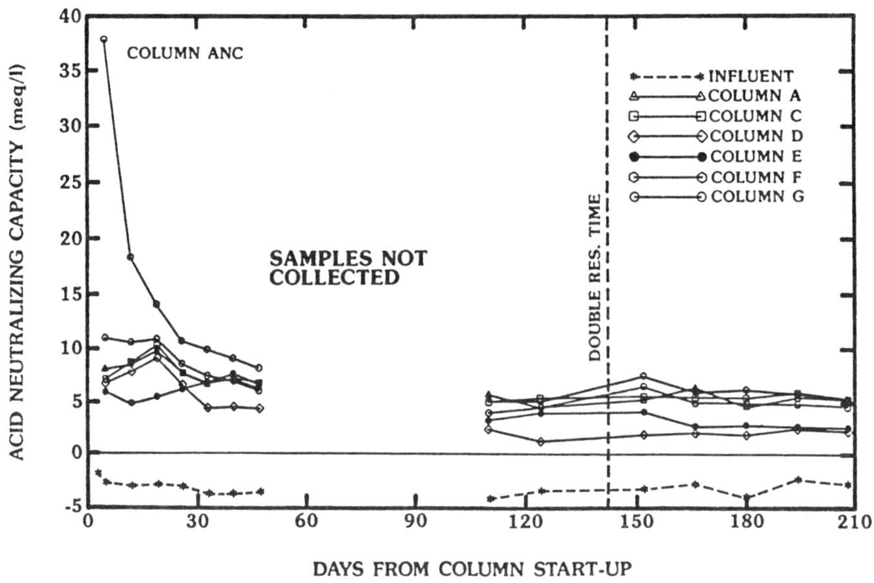

Figure 3. Column influent and effluent acid neutralizing capacity (ANC).

REFERENCES

1. McIntire, P. E., H. M. Edenborn, and R. W. Hammock. Incorporation of bacterial sulfate reduction into constructed wetlands for the treatment of acid and metal mine drainage, in *Proc. Natl. Symp. on Mining*. D. H. Graves, Ed. Knoxville, TN, 1990, 207–213.
2. Postgate, J. R. *The Sulphate-Reducing Bacteria*. 2nd ed. Cambridge University Press, Cambridge, 1984.
3. Hedin, R. S. Treatment of acid coal drainage with constructed wetlands, presented at *10th Annu. Conf. Assoc. Abandoned Mine Land Programs*. Wilkes-Barre/Pocanos, PA, October, 1988, 10–14.
4. Machemer, S. D., P. R. Lemke, T. R. Wildeman, R. R. Cohen, R. W. Klusman, J. C. Emerick, and E. R. Bates. Passive treatment of metals mine drainage through use of a constructed wetland, in *Proc. 16th Annu. RREL Hazardous Waste Res. Symp*. EPA Doc. No. EPA/600/9-90037, 1990, 104–114.
5. Kepler, D. and E. C. McCleary. Personal communication. Clarion, PA, EADS Group, 1990.
6. Tuttle, J. H., P. R. Dugan, and C. Randles. Microbial sulfate reduction and its potential utility as an acid mine water pollution abatement procedure, *Appl. Microbiol.* 17(2):297, 1969.
7. Gandy, L. C., S. J. Formica, and M. A. Gross. An evaluation of vertical flow sulfate reducing wetlands to treat low pH, low sulfate acid mine drainage using column experiments, in *Proc. 1991 Natl. Symp. Mining*. D. H. Graves, Ed. OES, University of Kentucky, Lexington, 1991, 81–93.
8. American Public Health Association (APHA). *Standard Methods for the Examination of Water and Wastewater,* 17th ed. Washington, 1989.
9. U.S. EPA. *Methods for Chemical Analysis of Water and Wastes*. EPA 625/6-74-003a, Cincinnati, OH, 1983.

CHAPTER 19

Contaminant Removal Capabilities of Wetlands Constructed to Treat Coal Mine Drainage

R. S. Hedin and R. W. Nairn

INTRODUCTION

Over the last 10 years, hundreds of wetlands have been constructed for the treatment of contaminated coal mine drainage. Owing to conflicting reports of the effectiveness of these systems, the technology is controversial. Constructed wetlands are in operation in southern and northern Appalachia that markedly lower concentrations of metal contaminants and discharge water that satisfies regulatory effluent criteria.[1-3] Wieder's analysis of influent and effluent contaminant concentrations at 146 constructed wetlands, however, indicated no strong statistical relationships between changes in contaminant concentrations and measures of water quality or system design. This finding led Weider to question the feasibility of the technology.[4] His conclusions have had strong impact on the regulatory treatment of constructed wetland systems, which has generally been cautious and noncommittal.[5]

In this article, we report measurements of contaminant removal by 11 wetlands constructed to treat coal mine drainage in western Pennsylvania. In vivid contrast to Wieder's conclusion that wetland performance is highly variable and unpredictable, we find that the removal of contaminants occurs in a manner consistent with well-known chemical and biological processes. The findings differ from those of Wieder's, in part, because of different methodologies. First, the effect of dilution on contaminant concentrations has been arithmetically removed, allowing the determination of the effect of wetland processes. Second, contaminant removal is evaluated from a rate perspective. Third, sites are considered only when influent contaminant concentrations are high enough to ensure that contaminant removal rates are not limited by the absence of the contaminant. Our approach is applicable to studies of wetlands constructed to treat any water pollutant.

EXPERIMENTAL APPROACH

Contaminant concentrations decrease as water flows through constructed wetlands because of two distinct processes: (1) contaminants are removed from solution by chemical and biological mechanisms, and (2) contaminant concentrations are diluted by inputs of uncontaminated water. Biological and chemical processes should respond to environmental conditions in a manner that is predictable and consistent between sites and sampling occasions. Dilution, however, will vary depending on weather conditions and engineering efforts to include or exclude uncontaminated water from the wetlands. In order to recognize the effects of biological and chemical processes on contaminant concentrations in constructed wetlands, first it is necessary to remove the effects of dilution.

In this study, a *dilution factor* was calculated from differences in concentrations of magnesium (Mg) between influent and effluent water samples.[6] The dilution factor was then used to separate changes in contaminant concentrations due to dilution from those due to biological and chemical processes. Magnesium was considered a good indicator of dilution in these systems for several reasons. In northern Appalachia, concentrations of magnesium in coal mine drainages are generally greater than 50 mg/L, while concentrations in rainfall, surface runoff, and shallow groundwater are less than 5 mg/L. Magnesium appears to be chemically and biologically conservative in constructed wetlands. $MgSO_4$ does not precipitate in wetlands because the compound is highly soluble (710,000 mg/L).[7] $MgCO_3$ generally forms only under evaporative conditions.[8]

0-87371-550-0/93/$0.00+$.50

Processes that remove Mg from coal mine drainage are negligible relative to the high Mg loadings that most constructed wetland sites in northern Appalachia receive. The average Mg loading of sites described in this study is 7000 $g/m^2/year$. Uptake of dissolved Mg by plants likely accounts for only 10 $g/m^2/year$, while sorption of Mg by cation exchange processes may account for a total of 50 to 100 g/m^2.

The second important aspect of this study is the evaluation of wetland performance from a rate perspective. Since the success or failure of a constructed wetland is based typically on the concentrations of contaminants contained in the effluent water, it is tempting to evaluate the performance of constructed wetlands from this perspective alone. Evaluations of the performance of different wetlands based on concentration data, however, cannot be legitimately compared. Only rates of contaminant removal can be compared between systems.[9]

Contaminant removal rates are ideally calculated by comparing contaminant loads at wetland influent and effluent locations and adjusting for wetland size. Accurate flow rates are, in practice, rarely available for all sampling locations. An alternative rate calculation method uses a single flow rate measurement, the dilution-adjusted changes in contaminant concentrations, and the wetland size. At a site where flow rates and water chemistry are known for all sampling points, estimates of contaminant removal derived by the dilution-adjustment method correlate closely with estimates derived from comparisons of measured contaminant loadings at each sampling point.[10]

A third feature of this study is our exclusion of sites that consistently lower contaminant concentrations to low levels (<2 mg/L). At completely effective sites, low rates of contaminant removal may result from contaminant loading limitations. For the purpose of determining contaminant removal capabilities of constructed wetlands, the most information is obtained from sites that are undersized with respect to contaminant loadings.

METHODS

Surface water samples were collected in 250-ml plastic bottles from the influent and effluent of each wetlands system. Both raw and acidified (2 ml conc. HCl) samples were collected. The pH of the raw sample was measured in the field with a calibrated Orion SA 270 portable pH/ISE meter. Field alkalinity was measured with the pH meter and the Orion Total Alkalinity test kit (Orion Research, Inc.). Concentrations of metals for the acidified samples were determined using inductively coupled argon plasma spectroscopy (Instrumentation Laboratory Plasma 100 model). Sulfate concentrations were determined for the unacidified samples by barium chloride titration using thorin as an indicator.[11] Net acidity was determined by boiling a 50-ml raw sample with 1 ml of 30% H_2O_2 and then either titrating acidic solutions with 0.1 *N* NaOH to pH 8.3, or titrating alkaline solutions (negative net acidity) with 0.2 *N* H_2SO_4 to pH 4.8.[10] Acidity and alkalinity concentrations are reported in milligrams per liter (mg/L) $CaCO_3$ equivalent. Cation and sulfate concentrations are reported in milligrams per liter (mg/L).

Flow rates were usually determined using a bucket and a stopwatch. Where flow rates were too high to measure with a bucket but the water flowed through a pipe of known diameter, slope, and composition, the depth of water in the pipe was measured and the flow rate was determined using the Manning formula.[11]

Contaminant Removal Rate Calculations

For each set of water samples from a constructed wetland, a dilution factor (DF) was calculated from changes in concentrations of magnesium between the influent and effluent station:

$$DF = Mg_{out}/Mg_{in} \quad (19\text{–}1)$$

Effluent contaminant concentrations were then adjusted upward by dividing the measured concentration (in mg/L) by the DF. Dilution-adjusted changes in concentrations of contaminants were calculated by subtracting the dilution-adjusted effluent concentrations from the measured influent concentrations. For example, the dilution-adjusted removal of iron (ΔFe_{DA}) was calculated:

$$\Delta Fe_{DA} = Fe_{in} - (Fe_{out}/DF) \quad (19\text{–}2)$$

Rates of contaminant removal, expressed as grams per square meter per ($g/m^2/day$), were then calculated from the dilution-adjusted removal, a flow rate measurement (L/min), and the surface area (SA) of the system (m^2);

$$Fe_{rem} = (\Delta Fe_{DA} \times Flow_{in} \times 1.44) / SA \tag{19–3}$$

where 1.44 converts minutes to days and milligrams to grams.

RESULTS AND DISCUSSION

The design characteristics of 11 constructed wetlands are shown in Table 1. The substrates used in the construction of the wetlands varied between mineral substances, such as clay and limestone rocks, to organic substances such as spent mushroom compost, manure, and hay bales. Cattails (*Typha latifolia* and, less commonly, *T. angustifolia*) were the most common emergent plants growing in the systems. Three sites contained few emergent plants. The sizes of the systems ranged from 643 to 8100 m^2. Most of the systems consisted of several cells or ponds connected serially. Three systems consisted of single long ditches.

The chemistry of the influent water at the sites varied widely (Table 2). Seven sites received water with circumneutral pH (6 to 8) and detectable concentrations of alkalinity. The contaminants at these alkaline mine water sites were iron and manganese.

At one site (Emlenton), low levels of alkalinity (<10 mg/L) were detected in the influent water, but the mineral acidity far exceeded the neutralizing capacity of this alkalinity. This site received water with a high net acidity, as did all other sites where the influent water contained no alkalinity. At these acid mine water sites, the drainage had low pH and was contaminated with iron, manganese, aluminum.

The mean influent flow rates of mine drainage at the study sites ranged from 8 to 7900 L/min (Table 1). The highest flow rates occurred where drainage discharged from abandoned and flooded underground mines. The lowest flow rates occurred at surface mining sites. Estimated average retention times ranged from 7 hours to more than 4 days (Table 1).

The average DF values for the 11 constructed wetlands are shown in Table 3. At 6 of the 11 sites, DF values ranged from 0.95 to 0.99 and dilution adjustments were minor. At 5 sites, DF values were less than 0.95. Sites where dilution effects were significant were characterized by low flow rates. When samples were collected from these sites during rainstorms, DFs were as low as 0.25, and adjustments of contaminant concentrations to reflect the effect of dilution were therefore as high as 75%. On those occasions where dilution was indicated, the calculation of removal rates without dilution-adjustment procedures would have overestimated actual contaminant removal.

Removal of Metals from Alkaline Mine Water

The removal of iron and manganese from mine drainage at the study sites (Table 3) was strongly related to the influent pH (Figures 1 and 2). At the Cedar, Keystone, and Morrison sites, where influent water had circumneutral pH and concentrations of alkalinity greater than 100 mg/L, iron was removed at rates of 21 to 24 g/m^2/day. The Blair and Donegal wetlands received alkaline water contaminated with iron, but a removal rate was not calculated because complete removal of iron occurred in an undetermined wetland area. The three sites for which iron removal rates were calculated each received water with markedly different iron concentrations and flow rates. These sites were also designed quite differently. The Cedar system is a shallow cattail wetland that received an average of 568 L/min of water contaminated with 95 mg/L Fe. The Keystone system is a deep-water ditch that received 7900 L/min water contaminated with 37 mg/L Fe. The Morrison system is a shallow-water ditch that received 10 L/min of water contaminated with 179 mg/L Fe.

Each of the three sites receiving alkaline, iron-contaminated mine water developed thick accumulations of iron oxyhydroxides (FeOOH). None contained an organic substrate to promote anoxic conditions and anaerobic biological or chemical processes. Removal of iron at these sites likely occurred primarily by oxidation and hydrolysis processes:

$$Fe^{2+} + {}^1/_4\, O_2 + H^+ \rightarrow Fe^{3+} + {}^1/_2\, H_2O \tag{19–4}$$

$$Fe^{3+} + 2H_2O \rightarrow FeOOH + 3H^+ \tag{19–5}$$

The proton acidity produced by iron hydrolysis (Equation 5) was neutralized at these sites by bicarbonate alkalinity,

Table 1. Construction Characteristics of the Study Systems

Site	Year	Design	Substrate[a]	Emergent vegetation	SA[b] (m^2)	Water depth (cm)	Flow rate (L/min)	Est. ret.[c] (day)
Morrison	1990	Ditch	Clay	None	70	1	10	0.5
Keystone	1989	Ditch	Topsoil	None	4200	100	7900	0.4
Cedar	1989	5 cells	Clay, LS	*Typha*	1360	15	568	0.3
Blair	?	Ditch	Manure	Mixed	1080	5	8	4.7
Shade	1989	2 cells	LS	None	880	10	21	2.9
Donegal	1987	8 cells	SMC, LS	*Typha*	8100	15	412	2.1
Piney	1987	1 cell	HB	Mixed	2500	50	388	1.3
Emlenton	1987	9 cells	Manure, LS	*Typha*	643	50	61	3.7
Somerset	1984	2 cells	SMC, LS	*Typha*	1005	15	27	3.9
Latrobe	1987	3 cells	SMC, LS	*Typha*	2800	15	86	2.9
FH	1988	6 cells	SMC, LS	*Typha*	667	15	15	4.6

[a] Substrate abbreviations are: LS, limestone; SMC, spent mushroom compost; and HB, haybales.

[b] SA is the surface area of the wet area. Water depths and flow rates are average values.

[c] Est. ret. is the estimated retention time, which was calculated from the water holding capacity and influent flow rate.

Table 2. Average Chemical Characteristics of Influent Water at the Study Sites

Site	n	pH	Fe	Mn	Al	Mg	SO_4	Alk	Net acid
Morrison	9	6.2	179	49	<1	115	1247	245	96
Keystone	13	6.5	37	<1	<1	14	331	106	–72
Cedar	9	6.3	95	2	<1	54	1260	300	–140
Blair	8	6.1	51	28	<1	68	493	163	–51
Shade	6	6.5	<1	34	<1	171	1258	95	–17
Donegal	12	7.7	5	8	<1	81	731	200	–182
Piney	16	6.1	<1	16	<1	221	1834	35	–6
Emlenton	23	4.5	96	85	9	272	2513	5	342
Somerset	24	3.7	202	59	3	228	2032	0	467
Latrobe	39	3.5	125	32	43	125	1655	0	617
FH	50	2.7	169	9	57	77	1688	0	956

Note: All anion and cation concentrations are in mg/L except pH (standard units). Alkalinity is mg/L $CaCO_3$ equivalent. Net acid is hot acidity. Negative net acidities indicate water with a net alkalinity. Flow rates are L/min.

$$HCO_3^- + H^+ \rightarrow H_2O + CO_2 \tag{19–6}$$

This buffering reaction is important in mine drainage treatment systems because it maintains circumneutral pH. The pH of the water influences both the mechanism and the kinetics of iron oxidation processes. For aerobic waters with pH values greater than 4, the removal of iron is limited by the oxidation process (Equation 4). At pH values greater than 6, abiotic oxidation processes dominate over bacterial oxidation processes, while at pH values below 5 the relationship is reversed.[13] In natural systems, the kinetics of the abiotic mechanism under circumneutral conditions can be 5 to 10 times faster than the biological mechanism at lower pH.

Significant removal of manganese occurred at four sites that received alkaline water; the rate ranged from 0.5 to 1.0 g Mn/m^2/day. Manganese removal rates at these sites were similar despite very different system designs. The Donegal wetland has a thick organic and limestone substrate and is densely vegetated with cattails. The Blair wetland was constructed with a manure substrate and is densely covered with volunteer vegetation. The Piney wetland was not constructed with an organic substrate and consists of a mosaic of deep open-water areas and shallow vegetated areas. The Shade treatment system consists of limestone rocks, no organic substrate, and few emergent plants.

The removal of manganese at sites with no organic substrate, coupled with the observation of black manganese oxide precipitates, indicated that removal likely resulted from oxidation and hydrolysis processes.

Table 3. Average Dilution Factors and Removal Rates for Manganese and Iron at the Study Sites

Site	Dilution factor	Removal rates Mn	Removal rates Fe
Morrison	0.60	0.1[a]	17.5[b]
Keystone	0.99	NA[c]	21.3[b]
Cedar	0.98	0.0[a]	24.2[b]
Blair	0.83	0.5[b]	NA[d]
Shade	0.94	0.5[b]	NA[c]
Donegal	0.99	0.5[b]	NA[d]
Piney	0.99	1.0[b]	NA[c]
Emlenton	0.94	−0.1[a]	10.4[b]
Somerset	0.75	0.1[b]	4.8[b]
Latrobe	0.97	0.0[a]	2.1[b]
FH	0.96	0.0[a]	1.4[b]

[a] Removal rate was not significantly greater than zero.

[b] The removal rate was significantly greater than zero (t test, $p < 0.05$).

[c] No rate is available because influent contaminant concentrations were <2 mg/L.

[d] No rate is available because complete iron removal occurs within an undefined wetland surface area.

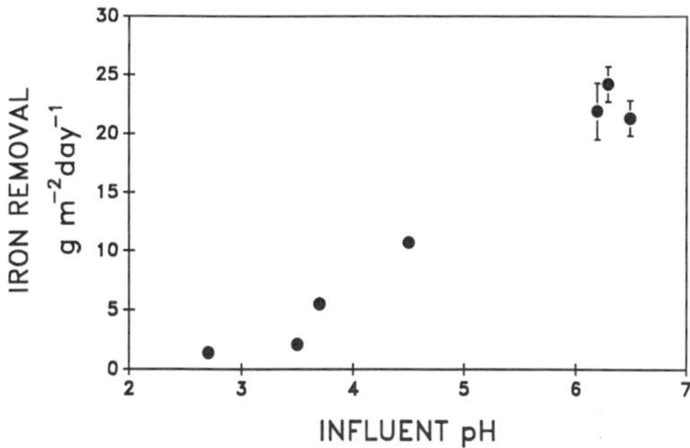

Figure 1. The relationship between influent pH and rates of iron removal at the study wetlands. Vertical bars indicate ± one standard error; where no bar is shown, it is less than the size of the symbol.

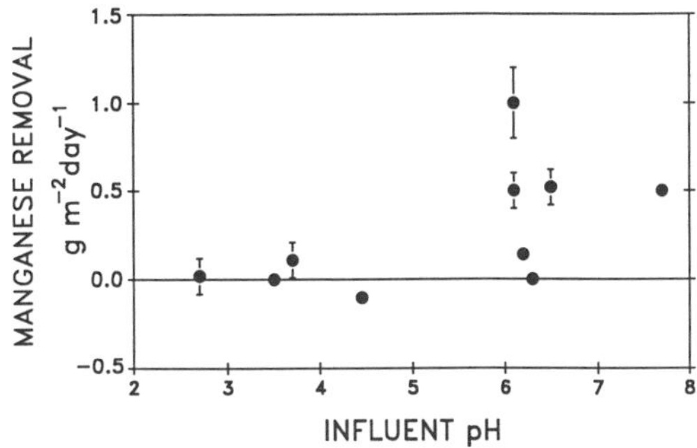

Figure 2. The relationship between influent pH and rates of manganese removal at the study wetlands. Vertical bars indicate ± one standard error; where no bar is shown, it is less than the size of the symbol.

$$Mn^{2+} + {}^{1}/_{2}\,O_2 + 2H^+ \rightarrow Mn^{4+} + H_2O \quad (19\text{–}7)$$

$$Mn^{4+} + 2H_2O \rightarrow MnO_2 + 4H^+ \quad (19\text{–}8)$$

Because the hydrolysis step (Equation 8) is extremely rapid, the oxidation step (Equation 7) is rate limiting.[14] Whether manganese removal at these sites is abiotic or is catalyzed by biological activity is unknown.

Although the processes that remove manganese and iron are mechanistically similar (both involve oxidation and hydrolysis reactions), the observed kinetics of the metal removal processes were quite different. Manganese removal rates were 20 to 40 times slower than iron removal rates for alkaline mine waters.

Coprecipitation of manganese with iron was not indicated at any site. Instead, the opposite relationship was observed; no removal of manganese occurred when dissolved iron was detected in the water, even if the water was highly alkaline (see the Morrison and Cedar Grove systems, Table 3). At the Blair and Donegal wetlands, where the raw mine water contained both dissolved iron and manganese, removal of manganese was not observed until after ferrous iron concentrations had decreased to less than 1 mg/L. These observations likely resulted from the reduction and resolubilization of manganese oxides by ferrous iron.

$$MnO_2 + 2Fe^{2+} + 2H_2O \rightarrow 2FeOOH + Mn^{2+} + 2H^+ \quad (19\text{–}9)$$

Dissolved manganese (Mn^{2+}) can reportedly be removed from solution by its sorption to charged FeOOH particles.[15] At the Morrison and Cedar sites, mine water within the treatment systems contained copious amounts of suspended FeOOH. After concentrations were adjusted to reflect dilution, no changes in manganese concentrations were detected at either site.

The presence or absence of emergent plants in the wetlands did not have a significant effect on rates of either iron or manganese removal at the alkaline mine water sites (see overlap of standard error bars in Figures 1 and 2). Bioaccumulation of metals in plant biomass is an insignificant component of iron and manganese removal in constructed wetlands.[15] The ability of emergent plants to oxygenate sediments and the water column[16,17] has been proposed as an important indirect plant function in wetlands constructed to treat polluted water.[18] Since the oxygen demand of metal removal processes is high in wetlands that receive alkaline mine water, the presence of plants was expected to increase metal removal rates. Either oxygenation of the water column is not a rate-limiting aspect of metal oxidation at the constructed wetlands or physical oxygen transfer processes that are promoted by waterfalls and sheet flow are more rapid than biological processes. Further research needs to be conducted to evaluate the role of emergent plants in these wetlands.

Removal of Contaminants from Acid Mine Drainage

Contaminant removal at sites receiving alkaline mine water is unambiguously evaluated through calculation of iron and manganese removal rates because no undesirable dissolved by-products are associated with the removal of these metals in a buffered chemical environment. In unbuffered waters, however, the oxidation and hydrolysis of iron is accompanied by a decrease in pH (Equation 5). Significant removal of iron can occur with little decrease in the net acidity of the water.

When mine water is acidic, wetland performance is best evaluated by calculation of acidity removal rates. Acidity, in this case, is the sum of mineral and proton acidity and represents the acidic characteristics of free hydrogen ion, and dissolved iron, manganese, and aluminum. Acidity is neutralized in acidic systems through the generation of alkalinity. Two common alkalinity-generating processes in aquatic systems are carbonate dissolution (Equation 10) and sulfate reduction (Equation 11):

$$CaCO_3 + H^+ \rightarrow Ca^{2+} + HCO_3^- \quad (19\text{–}10)$$

$$2CH_2O + SO_4^{2-} \rightarrow H_2S + 2HCO_3^- \quad (19\text{–}11)$$

where CH_2O represents organic matter. Both processes require the presence of reducing conditions that inhibit carbonate armoring by ferric hydroxides and promote the activity of anaerobic sulfate-reducing bacteria.

The rates of alkalinity generated from these processes in the constructed wetlands were estimated from dilution-adjusted changes in concentrations of dissolved calcium and sulfate, and stoichiometric adjustments that result in grams per square meter per day $CaCO_3$. These calculations assume that calcium concentrations

Table 4. Removal of Acidity and Generation of Alkalinity at Four Constructed Wetlands

Site	Acidity removal	Alkalinity generation	
		SR	$CaCO_3$
Emlenton	5.0[a]	3.3[b]	0.0[b]
Somerset	10.3[a]	6.3[a]	5.5[a]
Latrobe	6.7[a]	4.9[a]	2.3[a]
FH	7.2[a]	4.3[a]	2.5[a]

Note: SR = sulfate reduction; $CaCO_3$ = calcium carbonate.

[a] The removal rate was significantly greater than zero (*t* test, $p < 0.05$).
[b] Removal rate was not significantly greater than zero.

are only affected by carbonate dissolution and that sulfate concentrations are only affected by bacterial sulfate reduction. We currently consider these assumptions to be reasonable. Analyses of the water chemistry at the study sites using the WATEQ chemical equilibrium model have never indicated that any calcium-containing compounds should precipitate; even when calcium concentrations were artificially increased by 50 to 100 mg/L. The WATEQ model does indicate that sulfate-containing ferric oxyhydroxide minerals should precipitate, and the formation of these minerals has been reported for acid mine drainage environments.[19] Chemical analyses of iron precipitates collected from the constructed wetlands indicate that sulfate is incorporated into the precipitates collected from acidic environments at an average SO_4:Fe ratio of 0.11 (n = 5, sd = 0.02).[20] This incorporation of sulfate into ferric hydroxides could account for account for changes in sulfate concentrations of 10 to 15 mg/L at the acid mine water sites. Dilution-adjusted changes in sulfate concentrations at several of the acidic study sites were commonly 200 to 500 mg/L.

Rates of acidity removal and alkalinity generation by sulfate reduction and calcium carbonate dissolution were calculated for four constructed wetlands that received acidic mine water (Table 4). Significant removal of acidity occurred at all sites. The lowest rate of acidity removal occurred at the Emlenton wetland. This site consisted of cattails growing in a manure and limestone substrate (Table 1). No dissolution of the limestone was indicated and the sulfate reduction rate was highly variable.

The Somerset, Latrobe, and FH wetlands were each constructed with cattails growing in a spent mushroom compost and limestone substrate. Spent mushroom compost is a good substrate for microbial growth and also contains a high limestone content (10% dry weight). At these three wetlands, sulfate reduction and limestone dissolution both occurred at significant rates (Table 4). The summed amount of alkalinity generated by sulfate reduction and limestone dissolution processes correlated strongly with the measured rate of acidity removal at these sites ($r > 0.90$ at each site). At the FH wetland, where detailed studies of alkalinity generation have been conducted, 94% of the measured acidity removal could be explained by these two processes (Figure 3).[9]

Bacterial sulfate reduction was the dominant source of alkalinity at each of the wetlands. At the wetlands constructed with mushroom compost, sulfate reduction accounted for an average of 62% of the total alkalinity generated. The highest limestone dissolution rate occurred at the Somerset wetland, where the compost was supplemented with additional limestone before it was placed in the wetland.

SUMMARY

A method has been developed for the evaluation of the contaminant removal capabilities of wetlands constructed to treat coal mine drainage. Dilution effects are removed; contaminant removal is evaluated from a rate perspective, and sites which have low contaminant loading rates are avoided. When constructed wetlands were viewed in this manner, distinctive patterns of contaminant removal emerge. The alkalinity of the influent mine water is a major determinant of the kinetics of contaminant removal. Iron removal rates are highest under alkaline conditions; net removal of manganese only occurs under alkaline conditions. In this study, the presence of emergent plants in the treatment systems did not significantly affect either iron or manganese removal in alkaline systems. When influent mine water is acidic, contaminant removal is best evaluated by determining rates of acidity removal. Acidity is removed from mine water via bacterial sulfate reduction processes and carbonate dissolution. Both processes are stimulated by the use of organic substrates

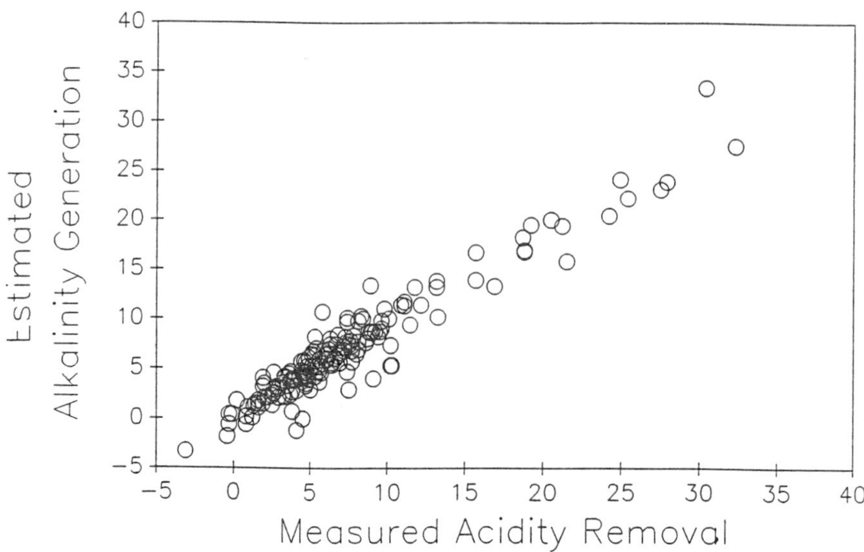

Figure 3. The relationship between measured rates of acidity removal and calculated rates of alkalinity generated from sulfate reduction and calcium carbonate dissolution at the FH constructed wetland. Units for both axes are g/m^2/day $CaCO_3$. The equation relating the parameters is: $Y = 0.91x + 0.33$, $r^2 = 0.92$, $n = 178$.

that have high limestone content. The contaminant removal rates reported are useful sizing criteria for the future construction of wetlands for the treatment of coal mine drainage.

REFERENCES

1. Brodie, G. A. Staged, aerobic constructed wetlands: Impoundment 4 case history and overview of the Tennessee Valley Authority acid drainage treatment wetland program.
2. Stark, L., E. Stevens, H. Webster, and W. Wenerick. Iron loading, efficiency, and sizing in a constructed wetland receiving mine drainage, in *Proc. Mining and Reclamation Conf. and Exhibition, Vol. 2.* West Virginia University, Morgantown, WV, 1990, 393.
3. Stilling, L. L., J. J. Gryta, and T. A. Ronning. Iron and manganese removal in a *Typha*-dominated wetland during ten months following its construction, in *Proc. Mine Drainage and Surface Mine Reclamation, Vol. I: Mine Water and Mine Waste.* U.S. BuMines IC 9183, 1988, 317.
4. Wieder, R. K. A survey of constructed wetlands for acid coal mine drainage treatment in the eastern United States, *Wetlands*, 9, 1989, 229.
5. Whitehouse, A. E. and H. V. Weaver, Jr. Office of Surface Mining Reclamation and Enforcement wetland policy update, in *Proc. 12th Annu. Abandoned Mined Land Conf.* Office of Surface Mining Reclamation and Enforcement, Breckenridge, CO, 1990, 345.
6. Hedin, R. S. The use of magnesium as a conservative tracer in studies of wetlands constructed to treat coal mine drainage, manuscript in preparation.
7. Weast, R. C. *Handbook of Chemistry and Physics.* CRC Press, Boca Raton, FL, 1968.
8. Krauskodf, B. K. *Introduction to Geochemistry.* McGraw-Hill, New York, 1967.
9. Hedin, R. S. and R. W. Nairn. Sizing and performance of constructed wetlands: case studies, in *Proc. Mining and Reclamation Conf. and Exhibition.* Charleston, WV, 1990, 385.
10. Hedin, R. S., D. H. Dvorak, S. L. Gustafson, D. M. Hyman, P. E. McIntire, R. W. Nairn, R. C. Neupert, A. C. Woods, and H. M. Edenborn. Use of a constructed wetland for the treatment of acid mine drainage at the Friendship Hill National Historic Site, Fayette County, PA, Technical Report to the National Park Service, 1991.
11. American Public Health Association. *Standard Methods for the Examination of Water and Wastewater.* American Public Health Association, Washington, D.C., 1985, 1268.

12. Grant, D. M. *Open Channel Flow Measurement Handbook*. Isco, Lincoln, NB, 1988, 227.
13. Singer, P. and W. Stumm. The oxygenation of ferrous iron, in *Water Polln. Cont. Res. Series 14010—06/69*, U.S. Dept. of the Interior, Federal Water Quality Admin., 1970.
14. Stumm, W. and J. J. Morgan. *Aquatic Chemistry*. 2nd ed. Wiley-Interscience, New York, 1981.
15. Sencindiver, J. C. and D. K. Bhumbla. Effects of cattails (*Typha*) on metal removal from mine drainage, in *Proc. Mine Drainage and Surface Mine Reclamation, Volume I: Mine Water and Mine Waste*. U.S. BuMines IC 9183, 1988, 359.
16. Armstrong, W. Aeration in higher plants, *Adv. Bot. Res.* 7:25, 1980.
17. Michaud, S. C. and C. J. Richardson. Relative radial oxygen loss in five wetland plants, in *Constructed Wetlands for Wastewater Treatment*. D. Hammer, Ed. Lewis Publishers, Chelsea, MI, 1989, 501.
18. Hammer, D. A. and R. K. Bastian. Wetlands ecosystems: natural water purifiers? in *Constructed Wetlands for Wastewater Treatment*. D. Hammer, Ed. Lewis Publishers, Chelsea, MI, 1989, 5.
19. Bigham, J. M., U. Schwertmann, L. Carlson, and E. Murad. A poorly crystallized oxyhydroxysulfate of iron formed by bacterial oxidation of Fe(II) in acid mine waters, *Geochim. Cosmochim. Acta.* 54:2743, 1990.
20. Hedin, R. S. Constructing wetlands for the treatment of acid mine drainage, Bureau of Mines Report, in review.

CHAPTER 20

A Peat/Wetland Treatment Approach to Acidic Mine Drainage Abatement

T. M. Frostman

INTRODUCTION

Research into the use of peat/wetland treatment for heavy metal removal from acid mine drainage (AMD) has increased over the past 10 years. Many of these projects incorporate a peat/wetland treatment system as an integral part of a reclamation and/or restoration operation. The majority of the work has concentrated on coal AMD, which has very low pH with high metal content. AMD from iron mining operations associated with Duluth Complex Gabbro generally has a moderate pH (5 to 6) with relatively low metal concentrations. These conditions are conducive to metal removal utilizing a peat/wetland system. The peat/wetland project addressed here is for treatment of AMD at an industrial site in the Upper Midwest.

DISCUSSION

A peat/wetland demonstration project was designed and constructed in 1988 to determine the ability of various wetland treatment configurations to remove heavy metals.[1] STS Consultants, Ltd. designated a demonstration plot and treatment system for a research project conducted by the Minnesota Department of Natural Resources (DNR). The design of the four test cells incorporated various flow distribution methods and vegetative types. The site selection process included surveys to determine peat availability and the wastewater composition based on chemical and physical analysis of pH, metal and nutrient concentration, and cation exchange capacity. The treatment efficiency of each cell and system life was evaluated to determine the type of application and/or design for a full-scale treatment wetlands. The operation was designed for use as a surface water (SW) treatment system.

The wetland treatment cells were constructed using peat with a low degree of decomposition. These types of fibrous peats have good hydraulic conductivity, allowing a uniform distribution of seepage water within the peat. One cell allowed for development of anaerobic conditions to encourage sulfate reduction for metal removal. Flow regimens within each cell were varied to evaluate the efficiency of removal with different flow regimens.

The wetland area used for the project was characterized by botanical description, measure of degree of decomposition (Von Post scale), pH, percent ash, metal content, nutrient concentrations, and cation exchange capacity. The criteria for selection of the demonstration plot location included low metals concentration and the high percentage of undecomposed fibrous peat within the top 8 in. of the material.[2]

Four wetland cells were designed to have a surface area of 2000 ft^2 each. The cells were separated by compacted peat berms approximately 3 ft high and 10 ft wide. The input into the demonstration cells was collected near the toe of a rock stockpile and transported to the plots through a gravity flow pipe system. Flow was distributed to each cell through a distribution trough that extends across the surface.[3] Influent flow to each cell was regulated by a valve to maintain the desired residence time of approximately 24 or 48 h. Cell effluent was collected in an effluent chamber and discharged to the surrounding environment. Influent and outflow volumes were recorded. Water samples were collected and analyzed in accordance with standard methods for copper, nickel, cobalt, zinc, calcium, magnesium, sodium, potassium, manganese, iron, sulfates, alkalinity, acidity, specific conductance, and pH.

0-87371-550-0/93/$0.00+$.50

Evaluation of the 1989, 1990, and 1991 data indicates an efficiency of approximately 90% for nickel removal. Effluent samples generally contained less than 0.2 mg/L of nickel. Over 90% copper and the other heavy metals analyzed were also removed. The chemical analysis data illustrate excellent removal of nickel and copper during the active vegetation growth period in the wetland (May through September). Reduction in the removal efficiency was observed during late fall.[4]

PEAT CHARACTERIZATION

Heavy metal removal efficiency is generally based on the characteristics of the peat such as pH, organic content, fiber characteristics, hydraulic conductivity, flow patterns, water level, residence times, and sulfate and organic availability. There are a number of possible mechanisms for metal retention and/or removal by peat. These include absorption, cation exchange, precipitation by sulfate reduction, and complexation with soil organic matter. In addition, the dynamics of the physical/chemical and biological components of peat/wetland systems utilize removal of suspended materials, biological uptake microbial oxidation, and absorption mechanisms. In anaerobic conditions, metal sulfide formation is the most effective metal retention mechanism because it forms an insoluble precipitant. Sulfide production is limited by the quantity of available sulfates and simple/available organics that provide the materials necessary for sulfate reduction activities.

Imported peat was used to supplement the peat/wetland treatment system to provide a high hydraulic conductivity and available organic content. This provides for more uniform contact between the organic absorption sites on the peat and the seep waters. The sulfate reduction process also requires a high organic content and soluble sulfates. The criteria established for the peat used in the treatment system includes:

- Hydraulic conductivity of greater than 1.3 m/day (5.5 cm/s)
- Fibrous peat with irregular length and split fibers
- Fiber decomposition; little or no fiber disassociation (Von Post H2 or H3)
- Organic content over 90% by dry weight
- Good cation exchange capacity
- pH above 5

Peat samples were examined for fiber content, extent of decomposition, wood content, and physical characteristics. To obtain the most effective peat material, a combination of reed sedge peat mixed with the screenings from a peat processing operation provided the best media to meet the criteria described above. The peat mixture was two parts process screenings and one part reed sedge peat. The two materials are mixed on-site prior to placement in the peat/wetland treatment system. The peat was placed loosely within the wetland to maintain an adequate hydraulic conductivity and minimize compaction of the peat. The peat mixture was placed with the least number of layers and minimal overlap.

PEAT/WETLAND TREATMENT SYSTEM DESIGN

An initial treatment system with limestone and peat mixture beds was designed for metals removal from the stockpile drainage. The peat/wetland treatment system design uses existing cattail marsh wetlands, which are enhanced by flow dispersal and ponding, coupled with a system of dikes for control of water levels. The peat/wetland treatment system designed for a seep is shown in Figure 1. The wetland enhancement will increase the interactions of biological and physiochemical activities to promote polishing of stockpile drainage from the limestone and peat mixture beds. The peat mixture contains reed sedge peat and peat process screenings to provide a medium with a high organic content and high permeability.

Limestone will be used to neutralize the pH of the input drainage water and provide for increased alkalinity and some precipitation of metal hydroxides. The increased alkalinity provides a more compatible water for sulfate reduction processes. Most of the stockpile drainage enters the treatment system with a pH greater than 6. Based on research, limestone removal efficiencies for copper of 50% and for nickel, cobalt, and zinc of 10% are expected.[10] The limestone beds will be located in easily accessible areas to accommodate maintenance or replacement as required.

The peat mixture beds are to be placed downstream of the limestone, with the flow dispersed laterally to increase the contact area and encourage subsurface water flow (SSW). The SSW flow through the peat mixture encourages the development of anaerobic conditions for sulfate reduction activities. The sulfate reduction

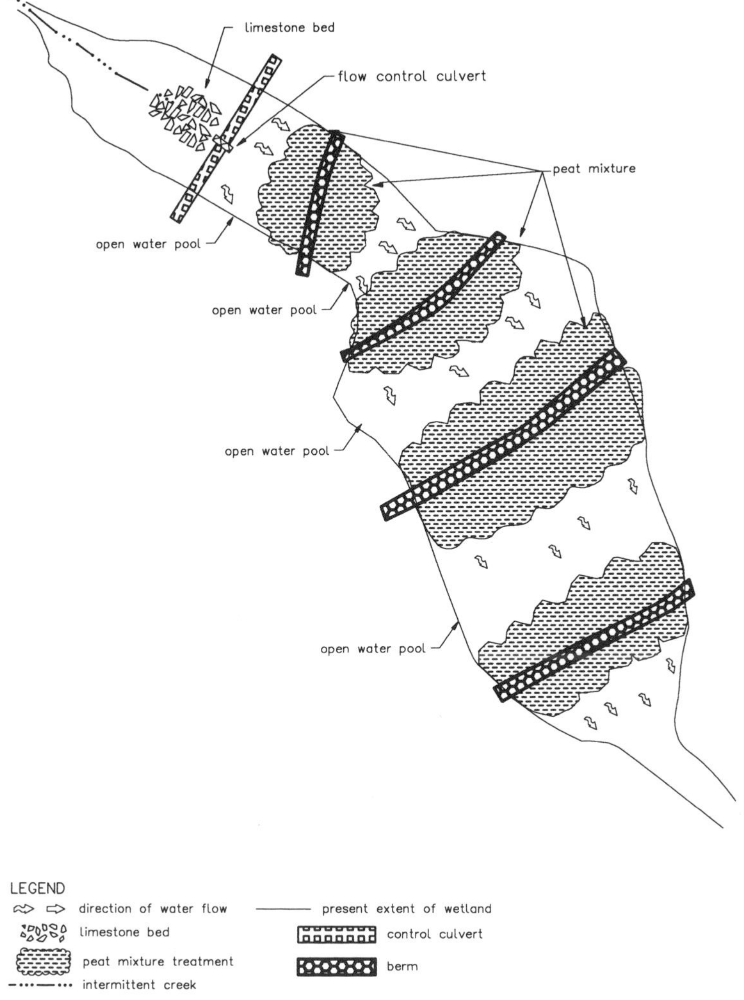

Figure 1. Typical peat/wetland treatment system, designed for heavy metal removal from stockpile drainage.

reactions result in formation of metal sulfide precipitates. A removal efficiency of 90% is anticipated, based on the performance of the wetland demonstration plots.

The peat mixture beds will be placed adjacent to existing berms to allow easy access for replacement of maintenance if required. Each peat bed will be seeded with cattails. The cattail seeds will be soaked in dish soap prior to hydroseeding. Additional limestone and/or peat mixture could be added if expected to perform reliably because of its ability to accommodate and attenuate flow and loading fluctuations in peat/wetland treatment systems.

Water Quality Following Initial Treatment

A series of till berms with peat mixture placed upstream and downstream sides will be constructed on the existing wetlands. Each berm system will form a water pool about 1 to $1^1/_2$ ft in depth. The berms will extend across the entire intermittent stream/wetland valley. The open-water pool is to encourage flow dispersal and movement of the water throughout the peat mixture.

The Peat/Wetland Treatment System is constructed to treat seep flows of 20 gpm. Metal removals are required to obtain water quality standards established by the regulatory agencies as they ultimately enter into the waters of the state. The acid drainage and metals are attributed to Duluth complex rock that has high sulfur and metal content.

The peat/wetland treatment system relies on significant metal removal from the stockpile drainage prior to entering the existing natural wetlands. Existing wetlands will be restored or enhanced to provide polishing of stockpile drainage flowing from the peat mixture treatment areas.

Enhancement will begin with an excavated lateral ditch. The ditch will be fanned and sloped to provide dispersal of drainage water across the entire ditch system. The ditch is designed to fill with water and eventually overflow across much of the existing wetlands. The existing stream channel will be blocked with the peat mixture to disperse flow and minimize short-circuiting. Several peat mixture channel blocks will be placed into the existing channel to further enhance or encourage a dispersed flow throughout the wetland area. The increased water levels in the wetlands will enhance biological activities, thereby producing more biomass. The annual biomass production and new growth replaces the majority used in metal uptake reactions. This provides additional organic matter for adsorption and sulfate reduction processes, thus increasing the life expectancy of the wetlands for treatment polishing.

The enhancement of the existing wetland systems design will include unused wetland areas for future use. In addition, the peat/wetland treatment system is designed to be compatible with the natural environment of the area, yet provide a subtle but effective passive treatment system.

CONCLUSION

The use of a peat/wetland treatment system has shown excellent results for heavy metal removal. The use of a passive peat/wetland treatment system requires little operational maintenance. The results of the demonstration project have illustrated that the dynamics of a peat/wetland treatment system are such that AMD can be successfully treated for pH adjustment and heavy metal removal. The design of a peat/wetland treatment system provides for enhancement of existing wetlands to provide increased vegetation diversity and habitat, thus developing dynamic ecosystems for increased wetlands quality.

ACKNOWLEDGMENTS

I wish to thank all the people and organizations that have provided assistance and encouragement to make this project a success. I especially appreciate the reviews and comments of Steve Gale, Peter Tiffany, and Leslie Stovring, STS Consultants, Ltd., and Paul Eger, MDNR. A special thanks to Debra Nevison who has had to make sense out of my dictations and scribblings and formulate this into a finished document.

REFERENCES

1. Eger, P. and K. Lapakko. Use of wetlands to remove nickel and copper from mine drainage, in *Proc. 1st Int. Conf. Constructed Wetlands for Wastewater Treatment*. D. A. Hammer, Ed., 1988.
2. Eger, P. and K. Lapakko. Nickel and copper removal from mine drainage by a natural wetland, in *Mine Drainage and Surface Mine Reclamation Conf.*, Vol. I, Bureau of Mines, Info Circular 9183, Pittsburgh, PA, 1988.
3. Wilderman, T. R., S. D. Machemer, R. W. Klusman, R. R. H. Cohen, and P. Lemke. Metal removal efficiencies from acid mine drainage, in *Big Five Constructed Wetlands — Proc. 1990 Mining and Reclamation Conf. Exhibition*. April 1990, 417–424.
4. Hedin, R.S., R. Hammack, and D. Hyman. Potential importance of sulfate reduction processes in wetlands constructed to treat mine drainage, *1st Int. Conf. Constructed Wetlands for Wastewater Treatment*. 1988, 508–514.
5. Frostman, T. M. Constructed wetlands for water quality improvement — heavy metal removal. Presented at University of Wisconsin, Green Bay, Wisconsin, 1988.
6. Steiner, G. R. and R. J. Freeman, Jr. Configurations and substrate design considerations for constructed wetlands wastewater treatment, *1st Int. Conf. Constructed Wetlands for Wastewater Treatment*. 1988, 363–377.

PART 4

SUBSURFACE TREATMENT

CHAPTER 21

The Use of Reed Bed Systems to Treat Domestic Sewage: The European Design and Operations Guidelines for Reed Bed Treatment Systems

P. F. Cooper

INTRODUCTION

This article is based upon the "European Design and Operation Guidelines for Reed Bed Treatment Systems".[1] These guidelines were produced by the European Community/European Water Pollution Control Association Expert Contact Group on Emergent Hydrophyte Treatment Systems. They were prepared by the group from work done over 5 years in 5 European countries, the U.K.,[1] Germany,[2] Denmark,[3,4] France,[5] and Austria,[6] and were presented at the International Conference on Constructed Wetlands in Water Pollution Control which was held at Cambridge, U.K. in September 1990. Performance details were also presented at that conference.

In Europe, the concept and use of constructed wetlands is generally different from those that exist in North America and Australasia. In Europe, they tend to be aimed at providing *secondary sewage treatment* for village-sized communities of up to 1000 population equivalent; while in North America, they tend to be used for *tertiary treatment* from larger populations. Another difference is that, in Europe, the systems are nearly always specifically excavated and planted, whereas in America they are often created from existing natural wetlands.

The use of reeds for the treatment of sewage was first investigated in Germany by Seidel and Kickuth in the 1960s. Since then, about 500 reed bed treatment systems (RBTS) have been constructed in Western Europe since 1984. In general, the experiences gained in the years since show that BOD (biochemical oxygen demand) removal is 80 to 90%, with a typical outlet concentration of 20 mg/L; that total N removal is 20 to 30%; and that total P removal is 30 to 40%. A general problem is surface flow of the feed caused by the hydraulic load exceeding the permeability of the substrate; that is, the wrong dimensioning of the systems.

The aim of this article[1] is to help the potential constructors/operators design and build safe plants and, in particular, help them avoid making mistakes which have produced poor results.

This article concentrates on *horizontal flow systems* since most of the European systems studied were of this type. However, late in the study, some data became available from vertical flow systems. The report mentions the vertical flow systems as being a promising design, but the EC group did not feel it could fully endorse this design by giving its "seal of approval" because there were not enough operating data.

The design guidelines were derived from experience with domestic sewage and are aimed at populations of up to 1000 Pe. (Pe, population equivalent). For BOD, this is 40 g/day for settled sewage or septic tank effluent. The mean flow for European populations is 200 L/day.

0-87371-550-0/93/$0.00+$.50

HORIZONTAL FLOW BEDS — GENERAL DESCRIPTION

The advantages claimed for RBTS systems include:

1. Low construction costs
2. Low operating costs
3. A possible treatment solution for very small flows previously untreated

The process depends upon the passage of the liquid (usually sewage or an effluent) horizontally through a bed of soil or gravel media in which common reeds (*Phragmites australis*) are growing. Figure 1 shows a cross-section through a typical RBTS.

The principles behind the process are claimed to be:

1. Rhizomes of the reeds grow vertically and horizontally, opening up the bed to provide a "hydraulic pathway".
2. Within the rhizosphere (the small area surrounding the rhizomes [underground stems]), large populations of common aerobic and anaerobic bacteria grow, which effect the biological breakdown of the organic components of the sewage.
3. It has been claimed that oxygen is passed to the rhizosphere via the leaves and stems of the reeds through the hollow rhizomes and out through the roots to provide some of the oxygen needed by the aerobic bacteria. *What is not clear yet is that rate at which oxygen is supplied by the reeds.*
4. Suspended solids in the sewage are aerobically composted in the above-ground layer of straw debris formed from dead leaves and stems.

DESIGN AND CONSTRUCTION — HORIZONTAL FLOW SYSTEMS

Pretreatment

Initially, it was thought that pretreatment was unnecessary and that screened or unscreened sewage could be applied to the reed bed. Over the past 10 years of experience, there has been a tendency for systems to be built that provide either use of an existing settlement tank or provision for a septic tank.

It is recommended that, where possible, RBT systems be preceded by screening and settlement; however, for the smaller systems (say <100 Pe) where this is not practicable, a well-designed septic tank should be sufficient. For systems larger than this, a conventional primary settlement tank is recommended; but if this is not possible, fine screening (≤6 mm) should be used.

Sizing of Bed

It is recommended that at least two beds be provided in all except the very small works to provide flexibility and allow the shut-down of one bed for maintenance.

Area

The surface area can be established by use of Equation 21–1.[3,7]

$$A_h = \frac{Q_d(\ln C_o - \ln C_t)}{K_{BOD}} \tag{21–1}$$

where A_h = surface area of bed, m^2
Q_d = daily average flowrate of sewage, m^3/d
C_o = daily average BOD_5 of the feed, mg/L
C_t = required daily average BOD_5 of the effluent, mg/L
K_{BOD} = a rate constant, m/day

The background to this equation is given in Reference 1.

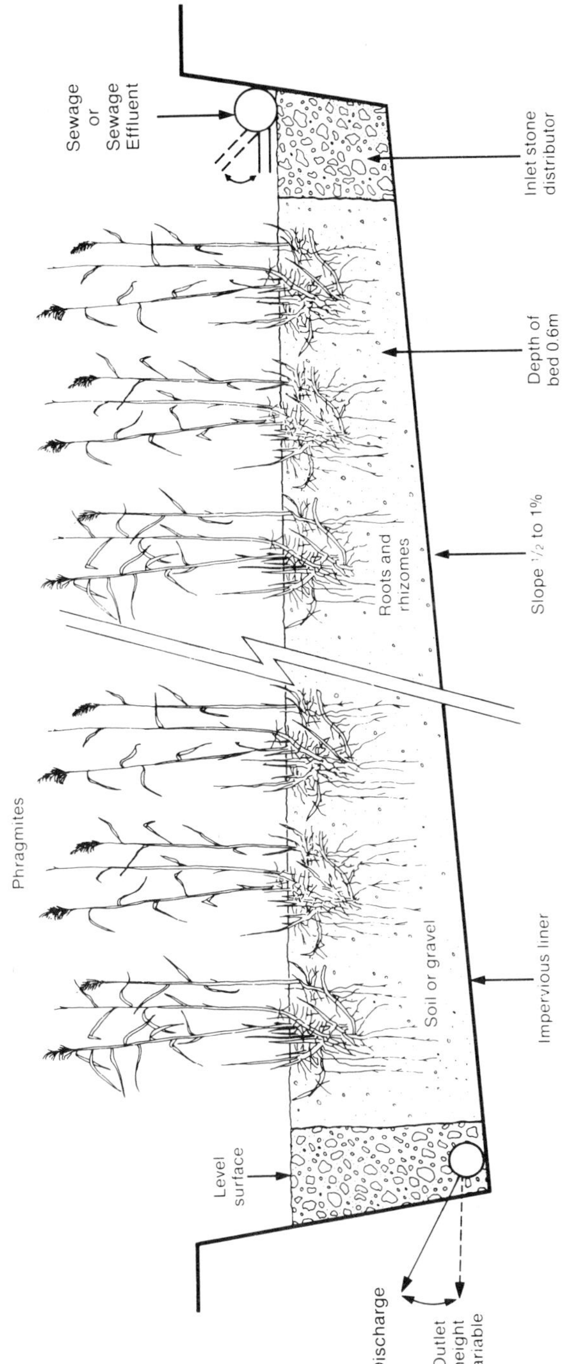

Figure 1. Typical arrangement for a horizontal flow reed bed treatment system.

For the present, the EC group recommended using a design of $5m^2$/Pe for settled domestic sewage or septic tank effluent of normal (European) strength (150 to 300 mg/L BOD). For stronger or weaker sewages, we suggest using Equation 21–1 with K_{BOD} equal to 0.1.

The second equation is a form of Darcy's Law:

$$A_c = \frac{Q_s}{k_f \frac{dh}{ds}} \qquad (21\text{–}2)$$

and is used to define the shape of the bed; that is length-to-width relationship.

where A_c = cross-sectional area of the bed, m^2
Q_s = average flow rate of sewage, m^3/s
k_f = hydraulic conductivity of the fully developed bed, m/s
dh/ds = slope of the bed, m/m

It has been stated[7] that the hydraulic conductivity of a fully developed reed bed built in soil would be about 3×10^{-3} m/s. This value is much higher than could be expected from the soil at the start, and its validity has been questioned.[8,9] One system has now been operated over eight growing seasons (planted June 1983), and there is no evidence that the hydraulic conductivity has increased.[6] *The EC group strongly recommended not using a design hydraulic conductivity of greater than the original media.*

If the soil on the site has a low conductivity, then it might be advisable to replace it with a more porous soil from elsewhere or to use gravel, as has been done in several beds.[10,11]

It is recommended that the bed slope only along the base and that the surface be flat. A sloping surface encourages overland flow and also makes it impossible to kill weeds (see later). It is recommended that the base slope by a maximum of only 1%, in order to prevent the inlet from drying out when the outlet level is dropped.

Depth

Most beds were constructed to a depth of 0.6 m, though some later beds were only 0.3 m deep. Where the recommendations on slope (see earlier) are followed, bed depth will not be constant. *It is recommended that the average bed depth be 0.6 m and that the depth at the inlet should not be less than 0.3 m.* Beyond 0.6 m deep roots, and rhizomes start to weaken.

It is recommended that the beds be built with at least 0.5 m head over the bed surface to the top of the bund walls. For small beds, this may be reduced. This is to allow for the anticipated growth of the bed due to reed debris, etc. and at least a 20-year life before the inlet end of the bed needs skimming to allow for prolonged use.

Slope

Many of the early beds were built with a surface slope. With these slopes, it is impossible to use flooding to effect weed control. *It is recommended to use a level surface which allows weed control and the minimum slope necessary for the base to allow the water to pass through the bed.*

The Bed Should be Constructed with Walls as Steep as Possible

Experience has shown that when the walls were sloped, the reeds did not grow near the more shallow (sloped) edges.

Sealing the Bed

If soil is available on-site, with a k_f value of 10^{-8} m/s or less, then it is likely to have a high clay content and may be puddled to seal the bed. If the conductivity is greater than this value, a plastic liner or membrane should be used.

Media

Most of the RBT beds built with soil as the growth medium have suffered from a large proportion of overland flow because of the low k_f value of the soil. There have been problems of channelling and scouring of the surface which has resulted in some areas being starved of water and, hence, poor reed growth. This phenomenon leads to by-passing and hence, reduced treatment performance. Gravel has been used in reed bed systems in several countries.[4,10-13] The idea was that if gravel were used, this would allow through-flow of water from the start and, if the bed gradually starts to block up with sewage solids, this might be counterbalanced by the growth of rhizomes and roots opening up the bed.

A number of gravel-based systems have been built and, thus far, the designers and operators have been pleased with the way they have performed. Soils will have a k_f value of 10^{-5} m/s or less, whereas a uniform gravel in the range 3 to 6 mm or 5 to 10 mm will have an initial $k_f = 10^{-3}$ m/s or higher. In some systems where there has been overland flow, the performance of the bed has been quite adequate because flow has passed through the "F Horizon", a layer of decaying reed leaves and stem debris. A good example of this is the bed described in Reference 6 where the flow is over the soil surface but in the debris layer for two thirds of the bed length before disappearing into the soil. If, as there, the outlet level is at its lowest, the hydraulic gradient over the last few meters of the bed can become very high.

A number of waste materials have also been used as media for RBT systems. Two of the experimental beds are built with waste pulverized fuel ash (PFA).[10,11,14,15] Those beds have allowed reeds to grow well. *When using a medium such as gravel or PFA which is lacking in nutrients, it is recommended that the seedlings be planted in enriched peat in order to avoid this problem at this critical time.* Once established, the plants should find enough nutrients in the sewage feed.

Claims have been made for RBT systems removing phosphate. It is not likely that this will happen unless a medium is used which is high in iron, calcium, or aluminium and the majority of the flow goes through the medium. Clays will contain high concentrations of aluminium, but will have low k_f values and hence are not recommended.

Inlet Flow Distributor

It is essential to try to get even flow distribution from the inlet to the bed in order to establish a good distribution of flow across the width of the bed. A number of different inlet designs have been used. *Using a weir for flow distribution is not recommended* since these are expensive to construct and maldistribution of flow is often caused by screenable material collecting on the edges. Also, the channel tends to act as a sedimentation tank and collects sludge and grit.

A simple pipe with tees or orifices which can be adjusted to produce even distribution is recommended (see Figure 2). The size of the pipe (and tees) will be dictated by the flow rate, but the pipe is likely to be 75 to 150 mm in diameter for small systems.

Evenly graded stones in the range 50 to 200 mm can be used. It has been found that wire mesh gabions are useful to retain the inlet zone stones in position while the media are being placed in the bed. The inlet zone stones serve as a secondary distributor and allow the level of water across the bed to be equalized.

Another alternative inlet arrangement is to pipe the feed into the "V-shaped" wedge formed by the sloping bund wall and the vertical wall of the gabion (see Figure 3). The method should only be used for settled sewage or effluent. Tipping troughs are another alternative that have been successfully used on small beds.

It is essential that the inlet distributor be easily cleaned (e.g., by flushing).

Outlet Pipe Collector and Water Level Control

The outlet collector will be similar in construction to the inlet. It is recommended that when using soil, a 0.5 m wide stone collector be used with a slotted pipe running along the base. As for the inlet, it is recommended to use a gabion to ease the retention of the stones while filling the bed. The stones should be the same size as the inlet. If gravel is used, it may not be necessary to provide a stone collector. The outlet pipe has to be passed through the liner to the outlet sump. The suppliers of the liner (if used) will make the necessary connections through the liner wall.

It is essential to provide a method for raising and lowering the water level in the bed. This may be done in the outlet sump in several ways:

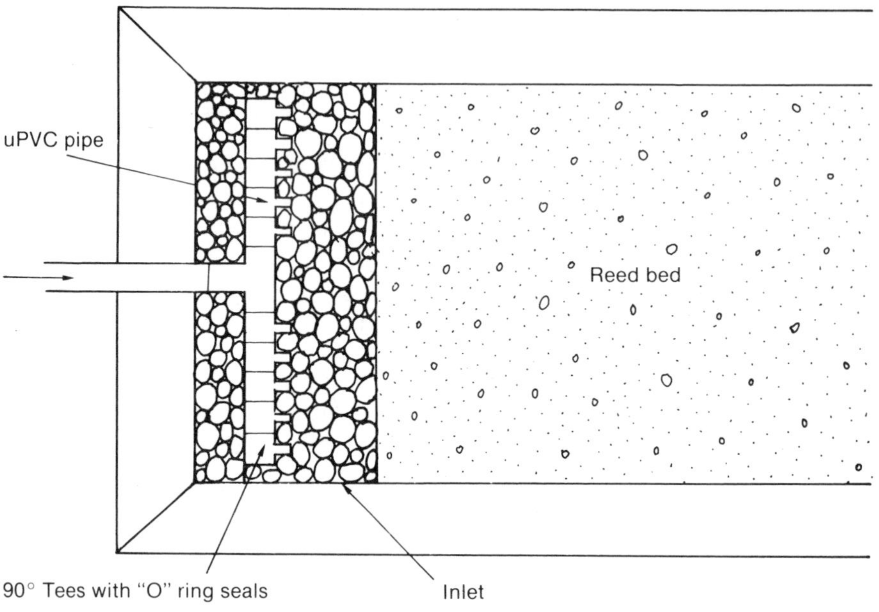

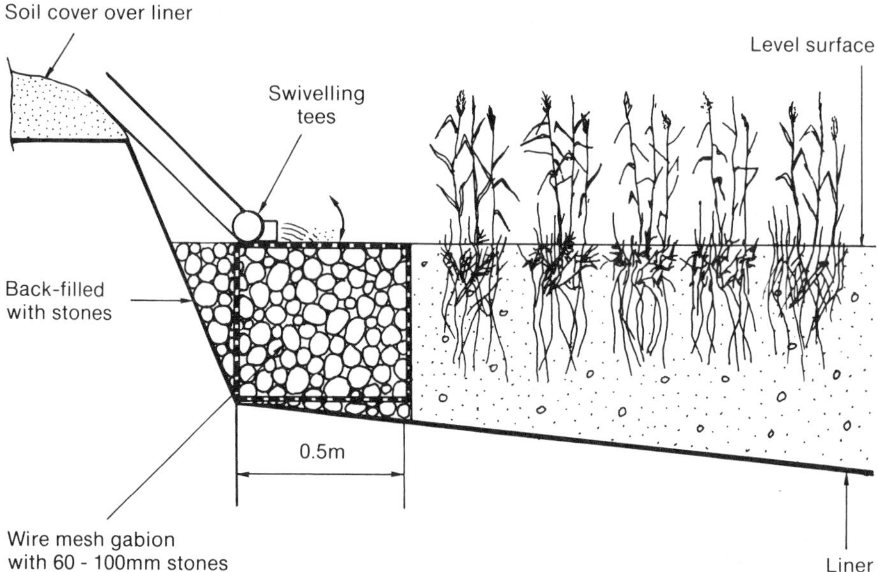

Figure 2. Inlet with swiveling trees.

1. The simplest and possibly cheapest method is to use a 90°-elbow with "O-ring" seals which allow the end of the pipe to be raised or lowered by swiveling at the elbow (see Figure 4).
2. The vertical outlet pipe can comprise a series of socketed sections which can be removed to drop the water level. *Stop-logs should not be used to achieve this because they tend to leak, are difficult to operate, and only give coarse control.* If they are used, the water level may have to drop a sufficiently large distance and at such a rate that scouring of surface debris may occur.
3. A piece of flexible pipe can be clamped in the required position. It is recommended that the outlet arrangement should allow the water level to be maintained at up to 20 cm above the bed surface and down to the base of the bed (i.e., liner level). Arrangements should be provided to allow for the outlet pipe to be flushed clean.

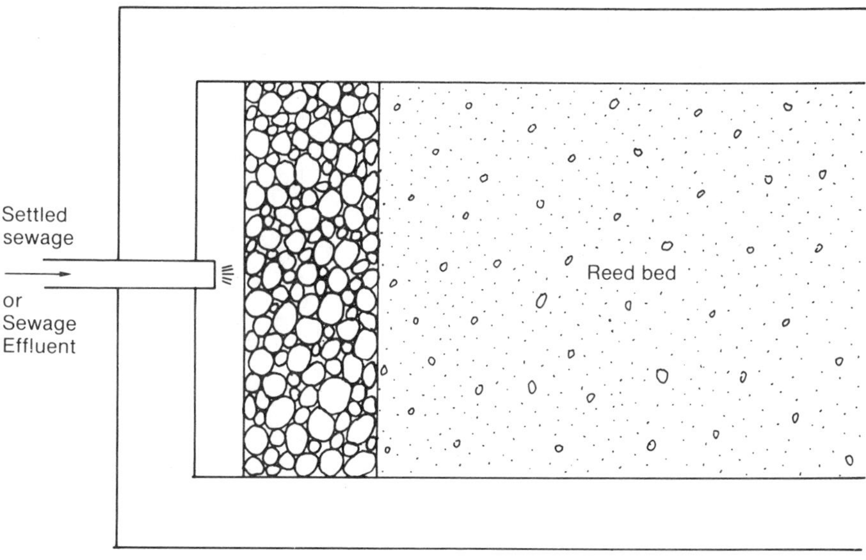

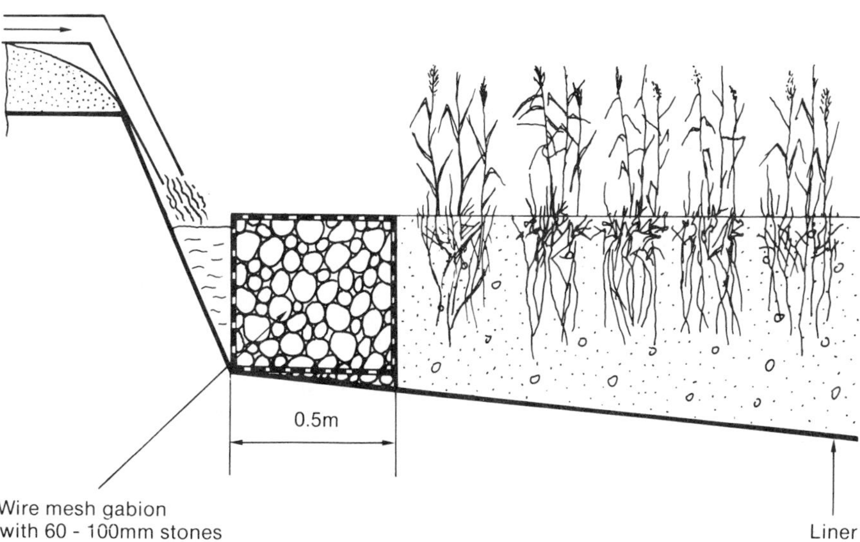

Figure 3. Inlet with feed behind gabion.

Planting and Establishment

The reed beds have been planted in a number of different ways:

1. Using rhizome sections
2. Using clumps of reeds
3. Using seedlings grown in a greenhouse

Several beds have used clumps of reeds (nominally 20 × 20 cm blocks) from existing natural beds at 1 clump per square meter. In all cases, the reeds have survived, but they have tended not to spread outwards to fill in the gaps as quickly as some other methods.

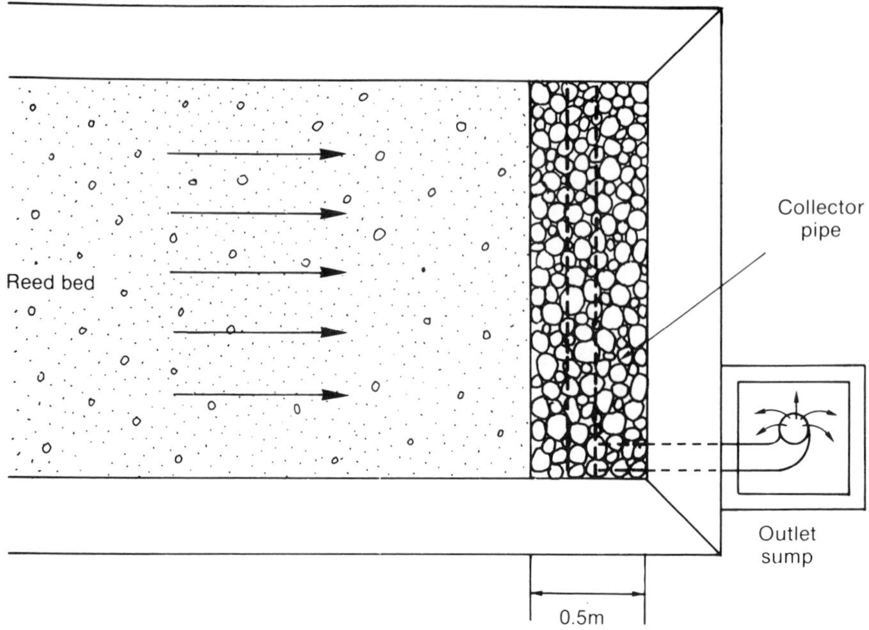

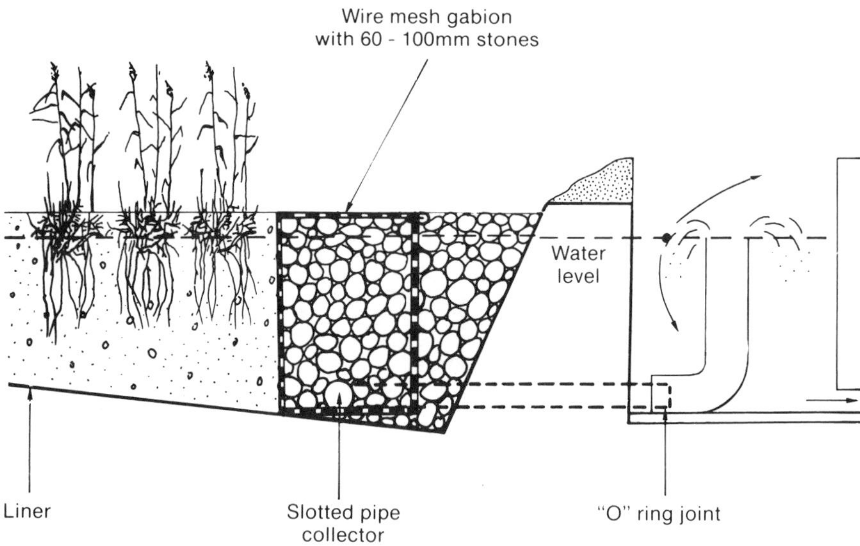

Figure 4. Outlet system with 90° elbow arrangement.

Seedlings are recommended because they tend to provide more rapid cover.[16] A bed planted with seedlings could be substantially covered after the first growing season, having a greater density of shoots and a more uniform cover than a similar bed established from rhizome sections.

Propagation of Seedlings

In Europe, seeds should be collected from fertile stands in November and December.[16] The seed should be germinated in February and March with the aim of planting in May or June. The seed should be sown onto a moist compost in seedling trays in a warm greenhouse. The daytime temperature should be kept between 25 and 30°C, with the nighttime temperature allowed to drop by 10°C. When the seedlings are 5 to 10 cm high, they should be pricked out into a peat compost in 7-cm pots. Planting at a density of 4 seedlings per square meter should be sufficient to obtain a good, even cover of reeds by autumn.

Planting of Rhizomes

It is recommended to plant rhizome segments with at least two nodes. It is recommended to plant two per square meter, and at a 45° angle. Planting in October or spring (March/April) is possible. It is recommended to plant rhizomes in late spring using small rhizome segments with one or two growing shoots.[16] Planting should be done obliquely, with some portion above the water level.

Planting Time

In Europe, it is possible to plant from March to the end of October, but the most successful planting has taken place with those planted in May and June. Seedlings should not be planted later than the end of August. Some of the beds planted in autumn had to be replanted the following spring. This has not been the case with those planted in the spring. Whatever time the planting takes place, it is essential that the seedlings be kept in a moist condition. It is also recommended to avoid periods when there might be frost; hence, the recommendation for late spring and before the beginning of November.

The Choice Between Rhizome Sections, Clumps, and Seedlings

Both seedlings and clumps have been successful and the users are very happy with the results. Rhizome segments have not been as successful. The use of *seedlings is preferred* because they will give more rapid cover and hence may not suffer from competition with weeds in the spring or the second growing season.

It has been noted that the appearance of a reed bed is often very important to how outsiders view a bed's performance. This perception of what a reed bed should look like often seems to be as important as the treatment performance and is something of which designers and operators should be aware.

Weed Control

Weeds have caused problems in several beds because of their excessive proliferation and the fact that, in spring, weeds started to grow earlier than the *Phragmites* and shaded the reed shoots, thereby retarding their growth. It is now clear that over three growing seasons, reeds will gradually win the competition with weeds. However, it is desirable from several viewpoints to establish a good cover of reeds as soon as possible and weed control is best effected by using a bed with a level surface and flooding to a depth of about 5 cm for the first 3 months of establishment for beds planted in spring/summer.

OPERATION

Water Level

It is essential that reeds have access to water at rhizome level at all times. For very permeable media with high hydraulic conductivity (such as gravel), it is recommended that the normal water level be kept about 2 to 5 cm below the surface of the bed. This may also be possible with some soils; but for most, it will be necessary to set the outlet control to the bottom setting (near the liner level) in order to produce the highest possible hydraulic driving force to encourage water to pass through the bed rather than over the surface.

Maintenance

The system should be inspected on, at least, a weekly basis to begin with; after or three growing seasons, this may be reduced.

1. The flow distributor should be looked at and, when necessary, adjusted to equalize the flow across the width.
2. The outlet flow control mechanism should also be inspected to ensure that it is set at the correct level. Where flow measurement is incorporated, this should be recorded.

Monitoring

The beds will need to be sampled at sufficient frequency for the operating authority to be confident that they will meet the effluent consent (standard) required. The collection of representative samples from the inlet and outlet should be easily accomplished.

Table 1. Performance of the MPIP System at Oaklands Park, U.K. August 1989 to September 1991

		Effluents				
	Influent	Stage I	Stage II	Stage III	Stage IV	Stage V
BOD, mg/L	258	57	14	15	7	11
SS, mg/L	169	53	17	11	9	21
NH_4-N, mg/L	50.5	29.2	14.0	15.4	11.1	8.1
Oxd-N, mg/L	1.7	10.2	22.5	10.0	7.2	2.3
o-PO_4-P, mg/L	22.7	18.3	16.9	14.5	11.9	11.2

Note: All values are means, but the number of samples vary slightly.

Table 2. Performance of the Plant at Saint Bohaire, France (mean values December 1983 to March 1989) Samples Collected During the School Period 170 days/year

		Effluents				
	Influent	Stage I	Stage II	Stage III	Stage IV	Stage V
BOD, mg/L	287	145	87	53	38	28
SS, mg/L	272	47	41	31	29	21
KjN, mg/L	64.0	50.5	42.0	40.0	35.7	31.1
NH_4-N, mg/L	43.6	39.5	35.0	34.0	31.8	27.6
NO_3-N, mg/L	0.11	0.20	0.20	0.11	0.15	0.17
Total-P, mg/L	16.1	20.6	22.5	18.1	21.9	19.7

VERTICAL FLOW SYSTEMS

General Description

Several papers presented at the International Conference on "The Use of Constructed Wetlands in Water Pollution Control", held in Cambridge, U.K., September 24–28, 1990, highlighted the promising performances of vertical flow systems (Tables 1 and 2).[5,17] These are based on systems developed by the Max Planck Institute in Germany (often called the MPIP, Krefeld, or Seidel system).

In the first two stages, the sewage to be treated passes vertically through the beds. There are several beds in parallel in the vertical flow stages, with each bed receiving flow in rotation. The flow then passes to two or three horizontal flow stages (or beds) in series. The vertical-flow stages are usually planted with *Phragmites australis*. The horizontal flow stages contain a number of other macrophytes. These may include *Iris, Schoenoplectus, Sparganium, Carex, Typha,* and *Acorus*. Diagrams of the general arrangement are shown in Figures 5 and 6. The media used in all beds is gravel, topped off with sharp sand in the vertical flow beds (Figure 7).

Average concentrations for the settled sewage influent and vertical flow bed effluents from the MPIP system at Oaklands Park in the U.K.[17] are shown in Table 1. These figures are for the first 2 years of operation (August 1989 to September 1991). The full design population had not been reached; thus, the first stage loading rate was a little below design at 0.74 m^2/pe.

Table 2 shows the concentrations of the influent and effluents at each stage of the Saint Bohaire plant in France.[5] This plant was designed for a boarding school in 1982. The loading rate is high (about 75 g BOD/m^2 on the Stage 1 beds) during the time of school activities (170 days/year). The results presented in Table 2 were measured during this period. The high concentrations of influent are explained by the fact that the plant is fed with raw sewage from a short, separate sewage system. The mean loading rate over the whole year was calculated to be about 42 g BOD/m^2 on Stage 1 beds and about 20 g BOD/m^2 based on the total area of the plant. In France, clogging problems occurred on the horizontal flow beds which are built in series and hence always in operation. Lack of oxygen and the absence of rest periods is the most likely cause of these problems. The results from Oaklands Park and Saint Bohaire, however, indicate the level of performance which may be expected from such systems.

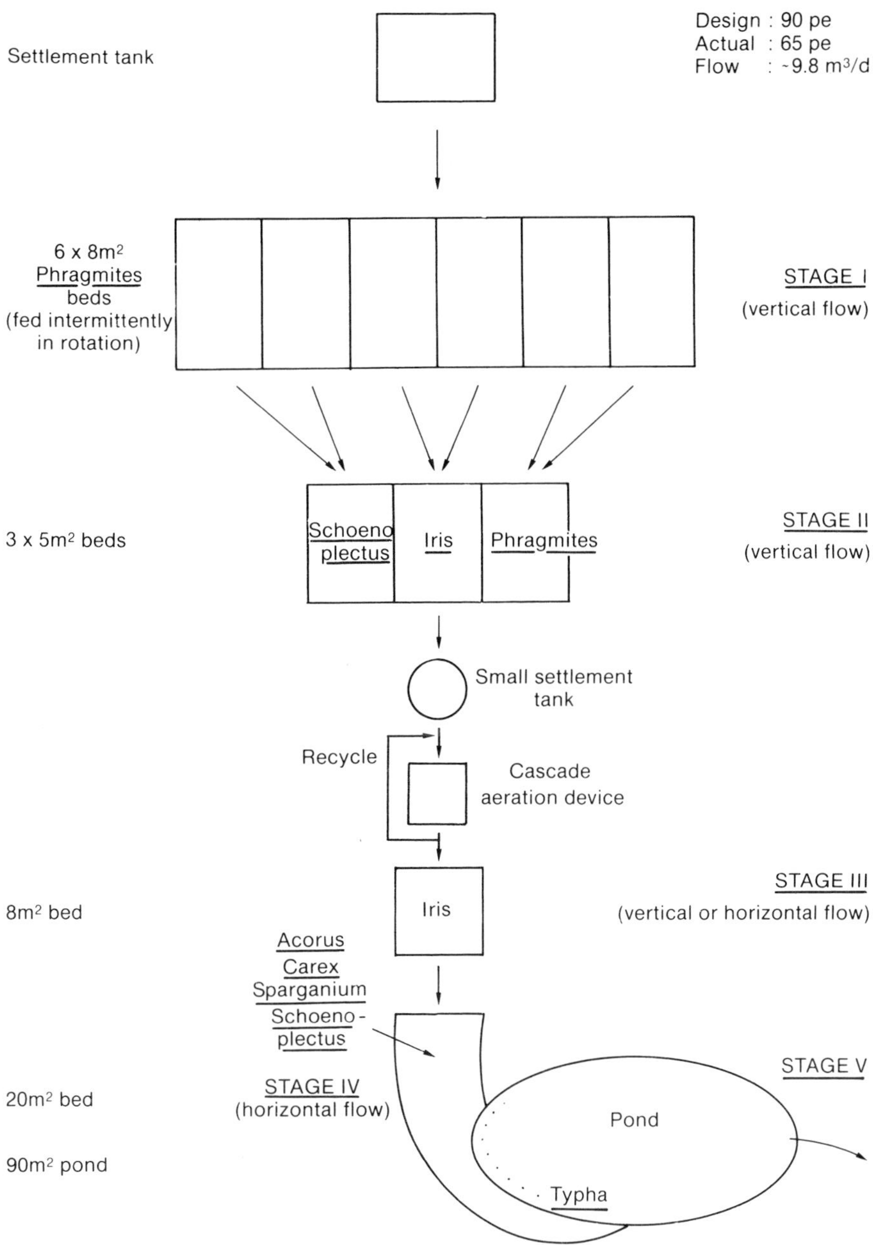

Figure 5. The vertical flow system used at Oaklands Park, in the U.K.

Pretreatment

Pretreatment is normally by fine screening (to <6 mm) and, if possible, settlement. However, raw sewage has been treated at three sites in France without any deleterious effect on the vertical flow beds.

Sizing of the Beds for Settled Domestic Sewage

The total area of the Stage I beds should be sized at 0.6 to 1 m^2/pe and the Stage II beds at 0.3 to 0.5 m^2/pe. The larger size is used when feeding with raw sewage. The Stage II beds are built with the same depth as the Stage I beds, but their number is half that of the Stage I beds. With big hydraulic loads, it might be necessary to increase the infiltration surface. The subsequent beds should be sized at 0.3 to 0.7m^2/pe.

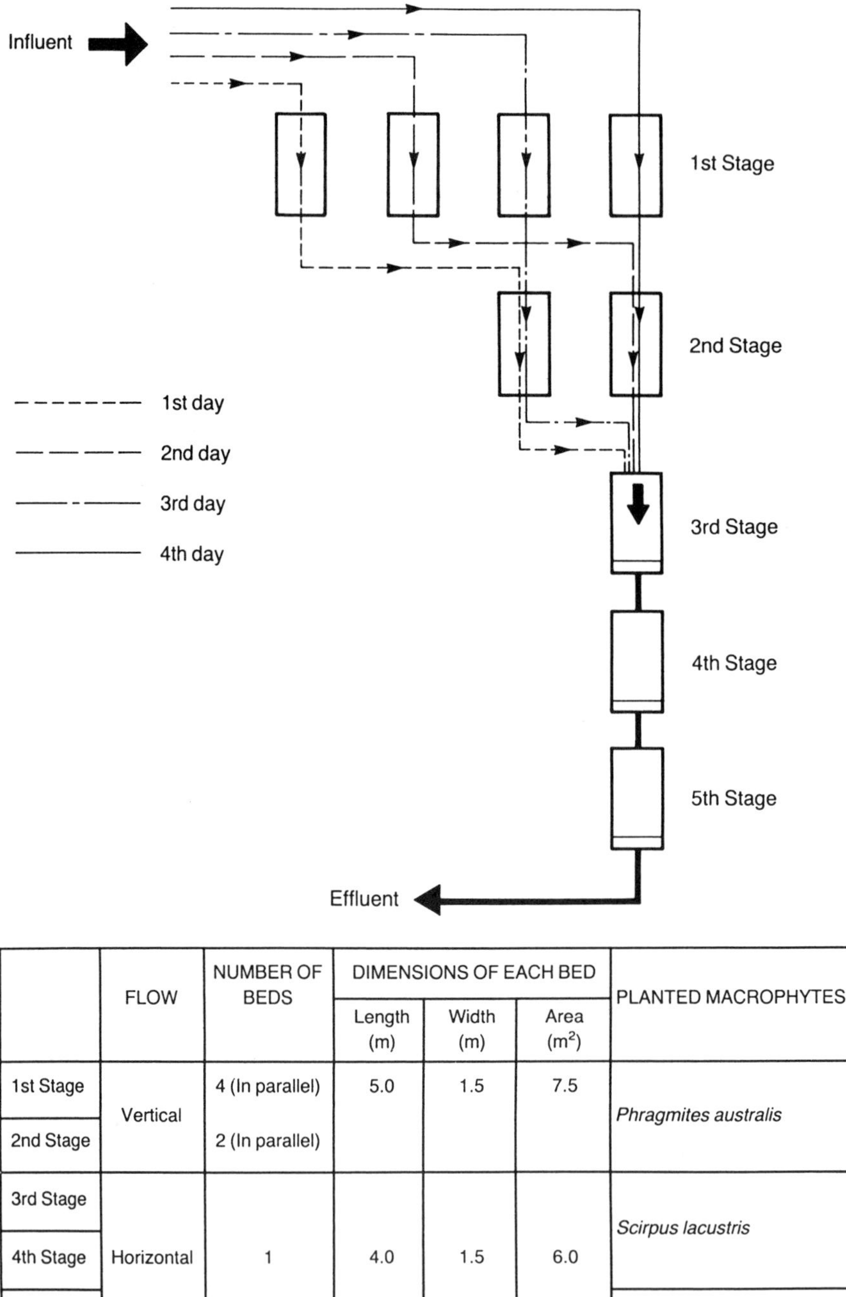

	FLOW	NUMBER OF BEDS	DIMENSIONS OF EACH BED			PLANTED MACROPHYTES
			Length (m)	Width (m)	Area (m^2)	
1st Stage	Vertical	4 (In parallel)	5.0	1.5	7.5	*Phragmites australis*
2nd Stage		2 (In parallel)				
3rd Stage	Horizontal	1	4.0	1.5	6.0	*Scirpus lacustris*
4th Stage						
5th Stage						*Iris pseudacorus*

Figure 6. General layout and characteristics of the Saint Bohaire Plant in France.

Dosing of the Beds

The beds are dosed for 1 to 2 days and then rested for 4 to 8 days. The beds should be dosed in rotation to the period stated above. The vertical flow beds being dosed should receive a minimum flow of 0.6 to 1 $m^3/m^2/d$ (0.6 to 1 m/d). The average of all the Stage I beds would be 0.16m/d.

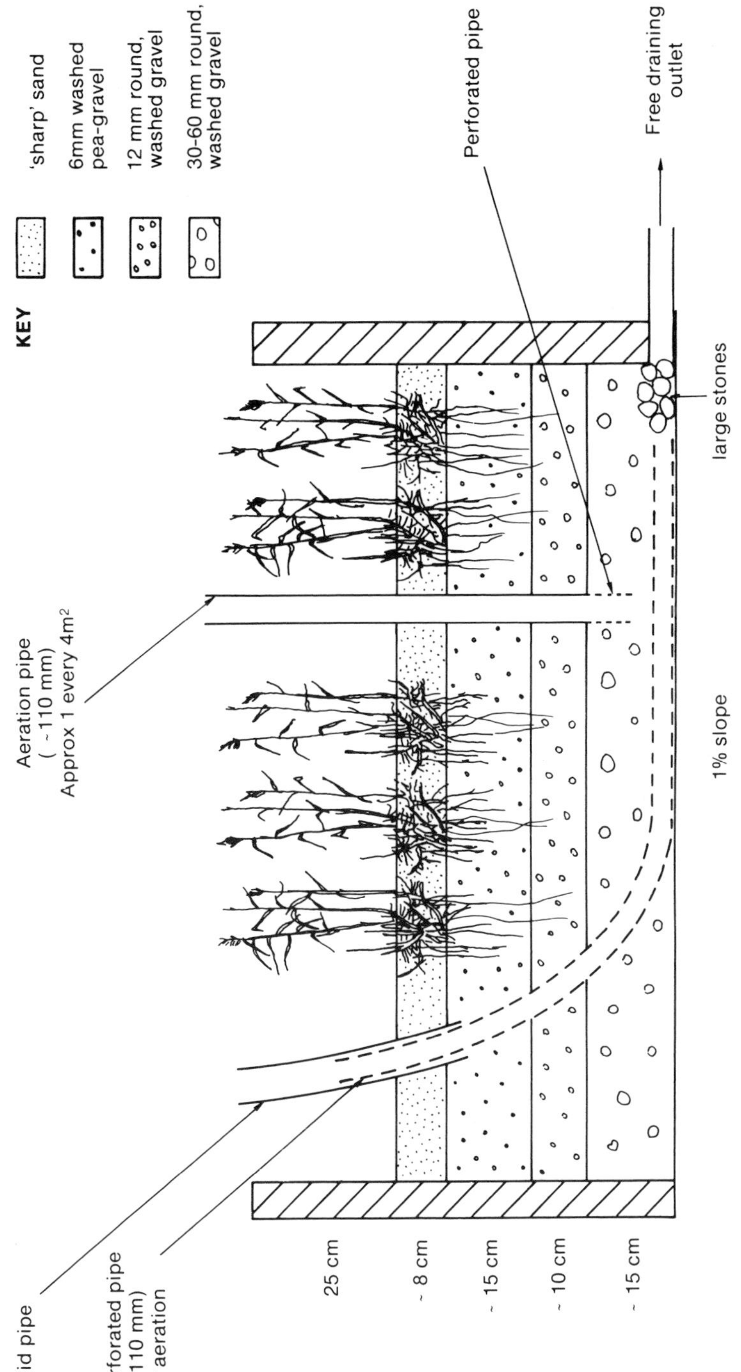

Figure 7. Cross-section of a vertical flow reed bed system — 1st and 2nd stages.

It is believed essential that the beds have time to dry out thoroughly before the next dosing period. This will assist in oxygen transfer and allow for biodegradation of organic matter and oxidation of ammonia, as oxygenation by the reeds alone is not sufficient. Oxygenation might further be improved by batch feeding.

Design of the Vertical Flow Beds

Figure 7 shows the cross-section through typical vertical flow beds. The beds should have a 1% slope on the base to allow free drainage. The surface should be flat to allow for even distribution of flow. If the hydraulic conductivity of the local soil is 10^{-8} m/s or less, it should be possible to puddle and seal the beds. If it is greater, then it will be necessary to use a plastic liner or membrane to seal the beds. It is recommended that 50 cm of freeboard is left above the bed surface.

Distribution of Flow onto the Vertical Flow Beds

Flow may be distributed onto the beds in a number of ways. The most popular has been a gutter channel suspended over the bed, from which flow either cascades over the edge or flows out of holes in the bottom. Where the flow from this distribution gutter hits the bed, splash plates are placed to prevent surface scouring.

Collection of Flow From the Beds

Flow from the first and second stage beds is collected in a perforated pipe (100 to 110 mm diameter) which runs along the base of the bed in the lower gravel (30 to 60 mm size). This pipe is connected to the surface to allow air to pass into the base of the bed.

REFERENCES

1. Cooper, P. F., Ed. European design and operations guidelines for reed bed treatment systems. WRc Report UI 17, September 1990 (revised December 1990), presented to the *Conf. Constructed Wetlands in Water Pollution Control.* Cambridge, U.K., September 1990.
2. ATV-Hinweis H 262. Behandlung von häuslichem Abwasser in Pflanzenbeeten. *Regelwerk Abwasser-Abfall, GFA.* 1989.
3. Brix, H., H.-H. Schierup, and B. Lorenzen. Design criteria for BOD_5 removal in constructed reed beds, paper presented at the *Conf. Small Wastewater Treatment Plants*, Trondheim, Norway, June, 1989.
4. Brix, H. and H.-H. Schierup. Root zone systems operational experience of 14 Danish systems in the initial phase. *Report to the Danish Environmental Protection Board.* July 1986.
5. Lienard, A, C. Boutin, and D. Esser. Domestic wastewater treatment with emergent hydrophytes beds in France, in *Proc. Int. Conf. Constructed Wetlands in Water Pollution Control.* P. F. Cooper and B. C. Findlater, Eds. Pergamon Press, Oxford, U.K., 1990, 183.
6. Haberl, R. and R. Perfler. Seven years of research work and experience with wastewater treatment by a reed bed system, in *Proc. Int. Conf. Constructed Wetlands in Water Pollution Control.* P. F. Cooper and B. C. Findlater, Eds. Pergamon Press, Oxford, U.K., 1990, 205.
7. Boon, A. G. Report of a visit by members and staff of WRc to Germany (FRG) to investigate to root zone method for treatment of wastewaters. WRc Report 367-S/1, Stevenage, U.K., August 1985 (revised February 1986).
8. Bucksteeg, K. Discussion of treatment of wastewater in the rhizosphere of wetland plants — the root zone method — H Brix. *Water Sci. Technol.* 19:1063, 1987.
9. Bucksteeg, K. Sewage treatment in helophyte beds. First experiences with a new technique. Paper presented to the *13th IAWPRC Conf. Post Conf. Seminar on The Use of Macrophytes in Water Pollution Control.* August 24–28, 1986, Piracicaba, Brazil.
10. Findlater, B. C., J. A. Hobson, and P. F. Cooper. Reed bed treatment systems: performance evaluation, in *Proc. Int. Conf. Constructed Wetlands in Water Pollution Control.* P. F. Cooper and B. C. Findlater, Eds. Pergamon Press, Oxford, U.K., 1990, 193.
11. Cooper, P. F., J. A. Hobson, and B. C. Findlater. The use of reed bed treatment systems in the U.K., *Water Sci. Technol.* 22:57, 1990.
12. Gersberg, R. M., B. V. Elkins, and C. R. Goldman. Wastewater treatment by artificial wetlands, *Water Sci. Technol.* 17:443, 1984.

13. Lienard, A., C. Boutin, and D. Esser. State of the art, the Max Planck Institute Process. CEMAGREF, Lyon, France, February 1990.
14. Bayes, C. D., D. H. Bache, and R. A. Dickson. Land-treatment systems: design and performance with special reference to reed beds, *J. Inst. Water and Environmental Management.* 3:588, 1989.
15. Wolstenholme, R. and C. D. Bayes. The evaluation of nutrient removal by the reed bed treatment system at Valleyfield, Fife, Scotland, in *Proc. Int. Conf. Constructed Wetlands in Water Pollution Control.* P. F. Cooper and B. C. Findlater, Eds. Pergamon Press, Oxford, U.K., 1990, 139.
16. Parr, T. W. Experimental studies on the propagation and establishment of reeds (*Phragmites australis*) for root zone treatment of sewage. Contract report to WRc. Institute of Terrestrial Ecology. Monks Wood Experimental Station, Huntingdon, U.K., November 1987.
17. Burka, U. and P. Lawrence. A new community approach to wastewater treatment with higher water plants, in *Proc. Int. Conf. Constructed Wetlands in Water Pollution Control.* P. F. Cooper and B. C. Findlater, Eds. Pergamon Press, Oxford, U.K., 1990, 359.

CHAPTER 22

Sustainable Suspended Solids and Nutrient Removal in Large-Scale, Solid Matrix, Constructed Wetland Systems

H. J. Bavor and T. J. Schulz

INTRODUCTION

Pollutant removal performance in solid matrix, subsurface flow constructed wetland systems depends upon the maintenance of hydraulic conductivity sufficient to allow interaction of effluent with the matrix. Required design criteria, critical to the use of such systems, relate to solids loadings which may be applied without loss of hydraulic conductivity or applied in conjunction with backwashing or other means of accumulated solids removal. Such criteria have been described for slow-sand, rapid-sand, and trickling filter systems, but are not well characterized for subsurface flow constructed wetland systems.[1,5]

Solids loading, removal performance, and in-system accumulation have been investigated in a number of large-scale, subsurface flow constructed wetland systems and in a rapid-sand filter system. Aspects of solids composition, loading, and accumulation will be discussed with respect to prediction of system performance and effective lifetime. Management options for maintenance of hydraulic conductivity and utilization of solid matrix systems in series with surface-flow wetland systems are evaluated.

MATERIALS AND METHODS

A large-scale investigation of constructed wetland systems for sewage effluent treatment was established in late 1983 and remains on-going at the University of Western Sydney-Hawkesbury, Richmond, New South Wales. During the study, aquatic macrophyte systems have been evaluated with respect to treatment performance for a number of effluent constituents, including nitrogen, phosphorus, BOD (biochemical oxygen demand), suspended solids, and bacteria. Details of the system layout, overall performance and methods of analysis have been described by Bavor et al.[1-3] and Roser et al.[6]

The systems were constructed in a trench format, each trench lined with an impermeable membrane, 100 m long, 4 m wide, and 0.5 m deep. The solid matrix systems, discussed below, comprised:

- A control system consisting of an unplanted gravel filled trench (5 to 10 mm crushed, graded, and washed local river gravel) and noted as "trench 2";
- A gravel-filled trench planted full length with *Schoenoplectus validus* (bulrush) and noted as "trench 3";
- Two "mixed-zone" systems consisting of alternating sections of gravel planted with *Typha orientalis*, open water, and unplanted gravel.

Estimates of particulate solid build-up in the systems were made by measuring the dry weight mass of particulate material which could be washed from cores of undisturbed gravel. For volatile solid determination, sediments were ashed at 550°C. The cores were taken using one of two methods. Surface samples were taken using a small (100-ml) core taken at intervals by gently pushing a PVC core tube into the surface layer of the gravel sections at the surface water level. The bottom of the core tube was capped, followed by removal of intact core and entrained solids. In addition, full-depth gravel cores (0.5 m) were secured in mesh socks (4-mm monofilament mesh) and left in submerged, perforated tubes inserted at intervals within gravel sections

0-87371-550-0/93/$0.00+$.50

of the trench systems. The mesh "plugs" were implanted at the commencement of the Richmond macrophyte trials in December 1983, and were removed at intervals over a number of years of system operation.

Hydraulic gradients in the systems were determined at 6- to 10-week intervals, using standard water table survey techniques. Water levels were measured along the length of the systems at 5- to 20-m intervals, depending upon the presence or absence of a gradient. Effluent flow and precipitation were monitored daily. Hydraulic conductivity was then calculated using the Darcy equation.[7]

RESULTS AND DISCUSSION

Initial samplings of the mesh socks were conducted at the end of the secondary treatment trial in July 1986. Visually, there was little evidence of clogging of the pores of this material, even in bottom samples. The only suggestion of solids build-up was a thin, microbial film associated with the gravel surfaces. Discoloration (blackening) increased in intensity with depth and proximity of the sample site to the inlet, reflecting sulfate reduction.

Solids levels of the order of 0.6 g/100 g of gravel were recorded, but there was no consistent correlation with distance from the inlet, which might have been expected from the pattern of solids removal apparent from longitudinal samplings (Figure 1A, B). There was some correlation between depth and solids, possibly indicating the introduction of dust/sediment from the surrounding paddock. Additional cores extracted in April 1987, after 6 months of application of primary effluent (following the secondary effluent trials), showed essentially the same pattern (Figure 1B, C).

Sampling of small surface cores also showed little evidence of solids build-up. Samples taken 5 m downstream from the system inlets were determined to have solids levels of 0.3 to 0.6 g/100 g (Figure 2A). Comparison of inlet levels with time showed initial accumulation of solids with time, leveling off after 6 months (Figure 2B).

Theoretical calculations of solids input based on influent levels and flow data indicated that the total primary and secondary solids applied to the system were each around 1.5×104 kg/ha, yielding a total solids input of 3.0×10^4 kg/ha for the study period. Given a gravel content of 10^7 kg/ha (assuming a density of 2.2 g/cc and an effective depth of 0.45 m), the total solids applied in each trial phase would have been on the order of 0.15 g/100 g gravel if spread throughout the systems. As noted previously, however, most deposition would have occurred in the first 10 to 20 m. If this had been the case, then levels of 0.75 to 1.50 g/100 g might have been expected (assuming no mineralization) after application of secondary effluent, and 1.5 to 3.0 g/100 g after application of both primary and secondary effluents.

Given a rate 0.2 g SS per 100 g gravel matrix per year, it would take approximately 50 years to add an additional 10% solids by weight to the Richmond system. This is a worst case projection, however, and assumes no decomposition of influent solids.

Quantitative accumulation is unlikely, given that 80 to 90% of both secondary and primary influent solids were volatile. Consequently, it is hypothesized that the gravel-based macrophyte systems, operated within suggested loadings, should remain functionally porous for a number of decades.

IMPLICATIONS OF RESULTS

Production of secondary pollution in the form of algal biomass is a major problem in oxidation pond operation. Two techniques used for removal have been high-rate sand filters and rock filters. Their operational features, summarized by Middlebrooks et al.,[4] have many features in common with gravel-based macrophyte systems.

Sand filters differ from the macrophyte systems in that flow is vertical, the flow paths are short (a few meters at most), and they require constant management in the form of intermittent back flushing. Sand filters also differ in that they use a particle size of <0.4 mm and hydraulic loading rates on the order of 5 ML/ha/day. As the filters directly trap particles in a shallow matrix, they cannot be operated with much larger sand size. A mean grain size of 0.7 mm passes excessive SS and BOD through such systems.

Rock filters are more comparable to gravel-based macrophyte systems. Two filter system designs, discussed by Middlebrooks et al.,[4] were configured in a very similar physical layout to that used in the Richmond systems (i.e., horizontal, gravity flow into a sloped bank of material, which was approximately 1 m deep).

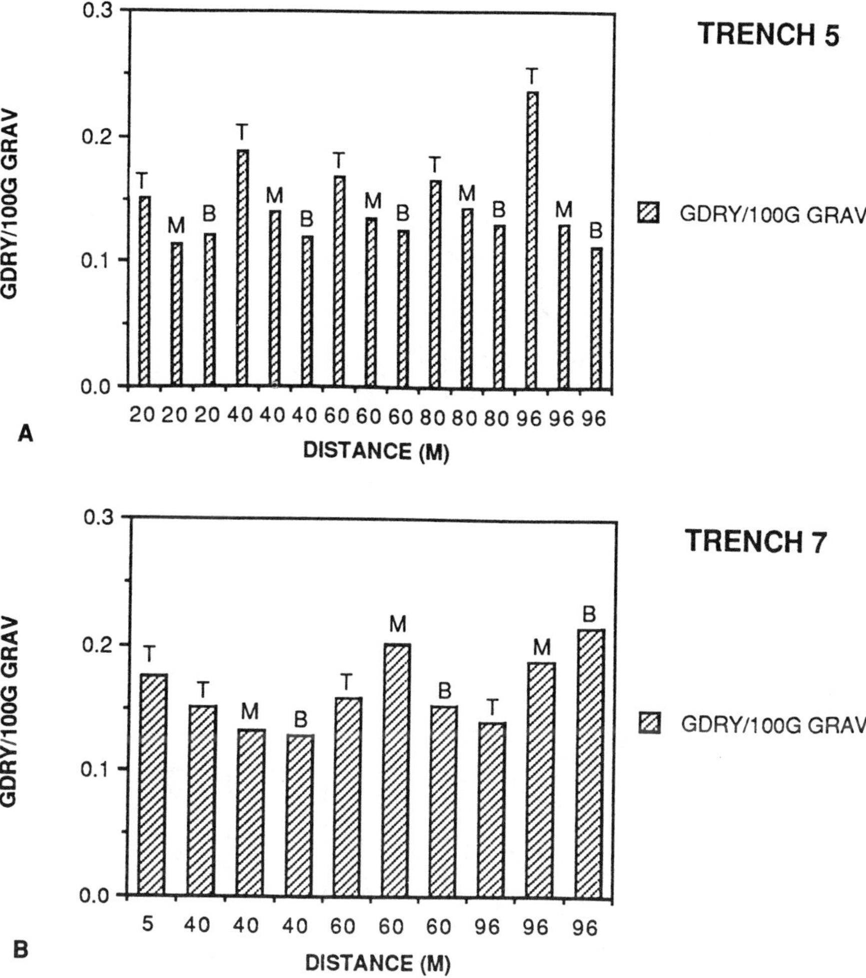

Figure 1. Solids levels from "Gravel Socks" extracted after 3 years of secondary effluent application (A and B), and after 6 months of primary effluent application (C and D). Solids values at the surface, mid-depth, and bottom of the core are noted as T, M, and B, respectively.

Three full-scale trial systems were constructed, one of which used a roughly comparable gravel particle size (9.5 to 25 mm) to that used in the Richmond systems. The other two were constructed of larger rock substrata (70 to 120 mm and 20 to 50 mm particles). The flow path in the rock filter studies, only 5 to 15 m, was shorter than that used for the Richmond studies. Loading figures were calculated on the basis of L/m^3/day, with peak loadings equivalent to retention times of only 0.2 days.

The macrophyte systems at Richmond utilized an intermediate particle size of 5 to 10 mm. The hydraulic loading rates applied to the Richmond systems were much lower than the above-noted rates, whether compared on a land area or retention time basis (0.15 to 1.0 ML/ha/day or 2 to 10 day' retention time for the Richmond systems, compared to 0.2 days' for the rock filters).

Clogging occurrences were characterized in all three system types (i.e., sand, rock, and gravel filters) and appeared at comparable solids loading rates as calculated in gram per square meter of infiltration area (rather than total volume). Also, the effluent solids level, achieved by the macrophyte beds, sand filters and rock beds were similar, being on the order of 5 to 20 mg/L.

In sand filters, solids accumulation leading to clogging problems appeared at a solids loading rate corresponding to 10 to 40 g/m^2/day (multiply by 10 to obtain equivalent figures in kg/ha/day), for sand grain sizes of 0.17 to 0.68 mm, with a tendency for larger sand to clog more slowly. Slightly below these loading

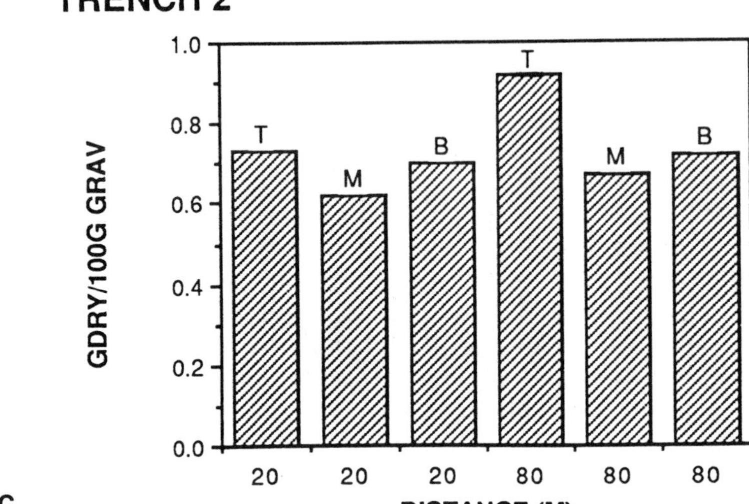

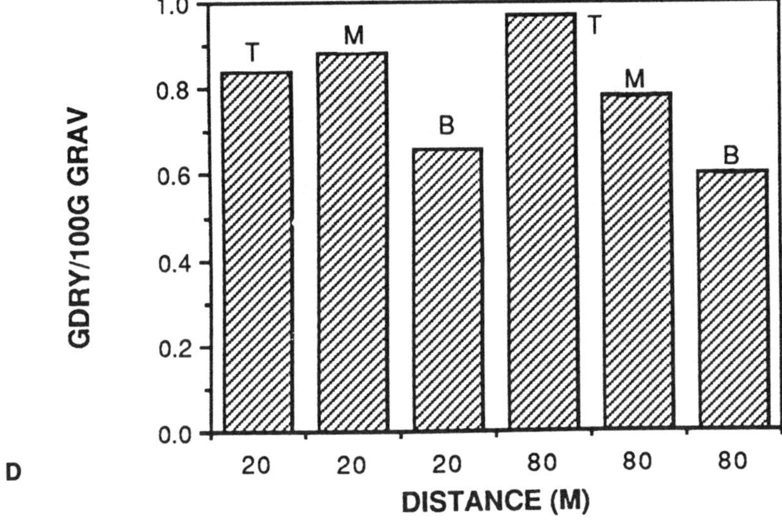

Figure 1. (continued)

rates, operation was shown to be possible for periods of more than 6 months, while slightly above it, clogging occurred in only days.

Large rock filters in a California study (Middlebrooks et al.[4]) also exhibited clogging at solids loading rates comparable to those applied to the sand filters and macrophyte systems. (Table 1). A rock filter with a particle size range of 9 to 25 mm and operated with a solids loading of 300 g/m^2/day at its interface clogged after 11 months. The filter with a marginally larger pore size rocks (20 to 50 mm) did not. A third filter, of larger gauge rocks, did not exhibit clogging problems but was also the least effective in reducing solids levels.

By comparison, the loading rate applied to the Richmond systems, total load on an areal basis (i.e., over 0.4 ha), was approximately 10 times lower than that used in the above study (2.0 g/m^2/day at a hydraulic loading of 0.4 ML/ha/day). The solids loading applied at the actual infiltration zones was, however, quite comparable. At the inlet regions, the infiltration area was estimated to be approximately 2 m^2 (inlets) and 4

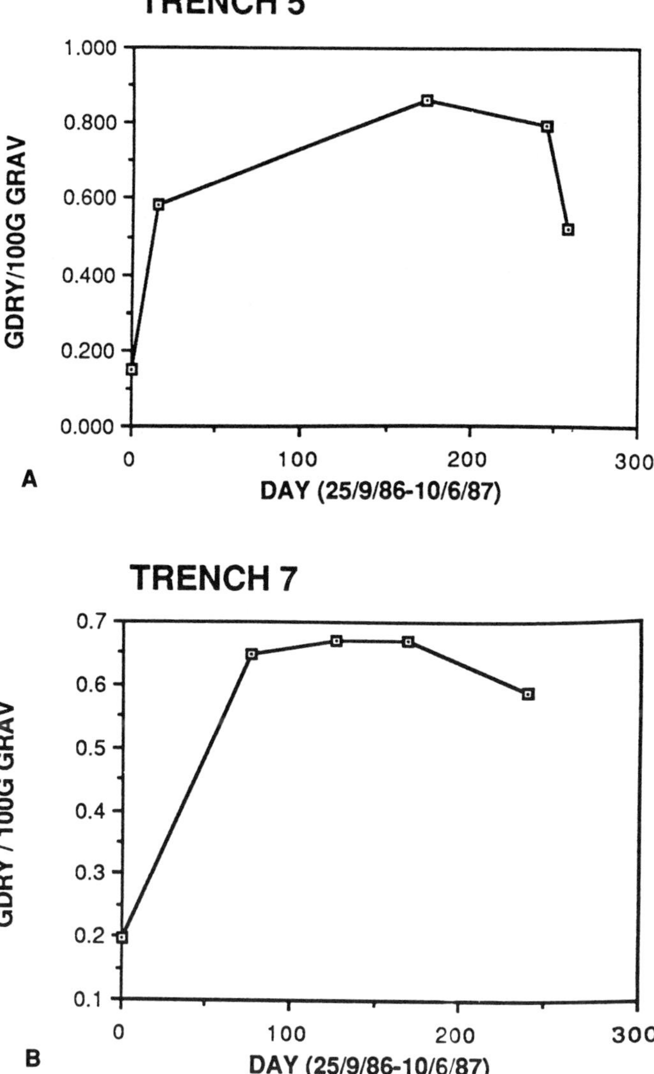

Figure 2. Solids accumulation in gravel at inlet end of trench systems during 9 months of primary effluent application (A and B), and after 1 year of operation (C, average for all trenches). Values are given as grams dry weight per 100 g gravel.

m^2 (open water/gravel interfaces in the mixed-zone systems). Assuming SS wastewater levels of 50 mg/L, the applied solids loadings would be in the range of 40 to 200 g/m^2/day.

Overall, comparison of our results and the behavior of studies noted in Table 1 on two different substrates suggests that porous filtration beds exhibit clogging at a solids loading rate around 10 to 400 g/m^2/day applied at the interface. There also appears to be a direct relationship between particle size and critical loading rate. We suggest that a sustainable solids loading rate for gravel-based macrophyte systems would be on the order of 40 g/m^2/day using domestic sewage effluent inputs. It is recognized that loadings may vary with composition of the effluent constituents.

When hydraulic conductivity estimates were made for the 0- to 20- and 20- to 100-m sections of Richmond trenches 2, 3, and 5, it was observed that the first 20 m had typical values of 0.1 m/s or much lower, whereas, the downstream ends consistently had values of 0.15 to 0.5 m/s. Regression analyses of K_p vs. weeks of system operation indicated that at the sensitive infiltration zone, permeability initially decreased rapidly and would

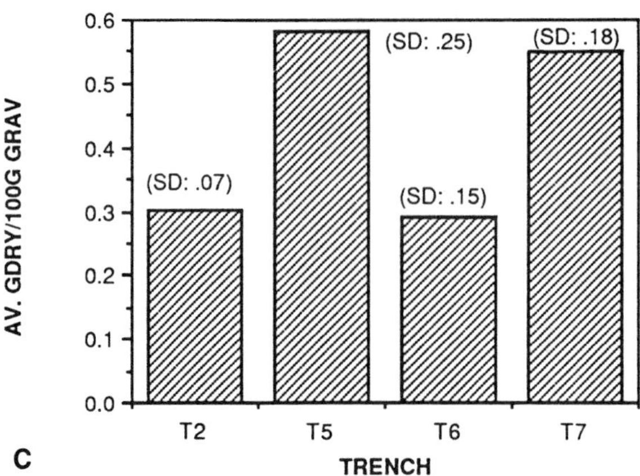

Figure 2. (continued)

Table 1. Comparison of the Behavior of Sand, Gravel, and Rock Filters Operated at Various Suspended Solids Loading Rates

Material	Typical particle size (mm)	Nominal SS loading rate (g/m²/day)	Performance
Sand	0.17	5	Clogging in >5 years
		10	Clogging in 50 days
		30	Clogging in <10 days
	0.40	10	Clogging in >0.5 years
		30	Clogging in 35 days
		70	Clogging in 10 days
	0.68	20	Clogging in >0.5 years
		40	Clogging in 50 days
		80	Clogging in 20 days
Gravel	5–10 (inlet)	40	Infiltration for 3+ years
	5–10 (w/g)	200	Clogging in 3 months
	40 (inlet)	18	Infiltration for 3+ years
	40 (inlet-primary)	80–160	Infiltration for 1+ year
Rock	9–25	13–464[a]	Clogging in 11 months
	10–50	113–629[a]	Infiltration for 17+ months but poor SS removal
	63–127	102[b]	Infiltration for 14+ months but poor SS removal

Notes: The loading rates were those estimated to apply per m² of surface available for infiltration. The data for sand and rock filters is adapted from Middlebrooks et al.[4] Gravel filters were at Eudora, Kansas and California, Missouri. Surface areas were estimated from the volumetric loading rates and estimates of the open surface in the illustrated designs. Gravel size at the water/gravel interface is noted as "w/g".

[a] Represents loadings with 50 mg/L algal solids.
[b] Represents loadings with 69 mg/L algal solids.

be negligible after 10 years. By contrast, similar regression analyses of downstream 80-m sections did not indicate significant reduction in hydraulic conductivity.

A number of regression analyses, with correlation coefficients of 0.39 to 0.41, indicated stabilization of gravel permeability, in the *Typha* system (trench 5), at approximately 0.15 m/s after 10 years. When the very small data set (10 points) was examined using linear or exponential regression; however, there was an indication of decreased permeability with time in the 20 to 100-m section. As noted above, the combined data for the gravel-based systems showed very little or no correlation with system age. We suggest that the systems stabilize at a hydraulic conductivity level of approximately 0.1 m/s, for the system as a unit (e.g., with a slightly lower K_p at inlet and higher K_p downstream).

A number of important inferences may be made from a comparison of the properties of the gravel bed macrophyte systems and rock and sand filters, which are particularly relevant to macrophyte system design criteria. These are as follows:

- Design criteria should include estimates of solids loadings which would be expected at infiltration zones and water//gravel interfaces.
- Infiltration and interface zones should have their cross-sectional area maximized and be large enough to limit solids loadings to a value on the order of 40 $g/m^2/day$. (Note that this figure is suggested during establishment and may vary with solids composition and other effluent characteristics; also, it does not allow for any prefiltration, SS reduction, which may be achieved using a larger rock filter).
- Effective presettling is vital for maintaining good hydraulic conductivity within and at infiltration interfaces of gravel-based systems.
- Sand or rock filters may be considered for pretreatment of effluent entering macrophyte systems.

The above considerations do not involve high levels of technology and/or maintenance and do not represent major deviation from our proposed design criteria.[1-3] It is suggested that in order to maintain appropriate hydraulic conductivity in similar solid matrix system, it would be prudent to address the above modifications.

Particular attention would need to be given to achieving an inlet solids loading of approximately 40 $g/m^2/day$. Higher loadings may be accommodated using an initial rock filter (or other methods) or an increased infiltration area system configuration, as suggested by Tchobanoglous.[8]

ACKNOWLEDGMENTS

Acknowledgment is given to Dr. D. Roser and Ms. S. McKersie for assistance during the research program. Funding by the Water Board, Sydney, and the Department of Public Works, NSW, is gratefully acknowledged.

REFERENCES

1. Bavor, H. J., D. J. Roser, P. J. Fisher, and I. C. Smalls. Performance of solid-matrix wetland systems, viewed as fixed-film bioreactors, in *Constructed Wetlands in Wastewater Treatment*. D. A. Hammer, Ed. Lewis Publishers, Chelsea, MI, 1989, 646–656.
2. Bavor, H. J., D. J. Roser, and S. A. McKersie. Treatment of wastewater using artificial wetlands: large scale, fixed-film bioreactors. *Austr. J. Biotechnol.* 3(1):67, 1987.
3. Bavor, H. J., D. J. Roser and S. A. McKersie. Nutrient removal using shallow lagoon-solid matrix macrophyte systems, in *Aquatic Plants for Water Treatment and Resource Recovery*. K. R. Reddy and W. H. Smith, Eds. Magnolia Publishing, Orlando, FL, 1987, 227.
4. Middlebrooks, E. J., C. H. Middlebrooks, J. H. Reynolds, G. Z. Watters, S. C. Reed, and D. B. George. *Wastewater Stabilisation Lagoon Design, Performance and Upgrading*. MacMillan, New York, 1982.
5. Middlebrooks, E. J. Research and development needs for utilisation of aquatic plants for water treatment and resource recovery: engineering status, in *Aquatic Plants for Water Treatment and Resource Recovery*. K. R. Reddy and W. H. Smith, Eds. Magnolia Publishing, Orlando, FL, 1987, 1009.
6. Roser, D.J., S. A. McKersie, P. J. Fisher, P. F. Breen, and H. J. Bavor. Sewage treatment using aquatic plants and artificial wetlands, *Water*. 14(3):20, 1987.
7. Tchobanoglous, G. and E. D. Schroeder. *Water Quality*. Addison-Wesley, Reading, MA, 1985.
8. Tchobanoglous, G. Aquatic plant systems for wastewater treatment: engineering considerations, in *Aquatic Plants for Water Treatment and Resource Recovery*. K. R. Reddy and W. H. Smith, Eds. Magnolia Publishing, Orlando, FL, 1987, 27.

CHAPTER 23

Hydraulics and Solids Accumulation in a Gravel Bed Treatment Wetland

R. H. Kadlec and J. T. Watson

INTRODUCTION

Subsurface flow wetlands are designed to provide contact between water, roots, and substrate. This type of system is variously called underground, gravel bed, or rock bed wetland, or rock reed filter. In the U.S., there are currently more than 130 such wetland systems, excluding tiny or individual on-site systems. The survey by Reed[1] documents their locations and parameters. A lagoon system in Benton, Kentucky (population about 4500) was modified by TVA to provide a gravel bed wetland (one of three wetland cells at the site). This facility has been described in detail by Steiner et al.[2] Distribution is via a perforated piping system, which allows the lagoon water to seep through a rock inlet region and then horizontally through gravel in the wetlands cell. There is a small gravel layer (25%) on top of a coarse gravel layer (75%). The wetland cell is 144 ft (44 m) wide and 1094 ft (333 m) long. The longitudinal cross section (Figure 1) shows the slight slope toward the outlet (about 0.1 %).

The maintenance of below-ground flow has sometimes proven difficult in this and other systems. The gravel cell at Benton has lost some of its ability to transmit water through subsurface strata. This set of studies was undertaken to learn the nature of the plugging. Field studies at the site were conducted during June 1990, and lab studies were conducted June to November, 1990. In addition, samples of rock and interstitial water and solids were analyzed for bacteria, algae, and fungi by the National Sanitation Foundation (NSF), Ann Arbor, Michigan. Supplementary information was obtained by TVA in an intensive field study in September 1990. The intent of the study was to identify probable causes of clogging, and consequently this work is to be considered a set of exploratory investigations rather than exhaustive research.

WATER BUDGET AND WATER ELEVATIONS

A survey of gravel and water levels was performed (closure to within 0.03 ft ≈ 1.0 cm). The depth to the clay was measured in each of four holes dug in the cell. Results show (Figure 2) the bed thickness to vary between 30 and 35 in (75 and 90 cm). Water was ponded to a depth of 4 in (10 cm) over the first 165 ft (50 m) during June 1990.

Spot determinations of water velocity in the gravel were made at several locations. A porous basket was imbedded in the gravel, and salt solution added. The elutriation of salt solution from the stirred contents was measured via electrical conductivity. The expected exponential change in concentration did occur, and the superficial (volume flow per unit cross-sectional area) velocities were calculated therefrom. Superficial velocities were also calculated from the measured depths and the measured total volume flow to the cell. The latter was adjusted at each distance to account for the measured evaporative loss. During the course of the field studies, the inflow was constant at 67,125 GPD (254 m^3/d) (orifice), and the outflow at 45,300 GPD (171 m^3/d) (H-flume). The evaporative loss was thus 0.22 in/d (0.56 cm/d), which is reasonable for this latitude and time of the year. Table 1 shows the averages of the spot checks at three distances compared to the mass balance velocities. While the spot averages are consistent with system mass balances, the range at any location is fairly wide; standard deviation was equal to the mean at the front end location. This indicates zonal blockage, or channelization, in the front end.

0-87371-550-0/93/$0.00+$.50

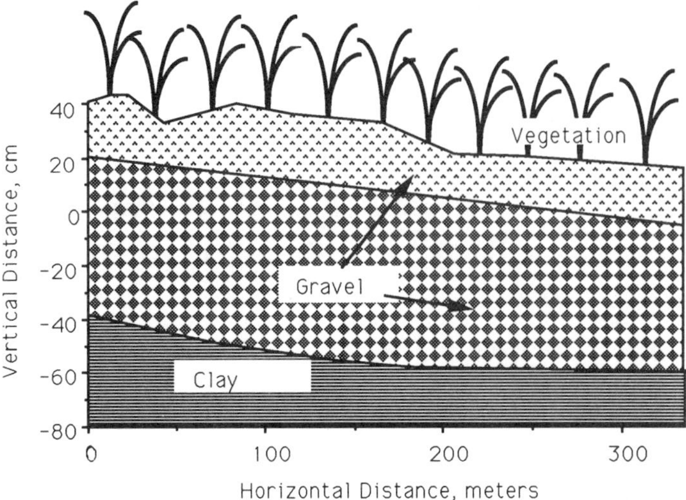

Figure 1. Cross-section of cell 3, Benton, Kentucky. The gravel is not uniform, a coarse fraction occupies the lower three quarters; a finer fraction the upper one quarter.

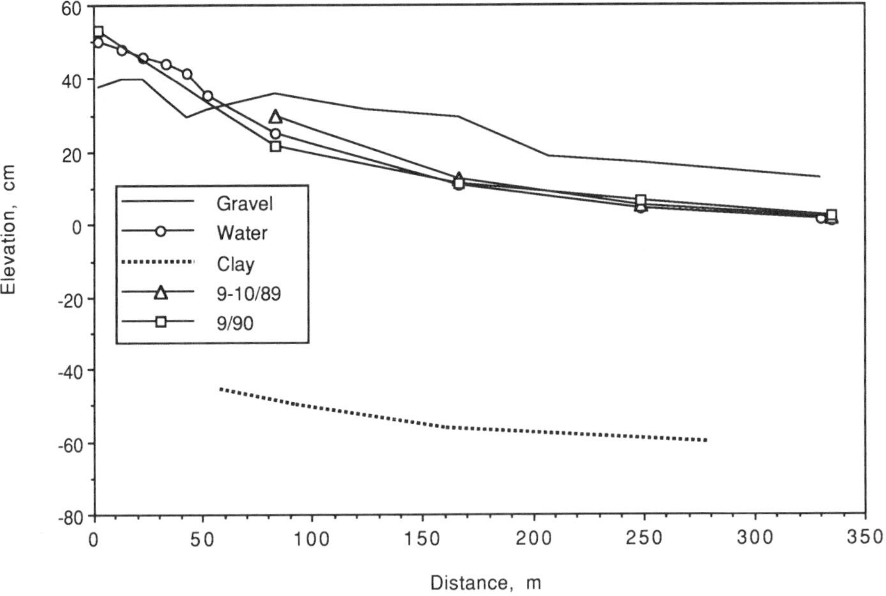

Figure 2. Elevations of soil, gravel, and water.

BED CONTENTS

The size distribution of both fine and coarse gravel were determined gravimetrically from about 1000 stones in each class. The surface area mean diameters were 0.55 and 0.89 in (1.40 and 2.25 cm), respectively. A wetting procedure was used to cross-check the surface area mean diameter. Stones were cut to rectangular parallelopipeds and the wetting water per unit area measured. Samples of rock were then wet, and the area determined. The sizes thus inferred were 0.50 and 0.81 in (1.27 and 2.07 cm), respectively. The density of the rock was measured to be 166 ± 5 lb/ft^3 (2.66 ± 0.08 g/cm^3).

Samples of gravel were taken from both the coarse and fine strata near the front end and near the middle of the bed. Triplicate measurements of porosity were made on compacted samples. Large and small rocks gave

Table 1. Velocities from Geometry and from Flow Tests

Distance (m)	Depth (m)	Flow (m^3/d)	From depths: velocity (m/d)	From salt tracer: velocity (m/d)
0	0.75	254	7.72	
33	0.70	246	7.96	
53				6.64 ± 5.88
67	0.63	238	8.57	
100	0.56	229	9.35	
133	0.51	221	9.87	
166				10.52 ± 2.77
167	0.49	213	9.94	
200	0.48	205	9.63	
233	0.49	196	9.10	
267	0.51	188	8.46	
289				8.08 ± 3.23
300	0.53	180	7.79	
333	0.55	171	7.12	
Mean			8.68	8.41

porosities of about 0.4 in the middle section. Near the front end, the large rock had porosity 0.375 and the small rock had porosity 0.3. The front end samples contained copious amounts of thick black "mud". The amounts of water and solids were determined for quantitative grab samples (Table 2). In the front end of the bed, the amount of adhering water is large; this represents the non-drainable gel. In the front end, 20 to 40% of the water and solids will not pour out of the rock, in contrast with 10 to 12% at downstream locations in the bed. The adhering waters contain 5 to 8% solids in the top front end. There are significant quantities of solids in all portions of the bed. Those in the front end retain water, so that only about half of the interstitial water is capable of easy movement.

The solids recovered on three recycle rinses of the large and small stones from the front end and middle of the bed were analyzed. At mid-bed in both upper and lower layers, these solids were 90 ± 5% mineral (ignition at 550°C). In the front end, they were 80 ± 5% mineral. These results were unexpected; consequently, the determination was repeated for front end solids. The rerun gave 78 ± 6% mineral.

These solids were settled for several days. The mid-bed solids settled readily to a thin compact layer. The front end solids settled only slightly, and then to a gelatinous mass which retained its character for 25 weeks. This gel contains about 17 ± 3% solids; the rerun analysis gave 17 ± 3%.

Holes were excavated in the bed at four locations down to clay and at many other locations into the top layer of gravel. Samples of the bed (water, solids, and rock) were taken for lab analyses by University of Michigan (UM) labs and by NSF. The NSF samples were air expressed to NSF on ice; those for UM labs were driven back in iced coolers. Samples of the surface sediment litter layer were also taken. Solids were rinsed from the rocks and placed in containers for lab analyses. Coarse and fine gravel samples were taken. The NSF bacteriological assay showed enormous numbers of bacteria in the front end of the bed, and a decrease toward the outlet end (Table 3). Algal assays show a similar pattern, but there is greater reduction in numbers as water passes through the bed. It should be noted that there is a size disparity between algae and bacteria, so algae are the dominant group on a mass basis. Fungal assays also show a similar pattern, but the numbers of fungi are three orders of magnitude smaller.

Samples of solids and water were examined microscopically. Three principal types of materials were identified: clay particles, algae, and amorphous solids of indeterminate origin. The algae were those normally associated with lagoon waters: *Nitzschia, Navicula, Synedra, Microcystis,* and *Scenedesmus.* The NSF also identified algae, and found diatoms to be the most common species. NSF found, in addition to the above, *Closterium, Chlorella, Cyclotetta, Palmella, Agmenellum, Carteria, Euglena, Ankistodesmus,* and *Rhodomonas.*

Surface litter in the front end of the bed averaged 21,490 ± 11,240 dry lb/ac (2410 ± 1260 dry gm/m^2). This is comparable to other FWS (free-water surface) wetlands. The rest of the surface of the cell was dry and contained much less litter. Of this, 62 ± 2% was fibrous plant leaf litter, the balance being small leaf fragments and microdetritus (sludge).

Table 2. Quantitative Grab Samples — Gravels

Distance (m)	Drainable voids % of total voids	Adhering gel % of total voids	Adhering volatile solids (mg/g air dry rock)
Small rock = Top			
2	57.01	43.81	11.77
58	88.98	11.47	1.94
93	91.53	8.92	2.35
156	89.48	10.68	0.86
279	87.52	12.75	1.44
330	84.70	15.57	1.48
Large rock = Bottom			
58	80.49	20.01	2.37
93	88.10	12.08	0.72
156	88.67	11.91	2.32
279	91.82	8.51	1.22

Table 3. Biological Assay of Bed Solids

	Distance, m =	Inlet 2	58	93	156	279	Outlet 330
Bacteria millions of cfu/ml	Litter layer	4.40	41.00				
	Small rock, top		1.10	0.47	0.26	0.14	0.39
	Middle depth		0.60				
	Large rock, bottom		4.30	0.85	0.25	0.18	
Algae millions of cfu/ml	Litter layer	5.20	16.00				
	Small rock, top		0.46	0.09	0.13	0.08	0.03
	Middle depth		0.57				
	Large rock, bottom		5.40	0.04	0.08	0.02	
Fungi thousands of cfu/ml	Litter layer	0.90	100.00				
	Small rock, top		2.00	3.20	1.00	0.80	10.00
	Middle depth		5.00				
	Large rock, bottom		80.00	7.00	0.80	4.00	

SOLIDS CHEMISTRY

Samples of the major solids involved were taken and processed for chemical analysis. Lagoon water (10 gal; 40 l) was filtered to collect suspended solids. These were dried and ashed at 550°C. Samples of interstitial clay (tan) and mud (black) from within the gravel bed were similarly processed to residue, as were samples of crushed gravel. Specimens were subjected to chemical analysis via a scanning electron microscope in the Department of Material Science Engineering at the University of Michigan. This procedure detects elements heavier than atomic mass 20, thus excluding carbon, nitrogen, and oxygen.

Results are summarized in Table 4. The proportions of the three principal elements — Ca, Al, and Si — in the mud indicate origins from all possible sources: rock, clay, and lagoon solids. One may speculate on this chemistry as being partially responsible for the clogging of the inlet zone. In the inlet zone of the bed, there is a low redox potential and the presence of algal silicate, coupled with calcium and aluminum. This combination yields the gel, or mud, which in turn traps organics until they undergo decomposition. This in turn keeps the redox low and the cycle becomes self-sustaining.

Records show that approximately 198,000 lb (90,000 kg) of solids have been introduced to the bed during its history, and 22,000 lb (10,000 kg) exited with water. If the balance, 176,000 lb (80,000 kg), remained in

Table 4. Solids Chemical Analysis: Principal Elements. Renormalized %.

Element	Interstitial mud	Lagoon solids	Interstitial clay	Rock
Aluminum	3.47	11.63	14.27	0.00
Silicon	21.60	53.51	85.73	2.63
Calcium	74.93	34.86	0.00	97.37
Other:	Sulfur	Phosphorus	Potassium	
		Potassium	Iron	
		Iron		

the bed, the average interstitial solids concentration would be 3.0%. It is only 1.36%, so only a fraction of the added solids are still present. Clearly, the organic fraction of the incoming solids has been greatly reduced by microbial activity: either aerobic, resulting in production of carbon dioxide, or anaerobic, resulting in the production of methane. Further, the solids now in the bed are mostly mineral, indicating that there has been quite a small accumulation of organic constituents.

HYDRAULICS

Coarse and fine rocks were placed in large diameter pipe and the hydraulic conductivity measured via pressure drop and flow rate. The fine gravel gave 50,000 ft/d (15,240 m/d); the coarse gravel 210,000 ft/d (64,000 m/d). These were clean gravels, probably not quite as compacted as in the field.

The material balance for evaporating flow through an inclined gravel bed is:

$$-kW\delta\frac{d(S+\delta)}{dx} = Q_0 - eWx \tag{23–1}$$

where k = hydraulic conductivity,
W = width,
d = water depth,
S = clay elevation,
Q_0 = pumped flow rate,
e = evapotranspiration rate,
and x = distance from the inlet.

For these conditions, it is not correct to simply use Darcy's (1856) law;[3] it does not deal with variable depth or evaporation. In this equation, a distance variable hydraulic conductivity must also be used in order to reflect the clogging. This was obtained from the measured gradients, depths, and flows (Figure 3). Note that the inlet region has some surface flow; hence, the conductivities in that region do reflect gravel conductivity only. Also note that there has apparently been little or no change in the hydraulic conductivity profile over the 9-month period from September 1989 to June 1990.

In turn, this equation may be integrated to produce the shape of the water surface (Figure 4). This figure may be compared with the survey of the water surface (Figure 2), with which there is reasonably good agreement. Indeed, this is only a consistency check since the k values were derived from the survey.The equation may be used to explore the consequences of different constant (unclogged) hydraulic conductivities on the water surface. At the current inflow rate and exit level settings, the bed overfloods in the front end for a conductivity of 20,000 ft/d (6096 m/d), but does not overflood at a conductivity of 50,000 ft/d (15,240 m/d). It may be shown that at higher flows, flooding will occur at higher hydraulic conductivities.

BLOCKAGE OF PORES

Correlations of the hydraulic conductivity with rock and bed properties exist in several forms. These predict the effect of reduced porosity and rock size on the hydraulic conductivity. These include the Ergun equation

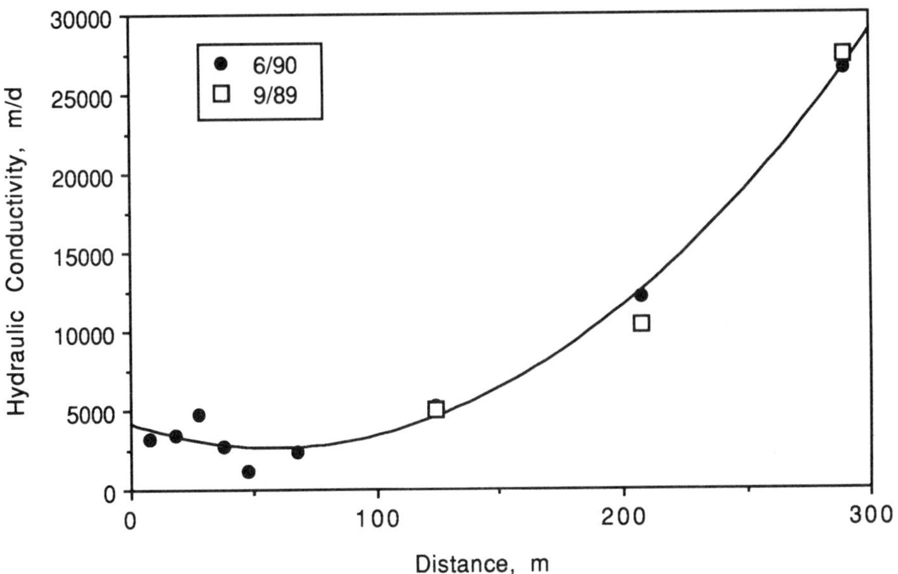

Figure 3. Hydraulic conductivity in the gravel bed.

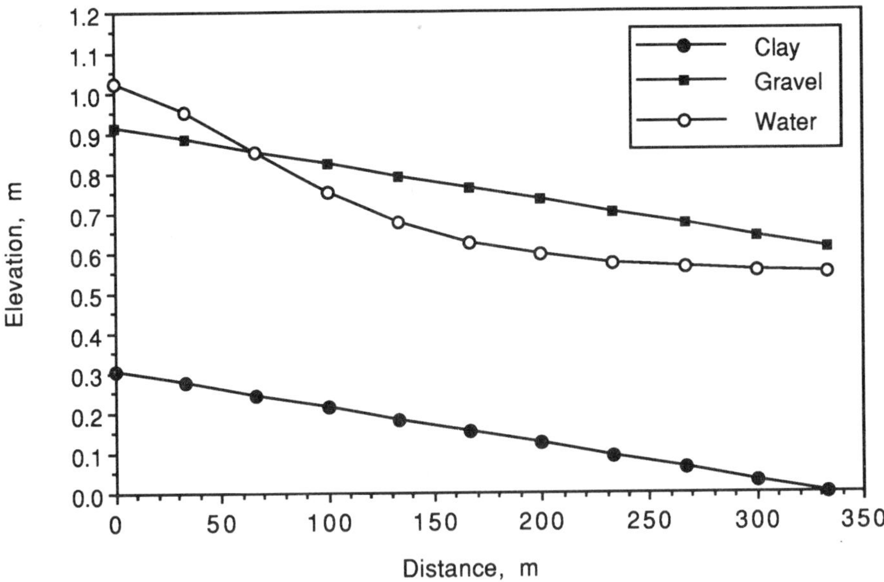

Figure 4. Calculated water elevation in the bed.

(Ergun,[4] 1956, reiterated by Kadlec,[5] 1989), friction factor methods (Brown and Associates,[6] 1956), and gravel-specific correlations (Idelchik,[7] 1986). The Ergun equation shows that the exit end of the Benton bed should have 88% of the porosity of the washed rock used in the lab hydraulic conductivity study; but the inlet end should have only 49% of the washed porosity at present. These calculations proceed as follows: hydraulic conductivity is proportional to porosity to the 3.7 power, as indicated by Ergun and Idelchik correlations. Hydraulic conductivity also depends on other factors, such as rock size, but those are not variable as the bed undergoes clogging. For the "clean" end of the bed, field data indicate a conductivity of 90,000 ft/d (27,432 m/d). The lab measurement gave 145,000 ft/d (44,196 m/d). Therefore:

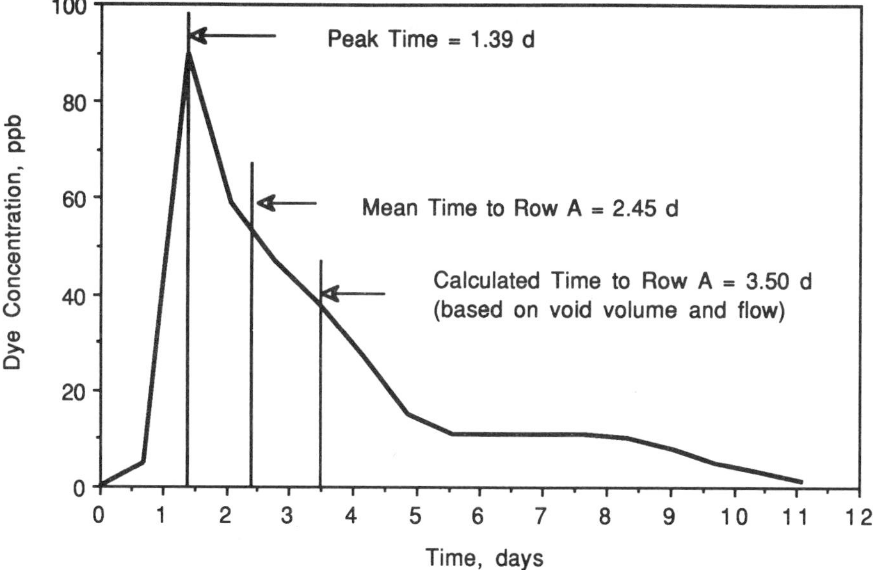

Figure 5. Tracer response curve for row A composite.

$$\frac{k}{k_{lab}} = \left[\frac{\varepsilon}{\varepsilon_{lab}}\right]^{3.7}$$

$$\frac{90{,}000}{145{,}000} = \left[\frac{\varepsilon}{\varepsilon_{lab}}\right]^{3.7}$$

or

$$\varepsilon = 0.88\varepsilon_{lab}$$

For the plugged end of the bed, this calculation may be repeated for field conductivity of 10,000 ft/d (3048 m/d), and lab conductivity of 145,000 ft/d (44,196 m/d), to yield:

$$\varepsilon = 0.485\varepsilon_{lab}$$

In contrast, note that Table 2 indicates that the front end is 57% open, the rest of the bed is about 88% open.

It should be noted that none of three correlations tested would have done an acceptable job of predicting the hydraulic conductivity of these particular gravels. The reason is not known.

The roots and rhizomes of the macrophytes do not at present contribute to porosity reduction, except in the inlet end, because they do not reach the water. The roots generally penetrate no more than 4 to 6 in (10 to 15 cm). In the flooded inlet region, they may block an additional 2 to 6% of the void volume.

In conjunction with the RWT tests, water level surveys were done in September 1990. Figure 2 shows that the elevation of the water surface has not changed over the period September 1989 through summer 1990. For a period of 1 year, there has been neither worsening nor improvement of the hydraulic conductivity of the bed.

DYE TRACER STUDY

Rhodamine WT was added in a pulse to the water entering the wetland, and detected at three rows of five stations and the effluent stream. Several items of interest are evident from the dye results.

First, the dye response data indicate a flow pattern which is not plug flow. Figure 5, the Row A composite from the September 1990 dye study, displays textbook characteristics of non-ideal flow. The response indicates tracer moving from a main flowing channel into dead zones and back out again. The peak occurs at the residence time for travel through the main channels, which is 1.39 days. The true mean residence time includes the tail of the response curve, and has the value 3.50 days for row A. An estimate may be made of the percentages of the dead and flowing volumes, following procedures given by Levenspiel.[8] The volume flow rate was 8900 ft^3/d (252 m^3/d), thus the total void volume is 31,440 ft^3 (890 $m^{3)}$ to row A, or 35%. But, the main channel void volume is only 12,360 ft^3 (350 $m^{3)}$ or 40% of the total.

The non-ideal flow has very important implications for data reduction for contaminant removal. Suppose that a process, such as BOD removal, follows first-order kinetics, which is the common presumption in the literature. For a half-life of four days, or a "k" of 0.1733 reciprocal days, the observed reduction for the actual residence time distribution would be 59.4%. Now, the intent of the Benton data collection was to infer k values. The procedure was to sample at the peak time, assuming this to be the average arrival time of the fluid sampled at the inlet. In this example, row A was sampled at 2230 min. For the hypothetical 59.4% reduction, the calculated k value would be 0.582 reciprocal days. That is 3.36 times too high.

Second, there are great differences in the dye responses at individual standpipes, both in timing and in magnitude. It is by no means clear that the five standpipes present a representative sample of all the water passing that longitudinal cut plane. In the case of row A, peak times range from 1000 to 4000 min; and peak concentrations range from 10 to 170 ppb. In terms of totals, nearly ten times as much dye passed standpipe #3 as passed standpipe #1.

Third, there are large diurnal fluctuations in the concentrations of dye leaving at the H-flume. The swings in concentration are 40% of the total. This implies potential errors in grab sampling of the outlet during normal working hours only.

On a more speculative level, there is an interesting observation to be made concerning flow distribution. In packed beds confined by vessel walls — an extremely common device in the chemical process industries — there are only minor deviations from plug flow behavior in long, slender geometries. A sharp pulse in the input results in a relatively sharp pulse in the output, although some concentration peak broadening occurs due to dispersive processes. That is definitely not true for the gravel cell at Benton, nor is it true for other gravel cells. There appear to be large zones of relative stagnation, with flow serpentining along preferential paths in the bed. This leads to the dye response curves observed repeatedly at Benton.

A possible explanation of this major departure from confined packed bed flow exists in the presence of the free-water surface. Very low surface water elevation gradients drive the flow in the gravel bed. Similar gradients in the transverse direction result in only minor surface elevation changes in the short transverse distance, yet they may influence crossflow in a major way. The measured cross gradients were up to 30% of the longitudinal gradients at some locations, indicating a strong cross flow component at those locations. Such transverse driving forces cannot occur in a confined flow because of the presence of the confining wall. The net effect is a highly non-ideal flow in a free surface packed bed; that is, a flow far removed from plug flow. A larger proportion of dead zones becomes possible, together with the necessary regions of bypassing.

CONCLUSIONS

This work produced the following information and tentative conclusions.

1. The gravel bed was designed close to its hydraulic capacity based on clean rock conductivity. A small amount of pore plugging pushed it to a flooding condition.
2. This mud is predominantly inorganic in nature (80%). The principal metals are calcium, silicon, and aluminum. These apparently derive from lagoon solids (silicon), substrate clays (aluminum), and the rock (calcium). Compounds of calcium and silicon are known to form gel precipitates. The organic fraction (20%) of the mud is dominated by bacteria and algae. On a mass basis, algae are of more significance. The numbers of these organisms decrease with distance from the inlet. Bacteria are probably associated with the organic substrate in the inlet end pore space. Algae are probably dying, settling, and being filtered; to form the organic substrate.
3. The front 20% of the gravel bed has been partly plugged with a dark gray, gelatinous material (mud). The material that is causing problems is most likely a clay-rock-organic gel, charge-stabilized by excess cation capacity. This material is mostly inorganic. The organic fraction is decomposing lagoon algae and

bacteria. The inorganic fraction results from mineral residues from lagoon algae and bacteria, as well as from minerals associated with clay and rock substrates.

4. The dry mud solids comprise an average of 2.3 to 3.3 weight percent of the total mass (water and mud) in the voids in the plugged zone. Due to the gel nature of the mud, it has blocked approximately two thirds of the void volume, based on dye study results.
5. The total solids fed to the gravel bed over its history total 80,000 kg, of which about 8000 kg were mineral ash. The total mud in the plugged zone total was about 20,000 kg. The mineral portion of the mud accumulation (80% of 20,000 = 16,000 kg) therefore resulted in part from other sources. The primary source is believed to be dissolution of limestone calcium and its subsequent reaction with silicon in the wastewater and the algae. The exit sections of the gravel bed also contain solids in the void spaces, but in lower amounts.
6. The hydraulic conductivity of the inlet zone of the gravel bed has been lowered substantially by the mud formation. Bed average conductivities are lower in the plugged zone by a factor of 10 (2800 vs. 28,000 m/d). However, most of this conductivity is in surface flow, so the situation is worse than indicated by averages. Conductivities in the gravel are probably reduced by two orders of magnitude.
7. The hydraulic conductivity of the bed has not changed over the past year; the clogging has apparently stabilized under the reduced flow operation used during that year.
8. The design of the gravel bed (fixed slope, fixed gravel depth, and uniform hydraulic conductivity in the longitudinal direction) permits only one flow rate if the water surface is to remain parallel to the gravel surface. In the presence of plugging, there is no possibility of maintaining water below and parallel to the bed surface without major bed alterations.
9. The flooded inlet zone contains healthy wetland vegetation, including a litter mat which is trapping lagoon solids.
10. The downstream zones of the gravel are dry to a depth of many centimeters. The roots of the plants do not reach water, and the vegetation is changing to a terrestrial mix.
11. The flow patterns within the bed are non-uniform. Preferential flow paths are causing bypassing. The fastest paths have velocities 6.3 times the average in the plugged zone, and 1.6 times the average in unplugged zones. This is manifested in early arrival times for the dye front at all locations, and by the unequal magnitude of the dye concentration response at points equidistant from the inlet. This is occurring because of the ease with which flow can be rerouted by very slight changes in the slope of the free-water surface.
12. The flow pattern is not close to plug flow. In fact, a more appropriate model would be a dead zone interchanging water with stirred tanks in series comprising the remaining void volume. Water quality samples taken at internal bed locations are not easily interpreted because flow patterns are not close to plug flow. Significant channeling and dead zones exist.
13. Design procedures should be changed for future systems. The hydraulic conductivity of the candidate gravel should be measured in advance. The hydraulics of the bed must be calculated during design, to produce the correct size and shape of the bed. This design should include enough flexibility to accommodate upsets, such as rainfall events, as well as changing average flows. The accumulation of interstitial solids is inevitable, and must be planned for.

REFERENCES

1. Reed, S. C. An Inventory of Constructed Wetlands Used for Wastewater Treatment in the United States. prepared for USEPA RREL, Cincinnati, OH, 1990.
2. Steiner, G. R., J. T. Watson, D. A. Hammer, and D. F. Harker. Municipal wastewater treatment with artificial wetlands — a TVA/Kentucky demonstration, in Aquatic Plants for Water Treatment and Resource Recovery. K. R. Reddy and W. H. Smith, Eds. Magnolia Publishers, Orlando, FL, 1987, 923.
3. Darcy, H. Les Fontaines Publiques de la Ville de Dijon. Dalmont, Paris, 1856.
4. Ergun, S. Fluid Flow Through Packed Columns, Chem. Eng. Prog., 48:89, 1952.
5. Kadlec, R. H. Hydrologic factors in wetland water treatment, in Constructed Wetlands for Wastewater Treatment. D. A. Hammer, Ed. Lewis, Chelsea, MI, 1989, 373.
6. Brown, G. G. et al. Unit Operations. John Wiley & Sons, New York, 1956.
7. Idelchik, I. E. Handbook of Hydraulic Resistance. Hemisphere, New York, 1986.
8. Levenspiel, O. Chemical Reaction Engineering, 2nd ed. John Wiley & Sons, New York, 1972.

CHAPTER 24

Gravel Bed Hydroponic Sewage Treatment: Performance and Potential

J. E. Butler, M. G. Ford, E. May, R. F. Ashworth, J. B. Williams, A. Dewedar, M. El-Housseini, and M. M. M. Baghat

INTRODUCTION

Gravel bed hydroponics (GBH) is a novel effluent treatment process that can provide cost-effective and aesthetic solutions to a number of environmental pollution problems.[1-3] GBH systems are sloping channels lined with an impermeable membrane and filled with gravel or an equivalent aggregate to provide a matrix in which hydrophytes such as *Phragmites australis* are planted. Feed water introduced at the top of the bed flows through the aggregate to emerge downstream as a final effluent. GBH systems constructed in Egypt and the U.K., are showing considerable potential for both secondary and tertiary treatment of domestic sewage. Furthermore, the treated effluent can be used to provide a nutrient solution for hydroponic crop production. Crops grown successfully in GBH systems include sugar beet, broad bean, sunflower, safflower, cotton, sorghum, maize, and napier grass.

TREATMENT MECHANISMS

Although GBH treatment mechanisms are still only poorly understood, many of the processes associated with percolating filters have been observed to occur within the gravel bed.[3] Establishment of a biofilm which can transform a variety of carbon and nitrogen compounds occurs within weeks of the introduction of feed water to the channels. Oxidation of reduced nitrogen species is then made possible by the increased concentration of dissolved oxygen which occurs as the sewage passes through the aggregate. This results in the establishment of an anaerobic region at the top of the bed and an aerobic zone towards the channel outlet; the GBH system therefore behaves as a combined anaerobic/aerobic digester.[3] Little treatment occurs in the absence of the biofilm.

Mechanisms of Oxygenation

Oxygen is supplied partly as a result of the convective movement of air from the aerial shoots of the common reed, *Phragmites australis*, through its aerenchyma channels, to reach the root surface where it is released into the surrounding media.[4] There is some controversy about the extent of this process and estimates of the rate at which the system can supply oxygen in this way range from as low as 0.02 g O_2/m^2 day for soil-based reed beds,[5] to as high as 12 g O_2/m^2 day.[4] The oxygen release rate is thought to increase with both plant density[4] and the hydraulic conductivity[5] of the bed. Oxygen may also be supplied as a result of air movement into the bed as the feedwater level falls during the flow-off period of an intermittent flow regime, and possibly as a result of flow around the gravel particles and through air-filled pores.[5]

Mechanisms Based upon Microbiological Processes

Analysis of the bacteria in the beds has demonstrated the presence of ammonifiers and organisms responsible for nitrifying and denitrifying activities. The numbers of ammonifiers and denitrifiers on the plant

0-87371-550-0/93/$0.00+$.50

root and gravel surfaces are usually higher than those responsible for nitrifying activity; the numbers of ammonia-oxidizing organisms are higher than those oxidizing nitrite; the denitrifying bacteria form the largest single group.[3]

The potential of the GBH system to provide effluents, either rich in nitrate or devoid of nitrogen species as a result of nitrate reduction to gaseous end products, such as nitrogen, therefore appears high. Work is underway at Portsmouth to measure the biochemical activities of the GBH microorganisms involved in nitrogen transformations, and to relate this to counts of the responsible bacterial groups present in the gravel beds. We already have evidence to suggest that the balance between the processes of nitrification and denitrification can be affected by the mode of operation of the GBH system.[3] Recent modeling studies based on unpublished data now suggest that this balance is likely to be influenced by the residence time of the effluent within the bed. If these early indications are confirmed, then it should be possible to operate the GBH system in either denitrification mode for discharge of the final effluent into waterways, where there is a requirement for an effluent low in nitrate, or in nitrification mode to produce a nitrate-rich effluent to act as a nutrient source for hydroponic crop production.

ESTABLISHMENT OF THE BEDS

Establishment of a healthy hydrophyte monoculture[1,2] is achieved by planting clumps of the reed, *Phragmites australis,* at a density of 5 per meter and maintaining a supply of sewage which ensures an average sewage depth greater than 5 cm above the channel bottom. Parr has recommended the use of clumps rather than rhizomes, which are particularly susceptible to drought and nutrient deficiency.[6] It is essential that the reed clumps are planted so that they contact the underlying membrane and maintain adequate immersion in the sewage. Plant height and density, and hence the supply of oxygen via the roots,[4] are critically dependent upon the depth of the sewage as illustrated in Figure 1A. Moreover, sewage depths below 5 cm result in unhealthy and uneven growth characterized by yellow chlorotic shoots and leaves (Figure 1B).

GBH PERFORMANCE

The performance of the GBH systems used for secondary treatment of settled sewage in the U.K. and Egypt, and tertiary treatment in the U.K. of mixed effluent from biological filters (60% of the effluent stream), or activated sludge plants (40% of the effluent stream), has been monitored for a period of 2 to 3 years. A large number of properties have been studied in an attempt to identify the mechanisms of treatment and how to improve performance by modified operation and design.[7] Some of the results of this study have already been published,[1-3,7-9] together with details of construction.[1,2,9]

Secondary Treatment in Mediterranean and Temperate Climates

Secondary treatment of domestic sewage in Egypt (Table 1; Figure 2) gave substantial reductions in suspended solids and BOD.[1,3] The oxygen concentration approached saturation at the end of the 100-m beds accompanied by a reduction in ammonia to very low levels, and a rise in nitrate to an average concentration of 5.6 mg/L. Over the same bed distance, counts of fecal bacteria and total aerobic bacteria were reduced 100-fold.[3] Figure 2 also shows a consistently satisfactory standard of treatment throughout the monitoring period, improvement in performance as the beds become established, a pronounced damping effect when outlet variability is compared to that of the inlet (Figure 2B), and the absence of seasonal effects on performance; the GBH system therefore provides a robust treatment.

For secondary treatment in the U.K., where average temperatures are substantially lower than those of Egypt, most of the performance criteria can be met using beds of 100 m or more in length (Table 1). However, ammonification was observed for the effluent discharged at the end of a 100-m bed for which unacceptably high concentrations of ammonia (25 mg/L) remained in solution. The ammonia concentration was reduced by further treatment which took place in the 40-m crop bed through which the effluent subsequently flowed during the summer months; this additional length of bed resulted in an overall proportional reduction (outlet concentration/inlet concentration) of 0.6; that is, an ammonia concentration at the crop bed outlet of 14.5 mg/L. Other determinants, including BOD, suspended solids, and soluble BOD, were also reduced by the additional treatment provided by the crop bed. Thus, increased bed length can produce more effective sewage treatment.[1,2]

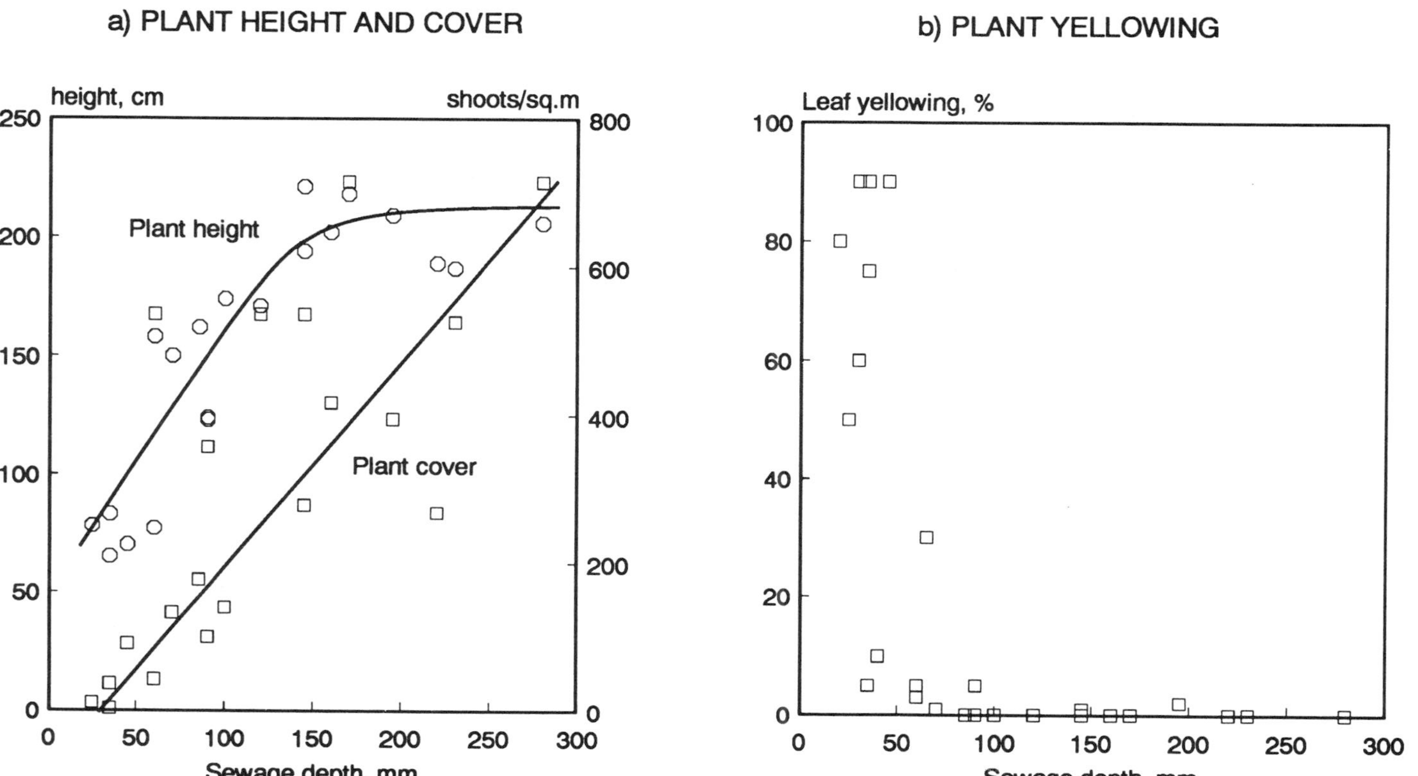

Figure 1. Growth characteristics of *Phragmites australis* in GBH systems in relation to sewage depth: A, plant height (○) and cover (□); B, plant yellowing.

Table 1. The Performance of Established 100-m GBH Channels Operated in Mediterranean and Temperate Climates

Determinant	Tertiary treatment U.K.	Secondary treatment U.K.	Secondary treatment Egypt
Biological oxygen demand (BOD)	0.09*	0.11*	0.23*
Suspended solids	0.56*	0.41*	0.30*
Ammonia	0.04*	1.16*	0.16*
Dissolved oxygen	6.5**	1.4**	4.9**

Note: Performance is based on the proportional reduction of feedwater concentration at the outlet* or the oxygen tension (mg/L)**. Hydraulic loadings were approximately 10 L/min/m bed width; residence times were typically 5h.

Tertiary Treatment in the United Kingdom

The treatment of effluent from biological filters and activated sludge processes leads to a further polishing to produce a final effluent with significantly lower values of BOD, suspended solids and ammonia, and higher concentrations of dissolved oxygen (Table 1). The reduced concentrations of the pollutants found in this final effluent should conform to the discharge consent standards which are now being set by government regulators. Furthermore, there is a substantial reduction in the variability of these determinant concentrations as shown, for example, for tertiary treatment of ammonia monitored over a 2- to 3-year period (Figure 3).

Bed Treatment Profiles

The effect of bed length on the effectiveness of treatment is probably most clearly demonstrated by inspection of the chemical profiles observed for increasing distance along the channel.[1,3] The profiles obtained for secondary and tertiary treatment are similar; examples are presented in Figures 4 and 5 for 100-m gravel channels used for tertiary treatment in the U.K.

Seasonal Effects

On average, a bed length of 50 to 60 m seems satisfactory for tertiary treatment of mineral nitrogen (Figure 4A), BOD and soluble BOD (Figure 4B), and organic matter (Figure 4d); significant oxygenation has also been achieved by 60 m (Figure 4C). However, seasonal effects can modify bed profiles; see, for example, the profile for ammonia which extends down a greater length of the bed during the winter months compared to the spring and summer (Figure 5A). This feature is also observed for mineral nitrogen (Figure 5B), which is lost from the bed more effectively during the summer time, presumably as a result of associated changes in the relative rates of nitrification and denitrification. Similar seasonal changes in bed profiles have been observed for secondary treatment in Egypt.

Removal of Organic Solids

The similar tertiary treatment profiles for dissolved oxygen and pH (Figure 4C) are of particular interest. At the top of the bed, the oxygen demand required for the oxidation of organic matter is high and anaerobic conditions are established; organic acids and ammonium ions are produced as a result of anaerobic fermentation. Thus, an initial consumption of oxygen is observed over the first 5 m of the bed and is associated with a decrease in pH to a value of 7.1, presumably as a consequence of the oxidation of carbon chains to fatty acids which are in turn substrates for methanogenic bacteria. The pH is maintained at around 7.1 to 7.2 for a distance of 30 to 40 m. This corresponds to the anaerobic (0 to 20 m) and predominantly facultative anaerobic (20 to 40 m) regions of the bed. By 40 m, both the BOD and the sediment and suspended solids have fallen to low levels (Figure 4B). The dissolved oxygen and the pH then start to rise to reflect the fall in oxygen consumption associated with the reduced availability of readily decomposable organic material and the neutralization of short-chain fatty acids to form salts, and the oxidation of these acids to carbon dioxide. There is evidence to suggest that progressive build-up of organic matter within the bed is prevented by movement of the solids and suspended organics from the anaerobic zone into the aerobic zone, where the process of humification can occur more efficiently. Thus, the porosity of the 100-m bed used for tertiary treatment remained more or less constant throughout the 2- to 3-year period of monitoring (Figure 6).

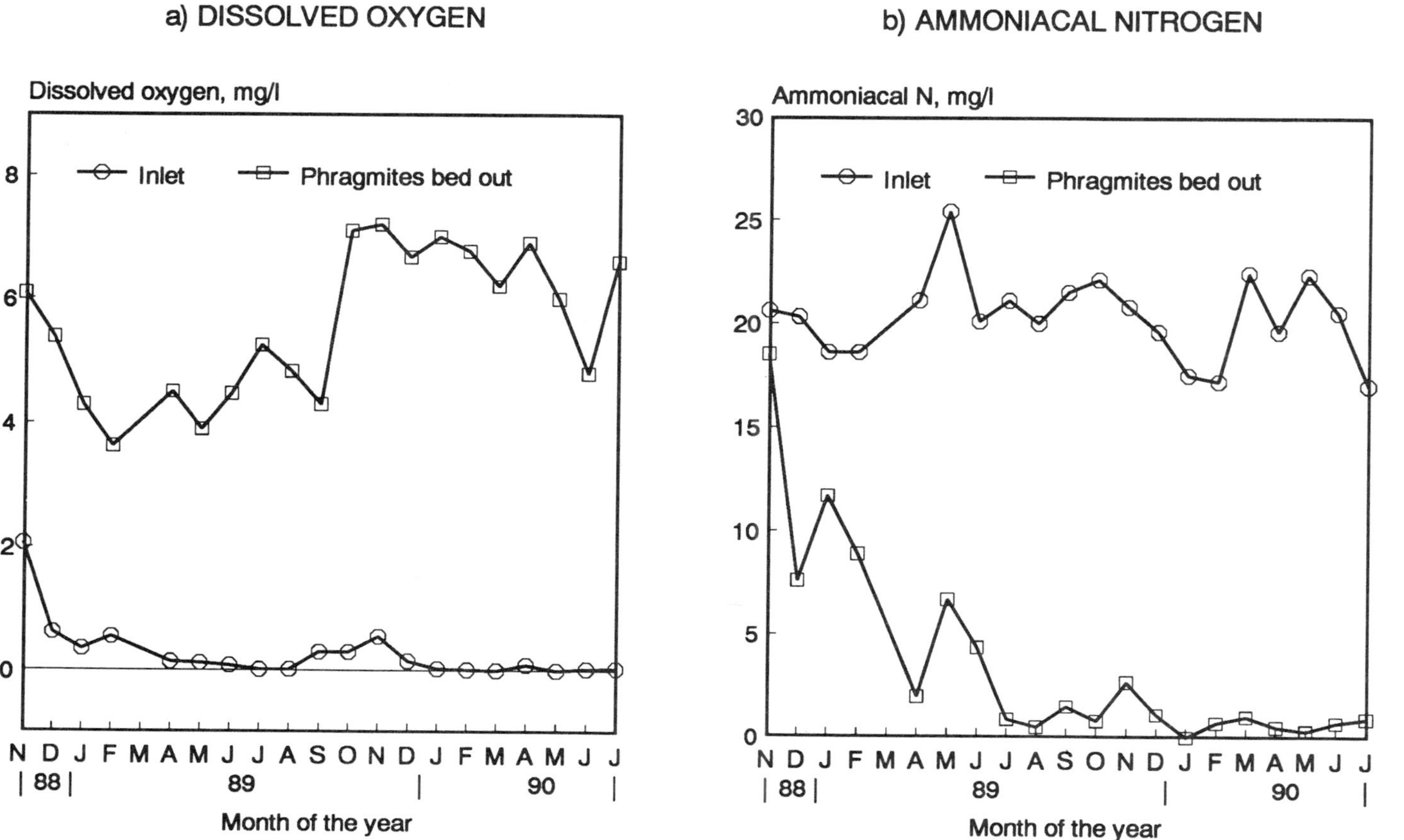

Figure 2. Performance of GBH systems in Egypt used for secondary treatment: A, dissolved oxygen; B, ammoniacal nitrogen.

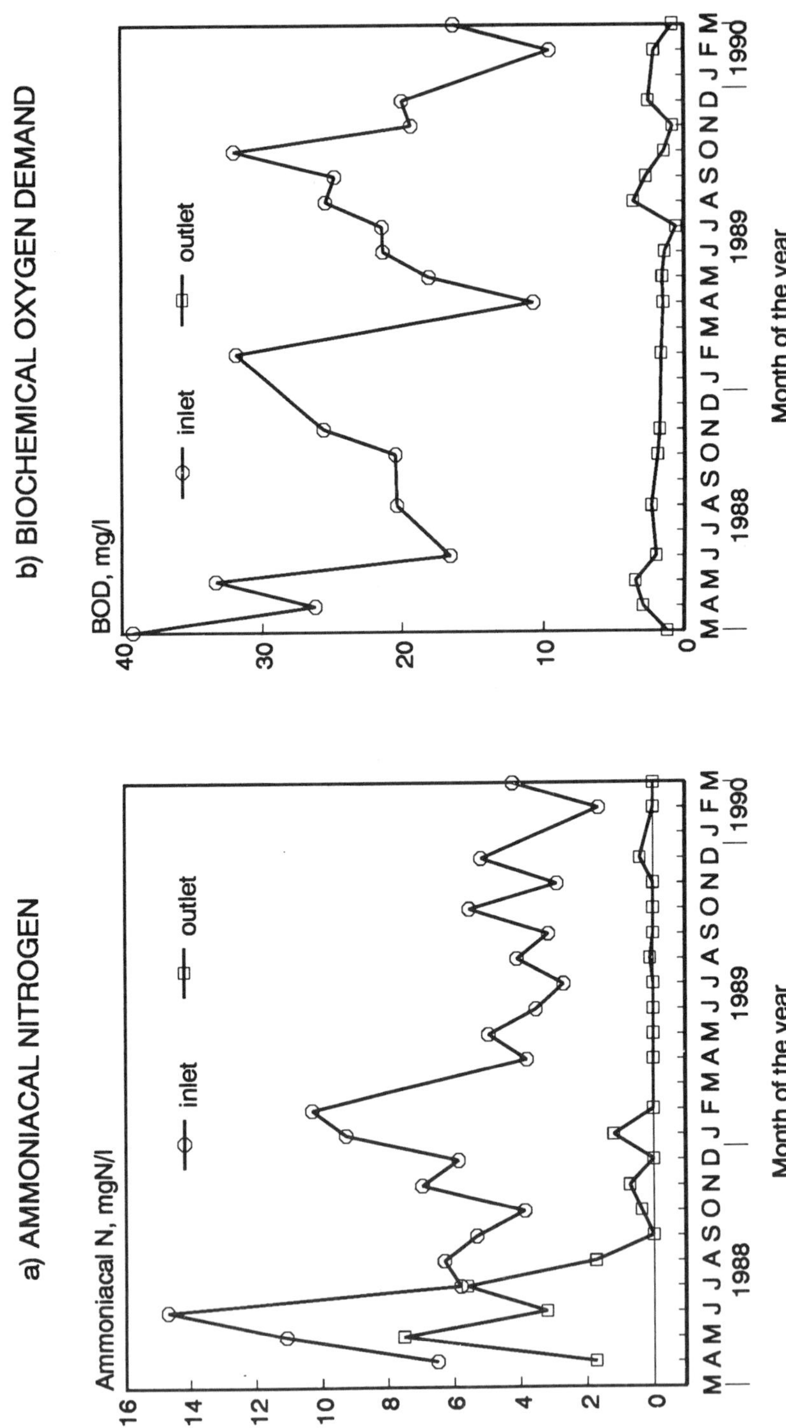

Figure 3. Performance of GBH systems in the U.K. used for tertiary treatment: A, ammoniacal nitrogen; B, biochemical oxygen demand (BOD).

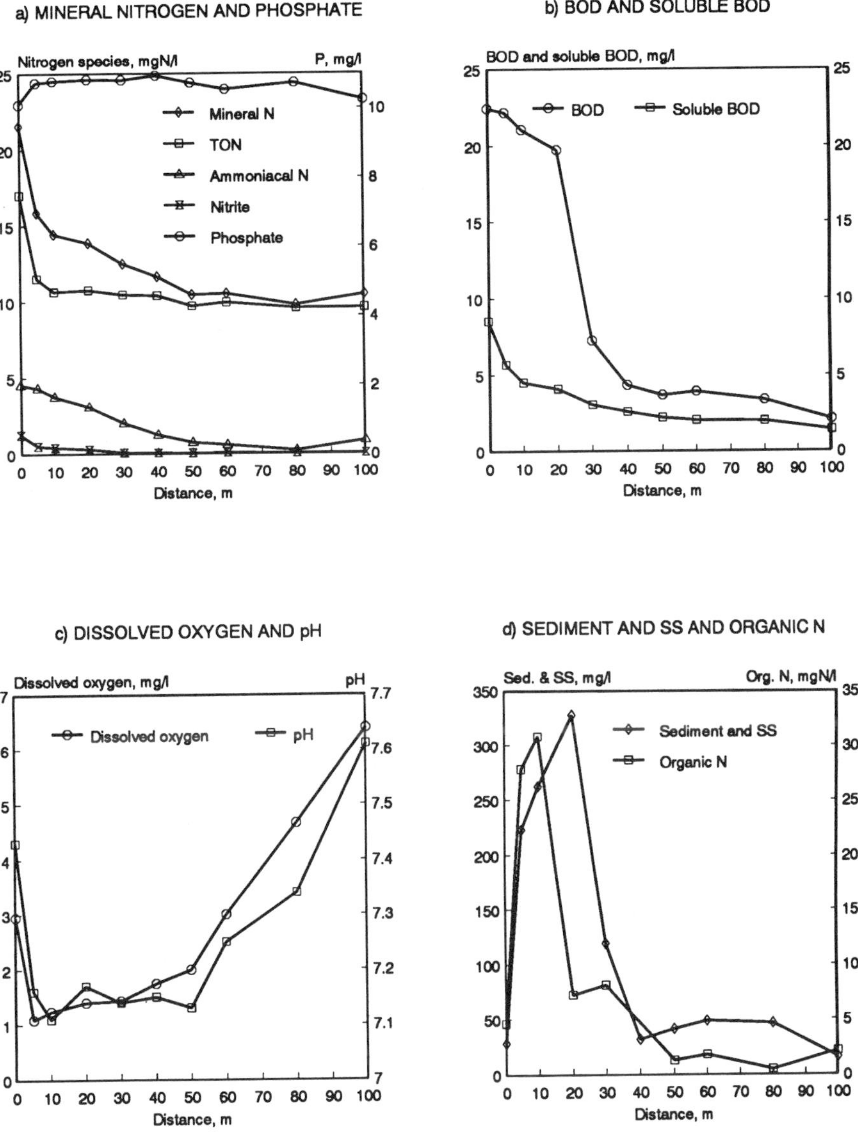

Figure 4. Averaged bed profiles for tertiary treatment of sewage in the U.K.: A, mineral nitrogen and phosphate; B, BOD and soluble BOD; C, dissolved oxygen and pH; and D, sediment and SS, and organic nitrogen.

Removal of Bacteria and Viruses

GBH systems are able to remove both bacteria[3] and viruses from sewage feed water, with exponential decay with respect to distance along the channel. The linear regression forms of the exponential equations describing the reductions in counts of aerobic bacteria (Equation 1) and fecal coliforms (Equation 2) are presented below; standard errors are shown in parenthesis.

Bacteria (tertiary treatment):

$$\log(\text{CFU/ml}) = 5.92 - 0.021\ (\text{bed length in meters}) \qquad (24\text{–}1)$$

$$(0.11)\ (0.002)$$

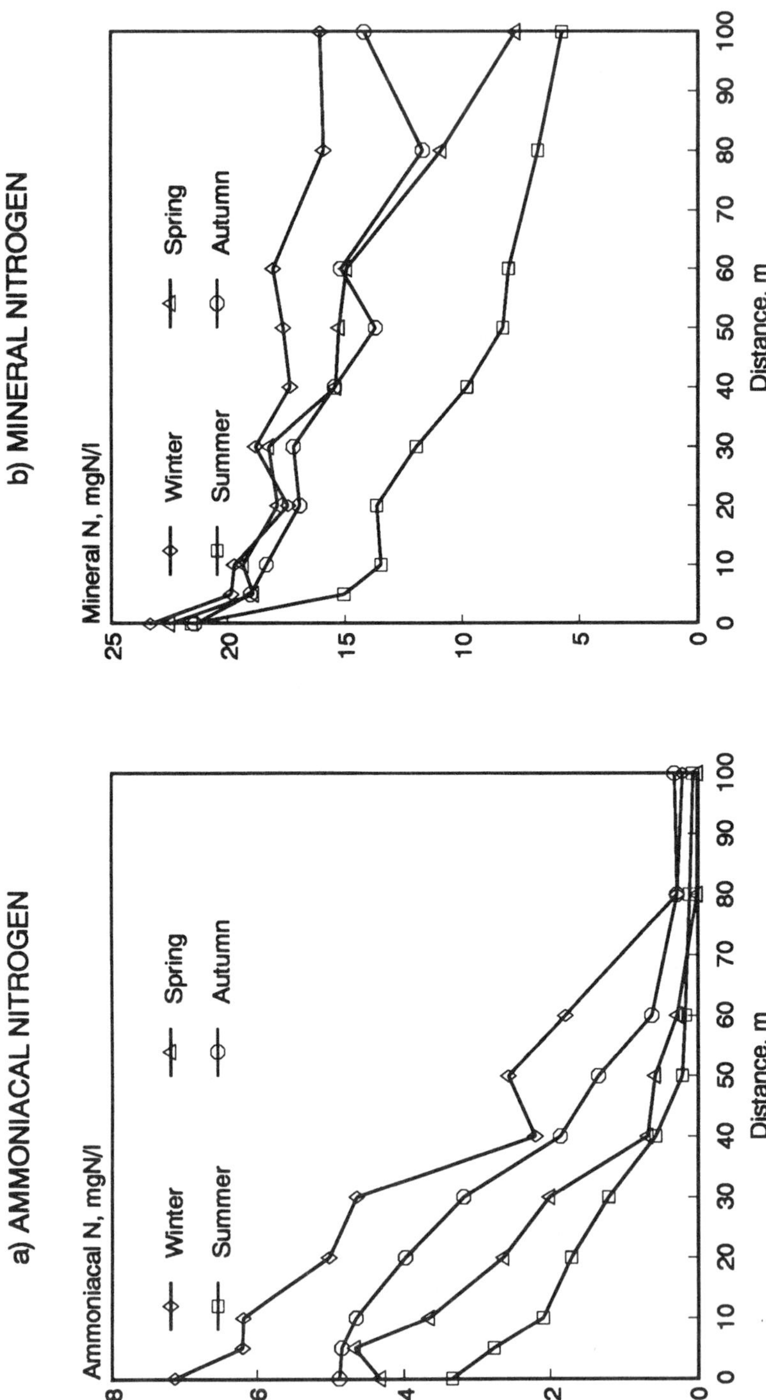

Figure 5. Seasonal effects on the bed profiles for tertiary treatment of sewage in the U.K.: A, ammoniacal nitrogen; B, mineral nitrogen.

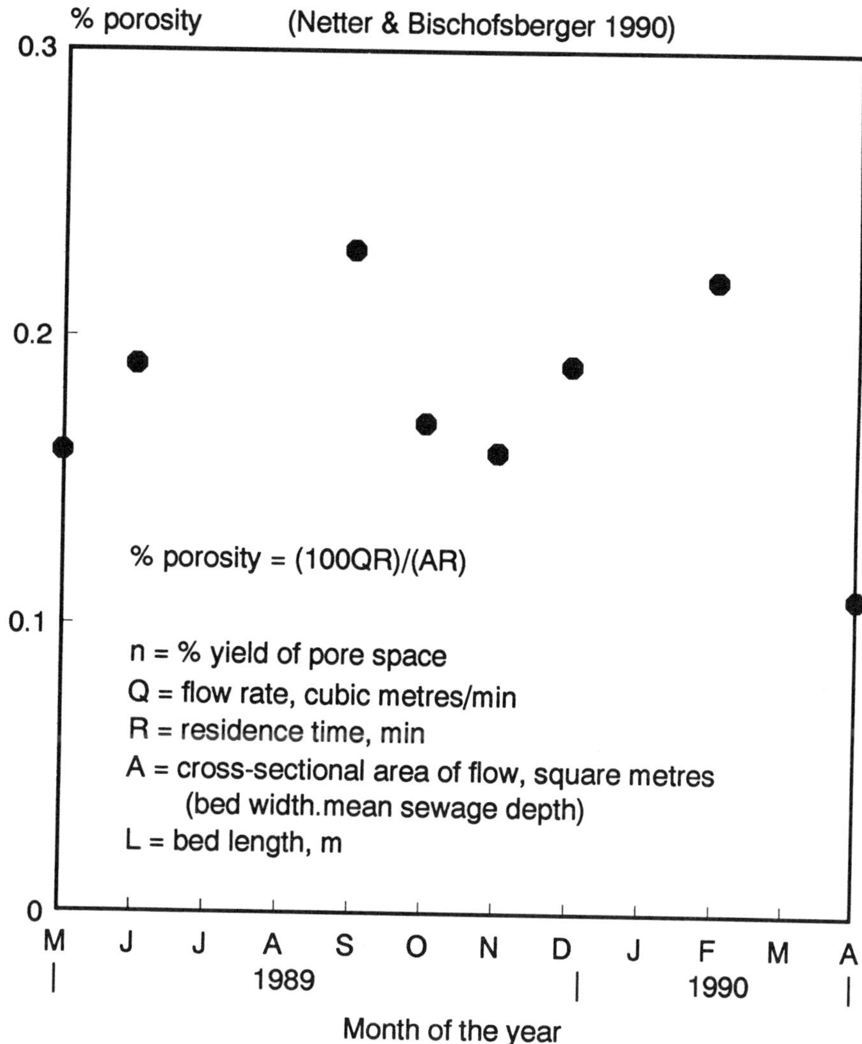

Figure 6. The effect of continual tertiary treatment on bed porosity.

$$\log(\text{CFU/mL}^{-1}) = 4.04 - 0.020 \text{ (bed length in meters)} \quad (24\text{–}2)$$

$$(0.10) \quad (0.002)$$

The reduction in numbers of virus particles has been estimated for populations of *Escherichia coli* infecting bacteriophage which are indigenous to the feed water (Equation 3), and for seeded populations of bacteriophage MS-2 applied to the top of the bed as a discrete dose of concentrated inoculum (Equation 4). Phage MS-2 has been recommended as a model organism for human virus inactivation.[10] The corresponding regression equations are:

Viruses (tertiary treatment).

$$\log(\text{PFU/mL}^{-1}) = 2.46 - 0.026 \text{ (bed length in meters)} \quad (24\text{–}3)$$

$$(0.12) \quad (0.002)$$

$$\log(\text{PFU/mL}^{-1}) = 6.58 - 0.058 \text{ (bed length in meters)} \quad (24\text{–}4)$$

$$(0.33) \quad (0.007)$$

There was no significant difference between the exponents of decay for aerobic and fecal bacteria; however, the viruses are removed more efficiently than the bacteria. Removal of indigenous phage is approximately half as efficient as that for the MS-2 phage, where dilution of the concentrated inoculum will have contributed to the reduction in counts. These results correspond to decimal reduction distances of 50 m for the bacteria and 17 to 40 m for the viruses; similar distances are observed for secondary treatment.

FUTURE PROSPECTS

GBH systems provide aesthetically pleasing, cost effective treatment of a variety of effluents containing unwanted pollutants. The system is available for commercial development as a treatment for domestic sewage, either as a secondary process fed settled sewage or as a tertiary polishing process. The final effluent from GBH treatment can provide a suitable nutrient solution for either hydroponic crop production or conventional irrigation in arid regions of the world where water is a scarce resource. We are now investigating the potential of the system to treat a variety of industrial wastes and concentrated biological wastewaters such as animal slurries and food processing effluents.

ACKNOWLEDGMENTS

This research was performed with grants from the Overseas Development Administration and the Science and Engineering Research Council. JBW was in receipt of a SERC case award in collaboration with Southern Water PLC; MMMB and M.El-H were in receipt of British Council travel awards. Technical assistance was provided by C. Hughes (School of Civil Engineering) and R. F. Loveridge (School of Biological Science).

REFERENCES

1. Butler, J. E., R. F. Loveridge, M. G. Ford, D. A. Bone, and R. F. Ashworth. Gravel bed hydroponic systems used for secondary and tertiary treatment of sewage effluent, *J. IWEM*. 4:276, 1990.
2. Butler, J. E., M. G. Ford, R. F. Loveridge, and E. May. Design, construction, establishment and operation of gravel bed hydroponic (GBH) systems for sewage treatment, in *Constructed Wetlands in Water Pollution Control*. P. F. Cooper and B. C. Findlater, Eds. Pergamon Press, Oxford, U.K., 1990, 539.
3. May, E., J. E. Butler, M. G. Ford, R. F. Ashworth, J. B. Williams, and M. M. M. Baghat. Comparison of chemical and microbiological processes in gravel bed hydroponic (GBH) systems for sewage treatment, in *Constructed Wetlands in Water Pollution Control*, P. F. Cooper and B. C. Findlater, Eds. Pergamon Press, Oxford, U.K., 1990, 33.
4. Armstrong, W., J. Armstrong, and P. M. Beckett. Measurement and modelling of oxygen release from roots of *Phragmites australis*, in *Constructed Wetlands in Water Pollution Control*. P. F. Cooper and B. C. Findlater, Eds. Pergamon Press, Oxford, U.K., 1990, 41.
5. Brix, H. and H.-H. Schierup. Soil oxygenation in constructed reed beds: the role of macrophyte and soil-atmosphere interface oxygen transport, in *Constructed Wetlands in Water Pollution Control*. P. F. Cooper and B. C. Findlater, Eds. Pergamon Press, Oxford, U.K., 1990, 53.
6. Parr, T. W. Factors affecting reed (*Phragmites australis*) growth in U.K. reed bed treatment systems, in *Constructed Wetlands in Water Pollution Control*. P. F. Cooper and B. C. Findlater, Eds. Pergamon Press, Oxford, U.K., 1990, 67.
7. Butler, J. E., R. F. Loveridge, and D. A. Bone. The use of hydroponics for sewage treatment in temperate and tropical areas, in *Proc. 7th Int. Cong. on Soilless Culture*. The Secretariat of ISOSC, Wageningen, The Netherlands, 1989, 103.

8. Butler, J. E., R. F. Loveridge, and D. A. Bone. Crop production and sewage treatment using gravel bed hydroponic irrigation, *Water Sci. Technol.*, 21:1669, 1989.
9. Butler, J. E. The design, construction and operation of a gravel bed hydroponic sewage treatment system in Egypt, presented at *IWEM Conf.*, Glasgow, September 2–4, 1990.
10. Havelaar, A. H. and W. M. Pot-Hogeboom. F-specific RNA-bacteriophages as model viruses in water hygiene: ecological aspects, *Water Sci. Technol.* 20:399, 1989.

CHAPTER 25

The Purification Efficiency of the Planted Soil Filter In See

R. Netter

INTRODUCTION

In Germany, a dichotomy exists with regard to enforcement of water pollution control regulations: in the western part of the country,[1] about 92% of the inhabitants are connected to sewerage systems; the connection rate is only 73% in the eastern part of Germany.[2] In rural areas, wastewater collection and treatment systems are inadequately developed in both the east and west. In the west only about 70% of the inhabitants of small villages are served by sewerage systems, and only about 63% of the sewage is biologically treated.[1]

Collection and treatment of sewage in areas with low population density is problematic, in general. Decentral wastewater treatment is mostly inevitable for economic reasons. The currently available technology for the treatment of wastewater from dwellings and small communities is in many respects unsatisfactory. "High-tech" constructions are often highly demanding with respect to both investment and operation and management. Alternative solutions, such as constructed wetlands, should be considered.

Constructed wetlands have been used for years for the treatment of sewage from dwellings or smaller drainage areas, up to 1000 population equivalents. They can be classified into three categories: hydrobotanical systems, aquacultures, and soil systems. In general, hydrobotanical and aquaculture systems should not be considered important for Germany for several reasons. Actually, they make up only 10% of constructed wetlands in operation.[3,4]

Soil systems can be differentiated by the granular composition of the filter material used. Seidel[3,4] proposed using mainly gravel, while Dafner[3-5] and Geller[6,7] prefer sandy and others cohesive soil materials.[3,4] The soil filter plant established at the village of See, described in the following text, can be classified as a soil system with sandy material.

DESCRIPTION OF THE SYSTEM

The planted soil filter at See was designed and constructed in 1984 by Dafner[5] for treatment of wastewater from a charity institution. Design capacity is 100 population equivalents. The plant is located about 50 km east of Nuremberg, Bavaria, on the plateau of the *Frankenalb*. Elevation is about 500 m above sea level. The bedrock is a very karstic limestone; this explains the lack of receiving water, a special problem in water pollution control.[5] High treatment standards are to be met. Figure 1 illustrates the system.

Pretreatment

The household wastewater is mechanically pretreated with septic tanks, holding a total capacity of 54 m^3. The average retention time in the septic tanks is 5 days. Until 1984, the mechanically pretreated sewage was utilized in agriculture. Due to the increase in the number of inhabitants, and the subsequent increase of the quantity of sewage, application to farmland became intolerable.[5]

0-87371-550-0/93/$0.00+$.50

Figure 1. Planted soil filter at See.

Construction of the Soil Filter

The constructed wetland consists of a planted soil filter with an area of 940 m^2 and a mean depth of 0.6 m. A natural loam layer (2 m) seals the soil filter from the karstic bed rock. Figure 2 is a plan map of the facility; all dimensions are in meters. Figure 3 is a cross-section of the planted soil filter. Wastewater is introduced into the system by subsurface flow distribution. An infiltration pipe which lies in a gravel pack distributes the sewage over the total width of the planted area.

The filter bed is divided by two dams into three regions. The first region, 410 m^2, is a submerged infiltration area that equalizes the variation of the hydraulic loading. The subsequent underground passage through the filter has a minimum length of 13 m. A deep-seated drain pipe on the effluent side of the bed collects the biologically treated sewage and transports it to a variable-height outlet. It is aerated by cascade flow (approximately 2 m in length) and directed thereafter to a pond in which storm water is also collected. Depending on the sewage quantity and the season, the water either seeps, evaporates, or is used for irrigation. Provisions for emergency overflow of the pond to a dry trench have not yet been needed.[8]

Soil Material

The soil material is a very ferruginous aeolian sand that is found in the neighborhood of the treatment plant. This fine-medium grading sand contains rounded iron ore particles of *limonit* (limonite) and *haematit* (hematite).[5] The uniformity coefficient of the soil is less than 5.

Plants

The plants which originally grew in the infiltration area (6 plants per square meter) are cattails (*Typha latifolia*). The non-infiltration area is planted with reeds (*Phragmites communis*).[5] Over the course of several years, the reed has pushed back the cattails to a small remainder. Figure 4 is a modern photograph of the facility.

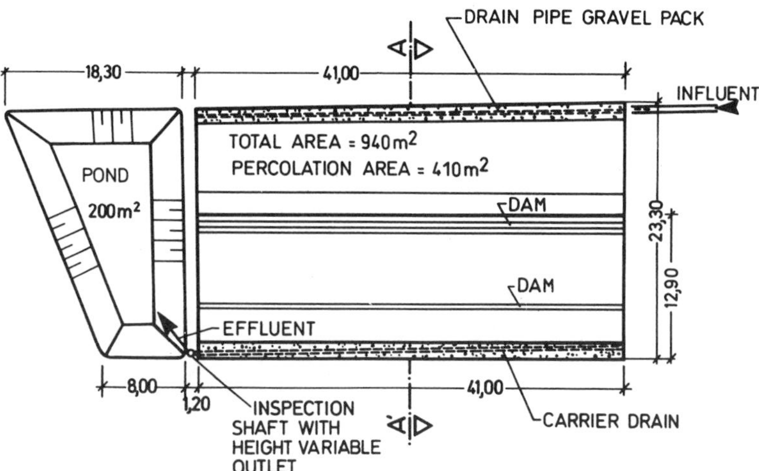

Figure 2. Planted soil filter at See: ground plan.

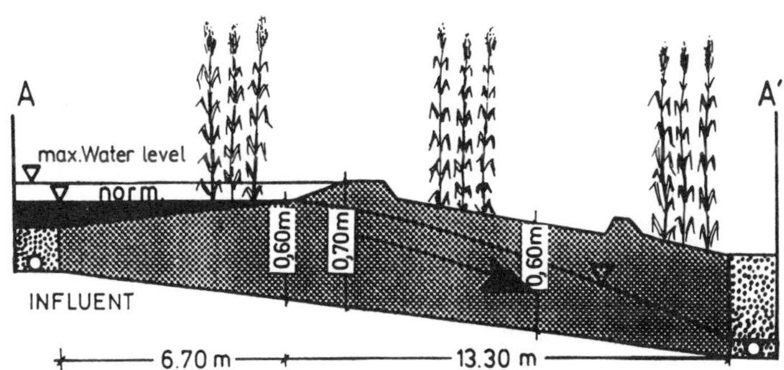

Figure 3. Planted soil filter at See: cross-section.

METHODS

Wastewater analyses, soil analyses, plant studies, and hydraulic investigations were carried out in the years 1984 to 1990 to study the performance of the system. Random samples were routinely taken, and nine intensive measurement were conducted, each lasting 10 to 14 days, with 24-h average samples taken for BOD_5, COD, and nutrient analysis. The continuous measurements allowed a better assessment of the purification efficiency of the system by mass balance studies. Microbiological studies included determination of parameter values such as total bacterial count, counts of coliforms, fecal coliforms and fecal streptococci. The flow measurements in the influent and effluent were conducted using magnetic inductive flowmeters (IDM).[9] Tracer tests to determine the hydraulic retention time were carried out with fluorescent dyes and the bromide anion.[9]

RESULTS

The relatively high inlet concentration of the wastewater treatment plant at See is due to the low specific water consumption of about 100 liters per inhabitant per day. The hydraulic surface loading varied during the

Figure 4. Planted soil filter at See.

continuous measurements from 4 mm/day to 12 mm/day. The BOD_5 mass loading was between 1 and 2.7 kg/day.

Continuous Measurements

The purification efficency was at least 99% with respect to BOD_5, and between 95 and 99% with respect to COD. The nutrient concentration in the effluent varied. Depending on the mass loading and the season, the total nitrogen removal varied between 71 and 97%. The elimination rate of total phosphorus was between 97 and almost 100% depending on the rate of application.[10-12] Table 1 reveals the concentration and purification efficiency at See.

Random Sample Measurements

The concentration in the effluent of the soil filter at See varied between 1 and 45 mg/L with respect to BOD_5, and 23 to 110 mg/L with respect to COD. An effluent concentration of 23 mg/L BOD_5 and 85 mg/L COD fell short by 90% of the samples (n = 50); see Table 2. The effluent concentration with respect to total phosphorus varied between 0.1 and 2.1 mg/L, whereas 90% of the samples showed the effluent values were kept below 1.1 mg/L.

Microbiological Investigations

The microbiological investigations showed significant removals of the total bacterial count (10^2 to 10^3), coliforms and fecal coliforms (10^3 to 10^4), and fecal streptococci (10^3 to 10^4).[12] The observed decrease of the counts (10^2 to 10^4 germs) lies well above the known results for ponds.

Hydraulic Investigations

Tracer tests which were carried out during 3 years showed that the mean hydraulic retention time was between 25 and 40 days with respect to hydraulic load, evapotranspiration, and the hydraulic gradient. Water-level measurements and tracer measurements at the planted soil filter showed that water flows homogeneously through the filter. No stagnant regions were detected.

Table 1. Planted Soil Filter at See — Average Concentration and Purification Efficiency during 9 Continuous Measurement Periods with 24-h Average Samples

Parameter	Influent (mg/L)	Effluent (mg/L)	Purification efficiency (%)
BOD5			
Minimum	141	2	>99
Maximum	283	6	>99
COD			
Minimum	320	28	95
Maximum	579	57	99
NH4-N			
Minimum	66	9	74
Maximum	116	44	98
NOx-N			
Minimum	<1	3	—
Maximum	<1	26	—
Tot N			
Minimum	92	19	71
Maximum	133	67	97
Tot P			
Minimum	13	<1	97
Maximum	21	<1	100

Table 2. Planted Soil Filter at See — Results of the Random Samples (n = 50)

Parameter	Influent (mg/L)	Effluent (mg/L)	90% probability[a] (mg/L)
BOD_5			<23
Minimum	305	1	
Maximum	440	45	
Median		2	
COD			<85
Minimum	519	23	
Maximum	734	110	
Median		49	
NH_4-N			<66
Minimum	39	22	
Maximum	142	73	
Median		46	
NO_x-N			—
Minimum	<1	<1	
Maximum	2.7	38	
Median		—	
Tot P			<1.1
Minimum	8	0.1	
Maximum	23	2.1	
Median		0.3	

[a] The effluent concentration is in 90% of the cases below these values.

CONCLUSIONS

The investigations conducted thus far demonstrate that the planted soil filter at the village of See is a practicable solution for sewage treatment in rural areas. The purification efficiency was above 99% with respect to BOD_5, and between 95 and 99% with respect to COD, provided subsurface water flow was maintained. The results with respect to nutrients varied. Depending on the pollution load and the season, the removal of total nitrogen varied between 74 and 98%. The elimination rate of total phosphorus was between 97 and almost 100%, depending on the rate of application.

ACKNOWLEDGMENTS

The investigation was carried out in context with the research project "Testing the Efficiency of a Planted Soil Filter for Wastewater Treatment", sponsored by the German Ministry of Research and Technology. We are very much obliged to G. Dafner who planned and constructed the planted soil filter, P. Riekes who operates the plant, and to A. Decker who assisted in preparing this paper.

REFERENCES

1. Gilles, J. Oeffentliche Abwasserbeseitigung im Spiegel der Statistik, *Korrespondenz Abwasser.* 34:414, 1987.
2. Zehrfeld, H. Abwasserentsorgung in den neuen Bundeslaendern — Situation und Zustaendigkeit, in *Dokumentation und Schriftenreihe ATV* 29. St. Augustin, Germany, 1991, 13.
3. Poepel, H. J. Uebersicht über Verfahrensvarianten, in *BMFT-Statusseminar Naturnahe Verfahren der Abwasserbehandlung — Pflanzenklaeranlagen.* Essen, Germany, 1989.
4. Poepel, H, J. Uebersicht ueber Verfahrensvarianten von Pflanzenklaeranlagen, *WAR.* 48:17, 1990.
5. Dafner, G. Betriebserfahrungen mit einer Pflanzenklaeranlage, *gwf-Wasser/Abwasser.* 128:241, 1987.
6. Geller, G. and A. Lenz. Bewachsene Bodenfilter zur Abwasserreinigung, *Korrespondenz Abwasser.* 29:142, 1982.
7. Geller, G., E. Englmann, W. Haber, K. Kleyn, A. Lenz, and R. Netter. Bewachsene Bodenfilter zur Reinigung von Waessern — ein von Abwasser beeinflußtes System aus Boden und Pflanzen, *Wasser und Boden.* 43:477, 1991.
8. Scheer, G. Bericht ueber die Anlage See, in *ATV-Seminar Pflanzenklaeranlagen.* Nuernberg, Germany, 1988.
9. Netter, R. and W. Bischofsberger. Hydraulic investigations on planted soil filters, in *Proc. Int. Conf. Constructed Wetlands in Water Pollution Control.* Cambridge, P. F. Cooper and B. C. Findlater, Eds. Pergamon Press, Oxford, U.K., 1990, 10.
10. Netter, R. and W. Bischofsberger. Erprobung der Leistungsfaehigkeit eines bewachsenen Bodenfilters zur Abwasserreinigung — Betriebserfahrungen und Bemessung, in *BMFT-Statusseminar Naturnahe Verfahren der Abwasserbehandlung — Pflanzenklaeranlagen.* Essen, Germany, 1989.
11. Netter, R. and W. Bischofsberger. Sewage treatment by planted soil filters, in *Proc. Int. Conf. Constructed Wetlands in Water Pollution Control.* Cambridge, P. F. Cooper and B. C. Findlater, Eds. Pergamon Press, Oxford, 1990, 525.
12. Netter, R. Leistungsfaehigkeit von bewachsenen Bodenfiltern am Beispiel der Anlagen Germerswang und See, *WAR.* 48:135, 1990.

CHAPTER 26

Constructed Wetlands for Wastewater Treatment in Czechoslovakia: State of the Art

J. Vymazal

INTRODUCTION

The need for wastewater treatment, together with new economic conditions in Czechoslovakia, have led to introduction of systems with low energy input, and low capital, operation, and maintenance costs. When land is available, constructed wetlands offer a simple and effective process design which is attractive for use by small communities. Recently, it has been demonstrated that constructed wetlands perform well, not only for municipal sewage treatment, but also for agricultural and industrial wastewaters, landfill leachate, acid mine drainage, and urban storm runoff treatment.[1-3] Since 1988, four full-scale constructed wetlands have been built and three systems are presently under construction.

PILOT PLANT EXPERIMENTS

The small reed bed model has been operating in the Prague Central Wastewater Treatment Plant since May 1988. The model was made of polypropylene and was 1.1. m long, 1 m high and 0.5 m wide. The slope of the bottom was 2.4%. The system was filled with a mixture of stony gravels (fractions 2 to 8 mm and 0.5 to 2 mm) and limestone gravel (0.5 to 3 mm) in the ratio 3:1:1. The porosity of the gravel bed was 38%.The bed was planted with common reed (*Phragmites australis*) from the locality in the form of rhizomes by planting 12 pieces. A detailed description of the model is given by Vymazal.[4,5]

Treatment of Sewage

Mechanically treated sewage from the Central Treatment Plant was applied to the reed bed system beginning in June 1988. The operational parameters were: average flow (Q), 19.9 L/s; hydraulic loading (L_h), 5.7 cm/d; organic loading (L_o), 114 kg BOD_5/ha/d; and theoretical retention time, 5.6 d. Composite samples were collected twice a week. The results from the first year of operation are shown in Table 1. The treatment system demonstrated effective removal of organics (BOD_5 and COD), suspended solids, and bacteria. The removal of nitrogen and phosphorus was less than 50%, which is typical for wetland systems. There was no significant difference between summer and winter treatment efficiencies.[4]

Treatment of Chicken Manure Wastewater

In the period October 1989 to August 1990, diluted chicken manure wastewater was treated in the same experimental system. Samples were taken in the periods of October to December 1989 (period 1), and August to September 1990 (period 2). Hydraulic residence time, measured by the dye method, was 13 days. Hydraulic loading rate was 2.5 cm/d. Loading rates in period 1 were 325 kg BOD_5/ha/day, 195 g P/m^2/year, and 2660 g N/m^2/year; in the period 2, 347 kg BOD_5/ha/day, 224 g P/m^2/year, and 2885 g N/m^2/year. In spite of very high loadings, the removal efficiencies were very high and steady during 1 year of operation (Table 2). The productivity of *Phragmites* was very high during this study. In September 1990, 480 stems were present, with an average height of 300 cm.

0-87371-550-0/93/$0.00+$.50

Table 1. Experimental Reed Bed Treatment System at Prague Central Treatment Plant

	Inflow	Outflow	Efficiency (%)
BOD_5	200 (80–490)	21 (4.5–61)	89.5
COD	380 (125–850)	64 (24–118)	83.1
Suspended solids	240 (30–630)	5.7 (1.0–12)	97.6
Total P	6.5 (1.9–15.4)	3.3 (0.5–10.6)	49.2
Total N	48 (17–99)	26 (9.8–39)	45.8
Coliforms	107×10^5 ($20–200 \times 10^5$)	3.6×10^3 ($10^3–12 \times 10^3$)	99.97
Fecal coliforms	79×10^5 ($5–140 \times 10^5$)	2.1×10^3 ($0.5–8 \times 10^3$)	99.97
Enterobacteria	18×10^5 ($4–40 \times 10^5$)	700 ($0–4 \times 10^3$)	99.96

Note: Average values in mg/L, bacteriological indicators in CFU per 100 ml, minimum and maximum values are given in parentheses, n = 84.

Table 2. Treatment of Chicken-Manure Wastewater

	Inflow		Outflow		Efficiency (%)	
	Period 1	Period 2	Period 1	Period 2	Period 1	Period 2
COD	2340	2500	480	310	79.5	87.6
BOD_5	1320	1550	230	190	82.6	87.7
Susp. solids	1060	1300	155	130	85.4	90.0
Total N	295	320	<95	64	>67.8	80.0
Total P	21.8	25	<4.3	5.5	>80.3	78.0

Note: Average values in mg/L.

FULL-SCALE TREATMENT PLANTS

Petrov

The first full-scale reed bed system was built in Petrov near Prague in 1989 and was originally designed for the treatment of runoff waters from an adjacent dung-hill. The total area of 1248 m^2 was divided into five separate beds, 17 m wide, in a cascade configuration. The average slope of the bottom was 5.6%. Bed 1 (166 m^2) was unvegetated and filled with a mixture of coarse gravel and stones. It was primarily designed as a mechanical filter for suspended solids removal. Beds 2, 3, and 4 (297, 333, and 308 m^2, respectively) were filled with a 40-cm layer of coarse slate-gravel ($k_s = 4.05 \times 10^{-5}$ m/s) in the bottom and a 30-cm layer of local soil on the surface. Beds 2, 3, and 4 were planted with *Phragmites australis* using rhizomes with a density of 4 pieces per square meter. Finally, basin 5 (144 m^2) was designed as a free-water surface accumulation reservoir with no soil or gravel. Water from this last basin was used for dilution of highly concentrated dung-hill runoff. All five basins were sealed by a 0.8-mm polyethylene liner covered on both sides by geo-texture fabric.

In 1989, the precipitation in Czechoslovakia was extremely low and the system suffered from the lack of runoff water. As result, basin 2 was dominated by two weed species, *Rumex* sp. (dock) and *Atriplex patula* (fat-hen saltweed), as well as 28 species of terrestrial plants. Beds 3 and 4 were completely covered with *Echinochloa crus-galli* (barnyard grass), which is a typical corn-field weed in Czechoslovakia. In fall 1989, the weeds were cut down and beds 2, 3, and 4 were mulched with a 30-cm layer of hay. Owing to the lack of runoff water, the treatment plant has been used for sewage from the village of Petrov since February 1990. As the village has no sewerage, sewage from local cesspools is brought by tankers, stored in the dung-hill accumulation basin, and continuously distributed to the reed bed system. The average flow is 50 m^3/d, which represents hydraulic and organic loadings of 4 cm/d and 220 kg BOD_5/ha/day, respectively.

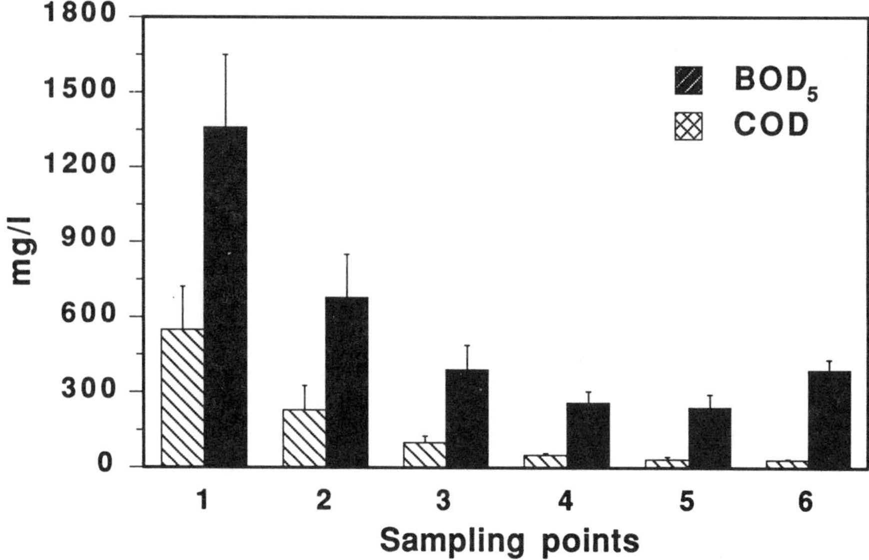

Figure 1. Reed bed system at Petrov: removal of organics.

Treatment Performance

Samples were collected in the period March to November 1990. The sampling points were chosen as follows: 1—inflow; 2, 3, 4, 5—outflows from beds 1, 2, 3, and 4, respectively; and 6-accumulation basin. It is important to note that in the accumulation basin, dense growth of planktonic algae appeared in early spring and persisted until November. Algae are known for their rapid uptake of nitrogen and phosphorus and the ability to store an excess amount of phosphorus ("luxury uptake").

BOD_5 removal amounted to 94% in sampling point 5, while COD removal amounted to 82.4% at this location (Figure 1). COD increased at sampling point 6 probably as a result of dense growth of planktonic algae. Removal of PO_4-P and total P at sampling point 5 amounted to 94.8 and 93.2%, respectively (Figure 2). Concentrations of orthophosphate decreased further in the accumulation basin, while total P increased. Uptake by planktonic algae and transformation of orthophosphate into polyphosphate as a storage "product" accounted for the increase in total P in the accumulation basin.

The reed bed system in Petrov exhibited very high nitrogen removal (Figure 3). Concentrations of ammonium nitrogen (NH_4-N) decreased down to 3.1 and 0.2% of the inflow concentrations at sampling points 5 and 6, respectively. The average NH_4-N concentration in the accumulation basin was only 0.3 mg/L. Nitrate nitrogen (NO_3-N) gradually increased from 6.7 mg/L in inflow to 47 mg/L in sampling point 5, suggesting nitrification within the system. In the last basin, the concentration of NO_3-N dropped to an average of 9 mg/L, probably as a result of uptake by algae. Removal of organic nitrogen was similar to NH_4-N. However, organic N concentrations increased in the last basin due to transformation of inorganic N into algal biomass. The reed bed system was very efficient, removing 99.7% of suspended solids at point 5.

Mulching with hay, together with continuous water supply, suppressed the growth of most terrestrial weeds and, in the fall, *Phragmites* covered about 60% of the area.

Other Systems

The reed bed system in Ondřejov is designed for common treatment of sewage from the village (population equivalent of 362) and the buildings of the Astronomical Institute of Czechoslovak Academy of Sciences. The system has been in full operation since May 1991.

The reed bed system in Kačice is designed as a tertiary step for the treatment of wastewaters from a dairy. It has been in operation since March 1991.

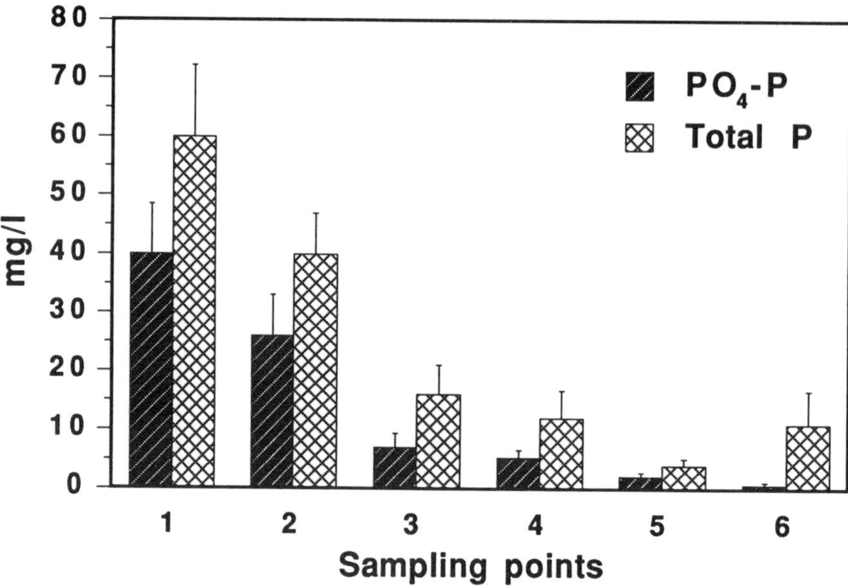

Figure 2. Reed bed system at Petrov: removal of phosphorus.

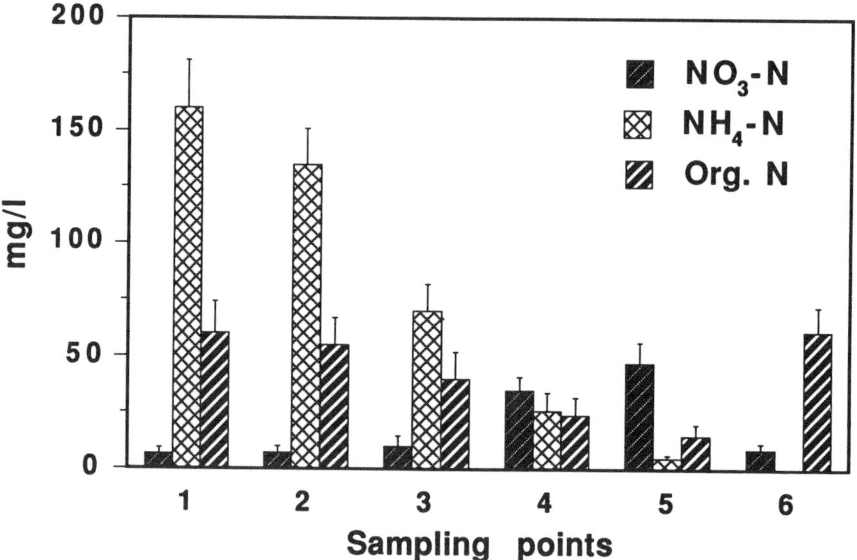

Figure 3. Reed bed system at Petrov: removal of nitrogen.

The constructed wetland in Rataje is designed for tertiary treatment of sewage from a recreational area. Design parameters for these constructed wetlands are given in Table 3.

CONSTRUCTED WETLANDS UNDER CONSTRUCTION

Zbizuby

The reed bed system is designed for treatment of village sewage (300, population equivalent). The treatment system comprises screening, primary settling, reed bed system, and stabilization pond which served as a single treatment unit.

Table 3. Design Parameters for Constructed Wetlands in Full Operation

Parameter	Ondřejov	Kačice	Rataje
Area (m^2)	800	600	200
Flow (m^3/d)	54.3	50	20.5
L_h (cm/d)	6.8	8.3	10.3
L_o (kg BOD_5/ha/day)	176	292	40.9
Average depth (m)	1.0	0.7	0.6
Slope (%)	2.5	1.0	2.0
Substrate	Gravel (8–32 mm)	Gravel (2–10 mm)	Gravel (8–12 mm)
Pretreatment	Primary settling	Primary settling	Primary settling
	Imhoff tank	Activated sludge	RBC[b]
Plants	Common reed[a]	Common reed[a]	Cattail[c]
Liner	PE (0.7 mm)	PVC (0.8 mm)	PVC (0.8 mm)

[a] *Phragmites australis.*
[b] RBC = rotating biological contactor.
[c] *Typha latifolia.*

Table 4. Design Parameters for Constructed Wetlands under Construction

Parameter	Zbizuby	Zásmuky	České Meziříčí	
Area (m^2)	1560	7760	10500	
Number of beds	1	4	4	
Flow (m^3/d)	130	604	432	263*
L_h (cm/d)	8.9	7.8	4.1	2.5*
L_o (kg BOD_5/ha/day)	68.6	24.2	74.1	200*
Average depth (m)	0.6	0.7	0.6	
Slope (%)	2	1	1	
Substrate	Gravel (5–20 mm)	Gravel (5–20 mm)	Gravel (8–16 mm)	
Pretreatment	Screening	Primary settling	Imhoff tank	
	Primary settling	Activated sludge	Settling ponds[a]	
Plants	Common reed	Common reed	Common reed	
Liner	PE (0.7 mm)	PVC (0.8 mm)	Clay (natural)	
Date of completion	1992	1993	1992	

[a] Data for treatment of "washing waters".

Zásmuky

The constructed wetland is designed for common treatment of landfill leachate and secondary treated sewage from the village. Since both leachate and treatment plant effluent (5.5. L/s) enter a small brook (1.5 L/s), the design was made for the brook itself.

České Meziříčí

The reed bed system is designed for treatment of sewage during sugar beet season (October to March) in the sugar refinery and "washing-waters" (soil needs to be washed from the sugar beets before processing) during the rest of the year. At present, sewage is treated in an Imhoff tank and "washing waters" are being treated in a series of 5 ponds, with a total area of 15 ha. The reed bed system has been used in place of the fourth pond. Design parameters for Zbizuby, Zásmuky, and České Meziříčí are given in Table 4.

ACKNOWLEDGMENTS

The author is grateful to Dr. Christopher Craft (Duke University Wetland Center) for his helpful comments on the manuscript, including language improvement.

REFERENCES

1. Reddy, K. R. and W. H. Smith, Eds. *Aquatic Plants for Water Treatment and Resource Recovery*. Proc. Int. Conf., Magnolia Publishers, Orlando, FL, 1987.
2. Hammer, D. A., Ed. *Constructed Wetlands for Wastewater Treatment*. Proc. Conf. Chattanooga, TN, Lewis Publishers, Chelsea, MI, 1989.
3. Cooper, P. F. and B. C. Findlater, Eds. *Constructed Wetlands in Water Pollution Control*. Proc. Conf. Cambridge, Pergamon Press, Oxford, U.K., 1990.
4. Vymazal, J. First experience with reed-bed systems in Czechoslovakia. *IAWPRC Spec. Group "The Use of Macrophytes in Water Pollution Control" Newsl. No. 2*. 51, 1989.
5. Vymazal, J. Use of reed-bed systems for the treatment of concentrated wastes from agriculture, in *Proc. Int. Conf. Constructed Wetlands in Water Pollution*. P. F. Cooper and B. C. Findlater, Eds. Pergamon Press, Oxford, U.K., 1990, 347.

CHAPTER 27

Subsurface Flow Wetlands at Mesquite, Nevada

C. C. Lekven, R. W. Crites, and R. A. Beggs

INTRODUCTION

Subsurface flow (SF) wetlands are used to provide effluent polishing following treatment of municipal wastewater in an aerated lagoon at Mesquite, Nevada. General SF wetlands design methodology is discussed in this article in addition to a discussion of the design of the wetlands facilities at Mesquite.

BACKGROUND

Mesquite is a small city in southern Nevada, along Interstate 15 at the Arizona border. The primary industries are agriculture and gambling. The wastewater treatment facility is located south of the city, immediately adjacent to the Virgin River. The wastewater treatment system previously consisted of flow measurement and coarse screening at the headworks, followed by treatment in a mechanically aerated facultative pond. Treated wastewater was stored in two storage ponds before being pumped approximately 1.6 km to a 25-ha alfalfa field for disposal by agricultural reuse. In addition to the treatment and disposal system described above, the facility also had two evaporation ponds located across a wash from the main facility.

The operator had been having problems with the wastewater treatment system. The Nevada Department of Environmental Protection (NDEP) permit for the system allowed effluent to average 30 mg/L 5-day biochemical oxygen demand (BOD_5). Daily maximum BOD_5 and suspended solids (SS) concentrations allowed in the effluent were 45 mg/L and 90 mg/L, respectively. Effluent BOD_5 and SS concentrations from the Mesquite facility often exceeded the levels allowed by the NDEP permit because of long retention times and growth.

The facilities plan recommendations were to increase the treatment capacity to 4540 m^3/day with a combination of "natural" systems consisting of an aerated lagoon, constructed wetlands, overland flow, and rapid infiltration disposal.[1] Figure 1 is a schematic diagram of the Mesquite facilities. The constructed wetlands portion of the plant expansion is discussed in this article.

A facilities plan, sponsored by Clark County Sanitation District (CCSD) in early 1989, was prepared by Nolte and Associates in association with Eckhoff, Watson and Preator.

CONSTRUCTED WETLANDS

Constructed wetlands represent a low-cost, low-maintenance form of wastewater treatment suitable for smaller communities which do not have the economic resources required to construct, operate, and maintain a conventional wastewater treatment plant.[2] Constructed wetlands are generally designed to operate in one of two different ways.

Free-water surface (FWS) wetlands operate in much the same manner as an overland flow system.[3] These FWS systems are generally shallow basins with emergent wetland vegetation growing in a soil medium.

0-87371-550-0/93/$0.00+$.50

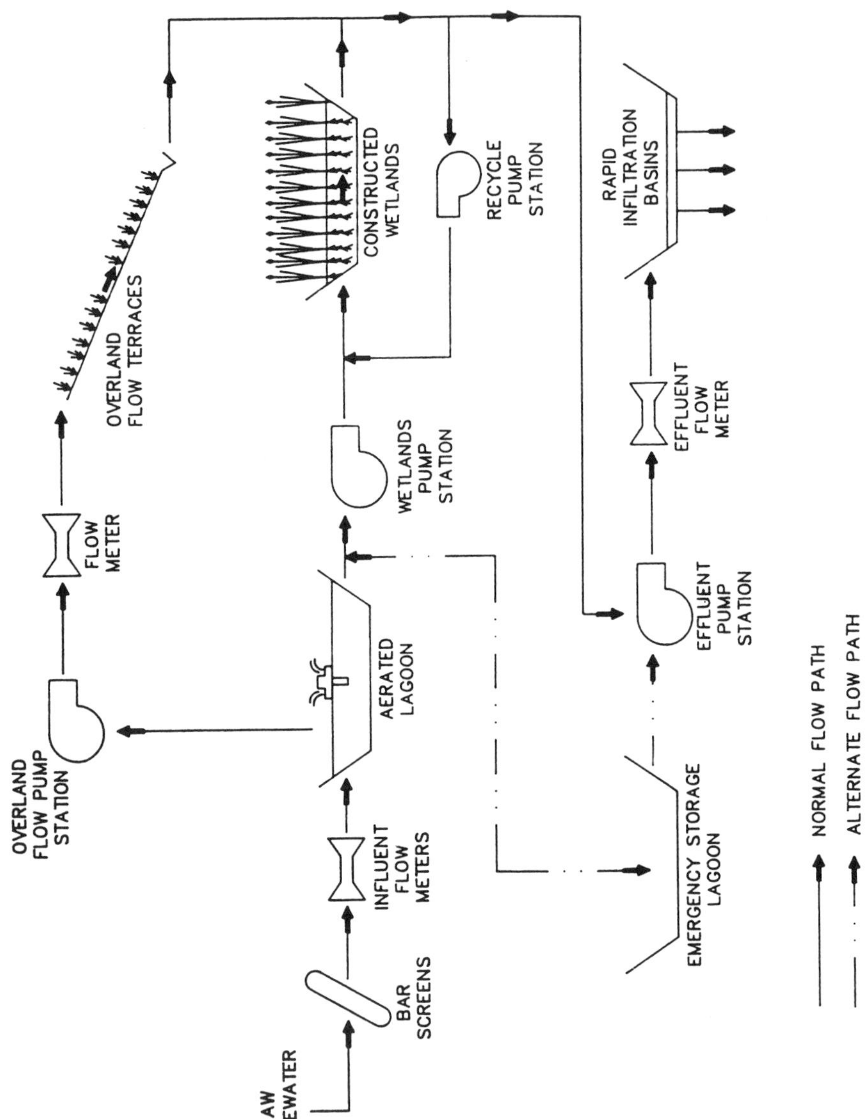

Figure 1. Schematic diagram of the Mesquite facilities.

Wastewater is allowed to flow at a shallow depth across the basin. The basins are typically designed with length-to-width ratios of 10:1 or greater to achieve plug flow conditions. Bacteria attached to the emergent vegetation provide much of the wastewater treatment. Quiescent conditions within the wetlands also allow suspended solids to settle out of the flow. Although algae are typically not produced within an FWS wetland when the emergent vegetation is dense enough to provide shade, efficient algae removal may not occur without long retention times. Mosquito breeding conditions can develop in FWS wetlands and the design of these systems should include ways of controlling vector populations, preferably without the use of insecticides. FWS wetlands were not considered to be a viable option at Mesquite due to the higher land areas required by this type of system (as compared to SF wetlands) and their relative ineffectiveness in removing algae.

Subsurface flow (SF) wetlands are essentially very low-rate horizontal trickling filters. Emergent wetland vegetation is grown in sand or gravel media within a lined basin. Wastewater is allowed to flow horizontally through the sand or gravel media, without surfacing. The roots of the emergent wetland vegetation supply oxygen to bacteria within the sand or gravel media matrix; hence, aerobic wastewater treatment processes occur even though the media are continuously saturated. Mosquito breeding conditions are avoided in SF wetlands since the wastewater flow is maintained below the surface of the media and thus, the stagnant water conditions required for vector reproduction do not exist.

SF WETLAND DESIGN METHODOLOGY

Reed et al. have suggested that BOD_5 removal in SF wetlands can be described using first-order plug flow kinetics.[3] The hydraulic residence time in the gravel or sand media is a function of media porosity, hydraulic gradient, wetland surface area, depth, and the flow rate. The equation used for design purposes is as follows:

$$C_e/C_o = \exp(-K_T A_s dn/Q) \tag{27–1}$$

where
C_e = effluent BOD_5, mg/L
C_o = influent BOD_5, mg/L
K_T = temperature dependent first-order reaction rate constant, day^{-1}
A_s = surface area of the system, m^2
n = media porosity, decimal fraction
d = depth of submergence, m
Q = average flow through system, m^3/day

As the entire flow must be maintained below the surface of the gravel or sand media, the required hydraulic cross-sectional area is determined using Darcy's law:

$$Q = KSA_c \tag{27–2}$$

where
K = hydraulic conductivity of the media, m/day
S = hydraulic gradient
A_c = hydraulic cross-sectional area, m^2

It is also important to check that the wetland vegetation will supply enough oxygen to the gravel matrix to oxidize the organic matter applied. To maintain primarily aerobic conditions, oxygen must be supplied at a rate of 1.5 times the BOD_5 loading rate to the wetland; hence,

$$\text{Required } O_2 = 1.5L_O \tag{27–3}$$

$$\text{Available } O_2 = (TrO_2)(A_s) \tag{27–4}$$

where L_O = BOD_5 loading rate, kg/day
TrO_2 = oxygen transfer rate for the vegetation
= 0.020 kg/m²/day
A_s = surface area, m²

ORGANIC LOADING CONSIDERATIONS

The aerated lagoon at Mesquite will be able to produce 70 mg/L BOD_5 effluent or better, based on the projected flow rate of 1510 m³/day and the existing aerated lagoon volume. The wetlands will further reduce BOD_5 to the 30-mg/L permit level during average January temperature conditions. Organic removal rates are expected to be much higher during the warm summer months, as the temperature-dependent rate constant will be higher.

The removal of suspended solids, particularly in the form of algae, was a major design consideration at Mesquite. SF wetlands have failed in the past when suspended solids caused a reduction in media permeability, resulting in surfacing and short-circuiting of the treatment process. Table 1 lists several SF wetlands systems and their design organic loadings to the gravel entry zones where problems with surfacing are most likely to develop. The loading rate at the Mesquite facility compares favorably with loading rates at other facilities which have not experienced surfacing.

Wastewater is introduced to the gravel medium at Mesquite through inlet trenches containing rocks 5 to 10 cm in diameter. The large diameter rocks will aid in the even distribution of wastewater along the length of the wetland. The gravel medium itself was sized to provide a large hydraulic safety factor to allow for the eventual partial clogging of the gravel matrix. The wetlands effluent is expected to contain less than 20 mg/L suspended solids on average.

MEDIA AND VEGETATION

The medium used at Mesquite is native river-run gravel 9.5 to 25 mm in diameter; a local quarry was the source of the gravel. No specific gradation was specified for the gravel other than the upper and lower diameters as listed above. This was done to lower gravel screening costs. Crushed rock is not suitable for SF wetlands as it will have a lower porosity than rounded gravel. Crushed limestone was used in SFS wetlands operated by the Tennessee Valley Authority, with poor results.[4] It is believed that fines from the limestone contributed to the eventual clogging of the media.

The gravel medium was covered with approximately 5 cm of a fine gravel/coarse sand mixture to aid seed germination. Some migration of fines from this surface layer down into the gravel medium is expected, but the quantities should be limited due to the gradation of the materials and no problems are anticipated.

Bulrush was selected for the wetlands vegetation at Mesquite because of its tendency to grow deeper roots than cattails or reeds. The depth of SF wetlands is generally designed to correspond with the rooting depth of the vegetation selected. The bulrushes at Mesquite are expected to create aerobic treatment zones throughout the full 81-cm bed depth. Alkalai bulrush (*Scirpus robustus*) was the species planted. This species of bulrush is frequently planted by duck clubs because of its high food value for waterfowl. As a result, bulk seed was available from a commercial seed supplier. The wetlands were hydro-seeded at a rate of 28 kg/ha of seed mixed with 2800 kg/ha of mulch fiber. The first wetland was hydro-seeded in July 1991. By September 1991, the wetlands had achieved approximately 20% germination. The hydro-seeded areas were watered by periodically flooding the wetland to levels above the gravel surface. It is believed that the sparse, delayed germination was a result of planting too late in the year and not keeping the hydro-seeded areas adequately moistened during the hot Nevada summer. Continuous flooding of the hydro-seeded areas is now believed the preferable method of irrigation for the bulrush species used, rather than the intermittent flooding technique. The remaining two wetland cells were hydro-seeded in early October 1991. Germination in these two cells is not expected until the spring of 1992. The wetland cells will be continuously flooded in the spring to a depth of approximately 7.5 cm to facilitate germination.

Table 1. Subsurface Flow Wetlands: Organic Loading (BOD_5) on Entry Zone

Location	Cross-section (kg/d/m²)	First 10 m of bed (kg/d/m³)	Surface flow?
Benton, KY	8	0.8	Yes
Great Britain (typ)	5	0.5	Yes
Denham Springs, LA	4	0.4	Yes
Haughton, LA	1.6	0.2	Yes
Mayo, MD	0.5	0.05	Yes
Pearlington, MS	0.2	0.02	No
Alexandria, LA	<0.2	<0.02	No
Mesquite, NV	0.1	0.01	No
Bear Creek, AL	0.02	0.002	No

Note: For comparison: typical loading, activated sludge = 0.8 kg/d/m³; low rate trickling filter = 0.25 kg/d/m³; and recirculating pea gravel filter = 0.04 kg/d/m³.

SYSTEM LAYOUT

The wetland cell layout at Mesquite is somewhat different from that of previous wetlands designs. Each of the three wetlands at Mesquite is divided into four cells, which operate in parallel. Each cell is approximately 15 m long and 116 m wide. The small length-to-width ratio is advantageous in SF wetlands for hydraulic considerations (a large hydraulic cross-sectional area) and for efficiency. Most of the treatment in SF wetland systems occurs in the first 20 m of bed length.[3] Figure 2 illustrates the layout of one of the three wetlands at Mesquite.

Flow Control

The flow to the wetlands is split proportionally by orifice plates so that each of the three wetlands receives an amount of wastewater proportional to its area. The wastewater is distributed throughout the wetlands by perforated pipe.

Treated wastewater is collected in a system of perforated drainage pipes. The water levels are maintained just below the surface of the gravel by an overflow weir in two of the wetlands, and by the effluent pump station float switches in the third.

Recycle

The wetlands at Mesquite were designed to operate with a continuous 1510 m³/day recycle flow (100% of forward flow). The recycle increases treatment efficiency by achieving better plug flow conditions than if there were no recycle provision. The recycle flow is also important to maintain healthy wetland vegetation during periods of high heat and low treatment plant influent flows.

Construction Costs

The low construction bid for the Mesquite wetlands was $515,000, or about $271,000 per hectare. The gravel medium is a major cost item for any SF wetland. An SF wetland system will only be cost effective if an affordable source of gravel media is located close to the project, as it was at Mesquite.

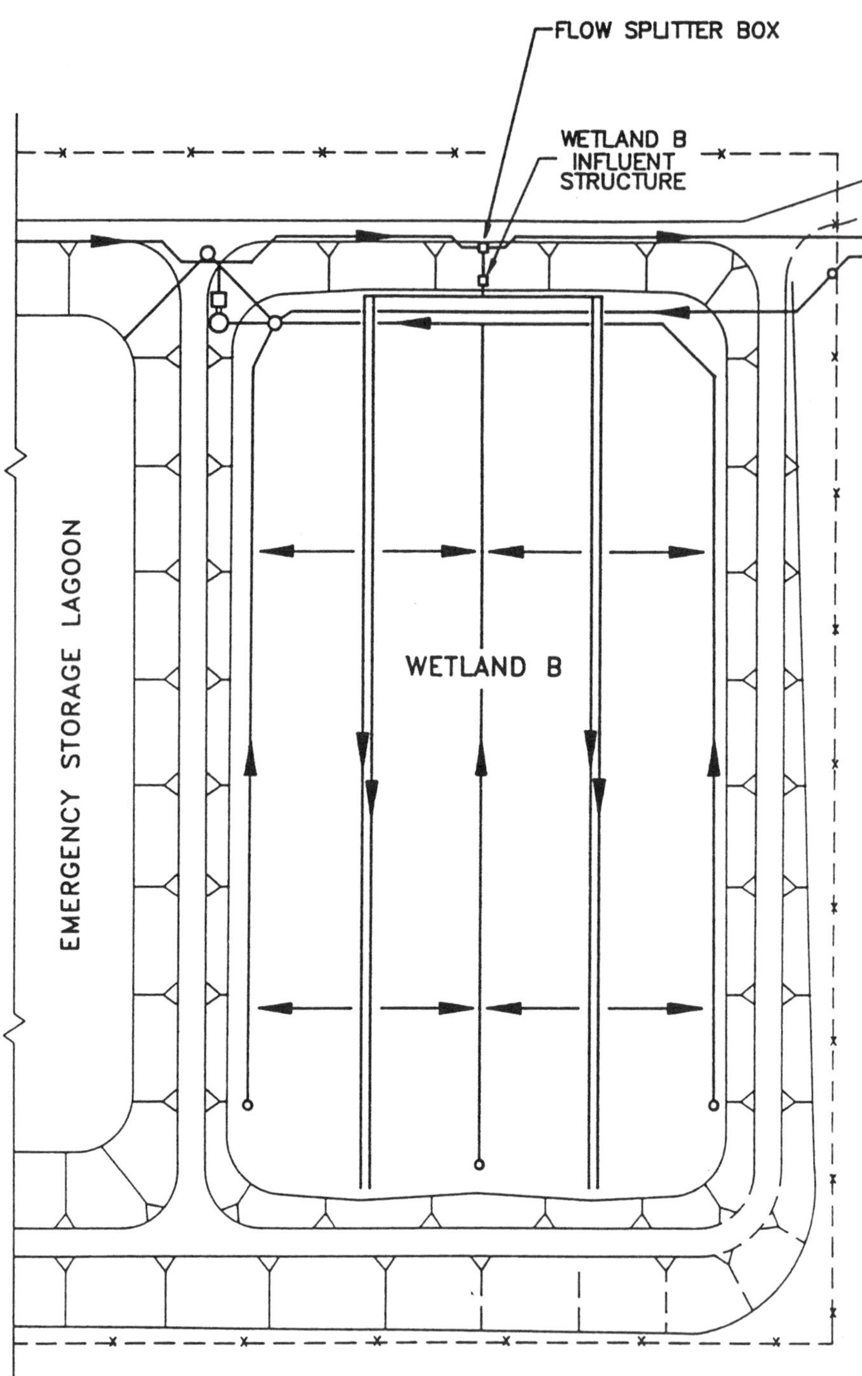

Figure 2. Typical wetland cell at Mesquite.

REFERENCES

1. Nolte and Associates and Eckhoff, Watson and Preator Engineering. Facility Plan, Mesquite Wastewater Treatment and Disposal Facilities, Clark County Nevada. May 1989.
2. EPA. Design Manual, Constructed Wetlands and Aquatic Plant Systems for Municipal Wastewater Treatment. EPA/625/1-88/022, September 1988.
3. Reed, S. C., E. J. Middlebrooks, and R. W. Crites. *Natural Systems for Waste Management and Treatment.* McGraw-Hill, New York, 1988.
4. Watson, J. T., K. D. Choate, and G. R. Steiner. Performance of Constructed Wetland Treatment Systems at Banton, Hardin, and Pembroke, Kentucky, During the Early Vegetation Establishment Phase, in *Constructed Wetlands in Water Pollution Control.* Proceedings of the International Conference on the Use of Constructed Wetlands in Water Pollution Control, 24–28 September 1990. P. F. Cooper and B. C. Findlater, Eds. Pergamon Press, Oxford, U.K.

PART 5

CHEMICAL PROCESSES

CHAPTER 28

Effective Phosphorus Retention in Wetlands: Fact or Fiction?

C. J. Richardson and C. B. Craft

INTRODUCTION

There are only a few studies that outline the mechanisms controlling phosphorus retention capacity in freshwater wetlands.[1,2] Most of those studies have shown that sediment/peat accumulation is the major long-term phosphorus sink and that natural wetlands are not particularly effective as phosphorus sinks when compared with terrestrial ecosystems.[3] Wetlands may have some functional ability to remove diffuse phosphorus runoff at low concentrations (long-term sink capacity) or may function as a downstream source for nutrients depending on the loading rate of nutrients, season of the year, or wetland type.[2] Recent studies have also shown that constructed wetlands may be designed to be more effective at removing phosphorus if their internal removal mechanisms are understood and if these removal and storage processes are optimized by management techniques.[4,5] In this article, we review the mechanisms controlling phosphorus (P) retention in wetlands, present a conceptual model of P storage in wetlands, and quantify how this model applies to a case study of P storage in the Everglades of south Florida.

REVIEW OF MECHANISMS CONTROLLING PHOSPHORUS RETENTION

It is well documented that phosphorus adsorption and retention in fresh-water wetland soil is controlled by interaction of redox potential, pH, Fe, Al, and Ca minerals, and the amount of native soil P.[6,7] The major P pool in natural wetlands is the soil/litter compartment.[2,8] There are no valency changes during biotic assimilation of inorganic P or during decomposition of organic P by microorganisms. Soil P primarily occurs in the +5 (oxidized) valency state because all lower oxidation states are thermodynamically unstable and readily oxidized to PO_4^{3-} even in highly reduced wetland soils.[6] Phosphorus has no gaseous phase, and thus it has a biogeochemical cycle which controls P retention.

The problem of maximizing long-term high efficiency P retention is based on our lack of understanding of the most efficient pathways which can move P from the water column into a permanent sink. Faulkner and Richardson[7] reviewed the effects of physical and chemical properties on P retention. Both pH and redox potential control the mobility of P in the environment. In acid soils, inorganic P is adsorbed on hydrous oxides of Fe and Al and may precipitate as insoluble Fe-phosphates (Fe-P) and Al-P. Precipitation as insoluble Ca-P is the dominant transformation at pH values greater than 7.0.

Redox potentials below +250 mV will cause the reduction of Fe^{3+} to Fe^{2+}, releasing associated P.[7] On the other hand, the reduction of redox potential caused by flooding can cause the transformation of crystalline Al and Fe minerals to the amorphous form; and amorphous Al and Fe hydrous oxides have a higher P sorption capacity than crystalline oxides due to their larger number of singly coordinated surface hydroxyl ions.[9]

Both macrophytes and phytoplankton are capable of taking up large amounts of P, especially during the first few years of additions. Recent studies of plant uptake capacity of P shows that phytoplankton are more efficient than macrophytes at removing P from the water column.[1] Howard-Williams and Allanson[10] calculated that in low P environments, almost twice as much phosphorus added to the water was taken up by *Cladophora*, epiphytic on *Potamogeton pectinatus* L., than by *Potamogeton* itself, although the biomass of the latter was some 30 times greater. Emergent wetland vegetation also takes up P from the soil. After the plants

0-87371-550-0/93/$0.00+$.50

die, much of the P is released to the surface water, so the short-term net effect of rooted emergent vegetation is to transfer P from the sediment to the water column, while root and residual decomposition products result in long-term P storage via peat. Algae take up P directly from water.[11] If it is possible to harvest algae efficiently, it may be more efficient to remove P from water by phytoplankton than by macrophytes.

In conclusion, physical input rates and retention times, chemical reactions, as well as biological uptake are important mechanisms controlling P removal and storage by wetlands. A conceptual model of how these processes interact in terms of both short-term uptake and long-term permanent storage is presented below.

A PHOSPHORUS RETENTION MODEL FOR WETLANDS

Our model of P retention in wetlands is based on the concept that P nutrient storage compartments have an operational capacity that is analogous to reservoirs of varying sizes (shown as buckets and bathtub) with either short-term or long-term storage capacity (Figure 1).[8] The size of the short-term storage compartments are drawn on a relative scale (soil adsorption precipitation > plants > periphyton), with the long-term retention capacity being directly related to peat accretion rates. The size and uptake rate of each short-term storage compartment is finite. Once they have reached capacity, they will no longer function as effective storage areas. The fact that short-term storage is equated with long-term storage is often the basis for the confusion in the literature concerning the efficiency of wetlands to store P. Short-term processes do control water quality, have fast recycle or uptake rates, (i.e., minutes, hours, or days vs. years), but limited storage capacity. The storage and cycling processes are primarily biological and chemical in nature, although the physical factors of high water input rates, increased water depth, and lower retention times for water can override these two processes and decrease storage efficiency.

Phosphorus enters the system and is immediately taken up by the microbes and algae (compartment 1), or is adsorbed by the soil or chemically precipitated directly from the water column (compartment 3) (Figure 1). Adsorption initially can be considered short term storage, since desorption back to the water column often takes place. Once this material is buried with the peat, it becomes part of the long-term storage. Chemical precipitation from the water column results in movement of the complexed material to the litter or the long-term peat storage compartment. Rooted plant uptake (compartment 2) is slower than the other processes and only occurs if the P that reaches the soil pore water is in a soluble form. The movement of input P to pore water in the soil requires sufficient retention time (days to weeks) and shallow surface water depths. For example, it has been suggested that water loading rates to wetlands not exceed 2 to 3 cm per week if maximum removal is desired.[5]

The initial rates of uptake in the short-term compartments can be quite high until the reservoirs have reached their capacity, especially in a system that has not received nutrient inputs before. This initial high rate of adsorption and uptake is often confused with long-term storage capacity of the wetland. The initial size of plant or periphyton compartments may increase slightly as new species are added and plant biomass increases, but once these short-term storage reservoirs reach capacity, they no longer function as storage areas (as shown by the full buckets, Figure 1). These compartments, having reached saturation from the nutrient inputs into the system, will no longer significantly decrease P water concentrations. A moving phosphorus front in downstream waters will develop once the short-term uptake capacity is overloaded. It should also be noted that each year plants and periphyton die back and release 35 to 75% of the P back to the water column and litter compartment.[12] The only way to maintain adequate P uptake from the water column is to balance input levels to uptake rates. Long-term storage for the wetland is thus keyed to the short term capacity and the collective turnover rates from each process.

The largest reservoir (bathtub size Figure 1) is the long-term peat or soil compartment. This compartment usually stores more than 80% of the nutrients in the top 20 cm of the strata and is comprised of buried organic phosphorus from undecomposed litter from previous years, as well as annual additions from adsorption and precipitation (PPT) (3), periphyton (1), plant roots (2), and new material from the litter compartment (4). The efficiency of any wetland to store P on a long-term basis is thus determined by the peat or soil accretion rate times the net increase of P stored by these processes each year.

To retain as much P as possible, the input rates should be limited to the long-term storage capacity, which is controlled by peat/soil accretion. Research to date would suggest that permanent storage of P is below 1.0 $g/m^2/year$ and usually averages around 0.5 $g/m^2/year$.[3,8,13,14] The question is whether runoff water from agricultural lands or municipal treatment systems can be distributed at a rate that will maintain efficient long-

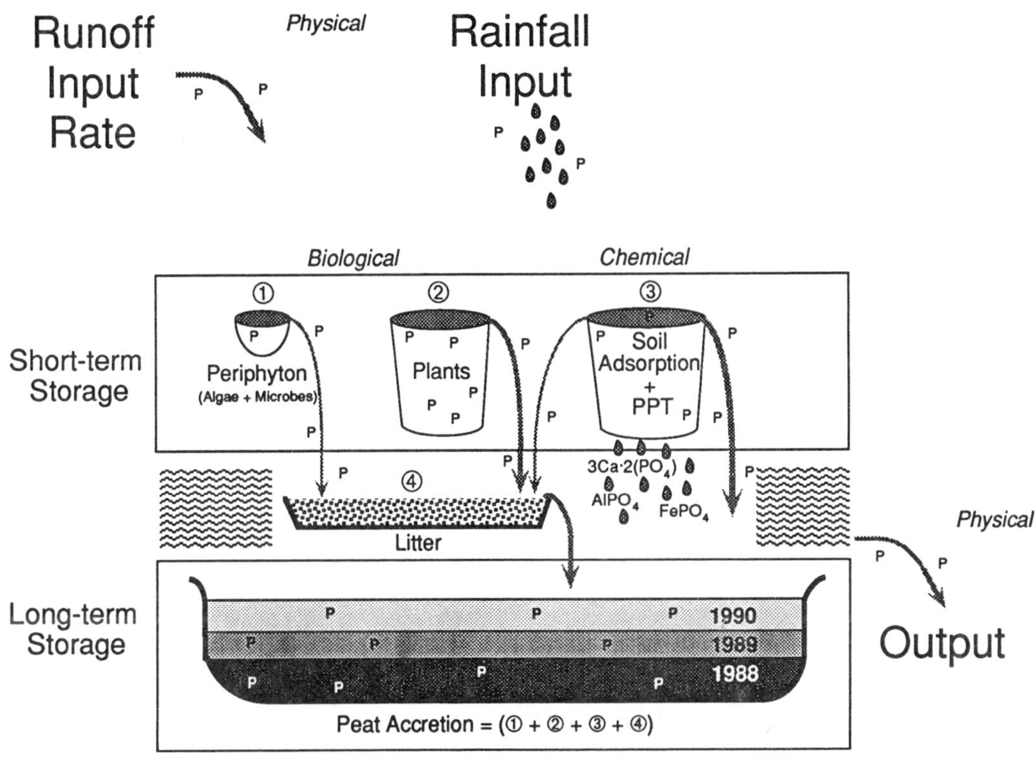

Figure 1. A conceptual model of phosphorus retention in wetlands. We have shown only the major reservoirs and no attempt was made to show a complete phosphorus cycle among the biotic and abiotic compartments.

term storage of P in newly constructed or natural wetlands and prevent downstream eutrophication. A case study analyzing agricultural runoff and the P storage capacity for P in the Everglades of south Florida gave us an opportunity to validate the retention model described in Figure 1.

EVERGLADES CASE STUDY OF P RETENTION FROM AGRICULTURAL RUNOFF

The Water Conservation Areas (WCAs) of south Florida encompass approximately 3500 km^2 of diked state wetlands which are managed by the South Florida Management District for flood control, water conservation, water storage and groundwater recharge, preservation of fish and wildlife resources, and recreational benefits, according to the 1948 federal law[37] (Figure 2). Historically, the WCAs were part of the 900,000-ha Everglades wetlands which, according to Davis[15] received water and nutrients mostly from rainfall. Nutrients, especially P, probably were in short supply[16] Nutrients and water in the WCAs have been altered significantly during the past few decades due primarily to the pumping of nutrient-laden water from the Everglades Agricultural Area (EAA) and Lake Okeechobee during certain portions of the year and withdrawals from the WCAs during drought periods,[11,17,18] According to estimates of the SFWMD[20] (see Reference 28a; Vol. III, Table B-16, p. B-133), the WCAs from 1979 to 1988 received an annual average loading of 429 metric ton of phosphorus (an areal loading rate of 0.14 to 0.25 g/m^2/year) and 12,170 metric ton of nitrogen (an areal loading rate of

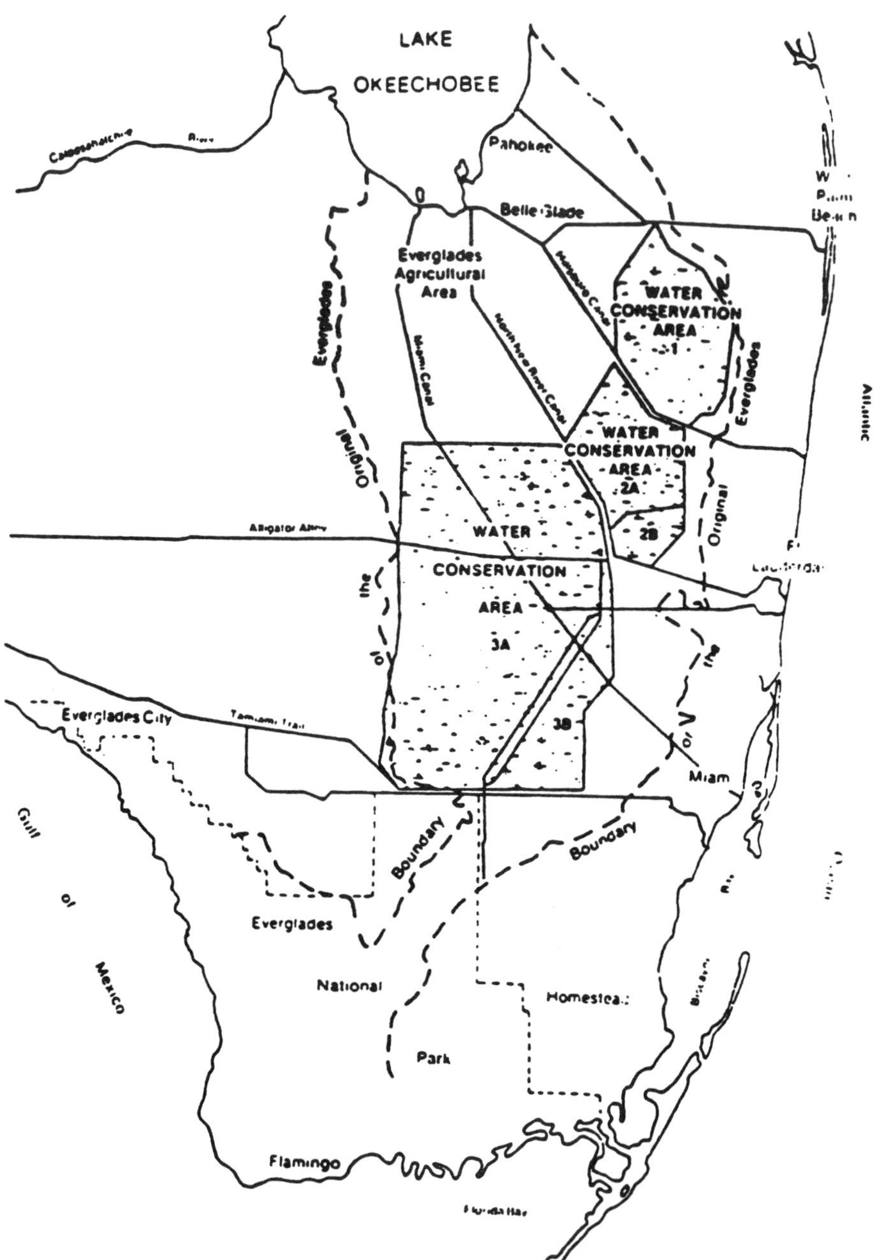

Figure 2. Location of Everglades Water Conservation Areas in South Florida.

3.51 to 8.56 g/m^2/year). Annual retention of P input ranged from 0.11 to 0.18 g/m^2/year (71 to 94%) in the WCAs, suggesting that the Everglades has some capacity to retain phosphorus. The input of water was 5.18 $\times 10^9$ m^3 (4.2 million acre feet). Approximately 39% of the P loading and 33% of the N enter via rainfall and are thus not controllable. Rainfall input comprised 66% of the water budget for the WCAs (see Reference 20, Vol. III, p. B-128).

A series of key questions related to this increase in agricultural runoff are directly related to the long-term nutrient storage capacity of the WCAs for N and P: (1) Are the northern Everglades areas that are subjected to increased P inputs able to store this nutrient permanently? (2) What are the mechanisms controlling phosphorus storage? To answer these questions, our research initially focused on the peat accretion mechanism as described earlier in Figure 1.

METHODS

Soils Methods and Collection Sites

Soil cores were collected from WCA 2A and 3A in June 1989 to assess nutrient loading effects and nutrient accumulation rates in the WCAs from areas receiving agricultural runoff and from unenriched areas. Cores were taken along a gradient from the nutrient impacted area in WCA 2A south through the unimpacted areas in WCA 3A to the northern boundary of the ENP.[21] Eight cores were collected from WCA 2A,[21] and five were taken from the nutrient-impacted area in the northern part of WCA 2A. This area was dominated by the cattail (*Typha domingensis*). Three cores were collected from the sawgrass (*Cladium jamaicense*) plains of western WCA 2A, 10 km south and southwest of the nutrient impacted area. Soils were sampled at three locations in WCA 3A (Alligator Alley), the north side of State Road 84, 6.6 km (4 mi) west of the Miami Canal, and approximately 100 m north of the S12-A and S12-C gates in the southern part of WCA 3A. Two cores were collected at each location in WCA 3A. One core was taken beneath a cattail stand, while the second core was collected from the sawgrass community.

Soil cores were sectioned into 1.5 cm depth increments and air-dried. Sections were weighed, ground with a mortar and pestle, and sieved through a 2-mm mesh diameter screen. Soils were analyzed at each depth interval for Cesium-137 activity (to estimate accretion rates). Cesium-137 was measured for 8 to 24 h using a high purity germanium gamma ray counter (EG & G Ortec, Oak Ridge, TN) coupled with a 2048 multichannel analyzer (The Nucleus, Inc., Oak Ridge, TN). Depth increments also were analyzed for bulk density total N and P, according to the methods described in Reference 21. Where appropriate, analyses were expressed on an oven-dried (75°C) weight basis.

RESULTS AND DISCUSSION

Soil Accretion Rates

Many wetland soils store N, P, and C through accretion of organic matter. Peat accretion, or peat build-up, is the difference between the rate of net primary production and the rate of decomposition. Cesium-137 (Cs-137) has been successfully used as a marker to estimate short-term (approximately 25 years) rates of accretion in estuarine[22-24] and freshwater wetlands.[25,26] Cesium-137 is a fallout by-product from above-ground testing of thermonuclear weapons by the U.S. and U.S.S.R. during the 1950s and 1960s. Production (and deposition on land surfaces) of fallout Cs-137 began in the mid-1950s and peaked in 1964. At this time, the U.S. and U.S.S.R. signed a treaty banning above-ground testing. With exception of the Chernobyl accident (which released measurable amounts of Cs-137 throughout eastern Europe), there has been no significant deposition of Cs-137 since 1964. Thus, the Cs-137 maximum in the soil profile is indicative of the location of the soil surface in approximately 1964. Accretion rates are estimated as follows:

$$\text{Accretion rate (mm/yr)} = \frac{\text{Depth (in mm) of Cs - 137 peak in soil}}{\text{Time (in years) since 1964}}$$

Figures 3A and 3B illustrate the use of Cs-137 as an accretionary marker. The Cs-137 maximum in Figure 3A occurs at 10.5 cm. The soil core was collected in 1989 so that the time elapsed since 1964 is 25 years. Therefore, the accretion rate for this core is 105 mm/25 year = 4.2 mm/year. In Figure 3B, the Cs-137 maximum is at the soil surface, indicating that there has been no accretion at this site in the past 25 years.

Accretion rates along the WCA 2A gradient ranged from 0 to 4.2 mm/year (Table 1). Accretion rates were greatest (2.0 to 4.2 mm/year) in the cattail-impacted portion of WCA 2A and lowest (0 mm/year) in the sawgrass plains of western WCA 2A. The higher accretion rates under cattail stands in WCA 2A may be due to increased net primary productivity of cattail (compared to sawgrass), or to a longer hydroperiod (and decreased decomposition) in the cattail-impacted area south of the S-10 structures. The average accretion rate in WCA 2A for the nutrient enriched areas was 3.23 mm/year (Table 1).

Accretion rates were less variable in WCA 3A, ranging from 1.8 to 3.7 mm/year (Table 1). Accretion rates in WCA 3A averaged 2.77 mm/year compared to 3.23 mm/year in the cattail-impacted areas of WCA 2A. Our

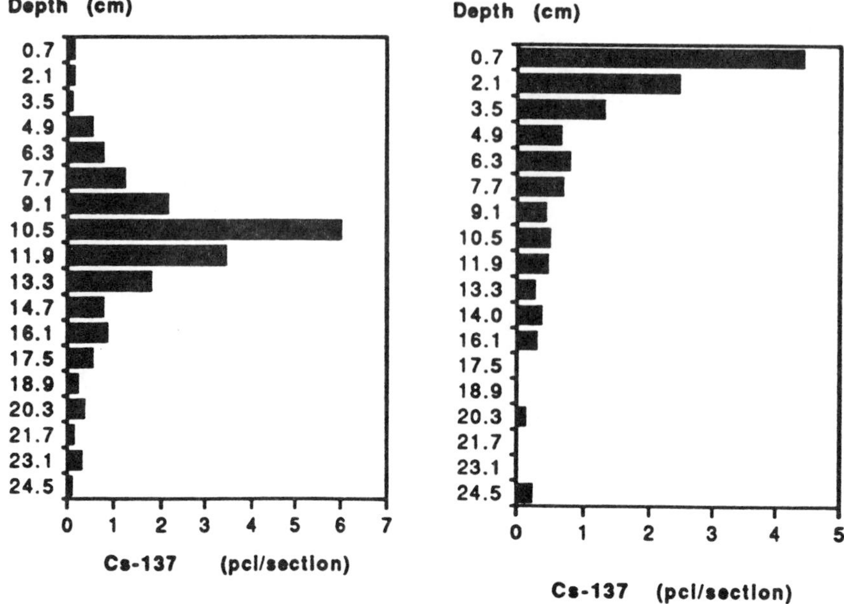

Figure 3. A: Cesium-137 activity in a soil core collected from the cattail impacted area in WCA 2A. (The accretion rate is 4.2 mm/year). B: Cesium-137 activity in a soil core collected from the sawgrass plains of WCA 2A. (The accretion rate is 0 mm/year).

estimates of peat accretion are higher than Everglades peat accretion rates calculated by Davis[27] (1.4 mm/year) and measured by McDowell et al.[28] (1.1 to 1.5 mm/year) using Carbon-14 dating (Table 1). The ^{14}C method is used to estimate peat accretion rates on millennium time scales (the half-life of ^{14}C is 5570 years). ^{14}C dating is usually performed on peat collected from the base of the profile (usually several meters or more below the soil surface). As succeeding layers of peat are deposited, the deeper (and older) peat is compacted by the weight of the overlying material. This results in underestimating more recent peat accretion rates if compaction is not taken into account. Our estimates of peat accretion in the WCAs also were higher than accretion rates in European (0.2 to 0.8 mm/year) and North American freshwater wetlands (1 to 2 mm/year).[29] However, most of these sites are located at cooler latitudes than south Florida. It is possible the higher accretion rates found in the WCAs may reflect increased plant productivity (and deposition of decaying plant material) at these warmer latitudes. The accretion rate in the cattail-impacted portion of WCA 2A (2.0 to 4.2 mm/year) was more than twice the world average of 1 to 2 mm/year (Table 1). This increase in peat accretion could increase the long-term nutrient storage capacity of the Everglades.

N and P Concentrations

Bulk densities were fairly uniform throughout the two WCAs with values ranging from 0.10 to 0.18 g/cm^3 (Table 2). There were no obvious differences in total soil N in the WCAs collected under cattail and sawgrass stands, although soils taken from sawgrass stands generally contained higher N levels. Soils collected from the nutrient-impacted area south of the S-10 structures in WCA 2A contained N levels similar to samples taken from areas of low nutrient inputs. These findings suggest denitrification may remove substantial amounts of N from surface waters entering WCA 2A.

In contrast to N, soil P concentrations were 1 to 5 times higher in the nutrient-enriched cattail-dominated zones compared to the sawgrass stands in WCA 2A (Table 2). Phosphorus-enriched water, which has been pumped through the S-10 gates[30] for the past 10 to 20 years, is responsible for high soil P concentrations south of the S-10 structures in WCA 2A. Soil P concentrations were much lower in WCA 3A (compared to the nutrient-enriched areas of WCA 2A) and probably are representative of soil P levels in WCA 3A prior to drainage and flood control. For this reason, we believe these values reflect the historical rates of P accumulation in the Everglades ecosystem.

Table 1. Mean Accretion Rates (Determined by Cs-137 Activity) of Peat Soils in Water Conservation Areas 2A and 3A in 1989

Location	Accretion rate (mm/year)
$\overline{X}$ (±1 SE) for WCA 2a (nutrient enriched)	3.23 ± 0.51
$\overline{X}$ n = 4	
for 2A (unenriched) n = 3	1.33 ± 0.66
$\overline{X}$ (±1 SE) for WCA 3A	2.77 ± 0.26
n = 6	
Everglades peat[a]	1.1–1.5
World average[b]	1.0–2.0

[a] From References 27 and 28.
[b] From Reference 29.

Table 2. Bulk Density and N and P of Soils in Water Conservation Areas 2A and 3A

Location	Bulk density (g/cm^3)	N (%)	P (mg/g)
$\overline{X}$ (±1 SE) for WCA 2A (nutrient enriched)	0.11 ± 0.01	3.35 ± 0.16	1164 ± 165
$\overline{X}$ (unenriched)	0.11 ± 0.01	3.68 ± 0.17	505 ± 93
$\overline{X}$ (±1 SE) for WCA 3A	0.14 ± 0.01	3.81 ± 0.21	722 ± 43

Note: Each number represents the mean values of the increments laid down during the past 25 years. In cores where the accretion rate = 0 (e.g., WCA 2A, sawgrass), the mean value in the upper 10 cm was used. Standard errors = SE.

Accumulation Rates of N and P

Rates of N and P accumulation over the past 25 years were estimated using accretion rates (Table 1), and soil bulk density and nutrient concentrations (Table 2). For example, the rate of P accumulation in a core collected from the cattail-dominated area in WCA 2A was calculated as follows:

$$\begin{aligned}\text{P accumulation} &= \text{P concentration} * \text{bulk density} * \text{accretion rate (g/m}^2\text{/yr)}\\ &= 1.503 \text{ mg P/g} * 0.10 \text{ g/cm}^3 * 0.42 \text{ cm/yr} * 10{,}000 \text{ cm}^2\text{/m}^2\text{/1000mg/g}\\ &= 0.63 \text{ g P/m}^2\text{/yr}\end{aligned}$$

Phosphorus accumulation rates were greatest in the cattail-dominated area of WCA 2A (0.63 g/m^2/year) and lowest on the sawgrass plains of southern WCA 2A (0.40 g/m^2/year) (Table 3). The highest rate of P accumulation we measured was (0.63 g/m^2/year) in the cattail-dominated area of WCA 2A. This value probably represents the maximum rate that can be stored under a high nutrient loading regime and extended hydroperiod. How long these systems can store P at this rate is unknown, but long-term storage only averaged 0.40 g/m^2/year during the past 25 years. Phosphorus accumulation rates in WCA 3A ranged from 0.17 to 0.35 g P/m^2/year, averaging 0.27 g P/m^2/year (Table 3). Average phosphorus accumulation in the nutrient enriched areas of WCA 2A (0.40 g/m^2/year) was nearly 1.5 times the mean rate of accumulation (0.27 g P/m^2/year) in the soils of WCA 3A (Table 3).

With the exception of the drier sawgrass stand in WCA 2A, accumulation rates of N were similar throughout the two WCAs. Nitrogen accumulation rates were 9.5 to 17.5 g/m^2/year, respectively. The similarity in N accumulation between nutrient impacted WCA 2A and WCA 3A suggests that denitrification may be important in regulating N accumulation in soils of the WCAs.

A Comparison of P and N Retention Capacities

The previously discussed peat accretion and P and N data from the cores in WCA 2A can be utilized to determine the long-term P and N storage capacities of the natural and fertilized wetlands over the past 25 years.

Table 3. Accumulation Rates of N and P in Soils of Water Conservation Areas 2A and 3A

Location	N (g/m²/year)	P (g/m²/year)
WCA 2A (cattail)	13.4	0.63 (maximum)
$\overline{X}$ for WCA 2A (nutrient enriched)	13.5	0.40
(±1 SE)	(1.15)	(0.09)
$\overline{X}$ (unenriched)	4.7 ± 2.4	0.08 ± 0.04
$\overline{X}$ for WCA 3A	14.2	0.27
(±1 SE)	(1.1)	(0.03)

Note: Accumulation rates were calculated from the accretion rate, soil bulk density, and nutrient concentrations.

The information obtained from this site is invaluable since it allows us to project the potential rates of nutrient retention under different loading rates, as well as determine the effects of cumulative loadings. Our analyses of the data suggests that the rates of permanent P storage that can be expected range from a low of 0.21 g/m²/year (1.9 lbs/acre/year) to a maximum of 0.63 g/m²/year (5.6 lbs/acre/year). The highest recorded rates are found closest to the S-10 structures along the Hillsboro Canal below the S-l0D gates, where the highest recorded loading rates (30 metric tons/year) were measured (Reference 20; Reference 20a, Vol. III, p. A-17, Figure A-2). The lower rates were found furthest from the canal and were in areas of lower P loadings. The mean P retention rate for the entire impacted area is 0.40 g/m²/year (3.6 lbs/acre/year).

Our long-term Cs-137 maximum P retention calculation (0.63 g/m²/year) is only 38% of the 1.67g/m²/year of P storage reported by Davis[31] and utilized in the 1990 Florida SWIM Plan to design constructed wetlands for P storage (Reference 32; Vol. III., Table IIIA-1 p. IIIA-8). These estimates are nearly three times higher than our figures. The reasons for this large difference is that the SWIM Plan erroneously equated plant uptake, below-ground storage data, and decomposition taken over several years, to be equal to P retention via long-term peat accretion, precipitation etc. Davis's upper finite level for retention is based on a 1.4 g/m²/year leaf uptake by cattail plus a 14% retention in below-ground biomass and litter bag storage. Unfortunately, this luxury uptake by cattails does not represent true P retention since between 35 to 75% of plant tissue P is rapidly released to the water column once the plant dies.[8,12,33,34] In addition, the litter bag decomposition study employed by Davis does not take into account release of P by long-term decomposition. Furthermore, plants placed under enriched-nutrient conditions reach a finite capacity of uptake and the annual uptake soon reaches a maximum[33] (Figure 1). Davis's estimate of P storage used in the design for constructed wetlands in the south Florida plan represents a realistic number for short-term uptake, but does not represent an accurate account of long-term storage of P. Thus, the only way the Davis estimate of P storage (1.67 g/m²/year) could even be partially reached is if the vegetation were harvested each year. This is possible, but it has proven to be an expensive and time-consuming process wherever it has been attempted.

In order to verify further our estimates of P and N retention in WCA 2A south of the Hillsboro Canal, we compared our accumulation rates with data from Gleason et al.[35] for the same region in 1973–1974. Gleason and co-workers calculated that 4848 ha (12,000 acres) had been affected by nutrient additions. Using data from Gleason et al., we calculated that 0.36 to 0.41 g/m²/year P and 10.5 to 10.8 g/m²/year N were retained by the area south of the Hillsboro Canal. Our historical Cs-137 profile analyses for the same period resulted in 0.21 to 0.49 g/m²/year estimates for P and 13.7 to 14.2 g/m²/year N retention during the 1973 to 1976 period.

Another estimate of the long-term P storage capacity is calculated from a generalized comparison of the areal extent (9696 ha) of nutrient enrichment in WCA 2A[21] in 1988 and the average loadings for the 1979 to 1988 period of 54 metric tons of P.[30] It is known that the loading rates are highest at the output point along the canals, and that loadings decrease southward (see section IV.A.2 of Reference 20; Reference 20a, Vol. III, Figure A-2, p. A-17). It is also known that once the loading rates exceed retention capacity, a front of increased P (i.e., above background levels) moves away from the saturated zone and into new wetland areas down gradient.[8] Thus, one can determine the total affected area, divide it by the average loading rate, and then obtain an approximate (order-of-magnitude) estimate of the storage rate, for the nutrients in question. This estimate is most accurate when the loadings have not shifted significantly. The size of the area affected will only be as large as the the difference between the loading rate and the rate of sustained uptake. Using this method our calculated rate of P storage from 1979 to 1988 was 0.56 g/m²/year of P storage vs. our earlier maximum estimate of 0.63 g/m²/year with Cs-137 analysis.

These two additional estimates of P retention support our position that the State of Florida's SWIM Plan has overestimated P retention capacity by two to three times the natural capacity of unharvested wetlands in the Everglades. An accurate estimate of long-term P storage capacity is essential to determine the size of the wetland necessary for removing P in the drainage water from the EAA. Unfortunately, short-term P uptake is being utilized by the State of Florida in their construction design.

Wetland Acreage Requirements for Permanent Storage of P

As mentioned in the previous section, natural Everglades wetlands that are fertilized with P can be expected to retain, on average, 0.40 g/m^2/year with a maximum value of 0.63 g/m^2/year. The long-term sustained capacity is probably closer to the mean value we found. We are currently analyzing additional cores to determine a better estimate of long-term sustained capacity, the mechanisms responsible for the retention, as well as loading retention relationships.

However, our preliminary data do allow us to estimate the amount of acreage that would be required to retain P. This P retention capacity is based on the data from WCA 2A and thus the loadings, mode of distribution, retention time, and plant communities would dictate that the P storage capacity found here only applies to these conditions. This means that constructed wetlands on peat soils may or may not reach this capacity if any or all of the above conditions are not simulated. It may, of course, also be possible to improve on these rates once one understands enough to optimize for P retention.

We utilized the previous information to calculate the number of acres that would be required to store P permanently based on an average and maximum P storage rate (Table 4). A sustained P storage at the maximum rate would result in a storage of 6.3 metric tons per 1000 ha (2740 acres). Thus, it would require 40,000 acres to remove approximately 100 metric tons and 108,000 acres to remove 275 metric tons. If one utilized the average P retention value we found (0.40 g/m^2/year), then 1000 ha would only remove 4.0 tons, or 63% of the maximum value. Thus, 20,000 acres would only remove 32.3 tons permanently, and 40,000 acres would sequester 64.6 metric tons.

These data indicate that the P storage capacity for the Everglades peatlands is very low and that vast acreages will be required to remove the proposed P tonnage on a permanent basis. Our data and findings are supported by previous reports[3,8,36] that wetlands containing peat soils are not efficient sinks for P, especially under high loading rates.

CONCLUSIONS

P Retention Model

- Micobial organisms and algae control the short-term uptake of P, not plants.
- The long-term (non-recycling) forms of P that control permanent storage are organic P (50 to 70%) and calcium phosphate in alkaline wetlands; Fe and Al control P adsorption and precipitation chemistry in acid wetlands.
- Constructed wetlands should be designed to maximize peat and soil accretion rates, soil adsorption, and sedimentation or precipitation processes for P.
- Output concentrations of P from wetland treatment systems vary according to hydroperiod, loading rates, retention time, and season.

Everglades Case Study

- Phosphorus accumulation rates were much higher in the nutrient-impacted areas of WCA 2A, compared to WCA 3A. The high P accumulation rates in WCA 2A resulted from increased peat accretion rates and increased soil P concentrations.
- Long-term sustainable P storage rates would probably be close to the 0.40 g/m^2/year average found for the nutrient-enriched areas in WCA 2A.
- Rates of N accumulation were similar in WCA 2A and 3A, suggesting that denitrification removes substantial amounts of N before surface waters reach the interior of WCA 2A.
- Based on our maximum (0.63 g P/m^2/year) and average (0.40 g P/m^2/year) rates of P accumulation, we estimate that 20,000 to 31,500 acres (8080 to 12,700 ha) will be needed to remove 51 metric tons of P. A 108,000-acre area will be required to remove 175 metric tons of P, assuming a storage rate of 0.40 g P/m^2/year.

Table 4. The Relationship Between Wetland Size and Phosphorus Storage Capacity Based on the Average (0.40 g P/m²/year) and Maximum (0.63 g P/m²/year) P Accumulation Rates

Area (acres)	Area (ha)	$\overline{X}$ 25-year P storage (g/m²/year)	Maximum P storage (metric tons)
2740	1000	0.40 (average)	4.0
		0.63 (maximum)	6.3
3000	1212	0.40	4.8
		0.63	7.6
3700	1495	0.40	6.0
		0.63	9.4
14,000	5657	0.40	22.6
		0.63	35.6
20,000	8080	0.40	32.3
		0.63	50.9
40,000	16,160	0.40	64.6
		0.63	101.8
64,470	26,045	0.40	104.2
		0.63	164.1
108,000	43,632	0.40	174.5
		0.63	274.9

- These data indicate that the P storage capacity of the Everglades is very low, and retention areas considerably larger than the 32,600 acres (13,170 ha) proposed in the State of Florida's SWIM plan will be required to remove P from agricultural drainage waters, providing the proposed flow-ways function as efficiently as the natural areas in WCA2A.

ACKNOWLEDGMENTS

This paper covers preliminary research completed by the Duke University Wetland Center for the period March 1989 to March 1990. Funding was provided by the Florida Sugar Cane League. Dr. Jerry Qualls helped with the review of this document; Eric Waldbauer and Paul Heine worked extensively on the laboratory analyses of plant, soil, and water chemistry; and Celia Best helped edit the manuscript. Song Quian provided background information and review material on constructed wetlands. Background information and reports on the Water Conservation Areas were provided by the South Florida Water Management District. The District also provided access to the Water Conservation Areas 2A and 3A, and the Duke Wetland Center greatly appreciates their cooperation on this project. We particularly wish to thank Steve Davis for help with background information and data. Data were not collected in the Loxahatchee National Wildlife Refuge nor the Everglades National Park, since access was denied by the judicial branch of the U.S. government.

REFERENCES

1. Howard-Williams, C. Cycling and retention of nitrogen a phosphorus in wetlands: a theoretical and applied perspective. *Freshwater Biol.* 15:391, 1985.
2. Richardson, C. J. Freshwater wetlands: transformers, filters, or sinks, in *Freshwater Wetlands and Wildlife*. 1989, CONF-8603101, DOS Symposium Series No. 61, R. R. Sharitz and J. W. Gibbons, Eds. USDOS Office of Scientific and Technical Information, Oak Ridge, TN, 1989.
3. Richardson, C. J. Mechanisms controlling phosphorus retention capacity in freshwater wetlands. *Science*. 228:1424, 1985.
4. Hammer, D. A., Ed. *Constructed Wetlands For Wastewater Treatment. Municipal, Industrial and Agricultural.* Lewis Publishers, Chelsea, MI, 1989.

5. Richardson, C. J. and J. A. Davis. Natural and Artificial Wetland Ecosystems: Ecological Opportunities and Limitations, in *Aquatic Plants for Water Treatment and Resource Recovery*. K. R. Reddy and W. H. Smith, Eds. Magnolia Publishing, Orlando, FL, 1987, 819–854.
6. Lindsay, A. L. Chemical Equilibria in Soils. Wiley & Sons, New York, 1979.
7. Faulkner, S. P. and C. J. Richardson. Physical and chemical characteristic of freshwater wetland soils, in *Constructed Wetlands for Wastewater Treatment: Municipal, Industrial and Agricultural*. D. A. Hammer, Ed. Lewis Publishers, Chelsea, MI, 1989, 41–72.
8. Richardson, C. J. and P. E. Marshall. Processes controlling movement, storage and export of phosphorus in a fen peatland. *Ecological Monogr*. 56:279, 1986.
9. Patrick, W. H., Jr., and R. A. Khalid. Phosphate release and sorption by soils and sediments: effect of aerobic and anaerobic conditions, *Science*. 186:53, 1974.
10. Howard-Williams, C. and B. R. Allanson. Phosphorus cycling in a dense *Potamogeton pectinatus* L. Bed. *Oecologia (Berlin)*. 49:56, 1981.
11. Swift, D. R. and R. B. Nicholas. Peliphyton and water quality relationships in the Everglades Water Conservation Areas: 1978–1982. South Florida Water Management District, Technical publication no. 87-2, 1987.
12. Richardson, C. J. and D. S. Nichols. Ecological analysis of wastewater management criteria in wetland ecosystems, in *Ecological Considerations in Wetlands Treatment of Municipal Wastewater*. P. J. Godfrey, E. R. Raynor, S. Pelczonski, and J. Benforado, Eds. Van Nostrand Reinhold, New York, 1985, 351–391.
13. Nichols, D. S. Capacity of natural wetlands to remove nutrients from wastewater. *J. Water Pollut. Control Fed*. 55:495, 1983.
14. Craft, C. B. and C. J. Richardson. Storage rates for nutrients, cations and metals, in *Effects of Nutrient Loadings and Hydroperiod Alterations on Control of Cattail Expansion, Community Structure and Nutrient Retention in the Water Conservation Areas of South Florida*. C. J. Richardson Ed. Duke Wetland Center Publication 91-08. Durham, NC, 1991, 128–144.
15. Davis, J. H. The natural features of southern Florida. *The Florida Geological Survey Bulletin, No. 25*. Tallahassee, FL, 1943.
16. Steward, K. K. and W. H. Ornes. The autecology of sawgrass in the Florida Everglades. *Ecology*. 56:162, 1975.
17. Davis, S. M. and L. A. Harris. Marsh plant production and phosphorus flux in Water Conservation Area 2, in *Environmental Quality Through Wetlands Utilization. Proc. Symp. Restoration of the Kissimmee River and Taylor Creek-Nubbin Slough Basin*. M. A. Drew, Ed. Feb. 28–Mar. 2, 1978. Tallahassee, FL, 1978.
18. Toth, L. A. Effects of hydrologic regimes on lifetime production and nutrient dynamics of sawgrass. South Florida Water Management District, Technical Publication no. 87-6, 1987.
19. Toth, L. A. Effects of hydrologic regimes on lifetime production and nutrient dynamics of cattail. South Florida Water Management District, Technical publication no. 88-6, 1988.
20. South Florida Water Management District. Surface Water Improvement and Management Plan for the Everglades. Nov. 8, 1989.

20a. Surface Water Improvement and Management Plan for the Everglades. (SWIMM) South Florida Water Management District. November, 8. West Palm Beach, FL, 1989.

21. Richardson, C. J. and C. B. Craft. Phase One: A Preliminary Assessment of Nitrogen and Phosphorus Accumulation and Surface Water Quality in Water Conservation Areas 2A and 3A of South Florida. Duke Wetland Center Publication 90-01, Duke University, Durham, NC, 1990.
22. DeLaune, R. D., W. H. Patrick, Jr., and R. J. Buresh. Sedimentation rates determined by Cs-137 dating in a rapidly accreting salt marsh. *Nature*. 275: 532, 1978.
23. Stevenson, C. J., L. G. Ward, and M. S. Kearney. Vertical accretion in marshes with varying rates of sea level rise, in *Estuarine Variability*. D. A. Wolfe, Ed. Academic Press, New York, 1986, 241–259.
24. Craft, C. B., E. D. Seneca, and S. W. Broome. Sedimentation and accretion in an irregularly flooded brackish-marsh, *10th Biennial Int. Estuarine Res. Conf.* Abstracts, 17, 1989.
25. Hatton, R. S., R. D. DeLaune, and W. H. Patrick, Jr. 1983. Sedimentation, accretion and subsidence in marshes of Barataria Basin, Louisiana. *Limnology Oceanography*. 28:494, 1983.
26. Kadlec, R. H. and J. A. Robbins. Sedimentation and sediment accretion in Michigan coastal wetlands. *Chem. Geol*. 44:119, 1984.

27. Davis, J. H. The peat deposits of Florida. Their occurrence, development and uses. *Florida Geological Survey Bulletin, No. 30*. Tallahassee, FL, 1946.
28. McDowell, L. L., J. C. Stephens, and E. H. Stewart. Radiocarbon chronology of the Florida Everglades peat. *Soil Sci. Soc. Am. Proc.* 33:743, 1969.
29. Mitsch, W. J. and J. G. Gosselink. *Wetlands*. Van Nostran Reinhold, New York, 1986.
30. Dineen, W. Presentation to the South Florida Water Management District. September 1989, unpublished.
31. Davis, S. M. Growth, decomposition and nutrient retention of sawgrass and cattail in the Everglades. South Florida Water Management District, Technical publication 90-03, 1990.
32. South Florida Water Management District. Surface Water Improvement and Management Plan for the Everglades. April 11, 1990.
33. Klopatek, J. M. The role of emergent macrophytes in mineral cycling in a fresh water marsh, in *Mineral Cycling in Southeastern Ecosystems*. ERDA Symposium series. F. G. Howell, J. B. Gentry, and M. H. Smith, Eds. National Technical Information Service, U.S. Department of Commerce, Springfield, VA, 1975, 367.
34. Davis, C. B. and A. G. van der Valk. Litter decomposition in prairie glacial marshes, in *Freshwater Wetlands: Ecological Processes and Management Potential*. R. E. Good, D. F. Whigham, and R. L. Simpson, Eds. Academic Press, New York, 1978, 99–114.
35. Gleason, P. J., P. Stone, and M. Rosen. Nutrient uptake and rates of nutrient deposition in conservation area 2A. Central and Southern Florida Flood Control District, Environmental Sciences Division, West Palm Beach, FL, Nov. 1974.
36. Kadlec, R. H. Wetland utilization for management of community wastewater: 1982 operations summary, Houghton Lake wetlands treatment project. Wetlands Environmental Research Group Report, University of Michigan, Ann Arbor, MI, 1983.
37. U.S. Congress. Comprehensive report on central and southern Florida for flood control and other purposes. 90th Congress, 2nd session, House document no. 643. U.S. Government Printing Office, Washington, D.C., 1949.

CHAPTER 29

Fate of Non-Point Source Nitrate Loads in Freshwater Wetlands: Results from Experimental Wetland Mesocosms

W. G. Crumpton, T. M. Isenhart, and S. W. Fisher

INTRODUCTION

In recent years, agriculture has changed from a labor-intensive to a chemical-intensive enterprise. Across the Midwestern Corn Belt, agricultural applications of fertilizers and pesticides have more than doubled since the mid-1960s, and agrichemical contamination of surface and groundwaters has become a pressing environmental problem.[1-4] Nitrate and pesticides are the agrichemical contaminants of foremost concern because of their potential impacts on both public health and ecosystem function, and because of the widespread use of nitrogen (N) and pesticides in modern agriculture. The total amount of N applied in fertilizers far exceeds that of any other nutrient, and annual application of fertilizer-N (FN) in the U.S. has grown from a negligible amount prior to World War II to approximately 9 million metric tons of N per year.[5] Not only crop acreage, but also the intensity of FN application has increased. Across the Corn Belt, average yearly FN rates on corn increased from 72 kg FN/ha in 1965 to 150 kg FN/ha in 1982.[6] Corn accounts for 21% of the total U.S. agricultural land use, but 43% of FN use and 42% of total pesticide use.[1,6] It is now clear that substantial amounts of the chemicals applied to farm fields are lost to surface and groundwaters in agricultural watersheds.[4,7] As much as 50% of the FN applied to cultivated crops may be lost in agricultural drainage water, primarily in the form of nitrate.[8]

The environmental impacts of agrichemical contamination of surface and groundwaters are a special concern in the Midwest. Non-point loads of nutrients to surface waters in this region are among the highest in the country,[7] and there is contamination of surface and groundwater by a variety of pesticides.[8,9] Despite our best efforts, it is unlikely that these contamination problems can be solved by chemical management alone. The best solutions will involve a combination of on-field and off-field approaches. In addition to improved farming systems which incorporate the best management practices and reduced inputs of nutrients and pesticides, the creation of buffer strips and other off-field sinks for chemical contaminants near their origin offers great promise for minimizing surface and groundwater contamination. In the Corn Belt, one of the most promising strategies for reducing agrichemical contamination of surface and groundwaters is the restoration of wetlands in agricultural watersheds specifically as sinks for agrichemical contaminants.[10]

Studies suggest that wetlands may act as sinks for a variety of compounds,[11,12] and wetlands may be especially effective as sinks for nitrate loads from cultivated fields. If wetlands are to serve as long-term sinks for nitrogen, differences in inputs and outputs must reflect net storage in the system through accumulation and burial in sediments, or net loss from the system through gaseous evolution of NH_3, N_2O, or N_2.[13,14] Most of the published papers dealing with wetlands and water quality note the probable importance of denitrification, with resulting gaseous loss of N_2O and N_2. In fact, with rare exception, denitrification is cited as the primary reason wetlands may serve as nitrogen sinks. However, there have been few actual measurements of denitrification in freshwater marshes.[8,12,15,16] As Neely and Baker[8] note, denitrification is assumed to be an important process in many freshwater wetlands based largely on circumstantial evidence; first, that conditions in the wetlands are suitable for denitrification (anaerobic conditions and a large base of organic carbon) and, second, that nitrate disappears rapidly from water overlying wetland sediments. Increased nitrate loading to wetlands in agricultural watersheds might be expected to stimulate denitrification, but there are no reliable

0-87371-550-0/93/$0.00+$.50

measurements of the effects of such loading or of the denitrification capacity of wetlands receiving agricultural loads.

A better understanding of the potential water quality functions of wetlands is especially critical at the current time. In 1986, with waterfowl populations at their lowest levels in 30 years, the U.S. and Canada agreed to the North American Waterfowl Management Plan to increase waterfowl populations 60% by the year 2000.[17] The plan calls for the protection and restoration of 1.1 million acres of wetlands in the U.S. In the past few years, over 5000 wetlands have been restored in Iowa, Minnesota, South Dakota, and North Dakota alone, mostly through the Conservation Reserve Program of the 1985 Food Security Act. The Food, Agriculture, and Trade Act of 1990 created the Wetland Reserve Program to assist farmers in restoring and protecting wetlands. This program has the goal of enrolling 1 million acres of wetlands by 1995.

These new initiatives in wetlands restoration offer a unique opportunity for water quality remediation in agricultural watersheds. Although the concerns motivating wetland restoration have related primarily to waterfowl and habitat loss, wetlands may have even greater value in agricultural watersheds for their water quality functions. However, because so little is known about the water quality functions of such wetlands, site selection criteria for wetland restorations have not generally considered water quality functions. Especially in light of the current efforts at wetland restoration in the prairie pothole region, there is a critical need for understanding the water quality functions of wetlands in agricultural watersheds.

In order to progress much beyond our current understanding of wetland function, we must be able to conduct controlled and repeatable ecosystem-level experiments. This is the rationale behind a major new research initiative of the Wetlands Research Laboratory at Iowa State University which specifically addresses the water quality functions of wetlands in agricultural watersheds using experimental wetland mesocosms. *Mesocosms* are experimental systems large enough to approximate the structure and function of at least some subset of critical processes and organisms in the ecosystem of interest, in this case wetlands. Mesocosms afford numerous advantages for wetland research: (1) they can be controlled with regard to physical forcing functions (hydroperiod, water inflow and flow rate, nutrient concentration, etc.); (2) they can be replicated in relatively large numbers and easily manipulated; (3) they can be dosed with radioactive and stable isotope tracers for studies of fate and flow; and (4) they can be sacrificed if necessary. In this article, we briefly summarize some of our initial mesocosm studies and discuss future applications.

METHODS

In 1989, a complex containing 48 experimental wetland mesocosms was constructed at Iowa State University's Hinds Irrigation farm, approximately 4 miles from main campus (Figures 1 and 2). This represents a significant and virtually unique facility for studies of agrichemical fate and effects in freshwater wetlands. The mesocosms were constructed using UV-stabilized polyethylene tanks which are 3.35 m in diameter and 90 cm deep, thus providing for approximately 9 m^2 of wetland in each mesocosm (Figure 3). The tanks were filled to a depth of 60 cm with sediment from a nearby drained wetland, planted with cattail rhizomes, and flooded. A deep irrigation well supplies the mesocosms with feed water having concentrations of anions and cations similar to those found in wetlands in glaciated terrain,[18] yet having concentrations of nitrogen and phosphorus which are low enough to allow experimental addition of these two elements. In-line Dosmatic Plus fertilizer injectors allow for the controlled addition of desired chemicals directly into the irrigation water. Mesocosms are individually valved and water is supplied to each unit through spray nozzles around its inside circumference. Bulkhead adapters for surface drainage are located 5 cm above the sediment level to prevent loss of all water in the event of a leak. Water level is maintained through the use of variable-height standpipes for drainage. Each mesocosm is also fitted with a drainage system which can provide a subsurface inlet or outlet for water. The mesocosms can be operated as either static or flow-through systems.

Studies using the mesocosms have been supplemented with stable and radioisotope tracer studies of nitrate and pesticide fate in smaller enclosures called *microcosms*. The microcosms used in these studies were developed in our laboratory for measurements of benthic microbial processes. The microcosms are designed for short-term experiments, but allow many more manipulations than are possible in the mesocosms. The microcosms are small enclosures consisting of 2-in. diameter, polycarbonate cylinders enclosing intact sediment and overlying water. The cylinders are milled to accept gas-tight closures with sampling and injection ports. This makes it possible to seal the microcosms for measurements of gas exchange or to control the partial pressures of gases in the water phase by injection of gas mixtures. We have used these microcosms

Figure 1. Aerial view of experimental wetland complex.

Figure 2. Ground-level view of experimental wetland complex.

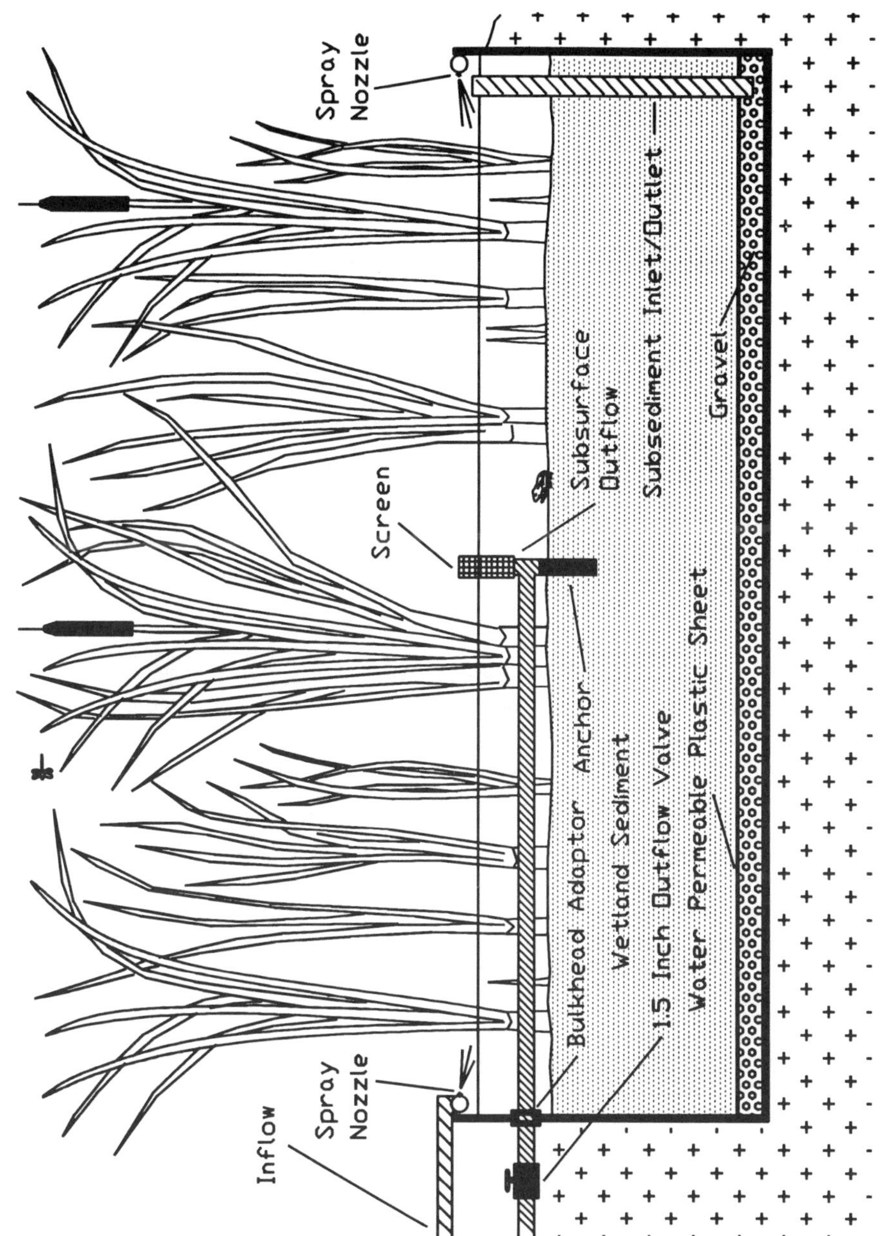

Figure 3. Cross-sectional diagram of a mesocosm.

for studies of ammonium and nitrate flux, for ^{15}N tracer studies of nitrogen assimilation, nitrification, and denitrification, for studies of oxygen profiles and flux rates using oxygen microelectrodes, and for studies of pesticide decomposition.

Our initial mesocosm studies have focused on defining the major factors affecting the assimilative capacities of natural and restored wetlands for nitrate, including the effects of nitrate loading, phosphorus addition, and sediment loading on nitrate loss. The effects of nitrate loading rates and patterns have been examined using a combination of batch dose studies, flowthrough mass balance studies, and direct measurements of transformations. In batch dose studies, mesocosms are dosed with nitrate solutions and the decline in nitrate concentrations in the overlying water is measured over time. In mass balance studies, mesocosms are configured as flow-through systems and mass balance budgets are constructed based on measured loading to and export from each mesocosm. Flow rates into and out of the mesocosms are controlled, so the mass balance budgets are calculated quite simply from flow rates and nitrate concentrations in the inflows and outflows.

RESULTS AND DISCUSSION

In order to be useful in agrichemical fate and effects studies, mesocosms must be large enough to approximate the structure and function of at least a critical subset of processes and organisms, yet small enough to allow the control and repeatability required to measure ecosystem-level responses. Although the mesocosms were constructed only $2^1/_2$ years ago, the establishment and growth of the transplanted cattails has been dramatic. By the end of the first growing season, cattail densities had increased from 1.7 to 31.3 shoots/m^2; and by the end of the second season, densities had reached 65.2 shoots/m^2, levels characteristic of field populations in the area. Based on measurements of macrophyte production, litter decomposition, benthic aerobic metabolism, and denitrification, the mesocosms now reasonably simulate nearby emergent wetlands. In addition, the mesocosms have proven to be remarkably replicable in both flowthrough and batch dose modes of operation. The coefficient of variation for macrophyte production and for nitrate loss are both generally less than 10%.

Our initial mesocosm studies have confirmed the considerable capacity of freshwater wetlands to transform nitrate. Even under highly aerobic conditions, nitrate concentrations decline rapidly in water overlying wetland sediments in all of our mesocosm and microcosm experiments. Results from batch dose studies illustrate the rapid decline of nitrate typically observed (Figures 4 and 5). These results are confirmed by longer-term, flowthrough mass balance studies. When mesocosms with residence times of approximately 1 week were loaded with 3 to 15 mg/L of nitrate nitrogen, percent retention for nitrate exceeded 80% (Figure 6).

Nitrate loss in mesocosm and microcosm studies can be described as a first-order process over a wide range of concentrations, and loss rates can be modeled based on factors controlling the rate of nitrate flux to anaerobic sites.[19] This is consistent with models for denitrification in agricultural streams which suggest that in the presence of high nitrate loads, denitrification rates are controlled by the NO_3^- concentration in the overlying water and the effective length of the diffusion path between the overlying water and the primary site of denitrification in underlying anaerobic sediments.[20] Studies using ^{15}N tracers confirm that denitrification is the dominant loss process in the experimental wetlands.[21,22] However, contrary to the assumptions of models for denitrification in agricultural streams, our data suggest that nitrate concentrations in the overlying water also significantly affect the activities and/or population densities of denitrifying bacteria.[21,22] In addition, our results demonstrate the effectiveness of decaying plant litter in providing anaerobic sites for denitrification.[21,22]

The overall objective of our mesocosm work is to determine the sustainable assimilative capacity of wetlands for agrichemical contaminants. Predicting assimilative capacities requires an understanding of the critical processes, their interactions, and the effects of system hydrology and loading patterns on their effectiveness. Our results to date have identified the most significant processes involved in nitrate loss in wetlands, as well as the major factors affecting rates of loss.[19,21,22] Research now in progress is designed to test the effectiveness of nitrate sinks of wetlands draining different size watersheds. In these studies, mesocosms configured as flowthrough systems are loaded to simulate different watershed/wetland acreage ratios; for example, a 1-acre wetland draining 100 acres of corn. Results should allow us to estimate the assimilative capacity of wetlands for agriculturally derived nitrate and thus to recommend site selection criteria for wetland restorations in agricultural watersheds.

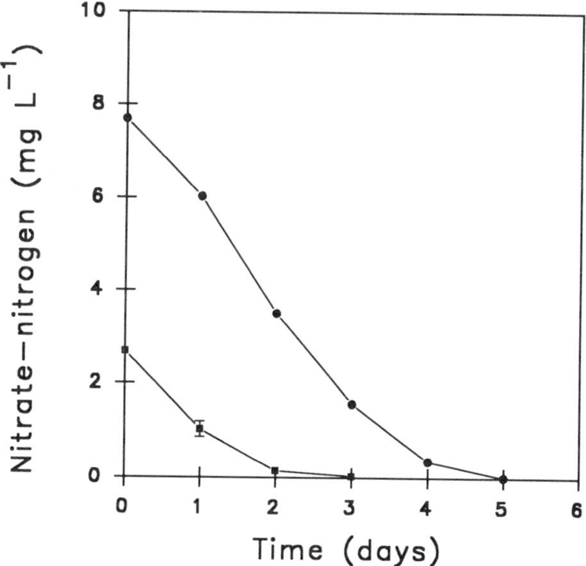

Figure 4. Nitrate-nitrogen concentrations in mesocosms during batch dose studies following experimental addition of nitrate-amended feed water. Bars indicate ± one standard error (n = 36 for higher concentration, n = 4 for lower concentration).

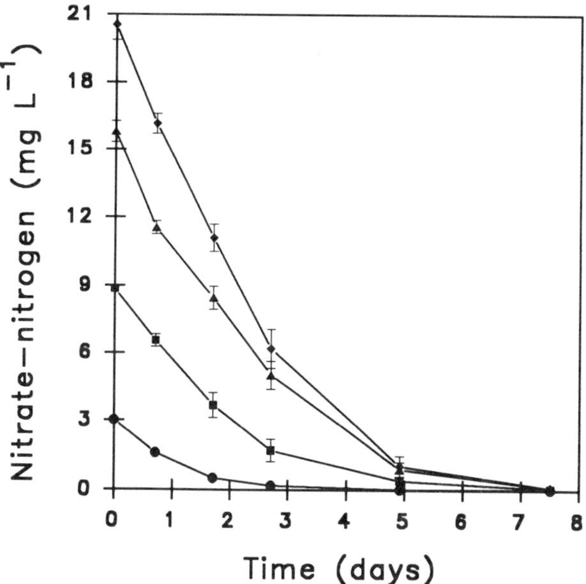

Figure 5. Nitrate-nitrogen concentrations in microcosms following experimental addition of nitrate. Bars indicate ± one standard error (n = 10 for each concentration).

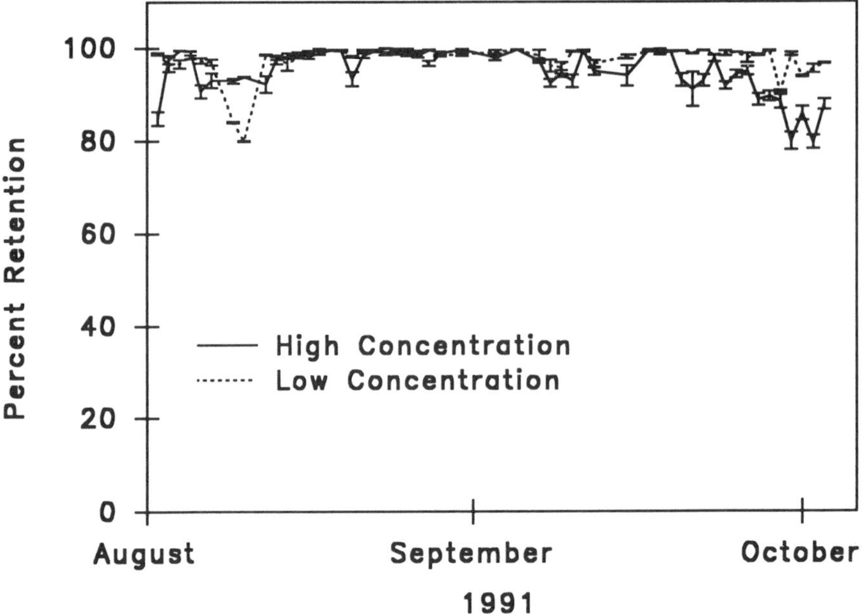

Figure 6. Percent retention of nitrate-nitrogen by mesocosms during flowthrough mass balance studies with 3 mg/L[1] nitrate-nitrogen (low concentration) and 15 mg/L[1] nitrate-nitrogen (high concentration). Bars indicate ± one standard error (n = 8 for each concentration).

REFERENCES

1. Gianessi, L. P., H. M. Peskin, P. Crosson, and C. Puffer. Nonpoint source pollution: are cropland controls the answer? Report prepared for the U.S. Environmental Protection Agency, U.S.D.A. Soil Conservation Service, and U.S. Geological Survey under E.P.A. Cooperative Agreement CR811858-01, Resources for the Future, Washington, D.C., 1986.
2. Gray, R. G., Sr. Agricultural chemicals: a growing concern for groundwater, *J. Freshwater*. 9:16, 1985.
3. National Research Council. Proceedings of the Colloquium on Agrichemical Management and Water Quality, National Academic Press, Washington, D.C., in press, 1986.
4. Keeney, D. R. Sources of nitrate to groundwater, *Crit. Rev. Environ. Control*. 16:257, 1986.
5. U.S. Department of Agriculture. 1983 Handbook of Agricultural Charts, U.S. Dept. Ag., Ag. Handbook No. 619., Washington, D.C., 1983.
6. Hargett, N. L. and J. T. Berry. 1982 fertilizer summary data, National Fertilizer Development Center, TVA, Muscle Shoals, AL, 1983.
7. Omernik, J. M. Nonpoint source-stream nutrient level relationships: a nationwide study, EPA-60013-77-105. U.S. Environ. Prot. Agency, Corvallis, OR, 1977.
8. Neely, R. K. and J. L. Baker. Nitrogen and phosphorous dynamics and the fate of agricultural runoff, in *Northern Prairie Wetlands*. A. G. van der Valk, Ed. Iowa State University Press, Ames, IA, 1989.
9. Kelley, R., G. R. Hallberg, L. G. Johnson, R. D. Libra, C. A. Thompson, R. C. Splinter, and M. G. Detroy. Pesticides in groundwater in Iowa, in *Proc. Agricultural Impacts on Ground Water Conf.* National Water Well Assoc., 1986, (8/11-13) Omaha, NE, 1986.
10. van der Valk, A. G. and R. W. Jolly. Recommendations for research to develop guidelines for the use of wetlands to control rural NPS pollution, *Ecological Engineering*. March:115–134, 1992.
11. Howard-Williams, C. Cycling and retention of nitrogen and phosphorus in wetlands: a theoretical and applied perspective, *Freshwater Biol.* 15:391, 1985.
12. Nixon, S. W. and V. Lee. Wetlands and Water Quality: A Regional Review of Recent Research in the United States on the Role of Freshwater and Saltwater Wetlands as Sources, Sinks, and Transformers of Nitrogen, Phosphorus, and Various Heavy Metals. Army Corp of Engineers, Waterways Experiment Station, Vicksburg, Mississippi, Technical Report Y-86-2. 1986.

13. van der Valk, A. G., J. L. Baker, C. B. Davis, and C. E. Beer. Natural freshwater marshes as nitrogen and phosphorous traps for land run off, in *Wetland Values and Functions: The State of Our Understanding*. P. J. Greeson, J. R. Clark, and J. E. Clark, Eds. American Waterworks Association, Minneapolis, MN, 1979, 457–467.
14. Sahrawat, K. L. and D. R. Keeney. Nitrogen transformations in flooded rice soils, *Fert. Res.*, 9:15, 1986.
15. Bowden, W. B. The biogeochemistry of nitrogen in freshwater wetlands, *Biogeochemistry*. 4:313, 1987.
16. Seitzinger, S. P. Denitrification in freshwater and coastal marine ecosystems: ecological and geochemical significance, *Limnol. Oceanogr.* 33:702, 1988.
17. Kasten, R. W. Legislative perspectives on wetland protection, in *Increasing Our Wetland Resources*. J. Zelazny and J. Feierabend, Eds. National Wildlife Federation Corporate Conservation Council, 1988, 32–34.
18. LaBaugh, J. W. Chemical characteristics of water in northern prairie wetlands, in *Northern Prairie Wetlands*. A. G. van der Valk, Ed. Iowa State University Press, Ames, IA, 1989, 56–90.
19. Crumpton, W. G. and T. M. Isenhart. in preparation, 1992.
20. Christensen, P. B., L. P. Nielsen, J. Sorensen, and N. P. Revsbech. Denitrification in nitrate-rich streams: diurnal and seasonal variation related to benthic oxygen metabolism, *Limnol. Oceanogr.* 35:640, 1990.
21. Isenhart, T. M. In preparation, Ph.D. dissertation, Iowa State University, 1992.
22. Isenhart, T. M. and W. G. Crumpton. in preparation, 1992.

CHAPTER 30

Mechanisms of Wetland-Water Quality Interaction

C. A. Johnston

INTRODUCTION

"Black box" (i.e., input–output) studies in natural wetlands receiving wastewaters have demonstrated that wetlands can be very effective in the short term at reducing suspended solids and nutrient concentrations in wastewater; but that over the long term, some of those functions can break down and the wetland can become a source of contaminants as it releases stored materials. In order to understand why, we need to break out of the black box of input–output studies and delve into the physical and biological mechanisms by which wetlands interact with materials in surface water. We need this understanding in order to design constructed wetlands that will serve a desired function over a long lifespan.

A wetland ecosystem consists of interacting biological and physical components that alter fluxes of materials. Once a substance enters an ecosystem, it may be stored, altered by chemical or biological action, or discharged via water or atmospheric fluxes. A generic model of these interactions contains seven different storage compartments and numerous pathways of flux between them (Figure 1). While some of these storage and discharge mechanisms are well understood, the difficulty in quantifying others has resulted in highly variable published results on the effects of wetlands on water quality.

The purpose of this article is to summarize findings about mechanisms of wetland-water quality interactions from studies in natural wetlands as a means of understanding how wetlands process inputs of sediments and nutrients. It does not deal with constructed wetlands *per se*, nor the impacts of water quality impairment on wetland health, but rather with the ability of wetlands to retain contaminants to the benefit of other surface waters. The article is based on a comprehensive review of over 300 literature sources.[1,2]

SEDIMENTATION

Excess turbidity in surface waters can reduce photosynthesis, decrease oxygen concentrations, impair respiration and feeding of aquatic animals, kill benthic organisms, destroy fish habitat, and stimulate the encroachment of exotic and undesirable plant species.[3] When water-borne suspended soilds enter a wetland, the decrease in water velocity causes them to settle out onto the soil surface, thereby benefiting downstream water quality. Although vegetation helps slow the water down and filters out particles to some extent, mineral sediment deposition is largely a physical settling process.

Mineral sediment deposition has dual benefits for water quality. Not only does it reduce turbidity, but also retains phosphorus and contaminants that are adsorbed to those particles.[4] In order for this mechanism to occur, however, there must be a source of sediment input to the wetland such as overland runoff, river flooding, or stormwater discharge. Mineral sediment deposition is a relatively irreversible mechanism: once deposited, the sediments remain part of the soil storage compartment indefinitely (Figure 1).

Sedimentation rates can be expressed in terms of vertical accretion (cm/year^{-1}) or mass accumulation (g/m^2/year). Accretion rates reported for wetlands range from 0 for wetlands receiving little or no sediment, to values ≥1.5 cm/year^{-1} for a shallow lake in the upper Mississippi River floodplain,[5] for rapidly subsiding wetlands of the Mississippi delta,[6,7] and for riparian wetlands receiving agricultural runoff.[8] Mass accumulation rates exceeding 5000 g/m^2/year^{-1} have been reported for floodplain wetlands[9] and wetlands receiving agricultural runoff.[10,11]

0-87371-550-0/93/$0.00+$.50

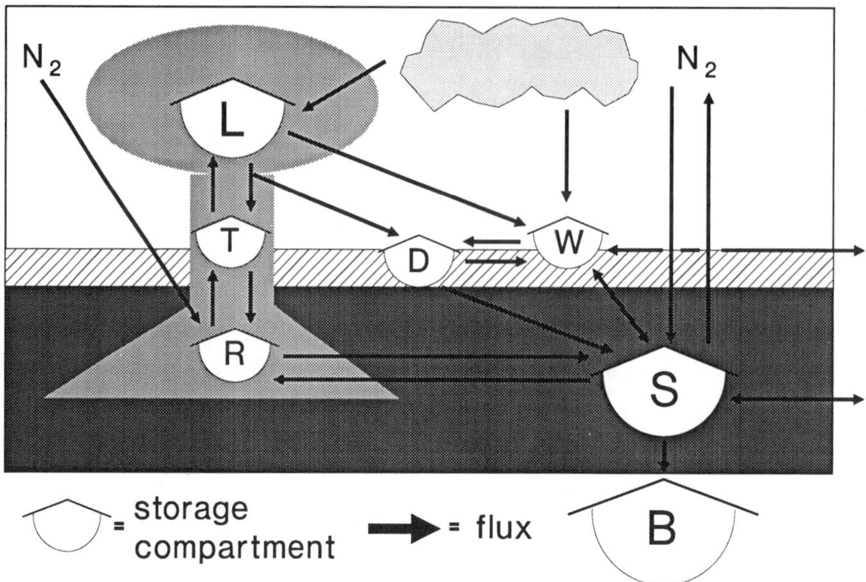

Figure 1. General model of material flux and storage in wetlands: L = leaves, T = woody stems, R = roots and rhizomes, D = litter, W = water, S = surface soil, and B = buried soil.

Organic soil ("peat") is formed by the accumulation of organic matter where biomass production exceeds decomposition rate. Unlike mineral sediment deposition, which depends on inputs of soil material from outside the wetland, organic soil accumulation may result from production of biomass within the wetland. In general, organic soil accumulation is a very slow process. Average annual accretion rates summarized from organic soil studies (0.12 cm/year) were only one sixth the rates summarized from mineral sedimentation studies (0.69 cm/year), and average annual mass accumulation rates were an order of magnitude less (96 g/m^2/year for organic soils vs. 1680 g/m^2/year for mineral sedimentation).[2] Organic soil accumulation is negligible in wetlands where most or all of the plant litter produced every year becomes decomposed.

When mineral sediments or organic matter are deposited in a wetland, they take with them associated nutrients and contaminants. Nutrient fluxes via sediment deposition were high, averaging 15 g N/m^2/year and 1.5 g P/m^2/year.[2] Nutrient fluxes associated with organic soil accumulation were about an order of magnitude lower than those associated with mineral soil deposition: 1.6 g N/m^2/year and 0.3 g P/m^2/year.

NUTRIENTS

Nutrient Standing Stocks

Once sediment and nutrients are added to the soil surface, they become part of its standing stock, the total amount of a substance per unit area of wetland storage compartment at a particular time. Standing stocks are computed as:

$$\text{standing stock} = \text{concentration} \times \text{mass}$$

where concentration is the amount of nutrient per unit of storage medium (e.g., milligrams N per gram of biomass) and mass is the amount of storage medium per unit area (e.g., biomass/per square meter). Nutrient standing stocks in soil are typically much higher than those in any other wetland compartment, with reported values ranging from 342 to 6900 g N/m^2 and 10 to 179 g P/m^2 (Tables 1 and 2).

Why are soil standing stocks so variable? One reason is the tremendous variation that exists in soil nutrient concentrations: 0.02 to 65 mg/gdw N, and 0.0001 to 7.0 mg/gdw P, differences of over three orders of magnitude.[2] Another reason for the variation in soil standing stock measurements is the difficulty in defining the soil mass. Soils are not discrete objects, so the depth of soil used in computing standing stocks is fairly arbitrary.

Table 1. Standing Stocks of Nitrogen in Wetland Storage Compartments

Compartment	Minimum (g/m²)	Maximum (g/m²)	Mean (g/m²)	n[a]
Soil	342	6900	2757	6
Below-ground				
Herbaceous roots and rhizomes	3.7	29	13	8
Tree and shrub roots	—	6	—	1
Green Tissues				
Emergent stems and leaves	0.6	72	21	30
Floating macrophytes	0.4	90	21	6
Tree and shrub leaves	4.7	13	8	3
Woody tissues	5.0	95	—	2
Litter				
Emergent stems and leaves	5.3	20	11	4
Tree and shrub leaves	4.5	24	12	4

Note: Values summarized by Johnston in Reference 2.

[a] n = number of studies summarized.

Table 2. Standing Stocks of Phosphorus in Wetland Storage Compartments

Compartment	Minimum (g/m²)	Maximum (g/m²)	Mean (g/m²)	n[a]
Soil	10	179	63	8
Below-ground				
Herbaceous roots and rhizomes	0.70	2.9	1.8	8
Tree and shrub roots	0.28	8.1	3.1	4
Green Tissues				
Emergent stems and leaves	0.10	6.8	2.2	28
Floating macrophytes	0.20	18.0	3.5	8
Tree and shrub leaves	0.08	1.7	0.9	9
Woody tissues	0.18	4.3	3.1	7
Litter				
Emergent stems and leaves	0.41	1.1	—	2
Tree and shrub leaves	0.21	1.9	1.0	8

Note: Values summarized by Johnston in Reference 2.

[a] n = number of studies summarized.

Average standing stocks in green tissues are one to two orders of magnitude smaller than soil standing stocks (Tables 1 and 2). Even in the Okefenokee Swamp, where nitrogen standing stocks in the mature cypress forest were the highest found in the published literature (99.5 g/m²), nitrogen standing stocks in vegetation were 1/50 those in the soil.[12]

Given the high variability in wetland soil concentrations, it is not really surprising that nutrient concentrations in green tissues generally fall within a narrow range of values, regardless of species: 1 to 3% N and 0.1 to 0.3% P. Free-floating aquatics such as *Lemna* and *Eichhornia* tend to have higher tissue concentrations, averaging 3.1% N and 0.5% P,[2] which is reflected in their higher P standing stocks (Table 2).

Biomass, the other half of the standing stock equation for green tissues, does vary substantially between species, and even within species. For example, above-ground biomass of *Typha* measured by a single investigator throughout North America ranged from 0.4 to 1.3 kg/m², a threefold difference.[13] An even wider range of values was reported for *Phragmites australis* in Scottish lakes.[14]

When nutrients are added to wetlands with low natural biomass production, there is often change to a more productive species. This occurred at a Michigan peatland used for municipal wastewater treatment, where large areas of a sedge-leatherleaf community were converted to a cattail stand in less than a decade.[15] Such shifts may result in greater changes in vegetation standing stocks in response to nutrient additions than would be possible by increasing tissue nutrient concentrations alone.

Above-ground woody plant tissues generally contain very low nutrient concentrations, averaging only 0.4% N and 0.01% P dry weight.[2] However, despite these low concentrations, this storage compartment can be primarily due to the large amount of woody biomass in wetland forests. In mature forests, woody biomass can contain up to 95 g N/m^2 and 4.3 g P/m^2,[12] ten times the amount of N, and five times the amount of P normally stored in leaves.

Below-ground standing stocks are rarely studied, especially in woody species, but studies indicate standing stocks that are often comparable to or higher than those in green tissues. Below-ground nutrient standing stocks were two to three times higher than above-ground standing stocks in three southeastern U.S. marshes,[16-18] while the ratio of below-ground-to-above-ground standing stocks in more northerly marshes ranged from 1:1 to 1:7 for N and 1:1 to 1:3 for P.[2] Less is known about below-ground standing stocks in forested wetlands because it is so difficult to quantify woody root biomass; however, values from three U.S. studies ranged from 0.3 to 8.1 g P/m^2 (Tables 1 and 2).

All green tissues eventually die and fall to the ground, where they become part of the litter compartment. Nutrient concentrations in litter are generally less than those in green tissues, although they can be higher if microbes living on litter surfaces take up nutrients from overlying surface waters. Litter standing stocks range between 4 and 24 g N/m^2 and 0.2 to 2.1 g P/m^2, slightly lower than standing stocks in green tissues (Tables 1 and 2).

Fluxes into Storage Compartments

Standing stocks give a snapshot view of the nutrients in place at a particular time. However, nutrient cycling in wetlands is extremely dynamic. Nutrients taken up by plant roots may be incorporated into below-ground standing stocks or translocated into above-ground tissues. At the end of the growing season, some of the nutrients are retranslocated from green tissues into perennial tissues, some are leached out of the green tissues into wetland surface waters, and the rest return to the wetland soil surface in litterfall. This process of nutrient uptake and return is known as *turnover*.

"Turnover times", estimated by dividing standing stocks by annual inputs, vary substantially among storage compartments. In contrast to P turnover in leaves and litter, which averages about 2 years, the turnover time for wood and soil is about 100 years. Soils in nutrient depauperate wetlands can have even longer turnover times, with one reported as 5600 years. A given flux into a short-term storage compartment, such as green tissues, is less important from a water quality standpoint than a comparable flux into a long-term storage compartment such as wood. If the flux rate into a long-term storage compartment is small, however, its net effects on water quality will also be small.

Given the large number of studies of nutrient concentrations and standing stocks in wetland plants, it is surprising that relatively few workers have attempted to quantify net annual retention of nutrients in herbaceous plant biomass (Table 3). A Wisconsin study indicated high annual uptake rates for both N and P, but only 26% N and 38% P retention, the rest being returned to the wetland surface in the form of leaching and litterfall.[19] A study of Wisconsin *Typha* marshes[20] revealed similar plant uptake and over-winter retention rates for P. Wetlands in an Alaskan wet tundra studied by Dowding et al.[21] were more efficient at retaining nutrients (55% retention), but their annual uptake was only 1.8 g N/m^2/year and 0.1 g P/m^2/year. A substantial portion of the retained nutrients is stored below ground over winter retranslocated into above-ground biomass the following spring.[20] This internal cycling of nutrients helps plants conserve nutrients, but reduces their net uptake of nutrients.

Note that percent retention is here expressed as percent of plant uptake, not percent of nutrients entering the wetland. Also, with the exception of floating plants such as *Lemna* and *Eichhornia*, wetland plants receive most (95 to 99%) of their nutrients from the soil in which they are rooted, rather than from overlying surface waters.[22] After senescing macrophytes fall to the soil surface, much of their nutrient content is reincorporated into sediments via litter decomposition, but some is leached into overlying surface waters. Therefore, vegetative uptake may not directly benefit surface water quality, and in some cases may even degrade it by transferring nutrients from soil to water. Rooted emergents may indirectly benefit water quality, however, by transporting oxygen below ground.[23]

Despite the large nutrient standing stocks in mature trees, uptake rates for forests are low in comparison to herbaceous plants (Table 4). Even though wood can contain large standing stocks of nutrients, the amount of P allocated to wood growth every year is generally less than 0.1 g/m^2/year. In all but one of the studies cited,

Table 3. Nitrogen and Phosphorus Cycling by Herbaceous Wetland Vegetation

Wetland type or species	Plant uptake	Leaching losses	Litterfall losses	Net annual retention	Percent retention	Ref.
Nitrogen						
Scirpus fluviatilis	20.8	7.3	8.1	5.3	26	20
Wet tundra	1.8	—	0.8	1.0	55	22
Phosphorus						
Scirpus fluviatilis	5.3	2.2	1.1	2.0	38	20
Wet tundra	0.1	—	0.05	0.06	54	22
Typha latifolia	4.3	—	2.5	1.3	30	21

Note: In grams per square meter per year (g/m²/year).

Table 4. Nitrogen and Phosphorus Cycling by Woody Wetland Vegetation

Wetland type or species	Plant uptake	Leaching & litterfall losses	Wood increment	Net annual retention	Percent retention	Ref.
Nitrogen						
Shrub fen	3.0	2.3	—	0.7	23	25
Sphagnum bog	3.8	3.8	0	0	0	26
Sphagnum bog	6.6	6.6	0	0	0	27
Phosphorus						
Shrub fen	0.17	0.10	—	0.07	41	25
Cypress swamp[a]	1.55	0.69	0.07	0.86	56	28
Cypress swamp	0.87	0.77	0.07	0.07	8	9
Cypress swamp	0.15	0.15	0.01	0.0	0	29
Cypress swamp	0.23	0.23	0.0	0.0	0	16
Nyssa swamp	0.53	0.34	0.06	0.19	36	30
Nyssa swamp	0.53	0.41	0.12	0.12	23	30

Note: In grams per square meter per year (g/m²/year).

[a] Annual root increment was 0.4 g/m²/year.

the majority of nutrients taken up by trees were returned to the wetland surface via leaching and litterfall. The study by Nessel and Bayley,[27] however, reported 56% of annual uptake retained when root allocation was included in their nutrient cycling model.

Denitrification

Denitrification, the process by which microorganisms convert nitrate into nitrogen gas, (a form of nitrogen unusable by most organisms) is commonly held to be a major pathway of nitrogen removal from wetlands. Unlike nutrient storage mechanisms, denitrification ultimately exports nitrogen out of the wetland system into the atmosphere, and apparently can proceed indefinitely without harm to the wetland.

Inasmuch as it involves a gas flux, denitrification is difficult to measure. The most common measurements are done as laboratory incubations on a small sample of soil for a period of days. Given the controlled conditions, short time period, and small amount of material used in laboratory incubations, extrapolations are rarely made to the annual, per area basis needed to compare this flux with other ecosystem fluxes.

Denitrification requires, among other things, a source of nitrate. In natural wetlands, the supply of nitrate is often small, limited by the rate at which ammonium is nitrified (i.e., converted from ammonium to nitrate). Therefore, most denitrification incubation studies amend soils by adding nitrate to determine denitrification potential. Denitrification rates in unamended soils are very low, about 0.1 g/m²/year, but denitrification potentials are two to three orders of magnitude higher.[2] Amended incubations provide a more realistic estimate of the potential denitrification performance of constructed wetlands that receive substantial amounts of N in wastewater.

Table 5. Flux Rates into Wetland Storage Compartment and the Atmosphere

Flux	N (g/m^2/year)	P (g/m^2/year)	Approximate turnover time (years)
Plant uptake (without harvest)	0–5.3	0.1–2.0	
Plant harvest	0.4–90.0	0.1–18.0	—
Wood increment	?	0–0.1	200
Litter	0.8–8.1	0.05–2.5	2
Peat accumulation	0.9–2.7	0.04–1.1	200
Mineral sedimentation	1.4–52.4	0.1–8.2	200
Denitrification	0–134	—	—

Note: Values summarized by Johnston in Reference 2.

In situ measurements of N_2O flux using acetylene block technique provide a more realistic estimate of actual denitrification losses, but are hard to do and are also subject to measurement error. As with laboratory incubations, much higher denitrification rates were obtained for soils with added nitrogen, up to 134 g/m^2/year.[30]

COMPARISON OF WETLAND FLUXES

The net effect of a wetland on water quality is the result of cumulative fluxes into storage compartments and the atmosphere (i.e., denitrification). When these different fluxes are compared on an annual basis, it becomes evident that some are more important than others with regard to water quality (Table 5). The highest annual vegetation retention rates in natural wetlands were those reported by Klopatek[19] for *Scirpus fluviatilis*: 5.3 g N and 2.0 g P/m^2/year. In harvested wetlands, however, much higher amounts of nutrient can be removed with vegetation, up to 90 g N and 18 g P/m^2/year. Wood constitutes a very large and long-term storage compartment in wetlands, but annual rates of P flux into woody tissues are too small (<0.1 g/m^2/year) to have much effect on surface water. Average nutrient concentrations retained via peat accumulation (1.6 g N/m^2/year, 0.3 g P/m^2/year) fall within the range of values reported for plant retention in natural wetlands.[2] Mineral sediment deposition, the physical settling out of suspended particles, can sequester up to 52.4 g N and 8.2 g P/m^2/year in a very long-term storage compartment.[31] This mechanism provides more phosphorus retention than any other in natural wetlands, but occurs only in wetlands receiving high particulate loadings. Denitrification constitutes the largest potential source of N removal from wetlands (Table 5).

These findings have important ramifications for constructed wetlands. In order to maximize nutrient removal rates, systems should be designed to promote sedimentation and denitrification. Harvesting vegetation in constructed wetlands will also increase nutrient removal rates. Additional studies in constructed and natural wetlands will increase our knowledge of the mechanisms affecting water quality, and allow us to design wetlands better for specific waste disposal needs.

REFERENCES

1. Johnston, C. A., T. Johnson, M. Kuehl, D. Taylor, and J. Westman. The Effects of Freshwater Wetlands on Water Quality: A Compilation of Literature Values. Natural Resources Research Institute Report NRRI/TR-90/15, University of Minnesota, Duluth, 1990.
2. Johnston, C. A. Sediment and nutrient retention by freshwater wetlands: effects on surface water quality, *Crit. Rev. Environ. Control.* 21:491, 1991.
3. Darnell, R. M. *Impacts of Construction Activities in Wetlands of the United States.* U.S. Environmental Protection Agency Ecological Research Series EPA-600/3-76-045, 1976.
4. Stall, J. B. Effects of sediment on water quality, *J. Environ. Qual.* 1:353, 1972.
5. Eckblad, J. W., N. L. Peterson, K. Ostlie, and A. Temte. The morphometry, benthos and sedimentation rates of a floodplain lake in Pool 9 of the upper Mississippi River, *Am. Midl. Nat.* 97:433, 1977.
6. Baumann, R. H. and J. W. Day, Jr. Mississippi deltaic wetland survival: sedimentation vs. coastal submergence, *Science.* 224:1093, 1984.

7. DeLaune, R. D., J. H. Whitcomb, W. H. Patrick, Jr., J. H. Pardue, and S. R. Pezeshki. Accretion and canal impacts in a rapidly subsiding wetland, I. ^{137}Cs and ^{210}Pb techniques, *Estuaries*. 12:247, 1989.
8. Cooper, J. R., J. W. Gilliam, R. B. Daniels, and W. P. Robarge. Riparian areas as filters for agricultural sediment, *Soil Sci. Soc. Am. J.* 51:416, 1987.
9. Mitsch, W. J., C. L. Dorge, and J. R. Weimhoff. Ecosystem dynamics and a phosphorus budget of an alluvial cypress swamp in southern Illinois, *Ecology*. 60:1116, 1979.
10. Johnston, C. A., G. D. Bubenzer, G. B. Lee, F. W. Madison, and J. R. McHenry. Nutrient trapping by sediment deposition in a seasonally flooded lakeside wetland, *J. Environ. Qual.* 13:283, 1984.
11. Lowrance, R., J. K. Sharpe, and J. M. Sheridan. Long-term sediment deposition in the riparian zone of a coastal plain watershed, *J. Soil Water Conserv.* Jul–Aug:266, 1986.
12. Schlesinger, W. H. Community structure, dynamics and nutrient cycling in the Okefenokee cypress swamp-forest, *Ecol. Monogr.* 48:43, 1978.
13. McNaughton, S. J. Ecotype function in the *Typha* community type, *Ecol. Monogr.* 36:297, 1966.
14. Ho, Y. B. Short development and production studies of *Phragmites australis* (Cav.) Trin. ex Steudel in Scottish lochs, *Hydrobiologia*. 64:215, 1979.
15. Tilton, D. personal communication, 1990.
16. Dolan, T. J., S. E. Bayley, J. Zoltek, Jr., and A. J. Hermann. Phosphorus dynamics of a Florida freshwater marsh receiving treated wastewater, *J. Appl. Ecol.* 18:205, 1981.
17. Boyd, C. E. Production, mineral nutrient adsorption, and biochemical assimilation by *Justicia americana* and Alternanthera philoxeroides, *Arch. Hydrobiol.* 66:139, 1969.
18. Zoltek, J., S. E. Bayley, T. Dolan, and A. Herman. Removal of Nutrients from Treated Municipal Wastewater by Freshwater Marshes, Center for Wetlands, University of Florida, Gainesville, 1978.
19. Klopatek, J. M. Nutrient dynamics of freshwater riverine marshes and the role of emergent macrophytes, in *Freshwater Wetlands: Ecological Processes and Management Potential*, R. E. Good, D. F. Whigham, and R. L. Simpson, Eds. Academic Press, New York, 1978, 195.
20. Prentki, R. T., T. D. Gustafson, and M. S. Adams. Nutrient movements in lakeshore marshes, in *Freshwater Wetlands: Ecological Processes and Management Potential*. R. E. Good, D. F. Whigham, and R. L. Simpson, Eds. Academic Press, New York, 1978, 169.
21. Dowding, P., F. S. Chapin, III, F. W. Wielgolaski, and P. Kilfeather. Nutrients in tundra ecosystems, in *Tundra Ecosystems: A Comparative Analysis*, L. C. Bliss, O. W. Heal, and J. J. Moore, Eds. Cambridge University Press, U.K., 1981, 647.
22. Wetzel, R. G. Land-water interfaces: metabolic and limnological regulators, *Verh. Internat. Verein. Limnol.* 24:6, 1990.
23. Gersberg, R. M., B. V. Elkins, S. R. Lyon, and C. R. Goldman. Role of aquatic plants in wastewater treatment by artificial wetlands, *Water Res.* 20:363, 1986.
24. Richardson, C. J., D. L. Tilton, J. A. Kadlec, J. P. M. Chamie, and W. A. Wentz. Nutrient dynamics of northern wetland ecosystems, in *Freshwater Wetlands: Ecological Processes and Management Potential*. R. E. Good, D. F. Whigham, and R. L. Simpson. Eds. Academic Press, New York, 1978, 217.
25. Hemond, H.F. The nitrogen budget of Thoreau's Bog, *Ecology*. 64:99, 1983.
26. Urban, N. R. and S. J. Eisenreich. Nitrogen cycling in a forested Minnesota bog, *Can. J. Bot.* 66:435, 1988.
27. Nessel, J. K. and S. E. Bayley. Distribution and dynamics of organic matter and phosphorus in a sewage-enriched cypress swamp, in *Cypress Swamps*. K. C. Ewel and H. T. Odum, Eds. University of Florida Press, Gainesville, 1984, 262.
28. Deghi, G. S. and K. S. Ewel. Simulated effect of wastewater application on phosphorus distribution in cypress domes, in *Cypress Swamps*. K. C. Ewel and H. T. Odum, Eds. University of Florida Press, Gainesville, 1984, 102.
29. Kuenzler, E. J., P. J. Mulholland, L. A. Yarbro, and L. A. Smock. *Distributions and Budgets of Carbon, Phosphorus, Iron and Manganese in a Floodplain Swamp Ecosystem*. Water Resources Research Institute, University of North Carolina, Report No. 157, 1980.
30. Lindau, C. W., R. D. DeLaune, and G. L. Jones. Fate of added nitrate and ammonium-nitrogen entering a Louisiana gulf coast swamp forest, *J. Water Pollut. Control Fed.* 60:386, 1988.
31. Johnston, C. A., G. D. Bubenzer, G. B. Lee, F. W. Madison, and J. R. McHenry. Nutrient trapping by sediment deposition in a seasonally flooded lakeside wetland *J. Environ. Qual.* 13:283, 1984.

CHAPTER 31

Pilot-Scale Nitrification Studies Using Vertical-Flow and Shallow Horizontal-Flow Constructed Wetland Cells

J. T. Watson and A. J. Danzig

INTRODUCTION

Constructed wetlands are effective in removing biochemical oxygen demand (BOD) and total suspended solids (TSS), but most systems have been relatively ineffective in removing ammonia nitrogen.[1-3] The primary limiting factor is believed to be dissolved oxygen. Systems that show the best potential for nitrification appear to be either shallow (12 in.) gravel horizontal-flow cells[4-6] or vertical-flow sand/gravel cells.[7-11]

Shallow horizontal-flow gravel cells are believed to be more effective than deeper cells for nitrification because flow is restricted to the most effective portion of the plant root zone. Most root biomass in gravel cells typically occurs in the top 6 to 12 in.[2,12,13] Oxygen is supplied to the nitrifiers from a combination of oxygen release from roots and rhizomes[7,14,15] and oxygen diffusion and convection directly from the atmosphere.[7] Vertical-flow cells have the potential to be more efficient than horizontal-flow cells because they can be operated as nonsaturated filters. Oxygen diffusion and convection rates are much higher for air-filled pore spaces than water-saturated pore spaces.[7]

EXPERIMENTAL DESIGN

Pilot Cell Details

Pilot-scale, shallow horizontal and vertical-flow cells were installed during the spring of 1991 at Benton, Kentucky, to develop design information for a full-scale constructed wetlands system for nitrification. The units consist of six modified septic tanks with a treatment surface area of 35 ft^2 for the five vertical-flow cells and 34 ft^2 for the horizontal-flow cell. *Phragmites australis* (reed) is planted in each cell except for one control cell. The cells are shown in Figure 1. Details on sand and gravel media used in each cell are provided in Tables 1 and 2. All media were locally available.

Reed rhizomes were planted on April 12 about 0.5-ft centers. The plants grew and expanded rapidly, achieving a mature (seed-bearing) height of 6 to 8 ft in 3 months.

The cells were dosed with effluent from an operational municipal wastewater constructed wetland. Operation began 5 weeks after planting and data collection commenced 1 week hence. A submersible pump was provided to feed each vertical-flow cell. Flow rates were controlled by a valved bypass for each pump. Dosing times and periods were controlled by a timer for each pump. An adjustable chemical metering pump was used for the horizontal-flow cell.

The wastewater was uniformly distributed to the vertical-flow cells by a pressurized pipe distribution system with drilled holes on top of the pipe network. Seven pipe laterals were evenly spaced across the top of the tank and holes were drilled at 0.5-ft intervals across the width. Initial hole diameter was 7/64 in., which provided a spray height from about 3 in. to 2 ft depending on the applied hydraulic loading rate (HLR). On July 25, the holes were enlarged to 9/64 in. in order to accommodate the larger HLRs. This system worked reasonably well for intermittent applications, but a large percentage of the holes clogged under continuous application (some of the Cell 5 test periods).

0-87371-550-0/93/$0.00+$.50

Figure 1. Pilot vertical- and horizontal-flow constructed wetland cells at Benton, Kentucky.

Table 1. Granular Media Used in the Pilot-Scale Cells at Benton

	Depth (Inches) of Media Layers					
	Sand		Gravel			
Cell	**Mortar**	**Concrete**	**Birdseye**	**Mixed**	**Large**	**Comments**
1		3	6	4	6	European type
2		12	3		10	
3	12		3		10	
4	12		3		10	Control (no reeds)
5			15		10	
6			12			Horizontal flow

Note: Characteristics of media are provided in Table 2. Surface areas are about 35 and 34 ft^2 for cells 1–5 and cell 6, respectively. Flow is uniformily distributed over the surface area of cells 1–5 by a pressurized perforated pipe system.

Effluent from each dose that was sampled was collected in a bottom storage tank. This tank allowed collection of a composite sample for each dose period and measurement of the total volume of water treated per dose.

Influent flows for the vertical-flow cells were measured using a Polysonics Doppler ultrasonic flowmeter. Influent flow for the horizontal-flow cell and effluents from all cells were determined using a calibrated bucket/graduated cylinder and stopwatch.

Table 2. Characteristics of Sand and Gravel Media

	Uniformity coefficient	Effective grain diameter (mm)
Mortar (fine) sand	2.0	0.20
Concrete (coarse) sand	2.0	0.28
Birdseye (pea) gravel	1.4	7.0
Mixed wash ($^1/_2$ in.)	1.3	9.0
Large wash (1 in.)	1.4	19

Loading Experiments

Since the cells functioned as polishing units with relatively constant, but light organic (2 to 40 mg BOD_5/L) and nitrogenous (7 to 14 mg NH_3-N/L) loads, vertical-flow experiments focused on determining hydraulic loading rates and drying cycles capable of consistently producing an effluent with ammonia nitrogen (NH_3-N) concentrations of less than 4 mg/L (the NPDES permit limit from May to October) without continuous ponding (ponding for the entire drying period between doses). Acceptable loading rates were expected to vary between the rates for intermittent sand filters for lagoon effluents (equal to or less than 12 gpd/ft^2)[16,17] and the rates used in Europe for second and third stage vertical-flow constructed wetland cells (16 gpd/ft^2).[18] Dosing cycles were expected to range from several (at least 4) per day to one per day.

Loading rates for the horizontal-flow cell were initiated at 0.765 gpd/ft^2 based on experience at Phillips High School, Bear Creek, Alabama.[4] Loading rates were gradually increased until effluent concentrations exceeded 4 mg/L.

Sample Locations and Types

Samples consisted of a grab sample of the influent to the cells (effluent from Cell 1 of the full-scale wetlands) and composite and/or grab samples of the effluent from each experimental cell. In addition, quality control (QC) samples typically included duplicates, spikes, and container blanks for each sample period.

Parameters

Parameters measured in the field were: temperature, pH, dissolved oxygen (DO), conductivity, oxidation-reduction potential (ORP), alkalinity, ammonia nitrogen, and flow (influent and effluent). A Hydrolab Surveyor 3 was used for temperature, pH, DO, conductivity, and ORP. Alkalinity was determined by pH titration to 4.5. Field ammonia concentrations were determined by an Orion ammonia electrode. All field instruments were calibrated before and after each sampling period and provided consistently reliable measurements.

Parameters measured in TVA's certified Environmental Chemistry Laboratory using EPA-approved analytical procedures were: BOD_5, TSS, total Kjeldahl nitrogen, ammonia nitrogen, nitrate + nitrite nitrogen, phosphorus, potassium, and fecal coliforms.

Sampling Frequency

Samples were collected weekly (except for the week of July 4) beginning May 22 and ending September 11.

RESULTS

The influent to the pilot cells is the effluent from constructed wetland cell 1 at Benton. It contains low levels of BOD_5 and TSS (average values were 13 and 13 mg/L, respectively). Ammonia nitrogen values ranged from 7.1 to 14 mg/L, with an average of 9.9 mg/L. Total Kjeldahl nitrogen ranged from 9.2 to 33 mg/L, with an average of 15 mg/L. There was more than adequate alkalinity for nitrification and pH ranged from 6.5 to 7.1. Dissolved oxygen was always low, less than 0.6 mg/L. The only factor believed to be restricting nitrification in the full-scale constructed wetlands is believed to be the low DO concentrations. Nitrification was attained within 2 to 4 weeks of operation of each pilot cell.

Hydraulic loading rates for the vertical-flow cells were extremely high (30 to 70 gpd/ft^2) during the first 2 weeks of operation to determine the clogging potential of the wastewater. Cells 1 to 4 failed hydraulically (water from the previous dose was still standing at the time of the next dose) within 2 weeks using 4 doses per day, 120 min per dose for the first week, and 60 min per dose for the second week. Reduction of hydraulic loading rates and dosing cycles immediately resulted in acceptable hydraulics, with the exception of cell 1 (discussed below).

Biological film growing on the sand reduced the hydraulic conductivity of cells 1 through 4 sufficiently to allow flooding over most of the top surface for most hydraulic loading rates, resulting in excellent distribution of the wastewater. Biological growth on the pea gravel in cell 5 was insufficient to cause surface flooding; consequently, flow distribution was the poorest in this cell. Also, clogging of the holes in the pressurized pipe distribution system was a continual maintenance problem that significantly affected flow distribution in Cell 5.

The vertical-flow scheme resulted in excellent aeration of the wastewater in all cells. Effluent from the sand cells was typically at 80% or higher saturation, with actual DO values typically in the 6 to 8-mg/L range. DO concentrations in the pea gravel cell effluent were typically in the 5 to 7-mg/L range (about 70% saturation). These high values indicate that unsaturated flow conditions predominated.

Objectable odors were not observed. The potential for any other nuisance or esthetic problems is considered low for sand cells.

Data presented in this paper are limited to ammonia nitrogen. However, significant correlations were not found between percent nitrification and organic loading, percent nitrification and BOD_5:TKN ratio, percent BOD removal and hydraulic loading, or between percent phosphorus removal and hydaulic loading rates. Under conditions of low organic matter ($BOD_5 < 20$ mg/L), hydraulic loading is a major parameter governing nitrification performance.[19] The ammonia data are listed in Table 3.

Cell 1

Cell 1 was designed according to the European Design and Operations Guidelines for Reed Bed Treatment Systems.[18] It contains 3 in. of "concrete" sand on top of several layers of gravel of increasing size, as shown in Tables 1 and 2. Loading rates and operation were similar to a European "Stage III" cell.

The hydraulic characteristics of cell 1 were the poorest of all cells even though the concrete sand was the largest of the two sand types used in the experiments. Clogging appeared to progress with time for all dosing cycles except once per day for 60 min. It is hypothesized that the drainage problem is caused by capillary action. The surface of the cell may stay wet due to capillarity. The drying time between doses may not have been sufficiently long to control the microbiological film on the sand particles. This film is believed to be the primary clogging factor. Although capallarity also exists in the other cells, it did not present a problem because its effect did not reach the top of the deeper sand layers.

Nitrification in the cell was good only during the first 7 weeks of sampling when the filter was clogging persistently. Following a week of drying, ammonia concentrations never dropped below the compliance value of 4 mg/L, even at very low loading rates (as low as 3.1 gpd/ft^2). The reeds were severely stressed after the 7-day drying period, but they appeared to recover fully within 3 to 4 weeks. There was poor correlation between ammonia removal and hydraulic loading rate, as shown in Figure 2.

Cell 2

This cell contains 12 in. concrete sand with an effective diameter of 0.28 mm. Clogging was not a problem within this cell using either two or three 60-min doses per day. Most application rates resulted in ponding that just covered the top surface.

Performance of the cell was good. Ammonia concentrations exceeded the target level of 4 mg/L only during high hydraulic loading rates (>28 gpd/ft^2). The relationship between loading rates and ammonia nitrogen reduction is shown in Figure 2. At the relatively high loading rate of 20 gpd/ft^2, reduction in ammonia nitrogen will be about 70%.

Cell 3

This cell contains 12 in. mortar sand with an effective diameter of 0.20 mm. Although clogging was not a problem with this cell using two or three 60-min doses per day, the pooling depth on top of the media surface was deeper than either cells 2 or 4 for equivalent HLRs. This probably reflects the plugging effect of the reeds.

Table 3. Ammonia Nitrogen Data for the Benton Pilot Cells

Date	Influent (mg/L)	Cell 1		Cell 2		Cell 3		Cell 4		Cell 5		Cell 6	
		mg/L	gpd/ft²	mg/L	gpd/ft²	mg/L	gpd/ft²	mg/L	gpd/ft²	mg/L	gpd/ft²	mg/L	gpd/ft²
05/22/91	10.0	10.0	56	8.4	66	7.6	49	7.0	56	10.0	65	8.6	0.91
05/30/91	7.6	1.0		2.0		1.5		1.6		7.1		3.3	0.91
06/07/91	6.8	3.8	9.9	1.7	8.6	0.86	8.6	1.5	8.6	4.8	7.2	0.25	0.33
06/12/91	7.8	1.9	9.3	1.1	9.6	0.40	8.6	0.88	11	5.8	8.6	1.3	2.5
06/19/91	7.1	1.3	9.9	0.64	9.9	0.25	8.2	0.47	9.3	5.1	23	0.07	0.85
06/27/91	7.4	0.14	9.3	1.5	14	1.2	14	1.3	13	5.4	25	0.83	1.82
07/10/91	10.0	0.23	5.4	2.9	15	2.2	18	3.4	20	6.6	62	[a]	0.25
07/17/91	9.9	8.2	4.4	3.7	21	2.2	20	3.0	21	7.5	280	0.01	0.42
07/25/91	10.0	5.9	6.9	4.2	29	3.5	28	4.5	35	7.1	330	1.1	3.5
07/31/91	10.7	5.4	13	4.7	32	2.3	23	4.3	31	6.7	330	1.6	2.8
08/07/91	11	5.5	12	3.2	16	1.7	15	3.3	17	9.0	59	10.0	13
08/14/91	8.1	4.3	11	3.1	13	1.3	14	5.9	14	7.7	64	7.5	6.7
08/21/91	11	4.0	3.6	2.9	17	1.4	13	1.8	13	6.1	43	3.1	2.7
08/28/91	14	6.0	3.4	3.4	21	2.8	23	2.5	21	9.9	99	4.5	2.4
09/04/91	14	4.9	3.1	3.6	23	2.7	23	1.6	20	9.8	97	8.0	3.3
09/11/91	13	5.9	3.8	1.5	19	1.8	22	2.2	24	8.4	74	6.0	3.2

[a] No effluent due to evapotranspiration.

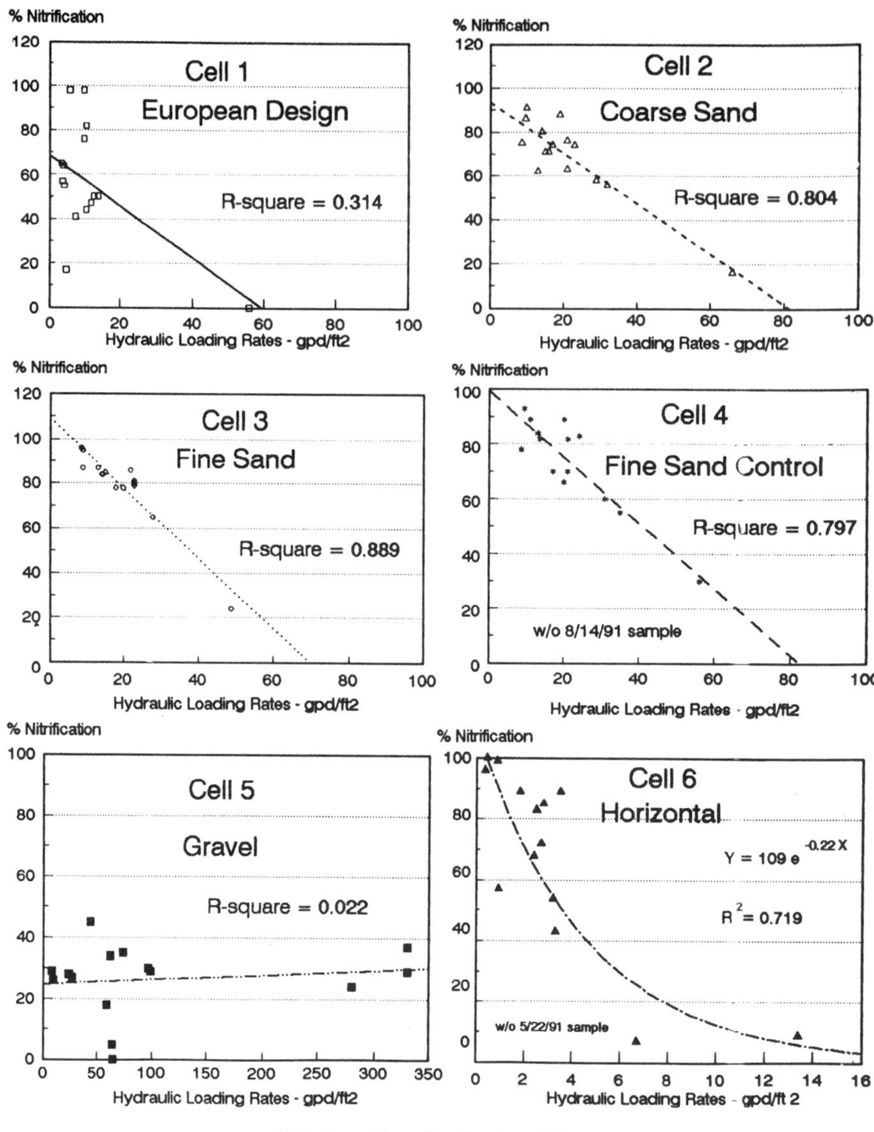

Figure 2. Percent nitrification vs. loading rate for the Benton constructed wetlands pilot cells.

Performance of cell 3 was generally the best of all cells. This is attributed to the slightly longer retention time resulting from the lower hydraulic conductivity of the top few inches of sand. The relationship between loading rates and ammonia nitrogen reduction is shown in Figure 2. This is the strongest relationship obtained, with a correlation factor of 0.889. Ammonia concentrations for the dose composite sample exceeded 4 mg/L only during the first sampling period when HLRs were very high (49 gpd/ft^2). At an HLR of 20 gpd/ft^2, reduction in ammonia nitrogen is estimated to be about 78%, which is adequate to achieve compliance at the maximum observed influent ammonia concentrations.

Cell 4

This cell is identical to cell 3 except it contains no reeds. No clogging problems were experienced for either two or three 60-min doses per day. Flooding depths on top of the media appeared to be very similar to those of cell 2 (containing larger sand) for equivalent HLRs, indicating the slightly lower hydraulic conductivities resulting from root growth.

Performance of the cell was good, generally falling between that of cell 3 and cell 2. Ammonia concentrations exceeded the target level of 4 mg/L only during high HLRs (>30 gpd/ft^2). The relationship between loading rates and ammonia nitrogen reduction is shown in Figure 2. At 20 gpd/ft^2, reduction in ammonia nitrogen is estimated to be about 75%.

Cell 5

This cell contains 15 in. pea gravel with an effective diameter of 7 mm. The media presented no clogging problems at any HLR. However, there was probably one very significant hydraulic problem: short-circuiting. Without ponding on top of the media, distribution of the wastewater depended on the pressurized pipe distribution system above the gravel. This restricted the majority of flow to a zone within a few inches of each drilled outlet hole. Distribution was better at higher HLRs because the pressure resulted in a spray height of more than 1 ft and bounced off of the reeds; however, clogging of the holes was a continual maintenance problem that contributed significantly to short-circuiting and the probable effective loss of much of the surface area. The growth of the reeds was also the poorest in this cell, probably due to the larger size of the media and the lower amount of moisture retained in the media between doses.

Performance of the cell was poor. Ammonia concentrations always exceeded the target level of 4 mg/L. There was very poor correlation between loading rates and ammonia reduction, as shown in Figure 2.

Cell 6

This cell contains 12 in. pea gravel, with an effective diameter of 7 mm. The operation of the cell differs from the others in that it is horizontal, saturated flow rather than vertical, unsaturated flow. Water levels were maintained just below the surface of the media. No clogging problems occurred.

Performance of the cell was excellent at HLRs below 4 gpd/ft^2 through July. Once cell 6 was forced out of compliance at high HLRs in August, it did not fully recover, even at HLRs less than 4 gpd/ft^2 in the time frame of the study. From Figure 2, ammonia nitrogen removals of 80% can be expected at HLRs of about 1.4 gpd/ft^2 (16.4 acres/MGD).

The efficiencies for this cell on a unit loading basis are not nearly as high as those of the vertical flow cells due to the lower dissolved oxygen resulting from saturated flow hydraulics and the lower surface area available for nitrifier colonization. The dissolved oxygen in the effluent was typically between 1 and 2 mg/L.

Dose Time Studies

Effluent ammonia concentrations for cells 2, 3, and 4 were monitored every 5 to 10 min during one dosing cycle on four separate occasions. Removal of ammonia nitrogen was not constant during the dosing cycle. Effluent concentrations during the first few minutes of the cycle were always very low (less than 1 mg/L) for these cells, but gradually increased during the cycle, as shown in Figures 3, 4, and 5. This type of trend has not been found in the literature for intermittent sand filters. It indicates that dosing periods of 2 h rather that 1 h would have resulted in composite samples with concentrations greater than 4 mg/L for most HLRs. This trend suggests that the population of nitrifiers was insufficient to utilize the available nitrogen. This may be caused by inadequate media surface area for the applied load or too long an interval between dosing cycles, resulting in excessive die-off of nitrifiers.

If the media surface area were lacking, there probably would be a significant difference between performance of the mortar sand (effective diameter of 0.20 mm) and the concrete sand (effective diameter of 0.28 mm). As shown in Figures 3, 4, and 5, on August 21, with two doses per day, concentrations rose quicker in the coarser sand of cell 2 than the finer sand of cells 3 and 4, supporting the hypothesis of insufficient surface area.

These cells received three doses per day during the other three sampling periods. On August 28 and September 4, curves for the finer mortar sand in cells 3 and 4 reflect a rapid increase in concentration, followed by a strong decline, and then by a gradual increase throughout the remainder of the cycle. The trend was strongest during the first week as loadings increased to three doses per day (August 28). This may indicate a rapidly multiplying nitrifier population that again is limited by the available surface area. On September 14, the up/down cycle has disappeared and performance is more stable, with overall lower concentrations under similar HLRs. The coarser sand of cell 2 did not exhibit the up/down cycles, but performance improved slightly when doses increased from two to three per day.

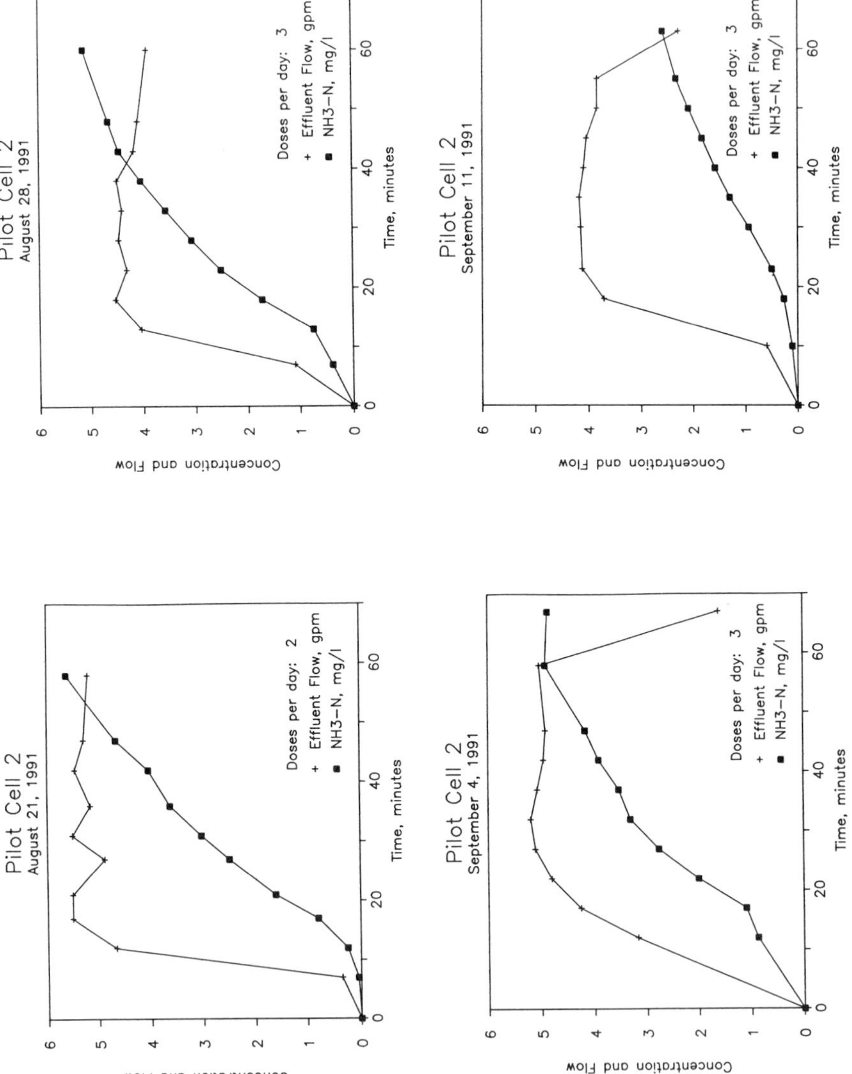

Figure 3. Dose time trend for ammonia nitrogen in cell 2.

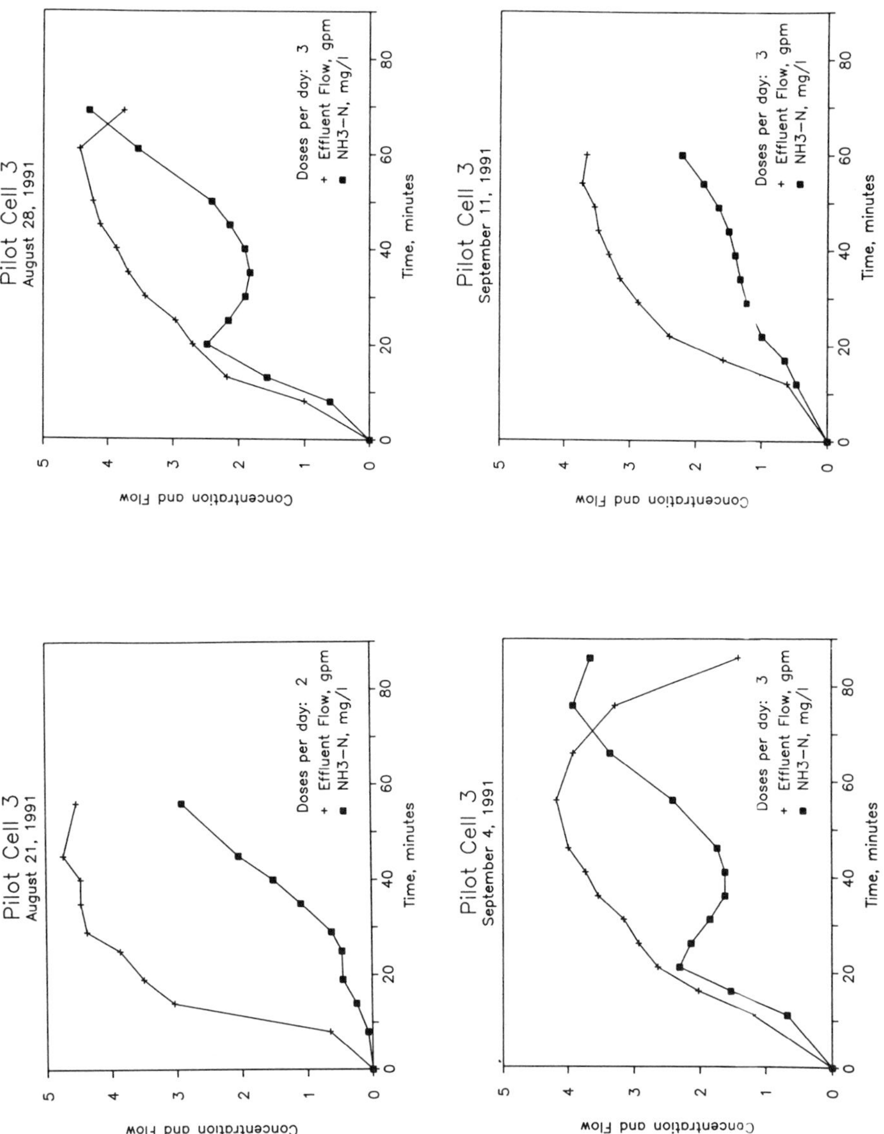

Figure 4. Dose time trend for ammonia nitrogen in cell 3.

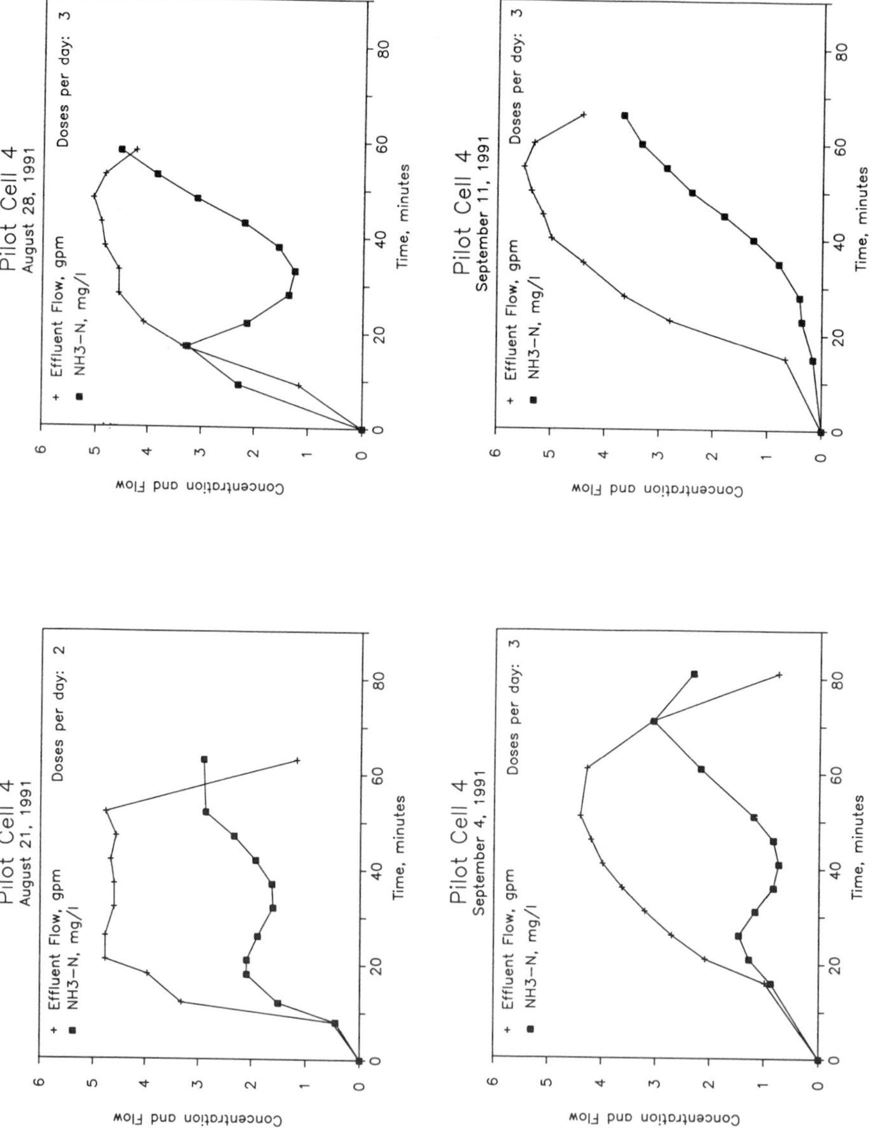

Figure 5. Dose time trend for ammonia nitrogen in cell 4.

These trends appear to support the benefits of greater sand depth (greater media surface area) and more frequent dosing to create a greater sustained population of nitrifiers. However, doses greater than three per day resulted in rapid hydraulic failure of all sand cells; consequently, there is a need to balance the biomass with acceptable hydraulics. Intermittent sand filter design criteria specify a minimum of 2 ft of sand,[20] which appears warranted based on this analysis.

IMPLICATIONS FOR CONSTRUCTED WETLANDS TECHNOLOGY

Benefits of Reeds

The issue of whether reeds improve treatment effectiveness of the vertical-flow cells is addressed by comparison of the performance of cells 2, 3, and 4. Cells 3 and 4 are identical except for the absence of reeds in cell 4. As shown in Figure 2, cell 3 provided slightly higher removals of ammonia nitrogen than cell 4. At the HLR of 20 gpd/ft^2, cell 3 is expected to remove about 78% of the ammonia compared to about 75% for cell 4. This slightly better performance is attributed to a reduction in hydraulic conductivity of the top few inches of sand, probably caused by the reed roots and/or poorer drying characteristics due to the reeds blocking wind and sunlight. The surface of cell 3 flooded to a deeper depth and drained more slowly than cell 4 for similar HLRs. The surface flooding of cells 2 and 4 were usually very similar for equivalent HLRs, indicating a similar clogging effect of the reeds on the coarser sand of cell 2. The difference in predicted ammonia reduction between cells 2 and 4 was also small, 70 vs. 75%, respectively, for a HLR of 20 gpd/ft^2. This difference would probably have been greater if cell 2 did not contain reeds.

Since the primary effect of reeds appears to be reduction of media hydraulic conductivity, its desirability in such systems is questionable. The primary technical basis for wetlands vegetation in granular-based treatment systems is oxygen transfer to aid the aerobic bacteria and stabilization of the hydraulic conductivity with time at a relatively high rate.[21] Since there is no need for additional oxygen in unsaturated vertical-flow systems, the issue focuses on stabilization of hydraulic conductivity. This would require a study of several years to address adequately. However, for polishing systems where the most significant clogging mechanism is microbiological film growing on the media, clogging appears to be easily controlled by the dosing/drying cycles. If clogging ever became so severe that the top few inches of media needed to be removed, a system without plants would have obvious advantages. The only clear benefit of using wetland species is esthetics. Most people would probably prefer the visual appeal of a filter with plants.

Use of Intermittent Sand Filters Following Constructed Wetlands

The issue is often raised as to the need or benefit of constructed wetlands if one has to utilize another technology for consistent nitrification. The answer lies in the cost effectiveness of each component when the design of each has been optimized to the extent practicable. This study has not attempted to provide a definitive answer to this issue, but it can be used to address it.

Intermittent sand filter design typically specifies a loading rate of about 5 to 10 gpd/ft^2 for wastewaters low in organics and a minimum depth of 2 ft.[20] This study indicates that the loading rate can probably be doubled using the relatively high-quality effluent from a constructed wetland. Although the depth of sand may also be decreased, the data suggest that 2 ft rather than 1 ft of sand may provide additional significant treatment efficiency. Consequently, the size and cost of a filter following a constructed wetland may be half of that otherwise needed. In addition, previous studies have shown that essentially all removal of BOD and TSS occurs in the front half of the Benton constructed wetlands, even though major short-circuiting occurs in these areas.[2] This indicates that the size and cost of the constructed wetlands for lagoon effluents may also be halved if the system is optimized. Consequently, for a design flow of approximately 1 MGD, to upgrade a lagoon system to achieve 70% nitrification, a hypothesized constructed wetland/intermittent sand filter system may consist of a surface flow constructed wetlands designed for about 5 acres/MGD for removal of BOD_5 and TSS and an intermittent sand filter (2 ft of relatively fine sand with a low uniformity coefficient) designed for approximately 20 gpd/ft^2 (50,000 ft^2 for 1 MGD), dosed at intervals of about 2 or 3 doses per day for 60 min per dose. Assuming such an optimized system performed as projected, its cost effectiveness should be relatively good compared to other alternatives. Another small constructed wetlands could follow the sand filter for denitrification if this were a concern. Also, recirculating sand filters may have the potential to be even more cost effective than intermittent filters for this type of application. However, these concepts still need to be demonstrated and refined before being adopted as an easy solution to a tough compliance problem.

SUMMARY

The following conclusions are offered for consideration.

- Vertical-flow, unsaturated sand, and pea gravel cells resolve the oxygen requirements for nitrification. This has been the primary problem in horizontal flow, saturated constructed wetlands systems.
- For wastewater with low BOD_5 and TSS, loading rates of 20 gpd/ft^2 will result in about 70% nitrification for a 1-ft deep fine sand filter operated with 2 or 3 doses per day for 60 min per dose. The efficiency will probably be much better with 2 ft of sand.
- Clogging appears to be caused by the microbiological film that develops on the surface of the sand particles. This is readily controlled by the dosing and drying cycles.
- The overall performance was poor for the cell based on European design information. Apparently, a greater depth of sand is needed to preclude adverse capillary action and provide additional surface area for higher populations of nitrifiers.
- The performance of the pea gravel, vertical-flow cell was also poor. Apparently, greater depths of gravel are also needed for this medium to provide adequate surface area to develop the nitrifier populations required for high efficiencies. A good distribution system is also critical for such larger media.
- Reeds increased treatment effectiveness only slightly by decreasing the hydraulic conductivity of the sand. Their primary benefit is esthetic since filters with reeds are considered to have a better visual appearance than those without reeds. However, their presence would preclude removal of the top layer of sand if clogging progressed to the point that drying could not reestablish acceptable hydraulics.
- The horizontal flow cell with 1 ft pea gravel performed very well for low loading rates. Ammonia nitrogen removals of 80% can be expected for HLRs of about 1.4 gpd/ft^2 (16.4 acres/MGD). This design is not capable of matching the performance of the vertical-flow cells on a unit basis because of the difference in oxygen transfer characteristics between saturated and unsaturated flow conditions. Nevertheless, the system is the simplest to construct, operate, and maintain, and may be a highly desirable design in certain situations.

REFERENCES

1. Watson, J. T., K. D. Choate, and G. R. Steiner. Performance of constructed wetland treatment systems at Benton, Hardin, and Pembroke, Kentucky during the early vegetation establishment phase, in *Constructed Wetlands in Water Pollution Control.* P. F. Cooper and B. C. Findlater, Eds. Pergamon Press, Oxford, 1990, 171.
2. Findlater, B. C., J. A. Hobson, and P. F. Cooper. Reed bed treatment systems: performance evaluation, in *Constructed Wetlands in Water Pollution Control.* P. F. Cooper and B. C. Findlater, Eds. Pergamon Press, Oxford, 1990, 193.
3. Choate, K. D., J. T. Watson, and G. R. Steiner. Demonstration of Constructed Wetlands for Treatment of Municipal Wastewaters, Monitoring Report for the Period: March 1988 to October 1989, TVA/WR/WQ—90/11, Tennessee Valley Authority, Chattanooga, TN, 1990, 43.
4. Watson, J. T. Design and Performance of the Constructed Wetland Wastewater Treatment System at Phillips High School, Bear Creek, Alabama, TVA/WR/WQ—90/5, Tennessee Valley Authority, Chattanooga, TN, 1990.
5. May, E., J. E. Butler, M. G. Ford, R. Ashworth, J. Williams, and M. M. M. Bahgat. Chemical and microbiological processes in gravel-bed hydroponic (GBH) systems for sewage treatment, in *Constructed Wetlands in Water Pollution Control.* P. F. Cooper and B. C. Findlater, Eds. Pergamon Press, Oxford, 1990, 33.
6. Butler, J. E., R. F. Loveridge, M. G. Ford, D. A. Bone, and R. F. Ashworth. Gravel bed hydroponic systems used for secondary and tertiary treatment of sewage effluent, *J. IWEM.* 4:276, 1990.
7. Brix, H. and H.-H. Schierup. Soil oxygenation in constructed reed beds: the role of macrophyte and soil-atmosphere interface oxygen transport, in *Constructed Wetlands in Water Pollution Control.* P. F. Cooper and B. C. Findlater, Eds. Pergamon Press, Oxford, 1990, 53.

8. Bahlo, K. E. and F. G. Wach. Purification of domestic sewage with and without faeces by vertical intermittent filtration in reed and rush beds, in *Constructed Wetlands in Water Pollution Control.* P. F. Cooper and B. C. Findlater, Eds. Pergamon Press, Oxford, 1990, 215.
9. Burka, U. and P. C. Lawrence. A new community approach to waste treatment with higher water plants, in *Constructed Wetlands in Water Pollution Control.* P. F. Cooper and B. C. Findlater, Eds. Pergamon Press, Oxford, 1990, 359.
10. de Zeeuw, W., G. Heijnen, and J. de Vries. Reed bed treatment as a wastewater (post) treatment alternative in the potato starch industry, in *Constructed Wetlands in Water Pollution Control.* P. F. Cooper and B. C. Findlater, Eds. Pergamon Press, Oxford, U.K., 1990, 551.
11. Rijs, G. B. J. and S. Veenstra. Artificial reed beds as post treatment for anaerobic effluents — urban sanitation in developing countries, in *Constructed Wetlands in Water Pollution Control.* P. F. Cooper and B. C. Findlater, Eds. Pergamon Press, Oxford, U.K., 1990, 583.
12. Edwards, M. E. *A Study of Soft-stem Bulrush (Scirpus validus) Growth in a Constructed Wetland, Hardin, Kentucky.* TV-81976V. Universit y of Tennessee at Chattanooga, Department of Biology, Chattanooga, 1990, 23.
13. Parr, T. W. Factors affecting reed (*Phragmites australis*) growth in UK reed bed treatment systems, in *Constructed Wetlands in Water Pollution Control.* P. F. Cooper and B. C. Findlater, Eds. Pergamon Press, Oxford, 1990, 67.
14. Armstrong, W., J. Armstrong, and P. M. Beckett. Measurement and modelling of oxygen release from roots of *Pharagmites australis,* in *Constructed Wetlands in Water Pollution Control.* P. F. Cooper and B. C. Findlater, Eds. Pergamon Press, Oxford, 1990, 41.
15. Reddy, K. R., E. M. D'Angelo, and T. A. Debusk. Oxygen transport through aquatic macrophytes: the role in wastewater treatment, *J. Environ. Qual.* 19:261, 1989.
16. Middlebrooks, E. J., J. H. Reynolds, C. Middlebrooks, R. W. Schneiter, and B. A. Johnson. Design Manual: Municipal Wastewater Stabilization Ponds. EPA-625/1-83-015, U.S. Environmental Protection Agency, Municipal Environmental Research Laboratory, Cincinnati, OH, 1983, 226.
17. Rich, L. G. and E. J. Wahlberg. Performance of lagoon intermittent sand filter systems, *J. Water Pollut. Control Fed.* 62:697, 1990.
18. European Design and Operations Guidelines for Reed Bed Treatment Systems. P. F. Cooper, Ed. EC/EWPCA Emergent Hydrophyte Treatment Systems Expert Contact Group, WRc Swindon, Swindon, England, A-5, 1990.
19. Grady, C. P. L., Jr. and H. C. Lim. *Biological Wastewater Treatment: Theory and Applications.* Marcel Dekker, New York, 1980, 802.
20. Clements, E. V., R. J. Otis, D. H. Bauer, R. L. Siegrist, E. J. Tyler, D. E. Stewart, and J. C. Converse. Design Manual: Onsite Wastewater Treatment and Disposal Systems. U.S. Environmental Protection Agency, Municipal Environmental Research Laboratory, Cincinnati, OH, 1980, 126.
21. Brix, H. and H.-H. Schierup. Danish experience with sewage treatment in constructed wetlands, in *Constructed Wetlands for Wastewater Treatment.* D. A. Hammer, Ed. Lewis Publishers, Chelsea, MI, 1989, 565.

CHAPTER 32

Phosphorus Removal from Wastewater in a Constructed Wetland

T. H. Davies and P. D. Cottingham

INTRODUCTION

In a constructed wetland using a high hydraulic conductivity matrix such as gravel, the wastewater will flow subsurface in the root and rhizome region of the aquatic macrophyte where bacterial growth on the surfaces (root, rhizome, and gravel) can absorb pollutants and break them down. There are several processes for phosphorus removal from wastewater, but some have a limited capacity for removal; once this capacity is exceeded no further phosphorus removal will occur.

As the aquatic plants grow, there will be an uptake of phosphorus into the plant cells which will continue until the plants are fully grown. At the end of the growing season, aquatic plants such as reeds die back and the leaves and stalks will eventually fall to the bed where they will slowly break down and return phosphorus back into the system. New growth will take up phosphorus again so that eventually an equilibrium will develop where the phosphorus take-up by plant growth in a year will equal the phosphorus return by dead plant breakdown. So, if no harvesting is initiated, the plants will bring about no net phosphorus removal.

The take-up of phosphorus into the plant cells is relatively small and the phosphorus content for plants such as reeds ranges from 0.9 to 1.35 mg/g (dry weight) for stems, 1.0 to 1.7 for leaves, and 0.9 to 1.63 for whole shoots.[1] A yield of 95 tons dry weight per year per hectare of a very vigorous stand of *Phragmites australis* up to 5 m tall was reported in Florida.[2] With this very high yield and using the highest plant phosphorus content of 1.7 mg/g, dry weight above, a yearly removal of less than 6% phosphorus by harvesting the reeds would be achieved from a typical constructed wetland with a loading of primary settled sewage with phosphorus content of approximately 8 mg/L delivered at a rate of 96 L/m^2/day. Hence, the use of macrophytes to remove phosphorus from wastewaters, even under optimum management conditions, is severely limited.

Bacteria can assimilate phosphorus in their cell structures; but once a steady state of biomass is reached within the bed, no net phosphorus removal will result thereafter.

If phosphorus is removed by means of the matrix material, it must be contained within the bed by adsorption, ion exchange, or chemical reaction to an inert insoluble form. As these factors have a finite capacity, phosphorus removal will cease when that capacity is reached. Clay-type bed media with its abundance of aluminium, iron, and calcium in its structure, and large surface area, has the greatest potential to trap and hold phosphorus, but its very low hydraulic conductivity in a bed designed for subsurface flow results in most of the water traveling across the surface and not making intimate contact with the bed proper to enable the containment to take place.

Gravel media, with is of high conductivity, permits all of the water to flow within the bed, but because of its impermeable nature has only a limited surface area for absorption, ion exchange, and/or chemical reaction to take place. Once the active sites are utilized, phosphorus removal would cease.

A study of almost 3 years of weekly monitoring of pilot size (30 × 5 m) reed beds, where there is no harvesting, has shown virtually no phosphorus removal in a basaltic gravel bed and an overall average of about 16% in a sandy soil bed of low hydraulic conductivity where most of the water flowed across the surface.[3]

Loss of phosphorus to the atmosphere as phosphine has been reported, but occurs only to a limited extent.[4] Hence, the use of constructed wetlands for significant phosphorus removal from wastewaters is very severely limited, even when regular harvesting is used, unless some major modification could be introduced which would enable large amounts of phosphorus to be contained within the bed.

0-87371-550-0/93/$0.00+$.50

From the foregoing, the conversion of the soluble phosphorus to an insoluble particulate form which would be contained within the bed, by chemical means, would be an efficient method of removal. The ability of the beds to remove suspended solids would ensure retention of the containment if the insoluble phosphorus particles could be produced. This thought led to the concept that chemical dosing within the bed, at a point where most of the natural solids had already been separated out, would be an effective and economical means of phosphorus removal.

Studies of the gravel bed above have shown that most of the wastewater solids had been removed in the first 10 to 15 m of the bed length,[5] so dosing at the beginning of the last 10 m of bed would be at a point where most of the void spaces in the gravel would still be available for collection of insoluble phosphorus particles. With a void space of 40%, the last 10 m of bed would have a void volume of approximately 12 m^3. The plant roots would take up some of this space, but since the roots are hollow, as they die and regrow this reduction would reach a constant value.

A reduction in phosphorus level from 8 to 1 mg/L by alum dosing for the pilot-size gravel bed treating 5.256 ML/year of primary settled sewage would produce about 145 kg of insoluble aluminium phosphate per year. If its density were 1.0 g/mL, it would take 83 years to fill the voids. If the density were 0.5 g/mL, it would take 41 years to fill. The presence of plant roots will reduce this time, but only to a limited degree. Hence, the beds have great capacity to contain the insoluble phosphorus, and if set up for chemical dosing would have a very long effective life. If some small localized blocking of voids did occur, a small amount of surface flow would not be a serious problem as it would flow in the plant litter layer out of sight, and the litter would remove any phosphate particles until the flow re-entered the bed further down.

When alum dosing is used in wastewater to remove phosphorus as the insoluble aluminium phosphate, at low levels of phosphate (<10 mg/L P), there is competition from the formation of aluminium hydroxide instead of the phosphate and this necessitates a higher Al:P ratio than the theoretical ratio of 1:1.[6]

$$Al_2(SO_4)_3 + 2Na_3PO4 \rightarrow 2AlPO_4 \downarrow + 6Na^+ + 3SO_4 = \qquad (32\text{–}1)$$

$$Al_2(SO_4)_3 + 6H_2O \rightarrow 2Al(OH)_3 \downarrow + 6H^+ + 3SO_4 = \qquad (32\text{–}2)$$

Formation of the hydroxide lowers the pH significantly and reduces the formation of the phosphate because of the rapid increase in solubility of aluminium phosphate as the pH reduces below pH 7. The addition of lime (calcium hydroxide) with the alum prevents the formation of acid conditions and thus improves the phosphorus removal.[7]

In this article we describe chemical dosing applied internally into the last section of constructed wetlands and its effectiveness.

EXPERIMENTAL DETAILS

Two experimental gravel matrix constructed wetlands, each 30 m long × 5 m wide, one planted with *Phragmites australis* and the other an unplanted control, were selected for the dosing experiments.

Each bed has a coarse rock distributor 1 m wide every 10 m, and the slotted pipes embedded in the distributor across the bottom of the bed are ideal for dosing experiments (Figure 1). The distributor at the 20-m distance from the influent was used for the dosing as it still left 10 m of bed for clarification and removal of the insoluble phosphate formed in the dosing. The dosing of the chemical solutions was accomplished by using an 8-channel variable-flow peristaltic pump, with each channel discharging into the slotted pipe at the bottom of one of the distributor sections of the bed at 0.5-m spacing across the bed (Figure 2). To facilitate mixing, a perforated air hose was placed into the slotted pipe across the bed and an air blower supplied low-pressure air across the bottom of the bed to give a steady stream of bubbles. Samples were taken of influent and final effluent. This enabled the effect of the dosing and the effectiveness of the bed to retain insoluble phosphates to be determined. The residual soluble phosphate and residual aluminium were also determined.

RESULTS AND DISCUSSIONS

The results for a range of alum dosing and combined alum plus lime experiments on the two gravel beds are shown in Tables 1 and 2. Residual aluminium was measured and for all samples was found to be less than

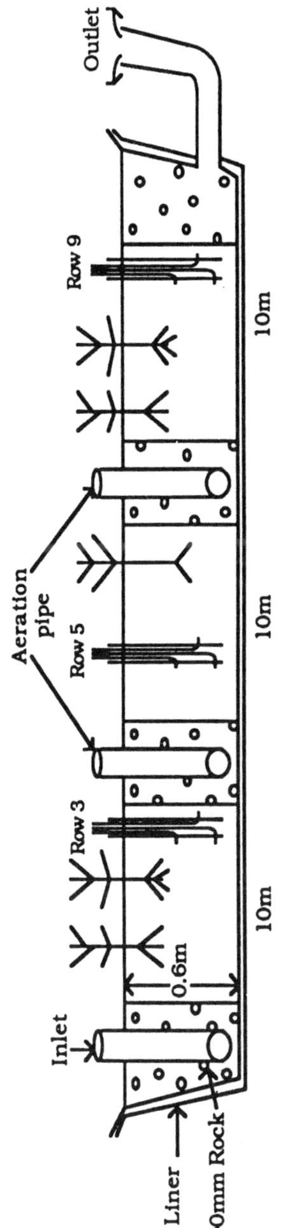

Figure 1. Longitudal section of bed.

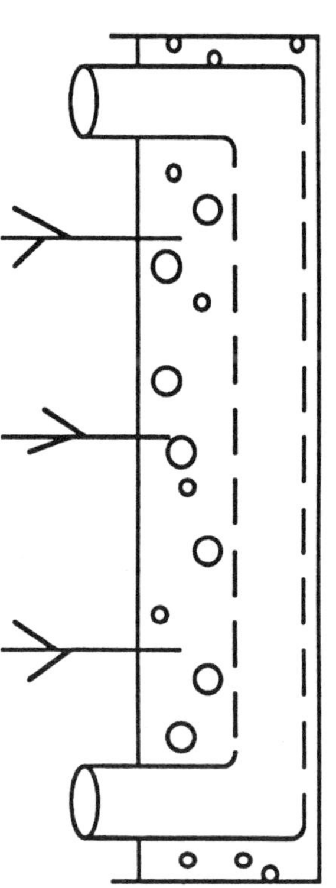

Figure 2. Cross-section of bed.

Table 1. Phosphorus Removal by Alum Dosing

	Planted gravel bed						Unplanted gravel bed					
	Total phosphorus		Soluble phosphorus				Total phosphorus		Soluble phosphorus			
Molar dosing ratio phosphorus: aluminium	mg/L	% removal	mg/L	% removal	pH	Number of samples	mg/L	% removal	mg/L	% removal	pH	Number of of samples
1:1.5												
Influent	8.0	—	6.1		7.4	10	10.6	—	9.6		6.9	10
Effluent	5.7	28.8	5.2	14.8	7.2	30	6.9	36.1	6.6	38.3	7.2	30
1:1.75												
Influent	8.9	—	7.5		7.3	10	9.4		7.5		7.2	10
Effluent	7.2	19.0	6.9	8.0	7.2	30	6.8	32.0	5.8	31.8	7.3	30
1:2.0												
Influent	11.1	—	7.2	—	7.3	10	10.4	—	9.0	—	7.2	10
Effluent	6.8	38.7	6.5	9.7	7.1	30	5.7	46.2	5.2	46.4	7.3	30
1:2.5												
Influent	9.1	—	6.2	—	6.4	10	10.3	—	9.6	—	7.1	10
Effluent	4.1	54.9	3.4	45.2	7.0	30	4.4	60.3	4.1	61.3	7.3	10

Table 2. Phosphorus Removal by Lime and Alum Dosing

	Total phosphorus		Inorganic phosphorus			
Dosing level	mg/L	% removed	mg/L	% removed	pH	Number of samples
100 mg/L lime + 100 mg/L alum						
Influent	7.07	—	5.93	—	7.0	10
Effluent	2.15	69.6	1.72	71.0	7.2	30
150 mg/L lime + 100 mg/L alum						
Influent	7.07	—	5.48	—	7.3	10
Effluent	1.58	77.7	1.21	77.9	7.1	30
150 mg/L lime + 50 mg/L alum						
Influent	5.39	—	4.31	—	7.3	10
Effluent	0.98	81.8	0.77	82.1	7.0	30

0.2 mg/L which is virtually the background value for undosed samples. Hence, the dosing contributed no extra aluminium to the solutions.

Although soluble alum is toxic to many plants, with the dosing being carried out in a 1-m unplanted coarse gravel section, there were no signs of any toxic effect or stressing of the plants adjacent to the dosing section; this suggests that the formation of the insoluble aluminium compounds was virtually complete before any of the dosed water reached any of the plants.

For the simple alum addition, phosphorus removal increased with increasing alum levels but was still relatively low at the optimum value, having phosphorus levels at more than 4 mg/L. Similar dosing in sewage treatment works easily reduces phosphorus levels to less than 1 mg/L, so obviously the wetland dosing was inefficient.

As stated earlier, there are two competing reactions for the aluminium and, if the insoluble phosphate is not quickly formed (Equation 1), the aluminium is then removed as the insoluble hydroxide with the reaction forming hydrogen ions (Equation 2), which makes the formation of the phosphate less favorable. Hence, it is essential to have rapid and efficient mixing of the alum with all of the phosphorus for the preferential formation of the phosphate, and the method of dosing within the bed does not allow this to occur efficiently. The presence of the gravel and the surrounding voids make the mixing quite difficult and better results would be obtained if the dosing area was free-water, with no gravel at that point (i.e., a small pond section).

Samples taken with a probe in the immediate dosing area showed low pH values between 4 and 5 which confirm the thesis of preferential aluminium hydroxide formation in the poorly mixed dosing area. This acidity soon disappears in the bed itself as it has a buffering capacity due to the presence of powdered limestone added during the construction of the beds.

The small difference between the effluent's total phosphorous and soluble phosphorus shows the effectiveness of the bed in retaining the insoluble phosphates within the bed.

The lime plus alum dosing, Table 2, resulted in increased efficiency in phosphorus removal with the optimum removal being produced by a concentration of 150 mg/L lime plus a relatively low 50 mg/L of alum reducing phosphorus levels to less than 1 mg/L despite the handicap of the poor mixing occurring in the dosing section.

The presence of the alkaline lime prevents the acid formation and thus increases the formation of insoluble phosphate at the expense of the insoluble hydroxide formation, this latter being a waste of the relatively expensive alum compound.

CONCLUSIONS

The use of lime (150 mg/L) plus alum (50 mg/L) is shown to be an economic and efficient dosing mixture for the removal of phosphorus as the insoluble phosphate within the bed itself.

Dosing directly into the bed entails mixing problems, leading to wasting of the alum by the formation of some hydroxide instead of the phosphate. This problem could be overcome by the use of three sections of installation, a planted bed for normal solids, BOD_5, and pollutant removal; a small pond section for chemical dosing, mixing, and sedimentation of the phosphate sludge; and a final planted bed for final treatment and removal of any insoluble phosphate carried over from the dosing pond.

REFERENCES

1. Hocking, P. J. Seasonal dynamics of production, and nutrient accumulation and cycling by *Phragmites australis* (Cav.) Trin. ex Stuedel in a nutrient-enriched swamp in inland Australia. II. Individual shoots. *Austr. J. Marine Freshwater Res.* 40:445, 1989.
2. Batterson, T. R. and D. W. Hall. Common Reed, *Phragmites australis* (Cav). Trin. ex Stuedel. Aquatic VI(2), Florida Aquatic Plant Management Society, 1984, 16.
3. Davies, T. H. and B. T. Hart. Reed bed treatment of wastewaters in a pilot scale facility. *Proc. Constructed Wetlands in Water Pollution Control Conf.* IAWPRC, Cambridge, U.K., 1990, 517.
4. Devai, I., L. Felfoldy, I. Wittner, and S. Plosz. Detection of phosphine: new aspects of the phosphorus cycle in the hydrosphere, *Nature.* 333:343, 1988.
5. Cottingham, P. D. and T. H. Davies. Progressive pollutant removal in a reed bed treating wastewater. *Biological Nutrient Removal Conf. Proc.* Bendigo, Victoria, Australia, 1990, 277.
6. Stumm, W. and J. J. Morgan. *Aquatic Chemistry.* 1st ed. Wiley-Interscience, New York, 1970.
7. Priestly, G. and S. Solin. Phosphorus (nutrient) removal from small module purification units. MMBW (Melbourne, Australia). Report, 1977.

CHAPTER 33

Nitrogen and Phosphorus Reduction in Secondary Effluent Using a 15-Acre, Multiple-Celled Reed Canarygrass (*Phalaris arundinacea*) Wetland

S. Geiger, J. Luzier, and J. Jackson

INTRODUCTION

This article provides a review of findings from the second year of operation of the Unified Sewerage Agency's (USA) Jackson Bottom Experimental Wetland (JBEW).[1] The JBEW was designed to treat secondary effluent during the period May through October from the Hillsboro Wastewater Treatment Facility, located west of Portland, Oregon in Hillsboro.

The JBEW is located in a 434-acre area immediately south of Hillsboro in a portion of the Tualatin River floodplain. Of the total area, 80% or 354 acres has been delineated wetland, with most of the wetland dominated by reed canary grass (Phalaris arundinacea). JBEW is located within the wetland.

Jackson Bottom is located approximately 1300 ft from the Tualatin River. The river, especially during the driest period of the year, is slow moving and eutrophic. The USA is faced with requirements to reduce phosphorus concentrations drastically during summer in four of its facilities that discharge into the river, two of which provide tertiary treatment. JBEW was developed to test natural means of providing tertiary treatment for phosphorus and nitrogen and achieve biodiversification in a monotypic wetland.

EXPERIMENTAL DESIGN AND MONITORING

Treatment Site Configuration

The JBEW is a 15-acre treatment wetland composed of 17 linear cells, each a combination of shallow and deep segments (Figure 1). The length of the cells are approximately 385 m. Flow into each cell was regulated at the inlet to achieve specified hydraulic loading rates. During the first year of JBEW operation in 1989, secondary effluent flowed through the facility from July to October. Results of the first year's monitoring were reported by Scientific Resources, Inc.[2] During the second year, effluent flowed into the wetland between June 25 and October 8.

Treatment Cell Design

Native soils of the area of JBEW were relocated in the fall of 1988 to form 3-ft high by 6-ft wide berms forming the linear cell configurations. Material for the berms was borrowed from excavated areas on the outside edges of the facility and along the lengths of the cells where deep areas were created to a depth of 3 ft below previous grade. This resulted in an alternation of reed canary grass-dominated "shallow" areas and the deeper areas where the grass, sod, and shallow soils had been removed to construct berms (Figure 1). Smaller deep areas were excavated at the head of each cell to facilitate water dispersion into first shallow cells.

The prevalence of reed canary grass in the shallow portions of the cells was of interest in different ways. In the first place, reed canary grass is not observed to grow in conditions of continuous flooding, which appeared to be the proposed setting of reed canary grass within the cells. There was only scattered and

0-87371-550-0/93/$0.00+$.50

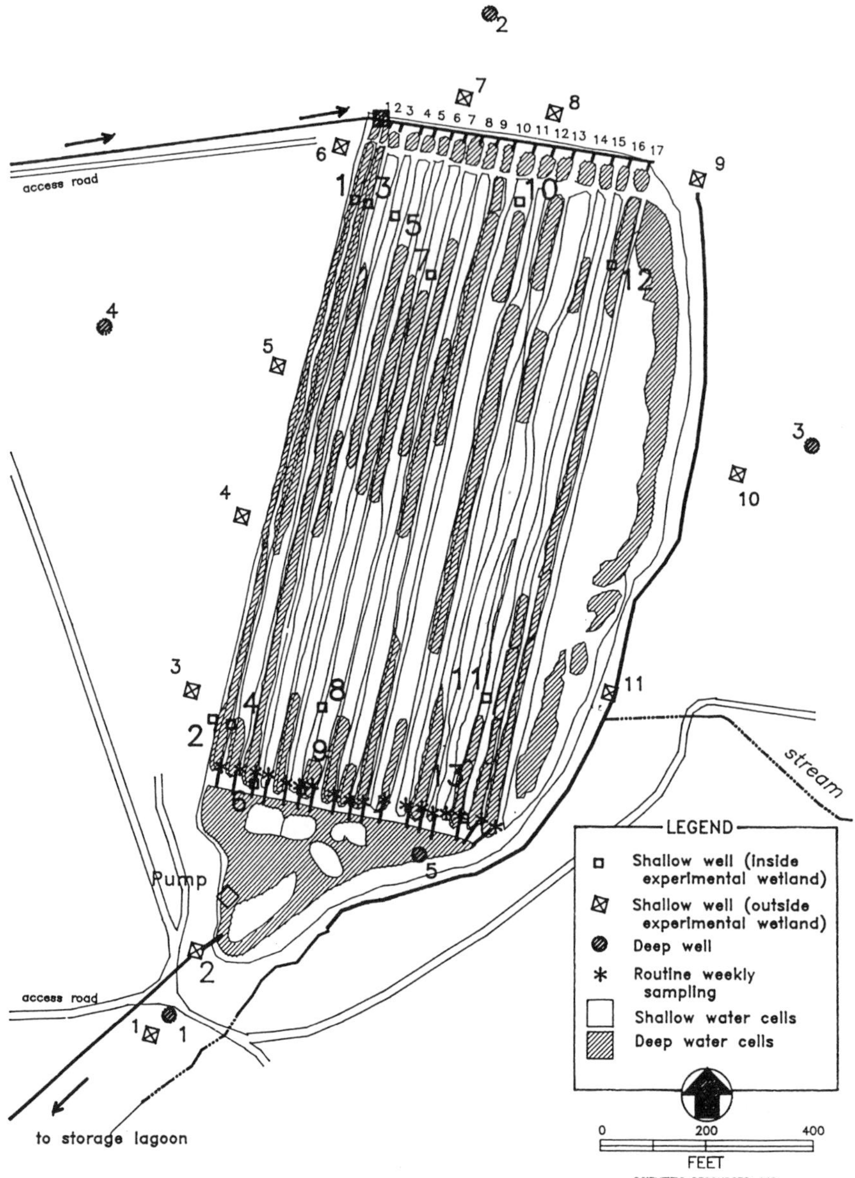

Figure 1. Jackson Bottom Experimental Wetland (JBEW) configuration and water quality monitoring locations.

nonregional investigation of the effectiveness of reed canary grass in influencing the water quality in wastewater treatment wetlands (e.g., Bell and Boll,[3] Giggey, Crites, and Brantner,[4] Pierce,[5] Ree[6]). The use of the wetland to test the effectiveness of reed canary grass in comparison with other plants on the treatment of wastewater was therefore of interest.

Sewage Treatment and Effluent Disposal Limitations

Wastewater treatment at the USA Hillsboro Facility affords standard secondary treatment through activated biofilter, aeration basin, and secondary clarification processes. During the summer (May to October), all plant effluent is disposed of by irrigation on land, in accordance with the requirements of the NPDES permit for the treatment plant. During winter, effluent is discharged to the Tualatin River. The JBEW is intended to function during the period of land effluent application. During the summer of 1990 (May 1 to October 31), approximately 370 million gallons (MG) of effluent (or approximately 2 MG per day) was produced and distributed to approximately 350 acres of bottom land, including wildlife ponds, a retention pond, open areas, and the experimental wetland.

Objectives and Monitoring Strategy

Objectives for the 1990 monitoring program at JBEW were to: (1) evaluate wetland performance at various loading rates; (2) evaluate impacts to JBEW ground water; (3) determine soil phosphorus assimilation potential; (4) evaluate performance in relation to soil types; (5) calculate a phosphorus nutrient budget for JBEW; and (6) characterize the biota of JBEW.

Surface water at the outlets of each cell was sampled weekly. Flow into and out of each cell was measured by meter and monitored twice weekly. A water budget was developed for the treatment system incorporating inflow/outflow data, precipitation measured at the treatment plant and, for an estimate of evaporation from the wetland, 0.7 × pan evaporation rates as measured at a nearby agricultural experiment station. Leakage out of the system into ground water was estimated by the difference between measured and estimated surface inflow and outflow. Five deep monitoring wells were installed in 1988–1989 prior to facility use. The 60-ft deep wells allow sampling of up to five depths. In addition, 24 shallow wells, 3 to 5 ft deep, were installed, 12 inside wetland cells and 12 outside the facility (Figure 1). The three soil series were sampled by augur to depths of 2 ft. Subsamples were analyzed for phosphorus species and phosphorus assimilation capacity.

RESULTS

Surface Water Hydrology and Quality

A total of 192,288 m^3 (50.8 mgal) of the Hillsboro plant secondary effluent flowed into the entire wetland between June 25 and October 8. This flow was 23.2% of the total plant effluent through that period. The total flow over the season of 108 days varied among the 17 cells ranging from 19,306 m^3 in cell 14 to 6,381 m^3 in cell 17. Mean hydraulic loading rates ranged from 0.013 to 0.068 m/da with an average of 0.040 m/da. Retention times ranged from 4.66 to 26.7 days with an average over all cells of 11.3 days.

A total of 52,962 m^3 (28.5 MG) flowed out of the wetland (27.15% of cell inputs). Of the total, 18,130 m^3 was used to fill the cells, and 16,258 m^3 (4.3 MG) was lost through evapotranspiration. Evapotranspiration values for the JBEW were based upon pan evaporation rates (× a factor of 0.7) from a nearby Oregon State University Agricultural Experiment Station. The remainder [107,742 m^3 (28.5 MG)] was leakage from the wetland. Leakage was estimated as the difference between measured flow and precipitation in and flow out, less evapotranspiration estimates. A summary water budget is provided in Table 1.

Inflow concentrations (Hillsboro West effluent) of total phosphorus ranged from 2.84 to 20.00 mg/L through the treatment period, with an average of 7.60 mg/L. Average soluble phosphorus concentrations in the inflow was 5.88 mg/L. Outflow concentrations of total phosphorus averaged 4.01 mg/L for all cells. Outflow concentrations were 52.8% of inflow on average (47.2% reduction in concentration). With respect to wetland performance in removing total phosphorus, the total mass of total phosphorus entering the wetland as plant effluent was 1461.4 kg. The total mass of total phosphorus leaving the wetland as surface flow was 211.8 kg (14.5% of the influent mass). This is a reduction in total phosphorus mass loading of 85.5%. Of the total soluble inorganic nitrogen (TSIN) entering the treatment cells as nitrate, nitrite, or ammonia, 99.4% was diverted by the treatment wetland into air, ground water soils, or biota.

In terms of a reduction in pounds per day, the JBEW reduced TSIN from 53.85 to 0.31 lb/day, total phosphorus from 29.82 to 4.32 lb/day, and soluble phosphorus from 23.08 to 3.53 lb/day.

Ground Water Hydrology and Quality

During the two summers of operation, the treatment wetlands have caused minor physical changes to the local ground water flow regime, such as slight mounding and increased ground water gradients. As noted in Table 1, leakage of treated water into the shallow ground water system constituted 55.2% of water entering JBEW.

Table 1. Summary Water Budget for the Jackson Bottom Experimental Wetland, 1990

	m^3	Mgal[a]	%
IN			
Inflow	192,288	50.8	98.56
Precipitation	2,805	0.7	1.44
Total	195,093	51.5	
OUT			
Evaporation	16,258	4.3	8.33
Outflow	52,962	14.0	27.15
Leakage	107,742	28.5	55.23
Cell volume	18,130	4.8	9.29
Total	195,093	51.5	

[a] Million gallons.

Background concentrations of ammonia-nitrogen measured in deep and shallow wells were several milligrams per liter, but nitrate-nitrogen was present at only trace levels. Soluble phosphorus concentrations were unusually high for natural ground waters, ranging between 1 and 4 mg/L. During the 2-year operation of JBEW, concentrations of ammonia-nitrogen and soluble phosphorus in shallow wells in the JBEW have been generally less than 0.5 mg/L (Figure 2).

Soils

Results from the phosphorus adsorption/desorption analyses indicated that soils within the wetland have a relatively high capacity to take up phosphorus from solution and maintain low phosphorus equilibrium concentrations (Figure 3). Extractable aluminum in the treatment wetland soils ranged from 371 to 2230 mg/kg, and extractable iron from 2451 to 27,820 mg/kg. In general, surface soils exhibited higher sorption potentials than did subsoil horizons. Evidence of increased phosphorus in soils between 1989 and 1990 implies that soil phosphorus is increasing through time as a result of treatment wetland operations. However, no discernible trend was evident with increasing soil phosphorus content and soil depth.

Biota

Vegetation within the JBEW relates to cell depth, with *Phalaris arundinacea* dominating shallow subcells and *Lemna minor* and *Potamogeton foliosus* dominating deep subcells. *Phalaris* appears to be increasing in coverage and it appears likely that it will eventually dominate all but the deepest subcells. Uptake of nitrogen and phosphorus by wetland vegetation relates to the quantity of biomass rather than the aerial coverage produced by each species (e.g., *Lemna's* extensive coverage of deep cells) with *Phalaris arundinacea* containing 91.5 and 90.8% of the total accumulated phosphorus and nitrogen in vegetation, respectively. Whereas macronutrient concentrations were highest in above-ground biomass, micronutrient levels (e.g., copper and zinc) were highest in the roots.

SYSTEM PERSPECTIVES

Phosphorus Model

A phosphorus model was developed for the JBEW system. Model results are provided in Figure 4. The ground water output estimates are provided in Figure 4. The ground water output estimates were made by multiplying mean total phosphorus in shallow wells by cumulative cell leakage. Soil loading was estimated by two methods, difference and estimation. The estimated soil loading value was based upon average bulk density for clay soils (1.6 g/cm^3), a 3-in. active soil horizon, and an average equilibrium phosphorus level of 4 mg/L, which gave sorbed phosphorus values of 230 to 256 mg/kg for the soils of the wetland.

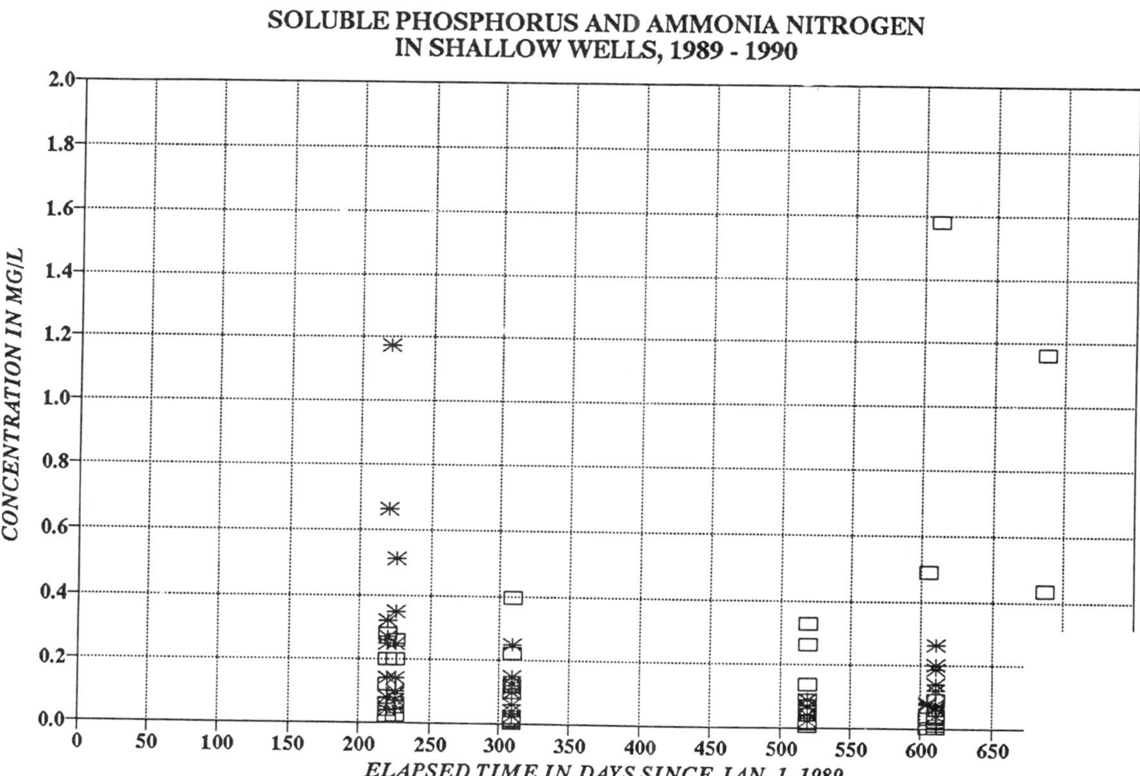

Figure 2. Concentrations of soluble phosphorus and ammonia-nitrogen in shallow ground water within the treatment wetland. □, soluble phosphorus; *, ammonia-nitrogen.

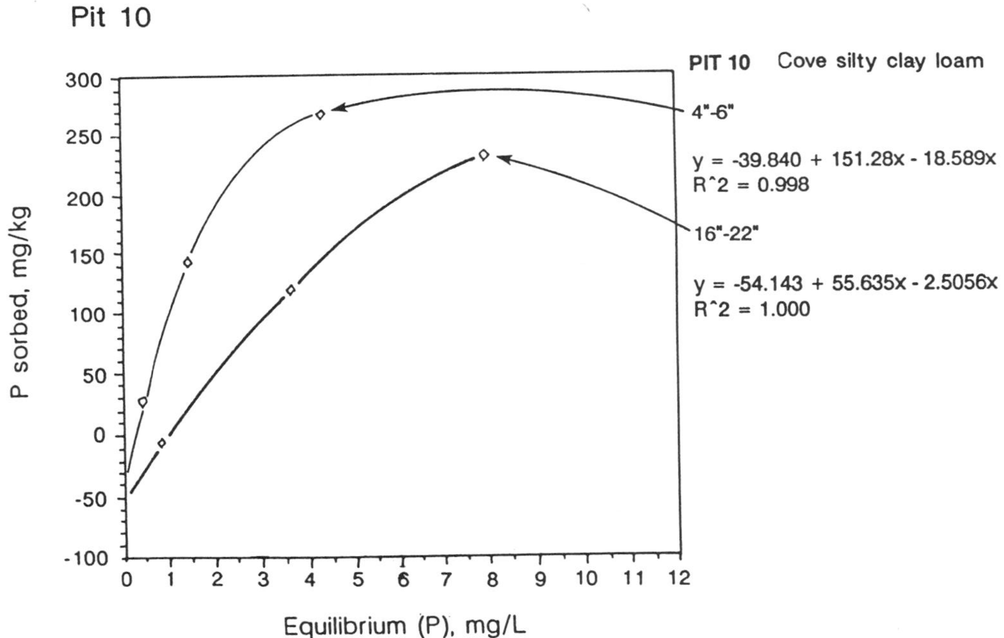

Figure 3. Phosphorus isotherms for Cove and Wapato soils series in the JBEW. Graphs show response of soils (5 g) during 72 h in deionized water and in solutions of 50-ml solutions of 3.1, 15.5, or 31 mg/L soluble phosphorus (KH_2PO_4).

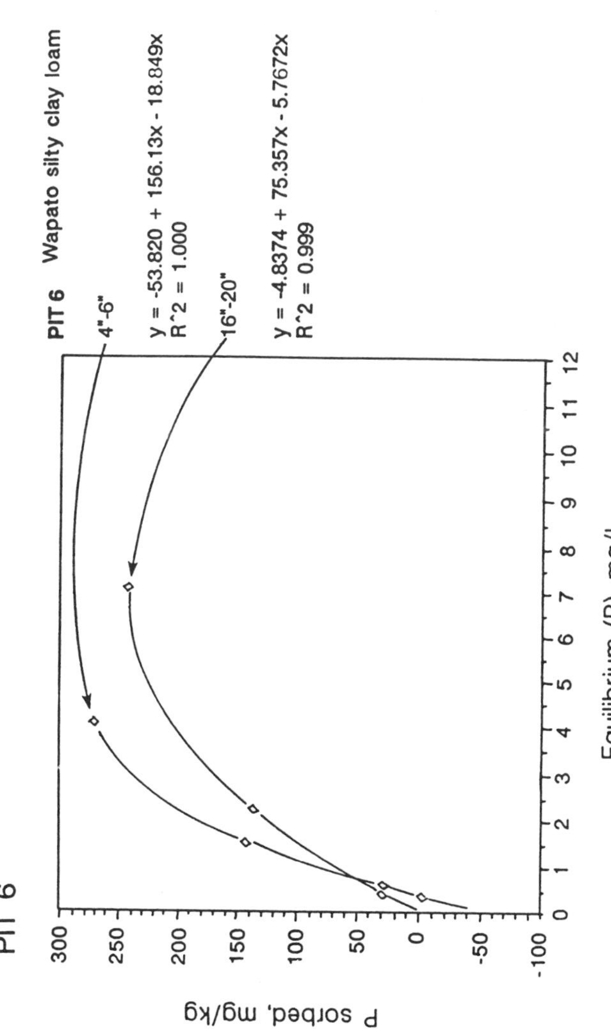

Figure 3. (continued)

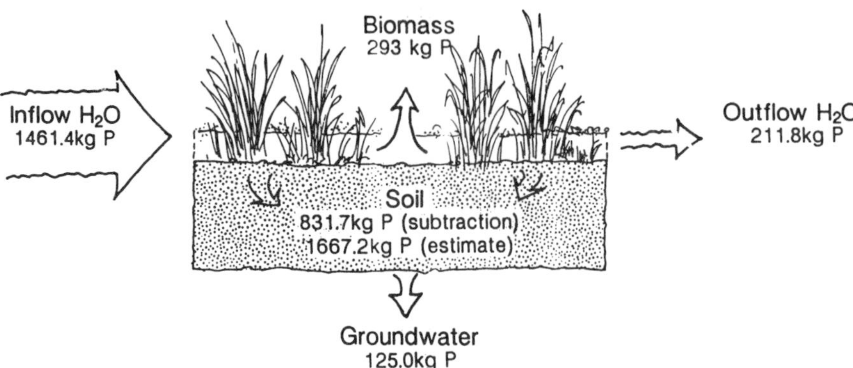

Figure 4. Model of phosphorus allocation for 1990. Soil compartment determined by measurement of soil phosphorus and by difference among other compartments.

The model results indicate 23.1% of the phosphorus left the system through ground and surface water routes (8.6 and 14.5%, respectively). Vegetation sequestered 20% of the total system phosphorus. The model indicates that the largest phosphorus pool exists within the wetland soil compartment (56.9%). Total phosphorus system efficiency (inflow minus outflow) was 85.5%.

Optimizing Phosphorus Reduction

It was found that leakage from the wetland was directly proportional to hydraulic loading. It appears that the highest total phosphorus, SRP, and TSIN reduction can be achieved through high mass loading. Phosphorus and nitrogen mass can be effectively removed through this system by incorporation into wetland soils, but ground water effects may not be tolerable at higher hydraulic loading rates.

The importance of total N and P mass reduction by the experimental wetland, rather than changes in concentration between inflow and outflow, confounds attempts to explore relationships between outflow concentrations and inflow mass loading rates, or even retention times. Plots of nitrogen and phosphorus reduction vs. mass nitrogen and phosphorus loading rates and retention times indicated no, or very weak, statistical relationships.

ACKNOWLEDGMENTS

SRI was assisted in the monitoring and data analysis by David McAllister (formerly with SRI, now with Oregon Department of Fish and Wildlife, Portland, Oregon) [biota] and Cascade Earth Science (CES) [soils] of Lake Oswego and Corbett, Oregon, respectively. Wes Jarrell and Mary Abrams of the Oregon Graduate Institute, Beaverton, Oregon performed soil adsorption/desorption analyses. We also acknowledge the valuable work of Ralph Vaga, formerly of SRI and now with the U.S. Geological Survey.

REFERENCES

1. Scientific Resources, Inc. Report on Monitoring of Jackson Bottom Experimental Wetland (JBEW), 1990, Final Report with Appendices. Prepared for the Unified Sewerage Agency, Hillsboro, OR, Lake Oswego, OR, 1991.
2. Scientific Resources, Inc. Report on Monitoring of Jackson Bottom Experimental Wetland (JBEW), 1989, Final Report with Appendices. Prepared for the Unified Sewerage Agency, Hillsboro, OR, Lake Oswego, OR, 1990.
3. Bell, R. G. and J. B. Bole. Elimination of fecal coliform bacteria from reed canary-grass irrigated with municipal sewage lagoon effluent, *J. Environ. Qual.* 5:417, 1976.

4. Giggey, M. D., R. W. Crites, and K. A. Brantner. Spray irrigation of treated septage on reed canarygrass, *J. Water Pollut. Control Fed.* 61:333, 1989.
5. Pierce, R. Reed canarygrass thrives on sewage, *U.S.D.A. Agric. Res. Sci. Education.* 28:12, 1979.
6. Ree, W. O. Effect of Seepage Flow on Reed Canarygrass and its Ability to Protect Waterways. #154, U.S. Agri. Res. Service Southern Region, 1976.

CHAPTER 34

Water Supply System Utilizing the Edaphic-Phytodepuration Technique

E. S. Manfrinato, E. S. Filho, and E. Salati

INTRODUCTION

In the majority of the cities in Brazil, water supply is derived from rivers or artificial lakes. Many of the existing water treatment plants were planned and built 30 to 40 years ago when the waters from rivers and dams were within the quality control required by the laws regulating water supply.

However, during the past few decades, due to increased industrialization and population growth in urban centers, numerous rivers and dams have become highly polluted, and for a large period of the year, their water quality is inadequate for use in conventional water treatment systems. The Piracicaba River Basin, for instance, which includes 44 municipalities and has an area of approximately 12,000 km^2, is one of the most polluted in the country.

A great number of cities developed without any regard to the implementation of systems for domestic sewage treatment. Of the waste load from $2^1/_2$ million inhabitants, only 5% of the load from the cities is removed by sewage systems, and the balance (76 tons BOD per day) is still discharged directly into rivers. The pollution potential from industrial organic load is on the order of 1440 tons BOD per day, or about 18 times the domestic load. However, approximately 95% of this is removed. In spite of this high removal percentage, the remaining 5% (77 tons/day) is still of the same magnitude as the discharged domestic load. Besides this contamination, the level of pollution is increased by agrotoxic residues transported along water bodies.

Solutions to problems of domestic water supply, which require changing the characteristics of the water treatment plants by using water from less polluted streams or rivers, also require large investments and are not always practicable.

In an attempt to find a solution to this problem, a technique was developed to clean river water using the purifying quality of floating aquatic plants and the filtering properties of soils cultivated with rice (*Oriza sativa*). This process is called "edaphic-phytopurification", as proposed by Salati in 1987 (patent requested #PI 8503030 in Brazil).[1]

OBJECTIVE

The present work is aimed at studying the efficiency of the edaphic-phytopurification process: in the decontamination of the waters of the Piracicaba River and analyzing the most relevant parameters (BOD, COD, total and fecal coliforms, pH, color, turbidity and inorganic compounds, NH_3^+, NH_4^+, Ba, Cd, Pb, Cu, Cr, Zn, SO_4^{-2}, P, Cl, Ca, Fe, Mg, Na, and PO_4^{-3}) for the characterization of the waters to be supplied to the public from the Piracicaba River, and was carried out during the period June 1986 to June 1987.[2]

0-87371-550-0/93/$0.00+$.50

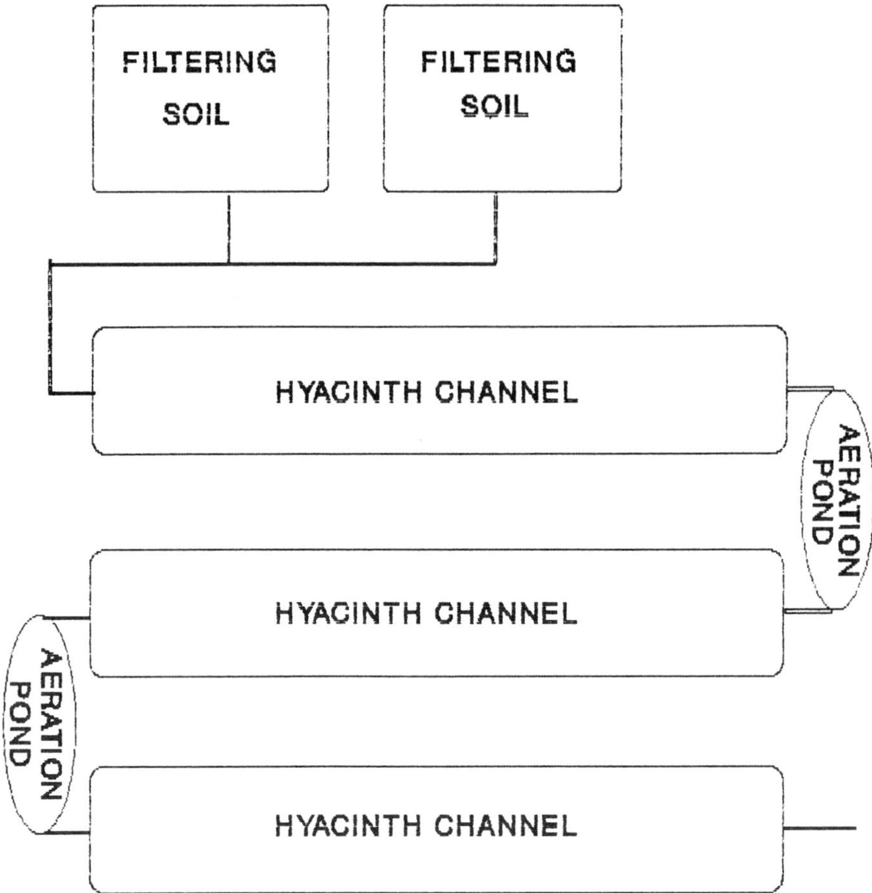

Figure 1. Hydric decontamination system using canals with aquatic plants and filtering soils.

MATERIAL AND METHOD

Location of Study

The study was conducted at "Terras do Engenho Central", located on the right margin of the Piracicaba River and centrally positioned in the city of Piracicaba (latitude 22° 42′ 30″ S; longitude 47° 38′ 00″ W; altitude 540 m), SP.

Description of the Decontamination System

Two decontamination systems were installed at the above location: (1) system I consisted of a channel cultivated with aquatic plants and a filtering soil bed. This system was built by the municipality and has been in experimental operation since 1984; (2) system II had two consecutive filtering soil beds.

System I — Channels Cultivated with Aquatic Plants and a Filtering Soil Bed

For channels of aquatic plants, the canals were cultivated with "water hyacinth" (*Eichhornia crassipes*). Three parallel channels were built, linked in series by two aeration ponds. Each channel was constructed with the following dimensions: 0.7 m deep × 15 m wide × 96 m long, occupying a total area of 4320 m^2. The aeration ponds are approximately 250 m^2 each (Figure 1).

The entering flow control in this system was made through a flowgate and the flow measurement through a Parshall measuring flow device of 9 in.; its calibration was determined after installation.

The filtering soil bed was 53 m long and 38 m wide, a total area of 2014 m^2 (Figure 1). This bed was a drainage system composed of 100-mm diameter PVC perforated pipes placed in a 40-cm deep ditch at every

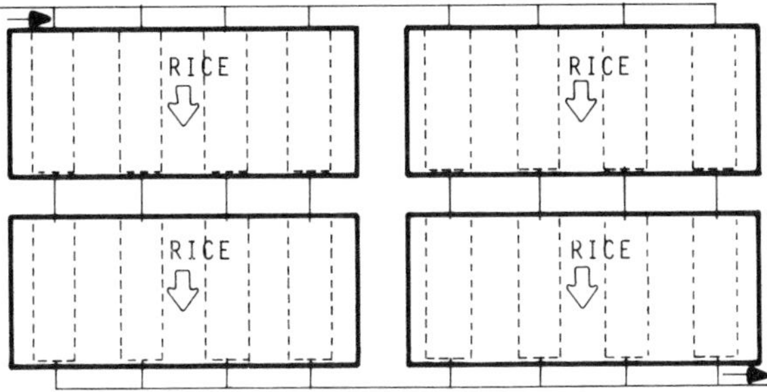

Figure 2. System consisting of a sequence of two filtering soils cultivated with rice.

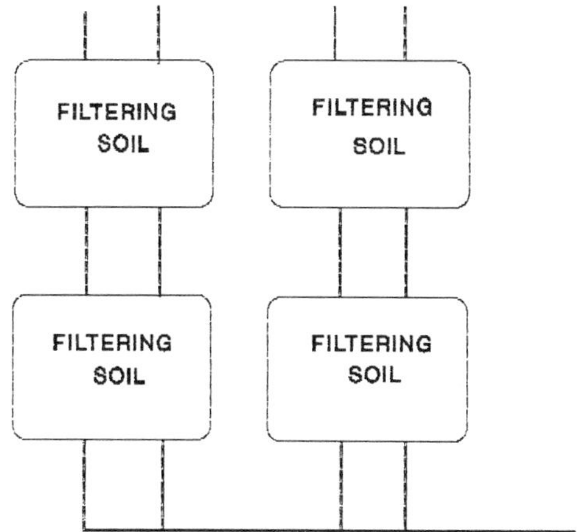

Figure 3. Hydric decontamination system using consecutive filtering soils.

2 m. These ditches were filled with gravel; a 15-cm deep lower layer of pebbles, and a 20-cm thick upper layer of soil (Figure 2). The soil (30 cm deep) was a mixture of clay soil (red-yellow latosol) from the Piracicaba area, and 20% of the mixture in volume of vermiculite.

System II — Consecutive Filtering Soils

Two modules with two filtering soil beds were built with an area of 410 m^2 each (Figure 3). The drainage system is similar to the one mentioned above (Figure 2). The entering water flow was controlled with the installation of a flow gauge in the system entrance.

Methodology

Handling of the Systems

System I — Channel with Water Hyacinth/Filtering Soil — During the experiment (June 1986 to June 1987), the entering flow in the channel with water hyacinth was kept at 40 L/s (93 L/s/ha of aquatic plants), while the residence time of the water on the canal was 27.5 h. In October 1986, 10% of the total mass of water hyacinth was taken out manually.

Part of the water flowing from the water hyacinth channel was diverted to the Piracicaba River. Therefore, only 20 L/s (99 L/s/ha of filtering soil) passed through the filtering soil plot.

System II—System of Consecutive Filtering Soils — The preliminary tests showed that it would be possible to keep an equivalent influx of 100 L/s/ha by using hydraulic pressure over the first 10 cm of filtering soil. However, a constant influx was kept during the entire phase of the rice development of the order of 2 L/s/ha module (491/s/ha). The constant water level was maintained by hydraulic pressure on the exiting pipes, located on the inverted position of the drainage system exit.

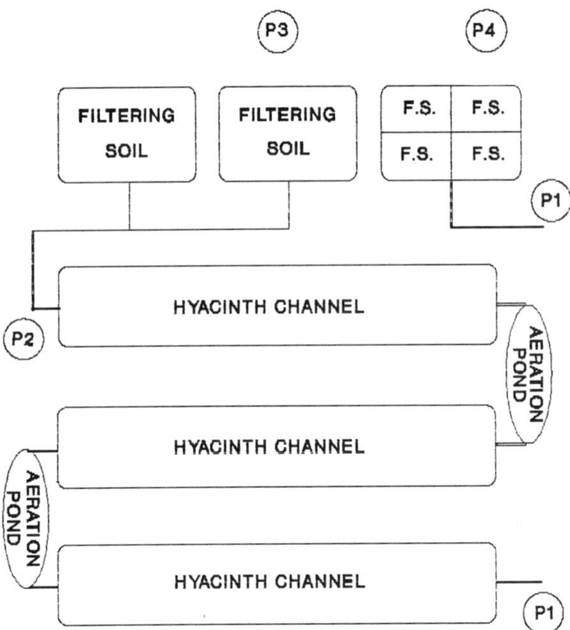

Figure 4. Localization of the collection points.

Water Sample Collection

Water samples were systematically collected during the period of June 23, 1986 to June 11, 1987. The methodology used consisted of a simple sample taken daily for five consecutive days and then repeated every 15 days. Samples were collected at four different points (Figure 4); one-liter samples were collected from each point.

Parameters Analyzed

- Biochemical oxygen demand (BOD)
- Chemical oxygen demand (COD)
- Total and fecal coliforms
- Color
- Turbidity
- Nitrogen compounds: nitrite-nitrate and ammonia

Other parameters include barium, zinc, cadmium, chromium, copper, lead, sulfates, chloride, calcium, potassium, magnesium, manganese, sodium, silica, and phosphorus.

RESULTS AND CONCLUSIONS

- Measurement of concentrations of the already-mentioned parameters indicated a substantial improvement in the quality of water after passing through the decontamination process and the soils cultivated with rice. The efficiency observed in the system using floating aquatic plants and then filtering soils was practically the same as that obtained when utilizing two consecutive filtrations with filtering soils.
- The monthly values obtained during the examination period for the most significant parameters, for both systems studied (water hyacinth/filtering soils and consecutive filtering soils), can be found in Graphs 1 to 10.
- The average efficiencies of both systems, obtained during the same period, for the most significant parameters are revealed in the table below.

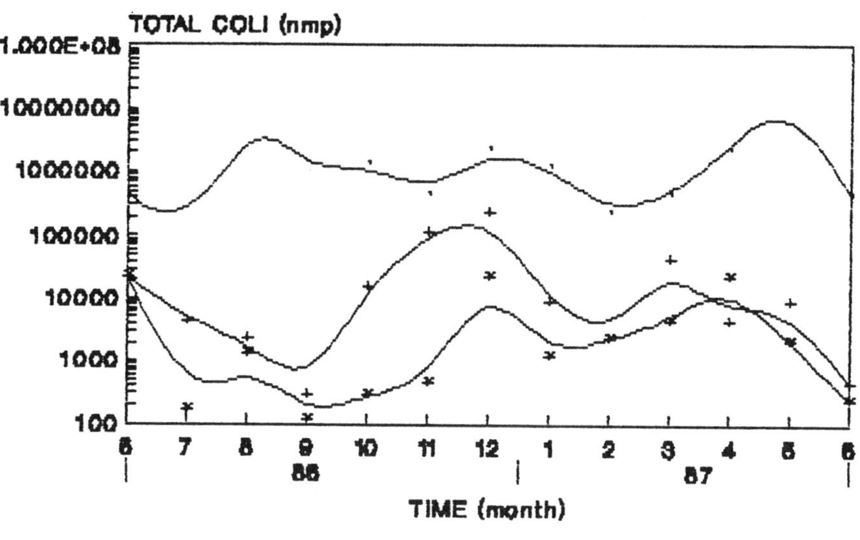

Graph 1. — INLET —+— OUTLET- H.C. —*— OUTLET- F.S.

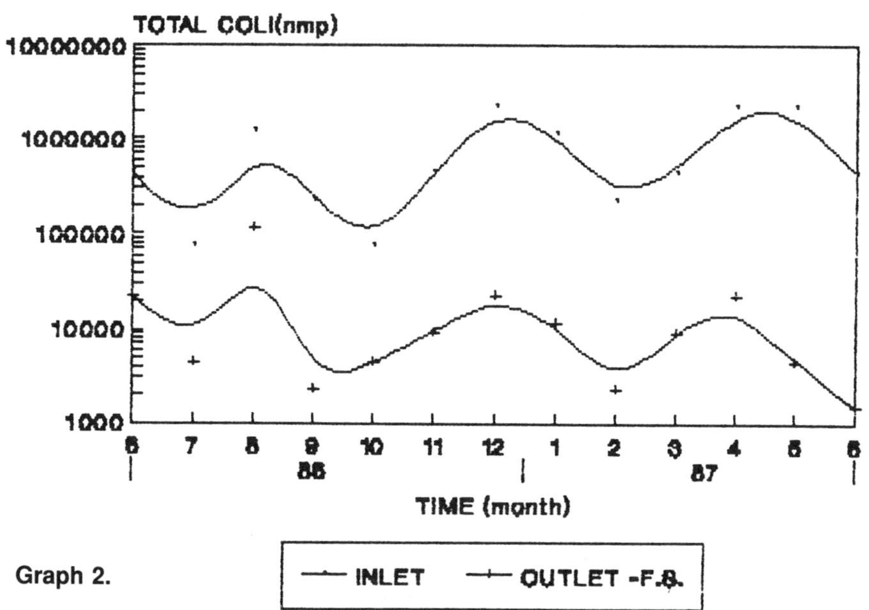

Graph 2.

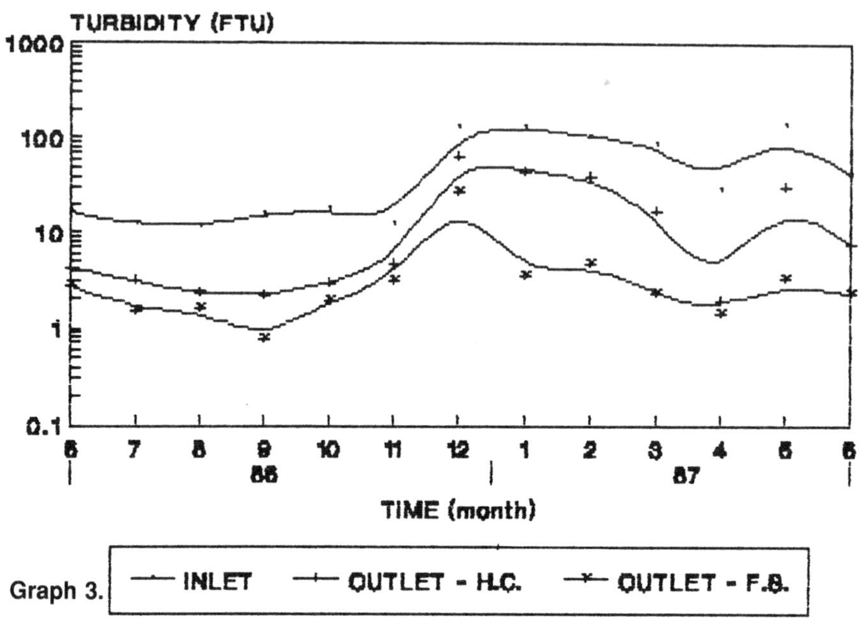

Graph 3.

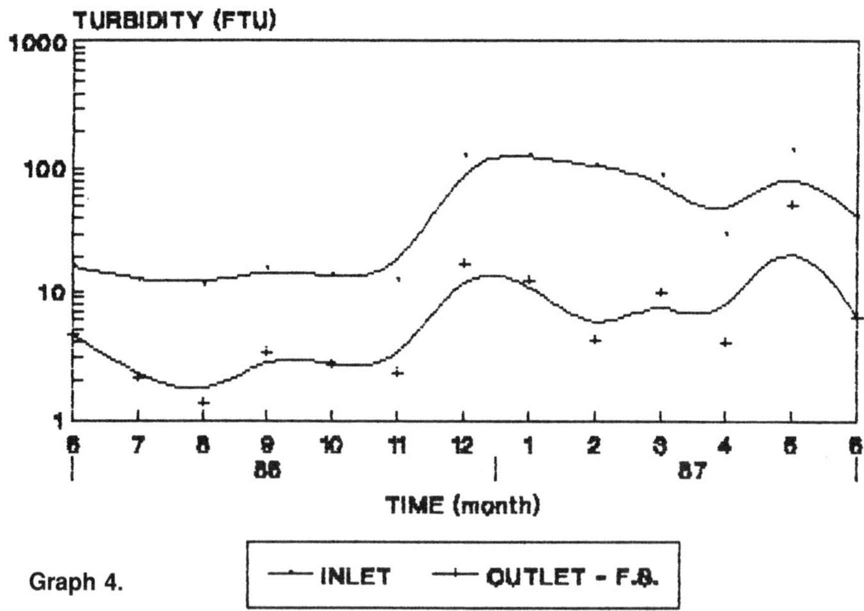

Graph 4.

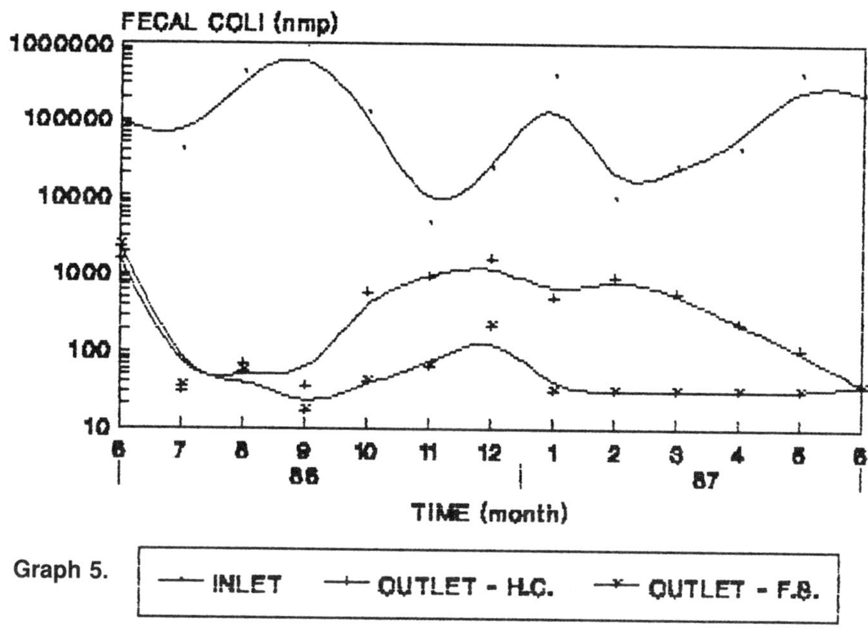

Graph 5.

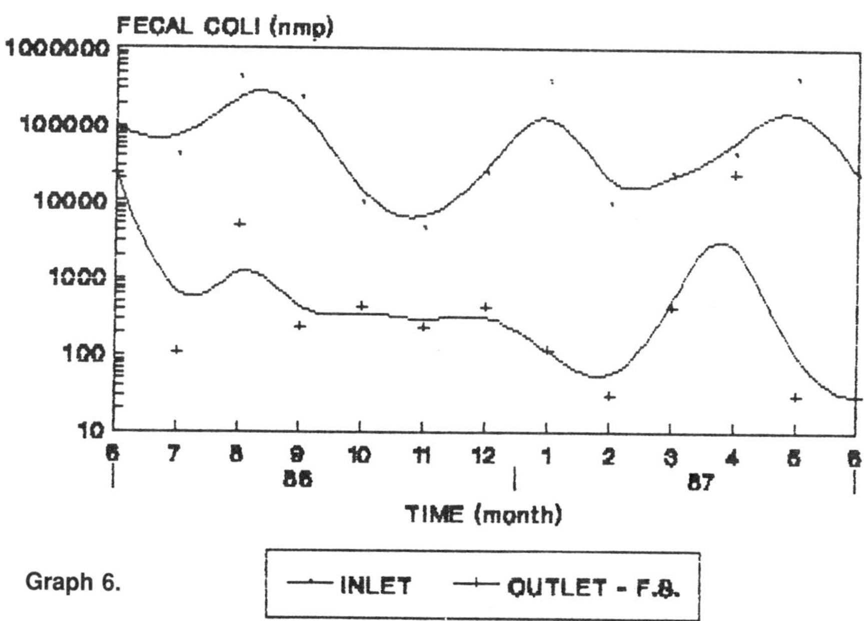

Graph 6.

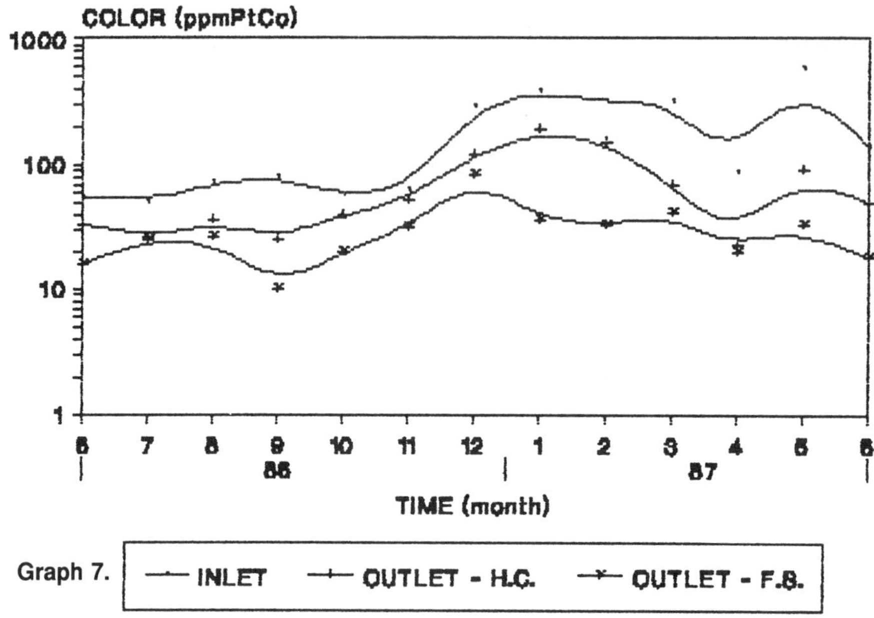

Graph 7.

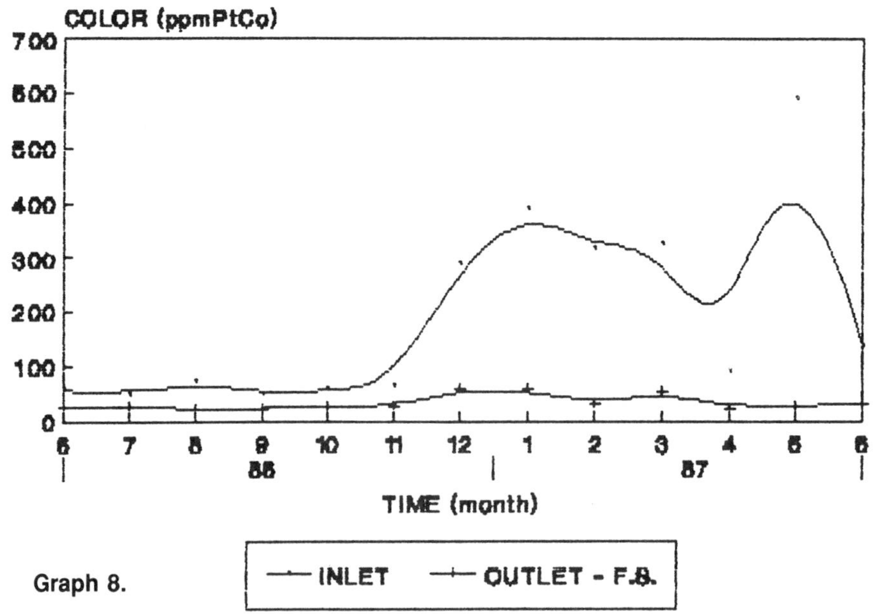

Graph 8.

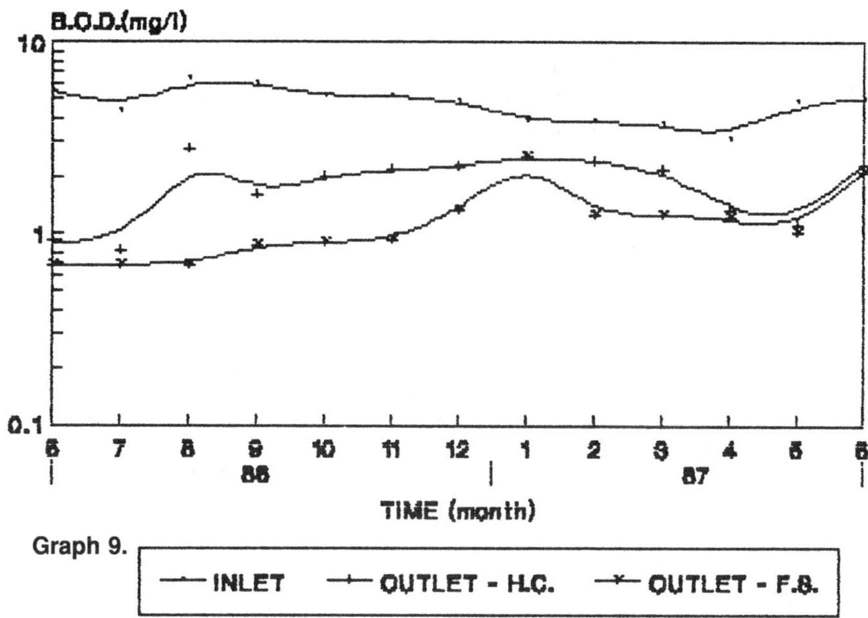

Graph 9.

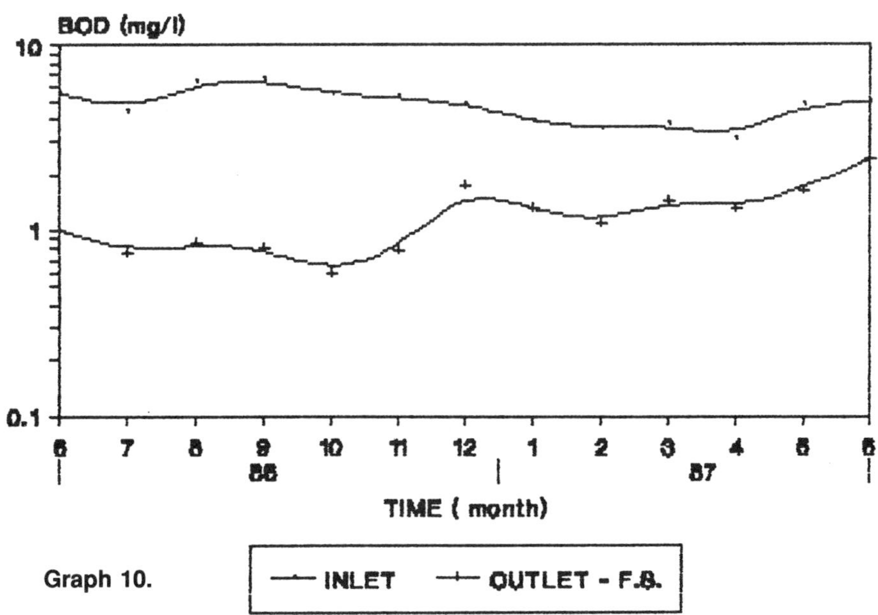

Graph 10.

Parameter	Average efficiency (%)
BOD	70
COD	70
Total coliforms	99
Fecal coliforms	99
Ammonium-nitrogen	60
Iron	80
Phosphorus	65
Aluminium	85
Nitrate	95
Color	90
Turbidity	95

- A variation in the efficiency of the system was observed during the year as a consequence of the biological activities based on climatic conditions and plant physiology.
- After treatment, water in the Piracicaba River showed characteristics appropriate for supply to conventional water treatment plants.
- During most of the 13-month study, the water coming out of the treatment system showed qualities which reached the levels required for potability; to be good for consumption, it would only need to be chlorinated. Further studies with varying flow indicated that when the hydraulic load varied on filtering soils, up to 200 L/s/ha outflow could be obtained, and the quality of the water for supply to treatment plants still remained good.
- The best result is obtained when the filtering soils are managed in an alternative way: the water passing through each system for 10 days at a time. A system based on this water supply experiment is under construction in the city of Parauapebas (40,000,000 persons on the Amazon). Another system, utilizing the water of a heavily polluted river, has been in operation for 8 months. This system is being used for water supply at one industry in Caruari, Pernambuco, Brazil, with a flow of 7 L/s.

REFERENCES

1. Salati, E. Edaphic-phytodepuration: a new approach to wastewater treatment, in *Aquatic Plants for Water Treatment and Resource Recovery*. K. R. Reddy and W. H. Smith, Eds. Orlando, FL, 1987, 199–208.
2. Manfrinato, E. S. Evaluation of the edaphic-phytodepuration method to pre-treatment of water. Piracicaba, 1989. M.S. thesis, Esalq/ USP. Piracicaba, SP.

PART 6

POINT AND NON-POINT SOURCES

CHAPTER 35

Treating Livestock Wastewaters with Constructed Wetlands

D. A. Hammer, B. P. Pullin, T. A. McCaskey, J. Eason, and V. W. E. Payne

INTRODUCTION

Polluted water has a detrimental impact on the health and environmental quality of human societies throughout the world. Public concern during the last 20 years has resulted in state and federal legislation that regulates wastewater discharges and provides for financial assistance for waste treatment facilities. Consequently, substantial progress has been made in treating point source discharges, especially from larger cities with major industries. Small communities which lack technical and financial resources for conventional waste treatment may benefit from the emerging technology of constructed wetlands.[1] However, anticipated improvements in the nation's waters have not been fully realized, and recent evaluations and legislation reflect a growing concern for non-point sources (NPS) of pollution, especially agricultural sources.

NPS pollution was reported as the principal remaining cause of water quality problems in six of ten EPA regions in 1982.[2] Agricultural activities (i.e., animal waste management and tillage practices) were considered the most pervasive problems in all regions. In 1984, officials from 49 states reported that 29% of lakes and reservoirs assessed were moderately to severely impaired by NPS pollution, principally from agricultural activities.[3]

Increased focus on agricultural activities has resulted from a reduction or elimination of other pollution sources and also from increased waste loading of receiving streams by intensified animal husbandry practices. Free-ranging livestock at relatively low population densities has had little impact on aquatic resources because wastes are widely dispersed and natural soil systems generally recycle nutrients on-site. However, the present practice of confining livestock in ever smaller areas to improve production efficiency also concentrates animal waste loading with subsequent run-off to nearby streams. Concurrent removal of riparian vegetation to increase available acreage, as well as vegetation depletion by livestock grazing and loafing activities, has eliminated the buffer strip that formerly protected streams from direct pollution. Unfortunately, reversing either trend is unlikely, given the economic pressures affecting farmers today.

Although regulatory agencies are aware of the potential magnitude of NPS pollution generated by agricultural practices, most regulatory actions are in response to public complaints. However, increasing attention to NPS pollutants is likely to generate support for revising present categories to include concentrated livestock operations within the provisions of NPDES in the near future; and permitted discharge limits could be more stringent in the future due to very high concentrations of certain pollutants from confinement facilities and the small size of many receiving streams in most cases. Affordable, effective wastewater treatment systems are needed.

METHODS

Auburn University's Sand Mountain Agricultural Experiment Station in DeKalb County, Alabama, was selected as the site to develop a constructed wetlands wastewater treatment system for swine waste. Most waste from this 500-animal farrowing and finishing operation is routed to a two-cell-in-series lagoon system, with a separate single cell lagoon collecting waste from one barn. Prior to installation of the constructed wetlands, lagoon overflow crossed a small meadow and entered the upper reaches of Bray Creek.

0-87371-550-0/93/$0.00+$.50

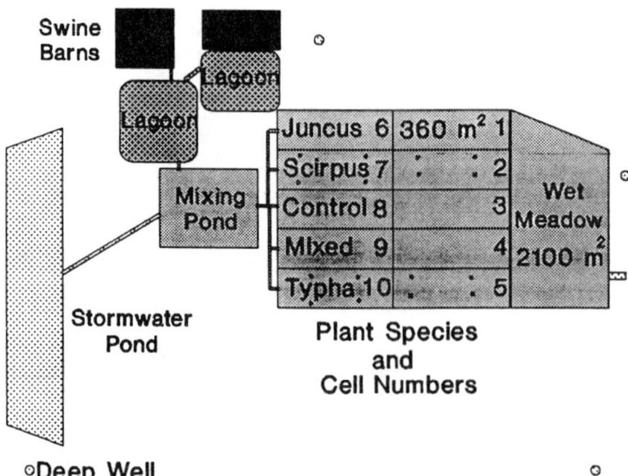

Figure 1. Schematic of the Sand Mountain Experiment Station's constructed wetlands for swine wastewater treatment. Lysimeter locations are depicted in wetland cells 2, 5, 7, and 10.

The wetlands, constructed in 1988, are a surface flow-type system located below the lagoons (Figure 1). Wastewater is discharged from the lagoon into a mixing pond which receives water from a farm pond located upstream from the wetlands. The mixture of swine effluent and clear water from the pond flows into five pairs of cells planted with marsh vegetation, then into a wet meadow for final polishing. Treatment area in the cells is 3600 m^2, with an additional 2100 m^2 in the wet meadow. Piping in the system provides for variable application rates and water level control within each cell. Substrate throughout the system is native soil. Initially, the cells were planted on 2-ft centers with cattail (*Typha latifolia*), soft-stem bulrush (*Scirpus validus*), giant cutgrass (*Zizaniopsis miliacea*), maidencane (*Panicum hemitomon*), common reed (*Phragmites australis*), and water chestnut (*Eleocharis dulcis*). However, monotypic stands were quickly invaded by other species.

Four ground water monitoring wells, two existing and two drilled, are located proximal to the wetlands. In addition, 16 lysimeters were installed at depths of 0.6 and 1.5 m in 4 of the wetland cells. Beginning in November 1990, water samples were collected monthly from the wells, lysimeters, mixing pond, wetland cells, and wet meadow, and analyzed for chemical oxygen demand (COD), biochemical oxygen demand (BOD), conductivity, pH, suspended solids, ammonia nitrogen, total Kjeldahl nitrogen (TKN), nitrate, phosphorus, fecal coliform, and fecal streptococci.

RESULTS

Influent BOD_5 levels ranged from 19.2 to 99.0 mg/L and averaged 63.7 mg/L during the 11-month sample period (Table 1). Average effluent BOD_5 values from the upper tier of wetland cells (numbers 6–10) varied from 4.9 to 17.6 mg/L, but were not significantly different. Likewise, effluent values from the lower wetland cells (numbers 1–5) were not different at $p = 0.05$. The BOD_5 removal rate by the entire wetlands system averaged 90.4% (Figure 2).

Suspended solids influent values varied by an order of magnitude (i.e., 21 to 210 mg/L). Nonetheless, average effluent levels from the initial receiving wetland cells (cells 6–10) did not vary. Average discharge levels of suspended solids from the lower tier of wetlands were also not significantly different (i.e., among cells 1–5). Average concentrations of suspended solids discharged from the lower cells were slightly higher than levels released from the upper tier of wetlands, but not significantly different ($p < 0.25$). However, removal of suspended solids was significantly enhanced during transit through the wet meadow ($p < 0.025$). The average removal rate of suspended solids by the treatment wetlands systems was 91.4% (Figure 2).

The average density of fecal coliforms entering the upper wetland cells was 1.75×10^5 CFU/100 ml. Discharge values from cells 6–10 ranged from 0 to 3.1×10^3 CFU/100 ml, with an average of 2.7×10^3 CFU/100 ml. The concentration of fecal coliforms did not change as a result of passing through the lower tier of wetland cells (mean = 2.7×10^3 CFU/100 ml). After transiting the wet meadow, the final effluent averaged 38% of wetland cells discharge levels. Overall, the system reduced fecal coliforms by 99.4%. Fecal strepto-

Table 1. Summary of Average Concentrations of Pollutants in Swine Treatment Wetlands Components During Nov. 1990 through Sept. 1991

Wetlands component	BOD_5 (mg/L)	TSS (mg/L)	Fecal col[a] (CFU/100 ml)	Fecal str[b] (CFU/100 ml)	Tot-P (mg/L)	TKN (mg/L)	NH_3-N (mg/L)
Lagoon eff.	110.8	346	817,500	118,750	48.8	116.0	84.0
Farm pond	32.5	51	1,022	679	2.9	3.8	1.1
Mixing pond	63.7	105	175,164	76,727	25.8	69.8	54.7
Upper wetlands							
Cell 6	17.6	31	3,005	3,780	11.9	18.9	13.2
Cell 7	4.9	21	2,530	5,008	7.3	7.8	4.9
Cell 8	15.6	21	3,798	4,634	10.9	25.2	20.3
Cell 9	16.4	35	2,438	3,403	11.9	18.5	12.8
Cell 10	15.3	19	1,894	5,873	10.6	18.9	13.8
Lower wetlands							
Cell 1	9.2	23	1,760	944	7.8	8.5	5.0
Cell 2	8.5	44	4,307	1,405	5.4	5.9	3.4
Cell 3	16.6	29	1,366	2,203	8.2	13.3	8.8
Cell 4	12.5	42	2,421	2,136	10.6	11.9	6.9
Cell 5	5.6	18	3,808	1,279	5.3	4.8	2.4
Final eff.	6.1	9	1,040	1,192	6.2	6.0	3.5
Loading rate (kg/ha/d)	3.5	2.8			1.4	3.0	
Removal (%)	90.4	91.4	99.4	98.4	75.9	91.4	93.6

[a] Fecal col = fecal coliform.
[b] Fecal str = fecal streptococci.

cocci densities entering the wetlands averaged 7.67×10^4 CFU/100 ml. Wastewater exiting upper wetland cells contained an average 4.5×10^3 CFU/100 ml, followed by a significant reduction to 1.6×10^3 CFU/100 ml from the lower wetland cells ($p < 0.025$). Compared to fecal coliform reduction, the average removal rate by the system for fecal streptococci was slightly lower, 98.4%.

Influent ammonia-nitrogen levels varied from 14.4 to 106.3 mg/L and averaged 54.7 mg/L. Similarly, influent TKN concentrations ranged from 25.3 to 121.3 mg/L, with a mean of 69.8 mg/L. If upper tier wetlands are compared, the bulrush cell (cell 7) discharged less ammonia nitrogen and TKN on average (7.8 and 4.9 mg/L, respectively); however, differences between cells were not significant. Average discharge concentrations of both pollutants were significantly higher in wetland cells 6–10 than in effluents from the second tier of wetland cells, indicating additional effective treatment by the lower tier wetlands. Similarly, total phosphorous levels were significantly reduced in the lower tier wetlands. Total phosphorous, TKN, and ammonia-nitrogen removal rates were 75.9, 91.4, and 93.6%, respectively (Figures 2 and 3).

DISCUSSION

Treatment performance by the five wetland cells within each tier was not affected by the type of vegetation present. Removal rates for BOD_5, total suspended solids, total Kjeldahl nitrogen, ammonia-nitrogen, and total phosphorous were consistent despite widely varying loading rates. Significant reductions of TKN, ammonia-nitrogen, total phosphorous, and fecal streptococci were accomplished by the replicate tier of wetlands; whereas, the removal of suspended solids, BOD_5, and fecal coliform bacteria was significantly enhanced by the wet meadow.

The constructed wetlands demonstrated effective and reliable treatment of swine lagoon effluent to acceptable wastewater treatment standards for BOD_5, TSS, nitrogen, and phosphorous during the first year of operation. Although fecal coliform and streptococci bacteria densities were reduced by 2 to 3 logs, discharge levels exceeded standards. Since bacteria are removed from wastewaters by filtration and absorption, performance by the wetlands should improve with succeeding growing seasons as plant densities increase. Monitoring will continue for another year.

Removal Performance
Swine Waste Treatment Wetlands

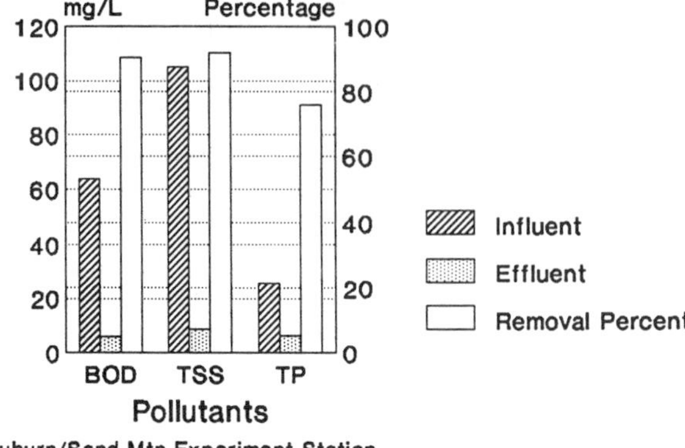

Figure 2. Mean concentrations of biochemical oxygen demand (BOD_5), suspended solids (TSS), and total phosphorus (TP) in swine lagoon wastewater entering and exiting treatment wetlands and resultant removal efficiencies.

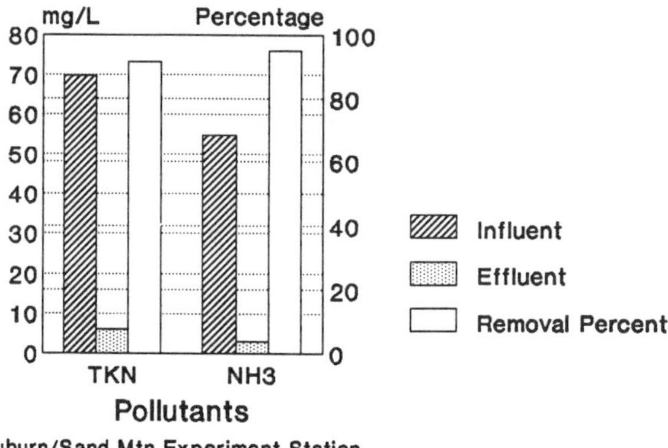

Figure 3. Average concentrations of total Kjeldahl nitrogen (TKN) and ammonia-nitrogen (NH_3) in swine lagoon wastewater entering and exiting treatment wetlands and resultant removal efficiencies.

Treatment effectiveness is dependent upon system design criteria. Swine waste at the Sand Mountain Agricultural Experiment Station receives primary treatment in three lagoons that are expected to reduce BOD_5 by 60%. The 500-animal swine operation is estimated to produce 90 kg BOD_5/day, reduced to 36 kg BOD_5/day in the final lagoon discharge. Minimum treatment area for 36 kg BOD_5/day at 150 m^2/kg BOD_5/day is 5400 m^2. The wetlands treatment system has 3600 m^2 within the wetland cells and 2100 m^2 in the wet meadow, for a total treatment area of 5700 m^2 or 158 m^2/kg BOD_5/day. Wetlands construction and planting costs are shown below:

Removal of old lagoon	$ 3,500
Grading and dikes	1,400
Pipe and materials	400
Labor for pipe installation	500
Plants and planting labor	7,000
Total	$12,800

Obviously, a farmer could construct a comparable wastewater treatment system (excluding the old lagoon removal) and reduce the direct planting costs by limited planting and appropriate water level manipulation to establish dense stands in two or three growing seasons. Total cost might approximate $3,000.

The advantages of constructed wetlands include (1) relatively low construction costs for grading, dike construction, piping and installation, and vegetation planting; and (2) low operating costs for monitoring water levels and plant viability, dike maintenance (i.e., mowing and repair of rodent damage as necessary) and, if required, water sample collections. Hopefully, simple efficient wetlands treatment systems will receive better acceptance and reduce the pollutant loading of our waterways from agricultural non-point sources.

REFERENCES

1. Gearheart, R. A., F. Klopp, and G. Allen. Constructed free surface wetlands to treat and receive wastewater: pilot project to full scale, in *Constructed Wetlands for Wastewater Treatment: Municipal, Industrial and Agricultural*. D. Hammer, Ed. Lewis Publishers, Chelsea, MI, 1989, 121–137.
2. USEPA. National Water Quality Inventory: 1984 Report to Congress. EPA 440/4-85-029. Washington, D.C., 1984.
3. ASIWPCA. America's Clean Water, The States Evaluation of Progress 1978–1982. Association of State and Interstate Water Pollution Control Administrators, Washington, D.C., 1984.

CHAPTER 36

Evaluation of Aquatic Plants for Constructed Wetlands

D. Surrency

INTRODUCTION

As the role of the Soil Conservation Service in water quality activities becomes more certain, vegetation (plant material) is likely to emerge as a major tool to use in soil and water conservation to address water quality problems. The plant materials program involvement will include the evaluation of promising plant materials for improving water quality and ultimately to provide the opportunity to expand the number of plant species that are suitable for agricultural and industrial uses for pollution control. Research information on the number and kind of plant species that are recommended for improving water quality is very limited. The utilization of plants for natural and constructed wetlands, filter strips, and land application of agricultural wastewater is a new concept for wastewater treatment. A fairly new technology of treating wastewater with constructed wetlands is being explored. Research has demonstrated that wetland ecosystems have a natural assimilative capacity to remove organic pollutants. If these systems can be constructed to provide a secondary treatment or final polishing effect, then overall water quality to receiving water resources may be improved and at lower construction, operation, and maintenance costs.

Limited information is available on sources of plant species to use in systems addressing water quality concerns. Plant species as sources for constructed wetlands for both agricultural and non-agricultural use are also limited. In most cases, the plant materials that are being used for constructed wetlands are collected from local natural stands, or purchased from nurseries and shipped to areas where the adaptation and performance of the plants is not known. Information on plant sources, planting and establishment methods, spacing, substrate, and water level requirements is not available to users.

Raw livestock wastewater is likely to have considerably higher concentrations of biochemical oxygen demand (BOD_5), 100 to 200 mg/L more of ammonia, and 75 to 150 mg/L more total phosphorus than municipal wastewater.[1] Ammonia is a factor that must be considered in plants that are used in constructed wetland cells, especially cell one in a multicell constructed wetlands.[2] The plants were evaluated in swine and dairy operations that had ammonia levels that have reached or exceeded 100 mg/L.

The low cost, yet effective, use of constructed wetlands in assimilating agricultural wastewater has shown much promise. There is concern, however, that the high concentration levels of nutrients, especially ammonia, in agricultural waste systems may exceed the tolerance level of most aquatic plants.[3]

Microbial populations in the water column, attached to vegetative and other substrates and within the soils, modify hydrocarbons, metals, pathogens, and nutrient loads causing precipitation of some pollutants and recycling, with subsequent settling out of others.[4] Not only does a dense stand of wetland vegetation provide a large surface area within the water column for microbial attachment, but the radial oxygen loss from wetlands plant root structures creates a substantial aerobic environment for microbial populations within the substrate.[1]

The vegetative component is a major factor in the treatment processes that occur in constructed wetlands. There is a complex symbiotic relationship among the plants (root system), microorganisms, substrate, soil, and the nutrients in wastewater. It is important to understand how a particular species reacts to wetland conditions before that species can be recommended for use in constructed wetlands. Wetland plants use a number of adaptive strategies to withstand varying edaphic and hydrologic conditions.[5] For example, wetland plants have air spaces (aerenchyma) in roots and stems that allow the diffusion of oxygen from the aerial portions of the plant into the roots.

0-87371-550-0/93/$0.00+$.50

An assembly of a large number of aquatic plant species was collected for testing and evaluation in typical swine and dairy operations, and a small community municipal constructed wetland in 1989 and 1990. Plant materials were selected based on criteria that were identified as very necessary characteristics for the operation of constructed wetland systems and for animal wastewater treatment.

There also exists the need for targeting best management practices (BMP) and monitoring the success of their implementation. The cooperative interagency demonstration projects in Putnam County, Georgia and Sand Mountain, Alabama are an evaluation of a selected BMP (constructed wetlands) utilized to capture discharge from overloaded lagoons. More than 20 different local, state, and federal agencies are participating in the project and each is responsible for implementing the project objectives that are considered common to the agency.

OBJECTIVES

The main objectives for conducting an evaluation of aquatic plants for constructed wetlands system (CWS) are:

1. Assemble and conduct initial evaluations (screening) of a large number of aquatic plant species that have characteristics that are essential for growth and survival in CWS
2. Determine the optimum plant spacing, establishment methods and techniques, and the initial loading and management requirements
3. Determine the plants that are adapted to the conditions and constructed wetland environment for agricultural wastewater treatment
4. Determine the symbiotic relationship that exists in the treatment of the wastewater and the role that the plants, substrate, microorganisms, root systems, and other parameters have on the treatment processes and retention time
5. Determine the fate of nutrients and other contaminants in the constructed wetlands
6. Determine the O and M requirements to maintain the vegetation in the CWS over time
7. Determine the effect of high nutrient concentration on survival and plant growth.

STUDY AREAS

The study areas for this evaluation includes three sites that are located in two states, Georgia and Alabama.

Sand Mountain, Alabama

The Sand Mountain area (Jackson, Marshall, and Dekalb Counties) of northeastern Alabama has been identified by the Alabama Agricultural NPS Task Force as the top priority watershed in Alabama. The 626-mile area has approximately 3000 farms, 12,000 cattle, 12,000 swine, 6.8 million poultry, and 65,000 acres of cropland eroding at greater than twice tolerance. Water supplies, local tributaries, and Guntersville Lake are severely impacted by sediment, nutrients, agrichemicals, and bacteria.

The Auburn University Agricultural Experiment Station on Sand Mountain operates a swine farrowing and feeding operation of 64 sows and 175 finishing pigs. Swine waste at the Sand Mountain Agricultural Experiment Station receives primary treatment in two lagoons that are expected to reduce BOD_5 by 60%.

The design and construction for the agricultural evaluation/demonstration constructed wetland system was performed by TVA in 1988–1989. The operation, monitoring, and evaluation of plant materials is conducted cooperatively with the Soil Conservation Service, Auburn University, Alabama Department of Environmental Management, and TVA. The CWS consists of ten cells, five primary and five secondary, that were ideal for conducting initial evaluations of many different plant species. The cells are 25 ft wide and 165 ft long. The large number of individual cells provides for test/evaluation plots with different emergent plant species.

Putnam County, Georgia

In 1987, the Piedmont Soil and Water Conservation District identified a potential problem with animal waste disposal around Lake Sinclair and Lake Oconee. Dairy farming is widespread in the Georgia Piedmont. Water use for a typical dairy (150 to 200 head) amounts to approximately 10,000 gallons per day for washdown of milking facilities. Wastewater from these operations is captured in lagoons which are intended to be a no discharge system. Although periodic pump-out to prevent solids overload is an integral maintenance feature necessary for lagoons, prohibitive costs (~$6000) for this service have precluded its occurrence. In addition to costs, inadequate pumping procedures have resulted in many lagoons becoming overloaded with solids, and thus, continuously discharging to adjacent water bodies.

The site of this demonstration/evaluation project lies in Putnam County near Eatonton, Georgia. Constructed wetlands have been installed at two dairy operations. One dairy, the McMichael Dairy, also has a newly constructed lagoon that was completed at the time the wetlands were installed.

The Key Farm, which is in close proximity to Lake Sinclair, has had a lagoon in place for several years. Both dairies have tributaries that drain into Lake Sinclair. Since the McMichael Dairy wetlands and lagoon are still in developmental stages, monitoring efforts are primarily directed toward the Key Dairy. As a result, this report will reflect findings of elevation efforts related to the Key Dairy.

Ochlocknee, Georgia

The small south Georgia town of Ochlocknee (population 588) located about 8 miles north of Thomasville, has had domestic waste water treatment problems. For many years the town relied on septic tank systems for sewage treatment for the houses, stores, mobile homes, and local businesses. However, septic tanks didn't work properly because of the poorly drained soils that are not suitable for septic tanks. Constructed wetlands have low energy requirements, compared to the other processes that require much larger initial investments to design and construct, and have higher operational costs (personnel and energy). They will also need either mechanical aeration or liquid distribution systems.

A constructed wetland system for Ochlocknee consists of two 2-cell systems that are 1 acre each, for a total treatment area of 4 acres. A 1-acre oxidation pond or lagoon provides initial settling of solids and some retention prior to being discharged into the constructed wetlands. Construction was completed in 1990 and the aquatic plants were established in June and July of 1990.

MATERIALS AND METHODS

Sand Mountain Agricultural Experiment Station

One complete cell system, primary (REP 1) and secondary (REP 2), was used to conduct the initial evaluation of 16 species or varieties of aquatic plants (see Table 4). Halifax maidencane *(Panicum hemitomon)*, cattail *(Typha latifolia)*, prairie cordgrass *(Spartina pectinata)*, smooth cordgrass *(Spartina alterniflora)*, common reed *(Phragmites australis)*, giant cutgrass *(Zizaniopsis miliacea)*, southern bulrush *(Scirpus californicus)*, giant bulrush *(Scirpus validus)*, four accessions of water chestnut *(Eleocharis dulcis)*, canna lily *(Canna flaccida)*, pickerelweed *(Pontederia cordata)*, elephant ear, or wild toro *(Colocasis esculenta)*, blueflag iris *(Iris virginica)*, giant reed *(Arundo donax)*, and arrowhead *(Sagittaria lancifolia)* were planted in 1989. Piping in this system provides for variable application rates to each plot and water level controls within each plot. Substrate throughout the system is native soil; 50 plants of each species were planted with hand dibbles on 3-ft centers in May 1989.

Monitoring

Ground water monitoring wells were installed up-gradient and down-gradient from the constructed wetlands. Prior to loading, samples were collected for determining background water quality. Auburn University and TVA are providing technical guidance to complete this project objective. Soil water samples (suction-cup lysimeters) were installed in two of the cells.

Putnam County, Georgia

Putnam County, Georgia contains many of the state's dairy herds. Owing to the great number of dairy operations and their subsequent pollution potential, Putnam County was chosen as a test site for evaluating aquatic plants for constructed wetlands.

A three-cell constructed wetland (surface flow) at the Howard McMichael Dairy was planted as described below in 1990.

Cell 1 — bulrush *(Scirpus californicus)*
Cell 2 — cattail *(Typha latifolia)*
Cell 3 — Halifax maidencane *(Panicum hemitomon)*
Cell Size: 42 × 225 ft
Spacing: 3-ft centers
Substrate: natural soil
Planting methods: tree dibble and a tractor drawn tree planter

Cell number two (2) at the McMichael Dairy contained a collection of cattails, 25 plants each from 12 states. The collection was made to determine if differences existed in cattails collected from a large geographic area. Cell size was 42 × 300 ft.

A three-cell constructed wetland was built in 1990 at the Richard Key Dairy. The planting plan for the cells are described below.

Cell 1 — cattail *(Typha latifolia)*
Cell 2 — bulrush *(Scirpus californicus)*
Cell 3 — Halifax maidencane *(Panicum hemitomon)*
Cell Size: 42 × 300 ft
Spacing: 3-ft centers

Cell 3 contained 100 plants each of canna lily, pickerelweed, elephant ear, arrowhead, prairie cordgrass, and blueflag iris planted along the edge in 3 in. of water for aesthetics and beautification. Animal waste lagoons were available for primary treatment and settling of solids at all locations.

Quarterly water chemistry sampling of the CWS and lagoon was begun in July 1990. Although not a task for the original plan of study, monitoring of non-point source run-off from the ungrassed feed lots and holding areas has been included as a project task by EPA. Water chemistry parameters for the constructed wetlands are ammonia-N, nitrate-nitrite-N, total Kjeldahl N, total phosphorus, total organic carbon, and total suspended solids. Water level recorders were employed at weirs to the wetland system influent and effluent.

Ochlocknee, Georgia

The Ochlocknee, Georgia CWS for the treatment of municipal wastewater is the first system to be demonstrated in south Georgia. Ochlocknee is a small town with a population of 588. The wastewater disposal system is a paired two-cell constructed wetland below a settling lagoon or oxidation pond. The cells are 1 acre each, for a total constructed wetland area of 4 acres.

The planting plan provided for giant cutgrass *(Zizaniopsis miliacea)* to be planted in the first third of the cell, cattail *(Typha latifolia)* the second third, and maidencane *(Panicum hemitomon)* in the last third of the cells. Wetland plants were obtained from USDA-SCS plant materials centers, USDA-ARS plant introduction stations, commercial nurseries, and collected from local stands. The wetland plants were planted on 3-ft centers with a tree dibble and a tractor-drawn tree planter in June 1990. Water level was designed to be maintained at 6 in.

Monthly water chemistry sampling of the constructed wetland was begun in 1990 by the Georgia Department of Natural Resources, Environmental Protection Division. Water chemistry parameters are ammonia, dissolved oxygen, BOD_5, suspended solids, pH, and flow. The data for the Ochlocknee constructed wetlands were obtained from GA.EPD.

Plant Materials

The aquatic plants were planted in the constructed wetland in May and June using vegetative material (whole plants), dormant rhizomes (maidencane), and corms (water chestnut). Whole plants such as cattails, giant cutgrass, and bulrush that were collected from local stands were topped to 12 in. and the roots pruned prior to planting.

Planting Methods

The constructed wetland sites in Putnam County, Georgia consisted of large cells that were located in the piedmont section on upland soils. The sites were dry and were easily accessible with a tractor-drawn tree planter. The planter worked extremely well and successfully planted cattail, bulrush, giant cutgrass, pickerelweed, canna lily, and elephant ear at the desired depth and spacing. Maidencane rhizomes were also planted with the planter. The wet sites located in Ochlocknee, Georgia and Sand Mountain, Alabama were successfully planted with a hand dibble. Plant materials were planted deep enough to prevent propagules from floating out of the planting hole.

Water Level Management

The plants were allowed to become well established before effluent from the animal waste lagoon was discharged into the constructed wetlands. Enough water from a pond or well should be used to maintain good saturation in the substrate, not flood it. As the plants become well established, then the water level may be raised to the designed elevation. When the plants were established (2 to 3 months), the water levels were raised to 6 in. The lagoon effluent was then applied to provide mixing of the water for the plants to become acclimated to the increased nutrient load. At no time should water levels overtop the plants. In contrast, water levels for emergent aquatic plants should never be lowered to the extent that the plants' roots become exposed. Dry cell conditions as a result of insufficient water will result in poor plant survival, growth, and development.

RESULTS AND DISCUSSION

Plant Evaluations

Table 1 shows that the percent stand was the best for giant cutgrass, maidencane, giant cutgrass, pickerelweed, and cattail at the Sand Mountain Experiment Station. Giant reed *(Arundo donax)* could not tolerate the 6 in. water level that was maintained in the constructed wetland cells.

Giant cutgrass, maidencane, and cattail stands were good at two other sites in Alabama and Georgia. Giant cutgrass is a very aggressive plant in the cells. It usually crowds out other species with prolific growth, spread, and a vigorous rhizomatous root system. This plant has never been evaluated for constructed wetlands but it appears to be extremely adapted to systems in the southeast. Water level adjustments up to 12 in. did not affect its vigor, growth, or survival. Based on the spread by rhizomes the first year, this plant can be planted on 6- or 8-ft centers and establish quickly. It also roots very easily and readily from the nodes when in contact with water and the substrate. Giant cutgrass survived winter low temperatures to –2°F recorded at the Sand Mountain Experiment Station in Alabama.

The plant materials dormancy (Table 2), shows that giant bulrush *(Scirpus californicus* and *S. validus)*, were not affected by low temperatures during the winter, as with other aquatic plants. In December and January 1990–1991, new growth or tillers were observed below the water level. The stems were green and vigorous. Based on the role that plants play in the constructed wetland environment, the winter dormancy of giant bulrush is a very important characteristic. Most of the other aquatic plant materials were burned back by the first or second heavy frost. The first killing frost usually occurs around October 15 to November 1 in the project areas. Pickerelweed, elephant ear, canna lily, and arrowhead are the first to be burned back from frost damage at these locations.

Table 3, which reveals the recommended plant spacing for constructed wetlands, is based on 3 years of data obtained by evaluating the plant growth and rate of spread in actual constructed wetland conditions in Alabama and Georgia. The initial plant evaluations in 1989–1991, Table 4, shows that the best aquatic plant materials at the Sand Mountain, Alabama site are giant cutgrass, Halifax maidencane, giant bulrush, and arrowhead.

Cattails *(Typha latifolia)* were not as vigorous the third growing season as were observed the first and second growing seasons. Cattails have been recommended for constructed wetlands in municipal systems, but it was not the best plant based on performance for agricultural constructed wetlands. Cattails were defoliated by an insect similar to army worms in 1991 at all three sites.

Water chestnut *(Eleocharis dulcis)* stands were significantly damaged by muskrats at the Sand Mountain, Alabama site. Water chestnut receives tremendous competition by invaders over time. It does not provide quick spread and its short growth is a major limitation.

Table 1. Percent Stand in CWS Sand Mountain, Alabama, 1989–1992

Species	1989	1990	1991	1992
Halifax maidencane	70	80	100	100
Giant cutgrass	100	100	100	100
Giant reed	50	0	0	0
Prairie cordgrass	80	70	65	65
Common reed	90	80	80	75
Water chestnut	90	70	50	30
Giant bulrush	95	100	100	100
Canna lily	90	80	70	70
Pickerelweed	95	100	100	100
Arrowhead	90	85	85	85
Cattail	95	100	95	70
Elephant ear	90	60	20	10
Blue flag iris	90	70	70	70

Table 2. Plant Materials Dormancy

Plant materials	October A B C	November A B C	December A B C	January A B C	February A B C
Cattail	1 1 1	2 2 1	3 3 2	4 4 4	4 4 4
Canna lily	2 3 1	4 4 3	4 4 4	4 4 4	4 4 4
Prairie cordgrass	2 2 2	3 3 3	4 4 4		
Common reed	2 2 2	3 3 3	4 4 4		
Water chestnut	3 2 1	4 4 4			
Giant bulrush (*S. validus*)	1 1 1	1 1 1	2 1 1	3 2 1	4 2 1
Giant bulrush (*S. californicus*)	1 1 1	1 1 1	2 1 1	3 2 1	4 2 1
Common bulrush	2 1 1	2 2 2	4 3 3	4 4 4	
Elephant ear	3 3 2	4 4 3	4 4 4	4 4 4	5 5 4
Blue flag iris	3 3 2	4 4 3	4 4 4	4 4 4	
Giant cutgrass	3 2 1	3 3 2	4 3 2	4 4 4	
Arrowhead	4 3	4 4			
Halifax maidencane	2 2 1	3 3 1	4 4 2	4 4 4	
Pickerelweed	4 3	4 4			
Giant reed	3	4			

Note: A, Sand Mountain Alabama (–2°F); B, Eatonton, Georgia; C, Ochlocknee. 1, Slight; 2, moderate; 3, frost damage (severe); 4, dormant; 5, winter kill.

Elephant ear *(Colocasis esculenta)* will not recover the second growing season in constructed wetlands when water levels are maintained at 6 in. during the dormant season. However, it will recover around the shallow edges where it is best suited to provide plant diversity.

Blue flag iris *(Iris virginica)* was planted to provide diversity and aesthetics to the constructed wetlands in urban areas. The plant grows well in 3 to 6 in. of water and can tolerate high nutrient levels that are usually common to the second cell in animal wastewater systems. Canna lily, arrowhead, and pickerelweed also produced beautiful yellow, red, orange, and white flowers.

Nutrient Tolerance

The Putnam County, Georgia constructed wetland sites provided the opportunity to determine the aquatic plants' (giant bulrush and cattail) response to high nutrient concentrations of wastewater that were recorded in the top cells. The discharge of effluent from the lagoon to the top cells had nutrient concentrations that averaged 160 to 170 mg/L ammonia. Giant bulrush plants did tolerate the high ammonia concentrations in the

Table 3. Recommended Spacing for Constructed Wetlands Systems

Plant materials	Spacing (ft)
Maidencane	3 × 3
Giant cutgrass	6 × 6
Prairie cordgrass	3 × 3
Common reed	3 × 3
Water chestnut	3 × 3
Giant bulrush	4 × 4
Cattail	3 × 3
Elephant ear	3 × 3
Blueflag iris	3 × 3
Canna lily	3 × 3
Arrowhead	3 × 3
Smooth cordgrass	3 × 3

Table 4. Plant Materials Evaluations, 1990–1991 Sand Mountain, Alabama

Plant material	Vigor	Tillers	Spread	Height	Disease resist.	Insect resist.	Basal areas (in.)
Giant cutgrass	1	60	1	9 ft	1	1	40
Halifax maidencane	1	50	1	22 in.	1	1	
Prairie cordgrass	3	10	3	50 in.	3	3	12
Common reed	3	10	5	80 in.	3	3	13
Water chestnut	3	20	3	12 in.	3	7	12
Giant bulrush (*Scirpus californicus*)	1	80	1	8 ft.	1	1	40
Giant bulrush (*Scirpus validus*)	1	40	1	5 ft.	1	1	40
Cattail (*Typha latifolia*)	3/5	30	3	45 in.	5	7	40
Elephant ear	1	5	5	5 ft	3	3	18
Blueflag iris	1	3	5	24 in.	3	3	12
Canna lily	1	10	3	38 in.	3	3	18
Arrowhead	1	25	1	72 in.	1	1	40
Smooth cordgrass	1	10	5	25 in.	3	3	12

Note: Evaluation scale for vigor, spread, disease, and insects: 1 = excellent; 3 = good; 5 = average; 7 = poor.

top cell. Cattails survived the high concentrations in the top cell, but the stands were stressed at the high ammonia levels.

It is important to know the nutrient concentrations that exist in the lagoon to be able to select the most tolerant plant species. This is an important planning consideration for developing plans for constructed wetlands that will be used to treat dairy and swine wastewater. It is important and essential to collect samples of the lagoon effluent and determine BOD, pH, ammonia, and suspended solids in older lagoons as part of the planning process. A multicell system of three or four cells in series provided the best reduction of effluent from the animal lagoon. Therefore, constructed wetlands must be a planned component of the total animal waste treatment system.

Wildlife Usage

Biologists with the Georgia Game and Fish Division and Georgia College monitored the constructed wetlands in Putnam County, Georgia to determine wildlife usage. Their reports indicate that more than 60 bird species and at least 10 species of both mammals and herpetofauna have been documented. Several broods of wood ducks were observed. The production of macroinvertebrates at the site appears to be important to avian species.[6]

Water Chemistry

Analysis of water chemistry data prior to installation of the constructed wetlands at the Key Dairy revealed elevated nutrient concentrations during active lagoon discharge from overloaded dairy lagoons. At that time, total nitrogen and phosphorus concentrations exceeded 160 and 35 mg/L, respectively. The constructed wetlands have dealt effectively with reducing frequency of discharge and solids and nutrient loadings to the receiving water. Comparison of water chemistry data from the lagoon and wetlands indicate a greater than 90 and 80% removal of total nitrogen and phosphorus, respectively.[2]

Sand Mountain, Alabama Experiment Station

The Sand Mountain wetlands have been efficient in the reduction of nutrients in swine lagoon wastewater. Ammonia and nitrogen reduction was more than 95%. BOD, total phosphorus, and total suspended solids reduction were 91, 78, and 92%, respectively.

Ochlocknee, Georgia

Analysis of water chemistry at the Ochlocknee constructed wetlands revealed a significant reduction of ammonia, BOD, total Kjeldahl nitrogen, and total suspended solids. Dissolved oxygen and pH levels from the wetlands were better than the receiving stream. Therefore, discharge from the wetlands improved the quality of water in the receiving stream.

CONCLUSIONS

Giant bulrush *(Scirpus californicus* and *S. validus)*, giant cutgrass *(Zizaniopsis miliacea)*, Halifax maidencane *(Panicum hemitomon)*, pickerelweed *(Pontederia cordata)*, arrowhead *(Sagittaria lancifolia)*, and cattail *(Typha latifolia)* are the best aquatic plants to use in constructed wetlands for treating waste water from dairy and swine operations and for municipal constructed wetland systems. Based on the tolerance of giant bulrush for ammonia, it is the best recommendation for the top cell in a 2- or 3-cell constructed wetland. Three-cell constructed wetlands in Putnam County, Georgia provided significant nutrient reductions.

Cattail *(Typha latifolia)* performed well as a standard of comparison for the other plant accessions, but it was susceptible to insect damage, and the vigorous stand appeared to be stressed when subjected to high ammonia levels in the top cell.

In general, a good plant spacing of 3 to 4 ft on center will result in a good stand and coverage the first growing season in constructed wetlands. Giant bulrush and giant cutgrass can be planted on 6- to 8-ft centers.

Although this paper reflects only interim results, it appears that constructed wetlands provide a positive approach to controlling overflows from dairy lagoons. Constructed wetlands for animal wastewater contributed to a greater than 90% reduction of total nitrogen and greater than 80% reduction of total phosphorus. In addition, a significant reduction in both nutrient and solids loading to area streams is occurring.

ACKNOWLEDGMENTS

I want to thank the U.S. EPA Ecological Support Branch in Athens, Georgia (Bruce Pruitt, Hoke Howard, and Don Shultz), for collecting cattails and providing technical assistance for the analysis of wastewater samples; Mack Hayes, RC&D Coordinator for his strong interest and support of constructed wetlands. The interest and support by Doris Griffin, Mayor, City of Ochlocknee is appreciated. Also to John Eason, Superintendent, Sand Mountain Substation, Auburn University for his technical support.

REFERENCES

1. Hammer, D. A. Constructed wetlands for treatment of agricultural waste and urban stormwater. Wetlands ecology and conservation: emphasis in Pennsylvania. S. K. Majumdar, R. P. Brooks, F. J. Brenner, and R. W. Tiner Jr., (Eds.), The Pennsylvania Academy of Science, 1989, 333–348.
2. Howard, H. S. Constructed wetlands for assimilation of dairy effluents: monitoring studies, *Proc. 1991 Georgia Water Resources Conf.* The University of Georgia, Athens, GA, 1991.
3. Wengrznek, R. J. and C. R. Terrell. Using constructed wetlands to control agricultural nonpoint source pollution, *Proc. Int. Conf. Use of Constructed Wetlands in Water Pollution Control.* Churchill College, Cambridge, U.K., 1990.
4. Wolverton, B. C. Aquatic plants for wastewater treatment: an overview, in *Aquatic Plants for Wastewater Treatment and Resource Recovery.* K. R. Reddy and W. H. Smith, Eds. Magnolia Publishing, Orlando, FL, 1987.
5. Wolverton, B. C. and R. C. McDonald. Vascular plants for water pollution control and renewable sources of energy, *Proc. Bio-Energy '80*, Atlanta, GA, 1980.
6. Harley, D. Wildlife Usage of Constructed Wetlands. Evaluations: Georgia Department of Natural Resources, Forsyth, GA, 1991.

CHAPTER 37

Controlling Agricultural Runoff by Use of Constructed Wetlands

M. J. Higgins, C. A. Rock, R. Bouchard, and B. Wengrezynek

INTRODUCTION

Protection of water resources has, until recently, focused on point source pollution as regulated by the Clean Water Act (CWA) of 1972. Point sources, such as municipal and industrial wastewaters, are easily identifiable and treatable based on their focused source. Non-point source pollution, mainly associated with storm water run-off from urban, agricultural, or mining land uses can be quite diffuse and difficult to treat. Non-point pollution is of sufficient quantity and diverse quality to have serious detrimental effects on regional water resources. The U.S. EPA's Report to Congress[1] stated that non-point source pollution contributed 76% of the pollution to lakes, and that the U.S. Agriculture was responsible for over half of the non-point source pollution.

In a survey of the Long Lake watershed in the St. John Valley of northern Maine, Bouchard et al.[2] found agricultural run-off to be the largest pollutant source to the lake. The survey documented an increased frequency of algal blooms in Long Lake, with agriculture-based activities contributing 64% of the total phosphorus load. As a result, the Soil Conservation Service, St. John Valley Soil and Water Conservation District (SWCD), and the Maine Department of Environmental Protection developed a watershed management plan for Long Lake that targeted the agricultural sources of phosphorus. The plan recommended that conservation techniques be applied by farmers to reduce run-off and pollutants. The St. John Valley SWCD estimated that conservation practices could reduce phosphorus loads to the lake by 10% and sediment loads by 25 to 30%.[3] In addition, the Soil Conservation Service designed treatment systems called Nutrient/ Sediment Control Systems (NSCS) to improve run-off quality further. Four systems, consisting of sediment basins, grass filters, and constructed wetland-pond components, have been constructed in the Long Lake watershed.

Most constructed wetlands are designed to treat domestic wastewater and focus on biochemical oxygen demand (BOD) and nutrient removal, whereas systems for mitigating agricultural (cropland) non-point runoff are directed at sediment and nutrients. While many of the basic principles are the same, functional considerations will vary the actual design. The application of constructed wetlands to treat agricultural (cropland) run-off differs from those used to treat wastewater in several significant ways. The hydraulic loading is intermittent and a significant organic load is usually absent. Also, agricultural run-off carries a heavy sediment load and constructed wetlands alone would have a limited capacity to treat agricultural run-off effectively. Agricultural run-off can carry high nutrient and pesticide levels which can create a shock load to any treatment system. Design of a treatment system must consider these problems. When combined with other treatment components, a constructed wetland could be effective in treating the run-off.

SYSTEM DESIGN

The constructed wetland-pond system has, in series, a sedimentation basin, grass filter strip, constructed wetland, and retention pond (Figure 1) which discharges to a final vegetated polishing filter. Run-off collected and diverted from cropland first enters the sediment basin where the water is slowed and detained to allow larger particles to settle and to reduce the hydraulic impact on downstream components.

0-87371-550-0/93/$0.00+$.50

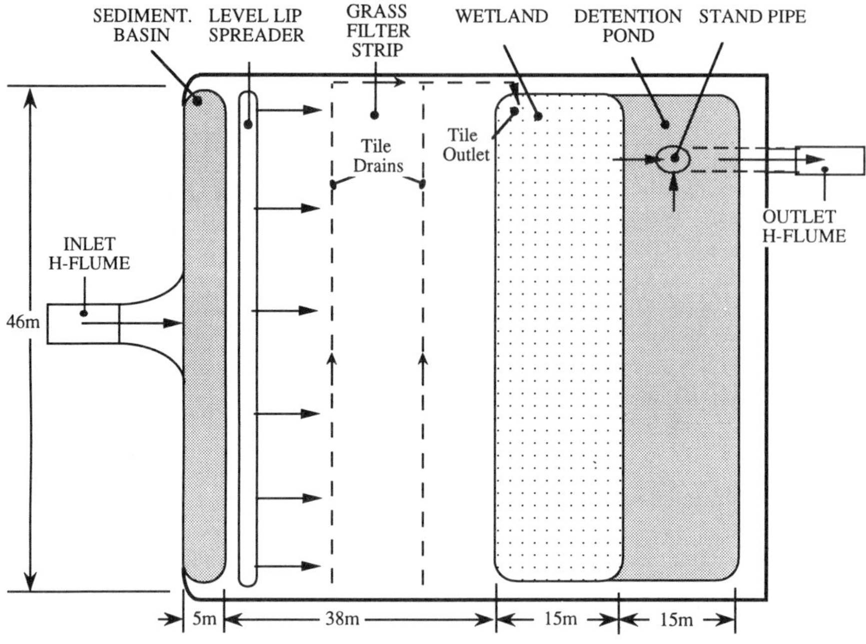

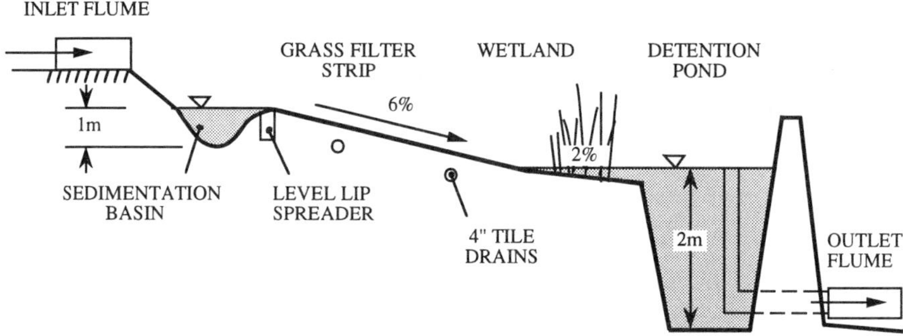

Figure 1. Plan-profile view of constructed wetland system to treat agricultural run-off.

Once the basin fills, the overflow enters a level lip spreader which is a 0.46-m deep, 0.6-m wide trench filled with crushed rock, The purpose of the level lip spreader is to distribute evenly the sediment basin discharge across the width of the filter strip, thereby reducing channelization and erosion of the grass filter strip. Drainage tiles were placed under the filter strip to promote infiltration of the run-off. Others have reported infiltration improved the efficiency of grass filters.[4,5] The water collected by the drainage tiles is discharged to the wetland in the southeast corner.

After flowing over the grass filter strip, the remaining run-off enters the wetland. The wetland vegetation further impedes flow, settling more particles; nutrients are adsorbed at the soilwater interface and are also taken up by plants and microbes. The floor of the wetland has a 2% slope. Finally, the water enters the retention pond. The pond, permanently flooded, gives greater retention times to settle smaller particles. The pond and wetland act as one integrated component because no barrier exists between the two components. The bottom slope increases to 2:1 at the wetland-pond edge, which defines the end of the wetland and beginning of the pond. The wetland-pond is stocked with small algaevorous fish and freshwater mussels. The diverse biological community creates a well-developed food chain which helps to remove nutrients from the water column. The water level in the wetland-pond is controlled by a standpipe which discharges into a 46-m long vegetated swale that drains to the lake.

The sequence of each component was carefully considered and the selected arrangement was developed specifically to address the unique problems associated with agricultural run-off. Recognizing that storm water run-off can be highly variable in magnitude and frequency, the system was sized to treat and contain many storm events in the first two elements. Since the run-off contains heavy sediment loads, the sediment basin was placed first to protect the downstream components from sediment overload. The grass filter strip was placed ahead of the wetland-pond to serve as an early indicator of the adverse impact of pesticides. In observing farming practices, it was found that the application of pesticides, herbicides, fungicides, and top killer often resulted in the killing of vegetation in run-off channels. A similar occurrence might jeopardize the integrity of the entire wetland/aquatic system. However, by locating the grass filter ahead of the wetland, the grass may be sacrificed to protect the downstream vegetation, depending upon how far through the system the toxic material moves. By placing the grass filter strip ahead of the wetland, an extra margin of safety is obtained. Further, it was reasoned that replacing the grass would be easier than replacing the wetland vegetation and other biota in the aquatic system.

The wetland-pond can also play a role in decreasing the impact of desorbed phosphorus. The soil pH in the potato fields is typically between 5 and 6.[6] At these pH levels, the soils have a greater ability to adsorb phosphorus.[7] When these soils are eroded to surface waters, which typically have a pH greater than 7, the soils' ability to hold phosphorus decreases, and phosphorus is desorbed. Originally, the SCS (Soil Conservation Service) planned to lime the pond and wetland to maintain the pH greater than 7, if necessary. However, the pH has remained greater than 7 without addition of lime. This allows desorption of phosphorus in the wetland-pond, where it can be assimilated by the diverse and abundant biological community.

The prototype system, denoted by the landowner's name (Tardiff), was built in 1988 and treats a 7-ha watershed cultivated mostly in potatoes. The land area covered by the system is 0.61 ha. The watershed has a 6% average slope, with moderately well-draining plaisted Howland soils, and is considered in good hydrologic condition according to an SCS field study.[8] Crop rotations usually contain 60% potatoes, with the remainder planted in small grains such as oats or millet. Diversions placed in the watershed direct flow to a culvert which flows underneath a road and into the system.

The sizing of the system was based in part on run-off calculations for the 7-ha watershed using the hydrologic method developed by the Soil Conservation Service[9] and on the capacity of a road culvert that carries water to the system. This culvert has a capacity of 0.9 m^3/s. If more flow results, it will continue past the culvert and bypass the system. The 0.9-m^3/s flow is approximately equivalent to the 10-year storm event for the watershed. The sizing of components is also partly based on similar constructed wetland-ponds used for municipal wastewater treatment. The USDA/SCS[10] currently has interim minimum sizing criteria based on watershed area. The cost of construction was about $14,000 for the entire system.

Vegetation in the grass filter strip was well established by spring 1989, having been seeded the previous fall, while the wetland was not planted until June 1989. Monitoring of the Tardiff system began June 17, 1989.[8] H-flumes were installed at the inlet of the sediment basin and the outlet of the retention pond (see Figure 1). The flumes were equipped with flowmeters and automated samplers which collect flow-proportioned composite samples during run-off events. The composite samples were analyzed for total phosphorus (TP), total suspended solids (TSS), and volatile suspended solids (VSS). In addition to the composite samples, grab samples were taken throughout several storm events to determine changes in pollutant concentration as a function of run-off. These grab samples were analyzed for TP, TSS, VSS, and dissolved phosphorus (DP).

RESULTS AND DISCUSSION

The constructed wetland-pond system combines the attributes of retention ponds, filter strips, and wetlands. The system approach incorporates design ideas based on the ecology of natural wetlands, in addition to design parameters already reported in the literature on the individual performance of ponds, filter strips, and wetlands. Although the design concepts for individual pond-wetland system components are available, the application and validity of these design parameters have not been studied as a system. Reed[11] makes this point for all constructed wetland systems, stating that there is "... no generally accepted consensus regarding design ... also (there is) no consensus on system configuration, and other details such as aspect ratio, depth of water to media, type of media, slope of bed, inlet and outlet structures." Thus, the monitoring program focused on assessing the overall efficiency of the system in terms of nutrients, hydraulics, and sediment retention capacity.

While monitoring began with the 1989 season, the data are not discussed in detail because only summer and fall seasons were monitored. The 1989 performance of the system was also limited by the fact that the wetland did not become fully established until late summer. For the period monitored (June 17 to November 20, 1989), removals were 92% for TP, 95% for TSS, and 94% for VSS.[8] Monitoring for the 1990 season began May 1 and in 1991 began April 26. Monitoring for both years ended in mid-November when the system froze over. The 1990 and 1991 data showed annual removal efficiencies of 82 to 91% for TP, 96 to 97% for TSS, and 92 to 94% for VSS.

Although the annual removals were good, seasonal removals varied considerably, with spring (April to May) flows actually exporting more phosphorus and sediment from the system than was imported (Table 1). As can be seen from Table 1, spring outflow was greater than inflow due to the high groundwater table which puts the system in a discharge area. During this time, groundwater surfaced in the system beyond the inlet monitoring station, but it was measured at the outlet monitoring station. Thus, more flow was measured exiting than entering the system. This groundwater flow was mostly intercepted by the under-drain tiles. The under-drain flow was grab sampled in spring 1991 to determine the phosphorus and suspended solids concentrations in the groundwater. These samples were only collected when no influent surface run-off was entering the system because the under-drains also collected water which infiltrated from the grass filter strip during run-off events. Discharge from the under-drain continued through the spring. The average concentrations were 0.010 mg/L TP, 0.008 mg/L DP, 0.6 mg/L TSS, and 0.42 mg/L VSS. These concentrations remained relatively constant during the periods of no inflow.

In order to calculate the pollutant mass balance for the spring season, the data from the inflow monitoring station was supplemented by the groundwater data. Measurement of the flow from the under-drain during 1991 showed that the under-drain flow was equivalent to the difference between inflow and outflow. For example, on May 10, 1991, the flow measured at the under-drain was 0.0012 m^3/s; the flow measured at the outlet in May 10 was 0.0013 m^3/s. This followed a 4-day period of no inlet flow so that the under-drain discharge represented groundwater flow. Therefore, the groundwater volume was approximately the difference between system inflow and outflow (direct precipitation on the system was considered offset by the evapotranspiration from the system). Using the measured concentrations and the estimated groundwater volume as outflow minus inflow, the spring mass inputs due to groundwater were determined and added to the spring influent mass. System performance was still negative after the groundwater input was added to the inlet measured values, thereby suggesting an internal source of phosphorus and solids. One possible source is leached phosphorus from dead and decaying plants and biota in the wetland and pond.[12] Another possible source could be the anaerobic release of phosphorus from bottom sediment in the pond since no dissolved oxygen was measured at the sediment-water interface during ice cover. Also, the spring overturn in the retention pond, coupled with wind mixing, could resuspend bottom sediments and phosphorus, which could then be washed from the pond by spring run-off. The large influx of groundwater could exacerbate the loss from the system. By May 28, 1991, outflow from the system had ceased; and on June 5, 1991, flow from the under-drain also stopped.

During the summer (June to August) of 1990, storage and evapotranspiration in the system held all incoming flow without discharge, thus 100% retention of phosphorus and sediment was recorded. During the summer of 1991, two storm events created outflow, but excellent removals were still achieved for the summer (99% for TP, TSS, and VSS).

Removals declined during both fall seasons (September to November) as outflow increased. During fall 1990, removals were 73% TP, 96% TSS, and 94% VSS. For the fall of 1991, removals were 83% TP, 95% TSS, and 88% VSS. High groundwater levels developed during the fall of both years, and similar to spring, more flow was measured exiting the system than entering; however, unlike spring, positive removals were maintained.

Grab samples from three 1990 storm events were analyzed for TSS, VSS, TP, and DP. Six storms were also grab sampled during 1991, along with one snowmelt event. Results from only one event will be discussed, since similar patterns were observed for all the events sampled. The run-off and subsequent outflow during August 19–20, 1991 (Figure 2) was the result of a 2-in. rainfall. As can be seen from Figures 3 and 4, influent TSS and TP concentrations followed the pattern of the hydrograph. Despite the variance for the influent concentrations, outflow concentrations remained relatively constant. Approximately 1102 kg TSS entered the system and 32 exited, a 97% reduction in TSS mass. Similarly, 4.55 kg TP entered, while 0.07 kg exited, a 98% reduction in TP mass. Although not shown in the figures, dissolved phosphorus concentrations remained

Table 1. Results of 1990-1991 Composite Sampling Programs

Season 1990	Flow in (m^3)	Flow out (m^3)	TP in (kg)	TP out (kg)	Percent removal	TSS in (kg)	TSS out (kg)	Percent removal	VSS in (kg)	VSS out (kg)	Percent removal
Spring	648	1767	0.06	0.13	—	7.3	8.1	—	3.4	7	—
Summer	292	0	3.06	0.00	100	1144	0	100	113	0	100
Fall	7296	12,295	4.63	1.26	73	3884	144	96	546	35	94
Total	8236	14,062	7.76	1.38	82	5036	152	97	663	42	94
Season 1991											
Spring	1387	7685	0.30	0.76	—	54	107	—	7	26	—
Summer	2023	743	12.4	0.11	99	3505	10.8	99	393	4	99
Fall	1526	3102	3.9	0.70	82	644	33.8	95	84	10	88
Total	4936	11,530	16.6	1.57	91	4203	152	96	484	40	92

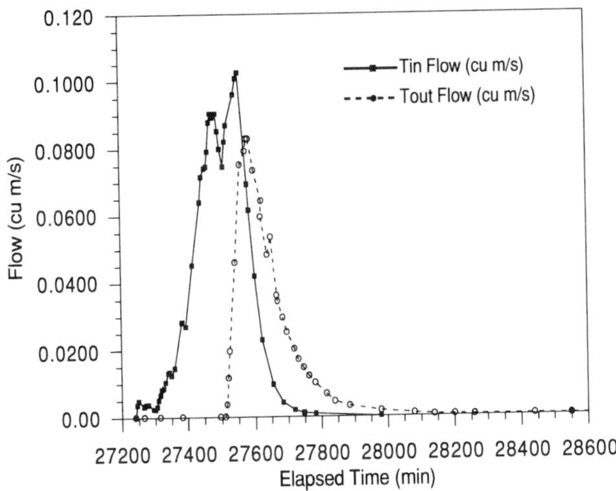

Figure 2. Inflow and outflow hydrographs for the storm of August 20–21, 1991.

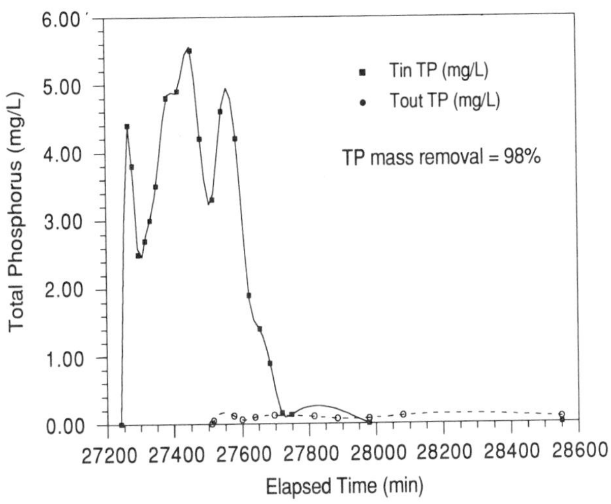

Figure 3. Inflow and outflow total phosphorus concentrations for the storm of August 20–21, 1991.

relatively constant in the influent and effluent. Mass reduction of dissolved phosphorus was 82%, and the flow weighted mean concentrations were 0.172 mg/L for the inflow and 0.047 mg/L for the outflow. The influent DP/TP ratio was 0.04, while the effluent ratio was 0.50. Similarly, the average influent VSS/TSS ratio was 0.11 and, at the outlet, the ratio was 0.16. Thus, a transformation of phosphorus and suspended solids species occurred within the system. Mitsch and Gosselink[12] described the function of wetlands to act as transformers.

Sediment accumulation was monitored annually in the sediment basin by measuring the depth of sediment overlaying the parent material placed during construction. Since the system began receiving flow (June 1989), 6.2 cm of sediment has accumulated in the sediment basin. The SCS[10] recommends sediment removal when accumulation reaches 30 cm to prevent overflow of solids into downstream components. To facilitate sediment removal by heavy machinery, an access ramp was constructed at the north end of the basin with a 10:1 slope. At the current accumulation rate, the basin will need to be dredged about every 10 years.

The accumulation of 6.2 cm sediment is calculated to represent 9500 kg sediment. The total mass of TSS measured entering the system since June 1989 was 17,000 kg. Therefore, the sediment basin retained about 56% TSS. The average TP concentration of the sediment was measured at 2300 mg/kg. Thus, the retained mass of sediment in the sedimentation basin is estimated to contain about 22 kg phosphorus. The total inflow of phosphorus to the system since June 1989 was 46 kg; thus the basin retained about 48% total phosphorus.

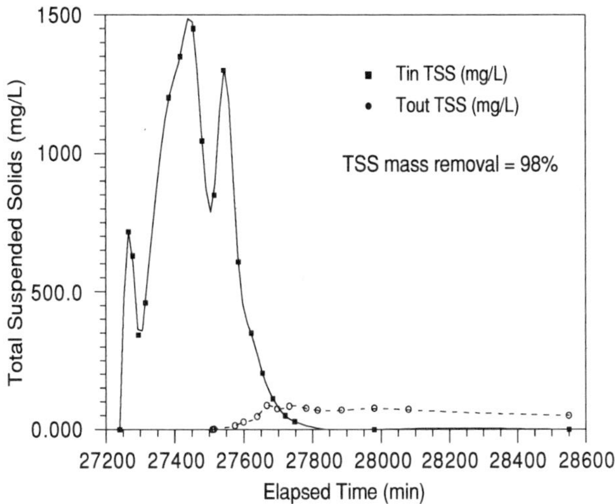

Figure 4. Inflow and outflow of total suspended solids for the storm of August 19–20, 1991.

Direct assessment of the grass filter strip efficiency was not performed because physical constraints did not allow for monitoring. However, observance of the filter revealed that channelization of flow occurred during large run-off events. This was due to the placement of the inlet relative to the sediment basin and grass filter. The inlet is perpendicular to the basin and once the basin is full, the flow has a minimal distance to travel before reaching the level lip spreader and grass filter (see Figure 1). Minimal reduction of the flow energy occurred in this distance, and the level lip spreader could not effectively distribute the flow across the width of the grass filter. The flow then channelized through the grass filter, directly across from the inlet. Aerial photographs of the system showed a noticeable enhancement of wetland growth along the flow path of this channel. Obviously, channelization reduces the effectiveness of the grass filter. Dillaha et al.[13] reported sheet flow was critical for adequate performance of a grass filter. Restructuring the flow path from the inlet would reduce channelization. The inlet could be placed at one end of the sediment basin. The flow path would then be parallel to the length of the basin instead of perpendicular, thereby increasing the travel distance of the flow and decreasing its energy.

The wetland was planted in the summer of 1989 with *Typha latifolia* at a density of 9 stems per square meter. In August of 1991, the mean density had increased to 31.6 stems per square meter. *Sparganium* sp. was also planted at the same time, but in 1 year had been mostly eliminated due to the vigorous growth of *Typha*. A stratified random sample of square meter quadrats (n = 15) revealed the mean tissue P density to be 805.5 mg/m^2 in the 980-m^2 wetland. Thus, the above-ground shoots and stems of the *Typha* in the wetland contained 0.8 kg phosphorus. This is only 5 to 10% of annual influent phosphorus mass measured during the study. Thus, recommendations for harvesting of wetland plants once a year during the fall or winter[10] would remove little TP from the system, especially during the recommended winter harvesting; by then, much of the phosphorus would likely have been translocated to the roots or leached from the plants.

The ability of constructed wetlands to remove phosphorus over long term appears to be limited, according to the literature.[11] Due to recycling of phosphorus from plants and other biota, long-term removal of phosphorus in constructed wetlands is limited to burial in sediments or accumulation in woody tissue. At best, constructed wetlands are typically able to remove 30 to 40% of incoming phosphorus.[11] The removal efficiency is important, though, when coupled with the removal of 48% TP by the sediment basin. The plants provide resistance to flow which will further reduce velocity and enhance sedimentation of smaller particles which have passed through the sediment basin. Settling of the smaller soil particles is important because they typically carry larger amounts of adsorbed material due to their larger surface area.[14] In addition, smaller particle size is indicative of cohesive soils such as clays and fine organic matter. Cohesive and organic soils have a much higher adsorption potential than noncohesive soils such as sand and gravel.[15,16]

Sediment collection cups placed on the bottom of the retention pond were annually monitored for sediment depth and TP concentrations. The sediment was typically fine, black, highly organic material with an anaerobic odor. Small leaf sections of *Typha* were found in the cups. An average of approximately 1.4 cm sediment was deposited per year in the pond. From the bottom area of the pond and sediment density, an estimated 6300 kg sediment was deposited on the pond bottom. However, a portion of this sediment was

Table 2. Cost Effectiveness of TP Reduction from Current Options

Treatment option	Cost/acre	Cost/kg TP removed
Conservation Reserve Program	$1250	
NSC System	$190	$500
St. Agatha POTW Diversion		$4350

biomass created within the wetland and pond, and does not represent only incoming sediment. Kadlec[17] reported that the generation of fine detritus litter occurs faster in wetlands receiving wastewater than in natural wetlands due to increased productivity and microflora and microfauna which die and accumulate in the bottom sediments.

The average phosphorus concentration of the sediment in the Tardiff pond was 913 mg/kg, which indicates the pond bottom contained about 5.7 kg phosphorus. This was 12% of the influent TP recorded entering the system during the 3 years of study. Unlike the sediment, phosphorus is not created within the system, although some was introduced by planting the *Typha*, but it is likely negligible. Some phosphorus was also present in the material placed during construction; but since only the material that had been deposited in the collection cups was sampled, it is representative of the influent material.

Bouchard et al.[2] compared two other alternatives of phosphorus reduction to Long Lake: a $3 million sewage treatment plant diversion and the Conservation Reserve Program (CRP). The CRP pays farmers to convert highly erodible cropland to grassland for 10-year periods. Bouchard et al.[2] estimated CRP land gave equivalent reductions of phosphorus as the NSCS. Between the three alternatives being used in the Long Lake watershed, the constructed wetland-pond system is the most cost effective in removing phosphorus, as shown in Table 2.

SUMMARY

It should be stressed that constructed wetland-pond systems are supplemental to conservation practices to be used on the cropland. These practices are important in reducing pollutants at their source and also protecting the resources of the farmer. The constructed wetland-pond system functioned well during the monitoring period, with minimal maintenance. The sedimentation basin provides about half the total removal. Most summer storms are contained within the sedimentation basin alone, and it is during these summer storms that high sediment and phosphorus loads occur. Thus, for the amount of land required and construction cost, the sediment basin appears most effective since the majority of phosphorus was associated with solids. Typically, outflow during the summer is minimal. As a result, removals are highest during the summer. Reducing the input of phosphorus to Long Lake during the summer will likely diminish the risk of algal blooms.

ACKNOWLEDGMENTS

The activities on which this publication is based were financed in part by the Department of Civil Engineering, University of Maine, Maine Department of Environmental Protection and the Department of the Interior, U.S. Geological Survey, through the University of Maine Water Resources Institute Program. The contents of this publication do not necessarily reflect the views and policies of the Department of the Interior, nor does mention of trade names or commercial products constitute their endorsement by the U.S. Government.

REFERENCES

1. U.S. Environmental Protection Agency. Report to Congress: National Water Quality Inventory. Washington, D.C., 1986.
2. Bouchard, R., M. J. Higgins, C. A. Rock, and J. W. Jolley. The role of constructed wetland-pond systems in watershed management, presented at the North Am. Lake Management Soc. 11th Int. Symp.

3. St. John Valley SWCD. Watershed Plan Long-Cross Lakes Watershed, Aroostook County, Maine. Ft. Kent, Maine, July, 1990.
4. Dillaha, T. A., J. H. Sherrard, D. Lee, S. Mostaghimi, and V. O. Shanholtz. Evaluation of vegetative filter strips as a best management practice for feed lots, *J. Water Pollut. Control Fed.* 60:1231, 1990.
5. Schellinger, G. R. and J. C. Clausen. Vegetative filter treatment of dairy barnyard runoff in cold regions, *J. Environ. Qual.* 21:40, 1992.
6. Erich. S. M. and G. Porter. Soil Phosphorus Levels and Phosphorus Fertilizer Recommendations for Potatoes Grown in Maine. University of Maine, unpublished, 1990.
7. Sposito, G. *The Chemistry of Soils.* Oxford University Press, New York, 1989.
8. Jolley, J. W. The Efficiency of Constructed Wetland-Pond Systems in the Reduction of Sediment and Nutrient Discharges from Agricultural Watersheds. M.S. thesis, Department of Civil Engineering, University of Maine, Orone, ME, 1990.
9. Soil Conservation Service. Urban hydrology for small water sheds, Tech. Release No. 55. U.S. Dept. of Agriculture, Washington D.C., 1986.
10. USDA/SCS. Nutrient/Sediment Control System (Interim Standard and Specification). USDA/SCS Orono, ME, Nov. 1991.
11. Reed, S. C. Constructed Wetlands for Wastewater Treatment, *BioCycle.* 32(1):44, 1991.
12. Mitsch, W. J. and J. G. Gosselink. *Wetlands.* Van Nostrand Reinhold, New York, 1986.
13. Dillaha, T. A., J. H. Sherrard, and D. Lee. Long-Term Effectiveness and Maintenance of Vegetative Filter Strips. Virginia Water Resources Research Center, Bulletin 153, Blacksburg, VA, 1986.
14. Faulker, S. P. and C. J. Richardson. Physical and Chemical Characteristics of Freshwater Wetland Soils, in *Constructed Wetlands for Wastewater Treatment.* D. Hammer, Ed. Lewis Publishers, Chelsea, MI, 1989.
15. Novotny, V. and G. Chesters. Handbook of Nonpoint Pollution Sources and Management. Van Nostrand Reinhold Company, New York, 1981.
16. Schreiber, J. D., D. L. Rausch, and L. L. McDowell. Callahan Reservoir. II. Inflow and outflow suspended sediment phosphorus relationships. Transaction of the ASAE Paper No. 74-2502, 1977.
17. Kadlec, R. H. Decomposition in wastewater wetlands, in *Constructed Wetlands for Wastewater Treatment.* D. Hammer, Ed. Lewis Publishers, Chelsea, MI, 1989.

CHAPTER 38

Constructed Wetlands R&D Facility at TVA's National Fertilizer and Environmental Research Center

C. Breed

INTRODUCTION

Since 1933, the TVA (Tennessee Valley Authority) has conducted fertilizer research and development at its facilities located in Mussel Shoals, Alabama. Approximately 80% of the fertilizer in the U.S. today is manufactured using TVA patents developed at these facilities. As the fertilizer industry has matured, the need for heavy emphasis on new process development has lessened, and TVA has refocused their research efforts more toward environmental issues dealing with fertilizer and related industries. This was evidenced by a name change for these facilities in 1989, from the National Fertilizer Development Center (NFDC) to the National Fertilizer and Environmental Research Center (NFERC).

As part of the refocusing effort, the NFERC staff participated in a suggestion, evaluation, and election process that identified potential "window of opportunity" (WOO) projects. These projects were considered those that, if successfully conducted, could result in major benefits (monetary and/or political support) to NFERC and the nation as a whole. One of the WOO projects was to expand the study and application of constructed wetlands to include wastewater and process water from the agribusiness and related industries, including fertilizer and agrichemical dealers, food processors, animal/poultry raising and processing, etc.

The initial strategy for the project was to utilize what appeared to be a wealth of existing data that had been generated on municipal wastewater systems and to modify them to meet the particular needs of the agribusiness industry. These modifications were to be tested first in the greenhouse, next in small- to medium-sized test cells, and then the modified systems were to be demonstrated at dealer sites as soon as possible. Information was gathered through a literature search, site visits, and personal contacts within TVA and with EPA, universities, and various consultants. As the information gathering progressed, it became evident that there was a lack of confirming data on such design and operating items as cell geometry, the role of the plants, plant species, cell media, hydraulic loading, ammonia removal, cell maintenance, operating costs, etc.

Based on this information, the primary mission and goals for this NFERC project were redefined (Table 1). The effort has been redirected mainly toward the development and operation of an R&D center that will still concentrate on the development of wetlands for the agribusiness industry, but will also address some of the major generic design and operating questions identified and facilitate the dissemination of data collected.

DESCRIPTION OF R&D FACILITY

The construction of the R&D facility was begun in summer of 1990 and currently is in its final stages. The major components of the facility are a greenhouse, an office/lab building, an influent building, 32 test cells, an effluent building, and two nursery cells. A description of each of these components follows:

Greenhouse

The greenhouse is being used for screening tests to evaluate a variety of variables in both single pot tests

0-87371-550-0/93/$0.00+$.50

Table 1. Statement of Mission and Goals

Mission

The mission of the constructed wetlands R&D project is to foster the development of the constructed wetlands technology using a scientific approach to generate reliable design data and apply that to commercial-scale demonstrations/operations.

Goals

- Design, construct, and operate a versatile R&D facility which will facilitate the gathering of reliable information
- Concentrate the major research work on the application of the technology for the treatment of a variety of agribusiness and chemical wastewaters
- Identify the types of agribusiness wastewaters that the constructed wetlands technique can effectively clean up
- Identify specific potential users of the technology, translate the R&D data into practical form, and apply it to field demonstrations
- Form alliances with other organizations, public and private, that are doing similar development work and provide research assistance when and where appropriate; research should complement rather than duplicate existing efforts
- Form an expert panel (inside and outside the TVA) to identify other potential work or problem areas periodically and to review and evaluate on-going research
- Obtain co-funding of the research effort from the public and private sectors

and mini-cells which can contain up to eight pots. The greenhouse, manufactured by Rough Brothers, is 40 × 100 ft and is capable of handling up to 800 single pot tests or 75 mini-cell tests. The climate in the greenhouse can be controlled using evaporative pads and electrical heat. A spray table for accelerating root development is to be added later. Data from the mini-cells will be taken using a portable data logger.

Office/Laboratory Building

The office/lab building is a 42 × 80 ft metal building which houses the headhouse for the greenhouse, a control lab for conducting routine measurements and for preparing samples, office space for on-site personnel, and an automatic data-logging system for the test cells. The headhouse contains drying ovens, a variety of plant tissue grinders, and a feed-water deionizer (for use in the greenhouse). The lab is set up to make field measurements for pH, BOD, DO, Cl^-, N, P_2O_5, NH_3, redox, etc. The automatic data-logging system is capable of continuously measuring and logging 100 different analog inputs throughout the test cells. Five different variables (pH, DO, level, redox, and temperature) will be logged on a routine basis.

Influent Building

The influent building, a 30 × 60 ft metal building, houses the feed system for the test cells. The composition of the contaminated water feed to each of the test cells is prepared and fed to the cells through magnetic flowmeters and totaling water meters. Since the original intent of the facility was to test agribusiness waste waters, the feed systems are designed for generally clear feed water. If a municipal or animal type wastewater is tested, the totaling meters will have to be supplemented. The building contains a control room where the analog signals from the field sensors/transmitters are routed through a patch panel to the data logger; this allows rapid changing of the data collection points.

Test Cells

There are 32 test cells of three different sizes and two construction techniques. Of the cells, 26 are of underground design and are double lined with 30- and 40-mil-thick ultraviolet-resistant PVC. The other six cells are constructed above ground using a modular plastic dike system; these are also double lined with the same type of material. A geometry of 20 × 30 ft. was selected as the standard size cell; there are 18 below-ground cells and 6 above-ground cells of this size. Four of the below-ground cells are half of that area and four are twice that area; these will be used to evaluate scale-up data. All of the cells are constructed with a slope of 1% from inlet to outlet and contain 2 ft of $^3/_8$- to $^3/_4$-in. river gravel. The influent to the cells will be pumped into the cells at a variety of inlet depths and the outflow from the cells will be controlled using a pumps and float switches with a $^1/_4$-in. sensitivity.

Effluent Building

The effluent building contains eight tanks with a capacity of 6400 gal. These tanks will be used to collect the wastewater after it has passed through the test cells. The water will be analyzed and, if it meets discharge requirements, it will be discharged to a nearby municipal sanitary sewer system. If the water quality is not good enough to allow discharge, it will be recycled through the cells or passed through other treatment processes (activated carbon, ion exchange, etc.).

TEST PLAN FOR INITIAL WORK

Greenhouse

The initial test work at the R&D facility will begin in the fall and winter of 1991–1992, with the first studies consisting of screening tests in the greenhouse. The work will be done in both individual pots and in mini-cells which contain up to eight potted plants. Results from the screening tests will then be used to guide the test work in the large outdoor cells beginning in the spring. The screening tests will be directed toward three main areas:

- Dilute nutrient/pesticide contaminated wastewaters typical of agrichemical dealers.
- Chloride-contaminated process wastewater typical of meat-packing operations.
- Plant tolerance to ammonia/ammonium concentrations.
- Five of the most commonly used pesticides (Atrazine, Alachlor, Metolachlor, Metribuzin, and Trifluralin) have been selected for the initial screening tests. Depending on the plant tolerance and contaminate destruction levels achieved in the screening tests, the two most promising pesticides will be then tested in the outdoor cells. If the greenhouse work proves that it is impractical to destroy the pesticides using the wetlands technique, then the spring work will focus more along the lines of the removal of nutrients (N and P_2O_5).

The effect of the ammonia/ammonium ion concentration in the wastewater will also be studied in the greenhouse, with the results being translated to the larger cells in the spring. The development of techniques to oxidize the ammonia will also be studied.

The chloride tolerance testing will be conducted in the greenhouse work, but actual large-scale confirmation of the results will probably be done at a plant site where a large supply of process water is available. This will be a similar approach in future work where the wastewater is difficult to simulate in large quantities, such as where there are high levels of toxic materials or high BOD loadings. Current examples in which the TVA is involved are a poultry rendering plant and a shellfish cleaning operation.

Test Cells

There were 18 test cells planted in August and September of 1991 to allow the establishment of good root systems in preparation for the spring testing, 11 different plants were selected for the initial planting (Table 2). These plants were placed at three different depth (5, 10, and 15 in.) to test the development and movement of the root system.

If the greenhouse test work goes as planned, the 18 cells will begin in the spring of 1992 to receive a synthetic wastewater containing the nutrient elements needed for growth, in addition to either of two pesticides. The plan is to study the remediation of one herbicide as a function of five different plant compositions, and of the second herbicide in three replicated cells having the same plant composition. The change in nutrient composition of the wastewater will be monitored, as affected by plant composition and by the second variable (e.g., herbicide or high ammonium level). Whatever the treatment, basic chemical and plant physiological studies will be made in all the pots, mini-cells, and outdoor cells. For example, phosphorus precipitation and redox potential around the different wetland plants species are among the several studies planned.

Table 2. Type of Vegetation Used in Initial Planting of Test Cells

Plants	Plant mix depth (in.)	Cell no.
Cattail *Typha latifolia*	15	7, 14, 16
Water Plantain/Pickerel *Pontederia cordata*	15	2, 4, 8, 9, 13, 17
Hardstem bulrush *Scirpus aeutus*	15	1, 3, 5, 6, 10, 11, 12, 15, 18
Green bulrush *Scirpus americanus*	10	7, 14, 16
River bulrush *Scirpus fluviatilis*	10	2, 4, 8, 9, 13, 17
Marsh smartweed *Polygonum* spp.	10	5, 11, 15
Wild reed "canes" *Phragmites australis*	10	1, 3, 6, 10, 12, 18
Woolgrass *Scirpus cyperinus*	5	7, 14, 16
Green bulrush *Scirpus americanus*	5	1, 3, 4, 6, 9, 10, 12, 17, 18
Wapato duck potato *Sagittaria latifolia*	5	2, 8, 13
Sweet flag/Alpine rush *Acornus calamus*	5	5, 11, 15

FUTURE STUDIES

The facility at Mussel Shoals will permit study of treatment variables in statistically sound experiments. Wastewaters can readily be transferred to any single cell or combination of cells in the facility, recycled if desired, and passed through the cells at predetermined rates and depths. The cells are sufficiently large to serve as demonstration cells, as well as to permit continuous monitoring of common parameters, and taking special measurements. Participation of researchers within and outside of TVA will be encouraged. A research review group is being formed to establish research priorities and guidelines. Comments concerning the facility and potential test work are always welcome.

CHAPTER 39

Stormwater Runoff Retention and Renovation: A Back Lot Function or Integral Part of the Landscape?

D. L. Ferlow

INTRODUCTION

Stormwater run-off from roofs, parking lots, roadways, and landscapes impacts and degrades water quality and habitat values within existing wetlands and water courses. Many communities require stormwater management with retention systems. Professional engineers calculate the required retention volumes and establish the configuration of the inlet and outlet structures. Concepts are utilitarian and created to store and slowly release run-off waters with a minimum need for maintenance and repair. Therefore, stormwater control basins are often broad, gently sloped, excavated or bermed, mowed grass site features, usually located away from active use areas, often relegated to the status of a "back lot" function.

Stormwater run-off basins can be integrated with aesthetic features and designed to provide features for run-off renovation. These basins, termed "biofilters", may have quite distinct and different visual characteristics. Included are vegetation-fringed ponds, shallow water marshes, seasonally flooded basins and intermittently wet and dry meadow or scrub brush systems. When properly designed and constructed, biofilters are created wetlands.

METHODS AND MATERIALS

Biofilters are the final element of a site's stormwater run-off control program. Rainfall runs across roofs, pavements, grass areas, and swales to points of collection in drain inlets. The inlets (catch basins) are often fitted with deep sumps (2 ft) which trap and hold the heavier sand and gravel materials carried in stormwater. Hoods in the catch basin structures are sometimes used to increase sediment trapping efficiency. Piping connects the drain structures and conveys the stormwater to points of outfall into the biofilters. Although biofilters may be of different design and construction, all function on the same principle. Run-off water flows are slowed in deep-water pools or meander through vegetation, "dropping" silts, sediments, and water-borne pollutants before discharge into down-gradient wetlands and water courses.

Pooling of Water Within a Biofilter

For maximum water quality renovation, a biofilter should impound water, either on a long-term (pond) or short-term (marsh or scrub shrub wetland) basis. Pooling slows and stills the stormwater and a high percentage of the silts and sediments, often accompanied by attached pollutants, settle out and bind into bottom sediments. Ponds and shallow water marshes have a permanent water pool with high sediment settlement and run-off filtration potential. Intermittently, wet scrub shrub biofilters pool water for short periods during and following a storm event. As settling is the primary mechanism for the removal of sediments and particulate pollutants, extended impoundment periods function best. Soluble pollutants are best removed in a permanent pond system or through extended run-off retention, utilizing biological processes within a shallow marsh or scrub shrub brush habitat.

0-87371-550-0/93/$0.00+$.50

Retention time varies, dependent upon the authority establishing the "standard". In Maryland, field studies have documented pollutant removal within extended retention ponds.[1] Similar extended retention periods have been utilized in northern Virginia.[1] Somerset County, New Jersey has established a standard to detain run-off from a 1.25-in., 2-hour rainfall in such a way that no more than 90% of the run-off will be released from the basin at the end of 18 h for residential land uses and 36 h for other development.

The key factor is the extended pooling period. The exact length of time should reflect a balance between sediment and pollutant removal efficiency, site area available for the biofilter, and long-term maintenance costs. Field experience, with several biofilters, has determined that heavily vegetated, flat-gradient ($\pm^1/_2\%$), shallow marsh and seasonally saturated scrub brush systems, built within small (5 to 20 ± acre) contributory watersheds, have the physical characteristics to pool water to shallow depths for extended periods (18 to 24 + h) with simple outlet control.

Vegetation Within a Biofilter

While extended retention removes significant levels of sediment and particulate pollutants, biological removal of soluble pollutants can be achieved through the vegetation of created wetland biofilters. Marsh plants, algae, and bacteria that grow on shallow, organic-rich substrate take up soluble nutrients needed for growth from the run-off water. Further, the bottom sediments of a created wetland are an excellent location for pollutant sorption. The degree of pollutant removal is not easily determined, but appears to be dependent upon the size of the biofilter basin in relation to the pollutant loading from the run-off waters.[1] In the northeastern U.S., pollutant removal varies seasonally. Highest removal occurs during the early spring and summer, and reduced removal in the late fall and winter after the plants have ceased growth. In the fall and winter months, created wetland biofilters sometimes become a net source of nutrients.[1] However, the nutrients are released at a time of the year when they will have a minimal impact on the receiving waters.

Studies have shown that the capacity of created wetlands to assimilate pollutant loading is highest when they are exposed to relatively dilute nutrient loads.[1] One source suggests that nutrient removal can be achieved if loadings do not exceed 45 lb. of phosphorus and 225 lb. of nitrogen per surface wetland acre per year.[1]

BIOFILTER BASIN DESIGN CONCEPTS — SCRUB SHRUB BRUSH SYSTEMS

General design guidelines have been developed as a result of the implementation and monitoring of several created wetland biofilters. A basic design was originally derived from CONNDOT construction project sediment pool standards and modified to interface with the more natural created wetland attributes desired within a biofilter. Original planning criteria has been updated based upon first-hand field knowledge and recent published "standards". Reference is made to Figure 1 which shows a typical cross-section through a marsh and scrub growth biofilter and details the level of run-off renovation effectiveness. From field experience, a biofilter with outlet control is recommended to contain:

- A "working" pooling depth before overflow of approximately 2 ft above the "normal" elevation of the biofilter. More frequent, small storms will saturate and flood the basin floor. Major storms will flood vegetation to levels and time periods resulting in plant material decline.
- As an optimum, a storage capacity (volume below overflow) of $^1/_2$ in. of run-off water from the watershed tributary to the biofilter. This volume (in cubic feet) can be obtained by multiplying the contributory drainage area in acres by 1815.[2] At a minimum, a biofilter should have a storage capacity for $^1/_4$ in. of run-off water from the watershed. Where storm run-off from the biofilter will directly enter water supply reservoirs or other environmentally-sensitive waters, consideration should be given to increasing the biofilter storage volume.
- Where possible, a shape so that the length of flow through the biofilter basin will be at least twice the width. Inlet locations should be as far away from the outlet structure as is feasible or should be baffled in a manner to turn the run-off water so it will pass across a maximum area of the biofilter before discharge. Further, the biofilter design should have an irregular, amoeba-like edge rather than a uniform, bowl-like shape, as this configuration will provide the potential for maximum habitat diversity and visual interest.
- A minimum surface area, at the overflow depth, that conforms to the formula $A = 180Q$, where Q is the flow expected from the 5-year design storm in cubic feet per second, and A is the surface area in square feet.[2]

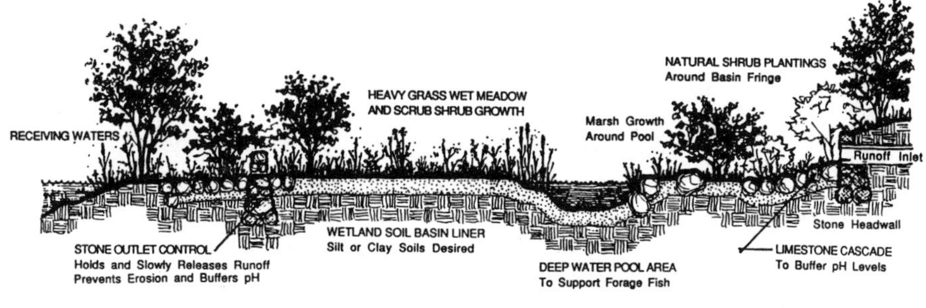

Stearns & Wheler

MARSH BIOFILTER SYSTEM

BIOFILTER/WETLAND WATER QUALITY TEST RESULTS

Tests from 1988 to 1991 show that biofilters function in much the same manner as a natural wetland toward storm runoff renovation. They quickly catch heavy sediments and remove water borne debris, buffer and help change pH levels toward neutral and generally reduce Nitrate, Phosphate, Iron and Chromium (hexavalent) levels across the system. In addition, biofilter visual values and habitat characteristics are positive environmental factors worthy of consideration for incorporation within development plans.

System	pH Level	Chromium	Iron	Phosphates	Nitrates
Test Results	pos. neg.	pos. neg.	pos. neg.	pos. neg.	pos. neg.
Biofilter (35/31)	46% 29%	35% 19%	77% 16%	84% 13%	45% 36%
Wetland* (21/21)	62% 28%	33% 28%	48% 24%	71% 5%	48% 19%

(1) The remaining tests showed either that no change occurred or the results were not determinable.
(2) Number of tests made in each system are shown thusly (pH/nutrients and metals).
(3) * indicates natural wetland area receiving development runoff and used as a "control" for testing.

Figure 1. A typical cross-section through a marsh and scrub growth biofilter, along with biofilter/wetland water quality test results.

- A design concept that enables long-term ponded biofilters to maintain a diversity of aquatic life to keep potential mosquito problems in check. An intermittently wet-dry biofilter might be integrated with a shallow water marsh or pond, or have several deep, permanent water pools.
- Invert elevations of inlet pipes or channels that discharge into the biofilter, optimumly set at or above the overflow spillway elevation. This will allow a biofilter basin to fill gradually with silts and sediments and continue to provide a filtration function without clogging drainage piping.
- An awareness of a proper landscape position for biofilters. Local surficial hydrology should be understood. Prior to design, seasonal and permanent groundwater elevations should be known to the extent possible. As a general guideline, biofilters should be excavated into existing grades without the need for extensive berming to impound run-off. Where berms are necessary, they should be constructed in accordance with accepted engineering practice.
- Silt loam or silty clay loam soils used in the construction of the biofilter basin. The characteristics of these soils will foster the extended water pooling and surficial saturation necessary for the created wetland to be a success. Where possible, soil material excavated from an existing wetland source should be used to line the biofilter bottom to provide the physical structure of a wetland soil and be a root stock and seed source for the almost immediate growth of wetland vegetation.
- Biofilter entrance systems to reduce entering water velocities to 1 to 1.5 ft/s when the water spreads out across the biofilter basin. Flow reduction is often provided through stone rip-rap. Use of limestone for the rip-rap will help buffer acid rain pH levels.
- Plant materials suitable for growth in hydric soil conditions established across the biofilter floor and plant species that grow in wetland fringe areas around the basin edges. Plant material may be allowed to generate naturally from the local root stock and seed sources in the biofilter basin topsoil. This is particularly true if wetland-based soils have been used in the biofilter's construction. Biofilter basins may be rough seeded for vegetation cover and erosion control until the more permanent species become established.

- Natural plant stock growth may be supplemented by planted materials. Plants may be obtained from nursery sources or gathered from the local area. Woody nursery stock should be planted at least 12 to 18 in. above normal water levels in the biofilter. Wildflower seeds may be collected and disbursed across the biofilter. Watering and weeding need only be rendered to levels necessary to insure the start of growth. From that point forward, biofilter systems should be allowed to grow without special care. Plants must be allowed to compete for survival and dominance within the biofilter. However, removal of undesirable species and hydrology adjustments will be needed in the initial growth period to maintain ecosystem viability.
- A long-term program for the removal of accumulated sediment and water-borne debris. When a created wetland biofilter "fills" and no longer serves its intended function, it should be excavated and reestablished.

BIOFILTER OBSERVATIONS AND CONCLUSIONS

Created wetland biofilters require a period of time to establish, stabilize, and grow into a viable functioning wetland. This may involve 2 to 5+ years, as each system is different and dependent upon the specific site conditions related to the hydrologic regime. For example, a pond or permanent shallow marsh biofilter may be an almost "instant" wetland success with dense growth achieved in one season; whereas an intermittently wet and dry scrub brush biofilter will require at least several growing seasons to become established and require several more years to present a natural stable visual character.

Pond biofilters, used as landscape features, have the potential to be impacted by run-off-carried debris and petroleum products which "float" across the surface and degrade general visual quality. These aesthetic impacts can be significantly reduced through construction of a stone-lined, heavily vegetated, bowl-shaped "trap" at the biofilter inlets. Entering run-off must pass through the rough stone layer and plant material growth. A significant percentage of the "floatables" will be caught and held at this point. "Trap" locations should be designed for easy access and periodic removal of the collected materials.

To ensure the potential for reasonable success, created wetland biofilters require detailed construction monitoring and maintenance during their time of establishment. Water levels must be checked, adjusted, and stabilized at optimum levels. Water elevation fluctuations of several inches can change the ecosystem type from wetland oriented to upland or drown desired plants. Undesirable pioneer plant growth, such as the reed (*Phragmites communis*) or purple loosestrife (*Lythrum salicaria*) must be controlled or weeded out until desirable higher-quality vegetation has had time to become established and has a reasonable chance for continuance. Once established, the desirable wetland vegetation appears to have the potential to compete reasonably with the more aggressive weed species.

In a newly created biofilter, wildlife impacts may have to be controlled to manageable levels, at least until the ecosystem can survive what might be considered normal wildlife predation impacts. For example, fine tuning of "normal" water level in a shallow marsh biofilter may be necessary to limit the habitat value of the created wetland for species, such as muskrat (*Ondatra zibethica*), which have the potential for devastating impacts on the vegetation.

As a created wetland biofilter grows and matures, it may change in visual character and habitat type from the original design. Rainfall cycles, water levels, natural plant growth patterns, and wildlife utilization will play a role in a potential kaleidoscope of visual features and the establishment of wetland functional values.

Though designed with the same basic standards, healthy functioning biofilters display quite diverse results. When correctly sited within the development plan, created wetland biofilters provide strong visual interest in the landscape and act as natural open-space buffers. Reference is made to Figures 2 and 3, which illustrate a stormwater management biofilter integrated with site landscape systems.

Since created wetland biofilters function as an integral part of stormwater management, they quickly show visual evidence of the collection of silts, sediments, and water-borne "debris" carried by the storm run-off water. Tests of three constructed biofilters made over a 3-year period have shown that biofilters also buffer run-off water pH levels, generally lower nitrate levels, and show mixed and inconsistent but, on the whole, positive results for phosphate and heavy metals reduction. Reference is made to Figures 1 and 4, which document biofiltration effectiveness and illustrate the aesthetic quality of a created wetland biofilter functioning as an open-space area.

The strong natural visual values, open-space elements, wildlife habitat characteristics, run-off control features, and water-borne materials reduction within created wetland biofilters represent positive environmen-

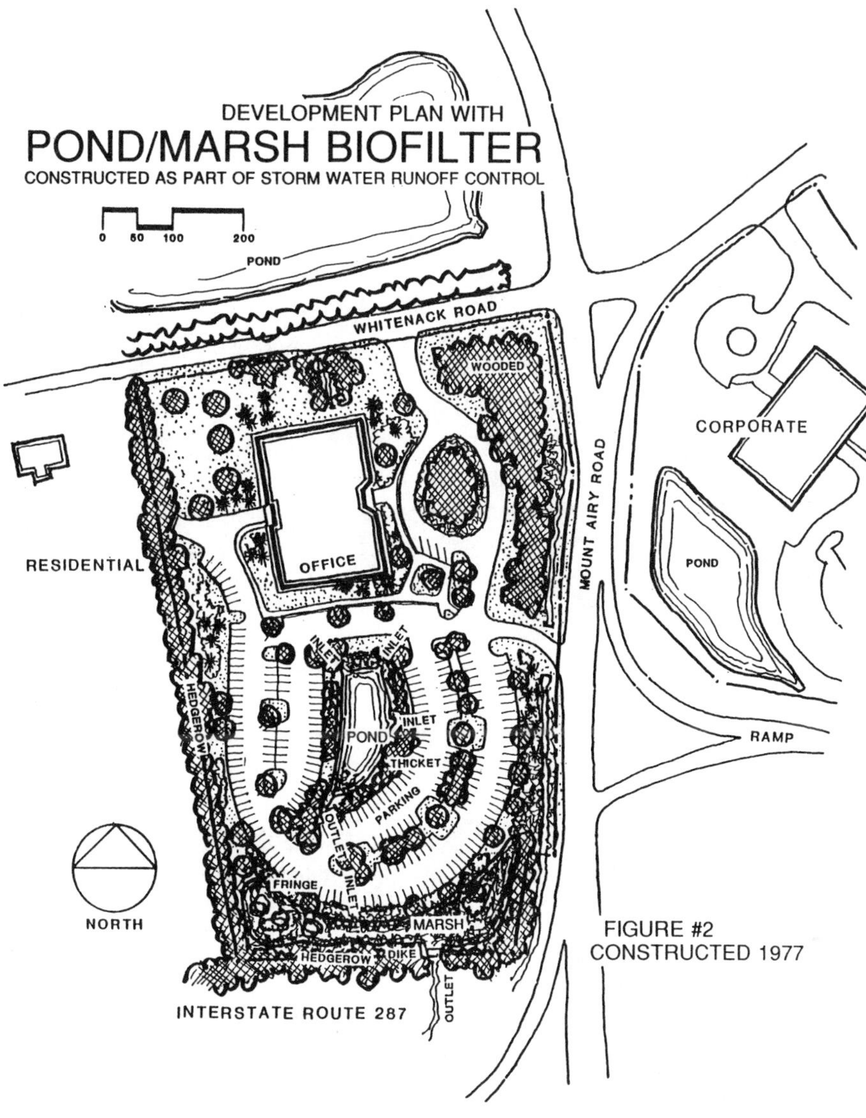

Figure 2. Development plan with pond/marsh biofilter constructed as part of stormwater run-off control.

tal factors. These are worthy of consideration for incorporation as site design features within development plans.

ACKNOWLEDGMENTS

The design criteria presented and the projects illustrated herein have evolved over a period of years. The original design and implementation of the biofilters was accomplished under my direction in my former firm, Environmental Design Associates, P.C., Wilton, CT, with more detailed on-going research and design updates carried forward into Stearns & Wheler.

Figure 3. View of marsh component of created wetland biofilter.

Figure 4. Visual quality of run-off biofilter open space area.

REFERENCES

The noted references have been used as design information sources for the planning and implementation of the author's latest biofilters. Original stormwater management basins constructed as created wetlands from 1977 to 1987 were accomplished without the benefit of established criteria and were, in essence, systems based upon an understanding of natural systems and a trial, error, revise, and update approach. The noted references have reinforced basic design concepts and have enabled the refinement of personal planning standards. Only a few specific reference sources have been noted in the test, as the research data has been utilized in the full context of the biofilter design process.

1. Schueler, T. R. Controlling Urban Runoff: A Practice Manual for Planning and Designing Urban BMPS. Department of Environmental Programs, Metropolitan Washington Council of Governments, Washington, D.C., 1987.
2. CONNDOT. Sedimentation Pool Design Standards — Excerpts from Guidelines, CONNDOT, Wethersfield, CT, undated.
3. USDOT, Federal Highway Administration. Retention, Retention, and Overland Flow for Pollutant Removal from Highway Stormwater Runoff, Research, Development, and Technology. Turner-Fairbank Highway Research Center, McLean, VA, 1988.
4. Somerset County, NJ. Water Quality Regulations, Somerset County Engineering Department, Sommerville, NJ, 1980.
5. Niering, W. A. Effects of Stormwater Runoff on Wetland Vegetation, draft. Connecticut College, New London, CT, 1989.

PART 7

HYDROPHYTE CONSIDERATIONS

CHAPTER 40

Wetland Systems and Their Response to Management

G. R. Guntenspergen, J. R. Keough, and J. Allen

INTRODUCTION

Constructed wetlands that have been designed for wastewater improvement have also been promoted for their wildlife habitat functions. Plans for multiple use of constructed wetlands should consider that, often, conflicting management strategies may be involved. When we manage constructed wetlands for wastewater improvement, one of the things we do is affect the natural hydrologic regime, and this may be incompatible with management for wildlife habitat. Increasingly, hydrology has been recognized as one of the most important variables controlling wetlands and their functions.[1]

Water level manipulation in wetlands is a commonly practiced wildlife management technique.[2,3] The objective is to encourage plant species and communities that will provide habitat and food for wildlife by manipulating the timing, duration, and depth of flooding.[4] An examination of wetland management techniques developed for wildlife populations can provide guidance to those seeking to use constructed wetlands for the dual purposes of water quality improvement and wildlife habitat development.

Our goal is to examine the relationship between hydrologic manipulation and the management of wetland habitats for wildlife. We first review the importance and role of the hydrologic regime in natural systems and examine processes underlying plant community response; then, we examine vegetation response to hydrologic management in green tree reservoirs and moist soil impoundments common to the southern U.S. that are managed for wildlife habitat.

NATURAL SYSTEMS

Water depth, flow, and the duration and frequency of flooding play major roles in controlling wetland structure and plant community dynamics. A conspicuous feature of many wetlands is the presence of zones of vegetation distributed along water depth gradients.[5,6] Water depth is correlated with the survival, distribution, and abundance of wetland species.[7-9] In addition, differences in hydro-period can have significant effects on species diversity[10] and plant productivity.[11]

Hydrology also plays a major role in vegetation dynamics. Van der Valk and Davis[12] have outlined wetland response to climate-induced changes in water level in prairie potholes. Fluctuations in the hydrologic regime result in changes in both plant species composition and structure. These changes in the plant community associated with water level fluctuations have also been documented for Great Lakes wetlands[13,14] and Carolina Bay wetlands of the southeast.[15]

Despite widespread recognition that natural communities are continuously subjected to environmental fluctuations, the effects of temporal variability (and the appropriate scale at which such variability is important) on wetland structure and function have received little study. Keough et al.[16] sought to determine the relationship between hydrologic regime and plant community composition in depressional wetlands in South Carolina. Long-term changes in plant community composition in these wetlands were attributed to the long-term climatic variation in precipitation patterns. They found no relationship between the seasonal pattern of water level and plant community structure. However, they did find that plant community composition was

0-87371-550-0/93/$0.00+$.50

related to residence time of water following major precipitation events. The vegetation in the sites that drained quickly following precipitation events was distinct from that in which drainage was slow or where water was lost primarily through evapotranspiration.

Changes in the hydrologic regime can affect which species will become established in a given site or which will decline in abundance. Wetland managers must understand the patterns and processes underlying shifts in species abundance and community change in order to sustain systems that will perform a variety of functions. However, few models exist to assess the impact of natural fluctuations in hydrologic regime on wetland plant communities.

Herendorf et al.[17] and Keddy and Reznicek[13] proposed community displacement models to assess changes in wetland vegetation types resulting from changes in water level. They emphasized the importance of seed banks and immigration of species along topographic gradients as major factors ameliorating periodic disturbance caused by high water. Van der Valk[18,19] proposed a species-oriented model which attempted to correlate plant life history traits and hydro-period. However, field tests using this model were only partially successful in making predictions about species composition in constructed impoundments subjected to drawdowns.[20]

The application of wastewater can alter the hydrologic regime of the receiving wetland. Wastewater application to a Minnesota wetland complex contributed 38% of the total inflow to the system affecting the natural hydrology of the system and resulted in changes in plant species composition.[21] Kadlec and Bevis[22] also concluded that hydrologic effects associated with the introduction of secondarily treated wastewater to a sedge-leatherleaf peatland contributed to changes in the composition of the plant community. Whigham[23] has proposed extending the van der Valk wetland succession model[18,19] to wetlands receiving wastewaters. He hypothesized that increases in the depth and duration of flooding could lead to the loss of species by eliminating established populations and suitable conditions for seed germination and regeneration.

GREENTREE RESERVOIRS

A common form of water level manipulation in forested wetlands is the establishment and management of greentree reservoirs (GTRs). Since their first introduction in the 1930s, at least 179 GTRs have been established in the U.S.[24]

GTRs increase local wintering waterfowl populations, at least over the short term, by providing food and resting/roosting habitat.[25,26] Because 89% of GTRs are used for hunting,[24] their ability to attract waterfowl is a great asset.

GTRs have also been established because of a perceived need to mitigate the loss of naturally flooded bottomland hardwood habitat. More than 100 of the 179 GTRs surveyed by Wigley and Filer[24] were located in the Lower Mississippi Valley, where less than 20% of the original 22 million acres of bottomland hardwood forest still exists.[27] By managing some of the remaining bottomland forests more intensively (i.e., as GTRs), managers can ensure that some flooded areas will be available at the appropriate time for waterfowl.

Techniques for the construction and management of GTRs are well established and at least three publications providing management guidelines are or soon will be available.[28-30] GTRs are most often established by constructing low levees around an area of bottomland forest that can be flooded either by diverting water from a source at a higher elevation or by pumping. Sites can also be passively flooded through overbank flooding of a nearby river or through the accumulation of rainfall. An essential feature is a water level control structure that is sufficient to drain all the surface water from the GTR. Generally, management guidelines call for flooding the GTR to a depth of no more than about 30 to 40 cm after the trees have become dormant in the fall and draining off the water prior to the growing season in the spring.

Despite the availability of management guidelines for GTRs, at least one major question about GTRs still remains, that is — do GTRs represent a truly sustainable form of management? GTRs generally have a hydroperiod that is substantially different from the natural flooding regime. In the dormant season, the water level in GTRs is more stable and generally higher than in non-impounded sites (Figure 1). Also, in practice, many GTRs remain flooded well into the spring and even early summer, either intentionally or due to problems such as beavers plugging up the water level control structure. The long-term impacts on vegetation of such hydrologic alterations are still unclear.

Early reports on the effects of GTR management indicated that they actually had positive impacts on existing vegetation. Broadfoot found smaller increases in the annual diameter growth of "representative" trees prior to establishment of a 160-acre GTR than for the same trees after establishment.[31] The site was flooded

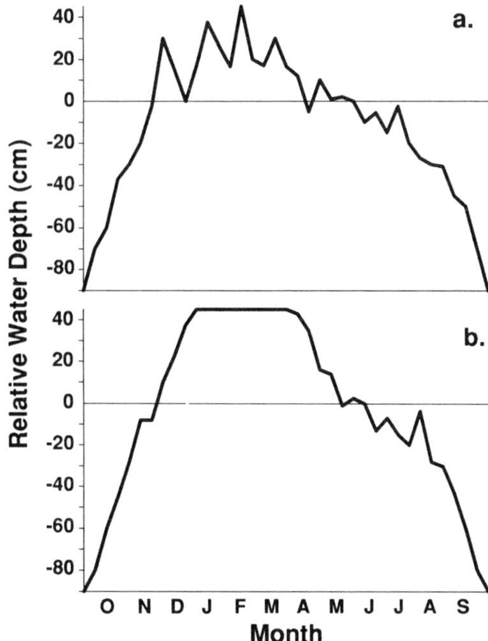

Figure 1. Depiction of A, "typical" hydrograph for naturally flooded bottomland hardwood forest; and B, hydrograph for same forest managed as a greentree reservoir (GTR).

annually to a depth of 15 to 30 cm. Broadfoot also reported increased radial tree growth for 12 species in two Mississippi Delta GTRs flooded as deep as 90 cm from February to early July.[32]

Mast production was also reported to be increased as a result of GTR management in the earliest reports. Minckler and McDermott compared acorn production in a 4-year study of a pin oak-dominated GTR in Missouri.[33] Acorn production within the GTR exceeded that of the control site in three of the years. They also found that a much higher percentage of the acorns were viable within the GTR, which they attributed to the effect of flooding on over-wintering populations of acorn weevils (*Curculio* spp.). However, this did not lead to increased oak reproduction. Winter flooding may have killed seedlings established the year before and waterfowl may have consumed a large proportion of the acorns before they could germinate.

Wigley and Filer's survey of GTR managers reported that 43% of the 179 GTRs had poor regeneration of desirable tree species and 40% had high levels of tree mortality.[24] A host of other problems was also cited by some of the managers, such as crown dieback, basal swelling, increased beaver damage, slow diameter growth, and poor acorn production. In general, more problems occurred on older (>20-year-old) sites.

The few studies of GTRs older than about 10 to 15 years have found a negative impact on the original vegetation. Nuttall oaks (*Quercus nuttalli*), flooded annually for 15 years in a Mississippi GTR, had a smaller rate of increase in crown area and had greater mortality than oaks on a control site.[34] This same population of Nuttall oaks also had reduced diameter growth and failed to increase their rate of growth in response to thinning when compared to trees in the control.[35] Newling studied this same Mississippi GTR and found no significant difference in overstory composition of the GTR and control site.[26] However, the understories of the two sites were quite different, with flood-tolerant species such as water hickory (*Carya aquatica*), overcup oak (*Q. lyrata*), and common buttonbush (*Cephalanthus occidentalis*) having much higher importance values within the GTR. Similar shifts towards more flood-tolerant tree, shrub, and herbaceous species have been reported at other sites.[36,37]

In at least one instance, changes in species composition within a GTR have been predicted using a computer simulation model. The model (FORFLO[38]) predicted a gradual shift in understory species composition towards wetter-site species in a large Arkansas GTR over the first 50 years.[39] Although the application of models for predicting the impact of water level manipulations on wetland vegetation is still in its infancy, results to date indicate that they may provide a useful tool for impact analyses and identification of sustainable water level management regimes.

The evidence suggests that the potential for gradual habitat degradation exists in a substantial proportion of GTRs. More work remains to be done, however, before the question of long-term sustainability can be answered with any certainty. In particular, long-term studies with replicated treatments and good controls are needed.[29] Further development and testing of models such as FORFLO should also prove valuable to GTR managers and others planning water level manipulations in forested wetlands. Given the substantial losses of bottomland hardwood habitat that have already occurred, managers can ill afford to degrade remaining forests for short-term benefits.

MOIST SOIL MANAGEMENT

Management of water levels in GTRs is critical to provide both suitable habitat for waterfowl without changing existing plant communities. In contrast, moist soil management is characterized by intensive manipulation of cleared habitat to promote specific plant and invertebrate populations beneficial to waterfowl and other wildlife.

Moist soil habitats — both managed and unmanaged — are usually found in flood plains, sloughs, and other shallow depressions. In the Lower Mississippi Alluvial Valley (LMAV), these habitats have alluvial origins. Their soils hold sufficient moisture for the growth of wetland and moist soil vegetation, yet are not flooded during the entire growing season. Typically, such habitat is found on sites of former bottomland forest, cleared in the past for timber harvest and agriculture. However, natural openings in forested sites and along water courses also support moist soil vegetation.

Moist soil habitat consists primarily of a community of herbaceous plant species managed to provide food for waterfowl. In addition, open water over food supplies is a critical requirement for wintering waterfowl habitat in the LMAV.[29] Thus, flooding during the waterfowl-use period in the LMAV is essentail in managing these sites.

Little is known about the natural hydrology of moist soil habitat; under management, either flooding and drawdown are completely controlled mechanically or the sites flood in response to rainfall. Water control structures are installed to provide predictable open-water habitat for waterfowl in the winter and spring. Occasionally, pumping facilities are available to add water; but generally, rainfall is sufficient to provide some standing water. Impoundments are typically allowed to flood in November, at about the same time that large numbers of migratory waterfowl arrive. Standing water is maintained through the winter and well into spring. Impoundments may be drawn down between March and July to encourage vegetation development.

Since the moist soil plant community is largely a managed one, and since the degree of management varies greatly between sites and managers, there is no characteristic species composition. Although many species lists have been generated for moist soil communities in the Mississippi Alluvial Valley,[4,40-43] few studies have concentrated on the dynamics of the plant communities.

The goal of moist soil management is to maintain the vegetation in a recurring state of early secondary succession dominated by species that provide abundant food for waterfowl and shore birds. The species that are desirable for meeting this goal are mainly invasive or weedy annual species characteristic of "old field" communities or aquatic communities. Characteristic species germinate and grow quickly on bare, moist soil. Most produce abundant seeds and some produce edible stems, leaves, rhizomes, tubers, or roots (Table 1). Perennial species such as *Cyperus esculentus* and *Sagittaria platyphylla* produce edible tubers. Their presence and abundance can be encouraged by "disking", a technique that appears to increase regeneration by tubers.[41] These tuber-producing perennials are treated as though they reproduce as annuals, leaving a "tuber bank" in the soil.

Since the moist soil plant community, as it is managed for waterfowl, is largely composed of annual and tuberous perennial species, species composition in any site can be highly variable from year to year. Community composition, growth, and seed and tuber production depend greatly on seasonal flooding and soil moisture conditions and the degree of soil disturbance. Common activities employed in managing moist soil vegetation include: manipulating the timing and rate of drawdown, reflooding, disking, and burning. These techniques are used to promote a community of desirable species and to eradicate or reduce undesirable species.

Management practices, such as drawdown timing, timing and depth of disking, burning, and summer reflooding, while generally giving desired results, are imprecise and often unpredictable. The surface propagule

Table 1. Range of Seed/Tuber Production Reported for Moist Soil Plant Species

Species	Production (kg/ha)
Seeds	
Bidens spp.	4.7–85.3
Cyperus erythrorhizos	0.2–0.6
Digitaria spp.	0.1–992.0
Diodia virginiana	0.4–14.9
Echinochloa crusgalli var. frumentacea	1680.0–2240.0
Echinochloa crusgalli	1.0–66.0
Eclipta alba	0.1–1.8
Eleocharis obtusa	1.3–96.2
Fimbristylis autumnalis	0.7–4.4
Leersia oryzoides	24.0–250.0
Lippia spp.	0.2–1.6
Panicum dichotomiflorum	1.0–192.0
Panicum virgatum	1680.0–2240.0
Polygonum densiflorum	490.0–745.0
Polygonum hydropiperoides	21.0–85.0
Rhynchospora corniculata	573.0–1205.0
Sida spinosa	0.2–5.5
Tubers	
Cyperus esculentus	12.0–330.0

bank, local moisture, temperature and nutrient conditions, and the temporal pattern of environmental conditions during the growing season control the sequence of germination, competitive interactions between species, and the final biomass of waterfowl food produced. Even minor alterations in the soil environment can cause year-to-year changes in the nature and production of plant species.[41]

Seed germination is the most critical factor determining the quality of moist soil habitats. Moist soil habitat is typically flooded during the winter and part of the spring to provide feeding and resting habitat for dabbling ducks. As climatic conditions begin to warm in the spring, aquatic plants begin to develop from over-wintering organs. However, the most desirable waterfowl species and many nuisance species will not germinate under water. It is only following drawdown that seeds of grasses, smartweeds, and many sedges germinate. Seed germination relative to the timing of drawdown determines the annual character and productivity of the moist soil community. Frederickson and Taylor describe the expected communities resulting from early and late and fast and slow rates of drawdown.[4] A slow drawdown early in the growing season provides the greatest opportunities for different species to germinate. Disking is an effective tool in stimulating seed germination. Soil mixing serves to bring dormant seeds to the surface. Some species are especially stimulated to germinate by after-ripening during burial.[44] However, disking also serves to aerate the soil and mix organic matter leading to increased decomposition and release of nutrients.

It is often difficult to predict the relative abundance of any particular species in a moist soil site; each has fairly specific requirements for germination, and the timing and balance of temperature change and rainfall vary from year to year. Random events, such as an early period of hot dry weather or a major rainstorm in mid-summer, may stimulate a species to grow and dominate the community.

Temperature and moisture act together to open and close "windows of opportunity" for seed germination by various species over the growing season. Dormancy and germination requirements are species-specific and may change over the growing season. Seeds of a given species may lose and regain dormancy over a period of a few weeks. Obligate winter annuals only germinate under autumn-like conditions, while "facultative winter annuals" germinate in spring or autumn. Some "summer annuals" germinate only in the spring, while others may germinate at anytime from spring through fall. A progression of species, germinating in sequence, is usually observed after a drawdown.[4]

CONCLUSIONS

Wetlands can be dynamic systems where the vegetation changes in response to changing hydrology.[45] An understanding of this relationship is important in recognizing the limits and consequences of management. An important goal for managing vegetation for wildlife is to produce consistently desirable habitat. When perennial vegetation is involved, such as in GTRs, management practices are aimed at maintaining the extant vegetation and discouraging undesirable species. Moist soil habitats dominated by annual plants require more intensive management because they must be regenerated from year to year. Water level manipulation is a powerful tool for managing wetland plant communities for waterfowl habitat. Different stages in the life history of plant species differ in their ability to tolerate flooding (depth, duration, and frequency) and the use of water level manipulation can influence plant species composition and dynamics.

Earlier studies have recognized the differential susceptibility of wetland species to flooding and the importance of topographic gradients which allow the displacement of species along water depth gradients. Numerous other processes are also affected by changes in water level, including germination, growth of seedlings, mortality and survival, and seed production. In addition, we must realize that the timing and degree of water level fluctuation can also be important determinants of the vegetation.[4]

It is clear that multiple use of wetlands constructed for wastewater treatment will often not be feasible because different management objectives will result in incompatible hydrologic regimes. Our understanding of the life history of wetland and aquatic species is the key to predicting the response of the wetland plant community to altered hydrology and the ability of sites to sustain management for multiple use.

ACKNOWLEDGMENTS

We wish to thank F. Stearns and G. Wein for reviewing an earlier version of this manuscript. GRG acknowledges support from the Wetlands Research Program (EPA-Duluth Environmental Research Laboratory) during the final preparation of this manuscript.

REFERENCES

1. Mitsch, W. J. and J. G. Gosselink. *Wetlands.* Van Nostrand Reinhold, New York, 1986.
2. Weller, M. W. *Freshwater Marshes: Ecology and Wildlife Management.* University of Minnesota Press, Minneapolis, 1981, 146.
3. Suring, L. H. and M. D. Knighton. History of water impoundments in wildlife management, in *Water Impoundments for Wildlife: A Habitat Management Workshop.* U.S. Department of Agriculture, Forest Service, General Technical Report NC-100, Washington, 1988, 15.
4. Fredrickson, L. H. and T. S. Taylor. Management of seasonally flooded impoundments for wildlife, U.S. Department Of The Interior, Fish and Wildlife Service, Resource Publication 148, Washington, 1982, 29.
5. Hutchinson, G. E. *Limnological Botany.* Academic Press, New York, 1975, 660.
6. Spence, D. H. N. The zonation of plants in freshwater Lakes, *Adv. Ecological Res.* 15:37, 1982.
7. Kadlec, J. A. and W. A. Wentz. State of the art survey and evaluation of marsh plant establishment techniques: induced and natural, 2 vols., National Technical Information Service, Springfield, 1974.
8. Kadlec, J. A., The effects of drawdown on a waterfowl impoundment, *Ecology.* 43:267, 1962.
9. Millar, J. B. Vegetation changes in shallow marsh wetlands under improving moisture regime, *Can. J. Bot.* 51:1443, 1973.
10. Wilcox, D. A. and J. E. Meeker. Disturbance effects on aquatic vegetation in regulated and unregulated lakes in northern Minnesota, *Can. J. Bot.* 69:1542, 1991.
11. Brown, S. L. A comparison of the structure, primary productivity, and the transpiration of Cypress Ecosystems in Florida, *Ecol. Monogr.,* 51:403, 1981.
12. van der Valk, A. G. and C. B. Davis. The role of seedbanks in the vegetation dynamics of prairie glacial marshes, *Ecology.* 59:322, 1978.
13. Keddy, P. A. and A. A. Reznicek. Vegetation dynamics, buried seeds, and water level fluctuations on the shorelines of the Great Lakes, in *Coastal Wetlands.* H. H. Prince and F. M. D'Itri, Eds. Lewis Publishers, Chelsea, MI, 1985.

14. Keough, J. R. Response of *Scirpus validus* to the Physical Environment and Consideration of its Role in a Great Lake Estuarine System. Ph.D. thesis, University of Wisconsin-Milwaukee, 1987.
15. Sharitz, R. R. and J. W. Gibbons. The Ecology of Southeastern Shrub Bogs (pocosins) and Carolina Bays: A Community Profile. U.S. Fish and Wildlife Service, Division of Biological Services, Washington, FWS/OBS-82/04, 1982, 93.
16. Keough, J. R., G. R. Guntenspergen, and J. Grace. Vegetation and hydrologic characteristics of Carolina Bays, unpublished report, 1990.
17. Herdendorf, C. E., S. M. Hartley, and M. D. Barnes, Eds. Fish and Wildlife Resources of the Great Lakes coastal wetlands within the United States, Vol. 1: Overview, Fish and Wildlife Service, Biological Services Program, Washington, FWS/OBS-81/02-V1, Washington, 1981, 469.
18. van der Valk, A. G. Succession in wetlands: a Gleasonian approach, *Ecology*. 62:688, 1981.
19. van der Valk, A. G. Succession in temperate North American wetlands, in *Wetlands: Ecology and Management*. B. Gopal, E. Turner, R. G. Wetzel, and D. F. Whigham, Eds. National Institute of Ecology and International Scientific Publications, Jaipur, 1982, 169.
20. van der Valk, A. G., C. H. Welling, and R. L. Pederson. Vegetation change in a freshwater wetland: a test of a priori predictions, in *Freshwater Wetlands and Wildlife*. R. R. Sharitz and J. W. Gibbons, Eds. USDOE Office of Scientific and Technical Information, Oak Ridge, 1989, 207.
21. Brown, R. G. and J. R. Stark. Hydrologic and water quality characteristics of a wetland receiving wastewater effluent in St. Joseph, Minnesota, *Wetlands*. 9:191, 1989.
22. Kadlec, R. H. and F. B. Bevis. Wetlands and wastewater: Kinross, Michigan, *Wetlands*. 10:77, 1990.
23. Whigham, D. F. Vegetation in wetlands receiving sewage effluent: the importance of the seedbank, in *Ecological Considerations in Wetlands Treatment of Municipal Wastewaters*. P. J. Godfrey, E. R. Kaynor, S. Pelczarski, and J. Benforado, Eds. van Nostrand Reinhold, New York, 1985, 231.
24. Wigley, T. B. and T. H. Filer, Jr. Characteristics of greentree reservoirs: a survey of managers, *Wildlife Soc. Bull.* 17:136, 1989.
25. Thompson, D. Q., P. B. Reed, Jr., G. E. Cummings, and E. Kivasalu. Muck-hardwoods as green-timber impoundments for waterfowl, *Trans. N. Am. Wildlife Resources Conf.* 33:142, 1968.
26. Newling, C. J. Ecological investigations of a greentree reservoir in the Delta National Forest, Mississippi. U.S. Army Corps of Engineers Miscellaneous Paper EL-81-5, Waterways Experiment Station, Vicksburg, 1981.
27. MacDonald, P. O., W. E. Frayer, and J. K. Clauser. Documentation, chronology, and future projections of bottomland hardwood habitat loss in the lower Mississippi Alluvial Plain, Vol 1., Basic Report, U. S. Fish and Wildlife Service, Vicksburg, VA, 1979, 133.
28. Mitchell, W. A. and C. J. Newling. Greentree reservoirs, Section 5. 5. 3, U.S. Army Corps of Engineers Wildlife Resources Management Manual, Waterways Experiment Station, Vicksburg, VA, 1986.
29. Rienecke, K. J., R. M. Kaminski, D. J. Moorhead, J. D. Hodges, and J. R. Nassar. Mississippi Alluvial Valley, in *Habitat management for migrating and wintering waterfowl in North America*. L. H. Smith, R. L. Pederson, and R. M. Kaminski, Eds. Texas Tech University Press, Lubbock, 1989, 203.
30. Fredrickson, L. H. and D. L. Batema. *Greentree Reservoir Management Handbook*. Gaylord Memorial Laboratory, University of Missouri-Columbia, Puxico, manuscript in preparation.
31. Broadfoot, W. M. Reaction of Hardwood Timber to Shallow-Water Impoundments. Mississippi State University Information Sheet 595, State College, 1958, 2.
32. Broadfoot, W. M. Shallow-water impoundment increases soil moisture and growth of hardwoods, *Soil Sci. Soc. Am. Proc.* 31:562, 1967.
33. Minckler, L. S. and R. E. McDermott. Pin Oak Acorn Production and Regeneration as Affected by Stand Density, Structure, and Flooding. University of Missouri Agricultural Experiment Station Research Bulletin 750, Columbia, 1960, 24.
34. Francis, J. K. Acorn Production and Tree Growth of Nuttall Oak in a Greentree Reservoir. USDA Forest Service Research Note SO-289, Southern Forest Experiment Station, New Orleans, 1983, 3.
35. Schlaegel, B. E. Long-Term Artificial Flooding Reduces Nuttall Oak Bole Growth. USDA Forest Service Research Note SO-309, Southern Forest Experiment Station, New Orleans, 1984, 3.
36. Karr, B. L., G. L. Young, B. D. Leopold, J. D. Hodges, and R. M. Kaminski. Effect of Flooding on Greentree Reservoirs. Technical Interim Report, Project Number G1571-03, Mississippi State University, Mississippi, 1989, 16.
37. Smith, D. E. The Effects of Greentree Reservoir Management on the Development of Basal Swelling Damage and on the Forest Dynamics of Missouri's Bottomland Hardwoods. Ph.D. dissertation, University of Missouri, Columbia, 1984, 126.

38. Pearlstine, L., H. McKellar, and W. Kitchens. Modelling the impacts of a river diversion on bottomland forest communities in the Santee River floodplain, South Carolina, *Ecological Modelling*. 29:283, 1985.
39. Allen, J. A., J. T. Teaford, E. C. Pendleton, and M. Brody. Evaluation of greentree reservoir management options in Arkansas, *Trans. N. Am. Wildlife and Natural Resources Conf.* 53:471, 1988.
40. Knauer, D. F. Moist Soil Plant Production on Mingo National Wildlife Refuge. M.S. thesis, University of Missouri, Columbia, 1977, 189.
41. Kelley, J. R., Jr. Management and Biomass Production of Selected Moist Soil Plants. M.S. thesis, University of Missouri, Columbia, 1986, 68.
42. Harrison, A. J. and R. H. Chabreck. Duck food production in openings in forested wetlands, in *Waterfowl in Winter*. M. W. Weller, Ed. University of Minnesota Press, Minneapolis, 1988, 339.
43. Wright, T. W. Winter foods of mallards in Arkansas, *Proc. Southeast Assoc. Game and Fish Commission*. 13:291, 1959.
44. Baskin, J. M. and C. C. Baskin. Physiology of dormancy and germination in relation to seed bank ecology, in *Ecology of Seed Banks*, M. A. Leck, V. T. Parker, and R. L. Simpson, Eds. Academic Press, New York, 1989, 53.
45. Niering, W. A. Hydrology, disturbance, and vegetation change, *Proc. Natl. Wetland Symp.: Wetland Hydrology*. J. A. Kusler and G. Brooks, Eds. Association State Wetland Managers, Technical Report 6, Berne, 1988, 50.

CHAPTER 41

Macrophyte-Mediated Oxygen Transfer in Wetlands: Transport Mechanisms and Rates

H. Brix

INTRODUCTION

In constructed wetlands with subsurface water flow, wastewater is believed to be purified during contact with the surface of soil/gravel particles and with the roots of macrophytes. Sediments in wetlands are continuously water saturated and therefore generally anoxic or anaerobic. Wetland plants are morphologically adapted to growing in a water-logged sediment by virtue of large internal air spaces for transportation of O_2 to roots and rhizomes. The extensive internal lacunal system, which normally contains constrictions at intervals to maintain structural integrity and to restrict water invasion into damaged tissues, may occupy up to 60% of the total tissue volume.[1] The internal O_2 movement down the plant serves not only the respiratory demands of the buried tissues, but also supplies the rhizosphere with O_2 by leakage from the roots.[2] This oxygen leakage from roots creates oxidized conditions in the otherwise anoxic substrate and stimulates both aerobic decomposition of organic matter and growth of nitrifying bacteria. Wetland plants are therefore believed to play an important function in relation to treatment of wastewater in constructed wetlands with subsurface water flow. This article summarizes the existing knowledge on internal gas transport mechanisms in wetland plants and oxygen leakage from their root system.

EFFECTS OF WATER-LOGGING

The primary difference between water-saturated and well-drained soils is the availability of oxygen for root respiration, microbial respiration, and chemical oxidation processes. In well-drained soils, the pore spaces are filled with air with a relatively high oxygen content. Microorganisms living in the soil and roots of plants growing in the soil are therefore able to obtain oxygen directly from their surroundings (Figure 1). As the soil pore spaces are interconnected to the atmosphere above the soil, the oxygen in the pore spaces is replenished by rapid diffusion and convection from the atmosphere. In a water-saturated soil, the pore spaces are filled with water. The rate of diffusion of oxygen through water is some 3×10^6 times slower than through air, principally due to the smaller diffusion coefficient in water, but also because of the low solubility of oxygen in water.[3] Consequently, water-saturated soils become anaerobic (oxygen-free or anoxic) except for a few millimeters at the surface. Therefore roots and rhizomes of plants growing in water-saturated substrates must obtain oxygen from their aerial organs via transport internally in the plant. The anaerobic conditions in water-saturated soils lead to chemical and biological transformations within the soil[4] and result in release to the soil solution of reduced substances such as Mn^{2+}, Fe^{2+}, H_2S, and organic acids. Some of these reduced substances can reach concentrations in the rhizosphere that are toxic to roots. Oxygen leakage from roots of wetland plants is important in relation to detoxification of these reduced substances within the soil solution.

GAS TRANSPORT MECHANISMS IN WETLAND PLANTS

Internal transportation of oxygen in wetland plants may occur by *passive molecular diffusion* following the concentration gradients within the lacunal system, and by *convective flow* (i.e., bulk flow) of air through the

0-87371-550-0/93/$0.00+$.50

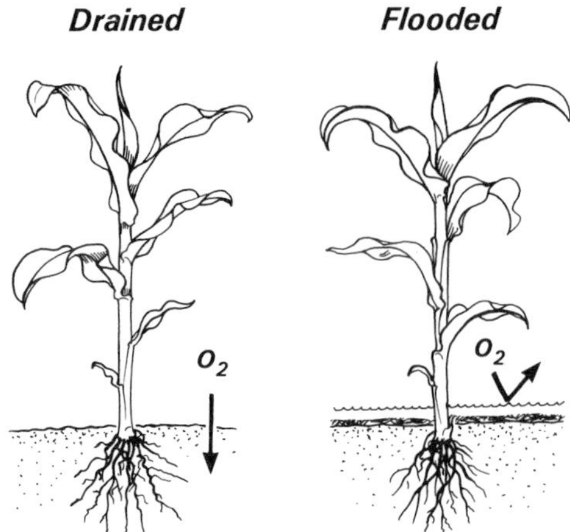

Figure 1. Sketch showing the main difference in soil oxygen status under well-drained and water-saturated conditions. In a well-drained soil, the pore spaces are filled with air with a relatively high content of O_2 which is readily replenished by rapid diffusion from the atmosphere. The plant roots are able to obtain respiratory O_2 from the soil atmosphere. In a water-saturated soil, the diffusion of O_2 through the water-filled pore space is slow and, as a consequence, the soil becomes anoxic. Plant roots must therefore obtain the O_2 needed for respiration from internal sources: that is, by transport from aerial plant parts.

internal gas spaces of the plants. Diffusion is the process by which matter is transported from one part of a system to another as the result of random molecular movement. The net movement of matter will be from sites with higher concentrations (or partial pressures) to sites with lower concentrations. The rate of diffusion of a gas depends on the medium in which the diffusion occurs, the molecular weight of the gas, and the temperature. These factors find expression in the diffusion coefficient, the magnitude of which for various substances can be found in standard tabular works. Diffusive transfer of matter can be mathematically expressed by Fick's laws.[5]

The internal gas transport mechanism in wetland plants was classically believed to be passive diffusion along the concentration gradients of the individual gases[6-8] (Figure 2). This hypothesis was based mainly on observations of reverse-order gradients of O_2 and CO_2 in the lacunae. In many cases, a steep gradient of O_2, with concentration decreasing with distance from the aerial plant parts, and gradients of CO_2 and CH_4 in the reverse direction, have been observed. In the common reed (*Phragmites australis*), the O_2 concentration is reported to decrease from close to atmospheric levels (20.7%) in the aerial stems to low levels (3.6%) in the lacunal air of the deepest-growing rhizomes.[9] At the same time, the concentrations of CO_2 increase from 0.07% in the aerial stems to 7.3% in the deepest-growing rhizomes. Such concentration gradients clearly show that internal diffusive transfer of O_2 plays a significant role for aeration of buried root systems.

The general belief that internal gas transport through the air spaces of shoots and roots of wetland plants occurred exclusively by gas-phase diffusion has recently been invalidated. In many species, *convection* (i.e., bulk flow of gas) plays a significant role for aeration of the below-ground tissues. Convective flow of air in plants can be *throughflow* or *non-throughflow*[10] and can be initiated by different physical processes.

At the end of the last century, observations on a convective throughflow in different species of water lilies and other floating-leaved plants were described.[11] This throughflow mechanism in water lilies was studied in detail in the early 1980s.[12-15] The convection in water lilies is driven by temperature and water vapor pressure differences between the inside of leaves and the surrounding air (Knudsen diffusion).[16] The air taken in from the atmosphere is vented through rhizomes and old leaves back to the atmosphere. The driving forces are two purely physical processes: *thermal transpiration* and *humidity-induced pressurization*. Both processes require a porous partition within the plant tissue, ideally with pore sizes less than the mean free diffusive path of the gas molecules (<0.1 μm), and consumption of energy in the form of heat. Humidity-induced pressurization requires, in addition, a constant supply of water inside the plant.

Thermal transpiration is the movement of gas through a porous partition when there is a gradient in temperature across the partition. Thermal transpiration leads to a pressure gradient across the partition, the pressure being higher on the warmer side. Humidity-induced pressurization is related to pressure differentials

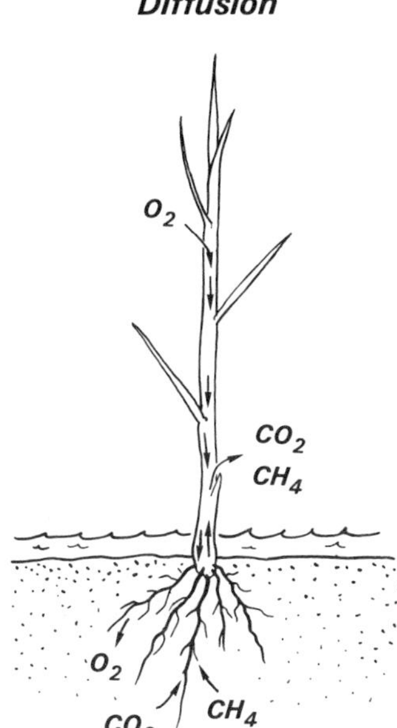

Figure 2. Passive diffusion of gases in the lacunal system of wetland plants. Oxygen diffuses along a concentration gradient from the atmosphere into the aerial plant parts and down the internal gas spaces to the rhizomes and roots. Conversely, CO_2 produced by respiration of the root system and CH_4 produced in the sediment diffuses along reverse concentration gradients in the opposite direction.

induced by differences in water vapor pressure across a porous partition. The result of humidity-induced pressurization is that the total pressure will be higher on the more humid side of the partition. In the case of water lilies, the pressurization is greatest in the youngest leaves. The slightly increased pressures in young leaves are released through the leaf petioles, the buried rhizome, and back to the atmosphere through petioles and leaf blades of older leaves (Figure 3). Pressurized ventilation of the root system is not restricted to water lilies and species with a similar morphology. The aeration of the rhizomes of the common reed (*Phragmites australis*) is significantly enhanced by a similar convective throughflow mechanism.[17,18] Furthermore, recent investigations have shown that internal pressurization and convective throughflow driven by gradients in temperature and water vapor pressure seem to be common attributes of a wide range of wetland plants, including species with cylindrical and linear leaves (e.g., *Typha*, *Schoenoplectus*, *Eleocharis*).[19]

Exchange of gases between the gas spaces of buried plant tissues and the surrounding water may also lead to convective air flow inside the plant. This mechanism is based on the different solubilities of O_2 and CO_2 in water, CO_2 being approximately 30 times more soluble in water than O_2. The first observation of convective flow based on *underwater gas exchange* was observed in the black mangrove (*Avicennia nitida*) which grows in the tidal zone in coastal subtropical areas.[20] One single tree or bush of *Avicennia* may have several thousand 20- to 30-cm high air roots (pneumatophores) which protrude vertically from the mud. The air roots have numerous lenticels (openings to the atmosphere) and are connected to the buried horizontal root system. The function of the air roots is to transfer O_2 to the root system. When the rising tide covers the lenticels on the air roots, the pressure in the air space tissue of the roots starts to drop. This drop in total gas pressure within the root system is caused by respiratory consumption of O_2 by the root system. In addition, part of the CO_2 produced by respiration dissolves in the interstitial water because of the relatively high solubility of CO_2 in water. The total gas pressure deficit in the root system persists until the lenticels are again exposed to the air by the falling tide, and atmospheric air is drawn into the root system (Figure 4).

The same mechanism of non-throughflow convection has been shown to function in deep water rice (*Oryza sativa*).[21] In this species, gas movement occurs in continuous air layers trapped between the hydrophobic surface of leaf blades and the surrounding water. In darkness, air is drawn down from the atmosphere through the air layers along a pressure gradient created by solubilization of respiratory CO_2 in the surrounding water. Solubilization of respiratory CO_2 has also been shown to produce some convective gas flow in other species

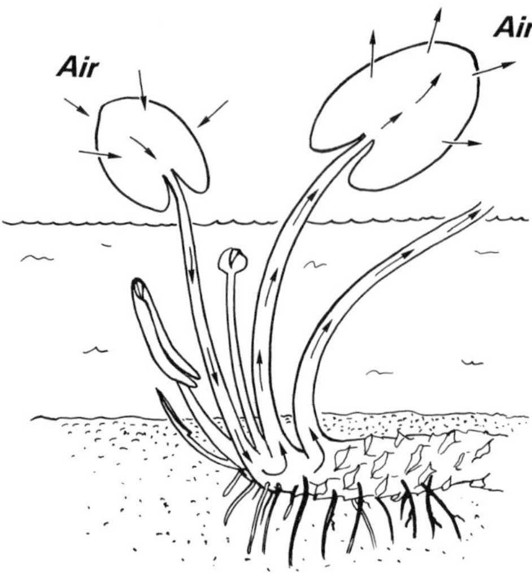

Figure 3. Convective throughflow of gas in water lilies. Air enters the youngest emergent leaves against a small pressure gradient as a consequence of humidity-induced pressurization and thermal transpiration, passes down their petioles to the rhizome, and up the petioles of older emergent leaves back to the atmosphere. (Modified from Reference 13.)

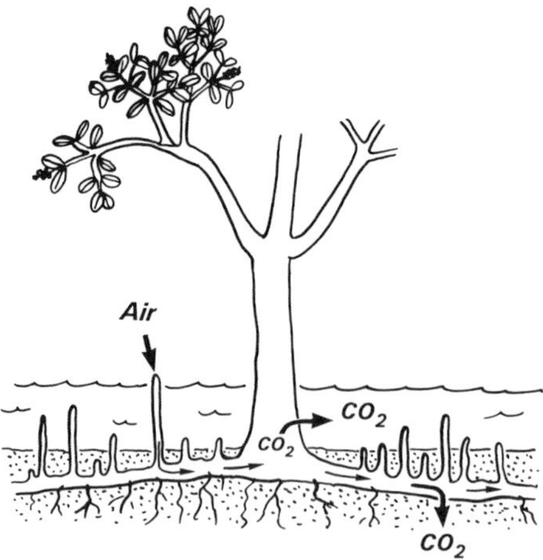

Figure 4. Non-throughflow convection in the black mangrove (*Avicennia nitida*) induced by underwater gas exchange. The black mangrove has air roots with numerous lenticels which protrude vertically from the mud. When the tide covers the lenticels on the air roots, the pressure in the air space tissue of the roots decreases as a consequence of consumption of O_2 in respiration and production and subsequent solubilization of CO_2 in the interstitial waters. When the tide falls and the lenticels are again exposed to the air, atmospheric air is drawn into the root system though the first emerging air root.

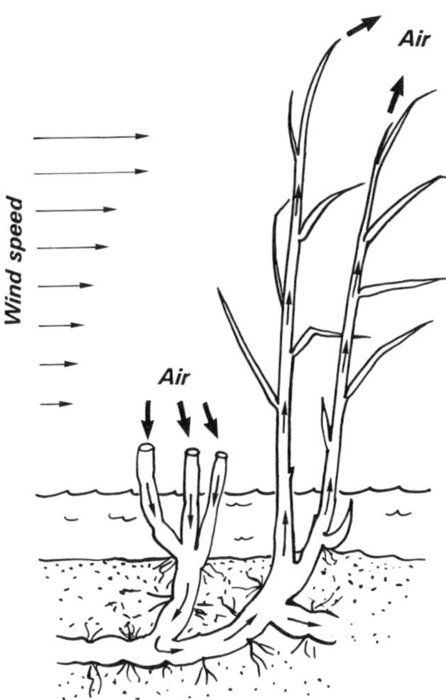

Figure 5. Venturi-induced convective throughflow in *Phragmites australis*. The taller shoots and old culms are exposed to higher wind velocities than broken shoots and stubbles close to the ground level. This induces a pressure differential which sucks atmospheric air into the underground root system.

of wetland plants (e.g., *Carex gracilis*).[22] Mathematical analyses have however demonstrated that this type of convection is subordinate to diffusion in the aeration process.[23] The cause of this is, that for every five O_2 molecules used by respiration, only one O_2 molecule is drawn in by convection, assuming all respiratory CO_2 is solubilized in the surrounding water.

Recently, another type of convective throughflow has been documented in *Phragmites*, namely *venturi-induced convection*.[24] This mechanism is based on the gradient in wind velocity around the plant, the velocity being higher at higher positions in the canopy. Venturi-induced convection is driven by pressure differentials created by wind blowing across tall dead culms which sucks atmospheric air into the underground root system via broken culms close to the ground level (Figure 5). The ventilation system is equivalent to the system used by prairie dogs to ventilate their tunnel systems. The prairie dogs maintain the openings of their tunnels at various elevations above the general surface of the prairie. The taller openings are exposed to higher wind velocities, and therefore lower pressures; as a result, air flows into the lower openings, through the tunnels, and out of the taller openings.[25] Venturi-induced convection in plants has until now only been documented for *Phragmites*, but it is likely that the mechanism also operates for other species of wetland plants. In contrast to humidity- and temperature-induced convection, which depend on specific porous structures within the plant tissue and gradients in water vapor pressure and temperature, venturi-induced convection can operate in damaged and dead plants and also during the night and winter, where water vapor and temperature gradients are small or lacking.

ROOT RELEASE OF OXYGEN

It is well documented that aquatic macrophytes release O_2 from roots into the rhizosphere. Most studies have been done using oxygen microelectrodes to measure radial O_2 losses from individual roots in oxygen-

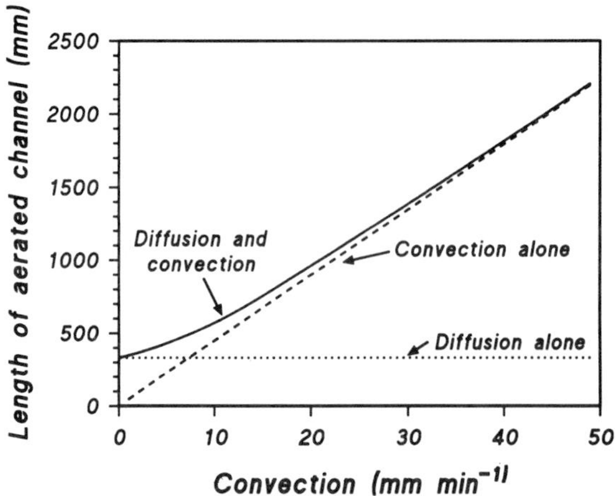

Figure 6. The length of a root which can be aerated by diffusion alone, throughflow convection alone, and combined diffusion and convection assuming a uniform respiratory O_2 demand of the root tissue of 100 pg/mm^3/s and no O_2 release. (From Reference 10. With permission.)

depleted solutions.[26-28] The O_2 release rates obtained by this technique vary from less than 10 to 160 ng O_2 per square centimeter root surface per minute, depending on species. The rates of O_2 release in wetland plants are generally highest in the subapical region of roots and decrease with distance from the root apex.[29] Decreased O_2 release rates with tissue age are probably caused by cuticularization and hypodermal suberization and associated decreased gas permeability of the root walls. The decreased gas permeability of the root walls minimizes radial leakage outward, allowing more oxygen to reach the apical meristem. Oxygen release from fine laterals at the base of roots can be significant, but generally, no release of oxygen from old roots and rhizomes are detected.[30] The heterogeneity of the O_2 release pattern of wetland roots makes it difficult or impossible to extrapolate from results obtained by the oxygen microelectrode technique to *in situ* release rates. Using different assumptions of root O_2 release rates, root dimensions, numbers, permeability, etc., Lawson[31] calculated a possible O_2 flux from roots of *Phragmites* to 4.3 g/m^2/day. Others, using different techniques, have estimated root O_2 release rates from *Phragmites* to be 0.02 g/m^2/day,[32] 1 to 2 g/m^2/day,[33] and 5 to 12 g/m^2/day.[34] The wide range in these values is caused partly by the different experimental techniques used in the studies, and partly by the seasonal variation in O_2 release rates. Root O_2 release rates from a number of submerged aquatic plants is reported to be in the range 0.5 to 5.2 g/m^2/day,[35-37] and from free-floating plants 2.4 to 9.6 g/m^2/day.[38]

DISCUSSION

Oxygen release rates from roots depend on the internal O_2 concentration, the O_2 demand of the surrounding medium, and the permeability of the root walls. Wetland plants conserve internal oxygen because of suberized and lignified layers in the hypodermis and outer cortex.[30] These stop radial leakage outward, allowing more O_2 to reach the apical meristem. This characteristic of wetland plants to conserve internal O_2 is inconsistent with the concept of wastewater treatment in subsurface flow constructed wetlands. The wetland plants attempt to minimize their O_2 losses to the rhizosphere, whereas the concept of wastewater treatment implies a high O_2 leakage from roots. Wetland plants do, however, release O_2 from their roots. Leakage occurs primarily at the root tip, and it serves to oxidize and detoxify potentially harmful reducing substances in the rhizosphere. Species possessing an internal convective throughflow ventilation system have higher internal O_2 concentrations in the rhizomes and roots than species relying exclusively on diffusive transfer of oxygen.[34] Furthermore, convective through flow significantly increases the root length that can be aerated, compared to the length by diffusion alone (Figure 6).[10] Wetland plants with a convective throughflow mechanism therefore have the potential to release more O_2 from their roots compared to species without convective throughflow. Thus, from the point of view of rhizosphere oxidation, these species are superior in constructed wetlands. However, other factors such as rooting depth, tolerance to high loads of wastewater, plant productivity, etc. also have to be taken into consideration when considering the suitability of different species in constructed wetlands.

REFERENCES

1. Studer, C. and R. Brändle. Oxygen consumption and availability in the rhizomes of *Acorus calamus* L., *Glyceria maxima* (Hartmann) Holmberg, *Menyanthes trifoliata* L., *Phalaris arundinacea* L., *Phragmites australis* Trin. and *Typha latifolia* L., *Bot. Helv.* 94:23, 1984.
2. Armstrong, W., P. M. Beckett, S. H. F. W. Justin, and S. Lythe. Modelling, and other aspects of root aeration by diffusion, in *Plant Life Under Oxygen Deprivation.* M. B. Jackson, D. D. Davies, and H. Lambers, Eds. SPB Academic Publishing bv, The Hague, The Netherlands, 1991, 267.
3. Drew, M. C. Plant responses to anaerobic conditions in soil and solution culture, *Curr. Adv. Plant Sci.* 36:1, 1979.
4. Gambrell, R. P. and W. H. Patrick. Chemical and microbiological properties of anaerobic soils and sediments, in *Plant Life in Anaerobic Environments.* D. D. Hook and R. M. M. Crawford, Eds. Ann Arbor Science, Ann Arbor, MI, 1978, 375.
5. Armstrong, W. Aeration in higher plants, *Adv. Bot. Res.* 7:225, 1979.
6. Evans, N. T. S. and M. Ebert. Radioactive oxygen in the study of gas transport down the root of *Vicea faba*, *J. Exp. Bot.* 11:246, 1960.
7. Barber, D. A., M. Ebert, and N. T. S. Evans. The movement of ^{15}O through barley and rice plants, *J. Exp. Bot.* 13:397, 1962.
8. Teal, J. M. and J. W. Kanwisher. Gas transport in the marsh grass, *Spartina alterniflora, J Exp. Bot.* 17:355, 1966.
9. Brix, H. Light-dependent variations in the composition of the internal atmosphere of *Phragmites australis* (Cav.) Trin. ex Steudel, *Aquat. Bot.* 30:319, 1988.
10. Armstrong, W., J. Armstrong, P. M. Beckett, and S. H. F. W. Justin. Convective gas-flows in wetland plant aeration, in *Plant Life Under Oxygen Deprivation.* M. B. Jackson, D. D. Davies, and H. Lambers, Eds. SPB Academic Publishing bv, The Hague, The Netherlands, 1991, 283.
11. Grosse, W., H. B. Büchel, and H. Tiebel. Pressurized ventilation in wetland plants, *Aquat. Bot.* 39:89, 1991.
12. Dacey, J. W. H. Internal winds in waterlilies: an adaptation for life in anaerobic sediments, *Science.* 210:1017, 1980.
13. Dacey, J. W. H. Pressurized ventilation in the yellow waterlily, *Ecology.* 62:1137, 1981.
14. Dacey, J. W. H. and M. J. Klug. Floating leaves and nighttime ventilation in *Nuphar, Am. J. Bot.* 69:999, 1982.
15. Grosse, W. and J. Mevi-Schütz. A beneficial gas transport system in *Nymphoides peltata, Am. J. Bot.* 74:947, 1987.
16. Schröder, P., W. Grosse, and D. Woermann. Localization of thermo-osmotically active partitions in young leaves of *Nuphar lutea*, *J. Exp. Bot.* 37:1450, 1986.
17. Armstrong, J. and W. Armstrong. Light enhanced convective throughflow increases oxygenation in rhizomes and rhizosphere of *Phragmites australis* (Cav.) Trin. ex Steud, *New Phytol.* 114:121, 1990.
18. Armstrong, J. and W. Armstrong. A convective through-flow of gases in *Phragmites australis* (Cav.) Trin. ex Steudel, *Aquat. Bot.* 39:75, 1991.
19. Brix, H., B. K. Sorrell, and P. T. Orr. Internal pressurization and convective gas flow in some emergent freshwater macrophytes, *Limnol. Oceanogr.* in press.
20. Scholander, P. F., L. van Dam, and S. I. Scholander. Gas exchange in the roots of mangroves, *Am. J. Bot.* 42:92, 1955.
21. Raskin, I. and H. Kende. How does deep water rice solve its aeration problem? *Plant Physiol.* 72:447, 1983.
22. Koncalova, H., J. Pokorny, and J. Kvet. Root ventilation in *Carex gracilis* Curt.: diffusion or mass flow? *Aquat. Bot.* 30:149, 1988.
23. Beckett, P. M., W. Armstrong, S. H. F. W. Justin, and J. Armstrong. On the relative importance of convective and diffusive gas flows in plant aeration, *New Phytol.* 110:463, 1988.
24. Armstrong, J., W. Armstrong, and P. M. Beckett. *Phragmites australis*: venturi- and humidity-induced pressure-flows enhance rhizome aeration and rhizosphere oxidation, *New Phytol.*, 120:197–207, 1992.
25. Dacey, J. W. H. How aquatic plants ventilate, *Oceanus.* 24:43, 1981.
26. Armstrong, W. The use of polarography in the assay of oxygen diffusing from roots in anaerobic media, *Physiol. Plant.* 20:540, 1967.

27. Laan, P., A. Smolders, C. W. P. M. Blom, and W. Armstrong. The relative roles of internal aeration, radial oxygen losses, iron exclusion and nutrient balances in flood-tolerance of *Rumex* species, *Acta Bot. Neerl.* 38:131, 1989.
28. Brix, H. and H.-H. Schierup. Soil oxygenation in constructed reed beds: the role of macrophyte and soil-atmosphere interface oxygen transport, in *Constructed Wetlands in Water Pollution Control.* P. F. Cooper and B. C. Findlater, Eds. Advances in Water Pollution Control, Pergamon Press, Oxford, 1990, 53.
29. Armstrong, W. Radial oxygen losses from intact rice roots as affected by distance from apex, respiration and waterlogging, *Physiol. Plant.* 25:192, 1971.
30. Armstrong, J. and W. Armstrong. *Phragmites australis* — a preliminary study of soil-oxidising sites and internal gas transport pathways, *New Phytol.* 108:373, 1988.
31. Lawson, G. J. Cultivating reeds (*Phragmites australis*) for root zone treatment of sewage. Contract report to the Water Research Centre, ITE project 965, Cumbria, U.K., 1985.
32. Brix, H. Gas exchange through the soil-atmosphere interphase and through dead culms of *Phragmites australis* in a constructed reed bed receiving domestic sewage, *Wat. Res.* 24:259, 1990.
33. Gries, C., L. Kappen, and R. Lösch. Mechanism of flood tolerance in reed, *Phragmites australis* (Cav.) Trin. ex Steudel, *New Phytol.* 114:589, 1990.
34. Armstrong, W., J. Armstrong, and P. M. Beckett. Measurement and modelling of oxygen release from roots of *Phragmites australis*, in *Constructed Wetlands in Water Pollution Control.* P. F. Cooper and B. C. Findlater, Eds. Advances in Water Pollution Control, Pergamon Press, Oxford, 1990, 41.
35. Kemp, W. M. and L. Murray. Oxygen release from roots of the submersed macrophyte *Potamogeton perfoliatus* L.: regulating factors and ecological implications, *Aquat. Bot.* 26:271, 1986.
36. Sand-Jensen, K., C. Prahl, and H. Stokholm. Oxygen release from roots of submerged aquatic macrophytes, *Oikos.* 38:349, 1982.
37. Caffrey, J. M. and W. M. Kemp. Seasonal and spatial patterns of oxygen production, respiration and root-rhizome release in *Potamogeton perfoliatus* L. and *Zostera marina* L., *Aquat. Bot.* 40:109, 1991.
38. Moorhead, K. K. and K. R. Reddy. Oxygen transport through selected aquatic macrophytes, *J. Environ. Qual.* 17:138, 1988.

CHAPTER 42

Control of Algae Using Duckweed (Lemna) Systems

S. J. Hancock and L. Buddhavarapu

INTRODUCTION

Wastewater treatment systems using ponds/lagoons are common in most countries of the world. Lagoon systems are inexpensive and are operated in areas where land is available. Generally, wastewater treatment lagoons require minimum operator maintenance. System designs are based on hydraulic retention time (HRT). With stringent effluent limits mandated by the U.S. EPA, lagoon systems are not able to perform adequately and therefore need upgrading. Reasons for the poor performances of these systems are short-circuiting and algae.

The problem of short-circuiting in a pond/lagoon has been reviewed many times and can be physically corrected with hydraulic baffles. Algal problems inherent to facultative ponds and subsequent control/elimination of algae need to be addressed. Most chemicals can control algae, but when released into open water without proper retention time, they are more harmful to the environment than the algae themselves. Here, we discuss control for algae using Lemna/duckweed systems which are inexpensive, easy to operate, and do not cause any damage to effluent receiving waters.

REVIEW OF NATURAL POND/LAGOON BIOLOGY

Normally, in a lagoon system, the primary pond/lagoon serves as a settling basin, where solids settle and gross BOD is reduced aerobically. Partially treated wastewater enters a secondary pond/lagoon for further treatment. BOD reduction leads to an increase in available nutrients and dissolved oxygen (DO). Temperature increases, especially in summer, coupled with available nutrients and DO creates an ideal condition for algae growth. This leads to the phenomenon known as *algae blooms*.

Algae have been used for many purposes, including wastewater treatment. These fast-growing plants uptake and incorporate available nutrients in the presence of light energy into plant tissue. The by-product of photosynthesis is oxygen, which is utilized by aerobic microorganisms to reduce BOD. Algae growth contributes to total suspended solids (TSS). Removal of algal biomass would remove nutrients and TSS from the water columns, eliminating recycling of nutrients from dead algae. Algae, however, cannot be easily harvested and the algae-laden water is released into the effluent receiving stream.

In certain instances, algae blooms die due to environmental conditions such as low temperatures, toxicity problems created by the population, increases in pH, release of toxic metabolites, low concentration for nutrients, etc. Dead algae sink to the bottom where their chemical constituents are transformed, solubilized, and recycled into the water. This contributes to increases in BOD and nutrients such as phosphorus and nitrogen. The fluctuation in the algal growth cycle affects the consistency of the effluent quality of the wastewater.

When greatly concentrated, algae can impart an unpleasant taste and odor to drinking water. In some cases, they release substances that can be harmful to aquatic animals. There have been instances where fish kills have occurred due to oxygen depletion by algal blooms.

0-87371-550-0/93/$0.00+$.50

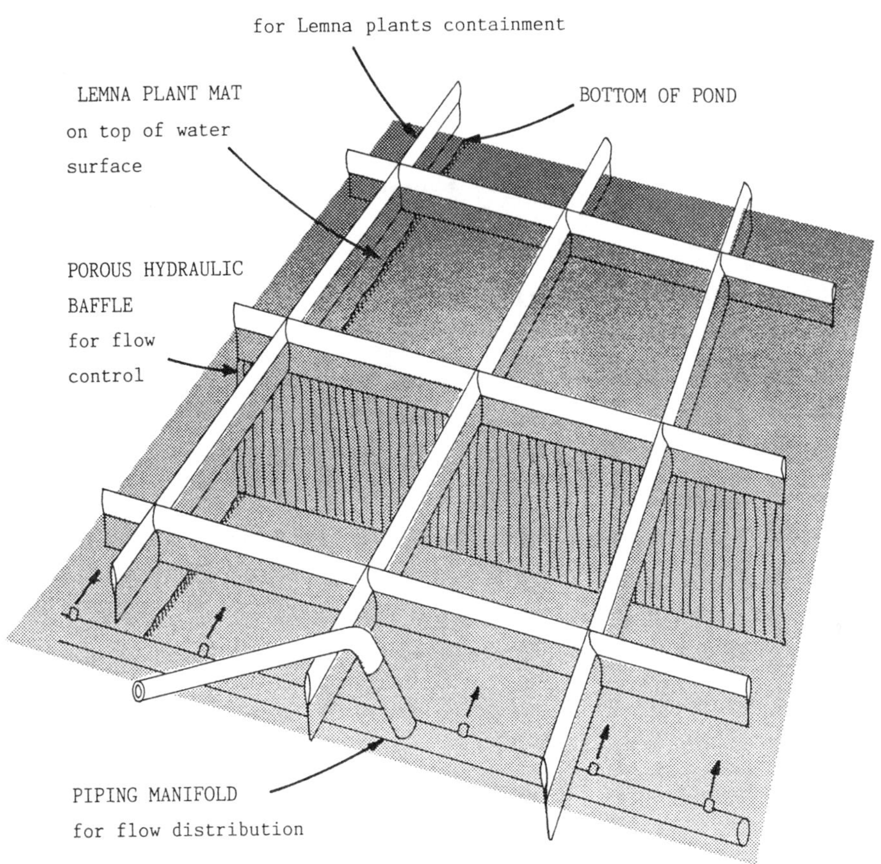

Figure 1. Portion of a typical Lemna treatment pond (a U.S. patented technology for wastewater treatment).

LEMNA POND BIOLOGY

The primary factor involved in photosynthesis is light. Elimination of light from a water body results in the death of algae. Lemna Corporation has incorporated this side in the treatment of wastewater in lagoons. Most of the research using Lemna for algae control has been done in the laboratory, with very little field data available on its performance. This study shows the control of algae (as TSS) at four different sites located in different climates in the U.S.

Lemna or duckweed is a small floating aquatic plant that grows very fast and covers the water surface. Due to its small size, the plant is susceptible to wind/wave movements. To overcome wind/wave movement, Lemna Corporation has patented a barrier system that forms a floating grid on the water surface (Figure 1). Each cell for the grid is 93 m^2 in area and creates a quiescent water surface that enhances the growth of the Lemna plants. At optimum conditions, Lemna can double in mass in 3 to 4 days and cover the entire water surface. This surface layer of plants forms a physical barrier and prevents light penetration into the water column, except the top few centimeters. The system stabilizes and becomes anaerobic. The top of the water column remains aerobic and helps in root respiration of the Lemna plants.

With the elimination of light, the algae die. Due to a lack of oxygen, the dead algae decompose anaerobically under the Lemna mat. Anaerobic bacteria enhance the rate of particulate conversion of organic matter to soluble organic and simple fatty acids. These acids lower the high pH created by algae to near-neutral conditions. Dead algae release lipids which remain in the particulate fraction until degraded by ethane fermentation. For methane fermentation to occur, an anaerobic environment with a pH >5.5 is required. Such

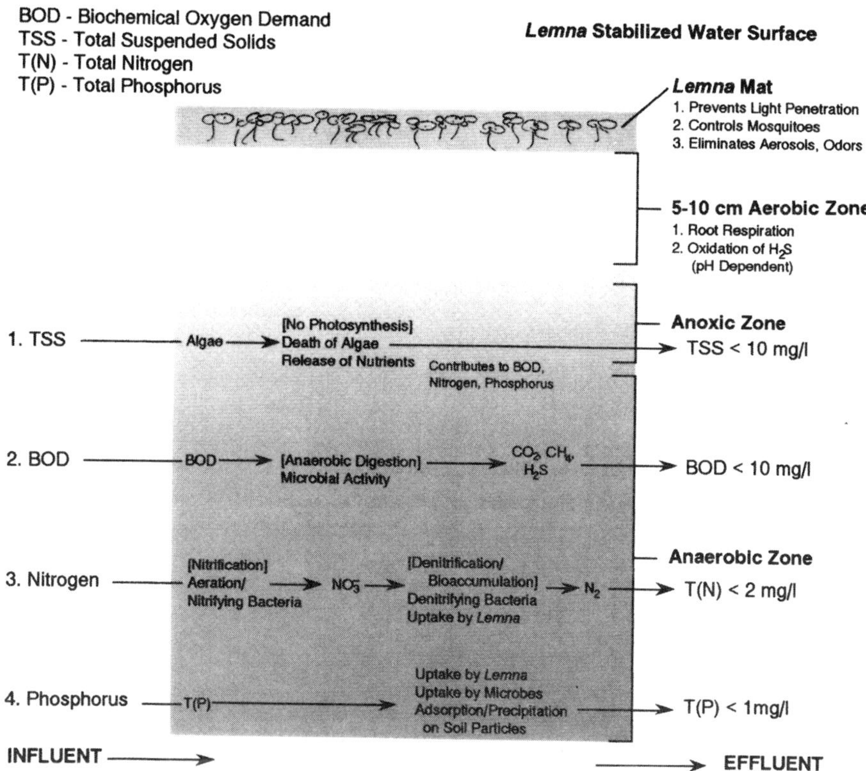

Figure 2. A vertical section showing Lemna pond biology.

an anaerobic environment is created by the Lemna cover. With appropriate conditions available under the Lemna mat, further biological decomposition leads to the release of volatile gases such as methane, hydrogen sulfide, and carbon dioxide. These gases can be transported by diffusion or dispersion to aerobic waters where they may be oxidized and taken up by the Lemna plants (Figure 2).

THE LEMNA SYSTEM

The Lemna system primarily consists of the following components:

- Floating barriers: A matrix of floating barriers divides the lagoon surface into sections. Made of durable synthetic material, these barriers extend 11.4 cm above the water and 11.4 cm below. A typical cell has 3.05-m sides. The cells' function is to shield Lemna plants from wind and prevent pond wave action. This network of floating cells is anchored to the shoreline. The grid barrier floats on the surface of the water and conforms to fluctuation in water level. It is permanently installed on the lagoon, thus stabilizing conditions. Sizes and configuration are designed to adapt to each lagoon based on pond geometry and wind conditions.
- Lemna harvesters: Harvesters capable of navigating over the matrix of floating barriers and hydraulic baffles are used to inoculate new ponds with Lemna plants, as well as harvest them. The watercraft utilizes hydraulic-operated conveyors to pick up biomass from the water surface. Conveyors located on shore transfer the biomass from the harvester to trucks.
- Lemna lagoons: Lemna lagoons are either specifically site-designed or retrofitted from existing ponds. Lemna system components can be installed on existing lagoons within a few weeks without interfering with on-going operations. New lagoons are designed to conform to existing topography and land use. A series of serpentine channels is built to minimize excavation, maximize retention time, and prevent short-circuiting. The lagoons can be designed as independent units, or in series, for maximum contact time and treatment.

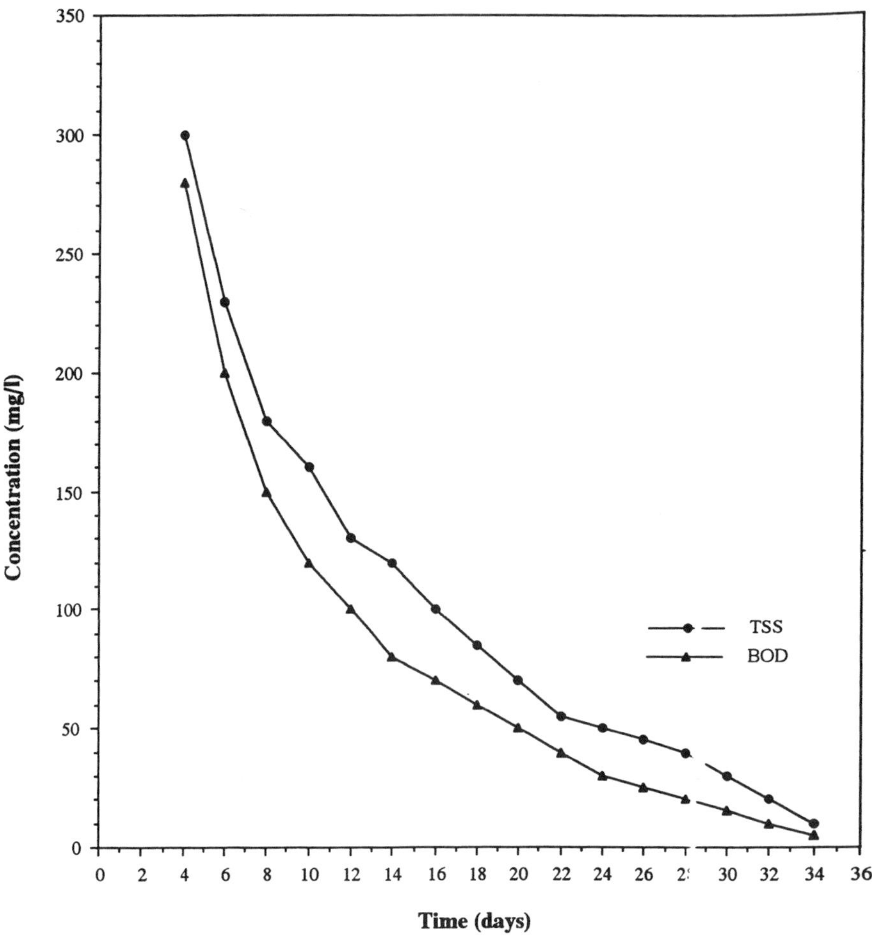

Figure 3. Typical BOD and TSS reduction by Lemna treatment.

LEMNA SYSTEM PERFORMANCE — CASE STUDIES

Four separate wastewater treatment facilities were upgraded with the Lemna system. The facilities are located in both northern and southern climates of the U.S. All facilities required TSS reduction prior to discharge of effluent to receiving waters. Furthermore, BOD and nutrients removal were necessary as well.

Each system was modified with a cover of Lemna plants according to design criteria retention times, as indicated in Figure 3. All facilities demonstrated dramatic improvement in effluent quality after installation of the Lemna plants. Each facility's design consideration and design summary are found in the following pages. Graphs of effluent BOD/TSS quality are included adjacent to the summary of design.

CONCLUSION

Modification of lagoon facilities with a Lemna aquaculture system dramatically improved effluent water quality and treatment reliability. Elimination of algal TSS, as well as polishing of BOD and nutrients removal, provided excellent effluent quality.

All modifications were incorporated into existing operating lagoons without disruption of wastewater flow. Hydraulic surge protection, low energy consumption, and operational ease were retained in all systems utilizing the described Lemna system improvements. The use of duckweed (lemna) systems was efficient and cost effective in both northern and southern latitudes in the continental U.S.

Devils Lake, North Dakota

Design Consideration

The Devils Lake has three stabilization lagoons in series, followed by a Lemna system which serves populations of 7500 in winter and 10,000 in summer. The primary and secondary lagoons cover 97.2 ha overall. The 20.2-ha Lemna system polishes water to advanced wastewater treatment standards.

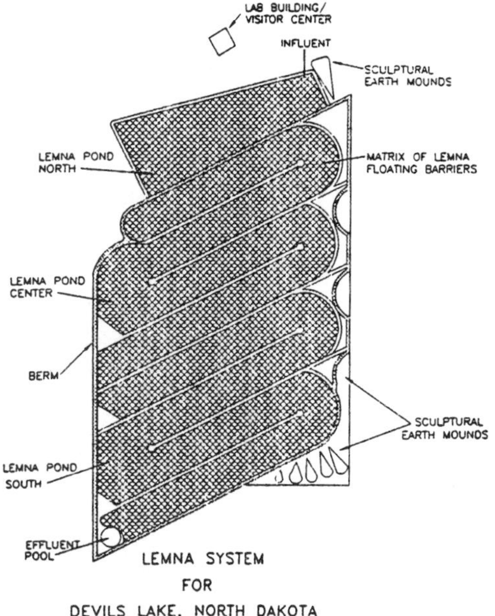

Design Summary

- Existing lagoons were upgraded by adding (raising) the berms an additional 2 ft and rip-rapping all three existing cells
- Design flow = 17,000 m^3/day
- Maximum daily flow = 20,800 m^3/day
- Lemna treatment: a new, three-cell, 20.2-ha Lemna advanced wastewater treatment facility now follows the preceding three lagoons; this facility also incorporates a visitors' center and is designed as an attractive Lemna Wastewater Treatment Park
- Active storage volume = 29,100 m^3
- Hydraulic retention time = 22 days
- Outlet: by gravity/pumping to Devils Lake
- Design effluent quality = <25 mg/L of BOD, <30 mg/L of TSS, and <1 mg/L of T(P)

Mean T(P) (mg/l)
0.3
0.2
0.1
0.0
Aug 90 Sep 90 Jun 91 Jul 91 Aug 91
Month

Total Phosphorus Levels from Devils Lake, ND

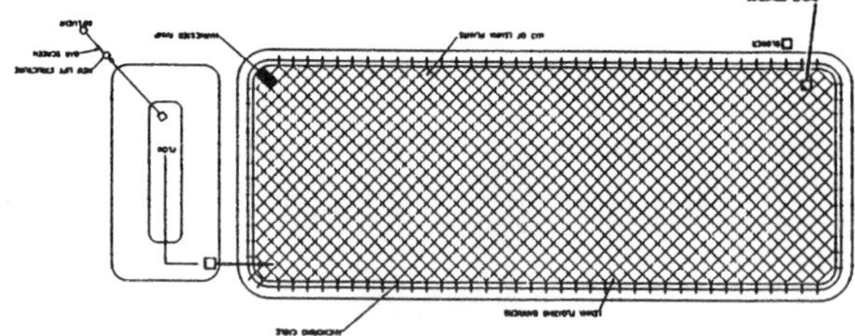

Ellaville, Georgia

Design Consideration

The city of Ellaville, Georgia has one primary aerated lagoon followed by a rectangular lagoon measuring 1.0 ha which serve a population of 1724. The current discharge limits are 30/90 mg/L of BOD/TSS.

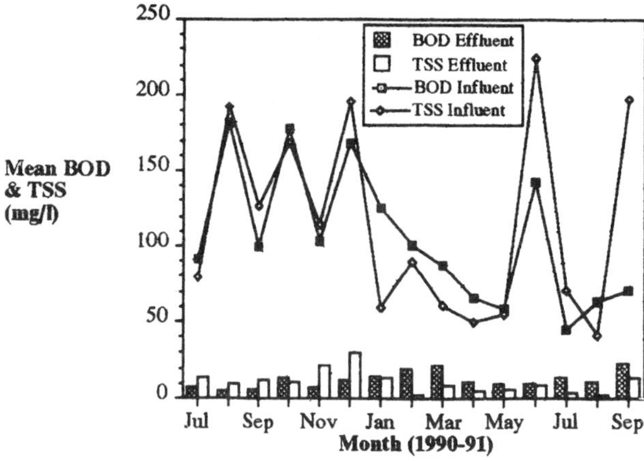

BOD and TSS Levels from Ellaville, GA (limits 15/30)

Design Summary

- Existing aeration cell remains as is
- Existing facultative pond = 1.0 ha dredged to a depth of 1.8 m (operating depth = 1.5 m, storage depth 3 m)
- Retain the existing inlet
- Design flow = 757 m^3/day
- Maximum daily flow = Design for up to three times average daily flow
- Lemna treatment: use existing primary aeration cell for initial BOD/TSS reduction; use second pond for polishing by a Lemna retrofit consisting of a superimposed matrix of Lemna floating barriers
- Active storage volume = 15,100 m^3
- Hydraulic retention time = 20 days
- Outlet: retain the existing effluent structure

Greenleaves Subdivision, St. Tammany Parish, Louisiana

Design Consideration

The Greenleaves subdivision of St. Tammany Parish, Louisiana has an aerated pond followed by a rock-reed polishing pond and a UV disinfection system. This facility is designed for 946 m^3 flow. Because of uncontrolled algae growth, the rock-reed cannot consistently produce the required 10/15 mg/L of BOD/TSS effluent quality. The Lemna System has been retrofitted between the initial aerated zone and the rock-reed filter bed to aid in TSS reduction. In June 1991, repairs were made to the existing hydraulic baffle to eliminate short circuiting. Additionally, the filter bed was trenched to reduce ponding.

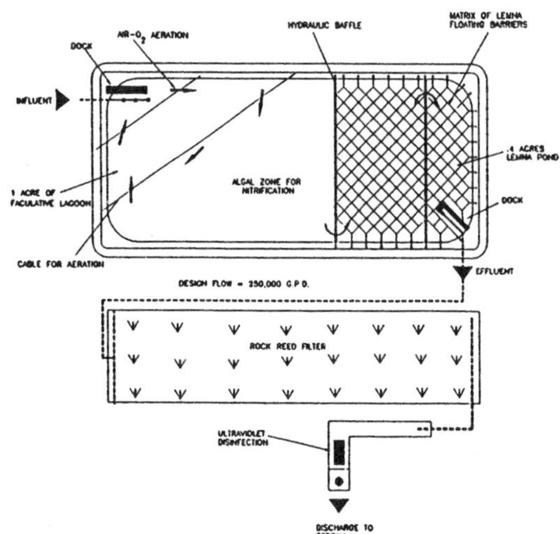

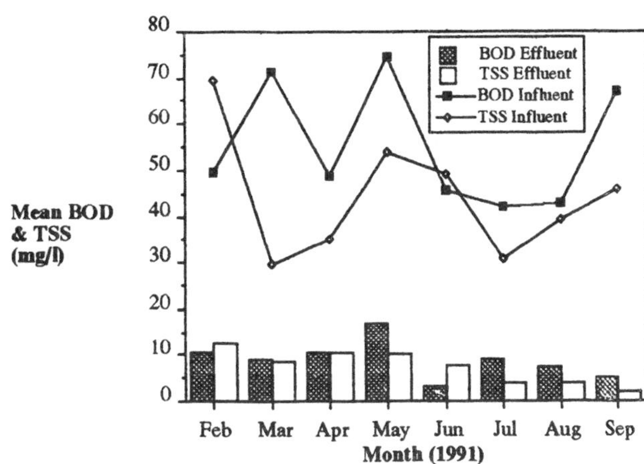

BOD and TSS Levels from Greenleaves, LA (limits 10/15)

Design Summary

- Existing pond area = 107 × 46 × 3 m deep
- Design flow = 946 m^3
- Lemna treatment volume = 40 × 46 × 3 m deep
- Lemna treatment: the Lemna retrofit consists of a matrix of Lemna floating barriers covering 0.45 acres of the aerated lagoon for TSS reduction; this process is upstream from the existing rock-reed system
- Existing influent and effluent structures retained
- Existing UV disinfection system retained
- Design effluent (entire system) = <10 mg/L BOD and <15 mg/L TSS

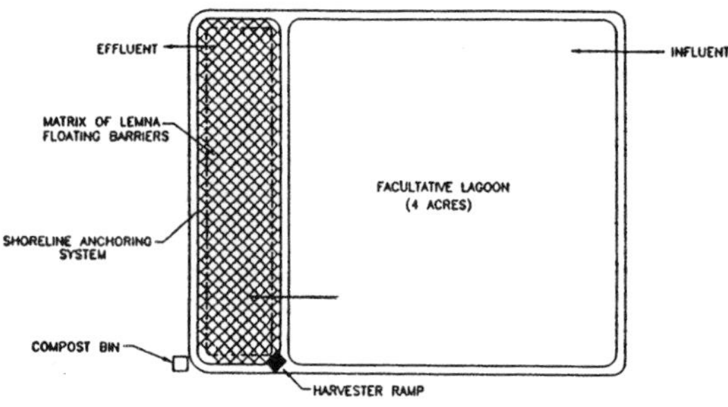

Ogema, Wisconsin

Design Consideration

The Lemna system at Ogema serves approximately 300 people. The facility comprises two stabilization lagoons in series. The primary lagoon has a volume of 19,380 m^3, giving a retention time of 146 days. The second lagoon has a volume of 5034 m^3, giving a retention time of 38 days. The Wisconsin Department of Natural Resources requires 150-day winter storage. However, the Lemna system increased treatment efficiency and produced secondary effluent quality even during winter months.

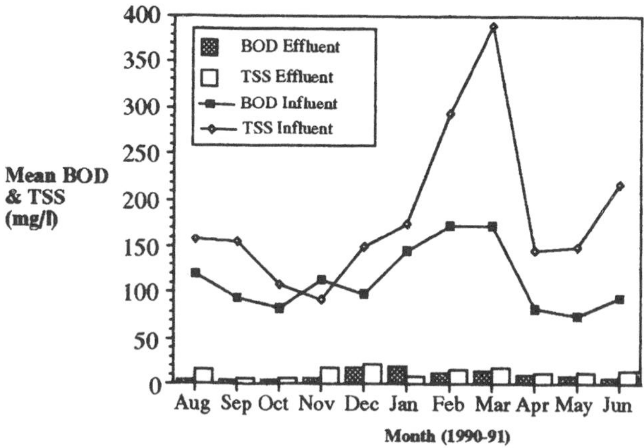

BOD and TSS Levels from Ogema, WI (limits 20/20)

Design Summary

The following is the design summary for the secondary cell of the two-lagoon system:

- Existing secondary cell pond = 0.4 ha (operating depth 1.5 m, storage 0.3 m)
- Retain the existing inlet
- Design flow = 284 m^3/day
- Maximum daily flow = designed for up to five times average daily flow
- Lemna treatment: the Lemna retrofit will consist of a matrix of Lemna floating barriers covering the pond
- Active storage volume = 5034 m^3
- Hydraulic retention time = 38 days
- Outlet: repair the effluent box to stop leakage
- Design effluent quality = <20 mg/L BOD and <20 mg/L TSS

CHAPTER 43

The Use of *Typha latifolia* for Heavy Metal Pollution Control in Urban Wetlands

R. B. Shutes, J. B. Ellis, D. M. Revitt, and T. T. Zhang

INTRODUCTION

The shock loads resulting from intermittent discharges of separately sewered stormwater and combined sewers can often seriously impair the quality of receiving waters, particularly in the case of lakes and semi-stagnant waters.[1,2] Although many of the effects may be highly localized and confined to the mixing zone, they need not always be transient and can, under certain conditions, persist for considerable periods of time.[3] Increased sediment oxygen demands and above-average levels of sediment bacteria for example, can last for several months.[4] In addition, high concentrations of heavy metals and organic micropollutants often persist for many years in the sediments close to sewer outlets and around lake inlet zones.[5] In the case of large receiving water bodies, dissolved oxygen levels usually drop for relatively short time periods, which may vary from 1 hour to several days, following stormwater discharges with high bacterial counts persisting for up to 36 hours.

The most common approach adopted in the U.K. to control urban stormwater run-off is through the use of flood storage basins. The primary, if not exclusive, objective of such retention basins however is to achieve flow attenuation and to maintain the downstream pre-development peak flow levels. Very little consideration in the past has been given to other secondary objectives, despite the fact that a large proportion of these basins have been, or are being developed for a variety of recreational and amenity uses. This attitude is now coming under critical review primarily through pressures to assess the likely public health risks associated with water-based activites on and around these freshwater sites. Additionally, there is a high probability that recreational freshwater standards will be introduced in the U.K. within the near future by the regulatory water agencies.

At the same time, it has become increasingly fashionable in recent years to consider the use of constructed wetlands to provide low-cost improvement for wastewater effluents.[6] While the large and highly variable stormwater run-off volumes, and associated pollutant loadings which discharge to urban storage basins, inevitably pose particular problems for biofiltration contact systems, they nevertheless do offer attractive stormwater management opportunities.[7] An urban wetland can in theory provide a simple purification system, by creating in effect a low-loaded biological filter. However, very little work exists which evaluates the feasibility of urban run-off quality enhancement using aquatic plants. This article describes the potential pollution control performance with regard to heavy metals of a macrophytic community within constructed urban wetland based on a comparison of laboratory and field studies.

BIOLOGICAL ASPECTS OF SECONDARY QUALITY ENHANCEMENT OF URBAN RUN-OFF

Table 1 compares the general performance characteristics and operational methods of typical wastewater stabilization ponds with a typical urban run-off retention basin facility. Both types of facility are similar in that they retain the fine particulate and associated pollutants of the influent wastewater for prolonged periods and thereby can facilitate the natural purification process. Both systems are relatively simple from an engineering

0-87371-550-0/93/$0.00+$.50

Table 1. Characteristics of Typical Wastewater Stabilization Ponds and Urban Flood Storage Basins

Wastewater stabilization ponds	Urban flood storage basins
Continuous flow	Intermittent flow
Permanent water body	Subject to sudden inundation and periodic drying out or fluctuations in level
Deep open water body without shoreline; no aquatic macrophytes in or around pond	Relatively shallow with well-developed shoreline supporting fringing growths of emergent macrophytes
Vertifical stratification develops in open water as a result of intense solar radiation; flow stratification occurs as warm influent flows over colder deeper water	Shallow ponds protected from intense sunlight by stands of emergents and floating mats; flow stratification prevented by barriers of emergent macrophytes
Anaerobic sludge builds up in sediments, contributing to oxygen depletion in the stratified pond	Sludge build-up is compacted and aerated during drying out and fluctuation in levels
Algal growth contributes to severe oxygen depletion at night; can substantially increase BOD during hot weather	Limited algal growth if good floating macrophyte cover
Efficient SS removal although algal/zooplankton blooms will cause deterioration	Efficient SS removal (Type I settling) and biofiltration by aquatic plants
Removal of ammonia-nitrogen through nitrification of little significance because low concentration of bacteria present in upper levels of pond water	Nitrification is important due to substrate provided by plants to microbial growth; in shallow ponds, nitrification also takes place at the sediment/water interface
Denitrification is directly related to pond area and probably takes place at the sludge-water interface under anoxic to anaerobic conditions	Denitrification probably occurs within the anoxic sediments
Removal of particulate metals by sedimentation is restricted by stratification; anaerobic conditions at the sediment/water interface may release additional metals to the soluble phase	Particulate metals removed by root sedimentation and macrophytic filtration; resuspension possible during storm events; soluble metal removal dependent on macrophyte uptake efficiency

viewpoint and are partially controllable although the prevailing physical and biochemical processes are complex. The main differences between the two types of receiving water body are that the urban flood storage basin is more liable to shock loadings and is obviously more likely to fulfill a multipurpose role.

Rhizomatous emergents such as *Typha latifolia, Phragmites australis,* and *Scirpus* spp., (as border marshes in water channels and lakes) are known to be capable of taking up and translocating macronutrients from the sediments to the aerial leaf tissues. However, studies on emergent macrophytes would indicate that the major pathway of nutrient as well as toxic metal and hydrocarbon removal in basins appears to be by root filtration and by settling of organic material rather than plant uptake.[6,7] From a management viewpoint, drainage and other modifications to urban wetlands to enable efficient harvesting would in any case be incompatible with the overall concept of the receiving basin as a balanced, seminatural ecosystem. In addition, harvesting and repeated cutting of emergent plants would have an effect of reduced growth as well as causing disturbance and mobilization of the contaminated substrate. It is therefore important to consider not only the uptake efficiency of aquatic macrophytes, but also to evaluate their filtration capacity, resistance to fluctuating levels of discharge velocity, and loadings. Identification of appropriate design and operational criteria is also necessary.

LOCATIONS AND METHODS

Typha latifolia and *Juncus effusus* plants for greenhouse-based experiments were collected from an unpolluted, spring-fed pond located in Trent country park in N London (site 1). Plants were also monitored at bi-monthly intervals during the period from October 1988 to December 1989, at the background site and at two polluted sites which included an ornamental pond receiving run-off from the M1 motorway in NW London (site 2), and a flood storage reservoir, the Welsh Harp, which receives stormwater inputs from both separate and combined sewer systems in NW London (site 3).

Separate greenhouse-based dosing experiments were carried out for *Typha* and *Juncus* over 8-week periods utilizing four undrained dosing tanks (diameter, 30 cm; height 35 cm) and one control tank. Irish moss peat was chosen as the growing medium because of its low background metal levels (Cd, 0.0; Cu, 2.0; Pb, 11.2, and Zn, 192 mg/kg dry wt) and each tank was planted with either 3 to 5 *Typha* plants or 10 to 15 *Juncus* plants. The control was treated with tap water, whereas the other plants were dosed twice weekly with a solution containing 10 mg/L of Cd, Cu, Pb, and Zn (as their nitrate salts). One *Typha* plant and five *Juncus* plants, together with associated surface and basal sediment, were removed sequentially from each dosing tank for weekly analysis. Samples from the control tanks were analyzed only at the end of the 8-week period.

Plant samples were carefully washed with tap water (leaves were washed with 2% nitric acid) followed by de-ionized water to prevent tissue damage and to remove all traces of attached sediment. The *Typha* plants were separated into three parts constituting leaf, rhizome, and roots; for *Juncus* plants, the stem and root sections were isolated. The plant materials (after thorough grinding) and sediment samples were oven dried at 100°C for 24 h and then digested with concentrated nitric acid. Duplicate samples were analyzed for Cd, Cu, Pb, and Zn by atomic absorption spectrophotometry using flame atomization. Water samples were extracted with concentrated nitric acid, followed by metal analysis using anodic stripping voltammetry.

RESULTS AND DISCUSSION

Metal Dosing Experiments

The dosing metal levels of 10 mg/L were considerably greater than the average soluble metal concentrations monitored at the most polluted field site (site 3: Cd, 8.9 μg/L; Cu, 53.4 μg/L; Pb, 36.2 μg/L; Zn, 136.6 μg/L), but neither the *Typha* nor *Juncus* plants suffered any observable toxic effects over the duration of the experiment. However, the short-term, high dosing level tests were not unrealistic compared to environmental field situations where the observed sediment metal concentrations (site 3: Cd, 12.4 mg/kg; Cu, 220.1 mg/kg; Pb, 841.2 mg/kg; and Zn, 778.9 mg/kg dry wt) were considerably higher for Cu, Pb, and Zn than those measured in the surface Irish moss peat at the conclusion of the dosing experiment.

Surface peat metal concentrations were substantially enhanced in both series of tests over the 8-week experimental period achieving average levels of 314, 198, 185, and 309 mg/kg for Cd, Cu, Pb, and Zn, respectively. The difference between the surface and the basal peat (15 cm depth) metal concentrations increased with dosing time, reaching a maximum enhancement factor of almost five-fold. Similar metal distributions have been demonstrated in previous dosing experiments and over shorter sediment depths, with Zn levels differing by a factor of 2.5 between depths of 0.2 and 1.0 cm.[8]

The increases in surface sediment metal concentrations have also been shown to parallel closely the metal uptake patterns of *Typha* root tissue, with the rhizome showing a reduced metal affinity and the leaf material being clearly the least important site with regard to metal concentrations.[9] This trend in metal concentrations (root > rhizome > leaf tissue) is consistent with previous results for Zn and was also demonstrated by *Juncus* tissue, although for this species there is less discrimination between root and stem metal concentrations (Figure 1).[8]

The *Juncus* tissue metal levels exhibited gradual and approximately linear increases with time and were similar to those exhibited by *Typha* leaf and rhizome material which do, however, indicate a greater tendency to achieve equilibrium levels within the time period of the experiment. The *Typha* root metal uptake patterns are less consistent, with Cd and Zn both exhibiting temporal exponential increases to values approaching 700 mg/kg. These metals show the greatest concentration enhancement in the root compared to the rhizome and

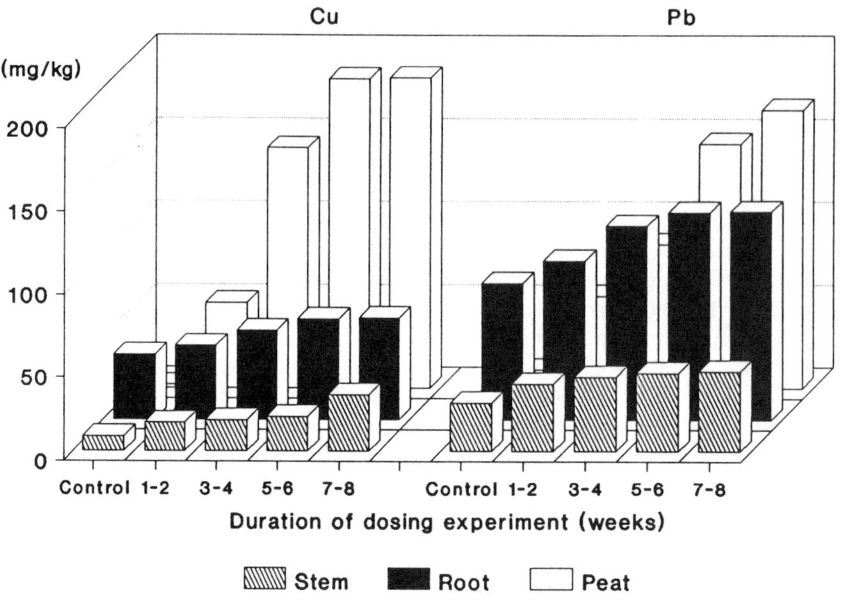

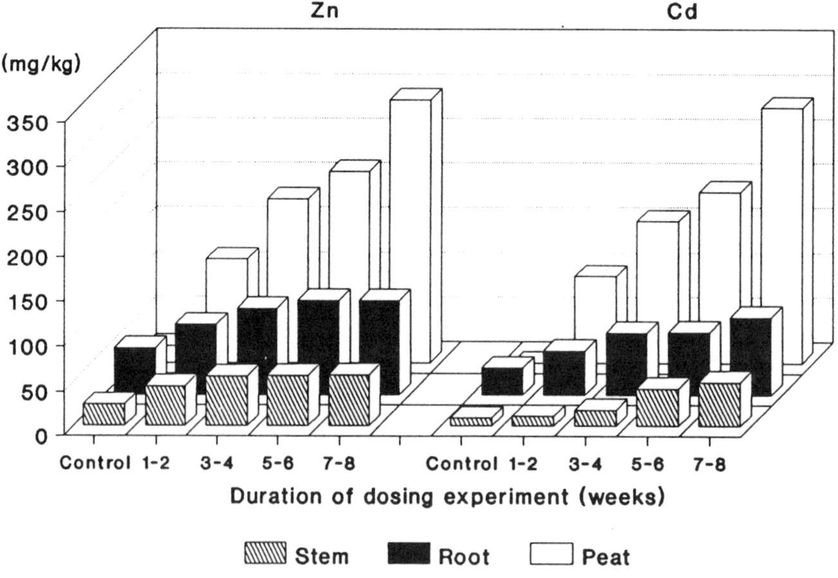

Dosing concentration 10 ppm
Dosing time 3/10/1989 - 29/11/1989

Figure 1. Temporal increases in surface peat and *Juncus* tissue metal concentrations during an 8-week dosing experiment.

Table 2. Tissue Metal Concentration Ratios After Dosing for 8 Weeks

	Typha latifolia		*Juncus effusus*
	Rhizome root	**Leaf root**	**Stem root**
Cd	0.13	0.07	0.59
Cu	0.37	0.13	0.55
Pb	0.25	0.16	0.40
Zn	0.15	0.05	0.55

leaf system, particularly towards the end of the exposure period. The final *Typha* root Cd and Zn levels are considerably elevated over those in the surface peat, which corresponds to their established existence in urban waters as predominantly soluble bioavailable metal ions or weak complexes.[10] However, the 4- to 5-week delay in rapid uptake suggests a superimposed influence due to sediment/water interactions which only release the bioavailable Cd and Zn for plant root uptake when a specific sediment saturation level has been reached, although the influence of reducing the number of plants in each tank may be a contributing factor. Previous experiments with *Typha* have shown that Zn concentrations in the plant root tissue can reach 1400 mg/kg after dosing with a 10-mg/L solution for 1 month.[8] *Juncus* root material also showed a higher uptake ability for Cd and Zn, with metal concentration factors after 8 weeks of 2.8, 2.0, 1.6, and 1.5 for Cd, Zn, Cu, and Pb, respectively, compared to the control plants. These ratios are much lower than the corresponding values for *Typha* root tissues (200, 7.8, 8.3, and 5.1 for Cd, Zn, Cu, and Pb), respectively, with the Cd concentration factor being artificially elevated due to the extremely low levels maintained in the control throughout the 8-week dosing experiment.

Although there are no large differences in the final root tissue concentrations for any of the metals in *Juncus*, the final accumulative Cu and Pb levels in *Typha* roots are much lower than those of Cd and Zn. The reduced or similar Cu root concentrations in both plants, compared to the peat Cu levels, are consistent with the established affinity of Cu for the humic and fulvic acids associated with soils.[11] The lower affinity of Pb for root tissues confirms the results of studies on other aquatic plants.[12,13]

The ratios of the metal concentrations in the different tissues of *Typha* and *Juncus* after dosing for 8 weeks are given in Table 2. These data indicate that the root to rhizome transfer mechanism in *Typha*, assuming negligible direct adsorption, is most efficient for Cu followed by Pb, with Cd and Zn showing lower but similar values. The tendency for Zn to remain concentrated in the root tissues of *Typha* has been demonstrated previously, as well as the high affinity of water hyacinth roots for Cd.[8] The leaf/root metal concentration ratios in *Typha* closely follow the trend of the rhizome/root values, although Pb demonstrates a greater ability than Cu with regard to translocation to the leaves. The *Juncus* metal tissue levels indicate relatively little discrimination between the four metals with regard to the root-to-stem transfer process which is consistently more efficient than for the corresponding *Typha* tissues.

Field Study

Average sediment Cu, Pb, Zn, and Cd concentrations exhibit exhanced levels typical of urban conditions at sites 2 and 3 compared to the rural site 1. All sediment levels show a high variability at the urban sites, with the greatest variation being observed at site 3 where average concentrations and standard deviations are Cu 219 ± 44.; Pb 841 ± 228; Zn 778 ± 139; and Cd 12.5 ± 7.8 mg/kg dry wt. Comparable sediment levels at sites 1 and 2, respectively, were 28.2 ± 7.9 and 119 ± 58.4 for Cu; 77.7 ± 15.6 and 496 ± 202 for Pb; 70.7 ± 30.6 and 506 ± 197 for Zn; and 4.2 ± 3.3 and 9.7 ± 4.8 mg/kg dry wt for Cd. The concentrations are undoubtedly elevated as a result of sediment flushing to the receiving basin following storm events, especially in the case of Cu, Pb, and Cd.

The temporal variability in *Typha* tissues, with peak levels of Cu and Zn in root, rhizome, and leaf coinciding with the growth period in June/July, and thus with an increase in demand for these associated plant micronutrients, has been previously reported.[9] There is bioaccumulation of Cu in root tissue during February/April, followed by a fall in concentrations from April/June when the rate of root growth exceeds the Cu uptake rate. Maximum Pb and Cd tissue levels occur during August/October, possibly due to reduced competition

Table 3. Metal Concentration Ratios of Sediment, *Typha* Root, and Rhizome to *Typha* Leaf (Site 3)

	Sediment	Root	Rhizome	Leaf
Cd	3.7	1.8	1.1	1
Cu	1.5	2.8	1.5	1
Pb	44.8	5.1	1.7	1
Zn	21.8	4.6	1.2	1

from Cu and Zn and to translocation from root tissue being augmented by atmospheric deposition as well as by increases in the permeability of the leaf during decomposition.

There is a progressive decrease in Cu, Pb, Zn, and Cd concentrations from sediment to *Typha* root and to rhizome and to leaf at all sites. This is generally consistent with the results of a previous study of Cu and Zn levels in *Typha latifolia*.[14] The ratio of sediment and *Typha* root and rhizome-to-leaf concentrations at site 3 are given in Table 3 which clearly reflect, for Pb and Zn, the extremely effective hydraulic and filtration barrier presented by the macrophytic vegetation in the flood storage barrier.

Although root-to-leaf mean concentration ratios increase for Pb and Zn at the urban sites, the ratio of rhizome-to-leaf concentrations is relatively constant for all metals at the three sites, ranging between 1:1 for Cd to 1.7:1 for Pb. This suggests that the ultimate tissue distribution is probably metabolically determined and is independent of root metal uptake rates. With the exception of Zn, there is less variability in leaf metal levels at the three sites than might be expected from the sediment concentration differences. Sediment-to-leaf mean Pb concentration ratios increase in the order 4.5:1, 27.1:1, and 44.8:1 at sites 1, 2, and 3, respectively, indicating the affinity of Pb for the particulate phase. However, the same trend is observed for Zn, which is known to occur predominantly in the more bioavailable dissolved phase.[10]

Comparison of Dosing Experiments and Field Study

The maximum metal concentrations in *Typha* root, rhizome, and leaf for the dosing experiments and field studies are shown in Figure 2. The maximum root Pb levels at the two urban sites are similar to the root Pb levels obtained after the 8 weeks dosing experiment, whereas Cd, Cu, and Zn levels are much lower than those of the dosed root. The maximum Cu, Pb, and Zn levels in *Typha* rhizome and leaf metal levels at the two urban sites are similar to those obtained from the greenhouse dosing experiments, whereas the Cd levels are much lower. The results suggest that the limit of uptake for Cu, Pb, and Zn in both rhizome and leaf is probably achieved at the urban sites.

Tissue metal load distributions have been determined for both dosing experiments and the field study and clearly indicate that the major metal bioaccumulation target area is the rhizome.[9] In the case of the field study, it has been shown that between 54 and 61% of the total metal load can be stored in this tissue area (Figure 3). The root and leaf tissue show similar bioaccumulation loads for all metals. This is in broad agreement with another study which showed that 58 to 75% Zn accumulated by *Typha* was being stored in root and rhizomes.[8]

CONCLUSIONS

Both the dosing and field results provide evidence of metal uptake, with target accumulation in the rhizosphere and root zones. If purification processes are largely confined within the latter, it is important to ensure good horizontal flows through this barrier zone and to avoid channeled short-circuiting. The use of a gravel substrate, which would allow adequate root growth but also support high hydraulic loadings, might provide a more suitable substrate for emergent macrophytes within urban retention basins. Additionally, the subsurface introduction of effluent into the basin through submerged inlets would also maximize the purification potential. The field and dosing studies suggest that metals will be taken up on the increased sediment-root surfaces and will be, at least to some extent, immobilized.

An obvious management concern is related to the fate of toxic contaminants and how often such macrophyte systems will need harvesting and/or removing. General experience to date does not show that emergent macrophyte systems need frequent harvesting. Metals, as in the case of nutrients, appear to be locked into

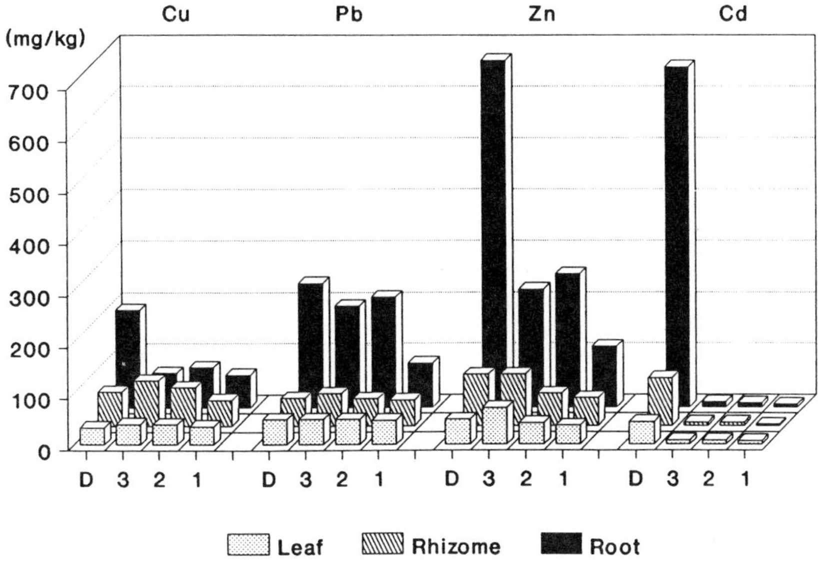

Figure 2. Comparison of *Typha* tissue metal levels in dosing experiments and field study.

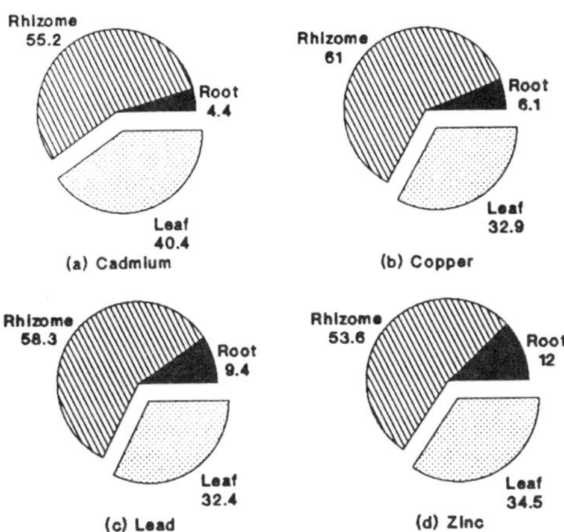

Figure 3. Field study metal load distributions in *Typha* tissue (%).

intractable organic material and are returned to the sediment when the plant tissue decomposes.[15] It is these sites of organic sediment that support the denitrification processes mentioned in Table 1. Although the sediment will attain high heavy metal levels, the superimposition of less contaminated incoming sediment will counteract the processes leading to saturation. Regular monitoring of the macrophytic community is certainly necessary, and at some stage harvesting may be required, although this is more likely to be needed to restore vigor to old macrophyte stands.

Wetland systems therefore do offer a cost-effective technology approach to the problem of controlling toxic stormwater inputs to urban flood storage basins. However, our present understanding of the uptake (and release) mechanisms, especially over time, is very limited and few accepted design guidelines exist. The possibility of failure and unsatisfactory performance must be accepted and urban wetland filter systems must still be regarded at the current state-of-the-art as comprising a promising, but unproven experimental approach.

REFERENCES

1. Ellis, J. B., Ed. Urban Discharges and Receiving Water Quality Impacts. Pergamon Press, Oxford, 1989.
2. Torno, H. C., J. Marsalek, and M. Desbordes, Eds. Urban Runoff Pollution. Springer-Verlag, Berlin, 1986.
3. Aalderink, R. H., L. Lijklema, and J. B. Ellis, Eds. Urban Stormwater Quality and Ecological Effects Upon Receiving Waters. Pergamon Press, Oxford, 1990.
4. Van Sluis, J. W., D. Ten Hove, and B. De Boer, Eds. Final Report of 1982–1989 Research Programme. Nat.Working Party on Sewerage and Water Quality. VROM, s'Gravenhage, The Netherlands, 1989.
5. Hall, M. J., D. L. Hockin, and J. B. Ellis. The Design and Operation of Flood Storage Reservoirs. Butterworths, London, 1991.
6. Cooper, P. F. and B. C. Findlater, Eds. Constructed Wetlands in Water Pollution Control. Pergamon Press, Oxford, 1990.
7. Schueler, T. R. Controlling Urban Runoff. Metropolitan Council of Governments, Washington, D.C., 1987.
8. Blake, G., J. Gagnaire-Michard, B. Kirassian, and P. Morand. Distribution and accumulation of zinc in Typha latifolia, in Proc. Seminar on Aquatic Plants for Wat. Treat. and Res. Recov. Orlando, FL, 1987, 487.
9. Zhang, T., J. B. Ellis, D. M. Revitt, and R. B. E. Shutes. Metal uptake and associated pollution control in urban wetlands, in Constructed Wetlands in Water Pollution Control. P. F. Cooper and B. C. Findlater, Eds. Pergamon Press, Oxford, 1990, 451.
10. Revitt, D. M. and G. M. Morrison. Metal speciation variations with separate stormwater systems, Environ. Technol. Lett. 8:373, 1987.
11. Luoma, S. N. and J. A. Davis. Requirements for modelling trace metal partitioning in oxidised estuarine sediments, Mar. Chem. 12:159, 1983.
12. Miller, J. E., J. J. Hassett, and D. E. Koeppe. Interaction of Pb and Cd on metal uptake and growth of corn plants, J. Environ. Qual. 6:18, 1979.
13. Chigbo, F. E., R. W. Smith, and F. L. Shore. Uptake of arsenic, cadmium, lead and mercury from polluted waters by the water hyacinth *Eichhornia crassipes*, Environ. Pollut. (Ser. A). 27:31, 1982.
14. Taylor, G. J. and A. A. Crowder. Uptake and accumulation of heavy metals by Typha latifolia in wetlands of the Sudbury, Ontario region, Can. J. Bot. 61:63, 1983.
15. Burgoon, P. S., K. R. Reddy, T. A. De Busk, and B. Koopwan. Vegetated submerged beds with artificial substrates, J. Env. Eng. ASCE. 17:(4), 1991.

CHAPTER 44

Growth of Soft-Stem Bulrush (*Scirpus validus*) Plants in a Gravel-Based Subsurface Flow Constructed Wetlands

M. E. Edwards, K. C. Brinkmann, and J. T. Watson

INTRODUCTION

Wetlands with aquatic plants are known to improve the quality of wastewater.[1-4] Emergent aquatic plants are thought to contribute to improved water quality through different means (such as oxygen transport to the root zone), thereby increasing aerobic sites for microbial degradation of pollutants and decreased nutrient concentrations.[5-8] Although much is known about plant growth in natural wetlands, limited information exists on plant growth in wetlands constructed for wastewater treatment.[9]

In cooperation with other government agencies, the Tennessee Valley Authority (TVA) undertook demonstration projects of constructed wetlands for wastewater treatment of small towns. One such project is located near Hardin, Kentucky.[10] Planned as part of an auxiliary system to polish effluent from a deteriorating sewage treatment plant, constructed wetlands were installed with emergent aquatic plants in two cells. Each cell was planted with a monoculture of either common reed (*Phragmites australis* [Cav.] Trin. ex Steudel) or soft-stem bulrush (*Scirpus validus* Vahl).

Initially, getting aquatic plants established was a problem; approximately a year after replanting, however, bulrush specimens had formed clones with multiple aerial shoots, and mature bulrushes were flowering throughout the constructed wetland. However, visual inspection suggested decreased growth along the direction of water flow. Specifically, at the influent end, clones had multiple long and wide aerial shoots, whereas clones 145 m away, near the terminus, possessed fewer, thinner, and shorter shoots. Clones located midway in the cell appeared to be intermediate in size between plants at either end. Thus, since no growth gradient of plants in constructed wetlands was identified from a survey of known literature, the present study was undertaken to: (1) determine if a linear growth gradient actually existed, and (2) establish a quantitative baseline for present and future growth comparisons.

METHODS

Site Description

As part of the constructed wetlands system, a hardwood-forested temporary storage area now manages excess inflow from rainstorms until the package treatment facility processes and relays pretreated effluent to the wetlands. Two equal-sized subsurface flow gravel cells (24 × 145 m) with emergent aquatic plants receive pretreated wastewater onto crushed limestone rock. Quality of water entering and leaving constructed wetlands cell 2 is described by growing season and overall averages of various parameters (Table 1). Gravel substrate in the wetlands cell is a mixture of limestone and river rock, with an average 1.3-cm diameter. Average depth is 0.61 m, ranging from 0.46 m at the inlet to 0.76 m at the outlet, and bottom slope is 0.2%. The intended subsurface water level is ca. 2.5 cm below the top of the gravel in the inlet area.[10-12] Influent flow rates fluctuate from 0.181 to 1.26 million liters per day.

0-87371-550-0/93/$0.00+$.50

Table 1. Water Quality Characteristics of Wastewater Entering and Leaving Constructed Wetland Cell 2, Hardin, Kentucky

	Influent		Effluent	
Parameter	Growing season averages 5/91–10/91	Annual overall average	Growing season averages 5/91–10/91	Annual overall average
H_2O temperature, °C	22.5	18.5	20.4	15.2
pH	7.2[a]	7.0[a]	7.0[a]	7.2[a]
D.O., mg/L	4.7	5.1	0.3	0.5
Conductivity, μmhos/cm	757	647	736	608
ORP	257	171	133	71
Alkalinity ($CaCO_3$), mg/L	147	133	235	189
Fecal coliform, no./100 mL	9994[a]	158,756[a]	203[a]	254[a]
TSS	218	171	14	10
BOD_5 day, mg/L	113	70	4.3	3.9
Total phosporus	8.7	5.8	3.3	2.5
Ammonia-N, mg/L	12.7	10.6	9.1	8.3
Nitrate-Nitrite, mg/L	0.19	0.48	0.01	0.21
Organic N, mg/L	21.5	14.4	7.3	2.9
TKN, mg/L	33.5	25.5	16.3	11.3
Total nitrogen, mg/L	33.7	26.0	16.3	11.5
Total potassium, mg/L	13.0	12.7	11.2	10.4

[a] Geometric mean values.

Originally planted at ca. 60-cm intervals in horizontal rows in October 1988, cell 1 received common reed (*Phragmites australis*) and cell 2 received only soft-stem bulrush (*Scirpus validus*). Survival percentages in both cells were low, probably resulting from incorrect management. Water depths occasionally topped new plants, and solids, flushed from the sewage treatment plant during high flows, accumulated (as deep as 15 cm) over the gravel surface of the front quadrant. Each cell was replanted in 1989 with the same species as before.[11] Again, in cell 1, survival of reeds was low (probably less than 5%); but in cell 2 (replanted in May 1989) numerous bulrushes were obviously established, although growth appeared to be less than uniform along the length of the cell. Thus, only vegetation in cell 2 (i.e., bulrushes) was investigated in this study.

A complicating (and potentially significant) factor should be mentioned. A muck layer (at least 8 cm deep), derived from organic solids flushed from the treatment plant when hydraulically overloaded, persists at the influent end and covers the gravel substrate. Although diminishing along the first quartile of the cell, muck has clogged the upper regions of the gravel layer and accounts for the surface flow of influent at the front end of the cell. Current operational practices allow periodic surface accumulation to extend along the first one third of the cell length, as evidenced by dried mats of algae and duckweed (*Lemna* sp.) in the litter layer of transect 2, 48 to 49 m from the front.

Four 1-m wide transects were established at ca. 44-m intervals along the constructed wetland cell perpendicular to the direction of water flow and to the length of the wetland, dividing the cell into three more-or-less equal and sequential sections. Transect 1 was 23 m long and was located 5 to 6 m downstream to exclude front row clones and any perimeter effect. Transects 2, 3, and 4 were 24 m long and crossed the cell at distances of 48 to 49, 94 to 95, and 139 to 140 m, respectively, from the input of influent. Bulrushes extended another 5 m beyond transect 4 to the terminus of the cell. For mapping and data collection, metric-scaled tapes were stretched from cornerposts to delineate 1-m^2 quadrats within transects. Field activities were conducted from July 1–15, 1990. All collections for biomass determinations, including fresh weight, were made between July 11–15, 1990.

From nearly equidistant transects along constructed wetland cell 2, bulrush clones were mapped and measured to provide data on: frequency; estimated canopy cover; basal diameters and basal areas of all clones;

mean diameters and mean lengths of representative clones; biomass (fresh and dry weights) of representative clones; and root depths (data not presented).

Mapping of Transects, Frequency, and Indexing of Bulrush Canopy

Ten-cm^2 maps were drawn to scale for each 1-m^2 quadrat of all four transects. Each map included the presence of bulrush clones, approximate area occupied by the base of a clone, approximate outer limits of expanding aerial shoots (i.e., maximal canopy), overhanging canopies of plants rooted outside the quadrat (regardless of whether derived from an adjacent quadrat or from outside the transect), and the presence of any species other than bulrush. Clonal frequency (number of bulrush clones) was counted from maps. All volunteer species were identified and found to be typical of the habitat and local flora (data not included).

Although canopy cover by bulrushes could be calculated from transect maps, a simpler alternative method, canopy indexing (similar to Oosting[13]), was adopted to evaluate the aerial density of bulrush. Direct visual estimates of the percentage of ground cover (or shade) resulting from bulrush shoots were made for each quadrat according to an index (*cf.* footnote to Table 2), which was developed on-site and applied consistently to all estimates. Mean index values for each transect were calculated from all 24 (23 for transect 1) quadrats.

Basal Diameters and Basal Areas

Diameter at breast height (DBH), a method from forestry, was adapted to measure bulrush clones and to provide size comparisons.[14,15] The sum of DBHs for a given area is defined as the basal area (BA).[14]

The diameter of bulrush clones at a height of 46 cm above ground level was measured (with the exception noted below) with a calibrated metal tape. Care was taken to prevent crushing, particularly of outer shoots, as all shoots were compacted during measurement. Ideally, shoots of a clone were compacted to the point that: (1) all shoots touched, (2) minimal space existed between shoots, (3) no shoots were damaged, and (4) equal tension was applied to tighten the tape before reading the diameter. The above procedures worked well for the smaller clones in transects 2, 3, and 4, which were measured first. However, when it came to the large clones of transect 1, levels of diameter measurements had to be raised to 61 (instead of 46) cm to prevent damage to outer shoots, while eliminating obvious space among interior shoots of the measured plane.

Basal diameters (BD) of all clones, rooted entirely or partially in a transect, were measured and recorded as to transect and quadrat. Basal areas were determined by summing all BD of a transect. For each transect, depending on average BD and dispersal along the transect, six or seven representative clones were identified for additional measurements of shoot frequency, diameter, length, and biomass.

Shoot Frequency, Diameter, Length, and Density

Having identified by basal diameters six representative clones in transects 2 and 4 each and seven representative clones in transects 1 and 3 each, living shoots were counted at the same level above ground at which BDs were measured. Similarly, diameters were recorded from a random sample of 50 shoots. Lengths of shoots included the distance from the gravel surface to the tapered tip, except in transect 1 where measurements were made from the muck surface. Densities were calculated by multiplying the number of clones per transect by the mean number of shoots in the representative clones and then dividing the product by the number of 1-m^2 quadrats.

Biomass

From the six to seven representative clones (identified for each transect by BD), three clones were selected from each transect and extracted for biomass determinations. Selection was based on (1) BDs similar to the median value of all clones in the transect, and (2) physical dispersal with selected clones coming from the middle and both sides of a transect. Clones were extracted by hand, washed to remove debris and embedded gravel, and divided for measurements into three parts: living shoots, dead shoots, and rhizomes and roots. Shoots were severed from rhizomes within 1 cm of attachment. Separated from any dead shoots, only living shoots were used for fresh weight measurements. Total shoot biomass included dry weights of both living and attached dead shoots. Impractical to impossible to separate, rhizomes and closely associated roots were kept together for weight determinations. All plant matter was dried to constant weight at 55°C.[9]

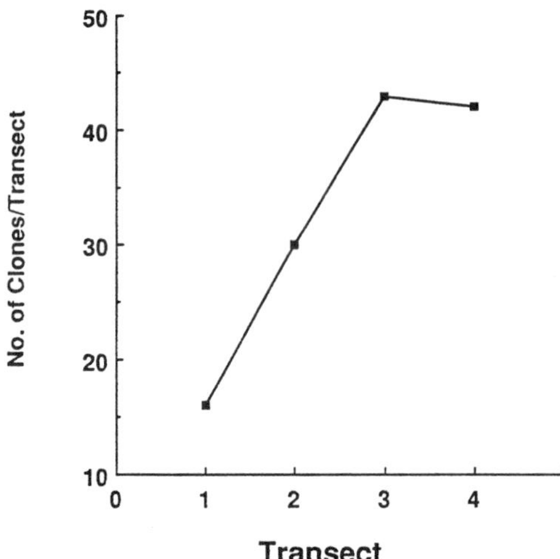

Figure 1. Clone frequency of *Scirpus validus* in constructed wetland transects along the path of water flow. Significant differences exist between transects 1, 2, and 3 ($p \leq 0.0001$), but not between transects 3 and 4.

Statistical Analysis

Statistical treatments (ANOVA) were computed by the StatView (1985 version) desktop computer program (Brain Power, Inc., Calabasa, CA).

RESULTS

Frequency of Bulrush Clones

The number of bulrush clones (i.e., frequency), rooted entirely or partially in each transect, increased along the constructed wetland cell and was highest in the last one third (Figure 1). No statistically significant difference existed in clone frequency for transects 3 and 4.

Bulrush Canopy Cover

As measured by canopy cover, bulrush density (or concentration per unit area) decreased along the cell. Located 5 m from wastewater input, transect 1 averaged more than 80% ground shade or overhang from bulrush plants, whereas transect 4 (positioned 5 to 6 m from the terminus of the 145-m cell) averaged less than 20% ground shade (Table 2). Intermediately positioned transects had intermediate index values. The absence of bulrush canopy did not always mean that substrate gravel or surface litter was exposed because volunteer species (e.g., *Polygonum* sp.) often provided ground cover. Nevertheless, although clonal frequency increased (Figure 1), a diminishing gradient of bulrush shade existed from input to terminus of the wetland cell.

Shoot Frequency, Diameter, Length, and Density

The mean number of shoots per clone (Figure 2), the mean shoot diameter, and the mean shoot length (Figure 3) decreased consistently from transects 1 through 4. Throughout the first three transects, the decreasing number of shoots per clone was significantly different ($p \leq 0.0001$). Likewise, along the first two thirds of the wetland cell, shoot diameters became significantly lower ($p \leq 0.0001$). However, shoot length was a less reliable indicator of bulrush growth because only two adjacent transects (transects 2 and 3) had statistically significantly different values. Calculated shoot densities for transects 1, 2, 3, and 4 were, respectively, 572, 413, 285, and 248 shoots per square meter.

Table 2. Bulrush (*Scirpus validus*) Mean Index of Canopy Cover

Transect	Mean index	SE	Median	Sample no.
1	4.35[a]	0.26	5	23
2	3.25[a]	0.26	3	24
3	2.08[a]	0.29	2	24
4	1.88[a]	0.20	2	24

Note: 0 = no overhang or ground shade by *Scirpus validus*/1-m² quadrats.
1 = some to 20% overhang or ground shade.
2 = 20–40% overhang or ground shade.
3 = 40–60% overhang or ground shade.
4 = 60–80% overhang or ground shade.
5 = 80–100% overhang or ground shade.

[a] One-way ANOVA (4 groups) gave significant difference at $p \leq 0.0001$.

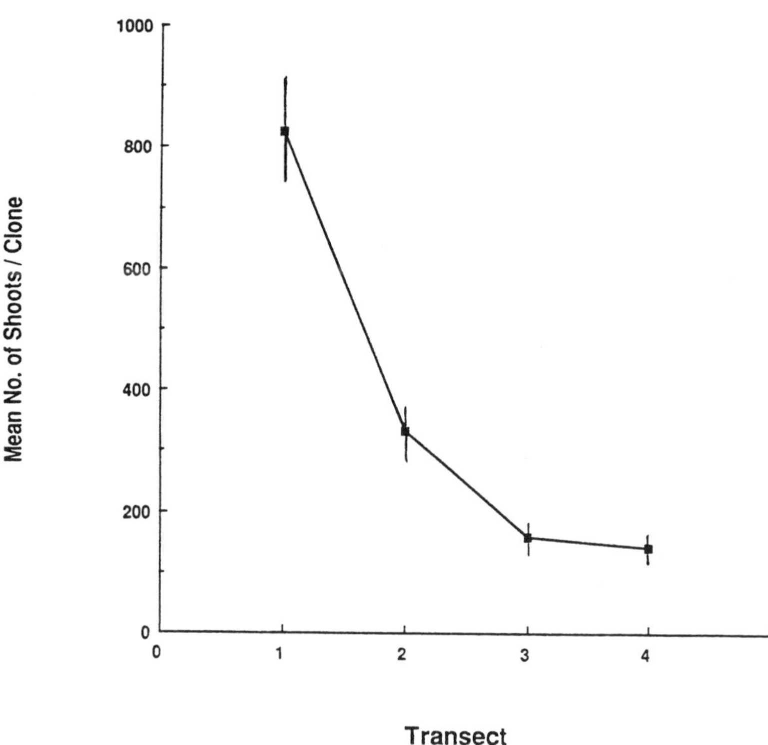

Figure 2. Shoot frequencies of representative *Scirpus validus* clones in constructed wetland transects along the path of water flow. Vertical bars indicate standard errors. Significant differences exist between transects 1, 2, and 3 ($p \leq 0.0001$), but not between transects 3 and 4. N = 50/clone; transects 1 and 3 = 7 clones each; transects 2 and 4 = 6 clones each.

Basal Diameters and Basal Areas

Although fewer clones per transect existed in the influent end of the cell (Figure 1), unadjusted mean BD and BA values for the influent end exceeded mean BD and BA values near the terminus (Figure 4). Keep in mind that BD in transect 1 had to be measured at 61-cm heights (and are unadjusted), whereas BD for other transects was at 46 cm. Thus, BA data are conservative since transect 1 measurements at 46-cm levels

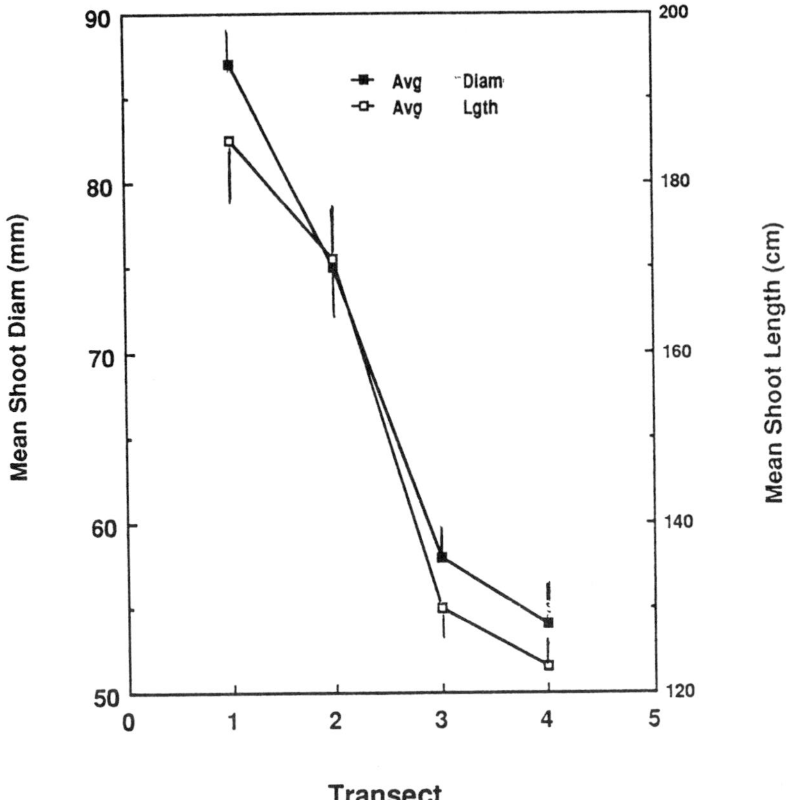

Figure 3. Shoot diameters and lengths of representative *Scirpus validus* clones in constructed wetland transects along the path of water flow. Closed squares (Avg Diam), mean diameters; open squares (Avg Lgth), mean length. Vertical bars indicate standard errors. N = 50/clone; transects 1 and 3 = 7 clones each; transects 2 and 4 = 6 clones each. Diameters in transects 1, 2, and 3 are significantly different ($p \leq 0.0001$). Lengths in transects 2 and 3 are significantly different ($p \leq 0.0001$). Other comparisions show no significant differences.

(impractical and counterproductive to constructed wetland objectives, since aerial stems near clonal perimeters would have been crushed) would have generated higher values. Since BD and BA represent a combination of bulrush number and size, basal parameters are more valuable indicators than frequency and density. Thus, based on BA data, more growth per unit area occurred in the first third of the wetland cell than in the last third of the cell.

Biomass

Shoot Fresh and Dry Weights

For representative clones, mean fresh weights of shoots decreased through the first three transects of the length of the constructed wetland (Figure 5). Similarly, mean dry weights of living and standing-dead shoots diminished through the same area of the wetland cell (Figure 5), with no statistically significant difference existing between mean values from transects 3 and 4.

Rhizome-Root Fresh and Dry Weights

Whether considering either fresh or dry weights of sampled rhizomes and roots, only transect 1 produced averages that were different (by one-way ANOVA) from other transects (Figure 6). From transect 1, mean fresh weight (5156 g) of the three clones, identified by BD as most representative of the transect, was ca. 2.5 to 5.5 times larger than average fresh weights of sampled clones from other transects. Mean dry weight (788

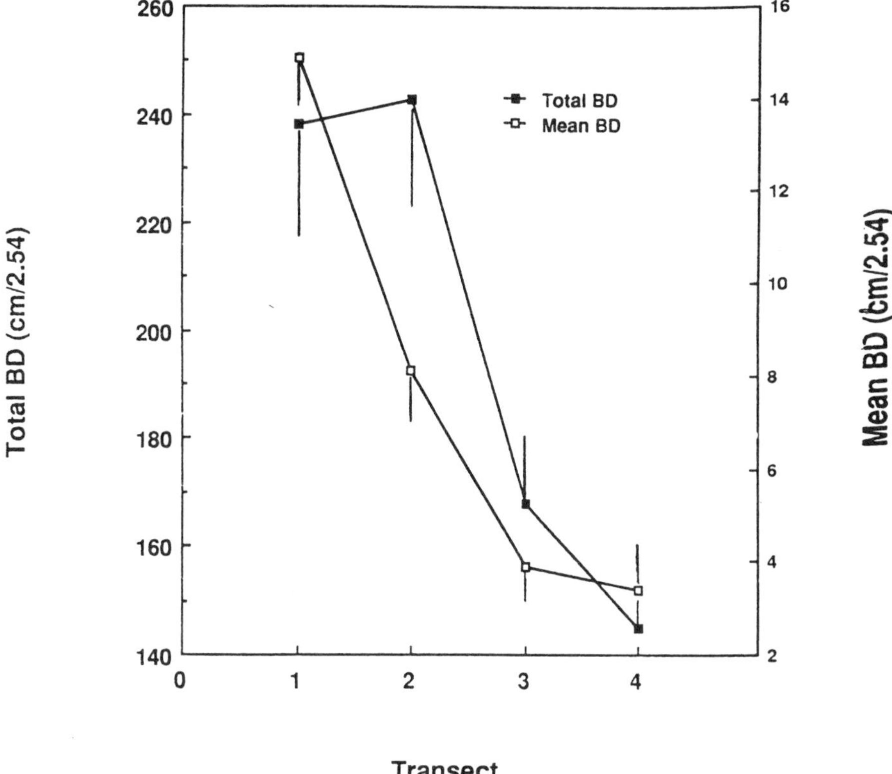

Figure 4. Shoot basal diameters (unadjusted) and basal areas of all *Scirpus validus* clones in constructed wetland transects along the path of water flow. Open squares (mean BD), mean basal diameter; closed squares (total BD), basal area. Vertical bars indicate standard errors. N = 16, transect 1; 30, transect 2; 43, transect 3; 42, transect 4.

g) of the same three samples of transect 1 was ca. 2.5 to 3.7 times greater than other dry weights of rhizomes and roots.

Total Bulrush Biomass

Mean total biomass (the sum of rhizomes and roots, living shoots, and dead-standing shoots) of three representative clones was greatest in transect l, intermediate in transect 2, and least in transects 3 and 4 (Table 3). For each transect, the ratio of aerial to subterranean growth was represented in the total shoot:rhizome-root ratios (Table 3), which was highest at the front end and lowest near the terminus of the wetland.

DISCUSSION

Undescribed previously for a constructed wetlands, an aerial growth gradient of bulrushes existed along the length of the constructed wetlands during the second growing season. A somewhat related study[9] at two other constructed wetlands in Kentucky reported no growth gradient. Differences between the earlier report and our findings may result from procedural differences, such as stage of vegetative development, sample size, and number and kind of examined parameters.

The question remains as to the cause for a growth gradient. One simple hypothesis is that the gradient results from a well-functioning constructed wetland. If plants at the front end effectively lower pollutant levels (and plant nutrients) from the wastewater stream, then bulrushes located downstream receive fewer nutrients, and a decreasing nutrient gradient would cause decreasing bulrush growth. Concentrations in effluent (Table 1) are consistent with this speculation, although nutrient concentrations along the cell are unavailable.

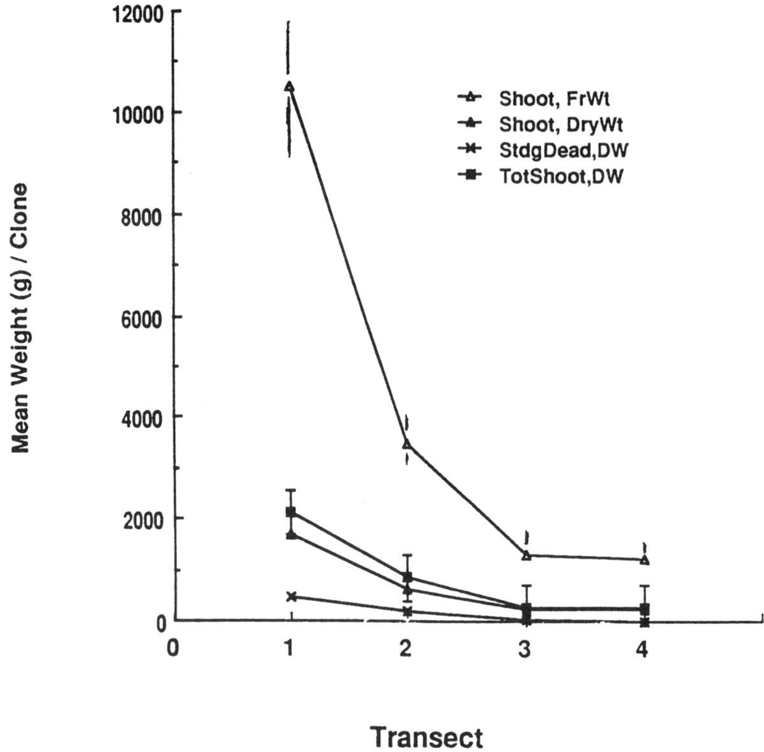

Figure 5. Shoot biomass of representative *Scirpus validus* clones in constructed wetland transects along the path of water flow. Open triangle (FrWt), fresh weight; closed triangle (DryWt), dry weight; × (StdgDead, DW), dry weight of standing dead; closed square (TotShoot, DW), dry weight of total shoot (standing dead + shoot). N = 3 for each transect. Vertical bars indicate standard errors. No significant difference exists between any mean values for transects 3 and 4.

The hydraulic gradient of the cell may also contribute to the growth gradient. Flooded conditions exist in the front end of the cell, while water levels were typically 8 cm or more below the top of the gravel in the lower end. This declining water elevation down the cell probably increased the proportion of roots in dry gravel. Consequently, the growth gradient may be due to water stress.

Related to the possibiltiy of water stress down the cell and also unresolved is the contribution of the muck layer to aerial growth near the front of this wetland cell. Possibly, the presence of muck affects the concentration of inorganics (e.g., essential macronutrients or trace elements) that influence plant growth.

An aerial growth gradient does not necessarily indicate the existence of a corresponding underground growth gradient. Based on much less data, a comparable growth gradient does not appear to exist underground, since only transect 1 exhibited a significantly higher rhizome-root biomass (Table 3). Relevant to the decreasing growth gradient aboveground is the decreasing ratio of aerial to subsurface biomass (Table 3). For many plant species in solution culture or soil, increasing nitrate concentrations favor aerial growth over that of the root system.[17,18]

A linear growth gradient may be a developmental phenomenon of a young constructed wetland, which eventually matures to outgrow its gradient. Continued studies are planned to evaluate this possibility. Until future growing seasons and additional data provide better answers, contributing factors to a bulrush growth gradient remain unestablished.

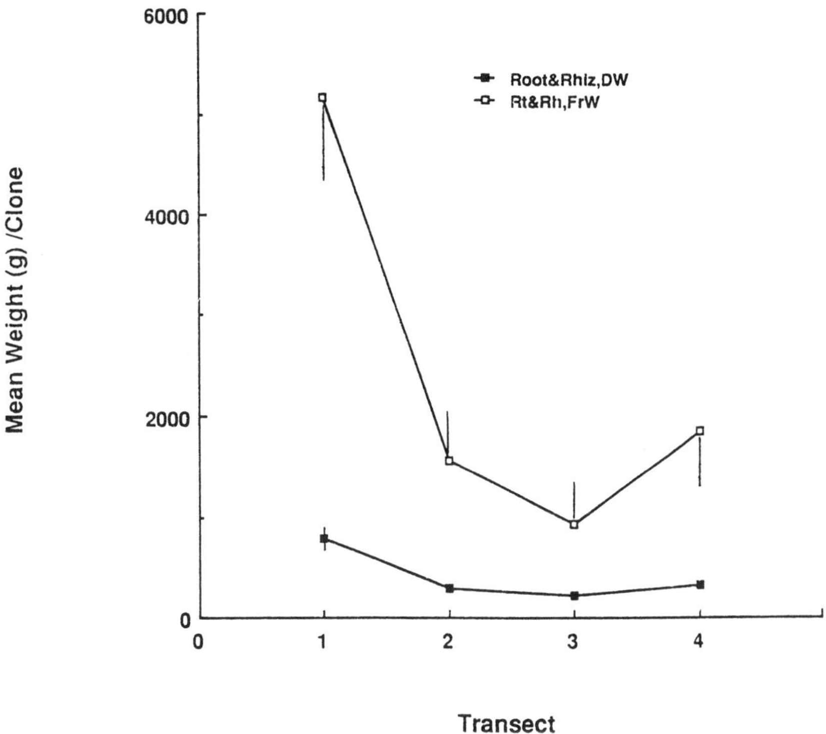

Figure 6. Root and rhizome biomass of representative *Scirpus validus* clones in constructed wetland transects along the path of water flow. Closed square (DW), mean dry weight; open square (FrW), mean fresh weight. N = 3 for each transect. Vertical bars indicate standard errors. Only transect 1 had mean values (both FW and DW) that were significantly different from other transects.

SUMMARY

Data consistently supported the conclusion that an aerial growth gradient of bulrushes existed along the length of the constructed wetland during the second growing season. First, aerial growth, as evaluated by canopy indexing (Table 2), was more than four times as dense near the influent end as near the terminus. Intermediate locations had intermediate degrees of canopy cover. Second, although transect 1 had the lowest frequency of individual clones (Figure 1), bulrush clones displayed: (1) the largest mean number of shoots per clone (Figure 2), (2) the widest mean diameter and the longest mean length of shoots (Figure 3), and (3) the largest mean basal diameter (Figure 4). Moreover, samples of the anterior transect exhibited the greatest fresh and dry weights, preferred indicators[16,17] of both shoot and rhizome-root growth (Table 2). All growth parameters of size and weight documented the conclusion that total growth in the front (ca. one third) of the constructed cell exceeded growth by the larger number of plants per unit area in the more distal regions. Continued studies are planned to follow bulrush growth and to identify contributing factors.

Table 3. Biomass of Representative Clones (n = 3) of Bulrush (*Scirpus validus*) from Transects along Cell 2

Clone source ratio	Rhizome & roots (g)	Living shoots (g)	Dead-standing shoots (g)	Total shoots (g)	Total biomass (g)	Total shoot: rhizome & root
Transect 1						
Sample A	647	1514	185	1699	2346	
Sample B	994	2460	699	3159	4153	
Sample C	722	1048	476	1524	2246	
Mean	788[a]	1674	453	2127[a]	2915[a,c]	2.70
Transect 2						
Sample A	373	631	373	1004	1377	
Sample B	275	645	130	775	1050	
Sample C	223	659	109	768	991	
Mean	290[a,b]	645	204	849[a]	1139[a]	2.93
Transect 3						
Sample A	117	108	15	123	240	
Sample B	165	282	24	306	471	
Sample C	364	374	41	415	779	
Mean	215[b]	255	23	281[a,b]	479[a,b]	1.29
Transect 4						
Sample A	228	247	20	267	495	
Sample B	270	202	21	223	493	
Sample C	464	276	15	291	755	
Mean	321[b]	242	19	260[b]	581[b,c]	0.81

[a] Values with this letter in the same column are significantly different ($0.025 < p \leq 0.05$).
[b] Values with this letter in the same column have no significant difference ($p > 0.05$).
[c] Values with this letter in the same column are significantly different ($0.025 < p \leq 0.05$).

REFERENCES

1. Brix, H. Treatment of wastewater in the rhizosphere of wetland plants — the root-zone method, *Water Sci. Technol.* 19:107, 1987.
2. Brix, H. Gas exchange through the soil-atmosphere interphase and through dead culms of *Phragmites australis* in a constructed reed bed receiving domestic sewage, *Wat. Res.* 24:259, 1990.
3. Wolverton, B. C. Aquatic plants for wastewater treatment: an overview, in *Aquatic Plants for Water Treatment and Resource Recovery.* K. R. Reedy and W. H. Smith, Eds. Magnolia Publishing, Orlando, FL. 1987.
4. Hammer, D. A. and R. K. Bastian. Wetlands ecosystems: natural water purifiers, in *Constructed Wetlands for Wastewater Treatment.* D. A. Hammer, Ed. Lewis Publishers, Ann Arbor, MI, 1989.
5. Kickuth, R. Degradation and incorporation of nutrients from rural waste water by plant rhizosphere under limnic conditions, in *Utilisation of Manure by Land Spreading.* Comm. of the Europ. Communities, EUR 5673e, London. 1977, 335–343 [cited in Brix, 1990].
6. Kickuth, R. Abwasserreinigung in mosaikmatrizen aus aerober und anaerober teilbezirken, *Grundlagen der Abwasserreinigung.* F. Moser, Ed. Schriftenreihe Wasser — Abwasser. 19:639, 1981 [cited in Brix, 1990].
7. Reddy, K. R., W. H. Patrick, Jr., and C. W. Lindau. Nitrification-denitrification at the plant-root-sediment interface in wetlands, *Limnol. Oceanogr.* 34:1004. 1989a.
8. Reddy, K. R., E. M. D'Angelo, and T. A. DeBusk. Oxygen transport through aquatic macrophytes: the role in wastewater treatment, *J. Environ. Qual.* 19:261, 1989.
9. Pullin, B. P. and D. A. Hammer. *Water Environ. Technol.* 3:36, 1991.

10. Steiner, G. R., J. T. Watson, D. A. Hammer, and D. F. Harker, Jr. Municipal wastewater treatment with artificial wetlands — a TVA/Kentucky demonstration, in *Aquatic Plants for Water Treatment and Resource Recovery*. K. R. Reedy and W. H. Smith, Eds., Magnolia Publishing, Orlando, FL, 1987.
11. Choate, K. D., J. T. Watson, and G. R. Steiner. Demonstration of constructed wetlands for treatment of municipal wastewaters, monitoring report for the period: March 1988 to October 1989. Tennessee Valley Authority, River Basin Operations, Water Resources. Chattanooga, TN, 1990.
12. Watson, J. T., K. D. Choate, and G. R. Steiner. Performance of constructed wetland treatment systems at Benton, Hardin, and Pembroke, Kentucky, during the early vegetation establishment phase, in *Constructed Wetlands in Water Pollution Control*. P. R. Cooper and B. C. Findlater, Eds. Pergamon Press, Oxford, U.K., 1990.
13. Oosting, H. J. *The Study of Plant Communities*, 2nd ed. Freeman, San Francisco, CA, 1958.
14. Odum, E. P. *Fundamentals of Ecology*, 2nd ed. Saunders, Philadelphia, 1959, 152.
15. Radford, A. E., W. C. Dickison, J. R. Massey, and C. R. Bell. *Vascular Plant Systematics*. Harper and Row, New York, 1974, 314.
16. Salisbury, F. B. and C. W. Ross. *Plant Physiology*, 3rd ed. Wadsworth Publishing, Belmont, CA, 1985.
17. Meyer, B. S., D. B. Anderson, R. H. Bohning, and D. G. Fratianne. *Introduction to Plant Physiology*, 2nd ed., VanNostrand, New York, 1973, 509.

CHAPTER 45

Species-Specific Aeration of Water by Different Vegetation Types in Constructed Wetlands

E. Stengel

INTRODUCTION

The "aeration function" of vascular aquatic plants, has only recently been investigated by attempts to quantify oxygen transport to the rhizosphere and the surrounding water with regard to natural situations[1,2] or (waste) water treatment.[3-9] These laboratory experiments (usually with single, small plants), *in-situ* measurements, and calculations based on this information may give fairly good estimates, but direct continuous measurements in technically relevant "mature" root horizons in order to obtain more integrated, "real" values are still lacking.

Surprisingly, no data under operating conditions can be found in the literature for the three helophyte species (*Iris pseudacorus* L., *Typha latifolia* L., and *Phragmites australis* (Cav.) Trin. ex Steudel), long used in practical water purification processes with respect to the aeration potential using direct measurements in flowing water.

The following contribution combines preliminary continuous recordings of (1) gas pressurization in detached plant parts, (2) pressurization in "intact" shoots of "polycormones" (see Reference 10) *in situ,* and (3) oxygen concentrations of water passing through the root horizon.

MATERIALS AND METHODS

Measurements were performed in outdoor constructed wetlands with subsurface water flow in U-shaped channels, 0.4 m deep, 0.6 m wide, 15 m long, and a total surface area of 9.1 m^2. The channels were filled with gravel (3 to 8 mm) and sampling sites were spaced between 1 and 1.5 m apart (Figure 1). The vegetated beds were dominated by *Iris pseudacorus*, *Typha latifolia,* and *Phragmites australis.* Local oxygen-saturated tap water was continuously supplied at a nominal dosage of 30 L/h = 79 L/m^2/day. Retention time was about 1 day. Isolated plant organs were connected to an electronic pressure transducer on an outdoor rack with an adapting glass tube, or *in situ* with needles sealed by Plasticine and grease and inserted into the second internode above the ground *(Phragmites*), or at one third of the leaf length above the ground (*Typha*).

Continuous recordings of oxygen concentrations were performed by Clarke-type electrodes (CONDUCTA electrode, type: 905, transducer type: OV). For technical reasons, the oxygen electrodes were situated at different lengths of the vegetated beds: *Typha*, *Iris*: 7.5 m (middle); *Phragmites*: 15 m (outlet). Oxygen profiles showed that after a distance of 5 to 6 m of all vegetated beds, the typical "steady-state" values of the outlet were already reached. Therefore, the results of the middle and the outlet electrodes are comparable.

RESULTS AND DISCUSSION

Pressurization in Above-Ground Plant Parts

Detached Plant Parts

Detached plant parts exposed under natural outdoor conditions (radiation, wind, humidity, etc.) showed pressurization for *Phragmites* and *Typha* and none for *Iris*. The fast responses to changing conditions, such

0-87371-550-0/93/$0.00+$.50

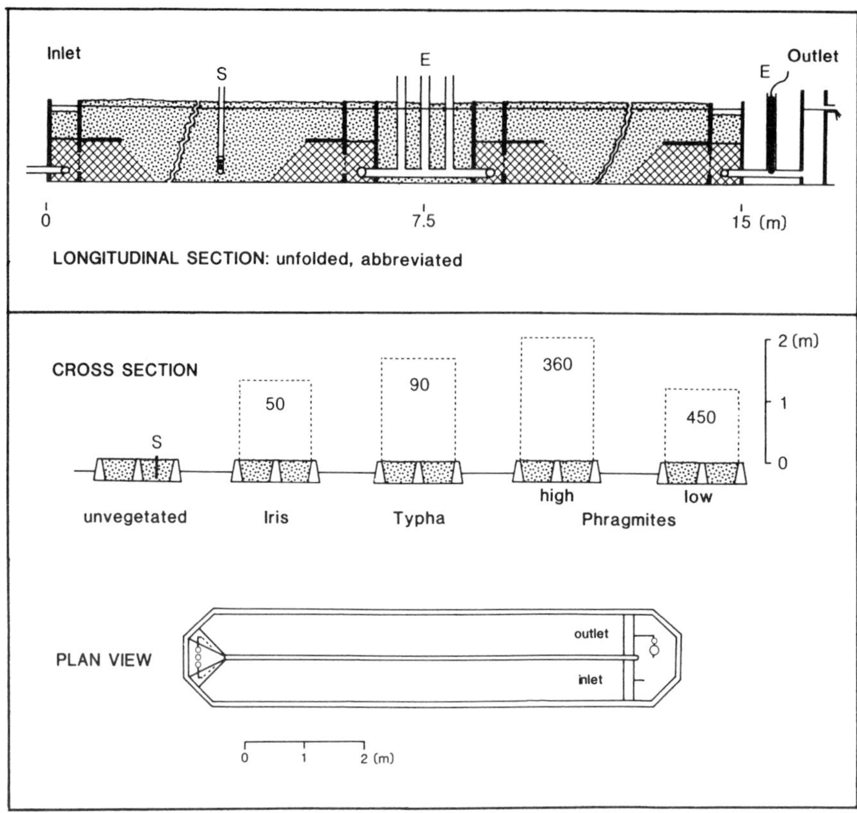

Figure 1. Basic features of the constructed wetlands at the Research Center in Jülich, FRG. Upper part: unfolded and abbreviated schematic longitudinal section. The ends of the two arms (7.5 m) of the U-shaped channel (total length, 15 m) are provided with water-distributing and -collecting areas with coarse gravel (16–32 mm diameter, see: cross-hatched area); the rest is filled with fine gravel (3–8 mm diameter, see: stippled area). Oxygen electrodes are situated in tubes at 7.5 m or 15 m. S = sampling tube; E = electrode tube. Lower part: approximate numbers of shoots per square meter (summer, 1991) are given inside the dotted lines showing the average plant height (summer, 1991).

as air movement and radiation, are identical for comparable shoots during simultaneous exposure (Figure 2, left graph). The pressurization potential was found to be higher in *Typha*. It seems that the *Typha* leaf represents a more integrating system compared to the *Phragmites* shoot which reaches dynamic equilibrium faster (Figure 2, right graph). Since the maximum theoretical value for "thermo-osmotically"-induced pressurization is about 800 Pa (5 K difference assumed[11]), the much higher values found clearly indicate a humidity-induced component (IDSO[12]). This is corroborated by an earlier observation that detached *Typha* leaves showed appreciable pressurization even in the shade under isothermal conditions ("indoor" situation) by partly emptying a U-shaped manometer against the hydrostatic pressure of a 90-mm water column (=900 Pa) protected from wind under a roof (12 August, 1988, ≈28°C, ≈30% rel. humidity).

In-situ *Continuous Recordings*

Roodenburg[13] published the first and (to my knowledge) only *continuous* pressure recordings of aquatic plants (*Nymphaea*). Our continuous recordings *in situ* agreed with individual observations with *Typha*[14] that the younger leaves (*Typha*) and the younger shoots (*Phragmites*) achieve a higher pressurization than the older parts (see Figure 3). The gas-conducting tissues that connect young and old leaves of the same shoot (*Typha*), or young and old shoots of the same polycormon (*Phragmites*, *Typha*), form the basic structures for a gas circulation by "volume flow".[15,16]

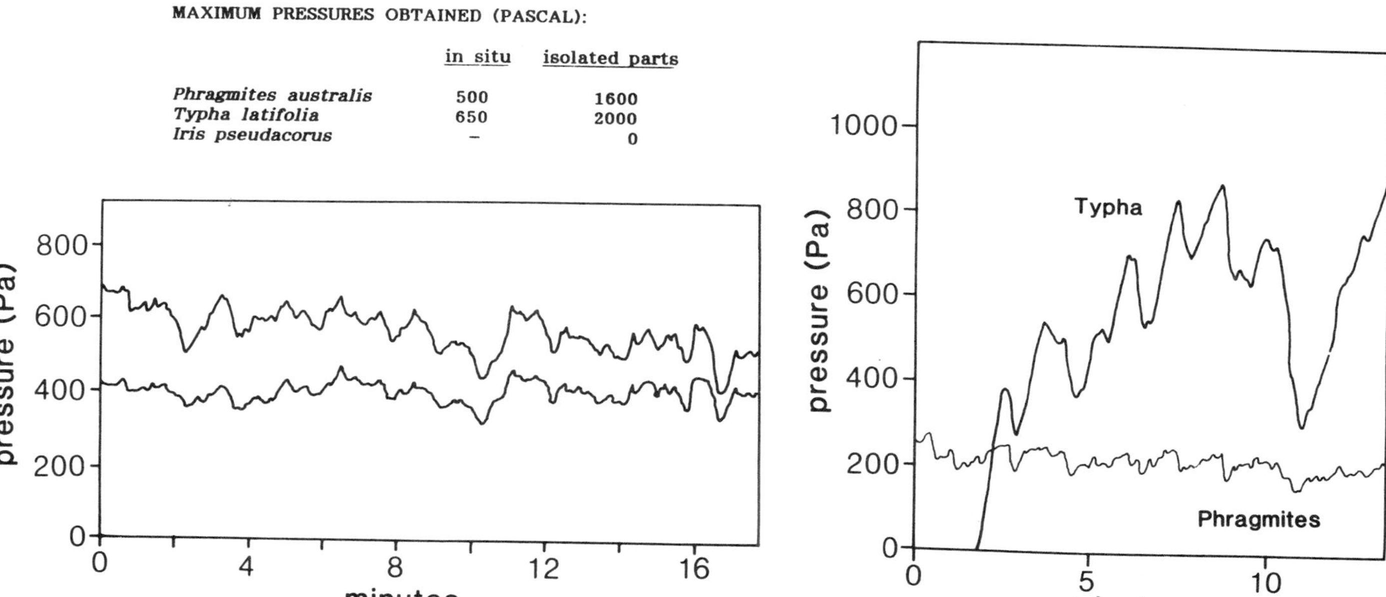
MAXIMUM PRESSURES OBTAINED (PASCAL):

	in situ	isolated parts
Phragmites australis	500	1600
Typha latifolia	650	2000
Iris pseudacorus	–	0

Figure 2. Gas pressures in helophytes exposed under natural conditions (radiation, air) measured both *in situ* and also at isolated parts of the plants (*Phragmites*: shoots; *Typha*, *Iris*: leaves). Simultanous exposure of plant parts on 8/24/89. (air temperature 21–22°C, humidity approx. 60%). Left graph: two similar *Phragmites* shoots (11.20–11.40 a.m.). Right graph: *Typha* leaf and *Phragmites* shoot (11.40–11.50 a.m.). Table: measurements *in situ* (9/20/89) and at plant parts (9/18/89), all values in Pascal (Pa).

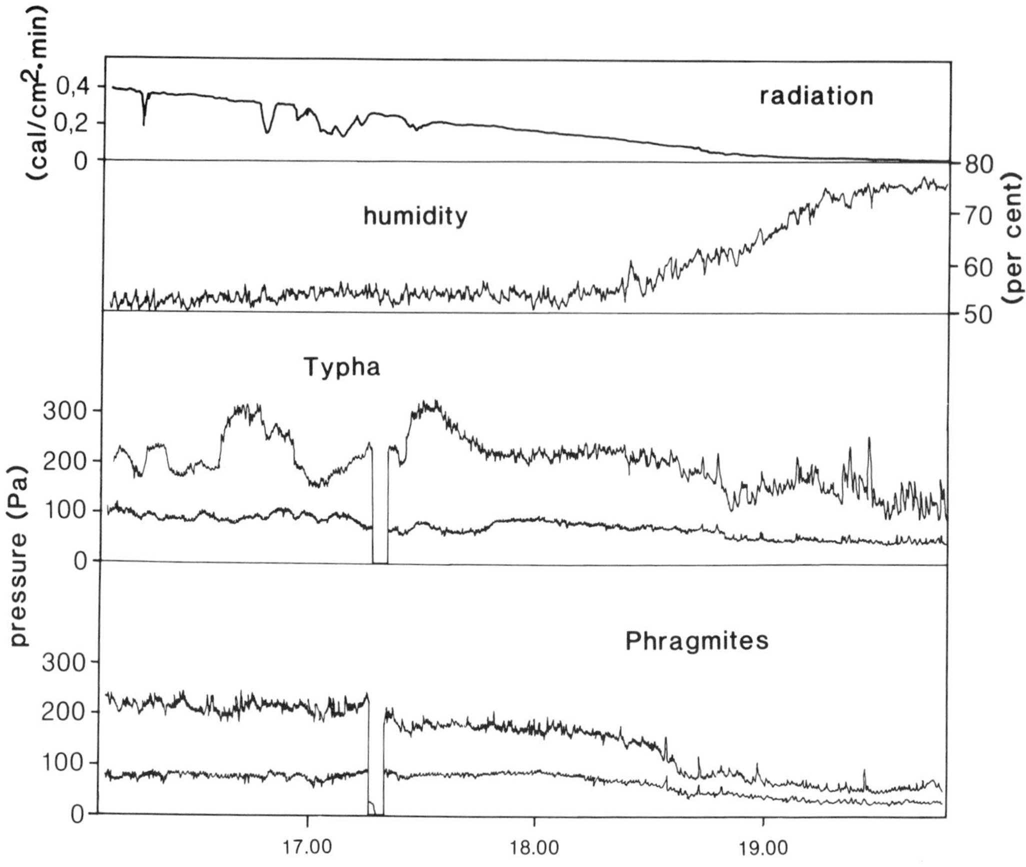

Figure 3. *In situ* measurements of gas pressures in intact plants as a part of a polycormon system (9/20/89). Air temperature was 24°C at 16.00 and 21°C at 20.00. The younger plants (shoots or leaves) show the higher pressures (upper lines). The interruption between 17.00 and 17.30 was caused by opening and reclosing of a valve to check the dynamic nature of pressurization.

Oxygen Regime in the Root Horizon

In summer 1990 (May to August), continuous oxygen recordings revealed three markedly distinct oxygen regimes in the root horizons of the three different types of helophyte vegetation (see Figures 4 through 7).

- *Iris*: No "aeration" in the sense of adding oxygen in measurable quantities to the water.
- *Phragmites*: Diurnal fluctuations with low amplitudes near "zero" oxygen during daytime. Kozuchowski and Johnson[17] reported on a similar rhythm of mercury emission by *Phragmites*.
- *Typha*: Very complex patterns of oxygen oscillations during day and night, often with large amplitudes. Only occasionally an overlying diurnal pattern has been found (see Figure 7). The daily oxygen amplitude vs. mean water temperature relationship in the root horizon showed three patterns (see Figure 8).

SUMMARY

- During summer of 1990, three different oxygen regimes were found in water passing through the root horizons of constructed wetlands vegetated by *Iris pseudacorus*, *Typha latifolia*, and *Phragmites australis*.
- Preliminary measurements of the ability of detached plant parts to generate pressurization enabling ventilation of underground organs revealed no pressurization by *Iris*; pressurization by *Typha* and *Phragmites*. The pressurization *in situ* measurements at polycormones of *Typha* and *Phragmites* suggest

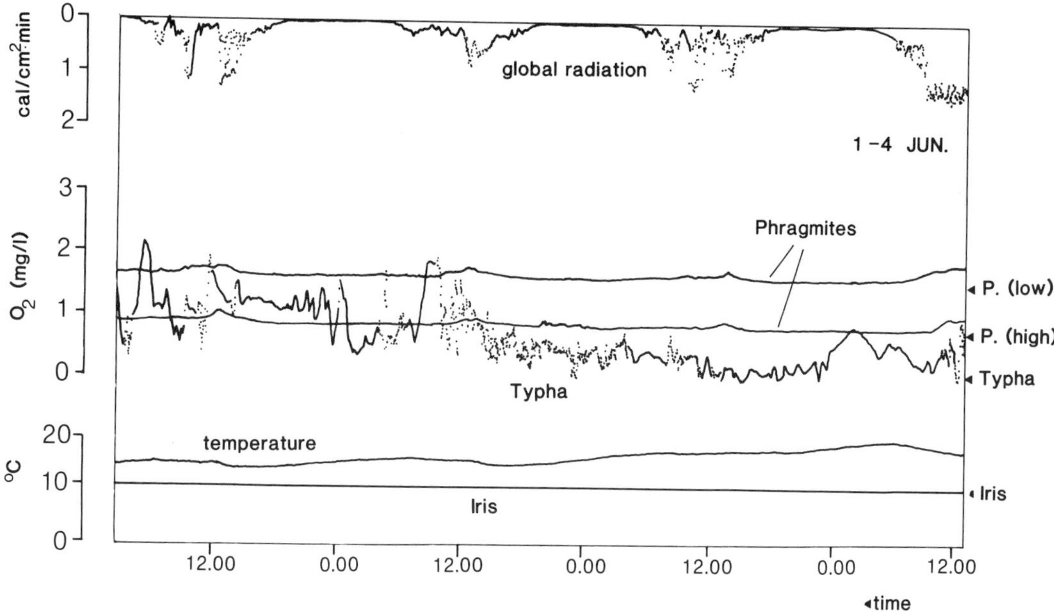

Figure 4.

Oxygen concentrations and temperature of water passing the root horizons of vegetation dominated by *Iris*, *Typha*, or *Phragmites* (summer, 1990). The "zero" positions of the different vegetation types are indicated by arrows. Graphically copied from the recording sheet.

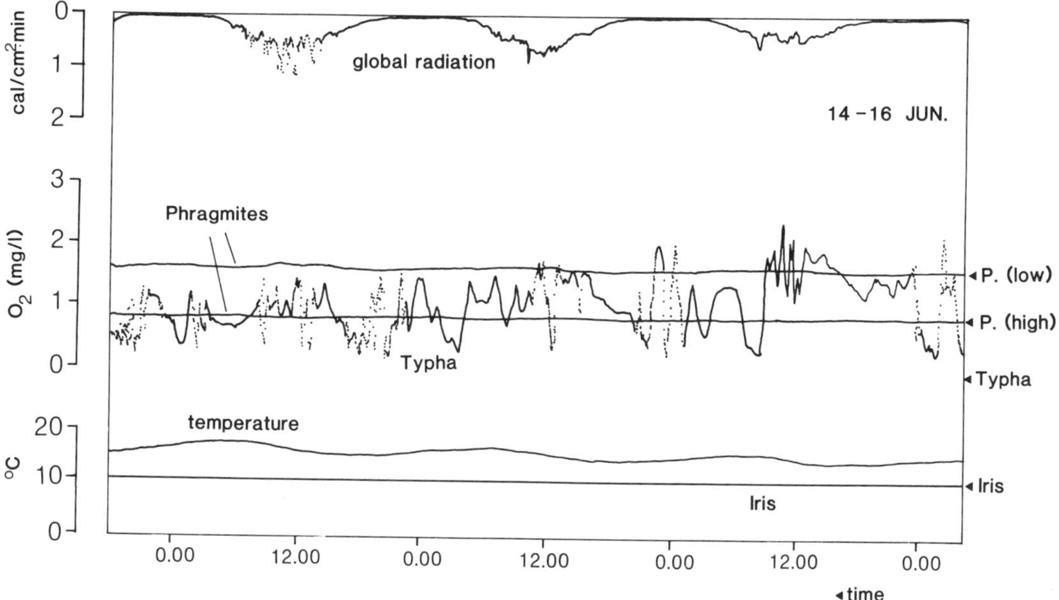

Figure 5.

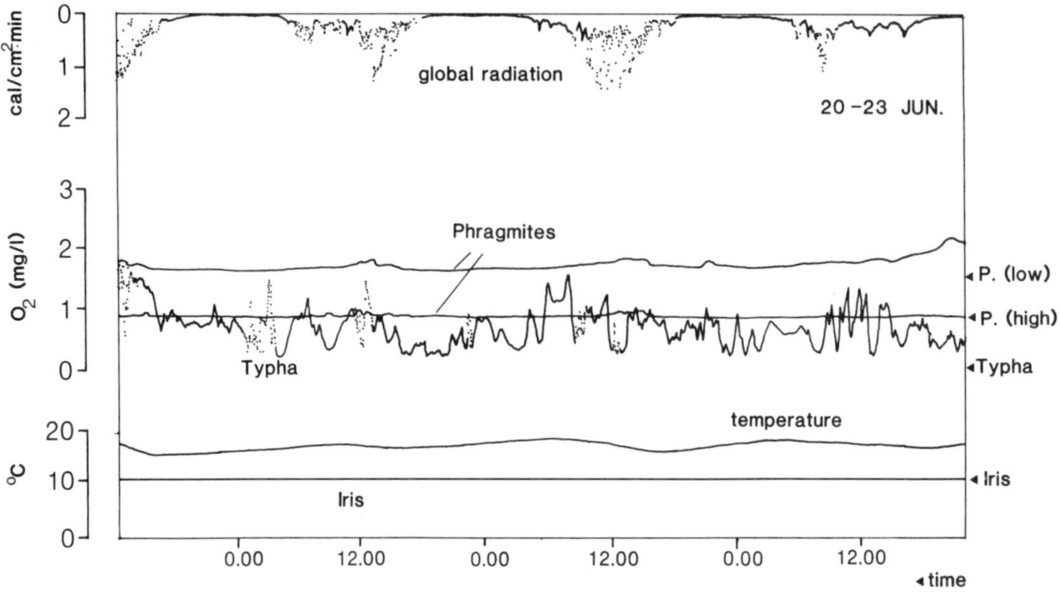

Figure 6.

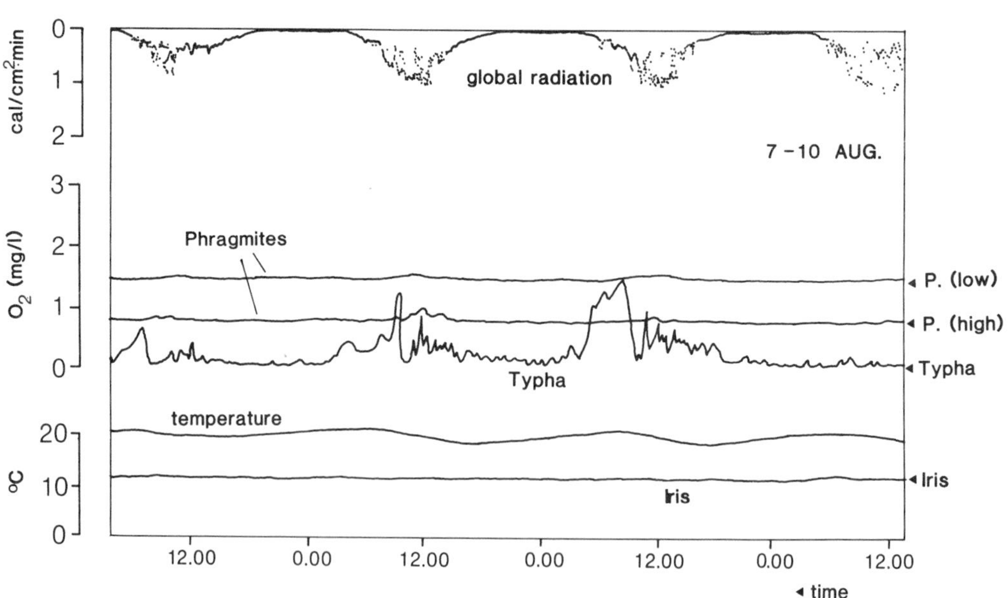

Figure 7.

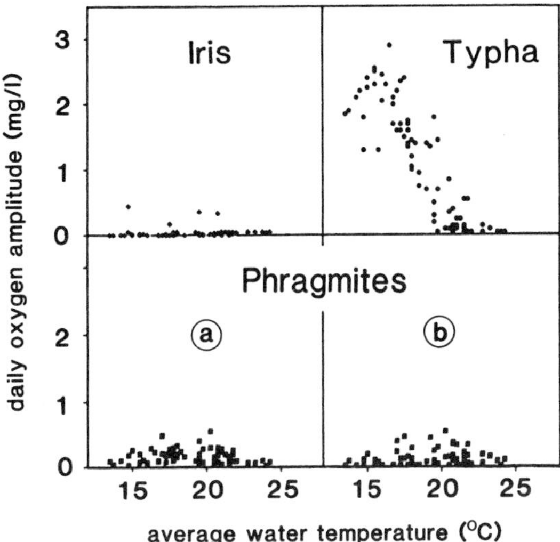

Figure 8. Relationship between water temperature and daily oxygen oscillations. The patterns for *Phragmites* are very similar, despite their differences in biomass and height (a << b).

that gas circulation by volume flow is possible and probably responsible for the observed oxygen patterns.

- The observed oxygen regimes are in agreement with the notion that volume convective gas flows inside the plants generated by purely physical mechanisms may have a special influence on oxygen transport into water passing the root horizons.
- The aeration of water flowing past the root horizons was found to be highest with *Typha*, much lower with *Phragmites,* and lowest with *Iris.*
- Disregarding the shortcomings of these preliminary observations with respect to a more extensive interpretation, the values are "realistic" since they are obtained from relatively matured "true" root horizons and therefore give an integrated, technically relevant signal from the whole root-rhizome-microbe community.

REFERENCES

1. Sand-Jensen, K., K. Prahl, and H. Stokholm. Oxygen release from roots of submerged aquatic macrophytes, *OIKOS*. 38:349, 1982.
2. Haldeman, Chr. and R. Brändle. Avoidance of oxygen deficit stress and release of oxygen by stalked rhizomes of *Schoenoplectus lacustris*, *Physiologie Végétale*. 21(1):109, 1983.
3. Armstrong, W., J. Armstrong, and P. M. Beckett. Measurement and modelling of oxygen release from roots of *Phragmites australis*, in *Constructed Wetlands in Water Pollution Control*. P. F. Cooper and B. C. Findlater, Eds. Pergamon Press, Oxford, U.K., 1990, 41.
4. Brix, H. Gas exchange through dead culms of reed, *Phragmites australis* (Cav.) Trin. ex Steudel, *Aquat. Bot*. 35:81, 1989.
5. Brix, H. Gas exchange through the soil-atmosphere interphase and through dead culms of *Phragmites australis* in a constructed reed bed receiving domestic sewage, *Water Res*. 24(2):259, 1990.
6. Dunbabin, J. S., J. Pokorny, and K. H. Bowmer. Rhizosphere oxygenation by *Typha domingensis* Pers. in miniature artificial wetland filters used for metal removal from wastewaters, *Aquat. Bot*. 29:303, 1988.
7. Moorhead, K. K. and K. R. Reddy. Oxygen transport through selected aquatic macrophytes, *J. Environ. Qual*. 17(1):138, 1988.
8. Michaud, S. C. and C. J. Richardson. Relative radial oxygen loss in five wetland plants, in *Constructed Wetlands for Wastewater Treatment*. D. A. Hammer, Ed. Lewis Publishers, Chelsea, MI, 1989, 501.
9. Reddy, K. R., E. M. D'Angelo, and T. A. DeBusk. Oxygen transport through aquatic macrophytes: the role in wastewater treatment, *J. Environ. Qual*. 19:261, 1990.

10. Penzes, A. Über die Morphologie, Dynamik und zönologische Rolle der Sprosskolonien-bildenden Pflanzen (Polycormone), *Fragmenta Floristica Geobotanica.* 6(4):505, 1960.
11. Grosse, W. Thermoosmotic air transport in aquatic plants affecting growth activities and diffusion to wetland soils, in *Constructed Wetlands for Wastewater Treatment.* D. A. Hammer, Ed. Lewis Publishers, Chelsea, MI, 1989, 469.
12. Idso, S. B. A technique for evaluating the potential for mass flow of gases in plants, *Plant Sci. Lett.* 23:47, 1981.
13. Roodenburg, J. W. M. Zuurstofgebrek in Den Grond in Verband Met Wortelrot, Ph.D. Dissertation Baarn, Hollandia Drukkerij (in Dutch), 1927.
14. Sebacher, D. I., R. C. Harriss, and K. B. Bartlett. Methane emissions to the atmosphere through aquatic plants, *J. Environ. Qual.* 14(1):40, 1985.
15. Dacey, J. W. H. Internal winds in water lilies: an adaptation for life in anaerobic sediments, *Science.* 210:1017, 1980.
16. Armstrong, W., J. Armstrong, P. Beckett, and S. H. F. W. Justin. Convective gas flows in wetland plant aeration, in *Plant Life under Oxygen Deprivation.* M. B. Jackson, D. D. Davis, and H. Lambers, Eds. SPB Publishing, The Hague, 1991, 283.
17. Kozuchowski, J. and D. L. Johnson. Gaseous emissions of mercury from an aquatic vascular plant, *Nature.* 274:468, 1978.

PART 8

INDUSTRIAL APPLICATIONS

Paper Pulp

Landfill Leachate

Petro Chemicals

CHAPTER 46

Performance of a Constructed Marsh in the Tertiary Treatment of Bleach Kraft Pulp Mill Effluent: Results of a 2-Year Pilot Project

R. P. Tettleton, F. G. Howell, and R. P. Reaves

INTRODUCTION

In May of 1989, personnel from The University of Southern Mississippi Department of Biological Sciences, and from Leaf River Pulp Operations of the Georgia Pacific Corporation, New Augusta, MS, began a cooperative study of the efficacy of tertiary treatment of bleach kraft pulp mill effluent by constructed marshes. The project was intended to generate information to develop a full-scale system, with emphases on appropriate plant species, hydraulic factors, cell configurations, and the potential for reduction in concentrations of various water contaminants. This article presents information on changes in concentration of pollutants in water, as monitored during a 2-year pilot study.

Constructed marsh systems have yielded good results in the treatment of domestic sewage, as exemplified by reductions greater than 50% that have been reported for BOD_5 (5-day biochemical oxygen demand) and TSS (total suspended solids) beyond the secondary treatment stage.[1] A pilot study by Weyerhaeuser Company represents the only prior application of constructed wetlands for tertiary treatment of pulp mill effluent.[2,3] Thut[2] performed his studies on small-scale systems (1.2 × 3.6 × 0.6 m) planted with either *Typha latifolia* (cattail), *Phragmites australis* (giant reed), or *Spartina cynosuroides* (cordgrass) in crushed marl submerged beds; these systems were monitored for 3 years. Average reductions in concentrations of water contaminants were: BOD_5 (32%), TSS (46%), ammonia (70%), organic-N (35%), and phosphorus (19%). Different plant species provided no statistically significant difference in providing improvements in water quality. Control systems without plants provided statistically equal treatment of mill effluents.

SYSTEM DESIGN

The system consisted of three 0.13 ha ($^1/_3$ acre) free-water surface wetland cells, placed in parallel and planted with torpedo grass *(Panicum repens)*. Each cell received influent through one pipe regulated by a hand-valve with a propeller-driven flow gauge. Outlet flow was regulated by a concrete weir with stacked stop logs. Cell influent was pumped continuously from the secondary clarifiers on the mill property. Flow rate was held constant, and equal retention times were maintained in all cells. Short-circuiting was minimized by careful grading of the cell basin before planting the vegetation.

Torpedo grass was sprigged on 30cm (12 in.) centers in cell 1 on June 5, 1989, cell 3 was completed on June 9, and cell 2 on June 12. Influent from the clarifier was introduced to the cells on June 30, 1989. *P. repens* grew as a monoculture and plant biomass increased throughout the first growing season, but did not stay rooted in the deeper areas of the three cells. During late December, a severe freeze (≈–19°C or 0°F) killed all emergent vegetation in the cells. On February 2, 1990, testing was stopped until vegetative cover was reestablished in the cells. The cells were replanted manually with clumps of grass obtained from edges of the cells. Second-year growth was less than the first year, and the system was successfully invaded by other plants. Reaves et al.[4] provide a more detailed description of the growth and performance of *P. repens* in this system.

0-87371-550-0/93/$0.00+$.50

WATER QUALITY PARAMETERS

Throughout all testing, water samples were taken from the secondary clarifier supplying the wetland cells and the outlet weirs of each of the three cells. The samples were analyzed for BOD_5, TSS, ammonia-nitrogen (NH_4^+-N), nitrate-nitrogen (NO_3^-), total Kjeldahl nitrogen (TKN), and total phosphorous (TP). All nutrient analyses were conducted by an independent laboratory. Routine monitoring of temperature, dissolved oxygen (DO), conductivity, true color, pH, and turbidity was also performed.

FIRST YEAR RESULTS: 1989–1990

From October 23 through November 27, 1989, and January 22 through February 2, 1990, tests were conducted for 1, 2, 3, and 4-day effluent retention times, all at a depth of 0.45 m. While occasional interruptions of influent supply occurred, they were of short duration and did not interfere with testing.

Data collected for the first year are summarized in Table 1. Due to small sample sizes, it was not possible to detect changes that may have resulted from different retention times. Outlet values for all parameters were pooled and compared against inlet values using either the t-test or z-test in MicroStat II.[5] A statistical significance in the difference between concentrations at the clarifier and those at the cell outlets was found for TSS ($p < 0.05$). While all parameters measured showed reductions in concentration, a variance ratio test indicated that variation in concentrations of TSS ($p < 10^{-14}$) and NO_3^- ($p < 10^{-8}$) was reduced across the system.

SECOND YEAR RESULTS: 1990

Water quality testing began in April. A constant depth of 0.30 m and a 1-day retention time were maintained until testing ended September 21, 1990. Data collected for the second year are summarized in Table 2. Either a t-test or z-test was used to compare sample means. A statistically significant difference between influent and outlets was found for the following: TSS ($p < 0.01$), and TKN ($p < 0.02$). All parameters measured showed reductions in concentration except TP, which increased, although not significantly. A variance ratio test indicated that variation was reduced for concentrations of NO_3^- ($p < 10^{-5}$) in the system.

DISCUSSION

Except for TP in the second year, concentrations for all water quality contaminants were reduced by treatment through the marsh system. Significant reductions in values for TSS in the first year, and TSS and TKN again in the second year, indicate that the process can effectively perform as a tertiary treatment system. While sample sizes were small for retention times of 2, 3, and 4 days, it appeared that longer retention times would be beneficial.

Some issues that should be considered in future projects are: (1) composition of the waste at the pulp mill varies due to production levels, wood type (softwood vs. hardwood), and treatment process changes; (2) the concentration levels of some water quality parameters change by orders of magnitude as the composition of the effluent changes; (3) accurate control of flow through the cells and short-circuiting (channelization of flow through the cells) both affect retention time in the system. Little can be done for the first two points stated above. However, an awareness of these issues provides insight into why the variation observed in the data may be so large. Measures were taken to ensure reliable retention time (e.g., flow gauges, control valves, and uniform weir logs), but the system design proved too crude for fine control of flow through the cells. Suspended solids, gravel, and other objects in the wastewater often fouled the flowmeters. Leakage around and between the stop logs allowed water to leak through the weir, reducing the expected retention times. Stagnant areas with no flow, and deep areas without vegetation, may have been indicators of short-circuiting. It is probable that retention times were shorter than the design specified, thus creating problems when matching the outlet sample to the inlet sample for analysis.

Table 1. Water Quality Data Collected the First Year of System Operation, 1989–1990

	n		Mean			S.E.		Variance ratio test	Range		
Parameter	In	Out	In	Out	Test statistic[a]	In	Out	F statistic[a]	In	Out	% reduction
BOD_5	7	20	29	27	$t = 01.62^{ns}$	5.3	2.2	0.95^{ns}	12–53	11–44	7
TSS	7	21	113	21	z = 1.40*	61.5	1.8	378.28**	22–480	9–38	81
NH_4-N	6	18	16	12	$z = 0.48^{ns}$	7.0	2.5	2.64^{ns}	0.35–48	2.5–32	25
NO_3-N	7	21	5	0.9	$z = 1.04^{ns}$	3.1	0.3	40.94**	0.1–23	0.1–5.5	80
TKN	6	18	21	18	$t = 1.00^{ns}$	6.5	3.1	1.43^{ns}	1–45	3–41	14
TP	7	21	7	3.5	$t = 0.98^{ns}$	2.2	0.9	1.95^{ns}	0.1–18	0.1–15	53

Note: All data is recorded in mg/L unless otherwise noted.

[a] ns, not significant; *, significant at 0.05; **, significant at 0.02.

Table 2. Water Quality Data Collected the Second Year of System Operation, 1990

	n		Mean			S.E.		Variance ratio test	Range		
Parameter	In	Out	In	Out	Test statistic[a]	In	Out	F statistic[a]	In	Out	% reduction
BOD_5	32	95	17	16	$z = 0.75^{ns}$	1.3	0.7	1.31^{ns}	7–36	3–35	6
TSS	36	108	55	37	z = 2.56**	6.4	3.1	1.36^{ns}	20–176	10–236	33
NH_4-N	7	21	9.3	7.6	$t = 1.38^{ns}$	1.8	1.3	0.67^{ns}	2–24	3–16	18
NO_3-N	7	21	1.4	0.5	$t = 1.15^{ns}$	0.9	0.1	13.7**	0.1–6.8	0.1–3.0	64
TKN	7	21	15	10	t = 2.80**	1.9	1.3	0.68^{ns}	9–23	2–24	32
TP	7	21	87	115	$t = 0.99^{ns}$	80	88	0.27^{ns}	4–570	3–1850	–32

Note: All data are recorded in mg/L unless otherwise noted.

[a] ns, not significant; *, significant at 0.05; **, significant at 0.02.
[b] Negative % reduction = increase.

RECOMMENDATIONS

Future projects should include better flow control structures. Solid weir plates at both inlets and outlets will be less likely to leak or become obstructed. This will allow for better control of both inflow and outflow. Careful grading of the cells should be made from side to side as well as front to back to reduce short-circuiting. Machine grading may be too coarse to achieve proper configuration of cell bottoms. Retention times should be monitored closely so that inlet and outlet samples match up properly. Tracer studies can be conducted regularly to verify retention time. The reduction in concentration of all contaminants tested warrants further investigation into tertiary treatment of bleached kraft mill effluent by constructed wetlands.

REFERENCES

1. Knight, R. L. Wetland systems, in *Natural Systems for Wastewater Treatment.* Manual of Practice FD-16, Ed. Reed, Sherwood C., Water Pollution Control Federation, Alexandria, VA, 1990.
2. Thut, R. N. Utilization of artificial marshes for treatment of pulp mill effluents, in *Constructed Wetlands for Wastewater Treatment: Municipal, Industrial, and Agricultural.* Proceedings from the First International Conference on Constructed Wetlands for Wastewater Treatment held in Chattanooga, Tennessee on June 13–17, 1988, D. A. Hammer, Ed. Lewis Publishers, Chelsea, MI, 1989.
3. Thut, R. N. Utilization of artificial marshes for treatment of pulp mill effluent, *TAPPI.* 73:93, 1990.
4. Reaves, R. P., F. G. Howell, and R. P. Tettleton. Performance of torpedo grass (*Panicum repens*) in a tertiary treatment system for pulp mill waste water, in *Proc. 21st Mississippi Water Resources Conf.* 26–27 March 1991, Jackson, MS, B. J. Daniel, Ed. Water Resources Research Institute, Mississippi State University, Mississippi State, MS, 1991, 105.
5. Ecosoft, Inc. *MicroStat II Interactive Statistical Software System.* Ecosoft, Inc. Indianapolis, IN, version 2.5, 1990.

CHAPTER 47

Feasibility of Treating Pulp Mill Effluent with a Constructed Wetland

R. N. Thut

INTRODUCTION

The use of constructed wetlands for a variety of treatment objectives has been underway for over 10 years. The earliest systems were used to treat wastewaters from small municipalities. Later, similar systems found a role in the treatment of storm water, agricultural wastes, and acid mine drainage. At present, there are over 150 systems in North America, ranging in size from a few square meters to over 500 ha.[1] For the largest facilities, wastewater volumes of up to 80,000 m^3/day are being treated (comparable to the volumes generated by a large pulp mill).

The application of this technology to industrial wastewaters has been limited. A chemical plant in Alabama recently constructed a 1.4-ha wetland of the subsurface flow type.[2] The system has proven effective in reducing the BOD_5 (5-day biochemical oxygen demand) concentration of a secondary-treated effluent. There appears to have been a beneficial effect on chronic toxicity as well. Several other facilities for industrial application are in the planning or early construction/implementation phases. These include installations of about 30 to 40 ha for the food processing (potato, meat, and sugar), petroleum, and pulp and paper industries. In the case of the pulp and paper industry, there are also a number of pilot studies underway. The results from some of these have been reported.[3-5]

CHARACTERISTICS OF PULP MILL EFFLUENTS

The chemical pulping of wood chips takes place under conditions of high heat and pressure in strongly alkaline solutions of sodium sulfide (the kraft process) or in mildly acidic solutions of a basic salt of bisulfite (the sulfite process). The primary intent of this pulping or cooking process is to solubilize the lignin in wood, thus isolating the cellulose fibers. Most of the lignin is recovered (burned for energy), but some enters the wastewater. Various mechanical pulping processes are also utilized whereby much of the lignin is incorporated into the finished product. If a bright white product is required, then several bleaching steps may be used. Most often, bleaching is accomplished with a combination of chlorine, chlorine dioxide, and sodium hydroxide. However, the use of oxygen and hydrogen peroxide has assumed increasing importance. Pulps produced by the mechanical method do not use chlorine bleaching agents.

Pulp mill effluents are complex mixtures of wood-derived organics as well as some inorganic ions and compounds. In untreated effluents, the BOD_5 is high (generally in the range of 200 to 800 mg/L). Secondary treatment (by aerated lagoon or activated sludge) can be quite effective, reducing the BOD_5 to about 10 to 100 mg/L. The compounds responsible for the BOD_5 of untreated effluents are primarily simple sugars, organic acids (e.g., acetic) and alcohols (e.g., methanol). After treatment, the residual BOD_5 is largely caused by biological solids and some more refractory organic compounds.

The dissolved lignin which enters a treatment system is composed of a mixture of breakdown products of enormous variety whose molecular weights range from a few hundred daltons to tens of thousands. Generally speaking, these compounds are not readily amenable to biodegradation, and although they are found in treated effluent, they contribute little to BOD_5. They do, however, have an impact on long-term BOD and COD.

0-87371-550-0/93/$0.00+$.50

If the mill uses chlorine to produce bleached pulp, some chlorinated organic compounds will be produced. Some simpler compounds that have been identified include chloroform, chloroacetic acids, and various kinds of chlorinated phenolics. The dissolved lignin material can also be chlorinated, producing large, complex molecules which cannot be readily identified using even the most sophisticated analytical techniques. A general measure of chlorine attached to organic molecules is used to monitor this type of material; the most commonly used technique at this time yields a measure of adsorbable (to activated carbon) organic halides or AOX. These chlorolignins are also difficult to biodegrade. The smaller chlorinated molecules (e.g., the chloroacetic acids and chlorinated phenolics) are readily removed by secondary treatment.

Well-treated secondary effluents demonstrate no acute toxicity and little chronic toxicity using standard bioassay protocols. Some studies have found pulp mill effluents to cause biochemical and histological effects on fish and invertebrates. Although the responsible chemical agents have not been identified, most attention has been focused on the chlorinated phenolics and resin and fatty acids extracted from the wood during the pulping process.

The most obvious characteristic of bleached pulp mill effluents is the often intense brown color. There is not complete agreement concerning the chemical origin of this color. Color is not removed in conventional secondary treatment facilities since the chromophores appear to be largely resistant to bacterial breakdown. At some mills, however, tertiary treatment, using coagulants or adsorbents, is used to reduce color.

PILOT STUDIES OF CONSTRUCTED WETLANDS FOR PULP MILL EFFLUENT TREATMENT

Two pilot studies have been conducted to determine the practicality of treating pulp mill effluents with constructed wetlands. Both utilized the subsurface flow/gravel matrix design.

Small-Scale Pilot Study

The earlier study utilized eight troughs, each 2.6 m^2 in area. The substrate was 40 cm deep and composed of a locally abundant marl-type gravel (1 to 2 cm). In addition to determining how effectively constructed wetlands remove pulp mill pollutants, it was also desired to gather information on the design parameters for a full-scale facility.

Three plant species were tested: the cattail (*Typha latifolia*), the giant reed (*Phragmites australis*), and a locally abundant species, the giant cordgrass (*Spartina cynosuroides*), all of which proved to be equally effective. The choice among them was made on the basis of local availability and how well the plant's growing season coincided with the period of lowest flow or greatest environmental concern in the receiving water. Much of the BOD_5 and TSS (total suspended solids) removal was accomplished by the gravel matrix, with little contribution by the plants (determined with two controls containing gravel but no plants). Nutrient removal, however, was dependent upon the presence of plants, either directly through root uptake or indirectly via bacterial denitrification mediated by the plant roots. All of the plant species grew prolifically in the pulp mill effluent. Biomass production of above-ground tissue in a single year was 4.4 kg/m^2, comparable to the highest rates of production reported in the ecological literature.

The effluent entering the troughs had already been effectively treated in an aerated stabilization basin. Suspended solids concentrations averaged only 6 mg/L; BOD_5 12 mg/L; ammonia-nitrogen and total nitrogen 2 mg/L and 3 mg/L, respectively; and total phosphorus 0.6 mg/L. Removal efficiencies were quite high for TSS, BOD_5, and ammonia-nitrogen, and were more modest for organic-nitrogen and total phosphorus (Table 1). There was no appreciable effect on color or AOX. Based on a small number of analyses, the chlorinated phenolics and resin and fatty acids were reduced by about 20%.

The studies were conducted at three relatively short retention times (6, 15, and 24 h) necessitated by the volume of effluent produced by a typical pulp mill. As expected, a retention time of 24 h was more effective than a retention time of 6 h. The retention time of 15 h was intermediate in effectiveness, but not a great deal less than that for 24 h (Table 1). If the primary intent of a constructed wetland of this type is to provide tertiary level treatment of TSS and BOD_5, a very short retention time could be sufficient depending upon the specific goals of the installation. For a bleached pulp mill of average size, a subsurface flow system of about 10 ha would provide a retention time of about 6 h.

Table 1. The Effect of Retention Time on Percent Removal Efficiency

	Retention time		
	6-h	**15-h**	**24-h**
TSS	54	61	65
BOD_5	36	38	41
Ammonia-N	21	62	70
Organic-N	18	26	26
Phosphorus	12	18	25

In the last of a 3-year study, the troughs were modified and replanted in order to study the effects of substrate depth (at 15 and 40 cm). The retention time was maintained at 15 h for all units; consequently, less water flowed through those units which had the lesser amount of substrate. Greater removal efficiencies were found in the troughs with the lesser amounts of substrate. There were, for BOD_5 (46 vs. 28%), for ammonia-nitrogen (96 vs. 42%), and total phosphorus (27 vs. 14%). Apparently, the greater amount of root mass and above-ground foliage per volume of substrate, and the higher levels of dissolved oxygen in the units with only 15 cm of substrate, were the factors responsible for these differences. The differences were insignificant for TSS (49 vs. 55% for 15 and 40 cm, respectively), and for organic nitrogen (31 vs. 33%). However, these pollutants are usually removed via filtration. The depth of substrate would not be expected to influence this mechanism. Although a constructed wetland with a shallower substrate would perform better in many respects, it might not be practical where land is at a premium. A decision in this regard would depend upon the relative costs of land and that of gravel.

Large-Scale Pilot Study

Based on the promising results from the small-scale pilot studies, a second investigation was initiated to test further the practicality of using constructed wetlands for tertiary-level treatment of pulp mill effluents. Design features incorporated the findings of the author's earlier studies, with adjustments made to account for constraints at the site if full-scale investigations were to be implemented.

The new pilot site had no slope, was 25 × 150 m (or about 0.4 ha), and was located at another pulp mill. Because of naturally occurring tight clay soils, lining of the shallow excavation was not considered necessary. The gravel matrix was about 50 cm deep at the input to the wetland and about 40 cm deep at the output. Due to flow resistance caused by the gravel, the surface of the water tends to be higher at the front end of this type of constructed wetland. After construction, the water surface height was measured and found, in fact, to be about 15 cm higher at the front end. The deeper gravel at this point reduces the likelihood of water short-circuiting over the top of the gravel. The gravel, which was of mediocre quality was purchased locally. It was small (1 to 2 cm) and poorly sorted, with an admixture of smaller material. As a consequence, the porosity of the gravel was only 30%.

Secondary-treated effluent was pumped from the final discharge of a thermomechanical pulp mill and distributed across the width of the wetland by a header with ports at 2.5-m intervals. After flowing through the gravel matrix, the effluent discharged into a headbox and over an adjustable weir, a structure which made it possible to vary the head and slope of the water surface. Between 550 and 600 m^3/day of effluent were delivered to the pilot, for a nominal retention time of about 20 h.

In November 1988, two kinds of vegetation were purchased and planted: *Phragmites australis*, which had been used with success in the earlier experiments, and *Scirpus californicus* (southern bulrush), which has been used in several constructed wetlands in the southeastern U.S. Only a few of these plants survived the winter. This may have been due to the effects of cold weather on the newly established plants. However, difficulties were also experienced in delivering effluent to the experiments on a regular basis. The pilot facility was completely replanted the following April with the bulrush, an iris, and *Panicum repens* (torpedo grass). In addition, several other plant species were collected on the mill site and transplanted. These included the cattail, *Typha latifolia*, the willow, *Salix* sp., *Ludwigia peploides*, and other as yet unidentified wetland plants. In addition, several other species (e.g., *Sagittaria* sp.) became naturally established before the conclusion of the study a year later. All of these plants, in particular the torpedo grass, grew well in the constructed wetlands.

Based on its rate of proliferation, growth form, and root structure, torpedo grass was selected as the plant of choice, in spite of the fact that this species is considered to be a noxious weed in parts of the southern U.S., it grows to a height of 0.5 to 1 m, forming a dense stand which shades the substrate, thereby discouraging the growth of algae and other unwanted plants. It belongs to the same genus as rye grass and assumes a similar growth form. It is salt-tolerant and can be used as cattle feed. A similar, and native plant, *Panicum hemitomon,* or maidencane, may also prove to be successful for this application, and may even be a more acceptable plant species in some areas.

The pilot facility proved to be very effective in removing BOD_5; during the first 5 months of operation, BOD_5 removal efficiency averaged 73%, and thereafter was consistently in the 80 to 90% range (see Figure 1). Inasmuch as the secondary-treated effluent delivered to the marsh was of a high quality (averaging about 13 mg/L), this resulted in BOD_5 values in the 1 to 2-mg/L range through much of the study. The background BOD_5 of the receiving water was generally 2 to 3 mg/L and consequently was not believed to have an impact on the BOD_5 or dissolved oxygen concentrations in the receiving water. Discounting the first 5 months, there was very little seasonal variation in BOD_5 removal efficiency. All values were in the range of 80 to 90%, with the exception of those for November and December which ran at about 75%. This was the time when many plants in the wetland began to senesce, and it is possible that some BOD_5 was added to the effluent in this way, thus apparently reducing efficiency.

Significant plant growth was not apparent until June, and BOD_5 removal was almost as effective before June as after. Month-to-month comparisons (e.g., January 1989 to January 1990) suggest some slight benefit due to the plants. After the wetland became established, the average BOD_5 removal efficiency was 83%. Perhaps 10% of this could be attributed to the plants, with the rest due to the action of the gravel and its associated bacteria. This is consistent with the observations from the earlier study.

It is encouraging to note that the constructed wetland was able to cope with a period of relatively high influent BOD_5. Due to maintenance dredging operations in the secondary treatment system, influent BOD_5 exceeded 30 mg/L in March and April. However, removal efficiencies of about 80% were maintained throughout this period.

The removal efficiencies for BOD_5 were much higher than the efficiencies observed in the earlier study (30 to 40%). The reasons for this are not clear. The small size of the pilot facilities in the earlier study may have introduced some experimental artifacts which were not operative with the larger pilot. The effluents came from different types of pulp mills, but there is nothing known to suggest that the BOD_5 of one should be more readily treated than that of the other, although the effluent used for the second pilot study had higher levels of suspended solids. Based on analyses of filtered and unfiltered BOD_5, only 27% of the influent BOD_5 was soluble, and much of the BOD_5 reduction may have been due to simple filtration of the solids. Pairs of standpipes were positioned at 15-m intervals along the length of the wetland, and samples from these, were collected for chemical analysis. Approximately half of the BOD_5 was removed in the first 15 m (or 10% of the total length), and three quarters of the BOD_5 was removed in the first 75 m (or 50%). It is expected that a much smaller wetland with less retention time would have been almost as effective as the one constructed.

TSS removal efficiency during the study averaged about 50% (see Figure 2). There was a period shortly after the inception of the study when removal efficiencies were quite low (10 to 20%). This was probably due to a sampling artifact whereby attached biological material was dislodged from the headbox at the rear end of the wetland. If these values are discounted, the average TSS removal efficiency was nearly 60%. This is similar to the values derived in the earlier small-scale pilot study.

The influent TSS was highly variable, ranging from about 20 to about 60 mg/L. During part of the year, much of the TSS was planktonic algae from the stabilization basin upstream. About two thirds of the influent suspended solids were volatile (i.e., organic). In the marsh effluent, 80 to 90% of the suspended solids were volatile, suggesting the gravel matrix was more efficient at capturing inert solids.

Influent ammonia concentrations were also highly variable, ranging from less than the detection limit of 0.1 to 6.6 mg/L (see Figure 3). Removal efficiencies were also highly varied. When the influent values were low, some small increase was often noted in transit through the wetland. However, during the summer and fall when plant growth was most apparent, ammonia removal efficiencies were very high. Between July and November, the average removal efficiency for ammonia was 92%.

The only form of phosphorus routinely measured during the study was orthophosphorus. Influent concentrations were low (in the range of less than the detection limit of 0.02 to 0.4 mg/L). The concentrations of orthophosphorus generally rose by a small amount (about 3%) in transit through the wetland. The increase was likely caused by the breakdown of other forms of phosphorus (for example, particulate phosphorus in the form of bacteria or algae). In the author's earlier study, and in most other studies of constructed wetlands, the ability

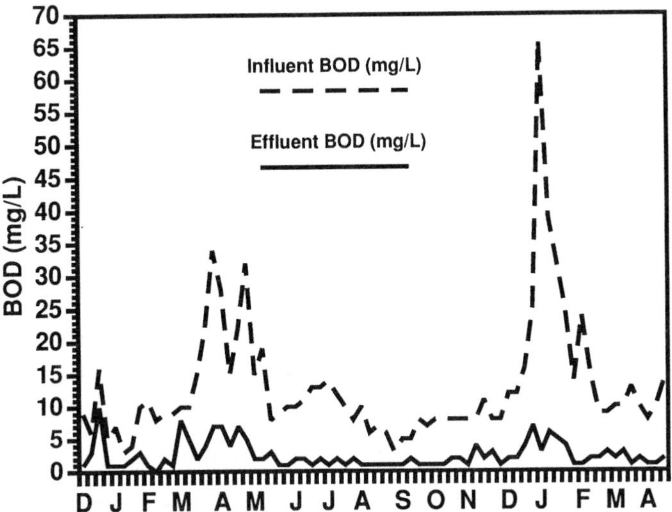

Figure 1. BOD_5 removal by constructed wetland.

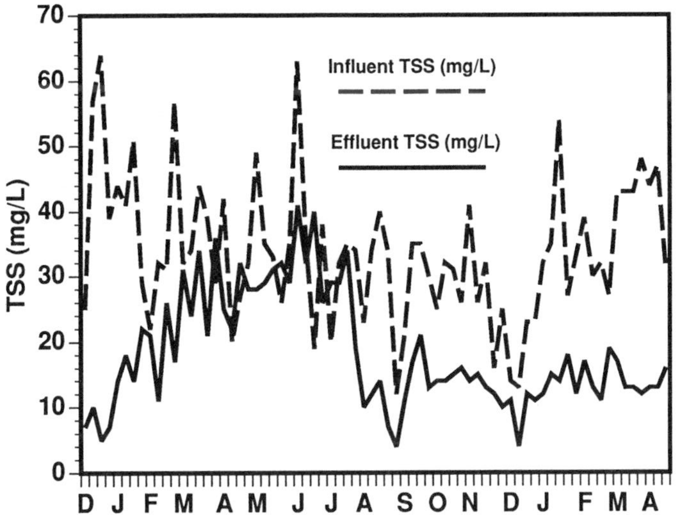

Figure 2. TSS removal by constructed wetland.

of these systems to remove phosphorus is modest (about 20%). The principal mechanism for phosphorus removal is plant uptake, although adsorption to soil or clay particles can also contribute.

DISCUSSION

The overall quality of the wetland effluent was very high and compared favorably to the water quality of the receiving water. Similar results could be expected for a bleached kraft effluent based on the earlier studies. However, a constructed wetland would have no effect on the color, and a small effect on the chlorinated organics found in such effluents.

Although this subsurface flow/gravel matrix system showed considerable promise, there are difficulties which should be recognized. First of all, the capital costs would be substantial. A full-scale system designed to treat 75,000 m^3/day would be 40 ha in size. The cost to construct such a facility would be about \$3 million,

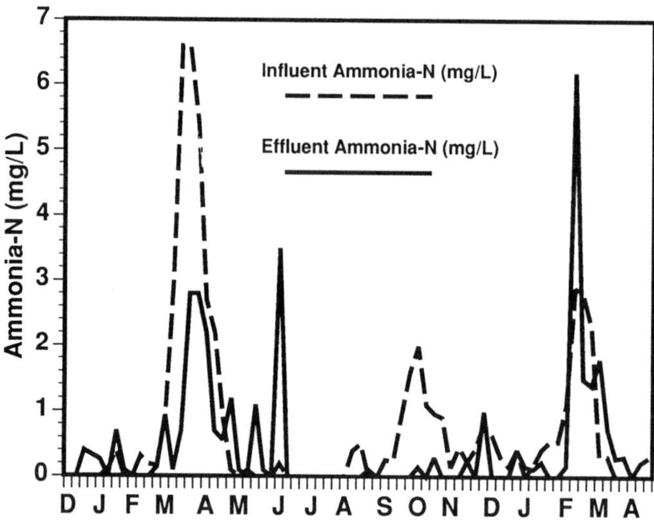

Figure 3. Ammonia-N removal by constructed wetland.

including a cost of $2 million for the purchase of gravel. Maintenance costs are unknown. The primary maintenance task would be to cleanse periodically or replace the gravel, which has a propensity for clogging.

Other systems of this type have had difficulties with clogging. This manifests itself when substantial volumes of effluent flow over the substrate rather than through the substrate. Evidence of this was shown by the stated pilot facility, even in the short period of time it was operative. Before this technology can be recommended for the large volumes of effluent generated by many industrial facilities, the problem of clogging will have to be satisfactorily resolved. It will not be possible to avoid clogging altogether; however, it will have to be managed so that maintenance costs are acceptable.

The flow of water through a gravel matrix is described by Darcy's law. It has several expressions. The most useful for our purposes is:

$$Q = K\frac{dh}{dl}A \tag{47–1}$$

where Q is flow, K is the permeability or hydraulic conductivity, dh/dl is the hydraulic gradient or slope, and A is the cross-sectional area.

Permeability is difficult to measure, particularly for gravels. It can be approximated by the Kozeny-Carmen equation:

$$K = \left(\frac{Pg}{\mu}\right)\left(\frac{n^3}{(1-n)^2}\right)\left(\frac{d_m^2}{180}\right) \tag{47–2}$$

where P is the density of the fluid (in this case, water), μ is the viscosity of the fluid, n is the porosity of the matrix (or void volume), and d_m is some representative grain size.

When gravels clog, the porosity (and therefore the permeability) is reduced and, consequently, less liquid can flow through the gravel matrix. In the stated earlier study, the initial porosity of the gravel was 44%. After 3 years, it had declined to 31%. The volume of flow that could pass through the gravel should have been reduced to less than one fourth (according to the middle term of the Kozeny-Carmen equation). A similar reduction in porosity (from 30 to 17%) in the stated latter pilot facility would have reduced intragravel flow to about one sixth.

Clogging can occur in three ways. The roots of the plants occupy a considerable volume; suspended solids become embedded in the gravel interstices; and the breakdown of BOD_5 generates biological solids (roughly one half unit weight of solids for each unit weight of BOD_5 broken down). Some of the suspended solids and the biological solids will partially break down in the gravel matrix, but there will still be some net accumulation of solids. Based on the earlier studies, root mass was the most important of the clogging mechanisms in the first year. By the third year, the solids caused by filtration of suspended material and breakdown of organic material were of equal importance.

There are several ways to extend the life of a wetland of this design.

1. Use a gravel of relatively large size. If the present study had used a gravel of twice the size of what was purchased, the flow capacity would have been quadrupled (according to the last term of the Kozeny-Carmen equation).
2. Use a well-sorted gravel of high porosity. The gravel used in the author's first set of experiments (44%) had almost five times the flow capacity of the gravel used in the second experiment (31%).
3. Slope the bed of the wetland as much as is practical. In the stated studies, the wetland had zero bed slope, and a hydraulic head of about 15 cm. A 1% slope (or 1.5 m) would have increased the head (and the flow capacity) by a factor of about 10. The amount of bed slope may be restricted to some extent by topography and the requirement that the velocity of water through the matrix not be too rapid.
4. If possible, the wetland facility should be used only when necessary to meet water quality standards in the receiving water. If it were used for only half of each year, its life would be extended by a factor of 2.
5. A settling pond could be situated before the wetland to remove some of the suspended solids. This will reduce the rate of accumulation of solids, thus extending the life of the wetland.
6. The plants could be left out. In essence, the facility would then be a gravel filter. A gravel filter would be almost as effective in removing BOD_5 and TSS, but not as effective in removing nutrients. The absence of plants (and their roots) would diminish the impact of clogging, at least in the first few years. Plants would have a tendency to develop in such a system due to natural seeding. They would have to be periodically removed.

If some combination of these alternatives were implemented, the result would be a constructed wetland which would not flood for several years after construction. At some point in time, the flow capacity would begin to decline leading eventually to flooding. After the flooding begins, the gravel would have to be reworked to dislodge the clogging substances or new gravel could simply be added over that which exists. It is easier to estimate the costs of this latter alternative. Assuming a reduction of about 5% per year in porosity, a corresponding amount of new gravel would have to be added for each year of operation. For a facility capable of treating 75,000 m^3/day, and with a capital cost of \$3 million, this would place annual maintenance costs for this particular item at \$100,000.

REFERENCES

1. Reed, S. C. Constructed wetlands for wastewater treatment, *Biocycle.* 32(1):44, 1991.
2. Gillette, B. Artificial marsh treats industrial wastewater, *Biocycle.* 30(2):48, 1989.
3. Allender, B. M. Water quality improvement of pulp and paper mill effluents by aquatic plants, *Appita.* 37:303, 1984.
4. Thut, R. N. Utilization of artificial marshes for treatment of pulp mill effluent, *TAPPI.* 73(2):93, 1990.
5. Reaves, R. P., F. G. Howell, and R. P. Tettleton. Performance of torpedo grass in a tertiary treatment system for pulp mill waste water, in Proc. 21st Mississippi Water Resources Conference, Jackson, MS, B. J. Daniel, Ed., Water Resources Research Institute, Mississippi State University, Mississippi State, MS, 1991, 105.

CHAPTER 48

Testing Color Removal from Pulp Mill Wastewaters with Constructed Wetlands

D. A. Hammer, B. P. Pullin, D. K. McMurry, and J. W. Lee

INTRODUCTION

Pulp industry effluents may contain small quantities of a variety of substances, but the most significant components by volume are the organic compounds that result in oxygen depletion (BOD) due to microbial decomposition. Conventional biological treatment systems, including aerated lagoons, aggregated precipitations, and activated sludges, provide favorable environmental conditions and adequate retention times mainly for bacterial populations to decompose organic substances in the wastewater. However, these methods provide poor control of water color. Lignins and their derivatives are abundant in pulp mill wastewater and, along with tannins, are principally responsible for the blackish-brown color.

Unlike cellulose and other carbohydrates, lignin is resistant to degradation by most microorganisms,[1] and only certain fungi and a few bacterial groups are capable of attacking this very complex polymer. Conventional wastewater treatment systems (i.e., lagoons dominated by bacteria) are likely to have few fungal species and thus low corresponding biomass. Consequently, potential for color removal is negligible.

Wastewater discharge limits at most pulp and paper mills are based upon a substantial dilution factor since most are located on major rivers. However, drought-induced low river flows have often prevented adequate dilution, and mill operations have been faced with closure and job losses in the absence of legal variances from or waivers for discharge limits.

Constructed wetland systems have been used successfully to treat wastewater from municipal, residential, and agricultural sources. These systems effectively remove BOD, suspended solids, bacteria, and macronutrients. Limited testing of constructed wetlands for treatment of pulp mill effluents has provided mixed results for color removal.[2] In most cases, substrates included have been various-sized rock under anaerobic conditions.[2,3] Obviously, a treatment system for tannins and lignins should be designed to optimize environmental conditions and retention times to enhance fungal decomposition of complex organics, and incorporate similar components for further decomposition by bacterial populations. Since fungal populations require an attachment substrate, a vegetated sand or porous soil substrate is likely to simulate natural soil conditions and provide the aerobic environment and hydraulic conductivity needed to enhance fungal growth.

METHODS

Six experimental marshes were constructed in galvanized steel water tanks measuring 2.4 × 0.9 × 0.6 m, and creating a surface area within each tank of 1.9 m^2. The tanks were lined with a 20-mil PVC fabric to prevent leaching of zinc. Substrates included a 0.3-m deep layer of native clay-loam topsoil obtained locally and topped with a 15-cm layer of decomposed wood mulch acquired from a tree cutting service. Cattail (*Typha latifolia*) taken from the pulp mill's wastewater retention pond margins were transplanted to the tanks on May 31, 1989, at the rate of 20 plants per tank on 0.1-m centers. By the end of the first growing season, new shoots had sprouted from rhizomes, and plants were 2 m tall.

Secondary-treated effluent was pumped from the main discharge channel of the bleached kraft mill to a 1134-l capacity head tank. A mercury float switch mounted in the head tank activated a submersible utility pump in the channel when the tank volume dropped to approximately 400 l. Flow from the head tank's

0-87371-550-0/93/$0.00+$.50

Table 1. Mean Values (mg/L) for True and Apparent Color of Pulp Mill Wastewater Received and Discharged from Wetland Cells from June 1989 to August 1990

True color	N	Mean	SE	Min	Max	GP
Head tank	22	1179.3	52.8	908.8	2048.0	a
Cell 1	21	1007.4	60.2	480.0	1628.0	b
Cell 2	21	998.8	51.9	620.0	1580.0	b
Cell 3	22	948.4	60.1	516.0	1600.0	b
Cell 4	22	1024.6	56.2	592.0	1588.0	b
Cell 5	22	986.0	54.9	520.0	1560.0	b
Cell 6	21	998.1	47.2	644.0	1432.0	b
Apparent color	**N**	**Mean**	**SE**	**Min**	**Max**	**GP**
Head tank	22	1727.3	120.1	1124.0	3104.0	c
Cell 1	21	1417.7	114.4	524.0	2640.0	d
Cell 2	21	1315.1	100.9	784.0	2508.0	d
Cell 3	22	1337.1	97.9	736.0	2392.0	d
Cell 4	22	1408.4	96.5	900.0	2420.0	d
Cell 5	22	1313.3	89.8	564.0	2316.0	d
Cell 6	21	1383.6	87.0	996.0	2580.0	d

Note: Application rates: cells 1.2 = 9.4 cm /day; cells 3,4 = 4.7 cm /day; and cells 5,6 = 3.1cm /day. GP: Groups identified by the same letter designation are not significantly different at $p = 0.05$.

distribution manifold into each replicated marsh was regulated by globe valves set at application rates of 9.4, 4.7, and 3.1 cm /day. Effluent entered the tanks through 8-mm (ID) tubing that dispersed the wastewater onto the mulch substrate. The liquid exited the opposite end of each tank through a port at the soil/mulch interface stratum. Wastewater from each tank flowed to a collection manifold and was ultimately returned to the main discharge channel.

The experiment was conducted through two growing seasons. Influent and effluent samples were collected weekly by Bowaters' environmental laboratory staff and analyzed for true and apparent color, total suspended solids (TSS), total dissolved solids, BOD_5, and pH.

RESULTS

True and apparent color measurements of effluent from all cells were significantly lower than the head tank influent. However, no differences were detected between the three application rates (Table 1). Other parameters (BOD_5, suspended solids, dissolved solids, and pH) did not change significantly in transit.

Removal of true and apparent color was consistent throughout the study period, with color removal efficiencies ranging from 2 to 59% over all sample dates. For all cells, the greatest removal of true color occurred during June and July 1989 (Figure 1). Cells 3 and 4 (application rate at 4.7 cm /day) were most efficient, reducing true color by an average of 36 and 28%, respectively (Table 2). The remaining cells averaged 20 to 25% removal. Removal efficiency slowed in September and October and was poorest during the winter. While true color removal improved into the second growing season, efficiency levels did not recover to the summer of 1989 levels.

In contrast, apparent color removal fluctuated for all cells during the first summer (Figure 2). Cell 3 outperformed the others, averaging 23% removal; while cells 4 and 2 reduced apparent color by 19% (Table 2). Cells 5 and 6 (with the lowest application rate) demonstrated steady improvement in apparent color reduction throughout the study period. All cells performed better during the second summer than during the initial growing season.

Although no significant differences were found between influent and respective cell effluent levels for BOD_5, suspended solids, dissolved solids, and pH, several trends were observed. (1) BOD levels in water discharged from all cells exceeded head tank values for the initial samples collected in June and July. The

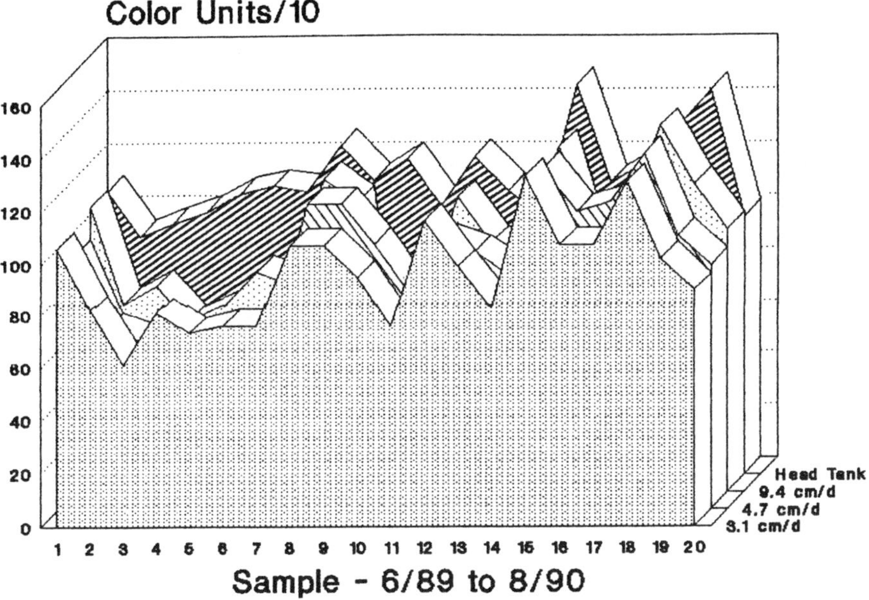

Figure 1. True color in pulp mill wastewater and wetlands cell effluents with three application rates.

Table 2. Average Removal Efficiencies by Wetland Cells for True and Apparent Color and BOD on a Seasonal Basis

	True color					
	AR = 9.4 cm /day		AR = 4.7 cm /day		AR = 3.1 cm /day	
	Cell 1	Cell 2	Cell 3	Cell 4	Cell 5	Cell 6
Jun–Jul, 1989	20	25	36	28	22	24
Sep–Oct, 1989	14	11	17	4	17	8
Jan–Feb, 1990	9	10	7	2	8	7
May–Aug, 1990	16	15	18	17	19	21
	Apparent color					
	Cell 1	Cell 2	Cell 3	Cell 4	Cell 5	Cell 6
Jun–Jul, 1989	13	19	23	19	11	11
Sep–Oct, 1989	26	26	10	14	23	19
Jan–Feb, 1990	20	27	23	14	25	22
May–Aug, 1990	18	26	28	26	35	29
	BOD_5					
	Cell 1	Cell 2	Cell 3	Cell 4	Cell 5	Cell 6
Sep–Oct, 1989	48	47	27	46	28	31
Jan–Feb, 1990	50	58	43	37	45	44
May–Aug, 1990	40	50	50	49	48	38

highest organic mulch substrate added to the native soil in the tanks contributed to the wastewater load. By September, the cattails had become established and the substrate stabilized, resulting in a range of BOD removal efficiencies from 0 to 69%. Average removal efficiency improved for five cells during the winter (Table 2). By the end of the second growing season, at least one cell from each wastewater application rate was reducing the BOD load by half. (2) Total suspended solids and dissolved solids in cell effluents often exceeded head tank levels. Despite settling and stabilization of the mulch substrate, tiny particles continued to flow through the tanks. (3) During the entire study period, influent pH was consistently lowered by 0.1 to 0.6 standard units after passing through the wetland cells.

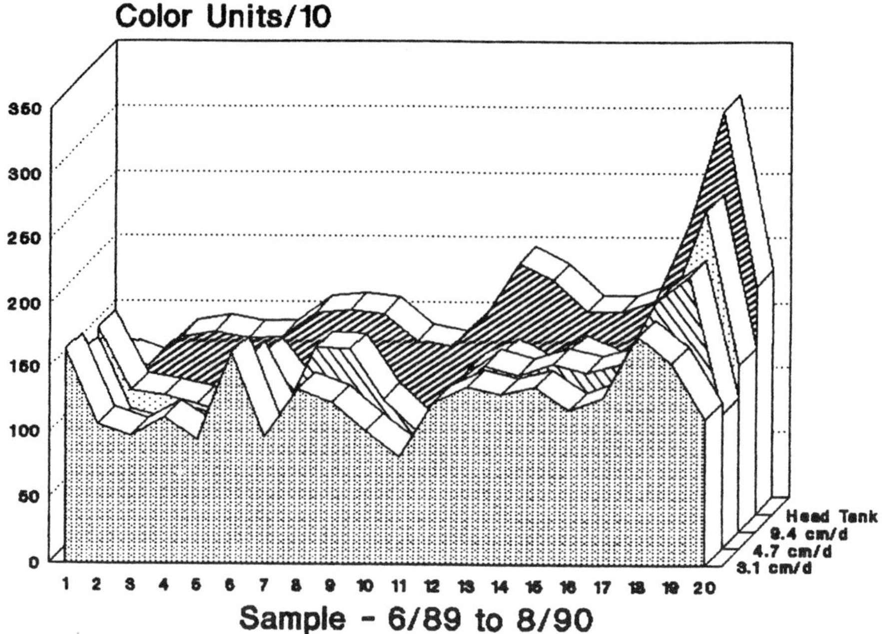

Figure 2. Apparent color in pulp mill wastewater and wetlands cell effluents with three application rates.

DISCUSSION

Initial color removal by all six cells was encouraging despite the concomitant export of BOD. The reduction of true color removal in the fall and winter was not unexpected since biological activity is temperature-dependent. Understandably, the levels of removal improved during the second growing season; however, efficiencies did not recover to previous levels. Other investigators suggest that additional carbon sources other than lignin are required by white-rot fungi to support cell growth and propagation necessary for the complete decolorization of pulp mill wastewater.[4,5] Supplementing treatment media with sucrose, glucose, ethanol, or other easily metabolizable energy sources may enhance decolorization or stabilize removal rates for prolonged periods. Further investigation is warranted.

The oscillating performance in apparent color removal by cells with moderate and high flow rates is difficult to interpret. Fluctuations in influent concentrations of BOD, suspended solids, and dissolved solids likely had an impact on the treatment on this experimental scale.

REFERENCES

1. Chang, H.-M., C.-L. Chen, and T. K. Kirk. The chemistry of lignin degradation by white-rot fungi, in *Lignin Biodegradation: Microbiology, Chemistry, and Potential Applications*. T. K. Kirk, Ed. CRC Press, Boca Raton, FL, 1980, 215–230.
2. Thut, R. N. Utilization of artificial marshes for treatment of pulp mill effluents, in *Constructed Wetlands for Wastewater Treatment: Municipal, Industrial, and Agricultural*. D. A. Hammer, Ed. Lewis Publishers, Chelsea, MI, 1989, 239–244.
3. Wolverton, B. C. Hybrid wastewater treatment system using aerobic microorganisms and reed, *Econ. Bot*. 36:373, 1982.
4. Fukuzumi, T. Microbial decolorization and defoaming of pulping waste liquors, in *Lignin Biodegradation: Microbiology, Chemistry, and Potential Applications*. T. K. Kirk, Ed. CRC Press, Boca Raton, FL, 1980, 161–177.
5. Marton, J., A. M. Stern, and T. Marton. Decolorization of kraft liquor with *Polyporous versicolor*, a white rot fungus, *TAPPI*. 52(10):1975, 1969.

CHAPTER **49**

Interfacing Constructed Wetlands With Traditional Wastewater Biotreatment Systems

J. Boyd, O. McDonald, D. Hatcher, R. J. Portier, and R. B. Conway

INTRODUCTION

Paper mills have traditionally used significant quantities of water for machine operations, resulting in moderate to high BOD (biochemical oxygen demand) discharges difficult to process on a daily basis. This article presents a case study on several innovative approaches for reducing organic loading rates in these discharges by focusing on all streams and sewers contributing to the problem, and developing strategies for mitigating problems at the point source. The effectiveness of treating paper mill discharges has been evaluated in the laboratory and in the field in a technology train consisting of a constructed wetland coupled to a pretreatment biotreatment system, a sequencing batch reactor (SBR), or an immobilized microbe bioreactor (IMBR). This pro-active approach resulted in significant BOD reductions in a facility that has been in operation for over 50 years.

Stone Container Corporation owns and operates a paper mill at Hodge, Louisiana. The mill currently produces approximately 1800 tons per day (TPD) of unbleached kraft paper and semi-chemical corrugating medium; 1400 TPD of paper and 400 TPD of medium. The Hodge mill first began operation in 1928, and over the years has increased production from 125 TPD to the current level. During these same years, water consumption and wastewater volume has increased from approximately 6 million gallons per day (MGPD) to the present 12.5 MGPD. Owing to the mill's location on the very small Dugdemona River, effluent treatment has always been an item of major concern. Consequently, a unique wastewater treatment system has evolved.

Wastewater from the manufacturing operations is currently segregated into two streams: one stream (10 MGPD) discharges to a continuous treatment system, and the other (2.5 MGPD) goes into an oxidation-retention (O-R) pond. The continuous system consists of two clarifiers in series, followed by aerated stabilization basins, color treatment, and continuous discharge to the Dugdemona River. The O-R pond is actually three ponds with a total surface area of approximately 600 acres. Discharges from the O-R ponds are seasonal and dependent on river flow conditions.

In a consent decree that became effective in May 1987, Stone Container was required to investigate cost-effective treatment alternatives to effect a reduction in BOD_5 discharged from the O-R pond. Data presented in this manuscript represent a capsulation of laboratory and field pilot tests conducted to reduce effluent discharge levels. The goal of the study has been, and continues to be, the reduction of BOD_5 such that discharges reflect 95% BOD_5 removal over influent concentration.

Biological treatment for BOD reduction was considered the most cost-effective process. However, which method and how to implement it remains a question. The following conceptual ideas were evaluated to support the biological treatment concept.

CONCEPT #1

BOD_5 to the oxidation/retention (O-R) pond originates from the primary clarifier, pulp mill, semi-chem, recovery, and tall oil/turpentine sewers. The treatment of this wastewater at the source was studied by selecting the treatability of only those sewers which contain elevated BOD_5. With 85% BOD removal of this wastewater stream and 70% BOD and removal of the remaining 15% by the existing O-R aerobic facultative process, a total treatment of 95% was hypothesized as being achievable.

0-87371-550-0/93/$0.00+$.50

CONCEPT #2

The combined BOD_5 source sewers entering the O-R pond can be segregated and isolated for study. The combined BOD_5 sewers average between 1.2 and 1.5 MGPD, with a BOD_5 of between 2500 and 3000 mg/L (ppm). Treatment resulting in a discharge to the O-R pond of BOD in the range of 375 to 450 mg/L can be realistically reduced by volume of other discharges to 188 to 225 mg/L BOD. Facultative treatment in the O-R pond could be expected to reduce this concentration further to between 56 and 68 mg/L at outfall 002.

SOURCE WATER PRETREATMENT

SEQUENCING BATCH REACTORS

Sequencing batch reactors (SBRs) have traditionally been used in domestic wastewater treatment as a single unit, multiple-cycle secondary treatment system to reduce organic loads in small community POTW operations. Based on a draw/fill, biological contact, and clarification/decant sequence, an SPR was considered a pretreatment step to constructed wetland system operations. Laboratory data[1,2,3] suggested that two SBR bioreactors in series or parallel having a capacity of 1747 ft^3 would be needed to achieve a BOD removal rate of 84% (flow rate of 2.5 gal/min [gpm]; hydraulic retention time [HRT] of 43.6 h). Air requirements would be approximately 1774 standard cubic feet per hour (SCFH). Figure 1 presents field pilot data for two source waters. The recovery sewer having a typical BOD feed concentration for most sewers in the mill showed 47 to 72% reductions for a 24-h contact cycle, while the combined plant sewer, having a higher percentage of readily biodegradable organics, gave a BOD reduction efficiency of 64 to 88% for a 24-h contact cycle.

Evaluation of the field pilot SBR data presented several possible scenarios for SBR utilization. Table 1 summarizes these findings. A 12-h cycle for SBR operations was preferable to 24-h due to the doubling of poundage removed on a per gallon per day basis. The more volume one introduces into an SBR, the greater the attractiveness of removal rate kinetics. However, with increased loading, the price that is paid is evidenced in the aeration efficiency, almost 50% with a 2-day cycle. It is also important to note that different streams generated different BOD removal rates. The composite O-R and the bag mill gave optimal results at a comparable 12-h cycle. As was expected from BOD data, the recovery sewer gave the poorest performance, although removal rates are adequate when considered with other management options. The O-R SBR test gave better results than the bad mill sewer, though not statistically significant, due to the sludge age or maturity of the biomass in the SBR when these tests were conducted.

General estimates for constructing 2.0, 1.0, and a 0.5-MGDP SBR systems to handle either bag mill or O-R sewer wastewaters are presented in Table 2. As can be shown, to handle the complete stream capacity from Hodge Mill, an SBR approaching 4 million gallons capacity would be required, employing a hydraulic retention time of 45 h. This assumes a 30% decant of the SBR every 12 to 14 h. BOD removal rates for such a large system would be approximately 80 to 85%. Reductions in reactor capacity, of course, occur when flow rates are reduced as shown.

To summarize pilot test data, an SBR was considered a viable treatment approach for the wastewater entering Pond A. Whether a 0.5-MGPD or higher flow rate is chosen, such a volume reduction in BOD prior to its introduction into Pond A would attain the following: (1) achieve the desired effect of reducing BOD at the river outfall; (2) allow for additional mill capacity in the future and remain within consent decree allowances; and (3) suggest mechanisms for preventing further sludge build-up in the oxidation pond network by reducing TSS (total suspended solids). This would greatly extend the operational life of the oxidation-retention constructed wetlands and pond network.

IMMOBILIZED MICROBE ANAEROBIC REACTOR PILOT

The applicability of using another variation of the microbe/substrate relationship, immobilized microbe column technology, was initially evaluated in Phase B laboratory studies. Tall Oil Sewer and Composite waste streams were evaluated anaerobically using a packed bed dual column laboratory bioreactor pilot. The 10-l reactor contained approximately 4.2 kg diatomaceous earth carrier (R635 Manville Corporation) in which a commercial inoculum currently being used at the Hodge Mill was immobilized using methods outlined previously.[4,5] Flow rates of approximately 4.5 mL/min were maintained through the reactor for all tests. BOD levels were determined for influent and effluent concentration throughout the course of the study. Specific

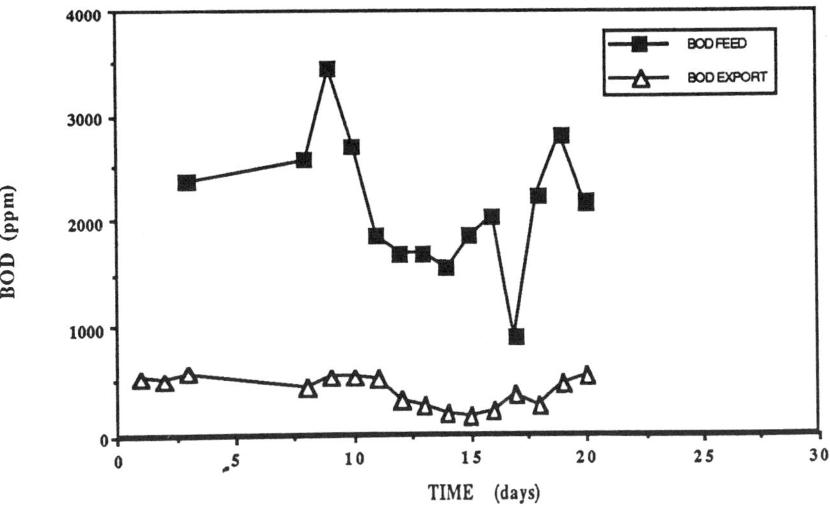

SBR Field Pilot: Recovery Sewer

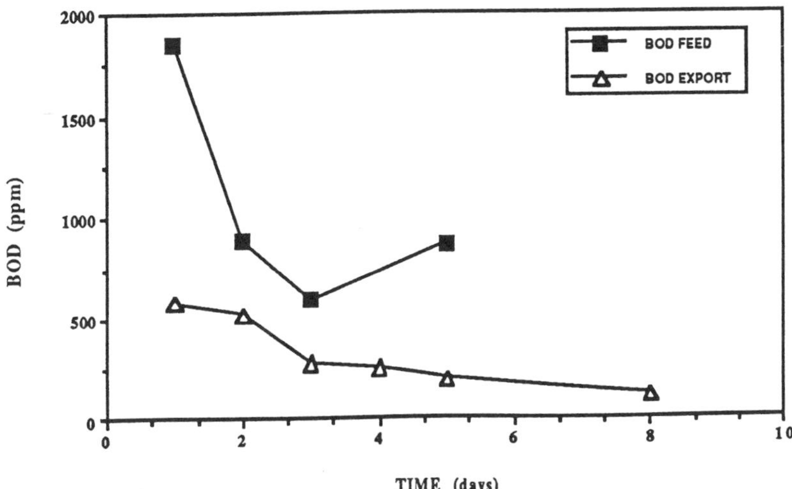

Figure 1. Performance of sequencing batch reactors as a source water pretreatment system.

Table 1. Assessment of Field Pilot SBR Parameters

Source (sewers)	Cycle	Rv[a]	O_2/BOD[b]
Bag/pulp mill	24	3.36	117.6
Bag/pulp mill	12	6.82	53.7
Recovery	24	1.983	49.8
Oxid/retent	24	5.13	75.0
Oxid/retent	12	8.74	42.3

[a] Lbs BOD/gallon/day $\times 10^{-3}$ removal rate.

[b] Efficiency of aeration system as expressed by lbs of O_2 for BOD removal rate.

Table 2. Estimated Capacities for a Full Scale SBR System

Flow[a]	HRT[b]	Reactor volume[c]	Sludge flow rate[d]
2.0	45.8	3.819	160,000
1.0	45.8	1.909	80,000
0.5	45.8	0.950	40,000

[a] Flow as measured in millions gallon per day (MGPD).
[b] Actual hydraulic retention time (HRT).
[c] Millions of gallons.
[d] Gallons per day.

waste streams were introduced on a continuous basis with no attempt made to replace and/or adjust the immobilized microflora and the column reactors themselves. Hydraulic retention time (HRT) through the reactors were approximately 36 h, equivalent to the draw/fill/treat steps for the sequencing batch reactor system.

Figure 2 presents data on the continuous treatment of a composite sewer waste water (drawn from the Bag Mill and the Tall Oil Sewer). As can be seen, variations and improvements in reactor performance and mineralization of BOD source carbon occurred with repeated application. The first batch steps resulted in greater than 91% reduction in BOD after 5 days retention. This was improved to less than 32 h at the conclusion of these tests. Figure 2 also presents similar data for the Tall Oil wastewater. Three batch tests were completed with improvement in HRT to 72 h after the third batch test. When one compares the treatment kinetics of both effluent sources in this anaerobic reactor, one sees improvements in removal rates for the composite stream as compared to the Tall Oil sewer.

CONSTRUCTED WETLANDS PILOT STUDIES

SPRAY HEADER SYSTEM (SHS)

Results generated in laboratory feasibility studies indicated that surfaces played a major role in the efficient reduction of BOD from Hodge Mill effluents. A spray header system (SHS) was installed in the O-R pond network so as to direct a BOD stream over a surface to achieve BOD reduction.[1,2] The surfaces in question here were the reed system as a constructed wetland. Rock reed treatment systems have been under development and operation by the Tennessee Valley Authority (TVA) for several years for municipal wastewater BOD reduction.[2] Reductions in feed BOD of 1000 to 200 ppm have been reported in extensive systems. The SHS approach used at Hodge is a variation on this theme and the overflow wier technology evaluated in earlier studies.

The SHS operated at a flow of 0.5 to 5.0 MGPD, distributing reclaimed water from the far end of the O-R pond. Flow was pumped using a 100-hp pump through a 12-in. PVC line and finally through a head spraying device. Bacteria inoculation tanks (a modified SBR) were located at the reclaim stream head so as to introduce viable biomass directly with the BOD effluent.

Figure 3 presents data for BOD reductions on the spray header/constructed wetland system over a 72-day period. Influent concentrations ranged from 955 to 1620 mg/L. Effluent discharges from the wetland system entering the remaining ponds of the OR system ranged from 374 to 410 m/L. The SHS reduced both BOD levels from 67 to 84% for the time period noted, as well as significant pulse events entering the system from mill operations. A notable increase in BOD was noted for days 14 to 18 with minimal loss in spray header efficiency.

DISCUSSION

Figure 4 presents data on the reductions in BOD concentration at the 002 outfall during 1988 and 1989 year to date. The 1989 data represent BOD values with the Spray Header/Constructed Wetland system in place. As can be seen in this data set, the level of pounds of BOD discharged in 1989 tracked at a rate much lower than the 1988 data set. It is believed that BOD reduction occurred for two primary reasons:

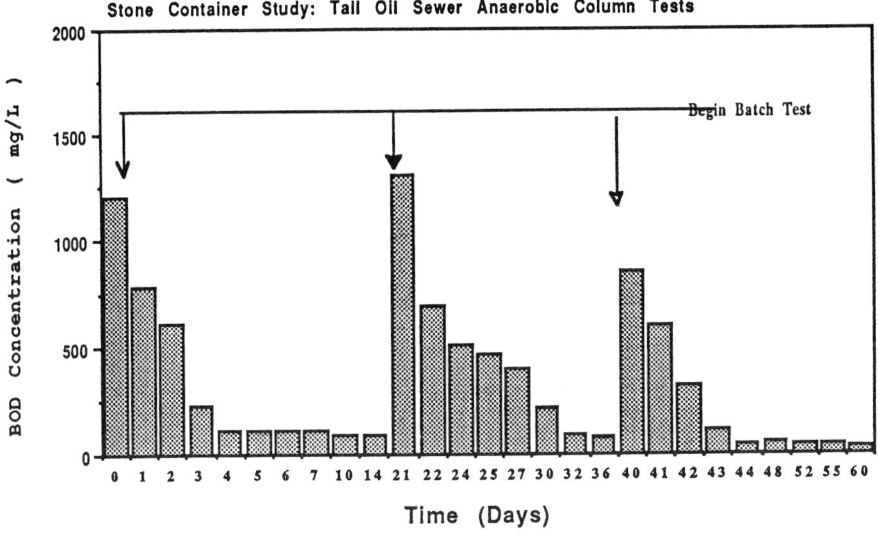

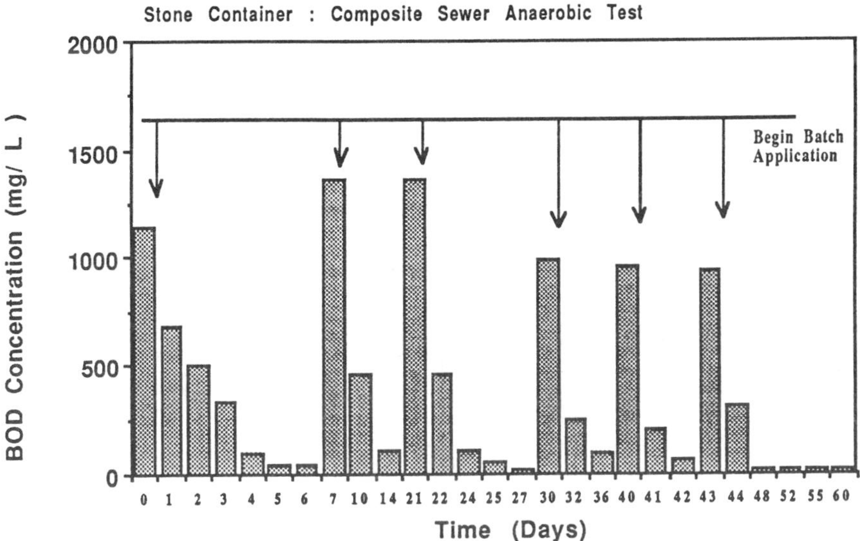

Figure 2. Performance of an anaerobic packed bed immobilized microbe bioreacter as a source water pretreatment system.

1. The cattail rock reed area at the head of the O-R pond is functioning as a surface for biomass colonization. This microbe colonization enhances the potential for BOD reduction in the Pond A network.
2. Excess biomass generated from this area, coupled with a lowering in the level of water in Pond A, has resulted in additional viable surface area for biomass colonization and, hence, treatment potential.

Additional investigations are continuing on the use of extruded surfaces and mats in the field to enhance further this surface/microbe phenomena.

Data shown in the previous section point to an important corollary to treatment technologies proposed for the Hodge Mill; namely, addition of allochthonous matter or addition of commercially available inocula. As mentioned, commercial inocula were added to the spray header so as to enhance colonization and utilization

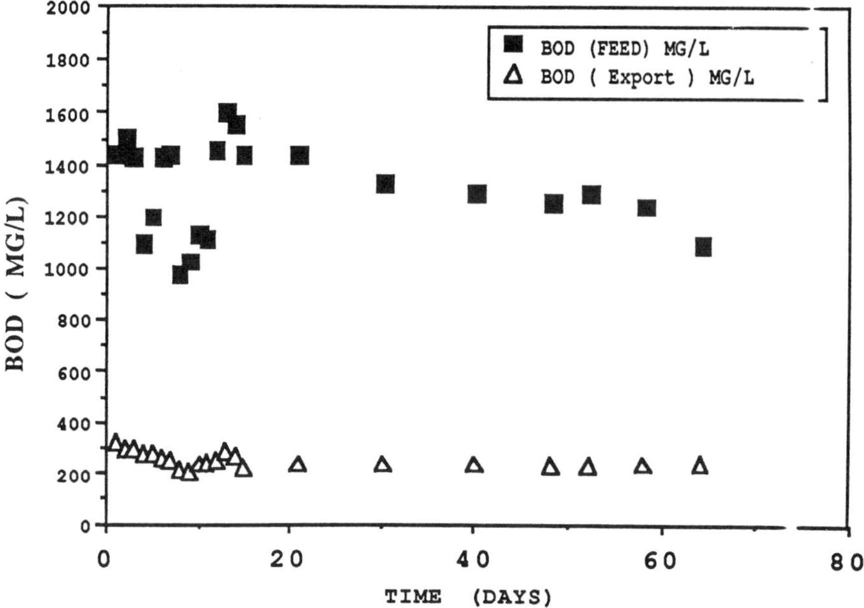

Figure 3. Reductions in BOD by a constructed wetland/spray header treatment approach.

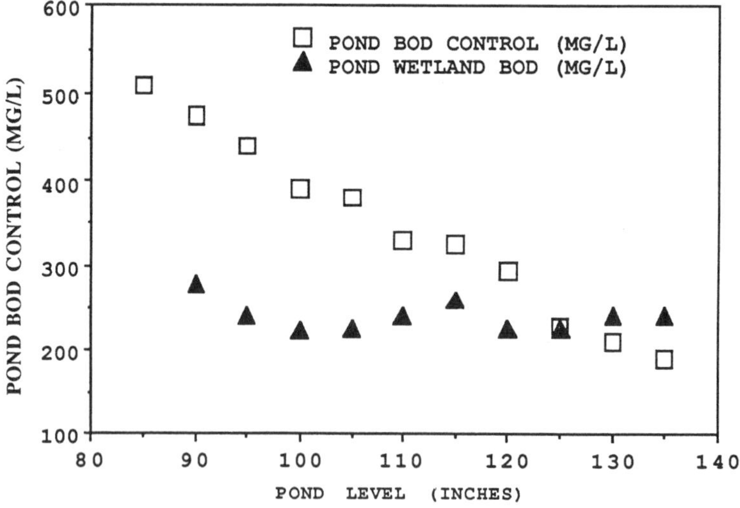

Figure 4. Annual BOD and pond level comparisons for control vs. constructed wetland operations.

of BOD waters at the header system. Additionally, this biomass addition promoted rapid colonization in the reed bed system. Subsequent tests of inoculation of standing sludge in exposed areas of Ponds A/B appeared to reduce overall sludge content in these areas. A reduction in sludge content in a pond results in reversal of the effects of eutrophication in that pond, promotes an increased O^2 presence, and extends the functional lifetime of the pond system.

CONCLUSIONS

As has been noted in previous sections, considerable field pilot demonstration efforts have been undertaken by Stone Container under the consent decree program. Each technology evaluated to date showed considerable promise for application at Hodge Mill. The SBR technology has the volume handling capability coupled with a desirable reduction in BOD. However, retention times are lengthy, necessitating reactor capacities which are quite large. Systems of this scale are, to our knowledge, not in commercial operation. The immobilized column technology shows promise for specific targeted streams at Hodge, but is also quite large when one considers handling of composite wastewaters. Bed volumes would be considerable, thus raising the concern of effective oxygen transfer for BOD reduction. For specific low-volume waste streams, this approach appears to be quite viable. More research data are needed.

The above technologies reflect an interest in treatment prior to the O-R pond network. The consent decree focused primarily on the pond network itself. The spray header wetland system gave the best indications for BOD reduction for the volume of wastewater to be treated. The system functioned as a primary/secondary treatment for significant BOD reductions as well as a biomass generator for seeding the remaining aerated pond network. In all aspects of its operation, the constructed wetland system allowed this mill to achieve the 95% BOD reduction target.

REFERENCES

1. Portier, R. J. and D. D. Friday. Modular bioreactor approaches to hazardous waste remediation, *Proc. Seventh Annu. Conf. Hazardous Materials Management.* Atlantic City, NJ, June 13–15, 1989, 322–327.
2. Portier, R. J. and S. Palmer. Wetlands microbiology: form, function processes, in *Constructed Wetlands for Waste Water Treatment.* D. A. Hammer, Ed. Lewis Publishing, Ann Arbor, MI, 1988, 89–105.
3. R. J. Portier. Evaluation of specific microbiological assays for constructed wetlands waste water treatment management, in *Constructed Wetlands for Waste Water Treatment.* (D.A. Hammer, ed), Lewis Publishing, Ann Arbor, MI, 1988, 515-523.
4. Miller, G. P., R. J. Portier, D. G. Hoover, and D. D. Friday. Biodegradation of chlorinated hydrocarbons in an immobilized bed reactor, *Environ. Prog.* 9(3):161, 1990.
5. Friday, D. D. and R. J. Portier. Development of an immobilized microbe bioreactor for VOC applications, *Environ. Prog.* 10(1)30, 1991.

CHAPTER 50

Effect of Season, Substrate Composition, and Plant Growth on Landfill Leachate Treatment in a Constructed Wetland

J. M. Surface, J. H. Peverly, T. S. Steenhuis, and W. E. Sanford

INTRODUCTION

The need for efficient collection and treatment of leachate from solid waste landfills to avoid surface-water and groundwater contamination is widely recognized. Traditional approaches (on-site treatment or transport to off-site facilities, usually municipal sewage treatment plants) are undesirable because (1) on-site plant operation and maintenance is costly and entails labor and services after landfill closing, and (2) transport of concentrated waste on public roads is costly and dangerous, and the addition of leachate to municipal sewage disrupts the normal biological processes upon which treatment depends.

A method of on-site treatment is needed that is inexpensive and requires little maintenance or power after landfill closure. Constructed wetlands have shown promise for treatment of sewage and municipal waste[1] and have several characteristics that are beneficial for leachate treatment, including high vegetation biomass, large adsorptive surfaces on sediments and plants, aerobic-anaerobic interfaces, and diverse, active microbial populations. The design and operating limits for treatment of landfill leachate in the humid temperate zone, however, are not well documented.

The treatment of landfill leachate by constructed wetlands is problematic because the high concentrations of biochemical oxygen demand (BOD; 42 to 10,000 mg/L), ammonium (NH_4^+; 0.01 to 1000 mg/L), iron (Fe; 0.09 to 678 mg/L), and manganese (Mn; 0.01 to 550 mg/L),[2] combined with the highly reduced state of the leachate, normally make the solution too toxic for most plants. A wetland system generally is most successful for wastewater treatment when only subsurface flow is allowed because this maximizes the contact between the wastewater and the substrate and rhizome or root surfaces. Wetland plants, such as the common reed (*Phragmites australis*) and cattail (*Typha latifolia*), have shown to increase aeration of the substrate by diffusing air from their above-ground shoots to the rhizomes and roots[3] and might also sustain the hydraulic conductivity of the substrate through eventual decay of rhizomes.[4] These plant functions would seem beneficial for the efficient treatment of landfill leachate because they provide the oxygen for breakdown of organic compounds and NH_4^+ and might offset plugging by large metal loads in leachate. Plants also might provide a more rapid decrease in leachate volume (through transpiration) than is provided by storage lagoons or ponds without plants.

In 1989, the U.S. Geological Survey (USGS), in cooperation with Tompkins County, New York Departments of Planning and Solid Waste, began a 3-year study at a municipal solid-waste landfill near Ithaca, NY, to test the efficiency of leachate treatment by constructed wetlands and to examine the associated treatment processes. Specific objectives of the study were to examine:

1. Treatment efficiency as function of substrate composition and grain size, degree of plant growth, and seasonal changes in evapotranspiration rates and microbial activity
2. Effects of leachate and plant growth on the hydraulic characteristics of the substrate
3. Chemical, biological, and physical processes by which nutrients, metals, and organic compounds are removed from leachate flowing through the substrate

0-87371-550-0/93/$0.00+$.50

This article addresses the first two of these objectives and briefly discusses the third. It (1) describes the substrate plots, leachate-distribution system, sampling methods, and analytical procedures; and (2) presents results in terms of (a) percent removal rates of selected constituents between inflow- and outflow-sampling points, (b) seasonal effects on constituent-removal rates, and (c) effect of substrate composition on constituent removal and plant growth.

MATERIALS AND METHODS

Four parallel constructed wetland plots were installed at the Tompkins County municipal solid waste landfill, 9 km south of Ithaca in western New York (Figure 1). Each plot is 3 m wide, 30 m long, 0.6 m deep, lined with 1.5-mm, high-density polyethylene, and is underlain by a leak detection system. The bottom slopes are 0.5% and the upper surfaces are horizontal. Bed I is filled with coarse gravel (2- to 4-cm average diameter), Bed IV is filled with pea gravel (0.5-cm average diameter), and Beds II and III are filled with a sand and gravel mixture (0.5- to 1-cm average gravel diameter, 0.05-cm average sand diameter). During the spring of 1989, *Phragmites australis* was planted in Beds I, III, IV; Bed II was left unplanted as a control.

A collection system beneath the perimeter of the landfill was installed to collect leachate, which was pumped to a 7.6-m^3 feed tank. A 5-cm polyvinylchloride (PVC) pipe runs from the feed tank to a distribution box (Figure 2), where it splits into four separate lines, one for each plot. Each line contains a valve to regulate flow, an inline water meter to record the quantity of leachate flowing, and a solenoid valve for emergency shutdown.

Leachate from the feed tank flows into a perforated pipe at the bottom of each plot. The treated leachate from each bed flows through an inline flowmeter to a 2.8 m^3 collection tank from which it is pumped back to the landfill 50 m away. The water level in each bed is regulated by the height of a flexible hose at the outflow. Five slotted Teflon* wells (slotted the full 0.6 m gravel depth) were installed at equal horizontal spacing in each bed for collection of water samples from the saturated zone.

Application of leachate at each plot began in July 1990 at a rate of 2000 L/day. Hydraulic controls at the head and foot of each plot maintain continuous sub-surface flow. A tipping bucket rain gage at the research plots was used for precipitation measurement. Evapotranspiration (ET) was calculated from Equation 1. Change in storage was assumed to be equal to zero (Table 1).

$$\text{ET} = \text{Inflow} + \text{precipitation} - \text{outflow} \tag{51–1}$$

Dissolved oxygen concentration, pH, temperature, and specific conductance of raw leachate and outflow from the four beds were measured bi-weekly in the field by standard methods,[5] and water samples from the five wells in each bed were analyzed monthly for the same constituents. Platinum redox electrodes[6] were placed at depths of 15 and 45 cm at two locations in each bed for biweekly measurement of pore-water redox potentials; pH was also measured bi-weekly.

All bi-weekly leachate samples were analyzed for nutrients, metals, and organic carbon by Cornell University's Soil, Crops, and Atmospheric Sciences Nutrients Analysis Laboratory. NO_3^-, NH_4^+, and total soluble phosphate were analyzed colorimetrically according to USEPA guidelines.[7] The Fe(III)/Fe(II) ratio was measured through a modification of Stucki's technique,[8] which allows determination of total Fe and Fe(II) in solution as an Fe-*ortho*-phenanthroline complex with photoreduction of Fe(III) to Fe(II). Organic carbon was measured with a carbon analyzer, where carbon compounds in the water are oxidized to CO_2 and then reduced to methane for detection.[5] BOD was determined by standard methods.[5] All other elements in water were determined by inductively coupled plasma (ICP) spectrophotometry.[5]

The standing crop of the rooted, submerged plants was sampled from each of five randomly spaced square study plots, 25 cm on each side, from each of the planted beds. A native stand of *Phragmites australis*, growing downslope from the landfill and receiving leachate from a seep, was also sampled as a control. All living plant material was removed from the plot, carefully washed, sorted by roots, shoots, and rhizomes, dried at 75°C for 48 h, and dry weights recorded. All samples were subsequently ground to a fine powder in a Wiley mill for chemical analysis.

Nitrogen in plant tissue was determined by the Kjeldahl method.[5] The other elements were determined as follows. Subsamples were ashed overnight in quartz crucibles at 500°C, treated with nitric acid, dried, heated

* Use of trade, product, or firm names is for identification purposes only and does not constitute endorsement by the U.S. Geological Survey.

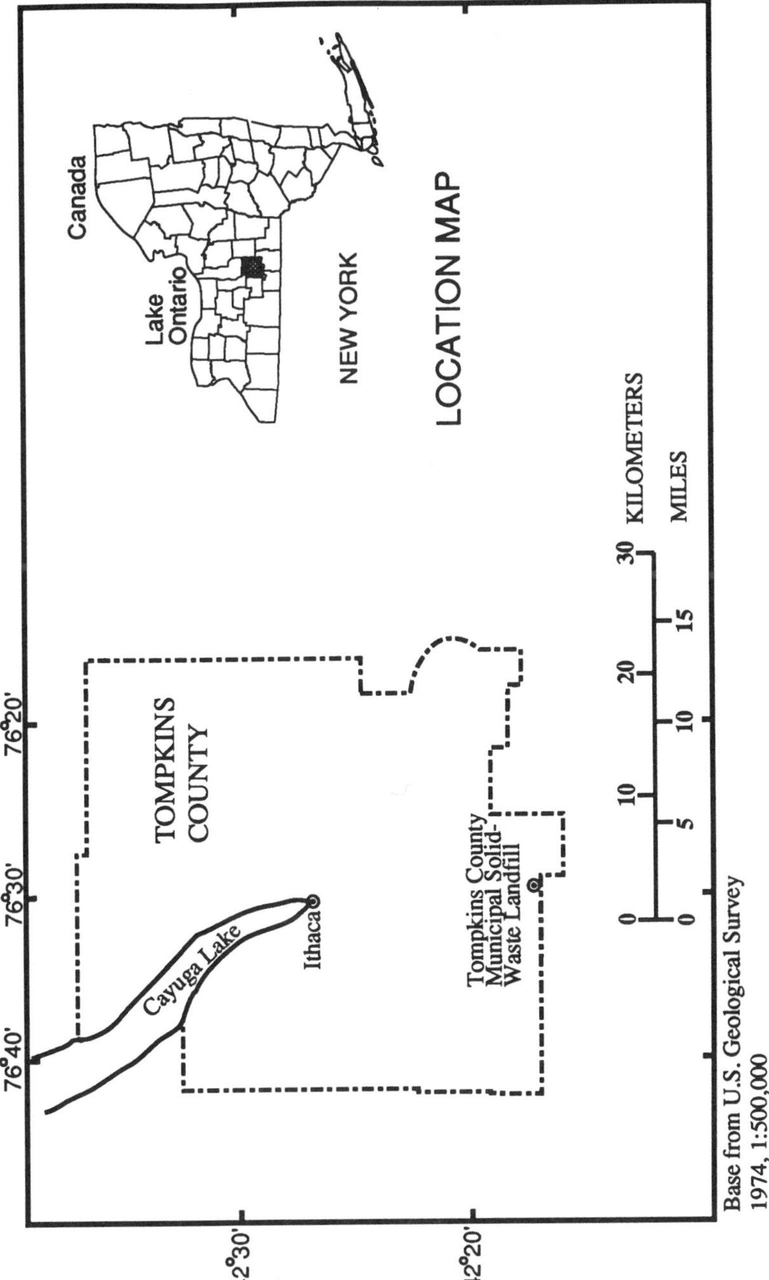

Figure 1. Location of Tompkins County municipal solid waste landfill.

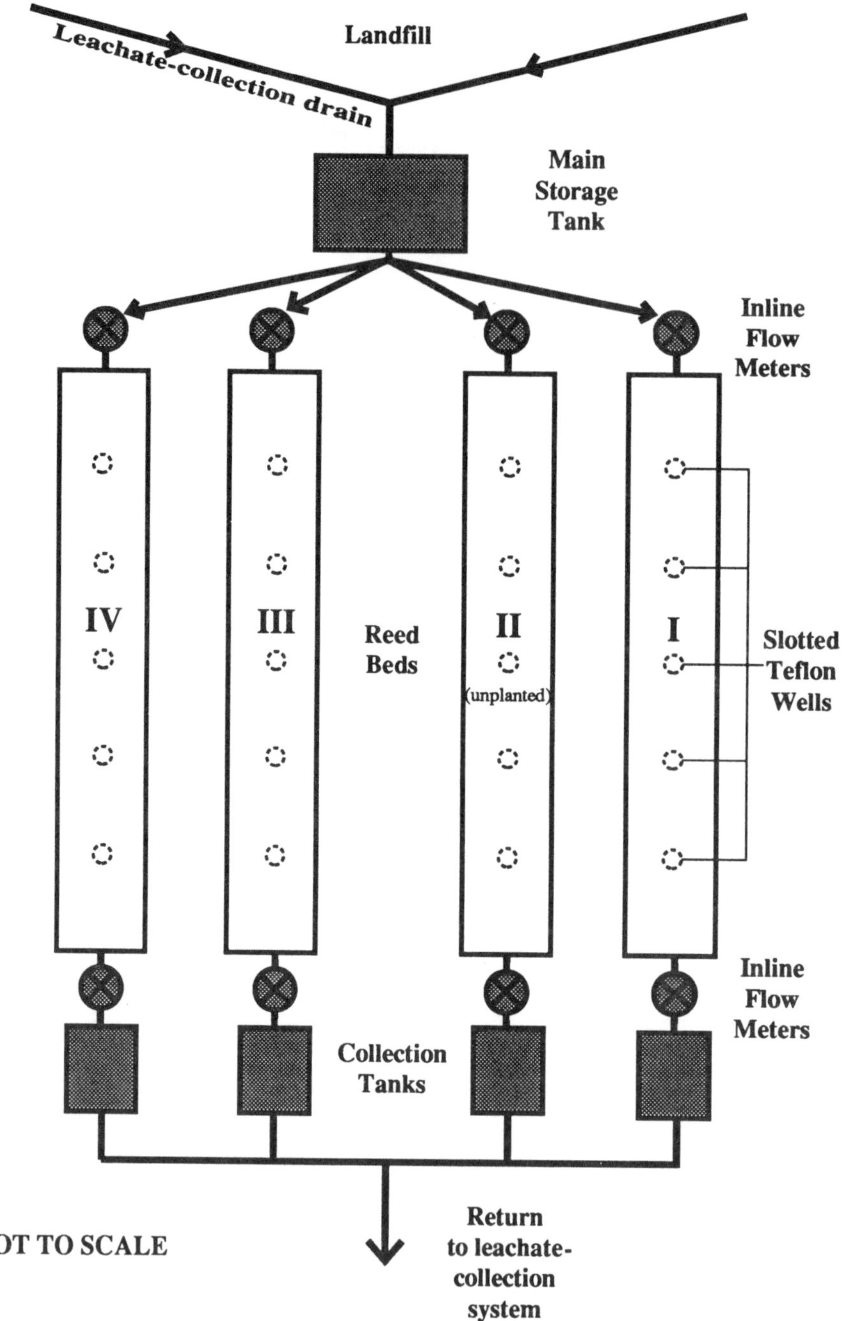

Figure 2. Plan view of constructed wetland research plots at Tomkins County municipal solid waste landfill.

to 400°C, and cooled. The residue was dissolved in 1 *M* HCl and diluted; tissue elements were analyzed by ICP.

The hydraulic conductivity of the substrate in each plot (Table 1) was calculated from Darcy's equation:

$$K_s = \frac{Q_i L}{A(H_i - H_o)} \tag{51–2}$$

Table 1. Air Temperature and Precipitation at Landfill, and Pore-Water Mean Temperature, Evapotranspiration, and Hydraulic Conductivity of the Four Constructed Wetland Research Plots, 1989–1991

Temperature, Precipitation, and Evapotranspiration

Characteristic	Season: Growing season (July 27– Oct. 9, 1990)	Non-growing season (Oct. 10, 1990– April 30, 1991)	Growing season (May 1– Aug. 1, 1991)
Landfill Vicinity			
Air temperature, mean high, °C	22.7	7.5	25.6
Precipitation, cm	23.8	64.3	26.9
Bed I (coarse gravel, planted)			
Pore-water mean temperature, °C	19.9	7.2	22.1
Evapotranspiration, cm	29.8	20.9	48.5
Bed II (sand and gravel, unplanted)			
Pore-water mean temperature, °C	19.5	6.4	22.0
Evapotranspiration, cm	28.7	4.2	29.2
Bed III (sand and gravel, planted)			
Pore-water mean temperature, °C	20.4	6.6	21.8
Evapotranspiration, cm	24.7	3.5	22.1
Bed IV (pea gravel, planted)			
Pore-water mean temperature, °C	20.2	7.3	22.8
Evapotranspiration, cm	11.1	20.9	28.1

Hydraulic Conductivity, cm/s

Location	Before leachate application (July 1989)	After experiment (September 1991)
Bed I	>16.7[a]	>16.7
Bed II	0.217	0.050
Bed III	0.167	0.033
Bed IV	6.33	4.00

[a] Hydraulic conductivity for bed I exceeds capacity of the outflow pipe.

where K_s = saturated hydraulic conductivity (cm per second),
Q_i = inflow rate (cubic cm per second),
L = length of the bed (cm),
A = cross-sectional area (square cm), and
$[H_i{-}H_o]$ = difference in water level between the inflow and outflow (centimeters).[9]

RESULTS

Concentrations of selected constituents in untreated leachate, and their total load and percent decrease in load after leachate moved through the four beds over a 15-day residence time, are given in Table 2. Comparison among the columns indicates the effects of the substrate type, the presence or absence of plants, and the season, on load decrease.

Effect of Precipitation

Concentrations of BOD (Figure 3) and Fe and Mn (Table 2) increased with recharge from precipitation, especially after periods of low percolation and slow leachate collection. The correlation between precipitation and BOD, K^+, NH_4^+, and organic C concentrations is illustrated in Figure 3. Relatively soluble constituents such as NH_4^+ and K^+ were diluted by increased flow (Table 2 and Figure 3).

Seasonal Removal of Biochemical Oxygen Demand, Ammonium, and Potassium

Maximum BOD removal from each bed ranged from 41.5% in bed I (coarse gravel) to 65.7% in bed II (sand and gravel, unplanted) (both in the 1990 growing season), and maximum NH_4^+ removal ranged from 32.6% in the 1991 growing season in bed I (coarse gravel) to 70.8% in the 1990 growing season in bed III (sand and gravel, planted) (Table 2 and Figures 4A and 4B). Beds II and III (mixed sand and gravel) produced the highest percent removal of these two constituents, and maximum removal was during the two warm periods (Table 2 and Figures 4A and 4B). In the second (1991) growing season, bed III (sand and gravel, planted) showed a higher percent BOD removal (44.2%) than bed II (sand and gravel, unplanted) (25.7%), although overall removal by both plots was smaller than in the 1990 growing season (60.3 and 65.7%) (Table 2 and Figure 4). The coarse gravel (bed I) showed the least BOD removal in each season, even though it supported the greatest *Phragmites* production (Table 3 and Figure 4A). The sharp decrease in BOD removal in all beds during the 1990–1991 winter (Table 2) is attributed to the decrease in microbial activity as bed temperatures dropped. Rates of BOD removal remained between 10 and 30% throughout the winter.

Seasonal changes strongly affected removal of NH_4^+ and K^+ in the two sand and gravel beds (II and III) (Figures 4B and 4C). Rates of NH_4^+ and K^+ removal in these beds were greatest during the two warm seasons, and both constituents were released during periods of low temperature (Figures 4B and 4C). Cation exchange could be a strong control on NH_4^+ and K^+ concentrations; if so, removal depends on the type of substrate used (Figures 4B and 4C).

Seasonal Removal of Iron, Manganese, and Phosphorus

Maximum removal of total Fe, Fe(II), Mn, and P from leachate during each season ranged from 76.3% (Mn) to 91.8% (P) (Table 2). The initial removal rate of total Fe in the coarse gravel (bed I) was low (44.0%) but increased in the second (1991) growing season (Figure 4D), perhaps in association with increasing plant biomass in that bed. Removal rates of total Fe and P fluctuated seasonally, even though the surface area of the substrate appears to have been the major control. Fe(II) and Mn removal does not appear to be seasonally affected, but Mn removal was increased by the greater adsorptive surface area in the fine-textured substrates (II and III) than in the coarse substrates (Table 2).

Biomass, Substrate Effect, and Hydraulic Conductivity

Plant biomass in the coarse gravel (bed I) increased during the study; the greatest increases were in bed II — 1.52 kg/m^2 for stems, 0.11 kg/m^2 for roots, and 1.15 kg/m^2 for rhizomes (Figure 4 and Table 3). Numbers of stems at the end of the study (October 1991) were greater in beds III and IV (156 and 125 per square meter, respectively) than in bed I (116 per square meter), but the small number in bed I was offset by the greater total biomass (Figure 5). The wide range in stem and rhizome mass contrasts with the relatively narrow range in root mass.

Substrate type seemed to have a distinct effect on growth form, but the reasons are unknown. *Phragmites* in bed I were taller than those in beds III and IV, had a greater total mass, and developed seed heads during the first year's growth. In contrast, leachate composition appeared to have no effect on tissue composition (Table 3) because it differed little among the beds or between the bed plants and the nearby control samples. In general, the plants were nutritionally normal and seemed undamaged by metal accumulation in the substrate or the anaerobic root zone. The accumulation of metals in or on the root tissue, but not in rhizomes or stems, suggests greater biological activity at root surfaces than elsewhere.

Plant biomass correlated positively with evapotranspiration in both growing seasons, and the coarse gravel (bed I) consistently showed the greatest decrease in leachate volume by this mechanism (Table 2 and Figure 6). Evapotranspiration decreased leachate volume by as much as 40%; even in winter, it decreased the volume in beds I and IV by 20%. Leachate evapotranspiration was directly related to stem biomass and underwent a maximum decrease of about 45% in the coarse gravel (bed I) in 1991.

Hydraulic conductivity of substrate materials was measured in July 1989 (before leachate was applied) and again at the end of the study (Table 1). The initial values at the start of leachate application (July 1990) were assumed to be equal to the pretest values measured in July 1989 (Table 1). Hydraulic conductivity of all

Table 2. Total Applications and Mean Percent Decreases in (A) Chemical Loads and in (B) Volume of Leachate Flowing Through Constructed Wetland Research Plots, 1990–1991

Constituents	Concentration in untreated leachate			Bed I Coarse gravel, planted		Bed II Sand and gravel, unplanted		Bed III Sand and gravel, planted		Bed IV Pea gravel planted	
	Max	Min	Mean	Total load[a] applied	Mean percent decrease	Total load[a] applied	Mean percent decrease	Total load[a] applied	Mean percent decrease	Total load[a] applied	Mean average decrease
NH_4^+											
Jul 27–Oct 9, 1990	386	292	313 (36)	34.1 (1.60)	6.16	41.6 (1.95)	65.1	38.3 (1.80)	70.8	17.3 (0.81)	10.4
Oct 10, 1990–Apr 30,1991[b]	228	143	185 (32)	34.4 (1.98)	22.4	35.4 (2.16)	–2.82	39.8 (2.29)	–5.02	40.4 (2.32)	31.2
May 1–Aug 1, 1991	376	0.7	301 (54)	38.7 (2.82)	32.6	37.7 (2.75)	51.7	31.1 (2.27)	54.3	38.3 (2.79)	33.2
K											
Jul 27–Oct 9, 1990	401	281	322 (42)	35.2 (1.86)	9.40	42.9 (2.27)	55.9	39.4 (2.08)	61.2	17.8 (0.94)	14.6
Oct 10, 1990–Apr 30, 1991[b]	214	125	180 (9.6)	33.4 (0.60)	17.7	34.4 (0.65)	–26.2	38.7 (0.69)	–19.1	39.3 (0.70)	34.9
May 1–Aug 1, 1991	—	—	—	—	—	—	—	—	—	—	—
P											
Jul 27–Oct 9, 1990	1.25	0.518	0.77 (0.27)	0.836 (0.01)	68.8	0.102 (0.02)	85.1	0.094 (0.01)	85.5	0.042 (0.01)	78.8
Oct 10, 1990–Apr 30, 1991[b]	0.462	0.135	0.28 (0.12)	0.052 (0.01)	32.2	0.053 (0.01)	56.7	0.060 (0.01)	18.6	0.061 (0.09)	24.1
May 1–Aug 1, 1991	1.84	0.214	0.70 (0.67)	0.090 (0.04)	71.7	0.087 (0.03)	78.6	0.072 (0.03)	91.8	0.089 (0.04)	75.4
Total Fe											
Jul 27–Oct 9, 1990	11.4	5.72	7.9 (2.5)	0.864 (0.11)	44.0	1.05 (0.14)	80.2	0.968 (0.14)	81.9	0.439 (0.06)	76.1
Oct 10, 1990–Apr 30, 1991[b]	40.2	10.6	26.4 (8.7)	4.92 (0.54)	43.6	5.06 (0.59)	71.1	5.09 (0.62)	69.2	5.77 (0.63)	59.3
May 1–Aug 1, 1991	64.1	6.60	21.7 (22.5)	2.78 (1.18)	78.6	2.72 (1.15)	90.8	2.24 (0.95)	80.4	2.76 (1.17)	81.3
Fe (II)											
Jul 27–Oct 9, 1990	—	—	—	—	—	—	—	—	—	—	—
Oct 10, 1990–Apr 30, 1991[b]	20.3	5.21	9.9 (5.3)	1.94 (0.33)	72.8	1.90 (0.34)	77.8	2.14 (0.38)	75.2	2.49 (0.39)	74.1
May 1–Aug 1, 1991	11.0	1.56	4.8 (3.6)	0.620 (—)	69.8	0.604 (—)	88.6	0.498 (—)	74.7	0.615 (—)	77.4

Table 2. Total Applications and Mean Percent Decreases in (A) Chemical Loads and in (B) Volume of Leachate Flowing Through Constructed Wetland Research Plots, 1990–1991 (continued)

	Concentration in untreated leachate			Bed I Coarse gravel, planted		Bed II Sand and gravel, unplanted		Bed III Sand and gravel, planted		Bed IV Pea gravel planted	
Constituents	Max	Min	Mean	Total load[a] applied	Mean percent decrease	Total load[a] applied	Mean percent decrease	Total load[a] applied	Mean percent decrease	Total load[a] applied	Mean average decrease
Mn											
Jul 27–Oct 9, 1990	1.87	1.00	1.56 (0.13)	0.170 (0.01)	20.6	0.207 (0.02)	76.3	0.191 (0.02)	70.2	0.086 (0.01)	5.81
Oct 10, 1990–Apr 30, 1991[b]	9.47	2.66	5.25 (0.82)	0.977 (0.05)	39.1	1.005 (0.06)	71.7	1.13 (0.06)	65.8	1.15 (0.06)	47.8
May 1–Aug 1, 1991	—	—	—	—	—	—	—	—	—	—	—
Biochemical Oxygen Demand											
Jul 27–Oct 9, 1990	143	89	118 (32)	12.8 (1.42)	41.5	15.7 (1.74)	65.7	14.4 (1.60)	60.3	6.52 (0.72)	48.8
Oct 10, 1990–Apr 30, 1991[b]	316	136	277 (55)	51.6 (3.44)	18.2	53.1 (3.75)	26.7	59.8 (3.98)	10.7	60.6 (4.04)	32.5
May 1–Aug 1, 1991	299	45.5	118 (94)	15.2 (4.93)	22.4	14.8 (4.80)	25.7	12.2 (3.96)	44.2	15.0 (4.88)	26.0
Organic Carbon											
Jul 27–Oct 9, 1990	—	—	—	—	—	—	—	—	—	—	—
Oct 10, 1990–Apr 30, 1991[b]	566	178	388 (40)	72.3 (2.48)	20.7	74.4 (2.71)	8.60	83.7 (2.87)	9.56	84.8 (2.91)	40.1
May 1–Aug 1, 1991	676	340	478 (112)	61.3 (5.89)	28.4	59.8 (5.74)	43.0	49.3 (4.73)	36.1	60.8 (5.83)	19.1
Leachate volume											
Jul 27–Oct 9, 1990					31		28		27		23
Oct 10, 1990–Apr 30, 1991[b]					21.1		11.2		1.9		20.4
May 1–Aug 1, 1991					42.7		26.2		23.0		24.9

[a] Total load is load (kg) of constituent entering the bed from July 27 through October 9, 1990; October 10, 1990 through April 30, 1991; or May 1 through August 1, 1991.
[b] No flow from January 10 through March 16, 1991

Note: Numbers in parentheses are standard error; standard error = standard deviation/$\sqrt{n}$. Concentrations are given in mg/L.

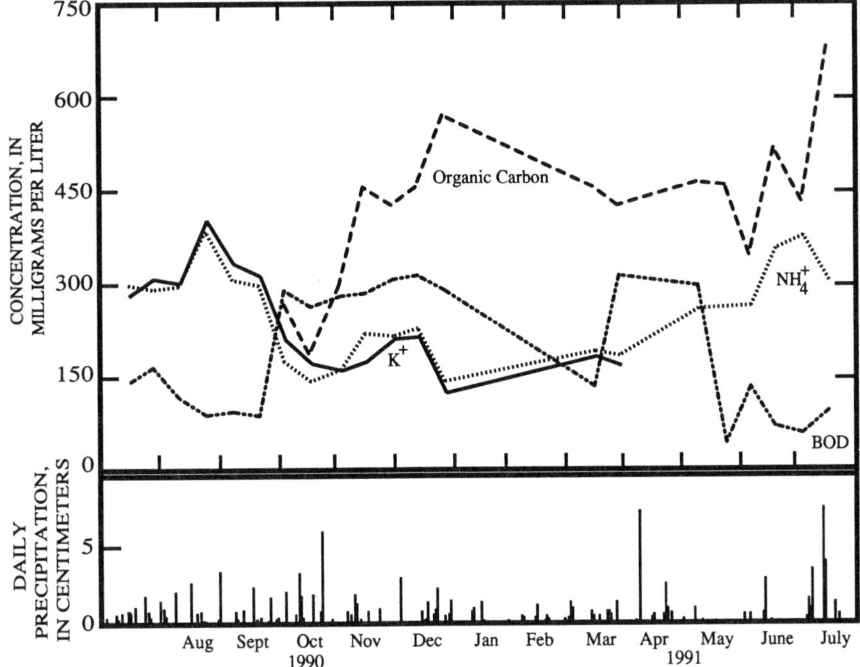

Figure 3. Precipitation amount and concentrations of biochemical oxygen demand, organic carbon, ammonium, and potassium in untreated landfill leachate at Tompkins County municipal solid waste landfill, July 1990 through July 1991.

substrates except the coarse gravel (bed I, in which it was unmeasurable) decreased throughout the study: by 77% in bed II, 80% in bed III, and 37% in bed IV. The 2-year-old plants had no mitigating effect on the decrease in hydraulic conductivity.

SUMMARY

Constructed wetlands may provide a treatment system for municipal solid waste landfill leachate that (1) decreases contaminant concentrations and total leachate volume and (2) requires little maintenance, management, or energy. Four constructed wetland research plots (3 × 30 × 0.6 m) were tested with various combinations of vegetation (common reed) and sand and gravel substrates during the summers and winter of 1990–1991 for removal of major leachate components (BOD, NH_4^+, P, K^+, Ca^{2+}, and selected metals). Hydraulic controls were installed to apply a continuous subsurface flow of leachate to each bed at a rate of 2000 L/day. Leachate residence time in each bed was about 15 days. Reeds grew vigorously and showed no ill effects in terms of tissue composition, production, rhizome extension, or evapotranspiration rate. Root tissue from all beds contained elevated metal concentrations, but stems and rhizomes did not.

Concentrations of BOD, Fe, and Mn in the untreated leachate increased as precipitation increased (October 1990), whereas K^+ and NH_4^+ underwent dilution. The two beds with a mixture of sand and gravel (one planted, one unplanted) were most effective in BOD and NH_4^+ removal and produced maximum removal rates of 65 and 70%, respectively. The planted sand and gravel bed showed no greater removal of BOD and NH_4^+ than the unplanted control bed. Removal of metals such as total Fe and Fe(II) in the two sand and gravel beds decreased by 71 to 91% during all season, and Mn in these beds decreased by 66 to 76%. Loads of BOD, NH_4^+, P, and K^+ decreased in winter, after bed temperatures dropped, although winter BOD removal remained between 10 and 30% in all beds.

Hydraulic conductivity of the substrate materials decreased between 37 and 80% during the study (in bed I it was unmeasurable). The 2-year old plants had no effect on the steady decrease in hydraulic conductivity, nor on the concentrations of most leachate constituents. Leachate evapotranspiration increased as stem

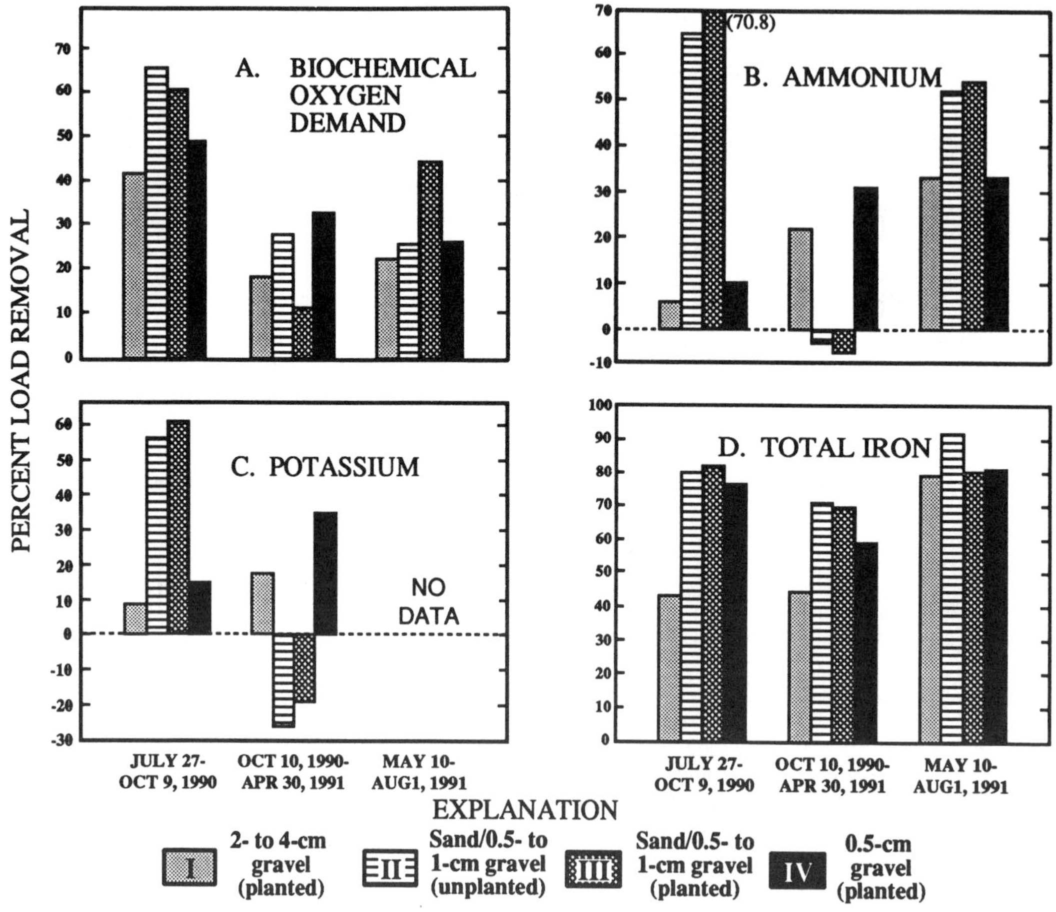

Figure 4. Removal rates of selected constituents in four constructed wetland research plots at Tompkins County municipal solid waste landfill, 1990–1991. A, biochemical oxygen demand; B, ammonium; C, potassium; D, total iron.

biomass increased; the maximum decrease in leachate volume, about 43%, was observed in the coarse gravel bed (I) in the second growing season.

CONCLUSIONS

- Constructed wetlands that contain *Phragmites australis* (common reed) can remove BOD, organic carbon, P, NH_4^+, Fe, Mn, and K^+ from landfill leachate to an appreciable degree.
- Differences in treatment effectiveness were related to type of substrate or bed temperature. The most efficient treatment was in summer in the beds containing a mixture of gravel and sand.
- Microbial activity and cation exchange are possible mechanisms for the removal of BOD, NH_4^+, and K^+, as indicated by decreased treatment efficiency and nutrient loss from the beds when bed temperatures dropped.
- Transpiration by plants can accelerate the leachate-evaporation process, especially where stem biomass exceeds 1.5 kg/m^2 (dry weight). Plants did not prevent a decrease in the hydraulic conductivity of the substrate, and no increase in root-zone aeration by plants was indicated, other than a greater removal of BOD from the planted sand and gravel bed (III) than from the unplanted sand and gravel bed (II) in the second growing season.
- Plant roots act as adsorptive surfaces for Fe and other metals, and effectively act as filters by limiting metal transport to the stems.

Table 3. Biomass and Tissue Analyses of *Phragmites australis* from Constructed Wetland Research Plots, I, III, and IV, and Adjacent Leachate-Seep Control Area at Tompkins County Municipal Solid Waste Landfill, 1991

	Constructed-wetland research plots			Adjacent leachate seep		
	Shoots	Roots	Rhizomes	Stems	Roots	Rhizomes
Biomass, in kilograms per square meter, 9/20/91						
Bed I (coarse gravel)	1.520 (0.160)	0.109 (0.023)	1.152 (0.212)			
Bed III (sand and gravel)	1.085 (0.230)	0.093 (0.013)	0.778 (0.184)			
Bed IV (pea gravel)	0.742 (0.171)	0.080 (0.007)	0.435 (0.083)			
Tissue Analysis						
Element, in percent by weight						
N	2.09 (0.10)	1.86 (0.14)	1.82 (0.18)	1.68 (0.24)	1.34 (0.20)	0.78 (0.39)
P	0.05 (0.005)	0.04 (0.005)	0.03 (0.003)	0.08 (0.005)	0.06 (0.02)	0.02 (0.005)
K	1.16 (0.08)	0.80 (0.06)	0.78 (0.10)	1.50 (0.41)	1.01 (0.38)	1.15 (0.24)
Ca	0.16 (0.02)	0.71 (0.16)	0.12 (0.03)	0.20 (0.03)	0.90 (0.56)	0.06 (0.02)
Metals, in milligrams per kilogram						
Fe	65.8 (8.6)	3709 (762)	287 (62)	92.0 (11.1)	9773 (3182)	328 (10.5)
Mn	68.9 (10)	289 (31)	40.8 (7.2)	60.2 (35.6)	270 (86)	40.2 (1.35)
Zn	17.4 (1.8)	35.0 (5.4)	12.4 (1.4)	21.4 (3.4)	91.6 (15.4)	4.0 (0.2)
Cd	0.09 (0.01)	0.39 (0.18)	0.15 (0.04)	0.05 (0.05)	0.1 (0.1)	0.05 (0.05)
Pb	0.21 (0.07)	8.02 (1.7)	0.45 (0.18)	0.30 (0.1)	11.8 (1.45)	0.25 (0.15)

Note: Numbers in parentheses are standard error (SE) values; SE = standard deviation/$\sqrt{n}$. Bed II was unplanted.

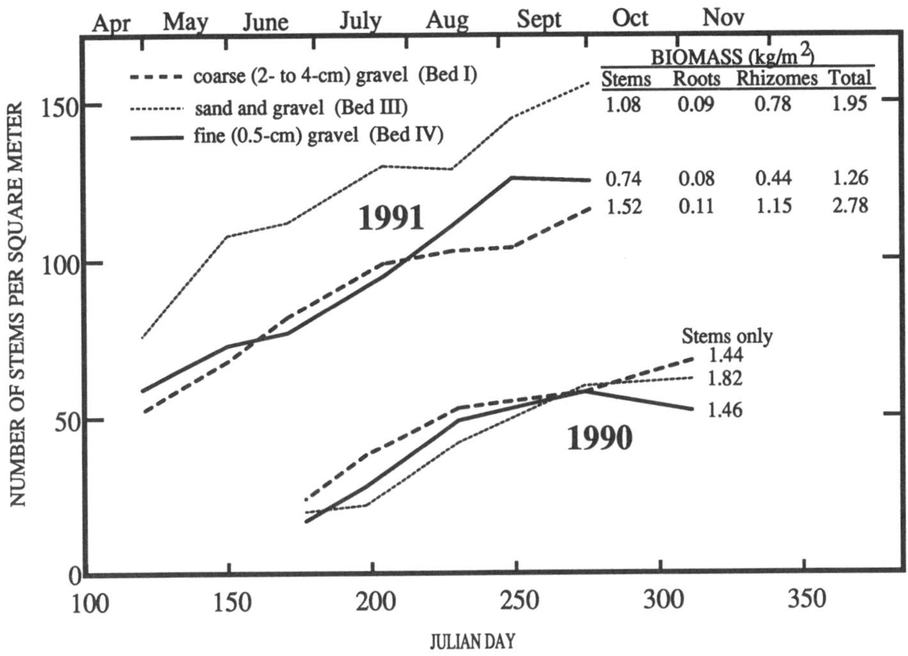

Figure 5. Biomass of *Phragmites australis* stems, roots, and rhizomes in three constructed wetland research plots at Tompkins County municipal solid waste landfill, June through November 1990 and April through September 1991.

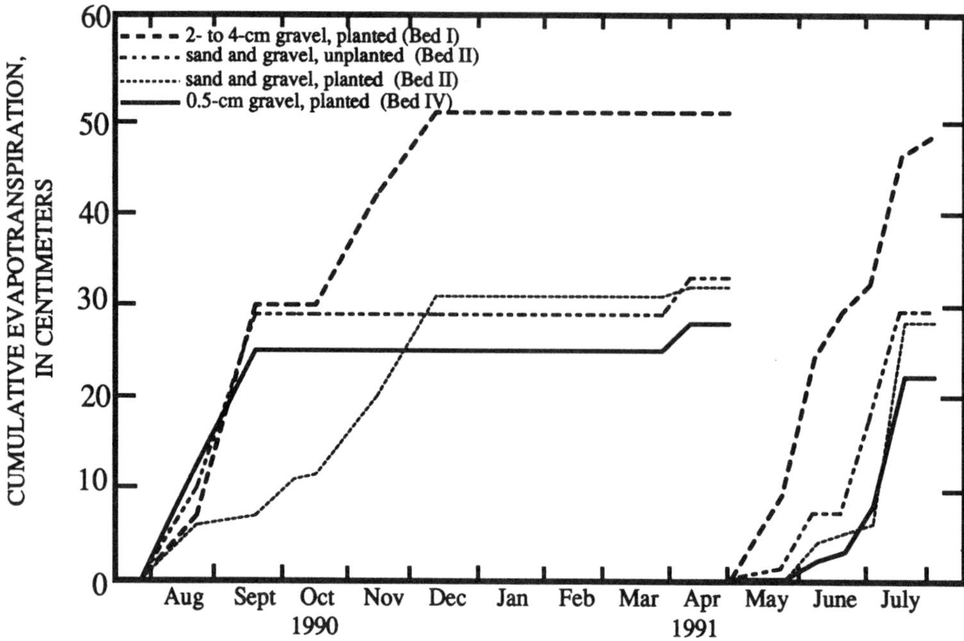

Figure 6. Cumulative evapotranspiration in four constructed wetland research plots at Tompkins County municipal solid waste landfill, July 1990 through July 1991.

REFERENCES

1. Hammer, D. A., Ed. *Constructed Wetlands for Wastewater Treatment: Municipal, Industrial and Agricultural.* Lewis Publishers, Chelsea, MI, 1989.
2. Shuckrow, A. J., A. P. Pajak, and C. J. Touhill. Management of hazardous waste leachate, *U.S. EPA Report* SW-871, 1980.
3. Reddy, K. R., E. M. D'Angelo, and T. A. Debusk. Oxygen transport through aquatic macrophytes: the role in wastewater treatment, *J. Environ. Qual.* 19:261, 1989.
4. Boon, A. G. Report of a visit by members and staff of WRC to Germany (GFR) to investigate the root zone method for treatment of wastewaters, Water Research Center Publication. 376-S/1, Processes. Stevenage, Hertfordshire, U.K., 1985.
5. Am. Public Health Assoc. *Standard methods for the examination of water and wastewater.* 16th ed. Am. Public Health Assoc., New York, 1985.
6. Mueller, J. C., L. H. Stolzy, and G. W. Fick. Constructing and screening Platinum microelectrodes for measuring soil redox potential, *Soil Sci.* 139:558, 1985.
7. U.S. Environmental Protection Agency. Methods for chemical analysis of water and wastewater. U.S. EPA Report 600/4-790-020, Washington, D.C., 1979.
8. Stucki, J. W. The quantitative assay of minerals for Fe^{2+} using 1,10 phenanthrolie. II. A photochemical method. *Soil Sci. Soc. Am. J.* 45:638, 1981.
9. Sanford, W. E., R. Kopka, J. Peverly, T. Steenhuis, J. Surface, and M. Lavine. An investigation into the use of subsurface flow rock-reed filters for the treatment of leachate from a solid waste landfill, in *Proc. Water Pollution Control Fed. Specialty Conf. Water Quality Management of Landfills.* Chicago, IL, 1990.

CHAPTER **51**

Mitigation of Landfill Leachate Incorporating In-Series Constructed Wetlands of a Closed-Loop Design

C. D. Martin, G. A. Moshiri, and C. C. Miller

INTRODUCTION

Constructed wetlands are a designed, man-made complex of water-saturated substrates with associated aquatic vegetation, animal life, and microbial communities. However, although these systems are designed to simulate natural wetlands, they are intended for direct human use and benefit in instances where the use of natural wetlands would not be practical. In recent years, constructed wetlands have been shown to be of particular value in the improvement of waters whose quality has been degraded through anthropogenic practices.[1] When properly designed and constructed, such systems will function as water purification systems as well as natural wetlands do. They are, however, less restricted by user conflicts, potential environmental concerns, and permit acquisition constraints than are natural wetlands.

In 1989, the Florida legislature enacted the Solid Waste Management Act which mandates certain practices to ensure groundwater integrity and other environmental matters at, and in the vicinity of, landfills within the state. This legislation required that all active landfills (1) be lined with an impervious membrane, (2) develop a system to collect and treat leachate, and (3) reduce the volume of incoming waste by 30% by 1994. Florida's Solid Waste Management Act was implemented in anticipation of the Federal Resource Conservation and Recovery Act (RCRA), Subtitle D regulations designed to regulate solid waste disposal practices nationwide.

Escambia County's plan to meet these new initiatives involves the staff of the Escambia County Solid Waste Management Department and the technical resources of the Wetlands Research Laboratory at the University of West Florida. This plan involves on-site recycling, composting, and leachate treatment using constructed wetlands (Figure 1). This article provides information on design characteristics. Publication of data on treatment efficiencies will follow at a later date.

LANDFILL DESIGN

Approximately 40% of all incoming solid waste is sorted for various recyclable materials, including ferrous metals, aluminum, glass, paper, and plastics. What remains after sorting is transferred to composting rows and composted. The remaining 60% is directly landfilled in cells lined with 3 ft of clay overlain by a 60-mil composite plastic liner. Each double-lined cell is also equipped with dual but separate leachate collection systems. The active cells are lined with a bottom layer of compacted clay covered with a 6-in. sand collection layer containing the below-liner leachate detection piping. This lower level of piping allows for sampling to ensure that the plastic and clay barriers have not been pierced. This lower level leachate detection system is then covered with 2 ft of compacted clay which is, in turn, covered with a 60-mil high-density polyethylene (HDPE) liner. The upper leachate collecting system is placed on top of the HDPE liner and consists of perforated piping covered in place with gravel and sand.

The resultant leachate filters through the sand and gravel-encased leachate collection system and flows into a 25-million gallon primary treatment lagoon lined with both clay and HDPE liner as already described. Septage is added, and the resultant wastewater is heavily aerated to reduce BOD and aid in the volatilization, oxidation, and mixing processes.

0-87371-550-0/93/$0.00+$.50

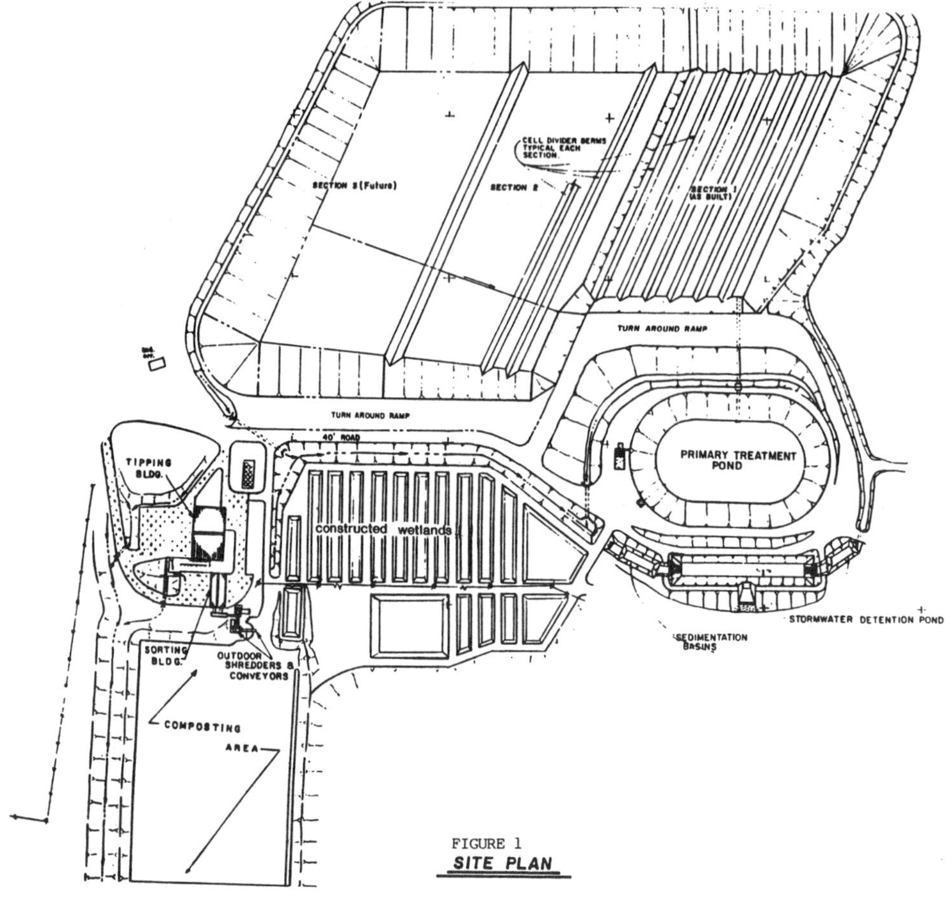

Figure 1. Site plan of Escambia County Solid Waste Facility.

Composting

Fixed spray nozzles are now used to broadcast the effluent onto rows of shredded solid waste placed on the 10-acre composting field. The site is lined with an impervious clay liner and graded so as to channel leachate generated from this field into a holding pond and ultimately into the constructed wetlands. The addition of the septage-augmented leachate provides the nutrients, moisture, and microorganisms essential for maximum composting activity to take place. The resultant leachate passes through a final sand filter as it is piped to the constructed wetlands.

Composting and recycling together result in a reduction in the volume of solid waste, as well as that of the leachate, through evapotranspiration and aerosol effects promoted by the leachate broadcast processes. The resulting noncommercial-grade compost has the consistency of soil and is used for daily landfill cover.

WETLANDS DESIGN

The basis for wastewater treatment in an aquatic plant system is the integrated growth of both the plants and their associated microorganisms. A major part of the treatment process for the degradation of organics is attributed to the microorganisms living in the rhizosphere associated with wetland plants. Periphyton and microbial communities translocate, metabolize, or utilize the various types of contaminants that are present in landfill leachate.[2] This final leachate treatment process involves a free surface flow marsh treatment system composed of ten wetlands (50 × 350 ft each) that terminate in a final sand filter. This surface flow approach is successfully used worldwide to treat various types of wastewater.

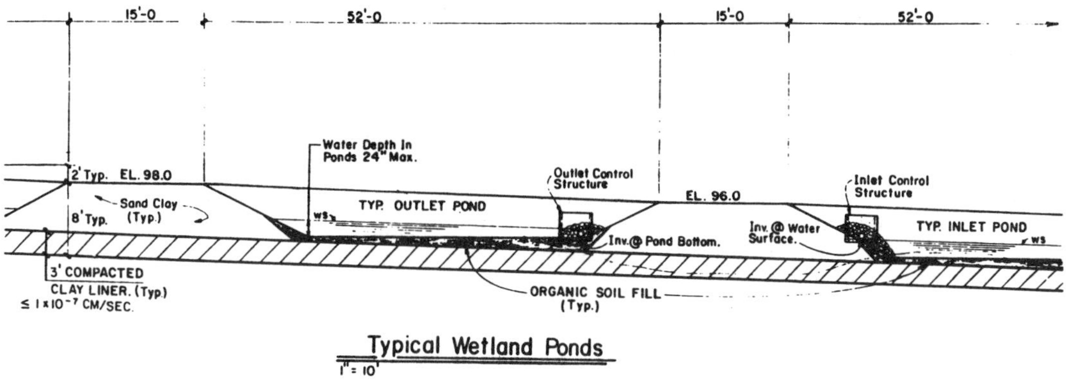

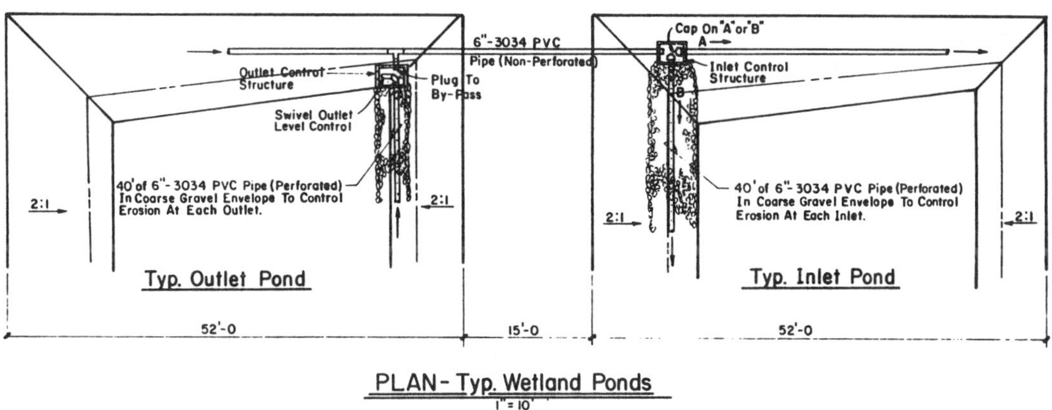

Figure 2. Typical wetland cell design at the Perdido Solid Waste Facility in Escambia County, Florida.

Construction of the treatment wetlands began with a 10-acre compacted clay basin designed to prevent seepage and thus help maintain desired water levels. The basin is divided into ten wetland cells as already described, with access roads constructed between cells. A suitable substrate composed of 74% sand, 17% clay, and 9% silt is used to provide support for the emergent vascular plants (Figure 2). The resulting series of cells provides the leachate with nearly 1 mile of contact in the wetlands complex.

Wastewater enters the system via a distribution network of piping designed to ensure an even flow distribution of leachate across the receiving lagoon, and help maintain aerobic conditions. Each wetland cell is constructed on a 2% grade for adequate drainage and maintenance purposes. At the opposite end of each treatment cell, water is collected from the underground bed via drain tiles. The effluent is then discharged through a swiveling standpipe into a concrete collection box. The standpipe serves to maintain individual basin water levels and ensure the desired retention times. Now, the effluent, aided by a 2-ft drop in elevation between each cell, exits the collection box and discharges into the following cell through its distribution pipe. Overflows that may result from heavy precipitation are prevented by the construction of a 2-ft high berm around each cell.

Vegetation

Plant species common to the region were selected for use in the wetland treatment system whose design depth was based upon the specific tolerance level of each species. *Eichornia crassipes*, *Typha latifolia*, *Scirpus* spp., and *Phragmites australis* were selected for use in the five initial receiving cells. These plants were chosen due to their demonstrated abilities to process nutrients.[3] The final five cells were planted with a mixture of species including *Sagittaria* spp., *Lemna* sp., and *Potendaria cordata*.

All individuals were obtained as young plants, seedlings, or rhizomes from a commercial nursery and were maintained in a greenhouse. To assure a dense, healthy stand of wetland vegetation, plants were set on 3-ft centers and cell water levels manipulated to encourage maximum rhizome growth. This procedure was based on the assumption that during the spring and summer months most of the growth is directed toward the emergent shoots. On the other hand, during the fall, most growth energy is directed toward the rhizomes, with vertical penetration aided by moist rather than saturated soil conditions.[4] Consequently, in September, the water level of each cell was lowered to approximately one third of the bed depth and kept at this level for 2 to 3 months. The plants used in these studies are adapted to such fluctuating water and nutrient levels and are more tolerant of high pollutant concentrations.[5]

Sampling

Sampling for bi-weekly water quality determinations included the influents and effluents from the first and final treatment cells. Parameters included volatile organic compounds, petroleum hydrocarbons, TOC, TKN, TPO_4, NO_2^--NO_3^-, TDS, and NH_3. Heavy metals assayed included As, Cd, Cr, Cu, Fe, Mn, Zn, Pb, and Ni. All of these characterizations were conducted against measurements of physical-chemical parameters including pH, conductivity, temperature, and BOD_5. Annually, sediment and plant tissue samples were also assayed for heavy metal accumulation.

Periodic harvesting was necessary when the accumulation of litter interfered with the hydraulic regime within each bed. At these latitudes, the frequency of such intervals are expected to range from 4 to 5 years.

CONCLUSION

Stringent discharge standards on wastewater effluent and National Pollution Discharge Elimination Systems (NPDES) requirements necessitate on-site management of wastewater. Proven technologies such as recycling, composting, and leachate treatment using constructed wetlands integrated into a closed-loop design reduce the need to discharge into state and federal waters. This integrated, relatively low cost, and easily operated approach achieves the present state and anticipated federal mandates, and has proven to be an environmentally sound approach to the resolution of routine landfill problems.

REFERENCES

1. Hammer, D. A., Ed. *Constructed Wetlands for Wastewater Treatment.* Lewis Publishers, 1989.
2. Trautman, N. M., J. H. Martin, Jr., K. S. Porter, and K. C. Hawk, Jr. Use of artificial wetlands for treatment of municipal solid waste landfill leachate, in *Constructed Wetlands for Wastewater Treatment: Municipal, Industrial, and Agricultural.* D. A. Hammer, Ed. Lewis Publishers, Chelsea, MI, 1989, 245–252.
3. Reedy, K. R. and W. H. Smith, Eds. *Aquatic Plants for Water Treatment and Resource Recovery.* Magnolia Publishing, Orlando, FL, 1987.
4. Reed, S. C., E. J. Middlebrooks, and R. W. Crites. *Natural Systems for Waste Management and Treatment.* McGraw-Hill, New York, 1987.
5. Feierabend, J. S. Wetlands: the lifeblood of wildlife, in *Constructed Wetlands for Wastewater Treatment: Municipal, Industrial, and Agricultural.* D. A. Hammer, Ed. Lewis Publishers, Chelsea, MI, 1989, 107–118.

CHAPTER 52

Innovative Marsh Treatment Systems for Control of Leachate and Fish Hatchery Wastewaters

R. Hunter, A. E. Birkbeck, and G. Coombs

INTRODUCTION

British Columbia Research Corporation (B.C. Research) has studied the application of engineered marshes for wastewater treatment in British Columbia over the past 4 years. Several wastewater sources have been tested in pilot-scale units including domestic sewage, pulp and paper effluent, landfill leachate, and woodwaste leachate. Full-scale systems for wastewater treatment have been designed and constructed for treatment of weak landfill leachate, woodwaste leachate, and fish hatchery wastewaters. Results to date suggest that locally available marsh plants can be used to create marshes which reduce wastewater oxygen demand and remove toxicity.

OBJECTIVES

This article describes some applications of engineered and natural marshes for treatment of different types of wastewater. General design of the marsh and associated retention times are described. Where possible, data showing treatment efficiencies have been included.

LANDFILL LEACHATE

Leachate control and treatment must be addressed at all landfills. To limit capital, labor, and maintenance costs for these facilities, the use of engineered marshes have been considered.

Pilot-Scale Studies for Landfill Leachate Treatment

B.C. Research has provided environmental services for several landfills over the last 15 years. In one study, leachate, partly treated in an aeration lagoon, was passed through six experimental engineered marshes. The objective was to determine if marsh treatment technology could replace the existing lagoon and spray irrigation treatment system at lower operating costs. Prior to discharging to a local fish-containing river, the leachate must meet disposal permit requirements. The wastewater characteristic of particular concern is toxicity, caused primarily by ammonia nitrogen, oxygen demand, and heavy metals content. The background, data, and assessment of the first year of the study were described previously.[1]

Ammonia Removal

Although the discharge permit does not specifically cite an ammonia-nitrogen level, earlier work demonstrated that treated leachate was not toxic to rainbow trout at ammonia-nitrogen concentrations below 7 mg/L.[2] This level was not achieved during the first year of the operation of the marsh cited in this study. It was anticipated that by increasing the population of active nitrifying organisms, ammonia-nitrogen removal would be increased.

Activated sludge from a local sewage treatment facility was added to the six experimental beds. Addition of active nitrifying bacteria caused marsh #1 to reduce the ammonia-nitrogen level from 32.2 mg/L in the

0-87371-550-0/93/$0.00+$.50

Table 1. Comparison of Six Pilot-Scale Engineered Marshes Treating Landfill Leachate

Sample	Retention time (days)		Ammonia-nitrogen (mg N/L)			Chemical oxygen demand (mg/L)		
	1991	1990 & spiked	1991	Spiked[a]	1990	1991	Spiked[a]	1990
Influent			32.7	32.2	32.3	179	176	176
Marsh #1	1.7	0.6	27.5	8.1	25.4	143	84	155
Marsh #2	—[b]	NA	NA	31.4	30.0	NA	190	183
Marsh #3	1.0	1.1	29.2	31.5	26.2	151	167	154
Marsh #4	2.6	2.6	30.4	24.4	28.0	146	147	158
Marsh #5	1.7	1.5	32.1	27.5	27.2	143	148	156
Marsh #6	1.5	3.1	30.5	23.4	27.2	144	120	150

[a] Spiked: data obtained following addition of waste-activated sludge to the influent.

[b] Unknown because outflow pipe not working. Sample was taken from ponding area close to outflow.

influent to 8 mg/L (Table 1). This marsh was constructed from gravel and sand and planted with cattails. Some improvement was also recorded in systems #4 and #6. Although visually the beds seemed to have suffered damage prior to monitoring during the second season, the ammonia levels were lower in the effluent as shown by the first year's data. However, it appears that the nitrifying bacteria did not survive the period between monitorings, as indicated by the generally higher values in the effluent during 1991.

Chemical Oxygen Demand

Like ammonia-nitrogen, the chemical oxygen demand was reduced more in the second year of monitoring than during the first year. Even greater reductions were observed when the marshes had been seeded with activated sludge.

Natural Marsh

In a second study, weak leachate generated at an active construction-demolition landfill was collected and passed through a natural marsh occupying approximately 13 ha. The landfill company desired to expand their operation into the marsh area — a plan that would reduce the available natural marsh treatment area and increase the leachate volume. Analytical results showed that the discharged effluent met permit levels except for chemical oxygen demand, dissolved oxygen, and occasionally ammonia-nitrogen. Therefore, the level of treatment in the natural marsh was considered unacceptable.

In order to use some of the natural marsh for fill area and to treat the expected increase in leachate volumes, the operators opted for a combination treatment system. Two thirds of the natural marsh was sectioned off for filling. The leachate passed through the remaining third of the natural marsh, then flowed through an aeration lagoon prior to discharge (see Figure 1). To compensate for the marsh treatment area lost, the natural marsh's plant population was enhanced by transplanting mature *Typha latifolia* to the site, followed by alterations to the flow pattern through the marsh. Since the landfill only discharges leachate during the winter, as yet no data have been collected comparing the natural marsh with the enhanced marsh and lagoon systems.

WOODWASTE LEACHATE

Woodwaste, referred to in the forest industry as hogfuel, is an abundant by-product and is commonly used as an inexpensive, light fill for road construction, horticulture, and animal husbandry. It is known to leach a highly coloured, odiferous, and toxic liquid. Because this leachate has a high oxygen demand and can be toxic to fish, it is essential to treat it prior to discharging to fish-containing receiving waters.

Pilot Scale System

A small-scale test unit, similar to the system described by Cooper and Hobson,[3] was used to examine the feasibility of treating woodwaste leachate in engineered marshes. One side of the unit was planted with *Juncus*

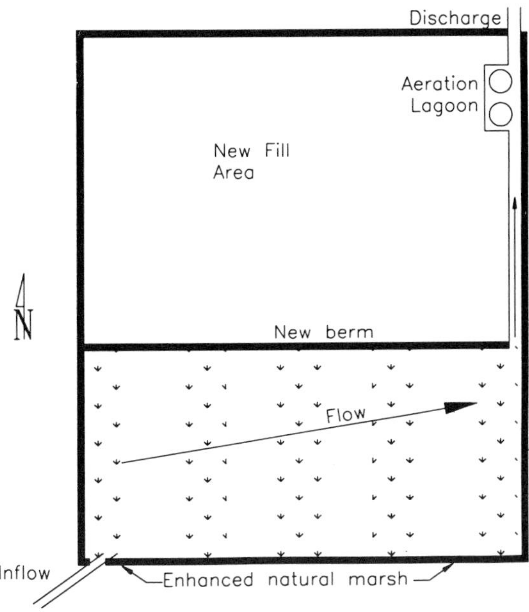

Figure 1. Modifications to natural marsh treating demolition landfill leachate.

Table 2. Treatment of Woodwaste Leachate in a Pilot-Scale Engineered Marsh with a 10-day Retention

	Biochemical oxygen demand (mg/L)	Chemical oxygen demand (mg/L)	Toxicity 96h-LC_{50}
Influent	463	1024	75
Effluent	263	721	>100
% Removed	43	29	

effusus, while the other side was left unplanted as a control. Woodwaste leachate was pumped into the two compartments. At a 10-day leachate retention time, toxicity was removed and biochemical and chemical oxygen demands were reduced by 45 and 30%, respectively (Table 2). Following these pilot studies, two full-scale engineered marshes were designed to treat woodwaste leachate.

Full-Scale Systems

Woodwaste Leachate from Highway Construction

Woodwaste has been used as a fill for entrances and off-ramps in construction of new highways over peat lands in the Fraser River delta. Owing to anticipated leachate production in the fill areas, a collection system was installed to collect the leachate generated. Based on the success of treating woodwaste leachate in small-scale marshes and the need to minimize the mechanical equipment and maintenance at the discharge site, a full-scale marsh was proposed as a leachate treatment method.

The average leachate flow rate in wet weather from the interchange area was estimated to be 20 L/min, and the desired retention time of 10 days required a marsh volume of 288 m^3. The available space limited the maximum width to 5 m and a design depth of 0.75 m was chosen based on the recommendations of Cooper and Hobson.[3] The substrate consisted of 0.5 m gravel, 0.2 m sand, and a layer of top soil. Figure 2 shows a schematic of the design.

The base of the marsh was local clay, lined with a geofiber. A subsurface inflow pipe was positioned at the bottom of the marsh as an extension of the leachate collection pipe. The marsh was planted with canary

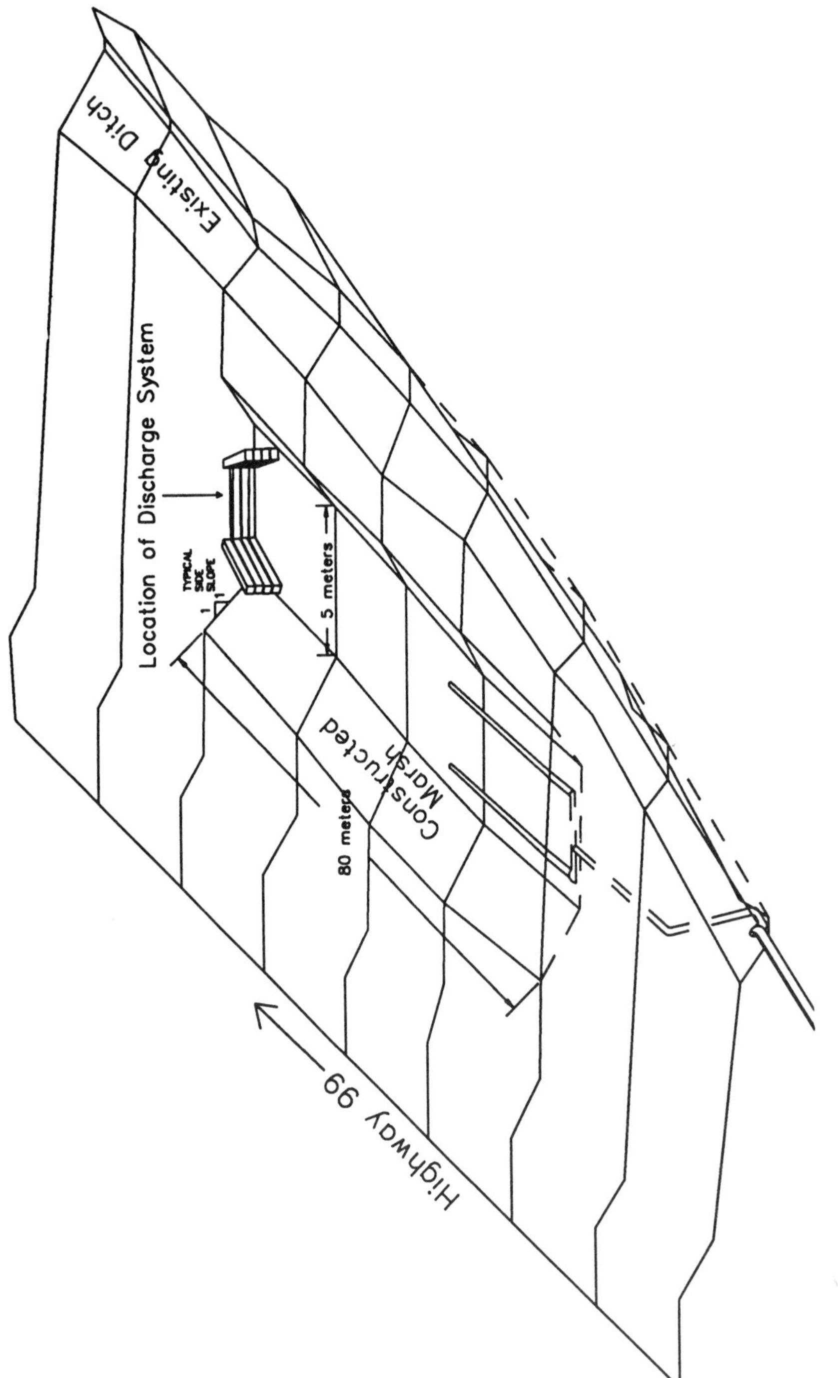

Figure 2. Schematic of highways marsh.

Table 3. Characteristic of Woodwaste Leachate Entering and Leaving a Full-Scale Engineered Marsh

Biochemical oxygen demand (mg/L)		Chemical oxygen demand (mg/L)		Toxicity 96h-LC_{50}	
Influent	**Effluent**	**Influent**	**Effluent**	**Influent**	**Effluent**
112	78	294	270	>100	>100
<20	48	95	321	<100	>100
109	150	242	338	>100	>100
184	193	411	459	>100	>100
144	154	335	359	<100	>100
160	<10	468	127	>100	>100
122[a]	106[a]	308[a]	312[a]		

[a] Average.

grass (*Phalaris arundinacea*) and soft rush (*Juncus effusus*). The marsh performance has been monitored for removal of toxicity (biochemical and chemical oxygen demands) to rainbow trout. There is no measurable flow through the marsh during the summer; therefore, limited data have been collected since the marsh's completion in January 1991. These data are presented in Table 3. The data show erratic oxygen demand reductions, causing some concern over the marsh's ability to treat the leachate. However, there is some reduction in toxicity levels.

Several reasons for the poor performance of these systems may include the immaturity of the marsh system; the present sampling technique (grab sampling); the irregular characteristics of the influent; and the dormant state of the plants during sampling events. In the late spring of 1991, the marsh was replanted with additional *Typha latifolia* in order to provide greater oxygen transfer to the root zones. The second year of monitoring will commence in the fall of 1991.

Woodwaste Leachate from Landfilling

In another instance, it has been proposed to landfill the woodwaste. A schematic of the proposed landfill is shown in Figure 3. The leachate is expected to be generated from winter precipitation onto the site and will be treated by means of a constructed marsh. The leachate will be collected at the base of the proposed fill area where it will pass into the 20 × 15 m marsh. A retention time of 13 days has been calculated, based on the estimated leachate volumes. The marsh base will be the natural clay-till and filled with sand to a depth of approximately 0.5 m.

Provisions have been made in the design to add a second marsh in-series with the first, if the 13-day retention time is insufficient to produce a nontoxic discharge.

FISH HATCHERY WASTEWATER

A Vancouver Island salmon hatchery has expanded its operation to a new site. At the fish hatchery, wastewater originates as two streams. The first is heated rearing tank water derived from small fish tanks used for rapid development of fingerlings from the swim-up stage of development. The larger flow, from the main fish holding tanks, is unheated water and is used to grow fingerlings up to smolt size. Both waste streams are provided preliminary treatment by screening through TRIANGEL filters. Backwash water from the TRIANGEL filters is settled and the supernatent passed into a small marsh. The screened waters are passed through a biological filter and then into a larger marsh (Figure 4).

The system has been developed to maintain the high-quality fresh waters of Coal Creek and protect an oyster fishery on the beach receiving the Coal Creek waters. Direct discharge of the fish hatchery waters has not been permitted. The treatment and disposal system maximizes removal of organic load, nutrients, and reduces the potential for transfer of diseases from the hatchery to wild stocks in the creek.

The backwash water flow rate is estimated at 111 m^3/day. This volume will require a marsh approximately 40 × 33 m, and the screened water will ultimately be discharged to a larger marsh. Discharge will be in a disperse manner to reduce impact of fresh water on the salinity levels of a commercial oyster beach. The

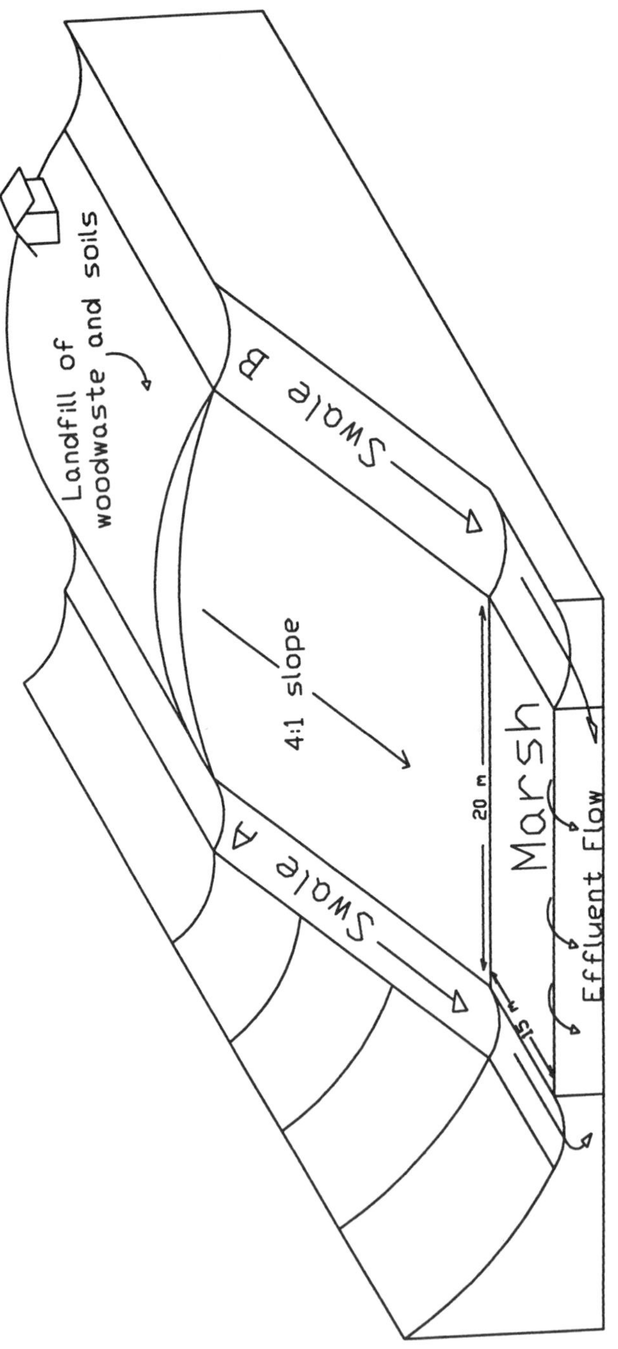

Figure 3. Proposed marsh for landfilled ravine.

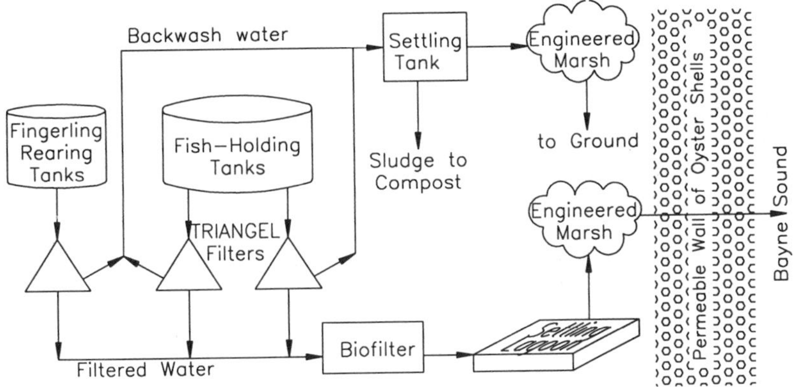

Figure 4. Fish hatchery wastewater treatment system at Coal Creek, Vancouver Island.

Table 4. Characteristics for Marsh-Treated Screened Fish Hatchery Wastwater at Coal Creek

	Marsh (mg/L)	Coal Creek (mg/L)
Total suspended solids	<1	<1
Nitrate-nitrogen	0.15	0.12
Nitrite-nitrogen	0.002	0.008
Ammonia-nitrogen	0.11	0.034
Phosphate	0.021	<0.02

Note: Analysis done by Cantest.

proposed system has been installed. Data (Table 4) show that there is no significant difference between the treated waters exiting the marsh and the Coal Creek waters.

CONCLUSIONS

Experience with construction of full-scale engineered marshes and pilot-scale studies has led to the following recommendations and conclusions.

- Spiking marsh systems with activated sludge organisms seems to enhance treatment efficiencies with respect to the removal of ammonia-nitrogen. Investigation of similar spiking for other targeted pollutants could also be considered.
- Full-scale marshes should be designed so that the natural hydraulics of the area supply and discharge the wastewater. This will eliminate the need for pumps and power and the need for maintenance which can be a problem in remote locations.
- Full-scale marshes should be designed to facilitate expansion.
- Locally available marsh plants in British Columbia, cattails (*Typha latifolia*), soft rush (*Juncus effusus*), and canary grass (*Phalaris arundinacea*) are suitable to construct engineered marshes which provide reductions in oxygen-demanding substances and remove toxicity.

ACKNOWLEDGMENTS

The authors acknowledge the Fraser River Harbour Commission, Eco Waste Industries Ltd., the British Columbia Ministry of Transportation and Highways, and Mr. John Holder of United Hatcheries Ltd.

REFERENCES

1. Birkbeck A. E., D. Reil, and R. Hunter. Application of Natural and Engineered Wetlands for Treatment of Low-Strength Leachate in Constructed Wetlands in Water Pollution Control. P. F. Cooper and B. C. Findlater, Eds. Pergamon Press, Oxford, 1990, 411–418.
2. Birkbeck A. E., L. J. Hunt, and H. Geiser. Characterisation and Treatment of Leachate from a Landfill on a Peat Bog, presented at the 6th Natl. Conf. on Waste Management. Vancouver, November 7–9, 1984.
3. Copper P. F. and J. A. Hobson. Sewage treatment by reed bed systems: the present situation in the United Kingdom, in Constructed Wetland for Wastewater Treatment. D. A. Hammer, Ed. Lewis Publishers, Chelsea, MI, 1989, 153–172.

CHAPTER **53**

Constructed Wetlands for Wastewater Treatment at Amoco Oil Company's Mandan, North Dakota Refinery

D. K. Litchfield

INTRODUCTION

To comply with stringent standards of the National Pollutant Discharge Elimination System (NPDES) promulgated by the United States Environmental Protection Agency (USEPA) in the early 1970s, a study was undertaken by Amoco's Mandan, North Dakota Refinery to evaluate alternate wastewater treatment systems. This evaluation reviewed carbon adsorption, activated sludge, and expanding the refinery's existing bio-oxidation system. The decision to expand the refinery's existing bio-oxidation system was based on availability of open land with natural drainage, past success of the existing system, and economics. Selecting this option was estimated to cost $250,000 vs. costs of alternate systems ranging from $1 million to $3 million (1972 U.S. dollars).

The Mandan Refinery covers 320 acres (130 ha) on a 960-acre (389 ha) tract of land north of the city of Mandan, North Dakota, located along the west bank of the Missouri River. The 640 acres (260 ha) surrounding the refinery are dedicated to wastewater treatment and wildlife management. The refinery has the capacity to process 60,000 barrels (7,592 MT/day) of crude oil per day, requiring 1.5 million gallons/day (3.7 million L/day) of water pumped from the Missouri River.

The Mandan Refinery's original wastewater treatment facilities consisted of an American Petroleum Institute (API) oil-water separator as primary treatment, and an aerated bio-oxidation lagoon as secondary treatment. This system offered limited flexibility. If a wastewater quality problem occurred from a process upset, wastewater could not be held or diverted while the problem was corrected. In addition, heavy rains and/or snowmelts reduced residence time significantly.

The additional bio-oxidation ponds, with average depths of 4 to 6 ft (1.2 to 1.8 m), were established by building dams across natural drainage channels. Dams were constructed based on the North Dakota Water Commission and local engineering guidelines. They are earth-filled, with clay keyways, overflow structures, and spillways. Upstream surfaces are rip-rapped for wave protection, and water levels are controlled by siphon lines, drain lines, and a series of culverts with slip gates. Cattail, bulrush, and a mix of other wetland plants have naturally invaded the lower ponds.

SYSTEM DESIGN AND OPERATION

Primary treatment of wastewater from the refinery process units is accomplished in a conventional API separator in which oil is separated and recovered. Water is then discharged into a 15-acre (6 ha) lagoon for initial secondary treatment and pumped to a high point for distribution among several routes through a series of cascading ponds and ditches before discharge to the Missouri River. Figure 1 shows a schematic of the wastewater treatment system and indicates the general flow route of the wastewater.

Normally, water flow is through 6 of the 11 ponds, representing a treatment area of 41 acres (16.6 ha). The five additional ponds, with an area of 47 acres (19.0 ha), are dedicated to wildlife habitats and their management, and provide diversion or holding capacity in the event of heavy rains, snowmelts, or other unexpected containment problems.

0-87371-550-0/93/$0.00+$.50

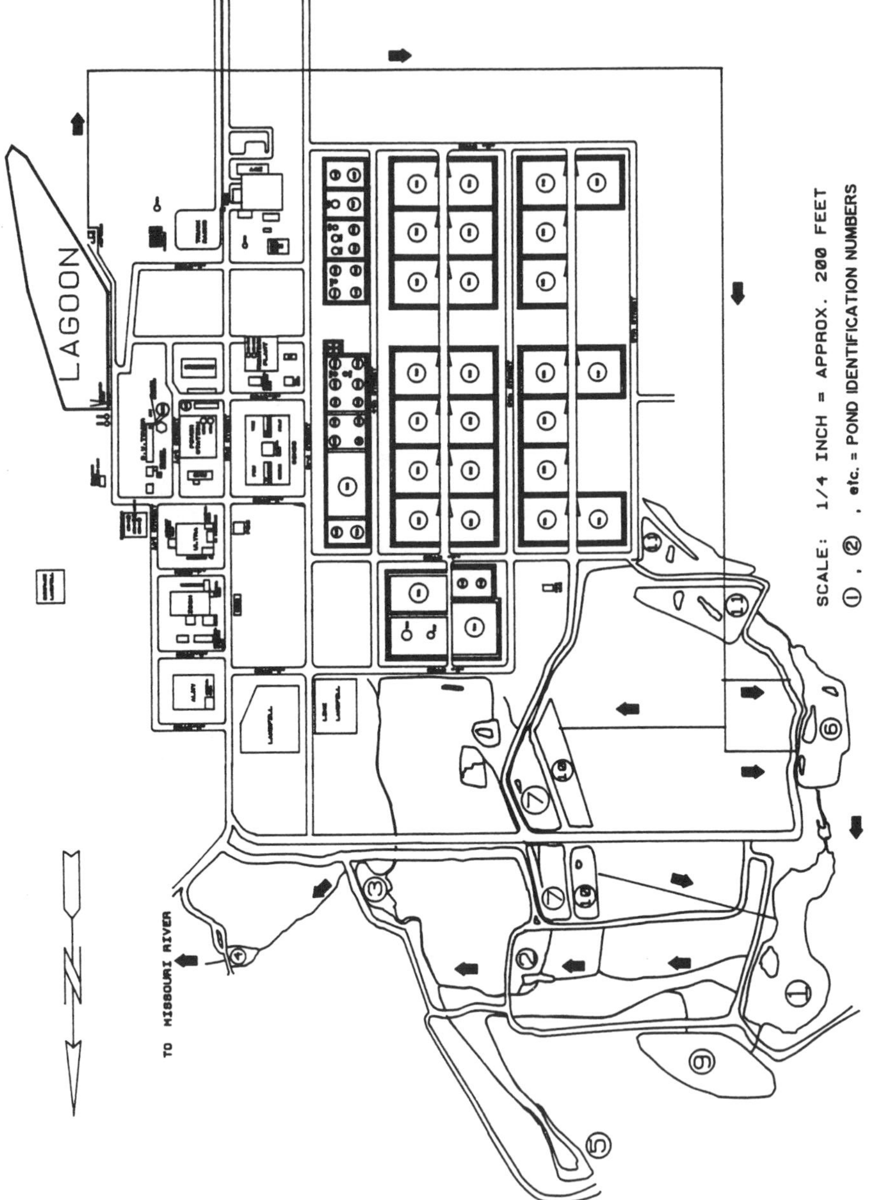

Figure 1. Cascade ponding system. Flow pattern from start (lagoon) to discharge (dam 4).

WASTEWATER QUALITY

Wastewater is routinely sampled at the API separator effluent (start of the bio-oxidation system), lagoon effluent (start of the cascade ponding/wetlands system), and the final discharge point (Dam 4 discharge).

Sample analyses indicate reductions in concentration of all pollutant species as the wastewater flows through the treatment system (Table 1). Hexavalent chromium, total chromium, and total suspended solids are not analyzed in the API separator effluent because of the oily nature of this water. The remaining pollutant concentrations, however, are reduced by 93 to 100% between the API separator effluent and the final discharge point. An exception is ammonia-nitrogen, which is reduced 84%. Reductions in parameter concentrations of 69 to 100% are obtained in the primary lagoon with the exception of ammonia-nitrogen, which is reduced 21%. Concentrations are further reduced by 79 to 100% in the cascade ponding system which, in effect, acts as a polishing system. Average discharge concentrations to the Missouri River are well below NPDES limits; however, there are occasions when a daily limit can be exceeded. This usually occurs during an extremely heavy rainfall or snowmelt, when surface run-off from surrounding drainage flushes out the lagoon and the cascade ponding system.

Although the average pollutant concentrations in effluent water from the primary lagoon are well below permit limits (Table 1), daily effluent limits would have been exceeded 27 times in 1990 had the lagoon discharge gone directly to the river instead of through the cascade ponding/wetlands system. Specifically, total suspended solids, biochemical oxygen demand, chemical oxygen demand, pH, and oil, and grease would have been exceeded at various times. In addition, at least once or twice a year, some operating upset increases the lagoon concentration of one or more of the pollutant species above normal levels, requiring longer retention times to complete the treatment.

In 1990, discharge from the API separator, lagoon, and Dam 4 averaged 530,000 gal/day (2,000,000 L/day); 664,000 gal/day (2,500,000 L/day); and 501,000 gal/day (1,900,000 L/day), respectively. The primary lagoon receives influent from stormwater run-off and discharges from surface impoundments, in addition to the API separator. As the wastewater flows through the cascade ponding system, its volume is affected by evaporation, percolation, stormwater run-off, and influent from natural springs in the drainage area.

In 1979, a limited survey was conducted on chromium, copper, iron, nickel, lead, and antimony reductions through the wastewater treatment system. Besides a general reduction, the samples indicated that most of the reduction occurred in a 0.5-mile (0.8 km) earthen canal between the primary lagoon and the first pond in the system. This canal had a heavy growth of aquatic vegetation.

WILDLIFE HABITAT

Superior wastewater quality and cost effectiveness are only two advantages of the refinery's constructed wetlands. Establishing this system has created a natural habitat for wildlife. Shortly after its development, a sharp increase in the number of wildlife species inhabiting the area was noted. In conjunction with various state and federal wildlife officials, refinery staff encouraged and expanded the propagation of wildlife and waterfowl by creating nesting islands within the ponds and planting 30,000 fruit-bearing trees and 20,000 other species of trees. Some 184 species of plant life have been identified in this area, including numerous wetland plants and grasses such as reed canary grass, wild rice, musk grass, cattails, and bulrushes. In addition, acres (hectares) of alfalfa, millet, flax, and corn are planted as food for waterfowl. Feeders and houses for wood ducks, swallows, and purple martins were installed, and ponds were stocked with rainbow trout, bass, and bluegill. Rainbow trout up to 5 pounds (2.3 kg) have been taken from these ponds for necropsy analysis. Results from annual necropsies have been normal.

Among the most popular residents are Canadian geese, which the North Dakota Game and Fish Department is introducing throughout the state. Since 1977, when state officials released 35 yearlings at the refinery, some 1246 goslings have fledged at the site; 1191 have been banded and 516 released at other locations in North Dakota. In addition, 201 species of birds have been observed, and 60 species nest in the area. Numerous other wildlife species, such as pheasant, grouse, partridge, deer, fox, badger, skunk, and raccoon also inhabit this site.

Table 1. Wastewater Quality (1990)

		Discharge Points		
NPDES parameter	NPDES limits (kg/day)	API Sep. (kg/day)	Lagoon (kg/day)	Dam 4 (kg/day)
BOD	197.7	456.9	61.9	7.2
COD	1477.6	1393.3	436.5	92.2
NH_3-N	131.8	35.5	28.0	5.6
Sulfides	1.3	42.8	0.1	ND
Phenols	1.5	6.4	0.2	0.012
Oil & grease	59.9	143.0	5.3	0.32
Hex Cr	0.24	—	0.04	0.009
Total Cr	3.00	—	0.10	0.012
TSS	137.8	—	88.9	11.4

Note: BOD (biochemical oxygen demand); COD (chemical oxygen demand); NH_3-N (ammonia-nitrogen); hex Cr (hexavalent chromium); TSS (total suspended solids); ND (not detectable).

CONCLUSION

The refinery's NPDES permit limits have not been exceeded in the last 3 years. Only two deviations have occurred in the past 5 years. These events happened during periods of heavy rainfall and/or snowmelt when so much water flowed through the cascade ponding system that kilogram-per-day limits were exceeded for either total suspended solids or biochemical oxygen demand. In addition, the refinery has received many awards during the past decade for the development of this wastewater treatment/wetlands complex. The most notable awards include the 1980 Citizen Participation Award from the Environmental Protection Agency for care of this planet and protection of its life; the 1986 Blue Heron Award from the National Wildlife Federation for use of a wastewater treatment wetlands as a wildlife sanctuary. Finally, in 1990, an Environmental Achievement Award was received from the Renew America Organization for a successful environmental program that reduces pollution and protects the environment. This system demonstrates that industrial and environmental concerns can exist harmoniously.

PART 9

SMALL SYSTEMS

CHAPTER 54

Small Constructed Wetlands Systems for Domestic Wastewater Treatment and Their Performance

G. R. Steiner and D. W. Combs

INTRODUCTION

Inadequate treatment of domestic wastewater in rural areas of the Tennessee Valley contributes to non-point source pollution of surface waters and contaminates groundwater supplies. Water quality of streams is impaired, public health is threatened, and the problems are growing. For example, in Alabama, 47% of the homes are on septic tank systems for household waste disposal, with 78% of these in rural areas.[1] An estimated 50% of the households depend on groundwater for drinking supplies; in 1989, 50% of the private well samples had high bacteria counts, an increase from the 35% in 1973. The Centers for Disease Control estimates that 43% of disease outbreaks (e.g., hepatitis, bacterial and viral gastroenteritis, and typhoid) can be traced to untreated groundwater which was invaded by sewage.[2] Septic tank systems that are improperly designed and constructed are suspected as the major cause of this serious problem.

Alternatives to conventional on-site wastewater treatment systems are needed to help provide solutions to this serious pollution problem. A small constructed wetlands (CW) system is believed to be one treatment alternative where conventional septic tank systems are ineffective due to poor or shallow soils, high groundwater table, Karst topography, or too small a lot size. The Tennessee Valley Authority (TVA), based on its experience obtained in demonstrating the CW technology for larger municipal systems, has developed and published guidelines for the design, construction, and operation of small systems.[3] Several systems have either been built or are being planned in the Tennessee Valley region using TVA information and guidance. Summarized below are the descriptions and performance information for several small systems.

DESCRIPTION AND PERFORMANCE

Home System, Signal Mountain, Tennessee

In August 1988, the first CW designed for a home by TVA in the Tennessee Valley was built in Signal Mountain, Tennessee. Due to poor percolating soil, the homeowner had a failed absorption field leaving the septic tank in his back yard. A CW was designed and approved by the local health department, and built to replace the absorption field.

The design flow is 2.27 m^3/day (600 gpd) as required by state criteria for a four-bedroom house. The CW area is 48.6 m^2 (523 ft^2) using an area loading rate of 21.3 m^2/m^3/day (0.87 ft^2/gpd). Since the lot is sloped, the CW is terraced into an upper first cell and a lower second cell, with 25 and 75% of the area put into the first and second cells, respectively. The TVA was not using Darcy's Law to calculate system dimensions when this system was designed. However, both the first cell and a third of the second cell were made trapezoidal in shape with wider inlet than outlet widths so the load could be distributed over a greater width. For example, the first cell has inlet and outlet widths of 3.6 m (12 ft) and 1.8 m (6 ft), respectively, with a 4.4-m (14.5-ft) length. The cell beds are flat (i.e, a 0% longitudinal slope). The first cell is lined to contain and treat the septic tank effluent. The second cell is lined on the sides only to prevent seepage out of the sides; its bottom is unlined to allow percolation. The principal substrate is 1.9 to 2.5 cm ($^3/_4$ to 1 in.) river rock, 45 cm (18 in.) deep in the first cell, and 30 cm (12 in.) deep in the second cell. For this initial system, 15 cm (6 in.) of topsoil covers

0-87371-550-0/93/$0.00+$.50

the gravel substrate in both cells. Wastewater is distributed in both cells with 10-cm (4 in.) PVC pipe headers with drilled holes and laid on the cell bottoms. Similar headers are the collectors in both cells. Water exits the first cell into a small chamber containing a standpipe of a fixed length which maintains the water level at about 2.5 cm (1 in.) below the surface of the soil. A valve on the collector at the end of the second cell is normally kept closed to prevent discharge through it.

A variety of vegetation has been planted in the system by the homeowner; but due to the tree cover, sunlight is limited. Arrowhead was eventually found to survive and grow well in the first cell. In the second cell, which is in deeper shade than the first cell, water iris, elephant ear, ferns, and other decorative species have been planted.

The system cost approximately $2000 (U.S.) to build in 1988, principally for materials and payment for part-time labor. Most of the work was done by the homeowner.

The system has now operated for over 3 years. There has been no discharge. The wastewater apparently is being cleaned to a degree which allows percolation into the soil. What was once a public health and aesthetic problem has been corrected by the CW. The homeowner once drained about 189 l (50 gal) from the second cell to drop some standing water below the substrate surface following a week of 33 cm (13 in.) of rain.

Three sets of samples have been collected from the system which provide some indication of the system's treatment effectiveness. Average concentrations for BOD_5 (5-day biochemical oxygen demand), total suspended solids (TSS), total nitrogen (TN), total phosphorus (TP), and fecal coliform (FC) are listed below. The cell 2 sample was collected from a standpipe about 75% down the length of the cell.

Sample location	BOD (mg/L)	TSS (mg/L)	TN (mg/L)	TP (mg/L)	FC (col./100 ml)
Septic tank effluent	247	60	64	9.3	280,000
Cell 1 effluent	57	25	35	4.9	260,000
Cell 2	27	13	26	3.3	61,000
% removal	89	78	59	65	78

Some operation problems have been noted for consideration in future systems. Many wetland plants do not grow well due to shortage of sunlight; systems would work best in a location having at least 6 h of full sun. The retaining wall made of landscaping timbers apparently is not supported well enough and has sagged a little. It may eventually need to be replaced. This sagging has caused the substrate to settle in one corner, in turn causing some standing water and resulting odor. Additional substrate has been added to level the surface of the bed. Also, the rigid water level control standpipe makes it difficult to adjust water level; a flexible hose is needed which can be adjusted easily.

Chattanooga Nature Center, Chattanooga, Tennessee

In July 1989, an on-site CW was designed for the Chattanooga, Nature Center (CNC). It is part of a Wetlands Education Center, a cooperative project between TVA and EPA Region IV.

The existing 3.8-m^3 (1000-gal) septic tank provides pretreatment and was cleaned of solids and inspected. The design flow for the CW is 4.92 m^3/day (1300 gpd) based on water use records. The CW area is 123 m^2 (1320 ft^2), using an area loading rate of 24.5 m^2/m^3/day (1.0 ft^2/gpd). Darcy's Law was applied to determine the dimensions. Using a 1.0% bed slope, a conservative substrate hydraulic conductivity of 122 m/day (400 ft/day), and an inlet substrate depth of 45 cm (18 in.), the effective width and total length of the system is 6.1 m (20 ft) and 20 m (66 ft), respectively. Two rectangular cells, each 6.1 m (20 ft) wide and 10 m (33 ft) long are used. The principal substrate is 1.9 to 2.5 cm ($^3/_4$ to 1 in.) river rock and the substrate surface is flat although the cell bottoms are sloped. The first cell is lined with 24-mil plastic sheeting to contain and treat the septic tank effluent. The bottom of the second cell is unlined to allow percolation. The substrate depths in the first and second cells are 45 and 30 cm (18 and 12 in.), respectively. Topsoil and mulch (8 cm, [3 in.] each) cover the gravel substrate in the second cell. The mulch/soil cover provides some water retention capacity for the vegetation if wastewater percolation through the unlined bottom limits water available to the vegetation.

In cell 1, septic tank effluent is distributed onto the surface of the gravel through a 10-cm (4-in.) PVC header pipe with adjustable swivel tees. The wastewater moves subsurface through the cell, collects in a 10-cm (4-in.) pipe with drilled holes, and discharges into a water level control structure through a fixed-length rigid standpipe which maintains the water level at about 2.5 cm (1 in.) below the surface of the substrate. Water exits the control structure and is distributed across the width of the second cell through a 10-cm (4-in.) PVC header with drilled holes and laid on the cell bottom. The second cell has a collector header and water level control structure similar to the first cell. The discharge pipe from the second cell is plugged since there is no discharge.[4]

A variety of vegetation has been planted. In cell 1, common reed (*Phragmites* sp.) has become the dominant species and the vegetation cover is 100%. Decorative flowering species are gradually being added to cell 2. This system was built by a contractor for approximately $8000 (U.S.).

Routine samples have not been collected from this system, but three studies were done using a dye to follow a plug of wastewater through the cell 1. These samples indicate the effectiveness of the lined half of the system and the quality of water entering the unlined second cell. Average concentrations for the inlet (septic tank effluent) and the cell 1 effluent are listed below.

Sample location	BOD (mg/L)	TSS (mg/L)	NH_3-N (mg/L)	FC, (col./100 ml)
Septic tank effluent	44	130	36	725,000
Cell 1 effluent	12	6	20	5800
% removal	73	95	44	99

Also, samples collected along the length of cell 1 for BOD and TSS indicate that most of the pollutant removal occurs in the first 25% of the cell, as shown in Figure 1.[5,6]

Coon Creek Science Center, Chester County, Tennessee

A similar but larger CW than the CNC system was designed for the Coon Creek Science Center, part of Memphis Museums, Inc. It was constructed and began operation in June 1989.

Three 3.8-m^3 (1000 gal) septic tanks and a grease trap provide pretreatment for the design flow of 12.4 m^3/day (3200 gpd). The CW area is 258 m^2 (2780 ft^2), using an area loading rate of 21.3 $m^2/m^3 \cdot$ day (0.87 ft^2/gpd). The dimensions were determined using Darcy's Law, a 1.0% bed slope, a substrate hydraulic conductivity of 244 m/day (800 ft/day), and an inlet substrate depth of 45 cm (18 in.). The effective width and total length of the system are 11 and 24 m (36 and 78 ft), respectively. Two rectangular cells, each 11 m (36 ft) wide and 12 m (39 ft) long are used.

The principal substrate is 2.5 to 5 cm (1 to 2 in.) gravel, similar to river rock. The substrate surface is flat, although the cell bottoms are sloped. The first cell is lined and the second cell is unlined to allow percolation. The substrate is 45 cm (18 in.) deep in cell 1 and 30 cm (12 in.) deep in cell 2, with 3 in. of topsoil. Mulch (8 cm, 3 in.) covers the gravel in the second cell. Wastewater distributes into cells 1 and 2 like at the CNC; however, less expensive corrugated drain pipe with slotted holes is the collector pipe in each cell. Also, a standpipe with a movable elbow is used in the water level control structure instead of a fixed-length pipe to maintain the desired water level. The discharge pipe from the second cell is plugged since there is no discharge.[7]

A variety of vegetation (e.g., cattail, bulrush, and arrowhead) has been planted in cell 1 and is growing and spreading very well. In the unlined cell 2, water iris survived and some volunteer species have established themselves.

The system was built by a contractor for about $7000 (U.S.), with an estimated 60% for labor and 40% for materials. This cost included about $2000 for septic tanks, making the wetlands cost about $5000.

No data have been collected from the system. However, it has now operated for over 2 years with no discharge from the unlined cell 2 (i.e., cleaned wastewater is percolating subsurface) and without any operational problems.

Home System, Canton, North Carolina

In Haywood County, North Carolina, the TVA developed designs for two residential CW for a cooperative project with Mountain Projects, Inc. (a community development agency), and the county health department. Construction of the first system, for a three-bedroom house, was completed in August 1990. The residence had been served by a septic tank–absorption field system. Wastewater was surfacing routinely from the absorption field due to poor percolating soils and a seasonal high water table. Ditches were dug by the homeowner to drain greywater off the yard to the drainage ditch at the street. A severe potential public health problem existed.

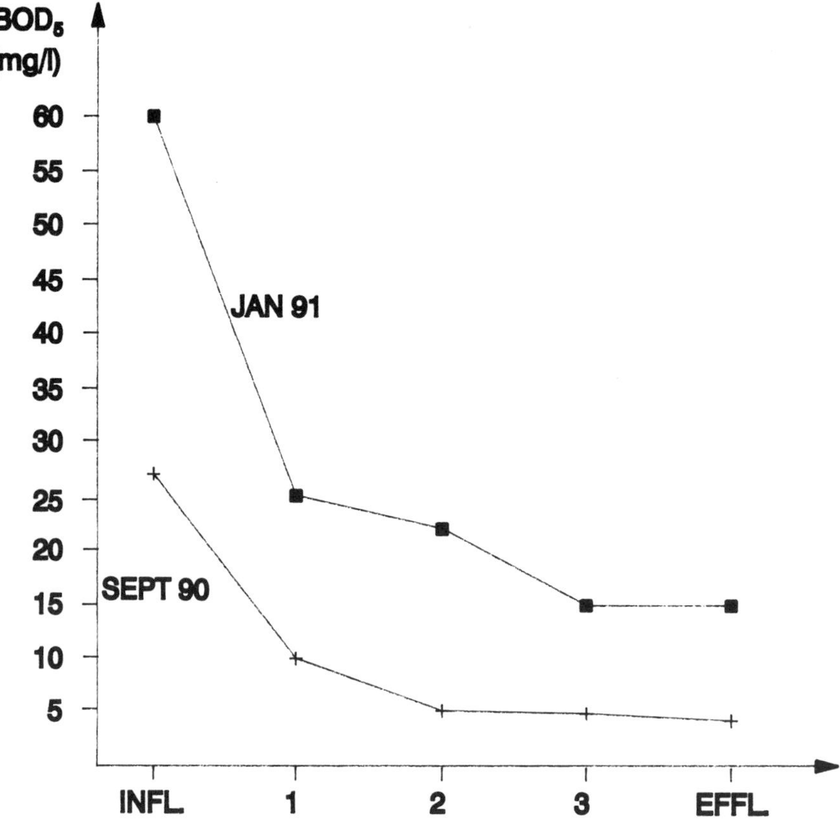

Figure 1. BOD reduction in cell 1 of the Chattanooga Nature Center constructed wetlands.

Since the existing septic tank was in poor condition, two new 3.8-m^3 (1000-gal) concrete tanks in series provide primary treatment. The CW is a subsurface flow type with porous substrate and two series cells. The septic tank effluent gravity flows about 18 m (60 ft) into the first CW cell.

The CW design flow is 1.36 m^3/day (360 gpd) based on guidance from the health department. The effective treatment area is 29 m^2 (314 ft^2), based on a selected area loading rate of 21.3 m^2/m^3/day (0.87 ft^2/gpd). Darcy's Law was used to determine the system dimensions. For a flat bed (0.25% slope for calculation) to minimize excavation, an inlet substrate depth of 45 cm (18 in.) and 244 m/day (800 ft/day) as the substrate hydraulic conductivity, the effective width and system length are 4.9 and 6.1 m (16 and 20 ft), respectively. Two rectangular cells, each 4.9 m (16 ft) wide by 3.0 (10 ft) long, are used.

Instead of earthen berms for cell sides, concrete block retaining walls are used. To prevent groundwater infiltration and wastewater seepage, the concrete blocks are coated with a waterproof sealant. Also, one layer of 6-mil plastic sheeting lines the inside walls. A 10-cm (4-in.) reinforced concrete slab is the bottom of cell 1, making it impermeable. Cell 2 is unlined to allow percolation when the water table was down. The substrate is #57 granite stone which has an average size of 2.5 cm (1/2 in.). It is 45 cm (18 in.) deep in cell 1. In cell 2, 30 cm (12 in.) of the same stone is used, but is covered by topsoil and mulch (8 cm [3 in.] each).

To distribute wastewater across the width of cells 1 and 2, and also collect the effluent in both cells, 10-cm (4-in.) PVC pipe with drilled holes is used. At the discharge of cell 1, a water level control structure with a swivel standpipe is used to adjust water level in cell 1. Any flow from cell 2 would travel through a 15 m (50 ft) long, 0.9 m (3 ft) wide, 0.6 m (2 ft) deep trench filled with 1.3-cm (1/2-in.) granite gravel before discharging into the street drainage ditch.[8]

Vegetation is principally soft-stem bulrush, with blue flag water iris planted around the perimeter of the cells.

The system was contractor built for $4200 (U.S.), of which $2400 was for materials (57%).

Only one sample has been collected from the system since it became operational in August 1990. The effluent BOD of cell 1 and cell 2 were 110 and 26 mg/L, respectively. The effluent of cell 2 had a TSS and

FC of 3 mg/L and 1000 col./100 ml, respectively. There was no significant reduction of nutrients. The sampling procedure may have caused concentrations to be higher than actual. There was no flow when the samples were collected. The water level standpipes were lowered to obtain some flow, and there was a flow surge which may have flushed solids from the cells.

In the first 15 months of operation, the system has operated well. Vegetation establishment is good. There has been no observed discharge from cell 2 into the gravel trench. There are no known operational problems.

Kentucky On-Site Systems

In March 1990, the TVA presented seminars on the CW technology to the Kentucky Department of Health Services. The Department is responsible for on-site wastewater treatment systems in the state and was searching for alternatives where soil conditions are too poor for conventional systems. As of October 1991, 121 CW have been installed and 31 others have been approved for installation. Monitoring is being conducted on a few systems to determine their effectiveness.

Home System, Washington, County, Kentucky

One small CW in Kentucky is for a three-bedroom house located in Washington County. It was designed and is being monitored by the registered sanitarian for the Lincoln Trail District Health Department. The estimated wastewater flow is 1.36 m^3/day (360 gpd). A 3.8-m^3 (1000-gal) septic tank is used for pretreatment. The septic tank effluent must be pumped to the CW. This CW consists of two series cells, each 1.8 m (6 ft) wide by 9.8 m (32 ft) long. The total area is 36 m^2 (384 ft^2), providing an area loading rate of 27 m^2/m^3/day (1.1 ft^2/gpd). Each cell has a 1.0% bed slope and substrate of 3.8 to 6.4 cm (1.5 to 2.5 in.) crushed limestone gravel, 45 cm (18 in.) deep at the inlet ends.

Wastewater enters the first cell under the gravel surface through a 5-cm (2-in.) header with 0.7-cm ($^1/_4$ in.) drilled holes across the cell width. Wastewater is collected in the first cell, and distributed and collected in the second cell with 10-cm (4-in.) headers. Adjustable standpipes at the end of both cells control water levels. Synthetic liners are not used since the soil has a high clay content, not allowing significant percolation. Discharge from cell 2 flows into a 1.2 m (4 ft) wide, 26 m (85 ft) long absorption trench with 30 cm (12 in.) gravel media covered by 30 cm (12 in.) topsoil.

Vegetation planted include cattail and pickerelweed in cell 1 and sweet flag and blue water iris in cell 2. After two growing seasons, there is about 75% vegetation cover in cell 1 and 80% in cell 2.

The system was built by a contractor for about $3300 (U.S.), which included septic tanks, pump, and electrical costs. Construction was completed in January 1990, at which time wastewater was routed to the system. Vegetation was planted in May 1990 and monitoring began in July 1990. There have been no operational problems to date, and the homeowner is pleased with the system.

For the first 16 months of operation, grab sample data for BOD and TSS in septic tank, cell 1 and cell 2 effluents are shown in Figure 2 and summarized below.

	BOD_5 (mg/L)			TSS (mg/L)		
Location	**7/90–10/91**	**7–10/90**	**11/90–10/91**	**7/90–10/91**	**7–10/90**	**11/90–10/91**
ST effl.	199	279	172	61	68	59
Cell 1 effl.	49	34	54	17	15	18
Cell 2 effl.	34	44	31	26	69	16
% removal	83	84	82	57	0	73

For monthly samples between July 1990 and October 1991, the effluent BOD ranged between 4 and 105 mg/L, averaging 34 mg/L. The effluent TSS ranged between 2 and 102 mg/L, averaging 26 mg/L. Cell 1 removed an average BOD of 75%, and cell 2 removed an additional 8%. The average TSS actually increased between cell 1 and cell 2 (from 17 to 26 mg/L). In 5 of 16 samples for both BOD and TSS, cell 2 effluent had a greater concentration than cell 1.

No definite reason is known for the occasional increase of BOD and TSS in the second cell, but at least two possibilities are recognized. First, it is noted that samples in the initial 4 months of operation had the most significant increases from cell 1 to cell 2. Whereas the 16-month average effluent BOD was 34 mg/L, the averages for the first 4 months and the last 12 months were 44 and 31 mg/L, respectively. The effluent TSS for the two periods had a more dramatic difference. The 16-month TSS was 26 mg/L, but the first 4 and final

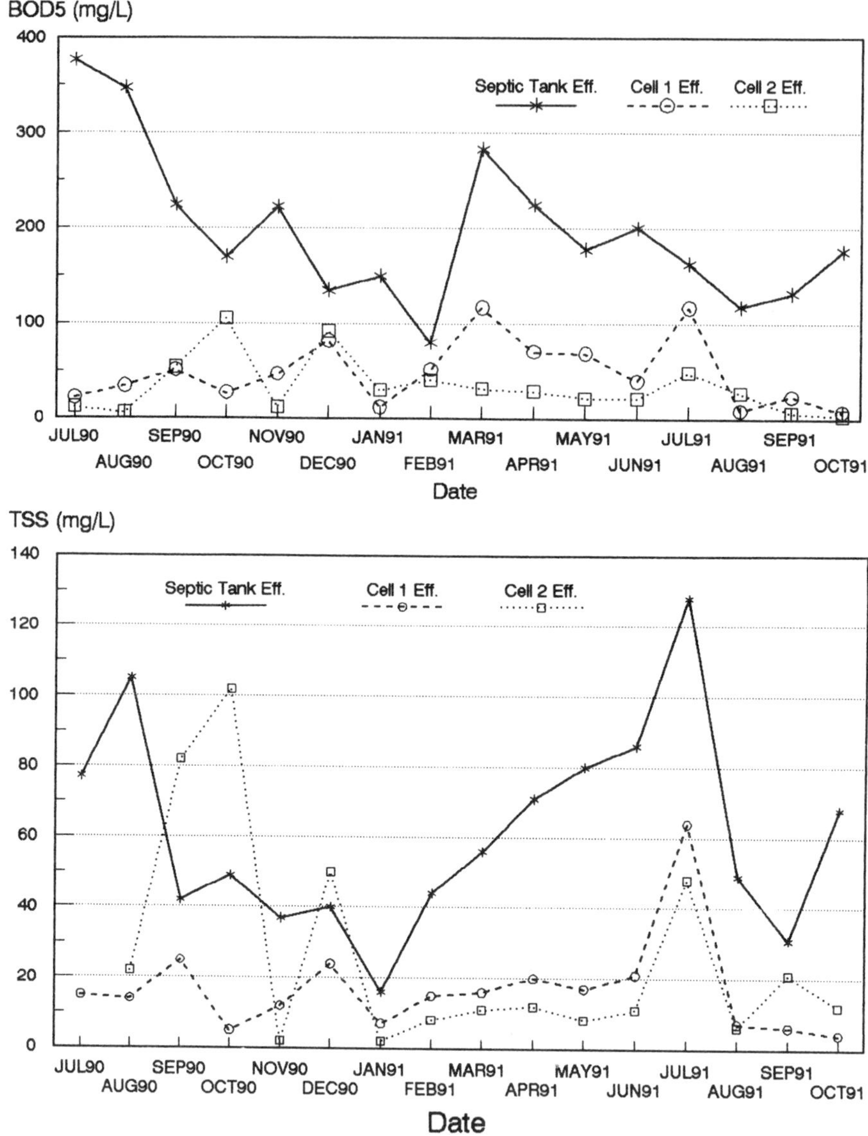

Figure 2. Home constructed wetlands, Washington County, Kentucky: BOD and TSS levels.

12 months were 69 and 16 mg/L, respectively. Therefore, factors during initial operation such as vegetation development and flushing of substrate fines may have caused the greater concentrations. This may emphasize the need to allow the vegetation to establish well in a small CW before wastewater is routinely discharged to it. Also, flow surges occur when the septic tank effluent pump cuts on, possibly causing settled solids to flush from the system.

The reduction of FC was significant; the geometric means of the septic tank, cell 1 and cell 2 effluents were 1500, 140, and 43 col./100 ml, respectively; an average fecal coliform reduction of 97%. For the 16 months, the effluent had 3 values above 200 col./100 ml; 13 values were below 200 col./100 ml.[9,10]

Samples collected for ammonia-nitrogen showed no or minimal reduction through the CW. In most cases, the effluent was equal to the influent.

Other Kentucky Systems

In addition to residences, Kentucky has small CW systems installed for schools, motels, truck stops, restaurants, mobile home parks, recreational areas, and a health department. Monitoring has been conducted for a few, but most are only being observed. Extensive monitoring is being conducted on a second system designed by the Lincoln Trail District Health Department. This system serves an elementary school and the initial results are also encouraging.[11]

FC monitoring on 5 to 10 other systems indicate a high degree of treatment. Most samples at the end of the CW are less than 100 col./100 ml.

Summary of Observed Problems

The major problems have been due to construction. To date, there is a lack of "quality" installers; inexperience may be the problem. Supervision is needed during installation to ensure that designs are strictly followed and "short-cuts" are not taken. In some cases, "dirty" substrate has caused partial plugging; adequate washing of the dirt and fines is mandatory. Leaking liners have occurred especially where inlet and outlet piping penetrate the liner; proper clamping and sealing have been the solution. Seepage under the berms of unlined cells occurred on some initial systems; a sidewall liner must be extended below the original soil surface (about 15 cm, or 6 in.). In some systems, die-off or nondevelopment of vegetation has been observed in the front few feet of the CW; this possible "toxicity" problem needs to be investigated and considered in designs. Cost of some initial systems are believed to be too high due to the installer billing for "experimental" systems; accurate reduced costs will probably occur with experience and time.

CONCLUSIONS

- Alternatives to conventional on-site wastewater treatment systems are needed to help solve and prevent non-point source pollution of groundwater and surface waters. A small CW system is a viable alternative.
- Simple design, construction, and operation guidelines for effective small CW systems are now available.
- The cost of a small CW system depends on local contractor costs and the amount of work done by the owner. The cost for a home CW has ranged from less than $2000 (U.S.) (owner constructed) to about $4000 (contractor constructed). Average costs will probably reduce as contractors gain experience.
- A small CW can produce a discharge with low BOD and TSS concentrations and also significantly reduce pathogen organisms.
- Ammonia-nitrogen reduction in a small CW is variable as in a large CW. More research is needed to enhance ammonia-nitrogen reduction and total nitrogen removal.
- Operators of a small CW are pleased with their performance and attractiveness.
- Wastewater cleaned significantly in a primary CW cell will percolate in an unlined secondary cell at a site having marginal soil percolation.
- Most "problems" to date have been due to construction errors. Construction guidelines are relatively simple but must be closely followed, and supervision is needed during construction.
- Use flexible instead of rigid water level control standpipes since they are easier for the system owner/operator to adjust.
- When retaining walls instead of earthen berms are used for cell sides, construct them properly to prevent sagging, leaks, and resulting problems.
- A small CW can be designed to handle pumped septic tank effluent, but additional work is needed to develop systems which manage flow surges so that solids are not washed from the system.
- The treatment effectiveness of a small CW improves over the first few months. It is advisable to allow good vegetation establishment before wastewater is routinely discharged to a system.
- Site a small CW to provide sufficient sunlight for optimum wetland vegetation growth.
- Use decorative wetland vegetation species in a small CW to enhance its visual appeal.
- Experience through monitoring and observation is needed with small CW systems to provide information on their long-range effectiveness.

REFERENCES

1. Kipp, S. Leaking septic tanks pose hefty health risk, *The Birmingham Post-Herald* (Birmingham, AL), March 19, 1991.
2. Loomis, J. Drinking water like gambling, *The Decatur Daily* (Decatur, AL), March 17, 1991.
3. Steiner, G. R., J. T. Watson, and K. D. Choate. General Design, Construction, and Operation Guidelines: Constructed Wetlands Wastewater Treatment Systems for Small Users Including Individual Residences. Tennessee Valley Authority, TVA/WR/WQ—91/2, March 1991.

4. Choate, K. D. and G. R. Steiner. Engineering Design, Chattanooga Nature Center Constructed Wetlands, Chattanooga, Tennessee. Tennessee Valley Authority, March 1989.
5. Watson, J. T. and T. H. Davies. Unpublished TVA data, 1990.
6. Von Uklanski, P. and S. Becker. Unpublished TVA data, 1991.
7. Steiner, G. R. Conceptual Engineering Design for Constructed Wetlands Wastewater Treatment System, Memphis Museum Coon Creek Science Center, Chester County, Tennessee. Tennessee Valley Authority, March 1989.
8. Choate, K. D. Constructed Wetlands Design Guidelines, Residential Septic Tank System, Canton, North Carolina. Tennessee Valley Authority, June 1990.
9. Terry, T. B. Designer Swamps, Lincoln Trail District Health Department, Hodgenville, Kentucky, February 1991.
10. Terry, T. B. unpublished data and personal communication, 1991.
11. Terry, T. B. Constructed Wetlands Wastewater Quality Improvement at Lynnvale Elementary School, presented at the *Int. Symp. Constructed Wetlands for Water Quality Improvement*. Pensacola, FL, October 21–24, 1991 (to be included in proceedings).

CHAPTER 55

General Design, Construction, and Operation Guidelines for Small Constructed Wetlands Wastewater Treatment Systems

G. R. Steiner, J. T. Watson, and K. D. Choate

INTRODUCTION

Overloaded and poorly operated wastewater treatment systems cause water quality problems and regulatory headaches, and often hinder community development. Also, unsatisfactory performance of conventional on-site systems in rural and non-sewered areas having poor soil conditions commonly contribute to non-point source pollution and groundwater contamination. To correct these problems small users especially need practical wastewater treatment alternatives that are affordable, effective, simple, reliable, and aesthetically pleasing.

Recognizing this need, the Tennessee Valley Authority (TVA) in 1986 began demonstration of the constructed wetlands technology as a treatment alternative for small communities. Demonstration results to date indicate that a constructed wetlands treatment (CW) system can be designed, constructed, and operated to provide a very high-quality, secondary-level effluent for BOD (biochemical oxygen demand) and suspended solids removal and fecal coliform reduction. Nutrient reduction can also be attained, but technology development is still needed on this aspect.[1]

As the process has been investigated, the TVA and others are realizing the potential of CW for treating various types and volumes of wastewaters. Besides large municipal systems, CW can be down-sized for small users, such as for schools, camps, and even individual homes.

Detailed, state-of-the-art, easy to follow guidelines for designing, constructing, and operating CW for small wastewater flows are needed for engineers, sanitarians, and developers. Based on its experience, the TVA recently prepared detailed guidelines.[2] These guidelines are summarized in this article.

GENERAL

CW is a relatively new technology. These guidelines will be modified as improved information is developed.

A CW is very site specific. Designs should be prepared and reviewed for each site. Design examples should not be used to "cookbook" a design. Designs must be coordinated with and approved by local and state officials.

Figure 1 is a cutaway perspective of one optional CW configuration, illustrating components and processes.

PRETREATMENT

Pretreatment prior to the CW is necessary to remove coarse and heavy solids. A septic tank(s), properly sized, built, and maintained according to appropriate state/county criteria will provide good pretreatment.

0-87371-550-0/93/$0.00+$.50

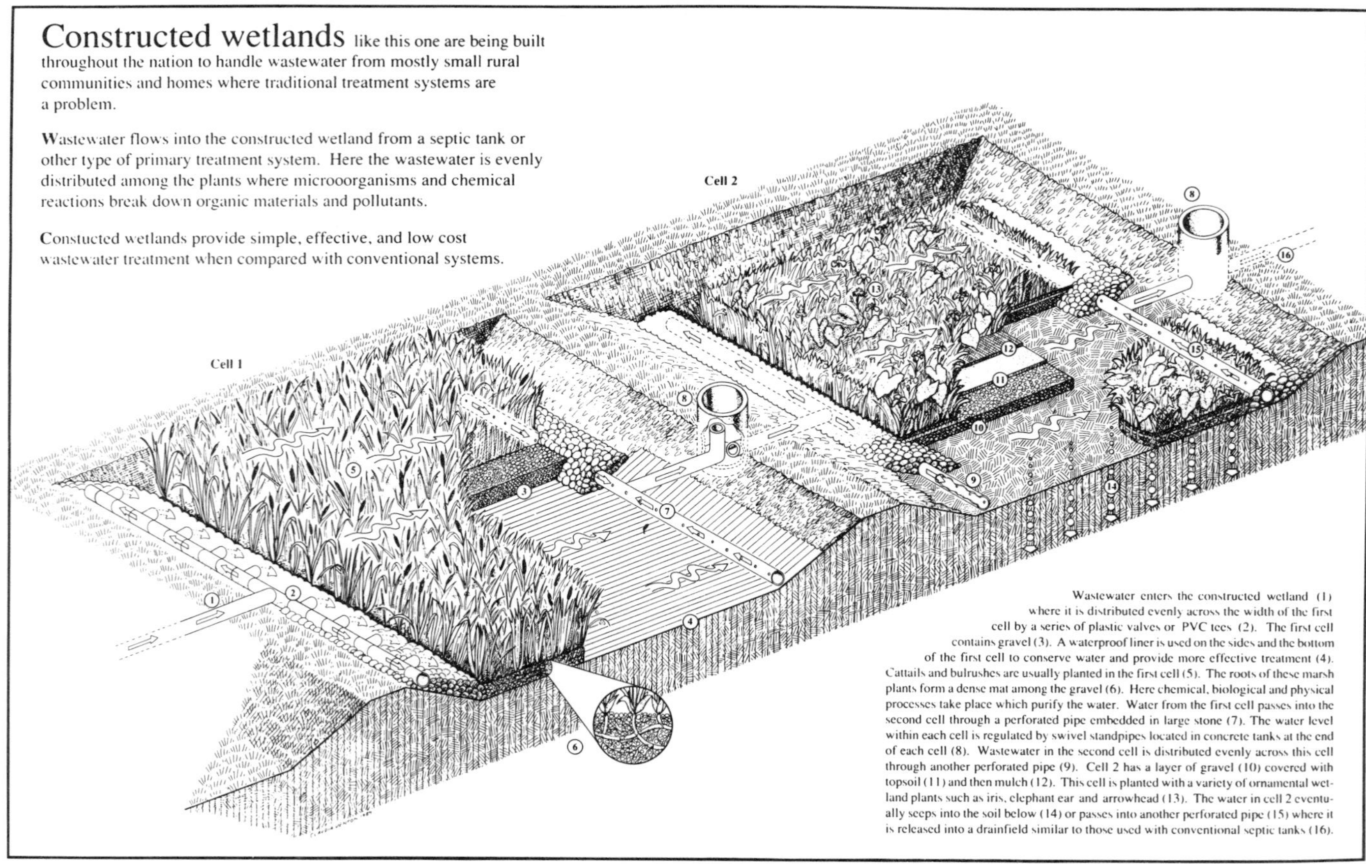

Figure 1. Cutaway perspective of a constructed wetlands system.

Water conservation measures, such as the use of low-flow plumbing fixtures, are recommended to minimize flow to the CW. For example, efficient ultra-low flush toilets using only 6 l (1.5 gal) per flush are available.

HYDRAULIC AND ORGANIC LOADING

Hydraulic Loading

Determine hydraulic loading in cubic meters per day (m^3/day), or gallons per day (gpd) based on the required flow per bedroom for home systems and flow per person or per fixture for other small systems. These rates are established by each state. For a home, a typical rate in the Tennessee Valley is 0.45 m^3/day (120 gpd) per bedroom.

Organic Loading

Determine the organic load in kilograms per day (kg/day) BOD_5 (or pounds per day [lb/day]) based on established criteria and assuming the amount decreased by pretreatment. For example, the average daily organic loading per person in a home is 0.077 kg (0.17 lb) BOD_5. A septic tank will remove about 40% BOD.[3] Therefore, a household would produce a BOD loading from the septic tank of about 0.046 kg/day (0.1 lb/day) per person.

BASIS OF CW DESIGN AND CONSTRUCTION

CW Type

Potential odor, mosquito, and public health problems must be minimized for a small CW, which would generally be built in close proximity to its user (e.g, residences, schools, or businesses). A CW can be either a surface flow system having visible standing water, a subsurface flow system containing porous substrate and designed and operated to prevent standing water, or a combination of the two. Since subsurface flow will eliminate or minimize the potential problems arising from standing water as discussed above, these guidelines provide criteria for subsurface flow CW only.

Configuration

A small CW can be a single cell, two cells in series, or multiple cells in parallel and/or series. A single cell system should be used on sites where the treated wastewater cannot be disposed of by percolation because of too high flows for the available site, a high groundwater table, shallow soil above impermeable rock, or very impermeable clay.

Two cells in series can be used at sites having soil that will allow percolation of treated wastewater. The first cell is lined, but the second cell is unlined to allow percolation of the treated wastewater and preclude or minimize a surface discharge.

A multiple-cell CW will allow operational flexibility. It can also be used to accommodate the natural topography (e.g., to prevent unrealistic cut and fill for a single cell).

Dimensions

Surface Area

Use a surface hydraulic loading criteria of 31.9 m^2 total surface area per cubic meter per day (1.3 ft^2/gpd) for a 30-cm (12-in.) deep bed to determine the surface area of a subsurface flow CW.

Cross-Sectional Area

The cross-section of the bed should be sized so that wastewater will not surface. Both hydraulic and organic loads should be considered, with the largest (most conservative) size used.

The substrate pore spaces must be able to pass the hydraulic flow physically. Use Darcy's Law, which considers flow through a permeable substrate, to calculate area for a given flow rate, a selected bottom bed slope of up to 1% which will approximate the hydraulic gradient, and assuming a conservative long-term hydraulic conductivity (259 m/day, or 850 ft/day). For a flat-bottom system (0% slope), assume a hydraulic gradient of 0.25%. For sloping lots, use bed slopes up to 2% to minimize cut and fill.

A too high organic load can cause substrate plugging, apparently due to formation of anaerobic slimes. An EPA contractor has observed that CW subsurface flow beds experience partial surface flow if the organic loading exceeds approximately 0.49 kg BOD per square meter cross-sectional area (0.1 lb/ft^2/day), or 2.04 m^2/kg/day (10 ft^2/lb/day).[4] A safety factor of 2 is recommended; therefore, use an organic loading criteria of 4.08 m^2/kg/day (20 ft^2/lb/day).

Depth

A 30-cm (12-in.) front end substrate depth is recommended. A shallow cell is believed to allow better wastewater reaeration than a deeper cell. Also, most root biomass will be in the top part of the substrate where more oxygen is available. If available surface area is limited, a 45-cm (18-in.) depth may be used, but a smaller surface area criterion (21.3 m^2/m^3/day [0.87 ft^2/gpd]) must then be used.

A sloped bed with a flat surface will have a deeper substrate depth at the effluent end. Limit influent and effluent depth difference to 15 cm (6 in.).

Width

Determine the effective width (cell bottom) from the cross-sectional area and depth. The cell surface will be wider if the cell berms are sloped.

Length

Determine the effective cell length (cell bottom) from the surface area and width. For two series cells, each cell length is half the system length.

Liner

Provide an impermeable liner to prevent seepage out of the cell and infiltration into the cell, and to allow water level control. Compacted clay, synthetic (20 to 30-mil polyvinyl chloride or polyethylene), or concrete and sealed concrete blocks can be used. Remove rocks, roots, and debris that might puncture a synthetic liner, and lay the liner on geotextile fabric or a layer of sand. Also, cover the synthetic liner to prevent UV deterioration.

For a single cell CW, install an impermeable liner on the bottom and sides of the cell. For a two-cell CW, line the first cell only; the second cell is unlined to allow for percolation. Seal liners where the influent/effluent pipe enters/exits the cell(s).

Substrate

Common substrate is sized, washed gravel. River rock which has rounded surfaces is preferred, in order to reduce substrate compaction. Crushed limestone is not recommended since it easily compacts. Use it only if it is the only available alternative. Sand or a gravel/sand mixture are other options.

Use pea gravel or small gravel about 1.3 cm (1/2 in.) diameter for the main substrate. (Graded gravel up to 5 cm (2 in.) diameter has been successfully used, but it is now believed that smaller substrate size will better enhance vegetation growth.) Sand should be coarse to very coarse.

Use at least 5 to 10 cm (2 to 4 in.) stone for a 60-cm (24-in.) length around influent distributor and effluent collector pipes to reduce clogging potential.

Use only washed substrate to eliminate fines which can plug substrate pore spaces and possibly cause surface flow.

Make the top of the substrate surface flat to facilitate water level control, vegetation planting and growth, and prevent formation of stagnant pools.

Mulch the substrate surface to help control potential odors, prevent reflective sun scalding of vegetation, and for visual aesthetics.

For the second cell of two series cells, use 30-cm (12-in.) coarse sand covered with 8 cm (3 in.) mulch. The sand/mulch is critical for plant survival during dry periods.

Pumping and Piping

Pumping to the CW is required when wastewater cannot gravity flow. Select a pump which will provide a minimal pumping rate and dosing volume which will prevent/minimize surface flow in the CW during pumping.

Use header pipes to provide uniform wastewater distribution and collection. Standard Schedule 40, 10-cm (4-in.) PVC water supply pipe with 2.5-cm (1-in.) diameter drilled holes or perforated drain pipe may be used. Place the influent header below the surface of the larger substrate and the effluent collector on the bed bottom, preferably in a shallow trench below the bed bottom.

Water Level Control

Water level control and adjustment is critical to establish and grow vegetation and prevent standing water. The cell discharge structure should incorporate water level control piping (swivel standpipe or flexible tubing) at the effluent end of the cell(s) for manipulating the water level. It should allow total draining of the bed for maintenance to flooding for intermittent weed control.

Berms

Earthen berms or a retaining wall around the cell(s) will retain wastewater inside and keep surface run-off out of the CW. Construct the top of the berm/retaining wall at least 15 cm (6 in.) above the CW substrate surface and at least 15 cm (6 in.) above the existing ground surface. Construct and line earthen berms to prevent seepage, and with exterior and interior slopes of 3H:1V and 2H:1V, respectively, and a 60-cm (24-in.) minimum top width. Retaining walls built with lined or sealed concrete blocks, crossties, landscaping timbers, or other strong, durable materials can be used to conserve space or for terracing.

Effluent Disposal

A small system (e.g., a house) may have no discharge due to evaporation, transpiration, and percolation water losses. However, where all water cannot be eliminated, excess water may be routed to a small absorption field, small sand or gravel filter trench, or used for landscaping, if approved by the appropriate agency. Larger systems having point source discharges must be approved and permitted by the state agency.

Miscellaneous

Divert surface water away from the CW. Install a french drain around the CW if necessary to drain away high groundwater. Enclose the CW with a fence if there are safety, trespassing, and sanitary concerns.

DESIGN EXAMPLES

Two design examples indicate how the recommended guidelines are to be used to determine CW dimensions. It is emphasized that these examples should not be used to "cookbook" a design for a specific system.

Example 1: Three-Bedroom House

Conditions and Assumptions

- System is for a three-bedroom house occupied by four people. The site has a relatively flat yard and area is not constrained. The existing absorption field failed due to poor soils having a slow percolation rate (exceeds 24 min/cm, or 60 min/in.). Pretreatment will be by septic tank (ST).
- Hydraulic load: Q = 0.45 m^3/day per bedroom × 3 rooms = 1.35 m^3/day (360 gpd)
- Organic load = 0.077 kg/day per person × 4 people × 0.6 (40% ST reduction) = 0.185 kg/day (0.4 lb/day)
- Hydraulic loading criteria = 31.9 m^2/m^3· day (1.3 ft^2/gpd)
- Organic loading criteria = 4.08 m^2/kg BOD per day (20 ft^2/lb/day)

- Use a two-series cell, with the primary cell lined and the secondary cell unlined to allow percolation of the treated water.
- Use a relatively flat bed slope, S, of 0.5%.
- Assume a hydraulic conductivity, K_s, of 259 m/day (850 ft/day).

Dimensions

- CW surface area, A_s:
  A_s = 1.35 m^3/day × 31.9 m^2/m^3/day = 43.1 m^2 (464 ft^2)
  Use one half, or 21.6 m^2 (232 ft^2), for each cell.
- Cross-sectional area, A_x:
  Based on hydraulic loading using Darcy's Law:
  $Q = A_x \times K_s \times S$
  where
  K_s = substrate hydraulic conductivity (long term), S = hydraulic gradient (assume equivalent to bed slope), and $A_x = Q/(K_s \times S) = 1.35/(259 \times 0.005) = 1.04$ m^2 (11.2 ft^2)
  Based on organic loading:
  A_x = 4.08 m^2/kg BOD per day × 0.185 kg BOD per day = 0.75 m^2 (8.1 ft^2)
  Use A_x = 1.04 m^2 (greater with hydraulic loading)
- Cell depth, D
  Use inlet depth of 30 cm for each cell. With the slope, the outlet depths will be slightly greater.
- Cell width, W
  $W = A_x/D = 1.04/0.3 = 3.5$ m (11.5 ft) each cell
- Cell length, L
  $L = A_s/W = 21.6/3.5 = 6.2$ m (20.3 ft) each cell

Note:

Construct the second cell at a lower elevation than the first cell to allow for independent water level control in each cell. Ideally, the top of the second cell should be at or below the bottom of the first cell so that lowering the first cell water level will not be restricted by the second cell water level.

The system's design is illustrated in Figure 2.

Example 2: Resort

Conditions and Assumptions

- System is for a resort with 10 cottages, a 16-room motel, and a 50-seat restaurant. Site is relatively flat and area is not constrained.
- Hydraulic load, Q = 37.8 m^3/day (10,000 gpd)
- Organic load = 8.2 kg/day × 0.6 (40% ST reduction) = 4.9 kg/day (10.8 lb/day)
- Hydraulic loading criteria = 21.3 m^2/m^3/day (0.87 ft^2/gpd)
- Organic loading criteria = 4.08 m^2/kg BOD per day (20 ft^2/lb/day)
- Use two parallel cells to provide operational flexibility.
- Use a bed slope, S, of 1.0%.
- Assume a hydraulic conductivity, K_s, of 259 m/day (850 ft/day).

Dimensions

- CW surface area, A_s:
  A_s = 37.8 m^3/day × 21.3 m^2/m^3/day = 805 m^2 (8662 ft^2)
  Use equal areas for each cell, 403 m^2 (4331 ft^2).
- Cross-sectional area, A_x:
  Based on hydraulic loading using Darcy's Law, each cell receiving 18.9 m^3/day (5000 gpd):
  $Q = A_x \times K_s \times S$, where
  K_s = substrate hydraulic conductivity (long-term)
  S = hydraulic gradient (assume equivalent to bed slope)
  $A_x = Q/(K_s \times S) = 18.9/(259 \times 0.01) = 7.3$ m^2 (79 ft^2)

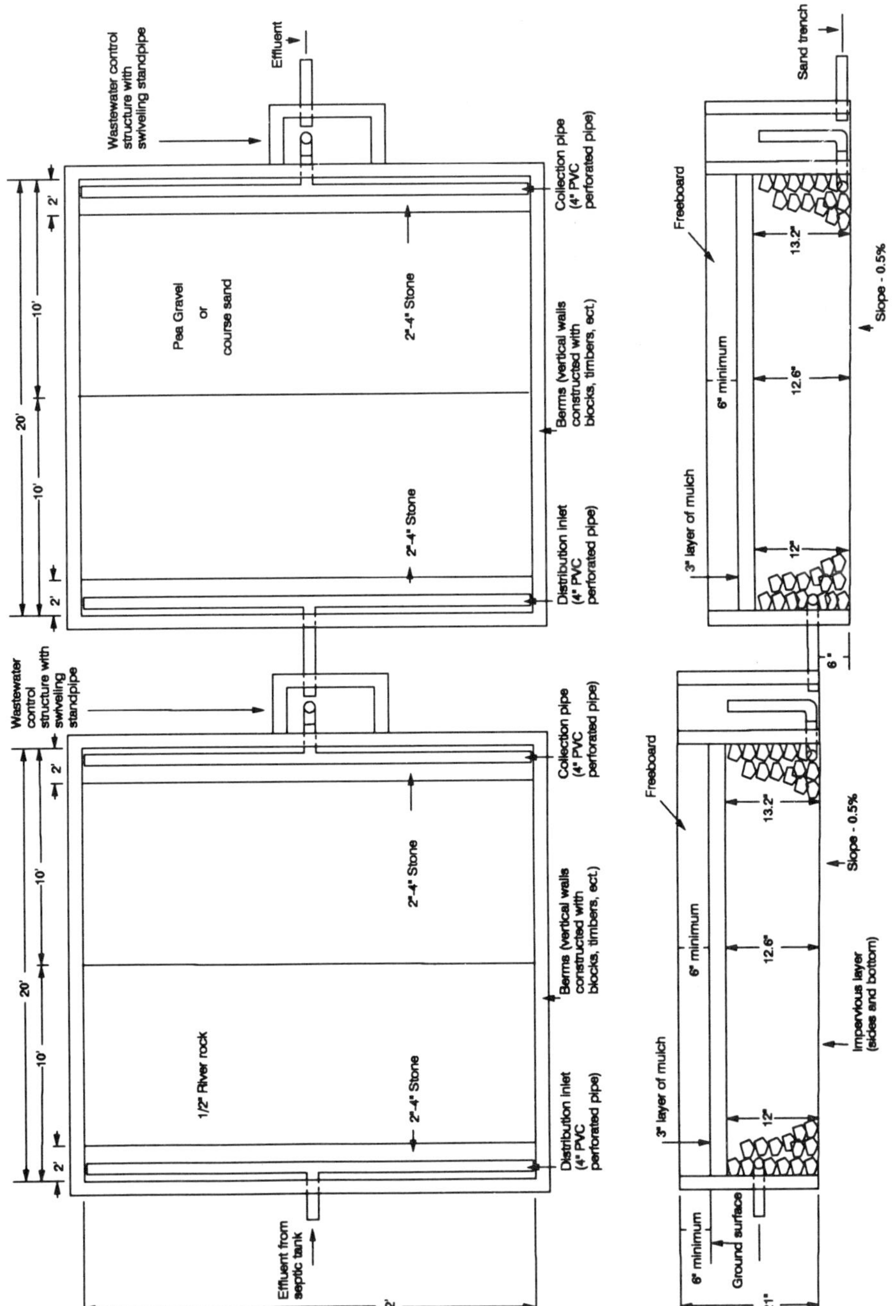

Figure 2. Design Example 1 — constructed wetlands for a three-bedroom house.

Based on organic loading, each cell receiving 2.45 kg BOD/d: A_x = 4.08 m^2/kg BOD per day × 2.45 kg BOD per day = 10 m^2 (108 ft^2)
Use A_x = 10 m^2 (greater with organic loading).

- Cell depth, D:
  Use inlet depth of 45 cm (18 in.) in each cell.
  Outlet depth, based on cell length (e. below) and 1.0% bed slope, will be: 45 cm + 0.01 × 18.2 m = 63 cm (25 in.) each cell
- Cell width, W:
  W = A_x/D = 10/0.45 = 22.2 m (73 ft) each cell
- Cell length, L:
  L = A_s/W = 403/22.2 = 18.2 m (60 ft) each cell

VEGETATION

Species

Use plant species that grow naturally within the region. Preferred species include, but are not limited to, *Typhaceae* (cattail family), *Cyperaceae* (sedge family), *Graminaceae* (grass family), *Junaceae* (rush family), *Scirpus validus* (softstem bulrush), and *Phragmites australis* (giant reed; considered "noxious" in some areas due to its aggressive growth).

Decorative, flowering species can also be used for visual appeal around the cell perimeter. These can include *Iris pseudacorus* (water iris), *Sagittaria latifolia* (arrowhead), *Acorus* sp. (sweet flag), *Safittaria lancifolia* (grassy arrowhead), *Pontederia cordata* (pickerelweed), *Canna flaccida* (canna lily), *Colocasia esculenta* (elephant ear), and *Zantedeschia aethiopica* (calla lily).

Most species require full sunlight for at least 6 hours. For shady locations, select shade-tolerant species such as ferns and *Sagittaria latifolia* (arrowhead). Some species may not be cold weather tolerant and will need to be replanted each spring.

Planting

Plant vegetation in early spring to early fall to obtain good root development which will reduce winter mortality. Plant as close as possible to obtain rapid vegetation cover. Preferably, use plants with a 15- to 30-cm (6- to 12-in.) stalk above the roots. Place roots in the water and keep the stalk above the water surface.

OPERATION AND MAINTENANCE

CW for small wastewater flows are not maintenance free. Casual observations and some care by the owner/operator are necessary to maintain an effective, attractive system.

Start-Up

If an existing septic tank is used for pretreatment, check the condition of sludge and scum accumulation and repair and clean as necessary.

If possible, allow at least 4 to 6 weeks before discharging wastewater regularly into a new CW. Fill the CW with water or wastewater, maintain the water level, and add liquid fertilizer for good plant growth.

Water Level

- Maintain the water level about 1 in. below the substrate surface. Adjust the water level using the water level control structure. Maintain the water level during extended periods of no flow (e.g., long vacations). Eliminate leaks from pipe joints in the level control structure.
- If surface ponding occurs and cannot be controlled by water level adjustment, check (1) clean-out in front of the CW for excessive solids carryover from the septic tank which may have plugged substrate; and (2) for excessive water flow which exceeds the hydraulic capacity of the substrate. Take necessary actions.

Berms and Liner

- Check for holes or erosion in earthen berms and repair. Check and repair leaks around berms/retaining walls by plugging, sealing, etc. Mow earthen berms or around retaining walls to maintain an attractive site.
- Maintain cover over the sides of synthetic liners which extend above the substrate and water level to prevent UV degradation.

Vegetation

- Check vegetation for signs of disease or other stress. (Natural seasonal changes will occur with temperature and length of day changes). If the water level is satisfactory, obtain horticultural guidance.
- Check for insect damage. Manually pick large insects. For serious insect infestation, apply a chemical agent after obtaining horticultural guidance.
- Add a balanced liquid fertilizer periodically.
- Replace dead plants as necessary to fill voids. Remove weeds, trees, and shrubs which will shade and crowd the desired plants. Remove mature wetland vegetation after the plants have browned in the fall by cutting about two thirds of the height of the plants.
- Encourage deep root growth by slowly lowering the water level over several weeks during the dormant vegetation period.
- Divide and replant decorative flowering species (e.g., iris) to enhance the system attractiveness.

Odor Control

- Standing water on the substrate surface is the probable cause of objectionable odor. Level low and high spots. Maintain the water level about 2.5 cm (1 in.) below the substrate surface.

Miscellaneous

- Maintain pump according to manufacturer's specifications.
- Do not empty household chemicals into the sanitary system, which can upset the septic tank and damage/kill the vegetation.
- Do not apply herbicides and pesticides near the CW which can damage vegetation.
- Maintain a 3-in. mulch layer on top of the substrate.
- Reroute surface drainage around/away from the CW.
- Prevent children from playing in the system to avoid contact with potentially infectious microorganisms.
- Prevent pets and animals from digging in the wetlands, which can damage vegetation and create holes in the substrate.
- Contact local county health department for guidance if any unusual problem is noted.

ADDITIONAL INFORMATION

Detailed guidelines may be obtained from the Tennessee Valley Authority, Water Quality Department, Haney Building 2C, 1101 Market Street, Chattanooga, Tennessee 37402-2801, or telephone 615/751-3164.

REFERENCES

1. Choate, K. D., J. T. Watson, and G. R. Steiner. Demonstration of Constructed Wetlands for Treatment of Municipal Wastewater, Monitoring Report for the Period: March 1988 to October 1989. Tennessee Valley Authority, TVA/WR/WQ—90/11, August 1990.
2. Steiner, G. R., J. T. Watson, and K. D. Choate. General Design, Construction, and Operation Guidelines: Constructed Wetlands Wastewater Treatment Systems for Small Users Including Individual Residences. Tennessee Valley Authority, TVA/WR/WQ—91/2, March 1991.
3. Design Manual — Onsite Wastewater Treatment and Disposal Systems, U.S. Environmental Protection Agency, 50, October 1980.
4. Reed, S. C. personal communication and unpublished data, 1990.

CHAPTER 56

TVA's Constructed Wetlands Demonstration

K. D. Choate, J. T. Watson, and G. R. Steiner

INTRODUCTION

The Tennessee Valley Authority (TVA) has implemented a municipal wastewater demonstration project in western Kentucky. Combined city, state, and TVA-appropriated funds were used to build three constructed wetland systems at Benton, Hardin, and Pembroke (Figure 1). Demonstration objectives include evaluating relative advantages and disadvantages of these three constructed wetland systems; determining permit compliance ability; developing, evaluating, and improving basic design and operation criteria; evaluating cost effectiveness; and transferring technology to others.

A demonstration monitoring project was implemented with a partnership of funds from the Environmental Protection Agency (EPA) Region IV, other EPA funds through the National Small Flows Clearinghouse (NSFC), and TVA appropriations. TVA is managing the 3-year project in cooperation with an inter-agency team consisting of EPA, Kentucky Division of Water, and NSFC.

This article describes the concept and status of each constructed wetland treatment system and summarizes the monitoring data for the March 1988 through August 1991 period. Recent operating revisions and activities are also discussed. Additional information on these constructed wetland systems is presented in other reports.[1,2,3]

SITE DESCRIPTIONS

Benton, Kentucky

A wetlands with a design capacity of 3785 m^3/day (1 mgd) was constructed at Benton to polish effluent from the existing two-cell (series) 10.5 ha (26-acre) lagoon system. The lagoon system was frequently overloaded hydraulically and occasionally exceeded National Pollutant Discharge Elimination System (NPDES) permit limits for various parameters. The 4-ha (10-acre) second lagoon cell was modified into a three-cell (parallel) constructed wetland (Figure 1) designed to receive effluent from the primary 6.5-ha (16-acre) lagoon. The three cells are equally sized. Cells 1 and 2 were designed to operate as surface flow systems, with each receiving 25% of the total flow, while cell 3 was designed to operate as a subsurface flow system and receive 50% of the total flow.

Each cell contains different combinations of substrates and plants. Cells 1 and 2 have substrates of native clay to provide surface flow. Cell 1 is planted mostly with *Typha latifolia L.* (cattail), along with a few decorative plant species. Cell 2 is planted with *Scirpus cyperinus (L.) Kunth* (woolgrass bulrush), *Scirpus validus Vahl.* (soft-stem bulrush), and cattail. Cell 3 has an average of 0.61 m (2 ft) crushed limestone to provide subsurface flow and is planted with soft-stem bulrush. Volunteer species also exist in each cell.

Plant establishment has been variable from year to year in all three cells. The woolgrass bulrush initially grew in clumps without spreading, and has mostly disappeared. Cattails in the lower end of cells 1 and 2, have also mostly disappeared. Contributive factors include depredation by muskrats, deep water, and a heavy clay bottom that hinders rooting. Soft-stem bulrush in the upper two thirds of cell 3 has also declined with time and is being replaced with other species.

0-87371-550-0/93/$0.00+$.50

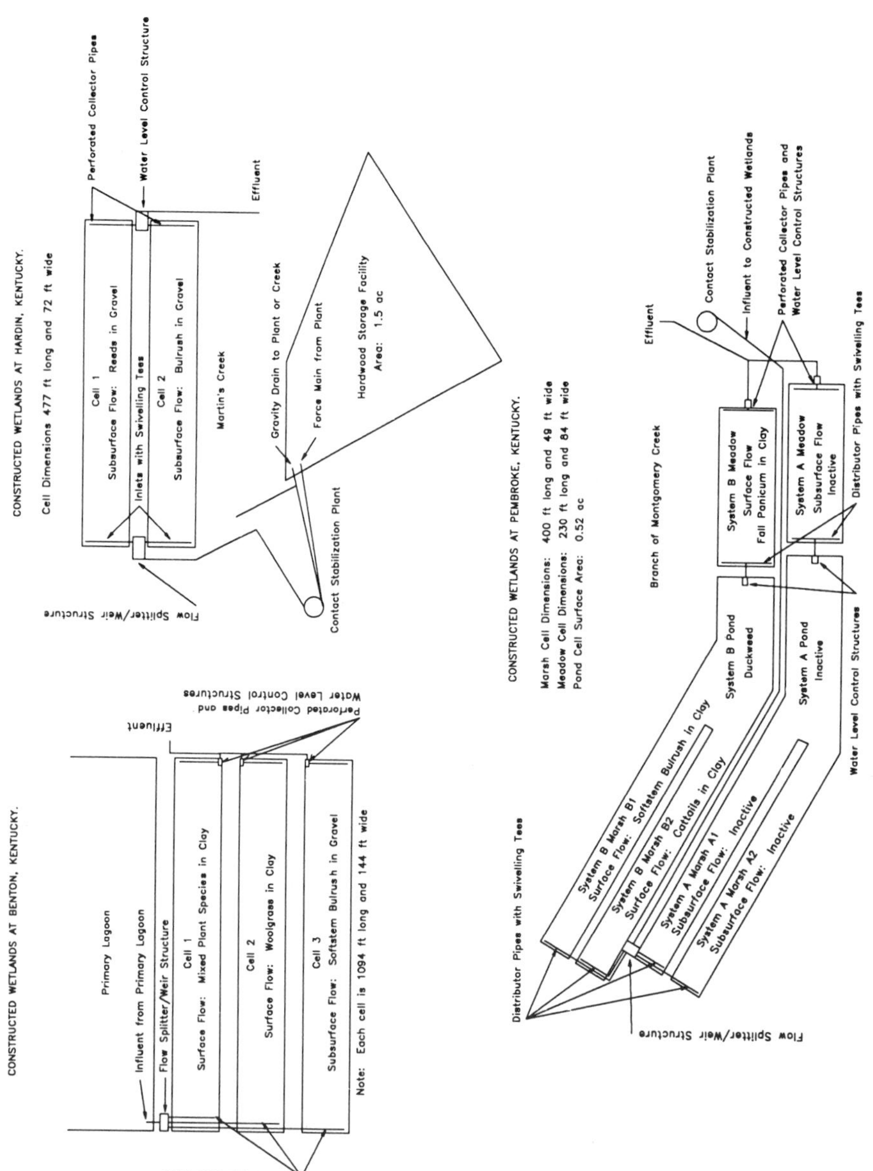

Figure 1. Constructed wetlands at Benton, Hardin, and Pembroke, Kentucky.

Hardin, Kentucky

A 379-m^3/day (0.1-mgd) gravel marsh system (Figure 1) was built to polish effluent from the package contact stabilization plant. The 265-m^3/day (0.07-mgd) package plant was overloaded hydraulically by high inflow/infiltration and was physically deteriorated.

The constructed wetlands consist of two equally sized subsurface flow cells that receive equal flow. The subsurface flow cells contain an average depth of 0.61 m (2 ft) of 1.3 cm ($^1/_2$ in.) diameter crushed limestone and river gravel. A diked hardwood storage area was designed to manage excess flow during high storm events. However, retaining water for extended periods has resulted in death of most of the trees.

Each cell contains different plant species. Cell 1 was planted with *Phragmites australis* (common reed), while cell 2 was planted with soft-stem bulrush. Volunteer species such as *Polygonum* (smartweed) have also established. Reeds were replanted several times before achieving good survival. At the present time, vegetation is very dense in both cells.

The package plant has not provided consistent influent concentrations to the constructed wetlands due to poor operation and high inflow/infiltration. Solids flushed from the plant have partially clogged the front portion of the two cells, resulting in surface flow in the inlet area. The solids are activated sludge floc carry-over from the final clarifier, resulting primarily from sludge bulking caused by several conditions, including inadequate oxygen, improper sludge recycle, and wasting operations. High flow during rainy periods exacerbates the carry-over problem. In order to prevent the wetland cells from being filled completely with solids, the package treatment plant has been extensively up-graded with the help of the Purchase Area Development District.

Pembroke, Kentucky

A wetland with a capacity of 341 m^3/day (0.09 mgd) was constructed to treat wastewater from the existing 265-m^3/day (0.07-mgd) contact stabilization package plant. The constructed wetland is a marsh-pond-meadow system, consisting of two parallel and independent systems, each having in series two parallel marsh cells, a pond, and a meadow (Figure 1). Designs of the two systems are the same except for substrates and marsh vegetation. System A contains gravel substrate in the marsh and meadow cells, while system B contains natural clay for surface flow.

Originally, only system B was planted and placed in operation. Cattails and soft-stem bulrushes were planted in the surface flow marshes. The first planting was successful in both marshes. Duckweed was seeded in the pond and multiplied rapidly. *Phalaris arundinacea* L. (reed canary grass) was sown several times in the meadow, but was never successful. Volunteer species became very dense.

The package plant provided the same problems as discussed for the Hardin constructed wetland. A sludge blanket exists throughout the entire length of the marsh cells, with the sludge depth decreasing with length of cell bed. Natural grasses cover the sludge deposits in the inlet area of both marshes, and bulrush has been observed sprouting in the sludge.

The system is currently being operated as a "no discharge" system. Recirculated meadow B effluent is pumped to pond A. Evapotranspiration and seepage losses have prevented any final discharge since the spring of 1991, with one exception following a heavy rain event.

RESULTS

Influent and effluent concentrations are summarized in Figures 2A through 2E for each constructed wetland system. Influent concentrations (effluent from conventional treatment plants) for parameters of concern were: 5-day biochemical oxygen demand (BOD_5), total suspended solids (TSS), total phosphorus (TP), ammonia-nitrogen (NH_3-N), organic nitrogen, total nitrogen (TN), and fecal coliform (FC). These varied considerably for Hardin and Pembroke, which have package treatment plants. Benton influent concentrations (effluent from lagoon) were lower and more stable than Hardin and Pembroke for all parameters.

All three systems experienced high inflow/infiltration, with total flows during the monitoring period ranging from 853 to 9832 m^3/day (0.2253 to 2.5973 mgd) for Benton, 168 to 1251 m^3/day (0.0444 to 0.3331 mgd) for Hardin, and 143 to 1376 m^3/day (0.0378 to 0.3636 mgd) for Pembroke. Although Benton had the largest total flow range, high inflow/infiltration affected the influent concentrations to the constructed wetlands more at Hardin and Pembroke.

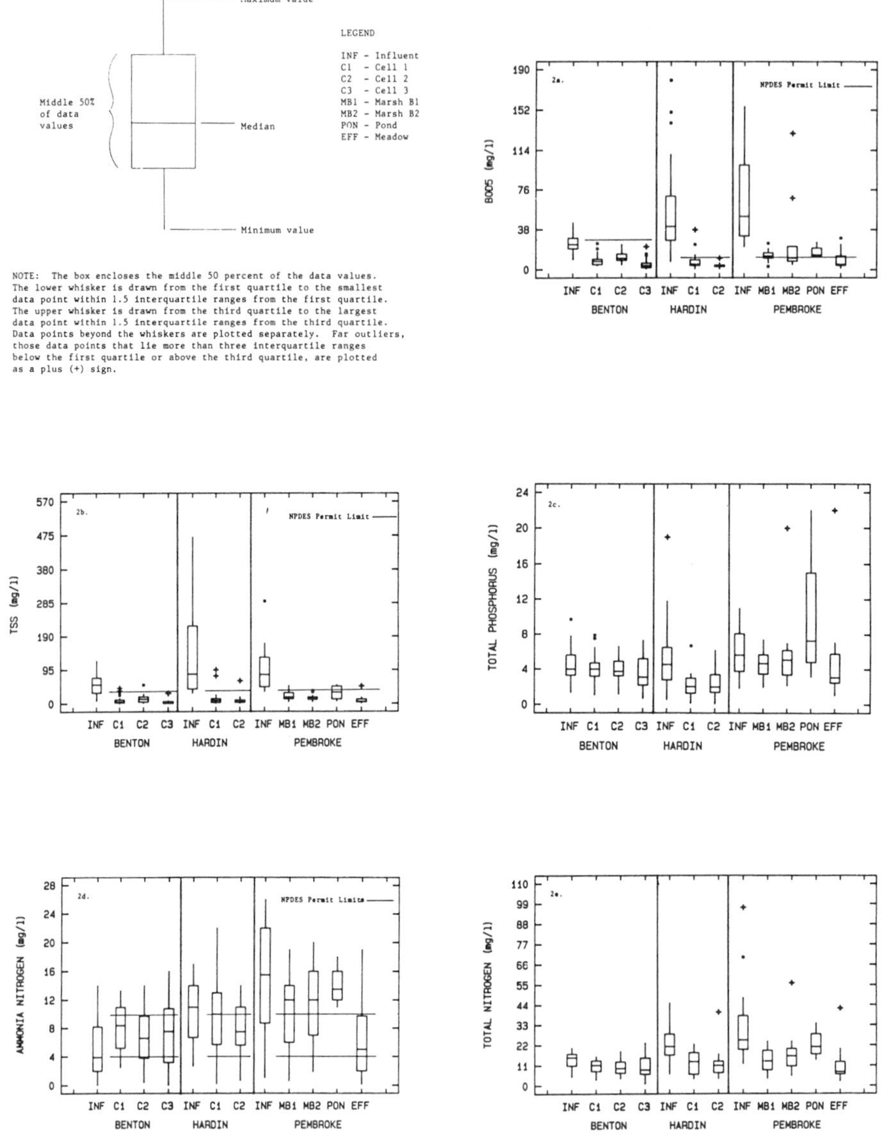

Figure 2. Constructed wetlands summary statistics: A, BOD_5; B, TSS; C, total phosphorus; D, ammonia-nitrogen; and E, total nitrogen.

All three systems effectively reduced BOD_5, with average BOD_5 effluent concentrations being below 15 mg/L (Figure 2A). BOD_5 effluent concentrations at Benton did not exceed the NPDES permit limit of 25 mg/L. A few concentrations at Pembroke and Hardin did exceed the NPDES permit limit of 10 mg/L. However, these exceedances generally occurred during unusual operating conditions such as during the initial start-up period when vegetation was still sparse and water levels not stabilized, or when water levels or flow proportions were being altered. Although hydraulic loading rates were similar, gravel cells had overall concentrations that were lower (averaging less than 10 mg/L) than surface flow cells.

Effluent TSS concentrations were reduced by all three systems, with average effluent concentrations being below 20 mg/L (Figure 2B). Like BOD_5, TSS deviations from the NPDES permit limit of 30 mg/L generally occurred during unusual operating conditions, as described above.

TP effluent concentrations varied considerably (Figure 2C). Performance of the two surface flow cells at Benton were similar. Cell 3 had a larger range of the middle 50% data values than cells 1 and 2, but had a lower median of 3.1 mg/L. Performance of the two subsurface flow cells at Hardin was also similar and had lower medians, 2.0 mg/L, than the other systems. Like the two surface flow cells at Benton, the two marshes at Pembroke had similar effluent ranges and medians. TP concentrations in the pond were generally higher than the influent, but concentrations in the meadow were below values measured in the marshes. Overall, the gravel cells produced lower TP concentrations than surface flow cells.

FC bacteria were effectively reduced in all three systems by natural processes (without chemical disinfection) (Table 1). Removals were very high for each, averaging more than 99%, with the exception of cells 1 and 2 at Benton. Removals for these two cells were 97 and 88%, respectively. The lower removal in cell 1 is probably caused by an increase in wildlife in this cell. Although removals were high in all three systems, values frequently exceeded the NPDES permit limit of 200 colonies per 100 ml.

Although more total nitrogen (TN) entered the Pembroke system, the system had a lower TN effluent concentration (median, 7.5 mg/L) than the other two systems (Figure 2E). Both marshes, which performed similarly, reduced the largest majority of TN. The pond increased TN, while the meadow reduced the concentration below values measured in the marshes. The two gravel cells at Hardin performed similarly. At Benton, the three cells produced similar effluent concentrations.

Nitrification was limited by low dissolved oxygen in all three systems, resulting in the systems frequently exceeding the summer and winter NH_3-N NPDES permit limits of 4 and 10 mg/L, respectively. Winter effluent NH_3-N concentrations were lower than summer concentrations, resulting in fewer permit exceedances. Organic nitrogen, which is converted to NH_3-N, was reduced in all three systems, having a median effluent range of 1.1 to 2.8 mg/L. NH_3-N effluent medians for all three systems were above or very close to the influent medians (Figure 2D).

Performance Within the Constructed Wetlands

Dye studies were performed at the Benton and Pembroke constructed wetlands using Rhodamine WT dye. Total amount of dye decreased with increasing distances from the inlet, indicating unexplained losses. Other tracers such as lithium have been recommended for use in future studies to provide more representative results.[4] Parameters associated with suspended material, such as TSS, BOD_5, FC, and organic nitrogen, were removed within the first half of the marshes at both sites. Dissolved oxygen concentrations generally dropped below 0.5 mg/L within the first quarter of the wetlands. No reduction pattern was detected for TP concentrations. NH_3-N concentrations generally increased in the first quarter of the wetlands length because most of the organic nitrogen was converted in this location. A gradual decrease followed the increase. Generally, TN gradually decreased as the wastewater moved through the wetlands. There was no significant difference in the patterns of each parameter between dye studies.

Short-circuiting occurred in the constructed wetland cells. Most of the wastewater traveled through the portion of the cells that had the deepest water due to not obtaining a zero lateral slope of the cell beds during construction. However, flow distribution improved greatly as the wastewater traveled through the cells.

Gravel Cells

All three gravel cells that were originally intended to operate as subsurface flow systems have experienced surface flow in the inlet areas. During startup, cell 3 received all of the total flow from Benton, thus causing voids in the limestone to clog quickly and resulting in partial surface flow by the end of the first year of operation. A specific hydraulic conductivity value at Benton was estimated to be two orders of magnitude less than the original value (28,042 m/day or 92,000 ft/day).

Clogging of the gravel cell at Benton was due primarily to an inorganic gel resulting mainly from dissolution of calcium in the limestone and its subsequent reaction with silicon in the wastewater and algae.[4] Approximately two thirds of the void volume in the inlet area is blocked.

Influent flow into cell 3 was reduced to only 12% of the total flow and the water level was lowered in order to reinstitute subsurface flow. This operation has succeeded in that cell 3 is currently operating as a subsurface flow system.

Specific hydraulic conductivities of the gravel cells at Hardin have apparently stabilized, which may be attributed to improvement of package plant performance. Surface flow continues to exist in the inlet area, but has not increased along the length of the wetland beds.

Table 1. Fecal Coliform Summary

	Benton	Hardin	Pembroke
Influent			
Geometric mean	6735	127,521	263,123
Range	100–250,000	1000–560,000	6100–850,000
Cell 1			
Geometric mean	822	682	1015 (MB1)
Range	20–27,000	5–25,000	10–40,000
% removal	87.8	99.5	99.6
Cell 2			
Geometric mean	207	266	1757 (MB2)
Range	3–3800	1–3400	100–48,000
% removal	96.9	99.8	99.3
Cell 3			
Geometric mean	87		344 (EFF)
Range	10–5450		10–10,300
% removal	98.7		99.9

Nitrification Activities

Several nitrification studies have been implemented in an effort to increase dissolved oxygen, resulting in achieving the desired nitrification. These are briefly discussed below.

Aeration platforms were installed in cell 1 at Benton in midsummer of 1990. Several combinations of pumps and platforms were experimented with in an effort to improve dissolved oxygen concentrations throughout the cell. None were successful. Dissolved oxygen decreased from near saturation at the platforms to less than 1 mg/L within a meter of the platform.

A recirculation pump and aeration system was also installed in the meadow at Pembroke, which has a low organic loading, to improve dissolved oxygen throughout the cell. No significant change in dissolved oxygen and ammonia-nitrogen was detected.

A pilot-scale study consisting of vertical flow and shallow horizontal flow constructed wetland cells is being conducted to determine if these concepts can achieve NPDES permit compliance. The findings are promising.[5]

Design and Construction

Performance, operation, and information on newer systems have provided better design criteria for surface and subsurface flow cells. Surface flow design should incorporate velocity considerations such as overflow velocities, as in sedimentation basin design.

Subsurface flow design needs to be based on Darcy's law or other appropriate hydraulic concepts, which relate flow to hydraulic conductivity of the substrate, hydraulic slope, and the area of the cell cross-section, resulting in no standard length-to-width ratio. A relatively low, long-range hydraulic conductivity value should be employed along with low organic loading (0.5 kg BOD_5 per day per square meter or 0.1 pound BOD_5 per day per square foot of cross-sectional area or less) for inlet areas.

Construction is relatively simple but is a critical factor in the performance of the constructed wetlands. Grading tolerances for the bottom of both types of cells and top of gravel need to be within 3.1 cm (0.1 ft) to minimize short-circuiting and ensure uniform water depths for vegetation management. Deep-water zones or internal dikes should be constructed throughout the length of surface flow systems to redistribute flow.

CONCLUSIONS

Several conclusions based on the three systems are presented below addressing compliance, performance, vegetation, operation, and design and construction.

Compliance

Constructed wetlands provide compliance for typical secondary and some advanced level limits for BOD_5 and TSS over a wide range of flow rates.

Compliance for FC and dissolved oxygen may be more efficiently achieved with supplemental unit operations following constructed wetlands.

Constructed wetlands have not been successful in achieving enough nitrification to meet NH_3-N permit limits. Dissolved oxygen added by recirculation and spray aeration is utilized almost immediately. However, research on vertical flow constructed wetlands has achieved high nitrification efficiencies.

Performance

Constructed wetlands remove suspended pollutants such as BOD_5, TSS, organic nitrogen, and FC within the first half of the wetland bed.

With time, there is no nominal change in concentration removal effectiveness.

Ammonification reduced organic nitrogen in all three systems. Nitrification also occurred, but was limited by available dissolved oxygen.

System Type

Surface flow cells are more adaptable and cost effective for handling high levels of suspended pollutants.

Both types of systems are effective in removing suspended pollutants; however, subsurface flow systems produce slightly lower concentrations, even when partially clogged.

Vegetation

Vegetation species do not appear to be a significant factor in overall performance under the loading conditions to date for surface flow cells.

Cattails, bulrush, and reeds are surviving and spreading in high solids deposition areas.

Cattails and woolgrass do not appear to be appropriate choices for deep-water depths.

Reed canary grass is difficult to establish by seeding.

Species that spread by rhizomes and roots are better for surface flow systems since they spread quickly and do not require water-level manipulation to germinate seeds.

Operation

Solids have accumulated massively at the front of the constructed wetland cells that have poorly performing contact stabilization plants as primary treatment. This results in clogging of the inlet area of gravel cells and the addition of long-term pollutant loads. The net effect is a significant reduction in both effectiveness and the life of the cell.

Controlling the water level in the wetland cells is critical for proper vegetation management.

Design and Construction

Overflow velocities should be considered in surface flow design.

Subsurface flow design should be based on Darcy's law or other appropriate hydraulic concepts.

Hydraulic conductivity values used in subsurface flow design should be relatively low, long-range values. Low organic loading for inlet areas should also be used.

Construction is a critical factor in the performance of constructed wetlands.

Grading tolerances for the bottom of both types of cells and top of gravel need to be within 3.1 cm (0.1 ft).

REFERENCES

1. Choate, K. D., J. T. Watson, and G. R. Steiner. Demonstration of Constructed Wetlands for Treatment of Municipal Wastewater, Monitoring Report for the Period: March 1988 to October 1989. Tennessee Valley Authority, TVA/WR/WQ—90/11, August 1990.
2. Choate, K. D., G. R. Steiner, and J. T. Watson. Second Monitoring Report: Demonstration of Constructed Wetlands for Treatment of Municipal Wastewater, March 1988 to October 1989. TVA/WR/WQ—90/11, Tennessee Valley Authority, Water Quality Department, Chattanooga, TN, 1990.
3. Watson, J. T., K. D. Choate, and G. R. Steiner. Performance of constructed wetland treatment systems at Benton, Hardin, and Pembroke, Kentucky during the early vegetation establishment phase, in *Constructed Wetlands in Water Pollution Control, Proc. Int. Conf. Use of Constructed Wetlands in Water Pollution Control*. Cambridge, U.K., September 24–28, 1990.
4. Kadlec, R. H. Analysis of Gravel Cell Number Three, Benton, Kentucky Wetlands. Final report presented to Tennessee Valley Authority, December 18, 1990.
5. Watson, J. T. and A. J. Danzig. Pilot-scale nitrification studies using vertical-flow and shallow horizontal-flow constructed wetland cells. Paper presented at *Int. Symp. Constructed Wetlands for Wastewater Quality Improvement*. Pensacola, FL, October 1991.

CHAPTER 57

Reed Bed Treatment for Small Communities: U.K. Experience

M. B. Green and J. Upton

INTRODUCTION

A high proportion of the population in the U.K. lives in communities provided with sewerage and sewage treatment. In England and Wales, 96% of households are connected to a sewerage system and 79% of these systems discharge to treatment plants.[1] Severn Trent Water Ltd. is fairly typical of the U.K. situation, with 97% of the contributory population connected to sewers discharging to treatment works, most of which (95%) include secondary treatment.

Some small communities within the Severn Trent region are classed as unsewered, but remain connected to "village drains". These systems are usually provided with a septic tank, the overflow from which goes into a ditch en route to the water course. Most of the village drains cause little concern and historically attention was restricted to emptying tanks and periodically pumping out the receiving ditches or "sewer dykes". The inception of the Water Act 1989 and the highly regulated private companies responsible for water has focused more attention on the small community systems and these discharges are now monitored and regulated by the National Rivers Authority.

Severn Trent Water Authority, the predecessor to Severn Trent Water, Ltd., carried out a full-scale trial in 1987, in which it replaced an existing sewer dyke which was causing pollution of the local brook with a reed bed treatment system. The design and early performance were described by Upton and Griffin.[2] This article will, in large part, extend the experience described in that paper and will put the work in the context of the company's larger commitment to reed bed treatment.

Little Stretton — Background

Little Stretton is typical of the "unsewered" small villages in its relative isolation from larger communities connected to public sewers. It comprises 15 dwellings and 2 farms. Of these, two have individual septic tanks and the rest connect to a short sewer ending with a single-compartment septic tank.[2] Washdown from a milking parlor and farmyard run off from one of the farms and surface water drainage combined with the septic tank outfall to the sewer dyke. Discharge from the sewer dyke to the local brook, the River Sence, caused unacceptable pollution which gave impetus to the replacement of the sewer dyke with a reed bed system.

OPERATION

The beds were planted with pots of *Phragmites* seedlings in July 1987, and there has been no replacement or replanting since that date. Initial concerns about relatively poor growth on the last terrace and at the front end of the first terrace proved groundless. In 1991, the reed stand was strong throughout the system despite damage caused by an incursion of cattle in the summer of 1990. Poorer reed growth on the first terrace, reported by Upton and Griffin,[2] is no longer apparent.

When Little Stretton was commissioned in July 1987, it operated as a series of gravel-filled lagoons with an impervious liner intact to the surface level at the outlet of each terrace. Liner flaps at the outlet were cut

0-87371-550-0/93/$0.00+$.50

down to 150 to 200 mm below the surface in September 1987 and then progressively reduced to the lowest practicable level (about 500 mm) below the surface by September 1988. The only operational/structural change since that date has been the fitting of a simple screening device, a "Copasac", for collection of plastic debris etc. at the inlet to terrace 1. This has increased the need for a monthly maintenance visit of 15 min to change the Copasac. When the septic tank is emptied each month, the inlet chamber in front of the V notch is also emptied, the vegetation on the banks is cut twice yearly, and the site is visited weekly to check the outlet effluent quality by simple operator testing in rotation with other small works in the District.

Problems caused by excessive input of heavily polluted washings from cattle standing areas were successfully overcome early in 1990 by diverting the run-off from cattle standing areas.

PERFORMANCE

Changes in Biological Treatment with Time

There was a dramatic improvement in the effluent discharged from the reed bed in 1990 (Table 1). The success in diversion of run-off from the cattle yard must have contributed to the improvement, but the pattern of spot sampling does not give clear evidence of this. In earlier years, all very high effluent BOD results were associated with high influent levels but effluent BODs ranging from 30 to 50 mg/L were relatively common.[2] In 1990 and 1991, the final effluent BOD improved dramatically and few samples had BODs in excess of 10 mg/L.

During the first 2 years of operation, there was little evidence of nitrification but this improved thereafter. During much of the first 2 years of operation, the individual terraces were ponded because of the outflow arrangements, and it is quite clear that much of the space in the matrix became filled with anaerobic material. Aerobic conditions would only prevail at the surfaces and in the lower terraces. With the lowering of the outlet flaps, it is likely that the distal ends of all terraces were aerobic, as were greater areas of terraces 7 and 8.

The degree of BOD and ammonia removal indicated in Table 2 would have required an oxygen transfer rate of about 13g $O_2/m^2/day$. This considerably exceeds the rates measured by Brix and Schierup in a reed bed treatment system in Kalo, Denmark.[3]

It may be that the root system of *Phragmites* was contributing to the oxygen supply, but it is at least as plausible that during periods of low flow, more of the sewage would have flowed through the gravel, under the influence of the hydraulic gradient between inlet and outlet, and thus have improved the oxygen transfer.

An intensive survey carried out in April 1991 (Table 3) showed that much of the BOD removal had taken place by the end of the third terrace. Diurnal variations in the input load were largely smoothed out through the system (Figure 1). During that survey, the overall average removal of BOD and COD was 86 and 60%, respectively (Table 4).

A feature of Little Stretton has been the highly variable composition of the sewage. It is likely that this is caused by the intermittent nature of the dairy parlor washings and untraced pollutant sources in the surface water. In a 7-day composite survey carried out in May 1991, the BOD and COD indicated a weak sewage, but the TKN and NH_3-N concentrations were relatively high (Table 5). During that survey, the BOD and COD removal rates were 95 and 68%, respectively.

There was an increase in ammonia-nitrogen between the inlet and the end of the third terrace in the April survey (Table 4). This was the result of the NH_3-N fraction of the TKN increasing by the hydrolysis of organic nitrogen. The TKN at terrace 3 often exceeded the TKN at the inlet (Figure 2) and it is reasonable to assume that such differences were due to a balancing effect in the upper terrace, with peaks being evened out. Nitrification becomes progressively more effective down the system, as shown by the change in ammonia concentration (Figure 3).

An interesting difference in the degree of nitrification showed up in the May survey (Table 5) when the removal of ammonia averaged 83%. Although nitrification would clearly be favored by the warmer weather of May, it is likely that the flow onto the bed also influenced the rate by affecting oxygen transfer. During the April survey period, surface flows predominated, with average flows of about 23 m^3/day. In the May survey, with flows of about 12 m^3/day, there was more opportunity for draw down toward the ends of the terrace, improving conditions for oxygen transfer.

A lithium trace was carried out during the April survey (Figure 4). The relatively short retention time and early appearance of the peak on terrace 3 suggest laminar surface flow. Peak concentration also appeared early,

Table 1. Performance of Little Stretton Reed Bed Unit in the Period 1987 to 1991

	BOD		SS		NH_3-N		TON	
	Inlet	Outlet	Inlet	Outlet	Inlet	Outlet	Inlet	Outlet
1987 (July to Dec)	147	29	132	19	10.0	10.0		
1988	112	33	85	24	12.2	13.8		
1989	162	34	127	43	14.9	11.3		
1990	112	3.9	93	28	24.8	12.1	2.2	6.2
1991 (Jan to Aug)	58	3.9	52	28	23.0	6.1	10.9	4.2

Note: Annual averages in mg/L.

Table 2. Performance of Little Stretton Reed Bed in the Period 06/09/90 to 14/09/90

	BOD		SS		NH_3-N		TON		oP	
	Inlet	Outlet	Inlet	Outlet	Inlet	Outlet	Inlet	Outlet	Inlet	Outlet
06/09	28	5.5	76	12	40.1	27.3	1.2	1.0	11.3	2.2
07/09	118	2.5	78	10	37.7	28.2	0.4	1.4	11.8	2.0
10/09	113	2.5	84	43	39.9	25.5	0.4	2.0	11.2	1.4
11/09	164	4.0	94	41	37.2	27.8	0.5	1.1	12.8	1.6
12/09	157	8.0	256	15	37.7	25.9	0.6	1.1	14.1	1.7
13/09	159	3.0	168	27	41.1	25.8	0.9	1.4	13.8	1.5
14/09	160	2.5	88	17	46.9	27.8	0.4	1.4	14.9	1.6
Mean	128	4.0	121	23.6	40.1	26.9	0.6	1.3	12.8	1.7

Note: Composite daily samples, in mg/L.

after 6 and 7 h, respectively, at the end of terrace 8 and at the outlet. Again, the retention times imply surface flow but the much flatter curves would have resulted from an increasing contribution from subsurface flow. At lower flows, more would be carried beneath the surface, thus offering greater opportunity for oxygen transfer.

Nutrient Removal

On most sampling occasions in 1991, there were significant concentrations of TON in the inflow to the reed bed unit. The concentrations were typically reduced to low levels by denitrification which, on the evidence of the April survey (Table 3), largely took place in the first few terraces. Oxidized nitrogen produced by nitrification of ammonia compounds was further removed within the bed matrix by denitrification. As a result of the April survey, the total nitrogen (TKN + TON) reduced from 30 to 10 mg/L. In the May survey (Table 5), the TON reduced from 17.6 to 2.8 mg/L, on average, and the total nitrogen from 76.6 to 12.5 mg/L.

There was some overall reduction in total phosphorus through the system (Table 1). At times of higher flow when the inputs were more diluted, giving relatively low total phosphorus concentrations, there was little change through the system. In the survey in April 1991 (Table 4), there was an overall 33% removal of total P; but on a day-to-day basis, the outlet was occasionally higher than the inlet (Table 3). In the survey period in May, the inlet concentration averaged 7.9 mg/L and the outlet averaged 2.8 mg/L, a removal rate of 65% (Table 5). Higher removal rates applied in the composite survey carried out in a dry period in September 1990 (Table 2) when the phosphate concentration was reduced on average by 87%.

There is no obvious mechanism for phosphorus retention in the system. It is likely that any surfaces in the gravels with available iron or aluminum would have been used up in the early months of operation, as would surfaces of the larger limestone rocks in the gabions.

The phosphorus concentrations in the solids accumulating in the system was higher than in the solids coming into the unit (Table 6). Although solids do accumulate in the litter layer, together with decaying leaves and stems of *Phragmites*, some solids are lost in the effluent.

Table 3. Performance of Little Stretton Reed Bed Unit in the Period 10/04/91 to 17/04/91 Based on Survey Samples Taken at Inlet, Terrace 3 (T3), Terrace 6 (T6), Terrace 8 (T8) and Outlet

	BOD (n)	COD (n)	SS (n)	TKN (n)	NH_3-N (n)	TON (n)	TP (n)
10/04							
Inlet	34.5 (13)	95 (13)	55 (13)	14.4 (13)	8.6 (13)	15.9 (13)	4.6 (13)
T3	7.8 (13)	63 (13)	22 (13)	13.2 (13)	11.1 (13)	1.1 (13)	3.0 (13)
T6	5.6 (13)	72 (13)	41 (13)	10.8 (13)	8.8 (13)	3.1 (13)	2.8 (13)
T8	1.9 (13)	28 (13)	8 (13)	6.9 (12)	6.4 (13)	3.6 (13)	2.3 (13)
Outlet	4.3 (13)	38 (13)	16 (13)	7.5 (13)	6.5 (13)	3.1 (13)	2.0 (13)
11/04							
Inlet	17.3 (24)	68 (24)	25 (24)	9.9 (24)	6.9 (24)	14.5 (24)	2.5 (24)
T3	5.7 (23)	45 (24)	8 (24)	10.5 (15)	9.5 (24)	1.1 (24)	2.6 (15)
T6	4.7 (24)	55 (24)	33 (24)	8.5 (24)	7.4 (24)	3.1 (24)	2.6 (24)
T8	2.0 (24)	39 (24)	10 (24)	6.5 (24)	5.6 (24)	3.6 (24)	2.2 (23)
Outlet	6.2 (24)	38 (24)	17 (24)	6.5 (23)	5.7 (24)	3.2 (24)	2.1 (23)
12/04							
Inlet	31.0 (13)	98 (13)	66 (13)	13.0 (12)	8.4 (13)	12.3 (13)	2.3 (11)
T3	3.8 (13)	44 (13)	8 (12)	11.0 (13)	10.2 (13)	0.7 (13)	3.6 (12)
T6	6.3 (13)	51 (13)	26 (13)	9.7 (13)	7.8 (13)	2.6 (13)	2.5 (13)
T8	3.5 (13)	53 (12)	13 (13)	6.4 (13)	5.6 (13)	2.6 (13)	2.1 (13)
Outlet	3.3 (13)	46 (12)	15 (12)	7.4 (13)	5.7 (13)	2.0 (13)	1.9 (13)
13/04							
Inlet	27.5 (4)	72 (4)	26 (4)	11.1 (4)	7.2 (4)	11.7 (4)	2.9 (4)
T3	7.0 (4)	66 (4)	5 (4)	13.5 (4)	11.4 (4)	0.3 (4)	4.6 (4)
T6	9.6 (4)	83 (4)	25 (4)	12.2 (4)	8.8 (4)	2.1 (4)	3.3 (4)
T8	6.8 (3)	65 (2)	15 (3)	7.8 (3)	6.3 (3)	2.2 (3)	2.4 (2)
Outlet	4.5 (4)	55 (4)	6 (4)	8.6 (4)	6.2 (4)	1.7 (4)	2.3 (4)
14/04							
Inlet	27.8 (4)	56 (4)	17 (4)	9.7 (4)	5.7 (4)	14.9 (4)	2.2 (4)
T3	7.1 (4)	63 (4)	8 (4)	13.5 (4)	11.7 (4)	0.4 (4)	4.7 (4)
T6	9.1 (4)	69 (4)	31 (4)	11.3 (4)	8.9 (4)	2.3 (4)	3.2 (4)
T8	7.5 (1)	52 (1)	11 (1)	7.3 (1)	5.7 (1)	4.0 (1)	2.3 (1)
Outlet	6.6 (4)	60 (4)	9 (4)	7.6 (4)	6.6 (4)	2.2 (4)	2.2 (4)
15/04							
Inlet	42.0 (8)	100 (8)	121 (8)	11.0 (8)	6.9 (8)	12.4 (8)	3.4 (8)
T3	10.6 (8)	65 (8)	11 (8)	14.1 (8)	12.2 (8)	0.2 (8)	4.8 (8)
T6	8.0 (2)	98 (2)	50 (2)	9.7 (2)	9.1 (2)	1.9 (2)	3.6 (2)
T8	6.7 (7)	41 (7)	12 (7)	8.2 (7)	6.9 (7)	2.8 (7)	2.5 (7)
Outlet	3.4 (8)	45 (8)	8 (8)	8.2 (7)	7.0 (8)	2.2 (8)	2.2 (8)
16/04							
Inlet	78.5 (8)	267 (8)	230 (8)	17.0 (8)	7.4 (8)	13.7 (8)	4.4 (8)
T3	7.7 (8)	45 (8)	9 (8)	13.3 (8)	11.3 (8)	0.3 (8)	4.4 (8)
T6	—	—	—	—	—	—	—
T8	3.3 (8)	38 (8)	13 (8)	8.3 (8)	6.8 (8)	3.0 (8)	2.6 (8)
Outlet	3.4 (8)	40 (8)	18 (8)	8.7 (8)	6.7 (8)	2.3 (8)	2.6 (8)
17/04							
Inlet	60.0 (4)	278 (4)	223 (4)	16.5 (4)	7.3 (4)	9.4 (4)	5.3 (4)
T3	9.5 (4)	71 (4)	57 (4)	15.9 (4)	12.2 (4)	0.6 (4)	5.0 (3)
T6	—	—	—	—	—	—	—
T8	2.7 (4)	43 (4)	27 (4)	8.0 (4)	6.3 (4)	3.7 (4)	2.8 (4)
Outlet	4.4 (4)	49 (4)	39 (4)	8.1 (4)	6.6 (4)	2.9 (4)	3.0 (4)

Note: Results given in mg/L.

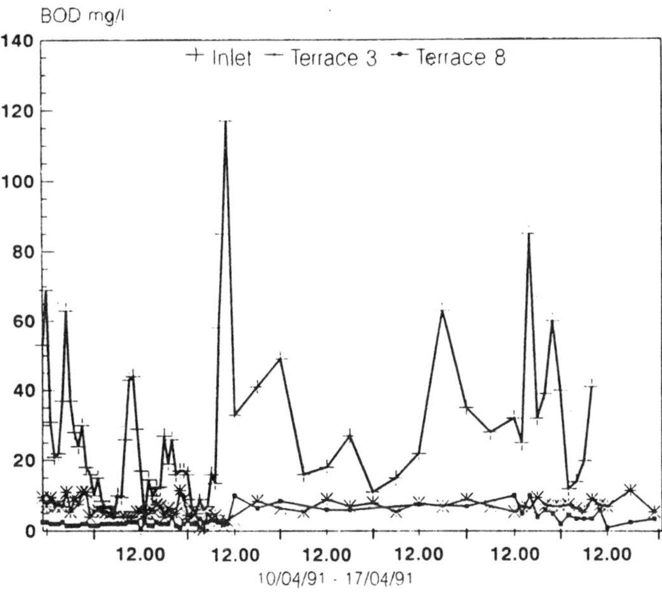

Figure 1. Little Stretton reed bed; variations in BOD concentrations at inlet, terrace 3 and terrace 8, in the period 10/04/91 to 17/04/91.

Table 4. Performance of Little Stretton Reed Bed Unit in the Period 10/04/91 to 17/04/91, Overall Average Results from Surveys at Inlet, Terrace 3 (T3), Terrace 6 (T6), Terrace 8 (T8), and Outlet

	BOD (n)	COD (n)	SS (n)	TKN (n)	NH_3-N (n)	TON (n)	TP (n)
Inlet	34.5 (38)	111 (79)	77 (77)	12.4 (78)	7.5 (78)	13.6 (79)	3.3 (76)
T3	6.4 (77)	53 (77)	13 (77)	12.4 (66)	10.7 (78)	0.7 (78)	3.7 (61)
T6	6.0 (60)	62 (60)	33 (60)	9.7 (55)	8.0 (60)	2.8 (60)	2.8 (56)
T8	3.2 (72)	41 (69)	12 (72)	7.0 (71)	6.1 (72)	3.2 (72)	2.3 (70)
Outlet	4.7 (78)	44 (76)	16 (77)	7.4 (76)	4.8 (76)	2.6 (78)	2.2 (77)

Note: Results given in mg/L.

Table 5. Performance of Little Stretton Reed Bed Unit in the Period 21/05/91 to 31/05/91

	BOD (n)	COD (n)	SS (n)	TKN (n)	NH_3-N (n)	TON (n)	TP (n)
Inlet	64.3 (6)	133 (7)	60 (7)	59.0 (7)	55.0 (7)	17.6 (7)	7.9 (7)
T8	2.4 (6)	44 (7)	22 (7)	10.0 (6)	9.2 (7)	3.8 (7)	3.5 (7)
Outlet	3.0 (7)	43 (7)	22 (7)	9.7 (6)	9.1 (7)	2.8 (7)	2.8 (7)

Note: Results are averages of daily composite samples, in mg/L.

LITTLE STRETTON DESIGN REVISITED

The reed bed unit was designed by the Water Research Center in 1986 using the best information available at the time, and some practical modifications were made on site during construction. In concept, it was assumed that the average flow would be 10.2 m^3/day. The area provided was what was available but at about 3 m^2/p.e. was believed to be adequate. An average effluent BOD of less than 10 mg/L would have been predicted from the formula used at the time.[4]

Eight terraces were built in a straight trapezoidal channel, each being about 12 m long and 2 m wide. The channel was lined with an impervious polyethylene membrane; the terraces were filled with 12-mm pea gravel

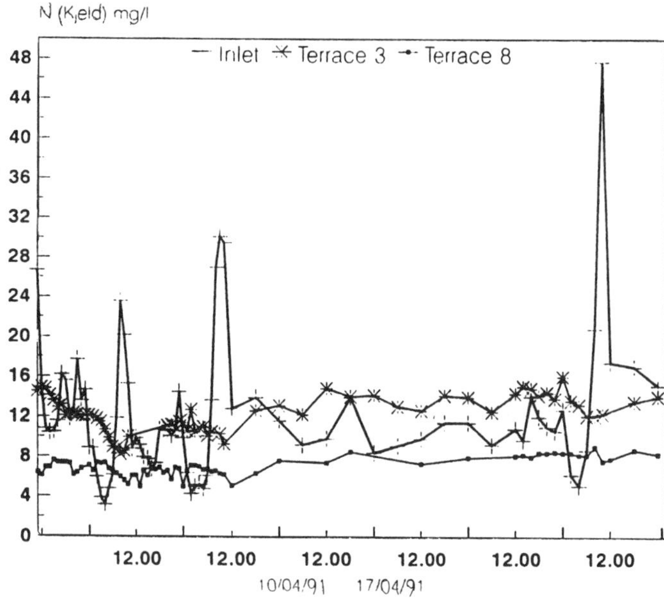

Figure 2. Little Stretton reed bed; variations in total Kjeldal nitrogen concentrations at inlet, terrace 3, and terrace 8, in the period 10/04/91 to 17/04/91.

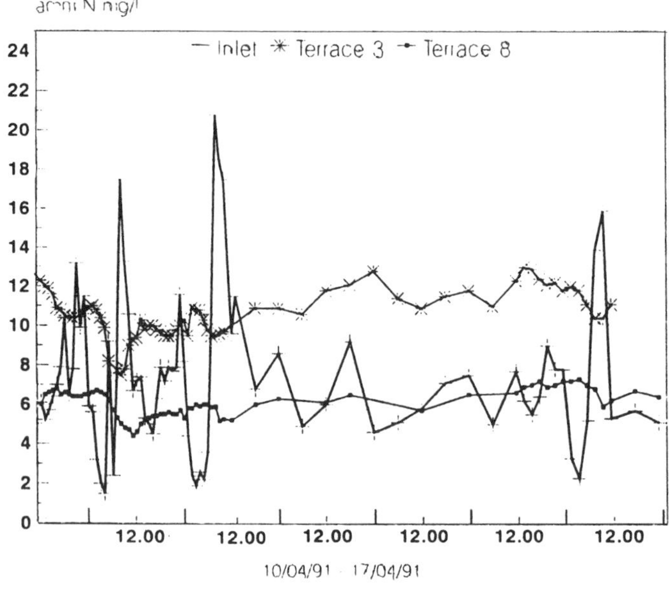

Figure 3. Little Stretton reed bed; variations in ammoniacal-nitrogen concentrations at inlet, terrace 3, and terrace 8, in the period 10/04/91 to 17/04/91.

and separated by gabions filled with limestone rocks. As the depth of gravel increased along the length of the terrace, the width increased, as indicated in the table of design details in Upton and Griffin.[1] The cross-sectional area required was calculated according to the equation derived from Darcy's law.[4,5]

For the design to work as intended the hydraulic conductivity (Kf) of the bed should have approximated to 3×10^{-3} m/s. It can be calculated that, as designed, there would be some overland flow on all terraces under

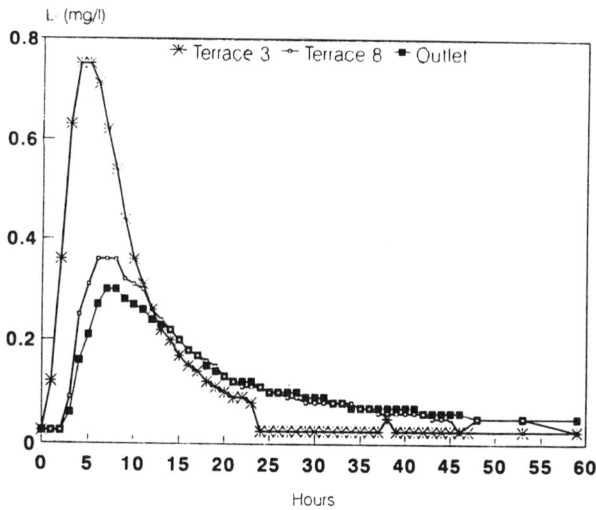

Figure 4. Little Stretton reed bed; Li survey; lithium concentrations at terrace 3, terrace 8, and outlet at time intervals from 11:00 h on 10/04/91.

Table 6. Analysis of Sludge Samples Collected from Little Stretton Reed Bed

	Dry matter, %		Loss on ignition, %		Total phosphorus, %	
date — 1991	**21/03**	**17/04**	**21/03**	**17/04**	**21/03**	**17/04**
Inlet	3.0	4.0	59.3	60.6	0.1	0.7
T3	2.2	5.0	43.9	36.3	0.8	0.7
T8	3.0	4.8	23.6	25.0	1.1	2.0
Outlet		5.4		28.1		1.3

average flow conditions until about half-way along their lengths, horizontal flow would thereafter continue through the gravel. In the event that surface-water drains connected to the system, average flows frequently exceeded the design value and hence obscured the error in the assumption of a Kf of 3×10^{-3} m/s. In the fifth season, the hydraulic conductivity has been measured at about 1×10^{-4} m/s in terrace 3 and 1×10^{-3} m/s in terrace 7. Current best practice, as described by Cooper,[5] recommends using a Kf of 1×10^{-3} m/s as a design figure for gravel-filled systems.

Experience in the U.K. and other European countries has caused a more cautious view to be taken of the loadings which can be accommodated on reed bed treatment units. A minimum area of 5 m^2 per p.e. has been advised. This has been adopted by Severn Trent Water as part of its design manual, as have other recommendations in the European Guidelines.[5]

If Little Stretton was to be built on the same site today, for the same assumed population, the terraces would be wider, at 4.8 m, but the overall length would be less, at 62.5 m. A free fall would be arranged between each of the terraces. Such arrangements would ensure that surface flow would only be significant at time of heavy rainfall.

Improvements to Little Stretton have been planned. These will provide:

1. A chamber for separating some of the surface flow inputs
2. A new septic tank discharging directly onto the first terrace
3. An outlet collection chamber with an improved sample collection facility

It is recognized that relatively inert solids are washed from the unit and that the outlet chamber will enable them to be removed.

FURTHER APPLICATIONS OF REED BED TREATMENT WITHIN SEVERN TRENT WATER

The company has chosen reed bed treatment as the preferred option in the treatment of sewage from small communities. A total of 140 village drains were identified in 1989. A significant proportion of these will be served by septic tanks followed by reed beds. In addition to these the company has 247 works serving populations of 50 or less, some of which qualify for replacement by septic tanks followed by reed beds. Reliability of performance is a key objective, and for very small communities this simple system has attractions over those of small package units such as rotating biological contactors (RBCs). The basic cost for a septic tank and reed bed unit to serve a population of 50 is about £35,000, but site difficulties and specific requirements for a "well engineered" appearance have increased this to as much as £80,000. For comparison, it is the company's experience that the total contract price for putting in small RBCs, to the standard it requires, is about £100,000 per installation. The great attraction of the septic tank/reed bed unit concept is the low requirement for operator intervention and low maintenance needs.

The company has a larger commitment to reed bed treatment for tertiary treatment. It has 465 sewage treatment works serving populations between 50 and 1500. About 50% of these have consent conditions more stringent than the standard 45/25/10 (SS/BOD/NH_3-N) and thus require effluent polishing. Eight works have been provided with reed bed units to date (September 1991); by the end of 1992, this number will rise to about 50. Over the subsequent 5 years, at least, the number will increase by about 30 per year. The design and performance of the first of these works will be presented elsewhere.[6]

SUMMARY

- Reed bed treatment at Little Stretton has progressed from being a qualified success.[2] It is believed that the improvement in performance in 1990 and 1991 is, in part, due to the diversion of farmyard run-off, but some improvements have come from the increased maturity of the system.
- The effluents from the system have consistently been of high quality in terms of BOD.
- Significant nitrification takes place in the system and it is believed that improved oxygen transfer caused by variation in the flow regime within the system is important for this.
- Nitrate from the inputs, together with some of the nitrate produced by nitrification, is removed by denitrification.
- No precise mechanism is advanced for the phosphate removal achieved.
- Reed bed treatment has been adopted by Severn Trent Water Ltd. as a preferred option for secondary treatment for populations of up to 50 and for tertiary treatment for populations of up to 1500.

REFERENCES

1. Turton, P. *Water Facts 1990.* Water Services Association Publications Sheffield, U.K., 1990, 20.
2. Upton, J. and P. Griffin. Reed bed treatment for sewer dykes, in *Constructed Wetlands in Water Pollution Control.* P. F. Cooper and B. C. Findlater, Eds. Pergamon Press, Oxford, U.K., 1990, 391.
3. Brix, H. and H.-H. Schierup. Soil oxygenation in constructed reed beds: the role of macrophyte and soil-atmosphere interface oxygen transport, in *Constructed Wetlands in Water Pollution Control.* P. F. Cooper and B. C. Findlater, Eds. Pergamon Press, Oxford, U.K., 1990, 53.
4. Boon, A. G. Report of a Visit by Members of Staff of WRc to Germany (FRG) to investigate the Root Zone Method for Treatment of Wastewaters, WRc Report 376-s, Water Research Centre, Stevenage, U.K., February 1986.
5. Cooper, P. F., Ed. European Design and Operations Guidelines for Reed Bed Treatment Systems, WRc Report U1 17, Water Research Centre, U.K., 1990.
6. Green, M. B. and J. Upton. Constructed reed beds: a cost effective way to polish wastewater effluents for small communities, in press.

CHAPTER 58

The Treatment of Septage Using Natural Systems

M. H. Ogden

INTRODUCTION

Natural treatment systems, which include aquatic systems, constructed wetlands, and overland flow systems, have been intensively researched as treatment options for the treatment of municipal, industrial, and mine wastewaters since the pioneering work by Seidel at the Max Planck Institute[1] and Small[2] at the Brookhaven National Labs, and many others, beginning in the early 1960s. Pond systems treating sewage have been used for at least 2000 years in China[3] for the cultivation of fish and vegetables.

Until 1988, natural treatment systems had not been applied to the treatment of septage. During the summer of 1988, Todd developed a natural treatment process for the treatment of septage using a combination of aquatic systems and constructed wetlands.[4] This process has subsequently been scaled up to a 5000 gallon per day system treating the septage in the community of Harwich, Massachusetts.

A design review of this system, as well as the City of San Diego's Aquatic System[5] for the treatment of sewage, the City of Austin, Texas's aquatic system for the treatment of sludge effluent,[6] and finally a review of reed bed systems for the treatment of sludge resulted in the design concept for a small-scale septage treatment facility. The design goal was to develop a small-scale treatment option that was self-regulating, self-maintaining, and inexpensive to build and operate.

Incorporated into this system is the fundamental design principal first articulated by Small at the Brookhaven National Laboratory[2] which is the idea of using the ecologies of the pond, marsh, and meadow to treat wastewater. These very different ecosystems provide habitats for various species of bacteria, protists, fungi, plants, and invertebrate populations.[7] This enormous species diversity results in self-regulating and self-regenerating systems capable of advanced levels of wastewater treatment using solar energy as the major energy source.

SYSTEM DESIGN

The Arroyo Hondo, New Mexico pilot septage treatment system consists of the following elements:

1. Receiving chamber with bar screens
2. Two-chamber aerated equalization tank
3. Reed bed
4. Water hyacinth lagoon, aerated (pond)
5. Two constructed wetlands, one within the greenhouse
6. Four irrigated hedgerows

The equalization tank, the hyacinth lagoon, and one of the constructed wetlands have been enclosed within a greenhouse to provide winter protection for the hyacinths and other cold-sensitive plants. The irrigated hedgerow has not been included in any of this original study for the purposes of determining treatment efficiencies. Sufficient data are available on overland flow, or meadow treatment systems to determine what will happen to the remaining nutrients leaving the external wetlands.

Figure 1 shows the various elements of the system, with the three pumps in their respective locations. Pump 1 (P1) is a sewage effluent pump used to pump septage from the equalization tank (ET) into the reed bed. This

0-87371-550-0/93/$0.00+$.50

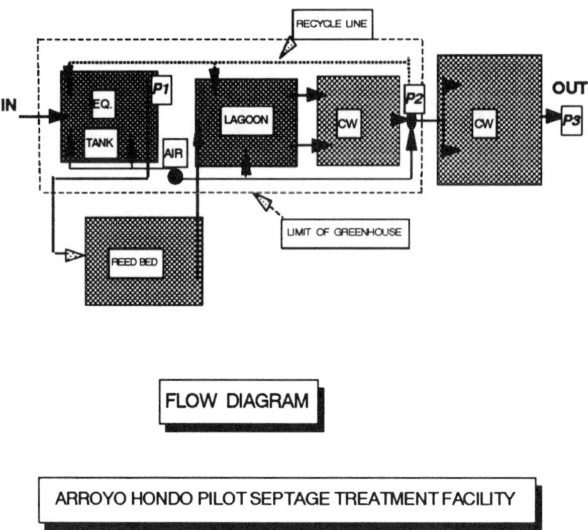

Figure 1. Flow diagram.

pump operates on a timed cycle during the day, and some portion of its flow is recycled to the front end of the ET. When the ET is not aerated, this recycle serves to mix solids.

The reed bed is designed to remove solids from the septage. After pretreatment in the ET, the solids float on the surface of the reed bed and eventually dry out, forming a crust-like solid that is easily removed from the surface by raking. Water and some very fine solids percolate through the gravel in the reed bed and drain into the lagoon.

The lagoon is a water hyacinth pond whose level ebbs and floods as P1 operates. As the pond floods, it overflows into a constructed wetlands. This concept was described by Teal of the Woods Hole Oceanographic Institute as a result of his studies of estuarine marshes,[8] and developed in this system. The importance of this ebb and flood is in the way in which the marsh can digest and assimilate the solids which are passed through the lagoon.

The constructed wetland contains numerous species of plants (Table 1) which have been previously used by other researchers.[1,7,9,10] As the solids move into the wetlands during the flood stage, they settle out onto the surface of the gravel and around the stems of the plants. As the water recedes during the ebb stage, the settled solids are exposed to the air and to drying. As the solids dry out, different organisms which have colonized the solid material can begin to digest and assimilate the sludge. Different bacteria, fungi, and algae are active during the wet and dry cycles.

After passing through the wetlands in the greenhouse, the water enters a sump in which pump 2 (P2) is located. P2's function is to recycle some portion of the wastewater to the equalization tank and the lagoon. The remainder of the water enters the external wetlands.

The external wetlands consists of two parallel beds, one constructed with pumice and one with gravel, which are also planted with numerous species of wetlands plants. These wetlands are designed to act as polishing marshes and to denitrify as much of the remaining nitrogen as possible, or have the nitrogen as nitrate assimilated by the marsh plants.

Water leaving this marsh enters a final sump where pump 3 (P3) pumps the water through a water meter and into an irrigation system for four hedgerows planted with a total of approximately 350 plants.The hedgerows consist of ashes, rosa rugosa, ponderosa pines, New Mexico privet, black locust, and Mexican sunflower.

MEASUREMENT AND DATA COLLECTION

Physical parameter data collection was accomplished with the simplest of instruments three times a week. Equipment included a YSI 51B dissolved oxygen and temperature meter, and Hach portable pH, TDS, and

Table 1. Plant Species Growing in Septage Treatment Facility

Common name	Latin name	Common name	Latin name
Water hyacinth	*Eichornia crassipes*	Daffodils	*Narcissus* spp.
Cattail	*Typha latifolia*	Foxglove	*Digitalis purpurea*
Bulrush	*Scirpus validus*	Winter wheat	*Triticum aestivum*
Cane grass	*Arundo donax*	Winter rye	*Lolium perenne*
Purple loosestrife	*Lythrum salicaria*	Lincoln broome	*Bromus* spp.
Reed canary grass	*Phalaris arundinacea*	Timothy	*Phleum pratenses*
Mint	*Mentha* spp.	Coontail*	*Ceratophyllum demersum*
Papyrus	*Cyperus alternifolius*	Duckweed*	*Lemmna* spp.
Watercress	*Nasturtium officinale*	Jasmine*	*Jasminum officinale*
Blue flag	*Iris versicolor*	Ragweed+	*Ambrosia psilostachya*
Yellow flag	*Iris pseudacorus*	Russian thistle+	*Salsola kali*
Daylily	*Hemerocallis dumortieri*	Elephant ears*	*Colocasia esculenta* var. *antiquorum*
Trumpet vine	*Campsis tagliabuana*		
Common reed	*Phragmities communis*	Siberian elm+	*Ulmus pumila*
Flat sedge	*Cyperus odoratus*	Tomato	Solanaceae
Jimson weed	*Datura meteloides*	Watermelon+	Cucurbitaceae
Narcissus	*Narcissus* spp.	Zucchini+	Cucurbitaceae
Variegated grass	*Phalaris arundinacea 'picta'*	Groundsel+	*Senecio douglasii*

* See text.
\+ See text.

oxidation-reduction potential (ORP) "Pocket Pal" meters. Water quality testing was performed once a week with samples returned to a small laboratory. Samples were taken from the equalization tank, from the lagoon, from P2 sump, and from the effluent sump (P3). These samples were analyzed for the following:

1. COD: Hach COD testing procedures, EPA approved
2. Ammonia: Hach Nessler Reagent/Colorimeter
3. Nitrate: Hach cadmium/colorimeter
4. Orthophosphate: Hach colorimeter

OPERATIONS

During the first year of operations, 492 m^3 (130,000 gal) of septage were processed. The system was designed to be self-regulating. If P1 failed, then flow ceased. No other piece of equipment is essential for operations. As long as P1 is pumping, the system was treating septage. In fact, the blower failed on three separate occasions. Surprisingly, treatment, when measured using COD as a basis, was not affected by a significant amount. This suggests a further experiment for the coming year.

The reed bed, added in the 11th month of operation, and the internal marsh required a program of regular sludge removal. This was done once a week, requiring approximately 1 man-hour of labor. The sludge was placed in a compost heap. Occasionally, water hyacinths were removed from the lagoon.

Growth of the water hyacinths was severely curtailed during the winter due to extremely cold temperatures. The greenhouse was designed for –18°C (0°F) night-time minimum temperatures; this past winter, the minimums were consistently in the –23 to –31°C (–10 to –24°F) range for several weeks. The hyacinths had leaves frost bitten; but because of the elevated water temperature, did not die. Some of other less cold-tolerant species were also affected, and did not survive. Dead plants were removed and composted.

It is important to stress that except for routine sludge removal and composting, this system operated without any operator involvement. Mechanical failures, not horticultural, or bacterial problems were the only problems that occurred. Of the mechanical failures, the clogging of P1 was the only problem that caused operations to cease.

BIOLOGICAL OBSERVATIONS

Plants

Some of the water hyacinths in the ET began to die, probably because of heat and environmental stress. Air temperatures at the water surface often exceeded 50°C, and ammonia levels in the water were routinely above 80 mg/L. Die-off began at the edges of the hyacinth leaves and progressed via increasingly larger areas of browning and brittle surfaces. No hyacinths survived in the ET. Hyacinths in the lagoon have continued to thrive and completely covered the entire surface. Harvest of some hyacinths is necessary. The hyacinths did not bloom. All of the marsh plants are doing well and are continuing to show new growth. Viable seeds and/or new shoots are common on all species.

There are several species of dry-land plants normally associated with arid environments that are growing in the external marsh. One of the more interesting species is *Datura meteloides*. There are numerous volunteers including Siberian Elm, Chinese Lantern, tomatoes, watermelon, Russian thistle, ragweed, as well as others listed in Table 1. Many different species of plants have been tried, and some have proven very successful. Those plants marked with an asterisk were planted, but did not survive. Those marked with a plus are volunteers.

When the air was turned off, large islands of sludge began to rise from the bottom of both the lagoon and the ET. These surfaces were colonized within 3 days by algae and snails. Bubbles of gas were observed to rise and break continuously over the entire surface of both lagoon and ET. Gas production was quite remarkable; it was similar to shaking a carbonated drink.

Insects

Insect populations are well controlled by predation. Since the sides of the greenhouse are open for all but the coldest days, birds and insect predators are free to enter the greenhouse. There are numerous ladybird beetles and various species of spiders preying on aphid and filter fly populations. During the winter months, both aphids and filter flies were very numerous and a nuisance.

The hyacinths are being eaten by two species of caterpillars. Occasionally, spider mites are a problem. Spider mites can be controlled by occasional spraying with fresh water. Both pests are also kept in balance by spiders, and in the case of the caterpillars, by birds. No chemical sprays are used. The plant ecophene has been developed for survivability and minimum care. Earthworms have colonized both the marshes and the sludge around the hyacinth roots. The only other significant insect is the larva of a drone fly. Honey bees are nesting in the ET room of the greenhouse.

Bacterial and Protozoan Communities

During the last 6 months of the pilot, attempts were made to document the various bacterial populations of each of the various stages of treatment. Since typing species is beyond the capabilities of this pilot project, a more limited approach was made using slide film and a microscope. Approximately 100 slides have been made of the various bacterial and protozoan populations in the various parts of the system. Cursory examination of these various populations show a highly diverse and very different microbial community at each stage of the process. The microbial community is presumed similar to that enumerated by Dinges.[7]

Perhaps the most noticeable difference is the absence of ciliates in the effluent and the appearance of copepods in the P2 sump. Slides of the ET and the lagoon show numerous populations of various different ciliates, ascarids (probably), and numerous encysted ciliates and/or egg cases of ascarids, toxascarids, etc. All of these populations are noticeably absent in the effluent. Another interesting phenomenon is the colonization or absence of colonization by cyanobacteria of various samples of water when left in the sun. Effluent and reed bed samples are readily colonized by cyanobacteria and turn green, while samples from P2 remain brownish gray in color.

Fecal coliform counts were begun on the effluent using prepared growth media. Preliminary indications show fecal coliform counts in the 5000 to 10,000 colonies per 100-ml sample. This represents approximately a four log reduction. This is consistent with other reported results for natural systems.[9-13] More testing is required.

Besides the obvious public health issues, the determination of what populations constitute a healthy treatment system is the ultimate goal of typing the various portions of the treatment system. The same techniques used by operators of activated sludge systems should be the basis of good operational procedures.

SUMMARY OF TEST RESULTS

ORP

The ORP (oxidation-reduction potential) levels are descriptive of the biological activities and amounts of oxygen being delivered to the septage. There is obviously some correlation between COD, total oxygen delivered, and ORP; but perhaps more important is the understanding of the relative bacterial activities and populations that are implied for a given ORP reading. Those parts of the system that are aerated have higher ORP values than anaerobic areas. This is consistent with the biological activity present in the four areas of the system. ORP readings are used as a confirming technique rather than as a specific performance reading. It is sufficient to know that each stage of the process has a different environment and of a desirable kind; and that each stage is sufficiently different to produce the desired results. Average values for elements in the system are ET = –72.9 mV, lagoon = –135 mV, P2 = –71.7 mV, and P3 = –110.5 mV.

Continuous monitoring of various portions of the system with ORP probes might provide some interesting insights as to the way in which ORP changes over time. There are perhaps some cyclic process present that are visible in some physical process, but which are not well understood at this point.

TDS

Total dissolved solids (TDS) have a tremendous variability with large standard deviations. However, the differences between the mean influent and the mean effluent are not significant. The most that can be said is that natural treatment systems do not increase the level of TDS by any significant amount. Influent TDS averaged 637 mg/L and effluent averaged 695 mg/L.

Nitrates

Nitrate is clearly produced by aerating the P2 sump, and subsequently removed in the external marsh. Nitrate is removed by both plant uptake and denitrification. Nitrate is not produced in measurable amounts in any other part of the system. Although the equalization tank is aerated, it must be assumed that the organic loading chemical oxygen demand at this point is too high to produce any nitrates. Other chemical reactions predominate in the ET and the lagoon. The pilot is a classic demonstration of biochemical activity.[7,9,11] See Figure 2.

pH

pH for the various elements of the system have remained relatively constant. Influent and effluent average values have remained within a 0.1 pH unit. The system has remained almost invariably alkaline. The range is from 6.5 to 8.2, with means at 7.6 (ET), 7.2 (lagoon), 7.4 (P2), and 7.5 (P3-effluent). Standard deviations are 0.3 to 0.4 pH units. Measurements have been taken at the same time during the day (i.e., 3–5 p.m.). The one noticeable point is the drop in pH from the ET to the lagoon, and the subsequent rise in pH. The lagoon generally has the lowest pH. This can be attributed to the ammonia generation at this point in the system.

Ammonia

Ammonia levels are significantly reduced in the ET by exposing the septage to high levels of aeration (1.3 m^3/min [45 cfm]) for an average of 7 days. Anaerobic conditions in the lagoon generate ammonia from the remaining solids which are then removed in the two marshes. The linear decline of ammonia supports previous work by Gersberg[14] and Gearhart[12] that ammonia removals are directly related to retention times.

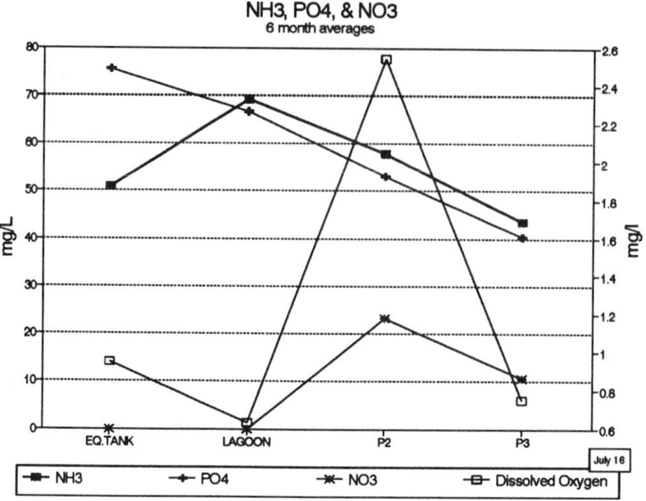

Figure 2. NH_3, PO_4, and NO_3.

Part of the apparent effectiveness of the internal marsh is related to the conversion of ammonia to nitrate in the P2 sump. Total nitrogen removed is therefore

Internal marsh Ce – Co = 1409 kg/ha/year (1300 lbs/ac/year)

External marsh Ce – Co = 1164 kg/ha/year (1074 lbs/ac/year)

where Ce is the influent concentration and Co is the effluent concentration. The apparent differences in removal rates are very likely due to temperature differences. These removal rates compare very favorably to overland flow systems which, depending upon the plants used, are about 40 to 50% of these removal rates. See Figure 2.

Phosphorus

There is a substantial decline in phosphorus throughout the system. If the EPA averages[9] are accepted as representative, the percentage reduction is 84%. This is extremely unlikely; based on experience with other wetlands systems, a more likely reduction is in the range of 50 to 60%.[10,11] Influent values are more likely to average 90 to 100 mg/L rather than 253 mg/L. Influent phosphorus levels are not measured by this program or by the City of Santa Fe. A sample program will be attempted during the second year. See Figure 2.

Chemical Oxygen Demand, COD

COD reduction is represented in a classical natural log reduction on the graph, although in fact, differential rates of reduction are occurring. The easiest and most volatile and chemically reactive substances requiring oxygen are oxidized first in significant amounts. COD is reduced by 18,260 mg/L in the ET and the hyacinth lagoon. The internal marsh and P2 sump (aerated) remove another 3400 mg/L and the external marsh a final 2537 mg/L. If COD removal is considered a time-dependent process, then we have the following removal rates:

COD removal rates (mg/L/day)

ET	Lagoon	Int. marsh	Ext. marsh
1714	1000	1360	265

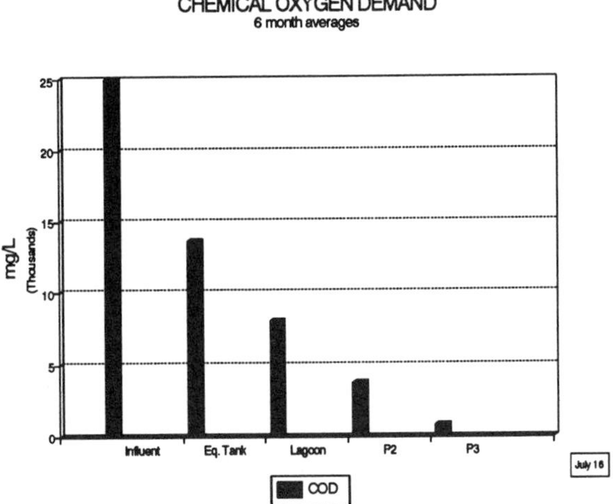

Figure 3. Chemical oxygen demand.

Both the ET and the internal marsh are more effective at removing COD because they are aerated or in the case of the internal marsh, the collection sump is aerated. The external marsh is treating water that has much of the solids and COD removed, such that comparative measures of performance are difficult. However, the lagoon is very likely more efficient because it removes COD without the addition of energy (in the form of air). The organic loading in the lagoon, in the internal marsh, and in the external marsh exceed previously reported loading rates for these types of natural treatment systems.

$$\text{Hyacinth lagoon removal rate} = 945 \text{ kg COD/ha/day}$$

$$(\text{Area} = 20 \text{ m}^2) = 315 \text{ kg BOD/ha/day}$$

The removal rates for the internal marsh are 1.36 times greater. These rates are about three to four times greater than generally reported elsewhere in the literature. Recommended loading rates are up to 110 kg/ha/day.[11]

COD reduction is of course not the only desirable end result. Conversion of amino acids and proteins to nitrates, phosphates, sulfates, carbon dioxide, water, and hydrochloric acid still requires the assistance of bacterial populations and plants that begin to do their work at the lower levels of COD. The last two marshes therefore perform the nitrogen and phosphorus removal functions without showing as noticeable reductions in COD, and without the requirement for additional oxygen. See Figure 3.

Retention Times

Treatment for any biologically based process is time-dependent. The average retention times for each portion of the systems are as follows: ET = 7 days; reed bed = 1 day; lagoon = 5 days; internal CW = 2.4 days; external CW = 9 days; and total retention time = 24 days, assuming that there is no short-circuiting.

ENERGY EFFICIENCIES

The energy cost of treating septage with aeration was approximately 0.0167 kWh/day/L, and 0.00132 kWh/day/L without air. At electric rates of 12.72 cents/kWh, the cost of treating 2385 L/day (630 gpd), which is the maximum flow rate, (3 kWh/day) without aeration was \$0.38/day or \$0.00016 per liter. With aeration, then the daily cost was \$5.05, and the cost per liter was \$0.002. Without air, an additional 11.3 g/day of ammonia was discharged from the P3 sump as fertilizer in the irrigation water. The cost of removing this 11.3 g was \$4.67, or \$0.41/g.

Table 2. Energy and Operating Cost

System location	Energy required (kWh/1000 liters)	Operating cost (cents/liter)	Ref.
Weyland/Sudbury, MA	9.5	5	15
Arroyo Hondo, NM	16.6	1	—
Yarmouth, MA	34.3 est.	7.9 proj.	15
Harwich, MA	39.1 est.	5.3	15
Orleans, MA	42.3 est.	6.6	16

Table 3. Water Quality Results

Parameter	Influent	Effluent	% Removed
COD	25,000	769	97
NH_3-N	157	44	72
TKN	677	55	92
NO_3-N		11	
Phosphate	253	40	84
Total Solids	38,800	20	99+

Clearly, the above calculations indicate that ammonia removal utilizing aeration is an expensive proposition; and since the effluent is being used to irrigate, the ammonia is better left in the effluent, provided of course that the ammonia is not at phytotoxic levels.

Using other septage treatment facilities for energy and cost comparisons, it is apparent that energy and treatment costs can vary over a wide range. Land application of the effluent as in the Arroyo Hondo system is apparently the key to low operating costs. Table 2 summarizes five systems.

While the Arroyo Hondo facility is not the most energy efficient (Weyland/Sudbury uses lime for solids removal), the cost savings of this small-scale system are significantly greater (i.e., 5 to 6 times greater than the other treatment technologies). It is reasonable to expect that efficiencies will be increased as the system is scaled up. The principal reasons for the cost savings are:

1. The reed bed is a very effective means of handling stabilized solids
2. Land application of treated effluent offers significant savings over any other treatment option for ammonia and nitrate removal

The use of air instead of lime to stabilize the solids allows the solids to be pumped out onto the reed bed. *In situ* de-watering and mineralization are important volume reduction processes that are passive, with zero mechanical energy inputs. After 8 to 10 years, the reed bed solids can be harvested and composted. Owing to the natural processes taking place in the reed bed which greatly reduce volumes, solids handling costs are minimized.

CONCLUSIONS

After 365 days of operation, generalizations are tempting to make and everyone would like to look at 1, 2, or even 3 years of data before reaching any major conclusions. However, having expressed the sentiments of caution, it is fair to say that the system is capable of: (1) treating septage equivalent to secondary treatment standards, and (2) treating septage for negligible amounts of dollars and energy.

An equally important consequence of this pilot is the demonstrated reuse and reclamation of a waste stream. Beneficial reuse at such low operational costs demonstrate conclusively the potential of natural treatment systems for the treatment of septage. The results summarized in Table 3 are even more significant when mass balances are calculated. Due to evaporation and plant transpiration, the effluent volume is 50% of the influent volume.

Finally, it is important to stress that effective treatment systems need to incorporate multiple ecologies; for example, aerated ponds, reed beds, constructed wetlands, and meadows (overland flow). The use of these various different ecologies allow a much greater range of biochemical, chemical, and physical reactions to occur. The designer must therefore understand the range of possibilities and incorporate the desired processes in the proper sequence.

ACKNOWLEDGMENTS

This project would not have been possible without the very generous financial support of "How on Earth", an environmental company and Mrs. Leslie Barclay. Thanks to Thomas Andrews, P.E. and Sherwood Reed, P.E. for their technical support and review of data and manuscript.

REFERENCES

1. Seidel, K., H. Happel, and G. Graue. Contributions to Revitalization of Waters. Max Plank Institute, Krefeld-Hulserberg, Germany, 1976.
2. Small, M. M. Wetlands wastewater treatment systems, presented at the *Int. Symp. State of Knowledge in Land Treatment of Wastewater*. Hanover, NH, August 1978.
3. Wang, B. The development of ecological wastewater treatment and utilization systems (EWTUS) in China. *Water Sci. Technol.* 19: 1987.
4. Crites, R. W. Final Report-Harwich Septage Treatment Pilot Study. Evaluation of Technology for Solar Aquatic Septage Treatment System. Nolte and Assoc., Sacramento, CA, April 1989.
5. Tchbanoglous, G. et al. Evolution and performance of City of San Diego pilot scale aquatic wastewater treatment system using water hyacinths, presented at the *60th Annu. Conf. of the Water Pollution Control Federation*. Philadelphia, PA, October 1987.
6. Doersam, J. The City of Austin experience with water hyacinths in wastewater treatment, presented at the *Annu. Meeting and Exposition of the Texas Water Pollution Control Assoc.* June 1987.
7. Dinges, R. *Natural Systems for Water Pollution Control.* Van Nostrand Reinhold, New York, 1983.
8. Valiela, I., J. M. Teal, and W. J. Sass. Production and dynamics of salt marsh vegetation and the effects of experimental treatment with sewage sludge, *J. Appl. Ecol.* 12(12):973, 1975.
9. Hammer, D. A., Ed. *Constructed Wetlands for Wastewater Treatment.* Lewis Publishers, Chelsea, MI, 1989.
10. Kreissl, J. F. *Septage Treatment and Disposal.* U.S. EPA, Cincinnati, OH, October 1984. Reed, S. C., E. J. Middlebrooks, and R. W. Crites, *Natural Systems for Waste Management and Treatment.* McGraw-Hill, New York, 1988.
12. Gearheart, R. A. *Nitrogen Removal in Wetlands.* Humbolt State University, Arcata, CA, 1989.
13. Herskowitz, J. Listowel Artificial Marsh Project Report. Ministry of the Environment. Toronto, Ontario. October 1986.
14. Gersberg, R. M., B. V. Elkins, and C. R. Goldman. Use of artificial wetlands to remove nitrogen from water. *J. Water Pollution Control Fed.* 56(2):152, 1984.
15. Giggey, M. Wright-Pierce Engineers. Personal communication, 1991.
16. John, T. Ocean Arks International. Personal communication, 1991.

CHAPTER 59

Constructed Wetlands Wastewater Quality Improvement at Lynnvale Elementary School

T. B. Terry, III

INTRODUCTION

In September 1989, the Hardin County Board of Education was informed that the discharge permit for the sand filter utilized for the treatment of wastewater at Lynnvale Elementary School would not be renewed by Kentucky Department for Environmental Protection. The Deputy Superintendent of Hardin County Schools, as well as other members of his staff, met with representatives of the environmental staff of Lincoln Trail District Health Department to discuss alternative on-site sewage disposal systems. A constructed wetlands (CW) was one of the alternatives presented to the group.

A number of existing CW systems were reviewed by the Board of Education before making the final decision to proceed with the design of a CW at Lynnvale Elementary School. In November 1989, the Hardin County Board of Education held a public meeting in White Mills, Kentucky to explain the nature of the CW sewage treatment process to members of the community surrounding Lynnvale Elementary School.

The CW system was completed in May 1990. A 3-year monitoring study was initiated in July 1990 by the Lincoln Trail District Health Department. This article will describe the design of the system and present the data collected through September 1991.

DESIGN AND CONSTRUCTION DETAILS

Formulas

At the time of construction, the only sizing criteria for CW systems in Kentucky came from a set of guidelines obtained from the Tennessee Valley Authority.[1]

The surface area is calculated from Equation 1.

$$A_s = \frac{Q_f}{Q_l} \qquad (60\text{–}1)$$

where A_s = surface area of CW, ft^2;
Q_f = average daily wastewater flow, gal/day; and
Q_l = hydraulic loading rate, ft^2/gal/day.

The cross-sectional area of the CW cells (A_x) is determined using Darcy's law.

$$A_x = \frac{Q_f}{K_s^* S} \qquad (60\text{–}2)$$

0-87371-550-0/93/$0.00+$.50

where Q_f = average daily wastewater flow, ft^3/day;
K_s = substrate hydraulic conductivity (long term) ft/s;
S = hydraulic gradient (bed slope), ft/ft.

The cell width (W) is calculated from Equation 3.

$$W = \frac{A_x}{D} \qquad (60\text{–}3)$$

where D = starting cell depth, ft.

GENERAL CONDITIONS

The maximum daily wastewater design flow was 1800 gal/day, based on actual water usage during the previous 12 months. The system (Figure 1) includes the original septic tank for primary treatment; three 1000-gal septic tanks in series for additional treatment and storage capacity; constructed wetlands for secondary/advanced treatment; a pump chamber with an effluent pump; and 420-ft of low-pressure pipe sewage disposal system for final disposal of any discharge from the wetlands.

CONSTRUCTED WETLANDS

The CW is of the subsurface flow type, with gravel as the principal substrate. It consists of two cells in series. Sewage from the last septic tank flows to the first cell, where a major portion of the suspended solids and carbonaceous BOD (biochemical oxygen demand) is removed. The first and second cell are both lined with a synthetic liner that is 30 mil. (30/1000 in.) thick.

Cell 1 is 27 ft wide by 22 ft long on the bottom of the bed. The bottom of the bed has a 1% slope, while the top surface is flat. The principal substrate is $^3/_4$- to 1-in. sized washed river rock with a starting depth of 18 in. Figure 2 shows how the wastewater enters the wetlands in a 4-in. PVC pipe connected to a 4-in. header that has five swivel tees spaced uniformly across the cell width. The header was leveled to insure equal distribution. The flow will enter the 4- to 6-in. stone under the header and distribute into the gravel substrate.

Wastewater is collected in a 4-in. PVC pipe placed level on the bottom of the bed at the end of the cell. Wastewater enters the collection pipe through $^3/_4$ -in. holes spaced uniformly 6 in. apart. The collection pipe connects by way of a tee to a 4-in. pipe leading to the water level control structure.

Normal water depth in the wetland cells is maintained about 1 in. below the gravel substrate. This is accomplished utilizing an adjustable standpipe. Discharge from cell 1 by way of the standpipe is distributed subsurface into cell 2.

Cell 2 is 27 ft wide by 44 ft long on the bottom of the bed. The bottom of the bed has a 1% slope, while the top surface is level. The principal substrate is $^3/_4$- to 1-in. sized washed river rock, with an average depth of 15 in. On top of the gravel, 3 to 5 in. top soil was placed, followed by an additional 1 to 3 in. shredded bark.

Wastewater enters this cell through a 4-in. pipe connected to a 4-in. header across the front bottom of the bed. The header is leveled and $^3/_4$-in. holes were drilled uniformly along its length. Wastewater exits the header into the gravel substrate. Collection and discharge is through a water level control structure like the one at the end of cell 1.

PLANTS

Cattail (*Typha latifolia*) and pickerelweed (*Pontederia cordata*) were planted in cell 1. Sweet flag (*Acorus calamus*), blue water iris (*Iris versicolor*), arrowhead (*Sagittaria latifolia*), and horsetail (*Equisetum hyemale*) were planted into cell 2.

FINAL DISPOSAL

Final disposal of any discharge from wetland cell 2 is into a pump chamber. From there, it passes through a water meter and into 420 ft low-pressure pipe lateral trenches.

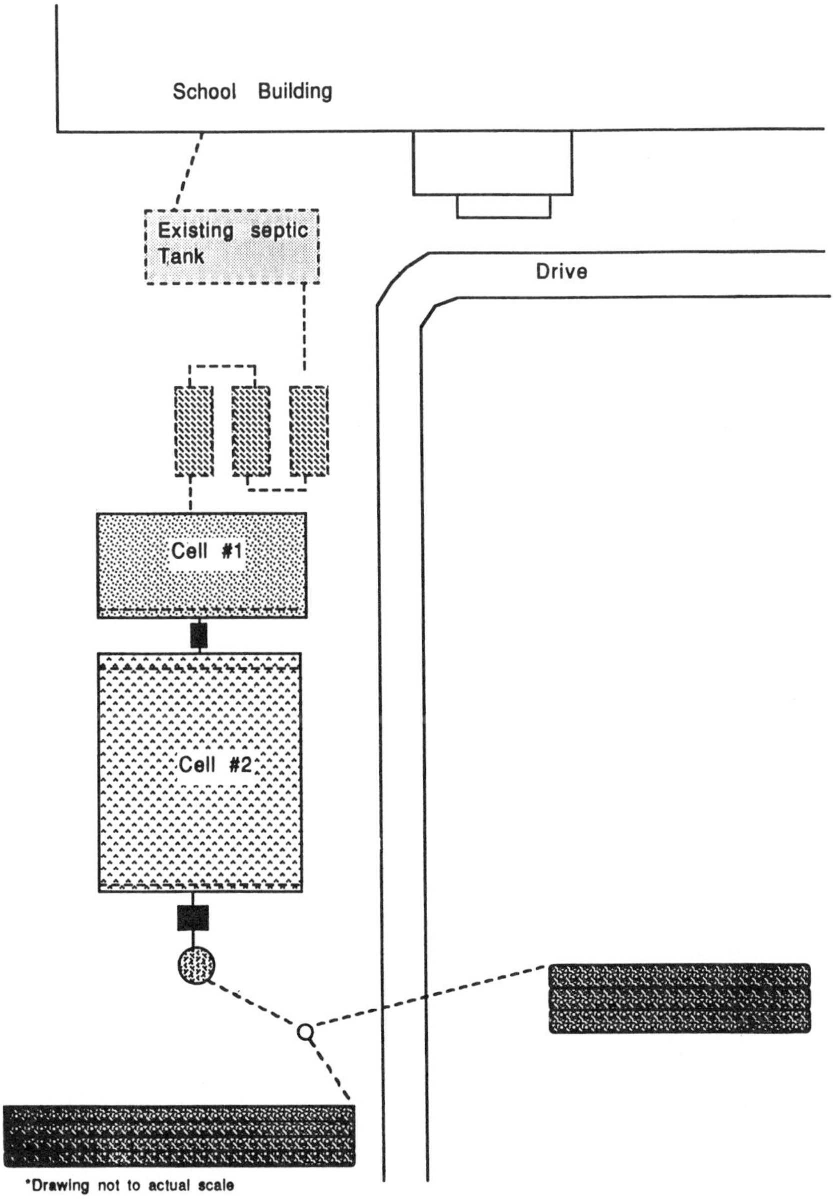

School Building
Existing septic Tank
Drive
Cell #1
Cell #2
*Drawing not to actual scale

Figure 1. Constructed wetlands at Lynnvale Elementary School, Kentucky.

BARK
TOPSOIL

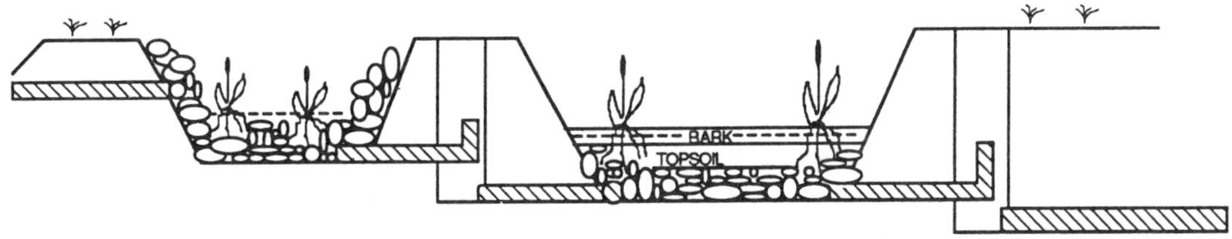

Figure 2. Profile view of constructed wetlands at Lynnvale Elementary School, Kentucky.

Table 1. Monthly BOD_5 Levels: Lynnvale Elementary School CW System

Sampling date	Septic tank (mg/L)	End of cell #1 (mg/L)	% removal	End of cell #2 (mg/L)	% removal	Cumulative % removal
July 31, 1990	60	17	72	21	—	65
Aug. 28, 1990	60	16	73	7	56	88
Oct. 1, 1990	221	38	83	112	—	49
Oct. 23, 1990	763	173	77	52	70	93
Nov. 14, 1990	265	130	51	75	42	72
Dec. 12, 1990	737	226	69	80	65	89
Jan. 29, 1991	682	247	64	247	—	64
Feb. 19, 1991	468	117	75	40	66	91
Mar. 19, 1991	163	39	76	37	5	77
Apr. 23, 1991	541	73	87	17	77	97
May 21, 1991	334	64	81	23	64	93
June 11, 1991	92	26	72	15	42	84
July 16, 1991	66	17	74	11	35	83
Aug. 20, 1991	16	10	38	6	40	63
Sep. 17, 1991	315	30	90	16	47	95

DESIGN SUMMARY

CW systems, as with most on-site disposal systems, are quite site-specific. At this site, the original design criteria specified the width to be 20 ft and the total length of the combined cells to be 89 ft in order to obtain the required surface area. This site was not suited to these dimensions; thus, the width was increased to 27 ft and the combined length was downsized to 66 ft. This allowed for placement of the two cells in series and still maintain the required surface area.

Also, the original plan specified that the second cell be unlined in order to facilitate percolation of some of the treated wastewater. Upon excavation of the second cell, a layer of sand was discovered that would lead to an increased percolation rate. This would have created problems, both in keeping the plants alive and in preventing ground water contamination. Thus, the decision was made to line the second cell as well as the first cell with a 30-mil liner.

RESULTS AND DISCUSSION

Performance results are shown in Tables 1, 2, and 3. The average effluent levels for BOD_5 were reduced from an average of 319 to 51 mg/L over the course of the first 15 months of monitoring. That is a cumulative removal rate of 85%. Even more encouraging is the fact that the average level of reduction for the first 3 months was 59%, but the level of reduction for the last 3 months increased to a reduction rate of 92%.

The average effluent levels for TSS (total suspended solids) were reduced from an average of 221 to 46 mg/L over the course of the first 15 months; that is a cumulative removal rate of 80%. The average level of reduction for the first 3 months was 42%; for the last 3 months, the rate of reduction increased to a level of 68%.

The average effluent levels of fecal coliform bacteria were reduced from an average colony count of 47,000 to 1043/100 ml over the course of the first 15 months; that is a cumulative removal rate of 98%. This level of reduction has remained consistent throughout the course of this study.

Results appear to be improving with time, but long-term studies must take into account seasonal variations, a decreased wasteload during school recesses (particulary during the months of June through August), and the effects of system maturity.

Table 2. Monthly TSS Levels: Lynnvale Elementary School CW System

Sampling date	Septic tank (mg/L)	End of cell #1 (mg/L)	% removal	End of cell #2 (mg/L)	% removal	Cumulative % removal
July 31, 1990	42	30	29	28	7	33
Aug. 28, 1990	56	12	79	23	—	60
Oct. 1, 1990	143	20	86	91	—	36
Oct. 23, 1990	289	44	85	37	16	87
Nov. 14, 1990	410	39	90	49	—	88
Dec. 12, 1990	327	58	82	41	29	87
Jan. 29, 1991	400	84	79	55	35	86
Feb. 19, 1991	460	46	90	44	4	90
Mar. 19, 1991	183	44	76	49	—	73
Apr. 23, 1991	388	40	90	43	—	89
May 21, 1991	167	31	81	61	—	63
June 11, 1991	60	19	68	42	—	30
July 16, 1991	293	16	95	60	—	80
Aug. 20, 1991	13	4	69	18	—	—
Sep. 17, 1991	86	23	73	51	—	41

Table 3. Monthly Fecal Coliform Levels: Lynnvale Elementary School CW System

Sampling date	Septic tank (mg/L)	End of cell #1 (mg/L)	% removal	End of cell #2 (mg/L)	% removal	Cumulative % removal
July 31, 1990	4000	<1	100	30	—	99
Aug. 28, 1990	<1	<1	—	<1	—	—
Oct. 1, 1990	10000[a]	5000[a]	50	130	97	99
Oct. 23, 1990	10000[a]	7000[a]	30	4000[a]	43	60
Nov. 14, 1990	20000[a]	1200[a]	94	5000[a]	—	75
Dec. 12, 1990	20000[a]	10000[a]	50	2080	79	90
Jan. 29, 1991	50000[a]	1000[a]	98	920	8	98
Feb. 19, 1991	60000[a]	10000	83	260	97	99.6
Mar. 19, 1991	40000[a]	2000[a]	95	72	96	99.8
Apr. 23, 1991	400000[a]	10000[a]	98	680	93	99.8
May 21, 1991	20000[a]	110	99	70	36	99.7
June 11, 1991	8000[a]	600	93	650	—	92
July 16, 1991	10000[a]	2000[a]	80	860	57	91
Aug. 20, 1991	3000[a]	585	81	890	—	70
Sep. 17, 1991	50000[a]	2000[a]	96	5	99.8	99.9

[a] Estimated value.

CONCLUSIONS

It is too early to draw conclusions, but the sampling data are encouraging. What can be concluded is that a failing sand filter that was polluting the environment has been replaced with an effective (short-term results) and attractive wastewater treatment system.

Weed growth in the second cell has been a problem from a visual standpoint, but does not appear to have affected the system's performance. The use of soil and shredded bark on the top of the second cell needs to be evaluated on any future systems. If a soil cap is needed, then some method of sterilization to destroy the weed seeds needs to be incorporated into the design process.

The amount of discharge from the CW system needs to be studied in order to determine the amount of subsurface lateral trenches required for final disposal. Continuous evaluation of data from this site and others throughout the state of Kentucky over the next few years may provide answers to these and other questions about the use of CW for on-site sewage disposal.

ACKNOWLEDGMENTS

This study was supported by a grant from the Kentucky Cabinet for Human Resources, Division of Local Health. The Lincoln Trail District Health Department also provided financial support.

REFERENCE

1. Tennessee Valley Authority, unpublished design guidelines, 1989.

CHAPTER 60

Microbial Populations and Decomposition Activity in Three Subsurface Flow Constructed Wetlands

K. Hatano, C. C. Trettin, C. H. House, and A. G. Wollum, II

INTRODUCTION

Constructed wetlands are efficient wastewater treatment systems characterized by low capital operating cost, low energy requirements, and many positive environmental attributes.[1] Microorganisms are the primary agent for removal of organic matter from wastewaters in constructed wetlands. The configuration and operation of these systems provide specific environments that can support microbe-mediated decomposition reactions.[2-5] The objectives of the research presented here are to characterize the microbial populations under three vegetation regimes in a constructed wetland receiving domestic wastewater, and to assess the significant contribution of individual types of organisms to organic matter degradation.

METHODS

Site Description

The constructed wetlands under study are located in eastern North Carolina near New Bern. This system combines an upland, wetland edge and a wetland (Figure 1).[6] The upland is a sand mound (9 × 3 × 0.75 m) capped with 20 cm of silt loam topsoil and planted with fescue (*Festuca arundinacea*). The entire system is underlaid by an impervious 20-mm PVC liner that slopes from the mound toward three gravel-filled wetland cells (3 × 3 × 0.6 m) containing either the cattail (*Typha angustifolia*), the common reed (*Phragmites australis*), or no vegetation. Septic tank effluent of domestic wastewater from a single-family dwelling is dosed into the mound through a pressure distribution network at a 40 L/m^2/day loading rate.

Sample Collection and Preparation

The wetland cells were sampled on January 21, April 2, June 10, August 5, and October 8, 1991. Duplicate gravel samples (ca. 100 g, wet weight) were collected at the inlet, middle, and outlet of each cell. The gravel samples were collected 15 to 20 cm below the surface of the gravel bed. Root samples from the cattails and reeds were collected near the outlet. An entire plant was extracted and the root portion removed. From these, root samples were obtained from fine roots and stored below 8°C until analyzed.

Gravel samples (1 g of dry weight; size, 2 to 3 mm) were suspended in 9 ml sterile water, and the suspension then treated for 10 min in an ultrasonic cleaner bath (45 KHz, FS5, Fisher Scientific, U.S.A.) to disperse the microorganisms attached to the surface of gravel. Thereafter, the treated suspension was serially diluted 10-fold with sterile water. Fine roots (0.5 g, wet weight) were placed in 9 ml sterile water, and processed as described for the gravel samples. Organisms enumerated in this manner were thought to represent those occurring in the rhizosphere.

Isolation and Enumeration of Microorganisms

Bacteria

Aliquots (0.10 ml) of the diluted suspension were spread on TY-agar plate (0.1% glucose, 0.5% Tryptone, 0.25% yeast extract, and 1.8% agar), and TY-CV agar plate (supplemented with 4 μg/ml crystal violet for

0-87371-550-0/93/$0.00+$.50

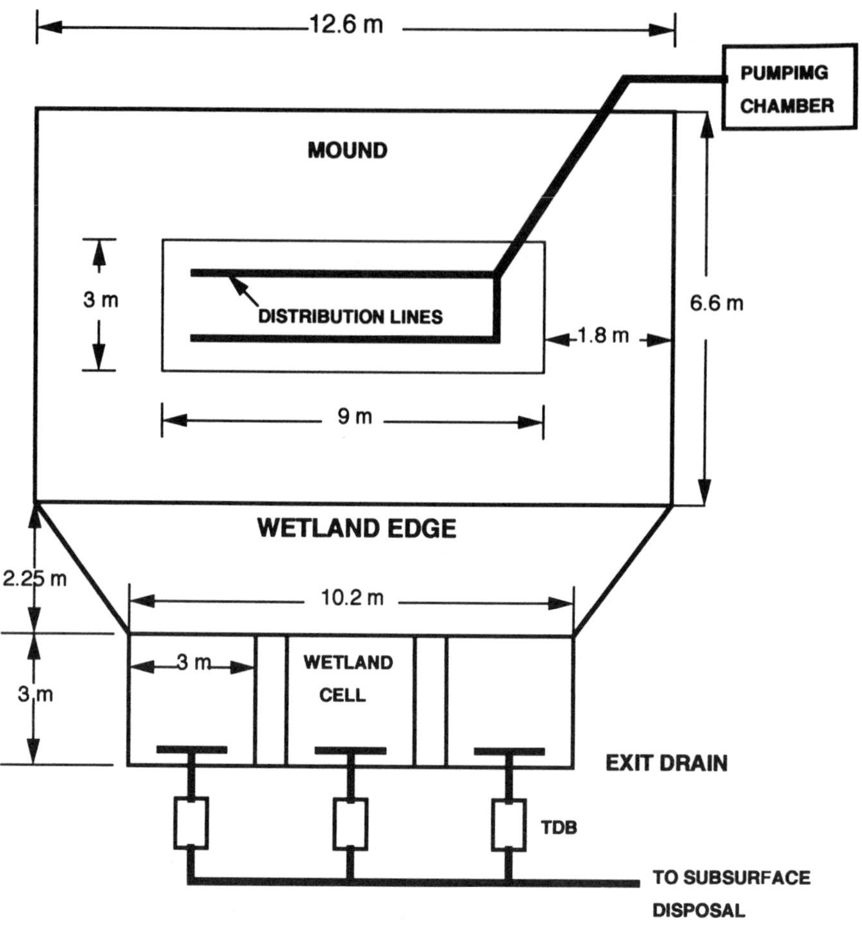

Figure 1. The constructed wetland system, combining an upland, wetland edge, and wetland treating domestic wastewater from a single-family dwelling.

inhibiting the growth of Gram-positive bacteria). The inoculated agar plates were incubated at 24°C for 3 days, and colonies of bacteria counted.

Actinomycetes

The soil suspensions were treated at 55°C for 5 min in a water bath and then immediately cooled in cold water. At this point, 0.1-ml aliquots of the treated suspensions were spread on starch-casein agar plates[7] whose medium consisted of 1% soluble starch, 0.03% lactic casein, 0.2% potassium nitrate, 0.2% sodium chloride, 0.005% magnesium sulfate, 0.002% calcium carbonate, 0.001% ferrous sulfate, 0.2% dipotassium phosphate, and 1.8% agar; at pH 7.5. After autoclaving the medium, cycloheximide is added to a final concentration of 40 μg/ml. The inoculated agar plates were then incubated at 24°C for 5 to 10 days, and representative actinomycetes counted.

Fungi

Similarly, 0.1-ml aliquots of diluted suspension were spread on Martin rose bengal agar plates (1% glucose, 0.5% peptone, 0.1% potassium phosphate, 0.05% magnesium sulfate, 0.003% rose bengal, and 1.8% agar; at pH 5.0). After autoclaving, streptomycin was added to a final concentration of 30 μg/ml in the medium. The inoculated agar plates were incubated at 24°C for 4 days and developing fungi counted.

Representative colonies from each medium were selected and inoculated on agar slants consisting of yeast extract-malt extract agar (1% malt extract, 0.4% yeast extract, 0.4% glucose, and 1.8% agar; at pH 7.2) for actinomycetes; Trypticase Soy Agar (BBL) for bacteria, and potato dextrose agar (Difco) for fungi. The inoculated slants were allowed to develop at 24°C for 7 to 10 days and then stored at 4°C.

Detection of Enzymatic Activity

Amylase, cellulase, chitinase, xylanase, and protease activities of microorganisms isolated from the gravel samples were characterized. Modified M9 minimal medium[9] (containing 0.05% of yeast extract) was used for the detection of amylase, chitinase, and xylanase. Semisynthetic agar medium[10] was used for the detection of cellulase. Skim-milk agar was used for detection of protease. Amylase was detected by the iodo-starch reaction. Xylanase was detected by the method of Treather and Wood[11] using 0.1% Congo red. Cellulase, chitinase, and protease activity were detected by the clear zone generated by hydrolysis of cellulose, chitin and milk casein, respectively.

RESULTS AND DISCUSSION

Microbial Populations

The gravel substrate of planted cells had greater microbial populations than the unplanted cell. Between the planted cells, the reed cell tended to have the highest microbial population. The average number of bacteria in the gravel bed for five sampling times was 3.9×10^5 colony-forming unit (cfu)/g of dry gravel in the unplanted cell, 8.6×10^5 cfu/g in the cattail cell, and 4.6×10^6 cfu/g in the reed cell. The average number of actinomycetes were 2.8×10^4, 6.0×10^4, and 1.3×10^5 cfu/g in the unplanted, cattail, and reed cells, respectively. Similarily, the average number of fungi were 1.4×10^3, 3.3×10^3, and 4.0×10^4 cfu/g in the unplanted, cattail, and reed cells, respectively. The Gram-negative bacteria population size ranked similarly to the total bacteria count, but the absolute population size in each cell was lower.

Seasonal Variation in Microbial Population Size

Variation in population size of aerobic and facultative anaerobic organisms depends on the amount and variety of available nutrients, temperature, and the oxygen supply. Since each cell within the constructed wetland received the same effective nutrient input from the wastewater influent, the primary factors affecting the microbial population are considered to be temperature, oxygen supply, and vegetation which represents a potential nutrient, energy, and oxygen source.

Time of sampling had little effect on the bacteria populations for the control and cattail cells (Figure 2). In contrast, the reed cell exhibited a significant increase in these populations for the June and October sampling times. The actinomycete populations also exhibited a significant increase in August and October for the reed cell and an increase in August for the cattail cell, while in the control cell the population was not different among the collection dates (Figure 3). Populations of fungi were not different among collection dates in the control and cattail cells (Figure 4); however, the reed cells exhibited a significant increase for the June and October collection date samples.

These results demonstrate that there is a vegetation interaction by season for microbial populations in the constructed wetland cells. The relationship between microorganisms and the root-soil environment is complex[12] and may, in part, be due to differences in plant growth, which occurs from January to October. The complexity of this relationship is further illustrated by the lack of a consistent seasonal response of fungi and bacteria in the control and cattail cells. Oxygen diffusion could be a factor regulating the microbial population;[13] however, it has been demonstrated that cattails have a higher capacity for oxygen transport than reeds.[14] The significant seasonal microbial response in the reed cells could also be affected by root exudates, accumulated organic matter, or root turnover.

A comparison of microbial populations in the rhizosphere of the cattail and reed plants with those in the gravel bed emphasizes the important effect vegetation has on microbial populations. Microbial populations in the rhizosphere are two to three orders of magnitude greater than those in the gravel bed (Table 1). Populations also varied by species, with the cattail rhizosphere having higher bacteria than the reed, and conversely, the reeds had a greater fungal population than the cattail cells.

Decomposition Activities

Decomposition activities of bacteria, actinomycetes, and fungi, as measured by their action on starch, cellulose, chitin, xylan and milk casein, were different among the three constructed wetland cells (Table 2).

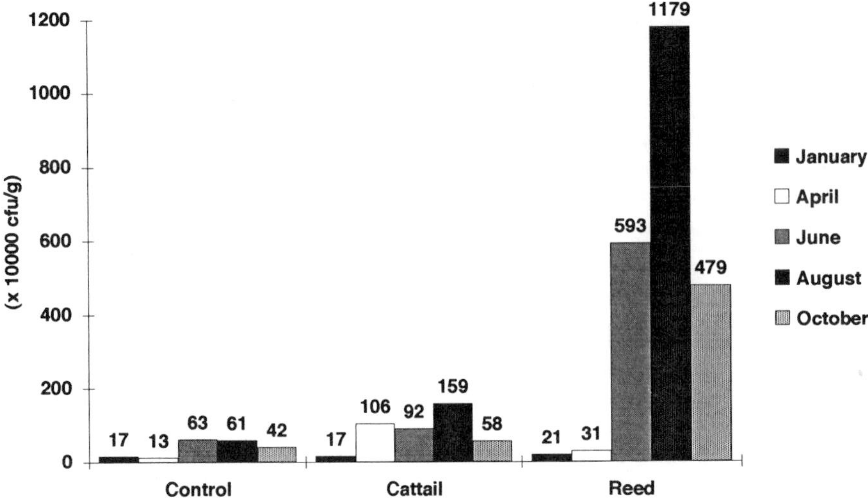

Figure 2. Populations of bacteria in gravel media from reed, cattail, and unplanted cells sampled on five dates in 1991. (Numbers indicate actual population size.)

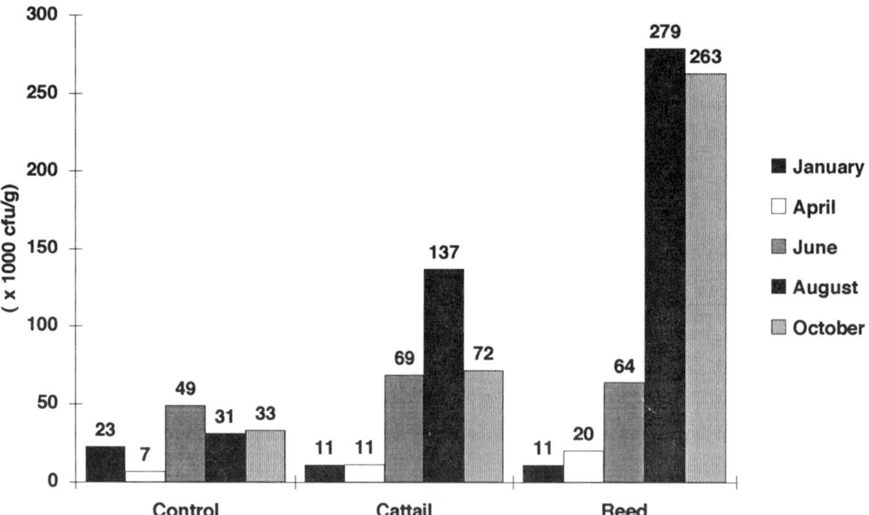

Figure 3. Populations of actinomycetes in gravel media from reed, cattail, and unplanted cells sampled on five dates in 1991. (Numbers indicate actual population size.)

Bacteria with the ability to degrade milk casein occurred in approximately 37% of total isolates, and approximately 27% exhibited the ability for starch hydrolysis. Only 6% bacteria were capable of decomposing xylan, which is a main component of hemicellulose. No bacteria having cellulase were detected in isolates from the three constructed wetland cells.

Actinomycetes had a wide range of hydrolysis activities. More than 70% of total isolates exhibited degradation of starch and chitin, and approximately 70% of isolates had protease activity. In addition, 20 and 40% of total isolates exhibited cellulose and xylan-decomposing activity, respectively.

The fungi also demonstrated a wide range of hydrolysis processes. They exhibited amylase and protease activity in approximately 50 and 70% of the isolates, respectively. More than 30% of the fungi showed chitinase and xylanase, while approximately 20% of the demonstrated cellulase activity.

These results suggest that actinomycetes, which have the ability to excrete a wide range of enzymes,[15] can produce relatively large populations and probably are important microorganisms for degrading organic matter

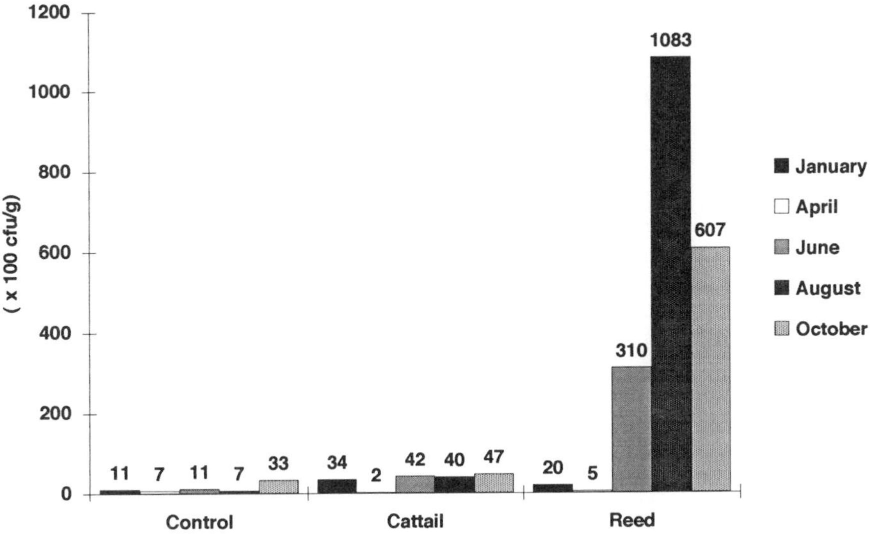

Figure 4. Populations of fungi in gravel media from reed, cattail, and unplanted cells sampled on five dates in 1991. (Numbers in indicate actual population size.)

Table 1. Comparison of Microorganisms Found in the Rhizosphere and Gravel of Three Subsurface Flow Constructed Wetland Cells in August 1991

Substrate	Bacteria	Actinomycetes	Fungi
Gravel[a]	0.6×10^6	3.1×10^4	1.0×10^3
Cattail gravel[a]	1.6×10^6	1.4×10^5	4.0×10^3
Cattail rhizosphere[b]	3.5×10^9	2.5×10^6	2.8×10^4
Reed gravel[a]	1.2×10^7	2.8×10^5	1.1×10^5
Reed rhizosphere[b]	0.6×10^9	1.3×10^6	1.6×10^6

[a] Expressed as cfu/g of dry weight.
[b] Expressed as cfu/g of wet weight.

Table 2. Proportion of the Bacteria, Actinomycetes, and Fungi Isolates Demonstrating Enzymatic Activity on Selected Substrates

Constructed wetland	Number of colonies	Amylase (%)	Cellulase (%)	Chitinase (%)	Xylanase (%)	Protease (%)
Control cell						
Bacteria	48	20	0	5	9	35
Actinomycetes	49	71	29	80	45	67
Fungi	12	63	44	38	25	80
Cattail cell						
Bacteria	47	38	0	11	3	39
Actinomycetes	47	74	17	74	43	64
Fungi	11	60	20	60	60	40
Reed cell						
Bacteria	47	29	0	16	7	36
Actinomycetes	39	87	23	78	33	66
Fungi	20	38	8	8	15	93

in wastewater. Bacteria, while high in biomass, have relatively low activities within the organic substrates tested. In spite of small populations, fungi exhibited a wide range of enzymatic activities, including cellulose hydrolysis. There appears to be a vegetation-fungi interaction in the cells tested. Fungal protease activity was lower in the cattail cells than either the control or reed cell, while chitinase and xylanase were significantly lower in the reed cell than in the cattail cell.

SUMMARY

This study demonstrated that plants have a significant effect on the microbial populations of constructed wetlands. These populations were dominated by bacteria; however, isolated actinomycetes exhibited a broader range of enzymatic activities. Fungal populations were relatively small, although they exhibited a wide range of enzymatic processes.

The reed cell tended to have higher microbial communities, particularly in the summer. Futhermore, the rhizosphere of the planted vegetation had a significantly greater microbial community than the gravel substrate, thus demonstrating the important role of vegetation for the development of microbial populations in wetlands.

Since most wastewater renovation in constructed wetlands is microbiologically mediated, it is important to understand the relationship of microorganisms to constructed wetland substrate types and vegetation. Results demonstrated that vegetation affects microorganism populations. Accordingly, the ability to manage the renovation process, particularly for a specific waste constituent, may be dependent on understanding the relationships between the organism, wastewater, substrate, and vegetation.

REFERENCES

1. Reed, S. C., E. J. Middlebrooks, and R.W. Crites. *Natural Systems for Waste Management and Treatment*. Mc-Graw Hill, New York, 1988.
2. Wood, A. Constructed wetlands for wastewater treatment — engineering and design considerations, in *Constructed Wetlands in Water Pollution Control*. P. F. Cooper and B. C. Findlater, Eds. Pergamon Press, Oxford, 1990, 481–494.
3. Portier, R. J. Evaluation of specific microbiological assays for constructed wetlands wastewater treatment management, in *Constructed Wetlands for Wastewater Treatment Municipal, Industrial and Agricultural*. D. A. Hammer, Ed. Lewis Publishers, Chelsea, MI, 1989.
4. Batal, W., L. S. Laudon, T. R. Wildeman, and N. Mohdnoordin. Bacteriological tests from the constructed wetland of the big five tunne Idaho spring, Colorado, in *Constructed Wetlands for Wastewater Treatment Municipal, Industrial and Agricultural*. D. A. Hammer, Ed. Lewis Publishers, Chelsea, MI, 1989.
5. Scheuerman, P. R., G. Bitton, and S. R. Farrah. Fate of microbial indicators and viruses in a forested wetland, in *Constructed Wetlands for Wastewater Treatment Municipal, Industrial and Agricultural*. D. A. Hammer, Ed. Lewis Publishers, Chelsea, MI, 1989.
6. House, C. H. and S. W. Broome. Constructed upland-wetland wastewater treatment system, in *Constructed Wetlands in Water Pollution Control*. P. F. Cooper and B. C. Findlater, Eds. Pergamon Press, New York, 1990, 567–570.
7. Kuster, E. and S. T. Williams. Selection of media for isolation of streptomycetes, *Nature*. 202:928, 1964.
8. Martin, J. P. Use of acid, rose bengal, and streptomycin in the plate method for estimating soil fungi, *Soil Science*. 69:215, 1950.
9. Maniatis, T., E. F. Fritsch, and J. Sanbrook. *Molecular Cloning, a Laboratory Manual*. Cold Spring Harbour Laboratory, New York, 1982, 150.
10. Eggins, H. O. W. and G. J. F. Pugh. Isolation of cellulose-decomposing fungi from the soil, *Nature*. 193:94, 1962.
11. Teather, R. M. and P. J. Wood. Use of congo red-polysaccharide interactions in enumeration and characterization of cellulolytic bacteria from the bovine rumen, *Appl. Environ. Microbiol*. 43:777, 1982.

12. Gunnison, D. and J. W. Barko. The rhizosphere ecology of submerged macrophytes, *Water Res. Bull.* 25:193, 1989.
13. Brix, H. Treatment of wastewater in the rhizophere of wetland plants — the root-zone method, *Water Sci. Technol.* 19:107, 1987.
14. McKee, K. L., I. A. Mendelssohn, and D. M. Burdick. Effect of long-term flooding on root metabolic response in five freshwater marsh plant species, *Can. J. Bot.* 67:3446, 1989.
15. Williams, S. T. Streptomycetes in biodeterioration — their relevance, detection and identification, *Int. Biodeterioration Bull.* 21:201, 1985.

CHAPTER 61

Biological Sludge Drying in Constructed Wetlands

S. M. Nielsen

INTRODUCTION

Traditionally, most of the sludge produced in Denmark is mechanically dewatered. Table 1 shows the dewatering capacities of the different dewatering methods at optimum operation.

Mechanical sludge dewatering involves conditioning with chemicals, usually in connection with the dewatering process itself. Either organic polyelectrolytes or inorganic conditioning substances are used. In biological dewatering, the process is governed by sun and wind energy, gravity, and vegetation. The dewatering process is facilitated by the natural conditioning of the sludge, either with root exudates or by freezing. The reject is more easily separated from the sludge, particularly after subsequent thawing.[2]

Water in sludge with a dry matter content of 5% can be divided into pore water (66.7%), capillary water (25%), and adsorption water and structurally bound water (8.3%).[3] Dewatering the pore concentrates the sludge to a dry matter content of approximately 15%. Further dewatering by removal of the capillary water could concentrate the sludge down to a dry matter content of approximately 50%. The rest of the water in the sludge can be removed by drying.[3] Figure 1 shows the physical quality of the sludge at different stages of dewatering.

REFERENCE PLANTS

The first sludge dewatering and mineralization systems were established in Denmark at Regstrup and Allerslev in the summer of 1988.[4] Since 1990, four new plants have been started, each with 3 to 10 beds with areas of 700 to 11,600 m^2. The systems consist of concrete or soil beds with drainage and aeration systems at the bottom.[4] Reeds (*Phragmites australis*) were planted in the beds; four plants per square meter are considered sufficient since it only takes about two growing seasons to cover the entire bed surface. Such systems often have an operating period of 8 to 10 years; the beds are emptied of sludge, reestablished, and a new operating period begins. See Table 2.

PRINCIPLES OF OPERATION

The efficiency of these sludge treatment plants is to a great extent dependent on their correct operation. Thus, it is important to match and adjust the amount loaded and the loading cycle according to the age of the plant and the quality of the sludge, especially as regards the dry matter content of the sludge. Sludge plant operation can be divided into four phases: loading (day 1–4); dewatering (day 5–7); aeration and mineralization (day 8–30).

RESULTS

Sludge Reduction

Sludge reduction consists of dewatering and mineralizing processes. Most of the reduction occurs during dewatering in the form of drain-off and evapotranspiration. For the volumes of sludge loaded and after

0-87371-550-0/93/$0.00+$.50

Table 1. Dry Matter Content Related to Dewatering Method

Dewatering method	Centrifuge	Filter belt press	Filter press	Traditional sludge bed	Biological sludge treatment
% dry matter	23 (15–20[a])	24 (15–20[a])	32	20[b]	40

[a] Normally observed values.

[b] Variable, dependent on the duration of the treatment period. Traditional sludge beds often dewater the sludge to a dry matter content of 10–20%.

From Reference 1, with modification.

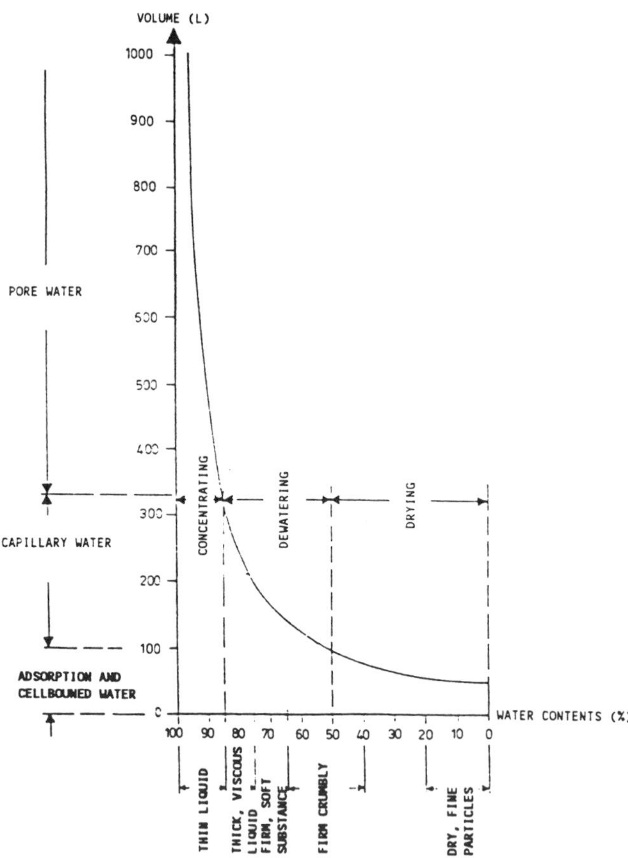

Figure 1. Variation in the water adsorption rate of the sludge compared to dry matter content. (From Reference 3, with modification.)

reduction, please see Tables 3 and 4. A reduction of approximately 98% (activated sludge) and 91% (concentrated sludge) was achieved with the sludge loaded onto the beds. The total sludge reduction for the entire monitoring period corresponded to a concentration of the activated and concentrated sludge of about 40, 10 (Regstrup), and 13 (Allerslev) times.

The relation between the loaded and dewatered sludge amounts was monitored. Figure 2 (A and B) shows the accumulated volumes of the loaded amounts of sludge. Also shown is the variation in volume of the resultant sludge residue during loading and rest periods.

At the Regstrup plant, the sludge residue increased by 21 and 16 m^3 after loadings of 831 and 991 m^3 in 1989 and 1990, respectively. Thus, the accumulation was 36% less in 1990 than in 1989 after adjustment for the difference in amounts of sludge loaded. Loading in 1989 did not follow the same pattern as that of other

Table 2. Reference Plants in Denmark — August 1991.

Municipality	Town	Date of opening	Load (pe)	Sludge (T TS)	Type of sludge	No. of beds	Area (total) (m^2)	Plantation
Jernløse	Regstrup	8/25/88	2000	25	Activated sludge (0.4% TS); anaerobically digested sludge (3% TS)	4	415	Reeds (*Phragmites australis*)
Præstø	Allerslev	9/29/88	1000	14	Activated sludge (1% TS); anaerobically digested sludge (4% TS)	2	215	Reeds (*Phragmites australis*)
Nakskov	Nakskov	11/1/90	30,000	870	Raw sludge (1–2% TS)	10	11,600	Reeds (*Phragmites australis*)
Galten	Galten	11/1/90	10,000	152	Chemical sludge; activated sludge (1% TS)	6	2377	Reeds (*Phragmites australis*)
Svinninge	Gislinge	10/11/91	3520	42	Activated sludge (1% TS)	3	700	Reeds (*Phragmites australis*)
Rudkøbing	Rudkøbing	1992	15,000	232	Chemical sludge; activated sludge (0.5–1% TS)	8	4000	Reeds (*Phragmites australis*)

years: a 5-month-long rest period caused the accumulation of sludge residue to be considerably less than the accumulation of other years. If the increase for 1989 is calculated from the time loading was resumed in May, the 1990 decrease in sludge residue growth is an estimated 54% less than in 1989.

At the Allerslev plant (Figure 2B), the loaded sludge volumes were substantially less; on the other hand, the dry matter content was about 10 times larger. The sludge residue accumulation in 1989 was relatively modest considering the amount loaded, which to a certain extent is due to the long rest period at the beginning of that year. Comparing the periods of operation in 1989 and 1990, the accumulation in 1990 was about 10% less than in 1989.

For both plants, the load — especially in 1988, but also in 1989 — caused a rapid accumulation of sludge residue. The increase during these periods was due to overloading and the early stage of development of the reed plants. The reed vegetation was planted in the systems in July (Regstrup) and September (Allerslev) of 1988. The vegetation had, at the most, reached a stage of development corresponding to $1^1/_2$ growing seasons and did not yet cover the entire surface of the beds. Thus, the reed vegetation had no influence in 1988 and in 1989 relatively little influence on mineralization, conditioning, dewatering, and thereby the entire process of sludge reduction.

In 1990 the vegetation had established itself to a much greater degree; if not 100%, then to an extent that considerably improved the efficiency of the treatment plants. This leads us to expect that the annual sludge residue growth will decrease even more once operations stabilize.

Sludge Dewatering

The dewatering rate depends on the quality and dry matter content of the type of sludge loaded (Figure 3, A and B). The figures show typical area-specific dewatering sequences for the sludge types used and the accumulated reject amounts relative to time and the sludge load. Measurements were taken in November 1989 (Regstrup) and July 1990 (Allerslev). Bed no. 1 at Regstrup was loaded with 21.6 m^3 activated sludge. Bed no. 1 at Allerslev was loaded with 18 m^3 concentrated sludge.

Recording of the dewatering sequence started about 24 h before the beds were loaded. The basic dewatering rate between sludge loadings was generally less than 0.01 L/m^2/min. The sludge loadings started at the time

Table 3. Sludge Loads and Reduction (Regstrup) 8/19/88–4/15/91

Bed no.	Sludge (loaded) (m^3)	Sludge residue (m^3)	Reduction (%)
1	2313[a]	58	97.5
2	2132[a]	51	97.6
3	685[b]	71	89.6
4	2154[a]	56	97.4

[a] Dry matter content of sludge: approx. 0.4%.
[b] Dry matter content of sludge: approx. 3%.

Table 4. Sludge Loads and Reduction (Allerslev) 9/14/88–4/24/91

Bed no.	Sludge (loaded) (m^3)	Sludge residue (m^3)	Reduction (%)
1	263[a]	18	93.2
2	253[a]	23	90.9

[a] Dry matter content of sludge: approx. 4.0%.

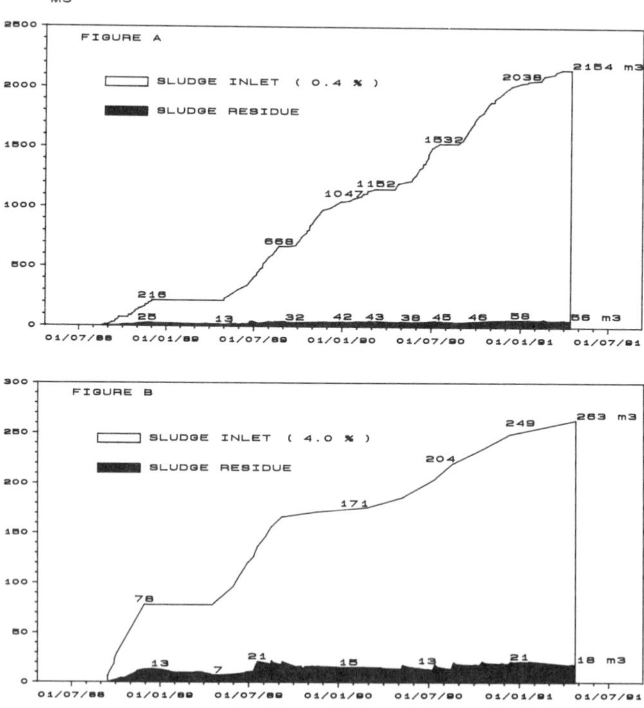

Figure 2. A: sludge reduction, bed 4 (Regstrup) — Operation period: 08/19/88 to 04/15/91 (activated sludge, 0.4%). B: sludge reduction, bed 1 (Allerslev) — Operation period: 09/14/88 to 04/24/91 (anaerobically digested sludge, 4% TS).

indicated as "0" on the graphs. The loading periods lasted from about 15 min (Allerslev) to about 40 min (Regstrup). The maximum dewatering rates of 0.30 and 0.053 L/m^2/min were achieved about 4 and 8 h, respectively, after loading commenced. As early as 24 h after sludge loading was begun, the basic dewatering rate was back to a level approaching the basic run-off rate (Figure 3A).

Approximately 17.2 m^3 (80%) and 5 m^3 (28%) of the sludge volume loaded was dewatered during the first 24 h after loading. After 144 h, about 98 and 55% of the pore water had been removed by dewatering (Figure 3B). Judging solely from the water balance of the dewatering of activated sludge, concentration to a dry matter content of about 15% takes an estimated 6 days. If evapotranspiration is also taken into account, then this period of time would presumably be shorter, particularly when the reeds are especially productive.

At the Allerslev plant, drainage from bed 1 ceased after a duration of 72 hours, in spite of the fact that the water budget calculations indicated that there should still be 10 cm^3 of water remaining within the sludge. Theoretically, at an evapotranspiration rate of 13 $L/m^2/day$, an estimated 3 cm of water should be left from the original 18 cm^3 of water within the sludge. Therefore, in the present studies, dewatering concentrated the sludge to a content of 24% dry matter within a six-day treatment period.

It must be concluded, especially for beds loaded with concentrated sludge, that the dewatering of the pore water involves, to a much greater extent, a combination of drainage and evapotranspiration, presumably with evapotranspiration as the most important element.

Sludge Residue Dry Matter Content

The primary indication of the dewatering efficiency of the treatment plants is the dry matter content of the sludge residue. The percentage of dry matter in the sludge residue varies considerably during the year and also depends on at what point in the loading period-rest period cycle it is measured. The dry matter content of the sludge residue from the activated sludge plant (Regstrup) and the concentration tank (Allerslev) varied between 13 and 24 and 25 and 35%, respectively (Figure 4). On average, the activated and concentrated sludges were dewatered to a concentration of 17 and 29%, respectively, which is a concentration corresponding to 43 and 17 times that of the loaded sludge.

The drainage and evapotranspiration levels are also shown in Figure 4. Field observations have shown that all drainage stops when the dry matter content of the sludge residue is greater than about 20%. Dewatering after this continues only in the form of evapotranspiration. Evapotranspiration can only dewater the sludge to an estimated maximum dry matter content of 50% (level of evapotranspiration). A higher dry matter content would most likely be "lethal" to the reeds.

In July 1990 (7/17 and 7/10), the dry matter content was 14 and 29%, respectively (Figure 4). Up to 8/29 and 8/16, the dry matter content increased to 22 and 36%, respectively. If we assume that the percentage of dry matter increases in a linear curve, and if loading had not been resumed, then the sludge residue would have been dewatered to 33 and 45%, respectively, if, for example, the period had been extended to October 1, 1990. In light of these theoretical considerations, these percentages are optimistic for the time period chosen. Had the time period started earlier, had lasted from June 1 to September 1, then the estimated levels would be more realistic.

Bed 1 at Regstrup had 4 periods during which the sludge was reduced in volume and concentration. In early spring and summer, 1990, the sludge depth decreased by 0.7 cm/week for spring and 1.2 cm/week for summer. Similarly, sludge thickness in bed 4 of the Regstrup plant showed an average reduction of 0.5 and 1.7 cm/week for the two seasons, respectively. For the two seasons, reductions per week were 20–50% greater in 1990 than in 1989.

Thus, the sludge reduction in summer is about three times greater than in winter. This discrepancy is presumably due to the favorable temperatures of summer, the increased evapotranspiration, and especially the vegetation. Furthermore, the vegetation was presumably the most significant cause of the increased reduction in 1990, as the reeds were much better established by then.

Water Budget

The water balances set up for the individual beds only apply to time periods starting in 1990 (Tables 5 and 6). In 1989, the vegetation was not yet fully mature and it had only spread to cover 60 to 70% of the bed area. By 1990, it had spread to cover almost 100% and had almost reached full height. Thus, 1990 was the first year when it could be assumed that the biological treatment systems were functioning at just about full capacity.

The system design made it possible to set up a simple water balance. Calculations were based on the water content in the sludge residue at the beginning and end of a time period (WIS_1 and WIS_2), sludge load (S), precipitation (P), and reject (R). The amount of water in the filter was considered to be constant for the entire period (Tables 5 and 6). Evapotranspiration can then be calculated as the only unknown in the formula:

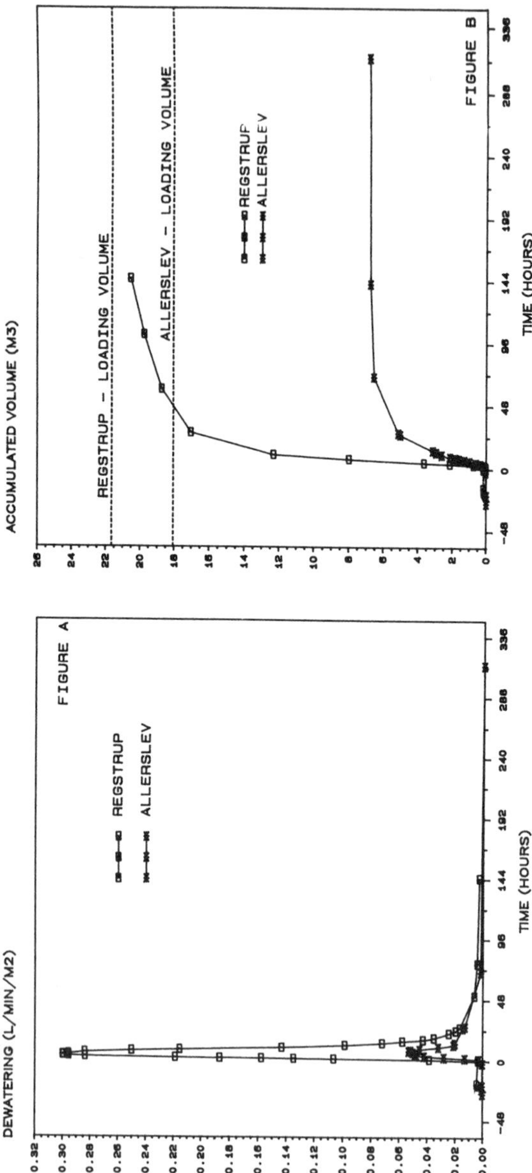

Figure 3. Area-specific dewatering rate measurement of Bed No. 1 at Regstrup and Allerslev. A: dewatering rate ($L/m^2/min$); B: accumulated reject (m^3).

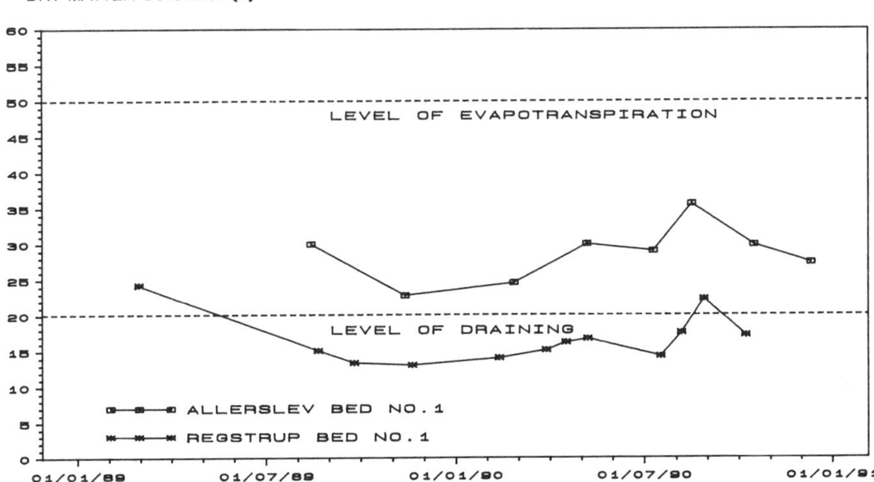

Figure 4. Variation in the average dry matter content in the sludge residue in bed 1 at both Regstrup and Allerslev.

$$WIS_1 + S + P = WIS_2 + R + E$$

On average, the evapotranspiration in 1990 was almost two times greater than in 1989, which further emphasizes the considerable influence of the vegetation on the total evapotranspiration and thereby on the water balance.

The variation in evapotranspiration from the individual beds is attributed to differences in the extent to which the vegetation has spread and its morphological amplitude. This study has shown that the dry matter content of the sludge residue was substantially greater — up to 10% greater — in areas with dense reed vegetation than in areas with no reeds.

In 1990, evapotranspiration was calculated to be about 1500 mm after 260 days at Regstrup and about 540 mm after 292 days at Allerslev. Water loss in the form of evapotranspiration accounted for approximately 23 and 80%, respectively, of the total dewatering (reject water + evapotranspiration) at the Regstrup and Allerslev plants.

The high levels of evapotranspiration at the Regstrup plant in 1990, which averaged about 1500 mm, are on the upper end of the scale (500 to 2000 mm) compared with earlier observations.[5] One explanation for the substantially greater amount of evapotranspiration at Regstrup could be that the much higher frequency of loadings caused the sludge residue to be generally much wetter and thus to be under conditions which caused the actual evapotranspiration to be close to the potential evapotranspiration. At Allerslev, which had a much lower loading frequency and where the sludge had a significantly higher dry matter content, it must be presumed that the actual evapotranspiration was far lower than the potential evapotranspiration.

DISCUSSION

During the loading periods, the activated sludge (0.4% TS) and the concentrated sludge (4.0% TS) were dewatered to an average of 17 and 29%, respectively. Before the basins are emptied of sludge, they are allowed to stand for 3 to 4 months so that the sludge becomes even more concentrated by means of drainage and evapotranspiration.

The reed transpiration in June, July and August was 8, 10, and 3.9 L/m^2/day,[6] respectively. The evaporation for one reed area was 400 mm: thus the daily evaporation amounted to a water loss of about 3 L/m^2/day.[6]

As regards the dewatering processes, it is assumed that the sludge can be dewatered to a maximum dry matter content of 50%, which means that the remaining water content is only adsorption water and structurally bound water.[3]

In a final sludge residue volume of 100 m^3 (Regstrup or Allerslev) with a dry matter content of 15, 20, or 25% at the beginning of the final 3 to 4-month rest period, the water content would consist solely of capillary

Table 5. Water Balance — Regstrup (1990)

Plant	Bed	Period	Water in sludge (start)	Load	Precipitation	Water in sludge (end)	Reject water	Evapotranspiration (calculated)	
				(m³)		(m³)		(m³)	(mm)
Regstrup	1	2/12–	38.7	710.3	50.8	42.1	592.3	165.4	1650
	2	10/8	34.4	666	50.8	59.2	587.3	104.7	1050
	4	2/12– 11/21	31.8	827.9	63.2	42.3	684.5	196.1	1700

Table 6. Water Balance — Allerslev (1990)

Plant	Bed	Period	Water in sludge (start)	Load	Precipitation	Water in sludge (end)	Reject water	Evapotranspiration (calculated)	
				(m³)		(m³)		(m³)	(mm)
Allerslev	1	2/26–	11.5	60.8	61.4	13.5	86.2	34.0	340
	2	12/11	12.4	58.1	70.7	14.9	52.9	73.4	640

water, adsorption water, and structurally bound water; the capillary water volume would then be, respectively, 64, 60, or 56 m^3 (calculated from the data in Figure 1), which must be considered to be the maximum volumes the natural dewatering processes can handle.

The sludge residue dry matter content values which theoretically can be obtained are calculated on the basis of the above-mentioned conditions, theoretical considerations, and selected values from the monitoring project. The correlation between the obtainable dry matter content, the dry matter content at the start of the final rest period, the total volume, and the theoretical evapotranspiration can be seen from the formula

$$TS(t) = TS(t_\infty) + \left[TS(t_o) \div TS(t_\infty)\right] \cdot \exp\left[-\left(100 \frac{Ep}{V_{tot} \cdot TS(t_\infty)}\right)t\right]$$

where $TS(t_\infty)$ = the maximum obtainable percentage of dry matter,
$TS(t_o)$ = the percentage of dry matter at the beginning of the drying period,
Ep = the potential evapotranspiration (m^3/day),
V_{tot} = the total volume (water + sludge) at the beginning of the drying period (m^3), and
t = the time in days.

If we insert theoretical values ($Ep = 1.0$ m^3/day, $TS(t_\infty) = 50\%$) and selected values ($V_{tot} = 100$ m^3, $TS(t_o) =$ 15, 20, and 25%) in the formula, the result is the theoretical dewatering of the sludge residue over the selected time (t) (Figure 5). The dry matter content basic level is about 15%, which is the minimum level the sludge residue is dewatered to as long as loading continues (Figure 5).

If the beds are allowed to stand in June, July, and August (90 days), the maximum concentration which can theoretically be achieved is 46% TS if the dry matter content in the sludge residue was 25% at the start of the drying period.

It must be taken into consideration that the beds are exposed to precipitation during the drying period, which will prevent the theoretical percentages of dry matter from being reached by the end of the period. The theoretical dry matter percentages should be considered to be overestimated — no allowance is made for any precipitation during the calculation period — and the potential evapotranspiration is considered to be constant. After a "life" of 8 to 10 years and a final 3-month drying period, the entire volume of sludge residue is expected to reach an average dry matter content of approximately 40%. Judging from the data collected thus far, this seems to be a reasonable expectation.

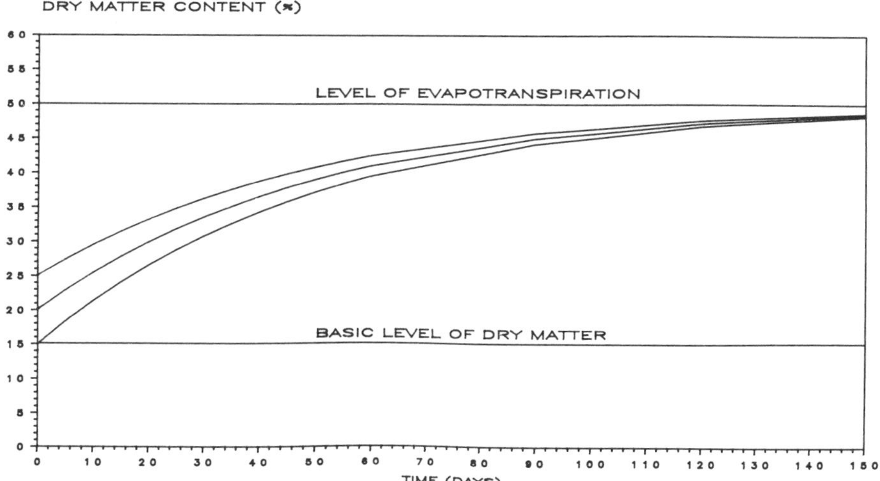

Figure 5. Theoretical dewatering of sludge residue over time.

CONCLUSION

Sludge Reduction. Most of the reduction in sludge volume is attributed to dewatering in the form of drainage and evapotranspiration. Just how much of the total reduction in volume was due to mineralization was not quantified. The activated and concentrated sludge underwent a total reduction of 98 and 91%, respectively, which is equal to concentrations of about 40 and 12 times.

Sludge Dewatering. The rate at which dewatering takes place depends on the quality of the sludge loaded and is inversely proportional to the dry matter content of the sludge loaded. About 80% (activated sludge) and 28% (concentrated sludge) of the volume loaded is dewatered within the first 24 h. The maximum area-specific dewatering rate is approximately 0.3 L/m^2/min for activated sludge and 0.05 L/m^2/min for the concentrated sludge.

The growing amount of sludge does not affect the dewatering efficiency. In 4 to 6 days, the loaded volume of activated sludge is dewatered to about 15% TS and the loaded volume of concentrated sludge to 24% TS. The evapotranspiration during the productive season of the vegetation was measured at about 1500 mm at Regstrup and 540 mm at Allerslev. An average dry matter content of up to approximately 35% was measured in the sludge residue; a dry matter content of approximately 40% is expected when the beds are emptied.

Thus, the degree of reduction is dependent on the type of sludge loaded and the condition of the vegetation and its stage of development. The load frequency and amount loaded each time were also operational parameters which had considerable effect on the efficiency of the systems and also on the increase in sludge residue.

The average dry matter content of the sludge residue depends on the length of the loading and rest periods and on the surface-volume ratios.

FINAL REMARKS

Judging by the results of this study, biological sludge treatment systems have proved that they can be quite useful. In comparison with existing sludge treatment systems, it has been shown that the sludge reduction efficiency of these biological systems is just as high as that of high-tech alternatives. This kind of biological system can treat a wide variety of sludge types: chemically and actively concentrated sludge with a dry matter content of up to approximately 10%, from municipal wastewater treatment plants as well as industrial ones.

Handling sludge in plant-based sludge systems is an entirely new sludge treatment method that distinguishes itself by being an inexpensive and environment-friendly alternative to traditional dewatering systems, and one that does not load wastewater treatment plants.

ACKNOWLEDGMENTS

The supervision project was carried out by Danish Land Development Service in cooperation with the technical administrations of the Jernløse and Præstø municipalities. The project was financed by the Danish Ministry of the Environment (30%), the Administration of Industries and Trade (20%), and Danish Land Development Service (50%).

REFERENCES

1. Orientering fra Miljøstyrelsen nr. 10, 1987. Spildevandsslam fra kommunale renseanlæg i 1987.
2. Thorsen, E., P. Rantala, P. Baamler, and B. Nordstrøm. Håndbog om drift af rensningsanlæg. Miljøstyrelsen of Spildevandsteknisk Forening, 1982.
3. Teknologistyrelsen. Biologiske metoder til behandling of genanvendelse af slamprodukter, 1987.
4. Nielsen, S. Sludge dewatering and mineralisation in reed bed systems, in *Constructed Wetlands in Water Pollution Control, Proc. Int. Conf. Use of Constructed Wetlands in Water Pollution Control.* Cambridge, U.K., September 24–28, 1990. P. F. Cooper and B. C. Findlater, Eds. WRc Swindon, Wiltshire, U.K.
5. Nielsen, S. Biologisk Slambehandling. Undersøgelse af tagrørsbeplantede slamafvandings- of mineraliseringsanlæg (in Danish, summary in English). *Miljøstyrelsen.* 1992, in press.
6. Burian, K. Primary production, carbon dioxide exchange and transpiration in *Phragmites communis* Trin on the Lake Neusiedler See, Austria, *Hydrobiologia.* 12:203, 1971.

PART 10

CASE STUDIES

CHAPTER 62

Constructed Open Surface Wetlands: The Water Quality Benefits and Wildlife Benefits— City of Arcata, California

R. A. Gearheart and M. Higley

INTRODUCTION

Small- to medium-sized communities have special problems regarding solutions to their wastewater management problems. These communities are required to have the same level wastewater treatment as larger communities. They are quite often limited in their tax base and thus in ability to pay the necessary service fees to sustain them adequately. EPA's construction grants program has assisted many medium to large communities in constructing wastewater treatment plants; however, many smaller communities have not participated in these programs. As the grants program has changed to one of state-managed revolving loans, the need to select low-cost, effective systems has become even more important.

The City of Arcata, California, has been involved with the testing and demonstration of open surface wetlands for treating domestic effluent for 12 years. These projects have been implemented by the Environmental Resources Engineering Department at Humboldt State University through the Public Works Department of the City of Arcata.

The city has supported effective, low-cost, environmentally sensitive wastewater treatment as a desirable goal in light of the energy issues, supply of public monies, and a variety of existing environmental policies.

The alternative Arcata proposed was based upon utilizing the natural capability of marsh plants, soils, and their associated microorganisms to be used to polish secondary effluent prior to disposal to Humboldt Bay. In engineering terms, this system is a free surface constructed wetlands as opposed to an existing natural wetland. The constructed wetland was to be designed to meet specific water quality objectives while affording a habitat for a wide variety of marsh plants, animals, and microorganisms.

THE SYSTEM AND SITE

In 1985, Arcata up-graded its existing 2.6 mgd primary and secondary treatment processes. Two treatment marshes (2.5 ac each) were built adjacent to the existing treatment plants, with a third treatment marsh added in 1989. The effluent from the treatment marshes is chlorinated prior to discharge to the effluent disposal wetlands (Arcata Marsh and Wildlife Sanctuary). The effluent disposal wetlands are comprised of three marshes, approximately 10 acres each, which were located a $^{1}/_{4}$ mile west across Jolly Giant Slough. The effluent from these three marshes is returned to the treatment facility where it is chlorinated and dechlorinated prior to discharge to Humboldt Bay. The restoration of a 10-ac log pond on the 150-ac site has afforded another freshwater swamp habitat. This swamp is watered with ground water available on the site. Recently, the city formally designated these coastal and aquatic habitats as the Arcata Marsh and Wildlife Sanctuary (AMWS). The AMWS is now about 150 ac, comprised of three effluent receiving marshes, one estuarine fishing lake, a fresh-water swamp, a closed-out landfill ("Mount Trashmore"), a restored estuary, and the open land areas in between.

The 150-ac AMWS located on the edge of Humboldt Bay encompasses marsh, estuarine, and upland habitats. Over 5 miles of foot trails are open for public access 21 h a day. Early estimates showed that over

0-87371-550-0/93/$0.00+$.50

130,000 people a year use the site for passive recreation, active recreation, educational, and relaxation. The AMWS is located near downtown Arcata (about $^1/_2$ mile), on the edge of the bay, which makes the marshes an attractive and readily accessible place commonly referred to as Arcata's bayview window.

WATER QUALITY BENEFITS

The City of Arcata has 12 years of experience with pilot- and full-scale constructed wastewater wetlands. Constructed wetlands are being used in Arcata to remove BOD, suspended solids, disinfection by-products, metals, nutrients, and organisms of significance in public health. Wastewater treatment systems utilizing aquatic plants (i.e., biological filtration) are very reliable since the rooted marsh plants of these systems show a surprising ability to withstand extreme shock load from toxic materials. It is emphasized that wetland treatment results must be interpreted in relation to the particular characteristics of the wetland system for which the results have been derived. Accordingly, treatment data presented in this article should be interpreted in the light of associated data on wetland characteristics.

BOD

The wetland system affords a complex microbial community which processes both particulate and dissolved organic material as it moves through the various communities. BOD (biochemical oxygen demand) removal is performed by both physical and microbiological processes in a wetland. Particulate BOD is removed by settling, impaction, filtration, and predation. Dissolved BOD is removed by planktonic and attached decomposers. The City of Arcata's pilot project showed that lower hydraulic loading rates produced higher BOD removals. Seasonal variations in effluent concentration were affected by vegetation type, density, and distribution. The removal efficiency rate varied from 41 to 65%. Those cells loaded at lower rates consistently produced BODs of 20 μg/L or less (Table 1). BOD in wastewater can arise from either dissolved material in solution or it can be associated with particulate and colloidal material. A wetland reduces BOD levels either through physical sedimentation of solids within the wetland channel system, through direct physical filtration of solids by wetland plants, or through biological processes associated with microorganisms in the wetland. Free surface flow wetlands BOD reduction (or dissolved oxygen enhancement) may be achieved by provision of a central open area in the wetland within which wind-induced reaeration can maintain dissolved oxygen levels. Epiphytes and phytoplankton are major oxygen contributors to the water column. Open areas provide habitat for phytoplankton, while macrophytes provide attachment sites for periphyton.

Constructed wetlands are able to process shock organic loads with little to no effect on effluent quality. The effluent BOD did not significantly increase until organic loadings exceeded 200 lbs/ac/day.[3] Suspended solids levels, on the other hand, were minimally affected, even at this higher loading. There are very few wastewater treatment processes that can produce an acceptable effluent quality over an order of magnitude increase in the BOD and suspended loading to the system. This ability to accept shock loads and to recover without any external process control is an important characteristic of wetland treatment systems.

A study was performed on the pilot project where the hydraulic load was increased a function of 10 in the summer months.[1] A BOD removal curve fits a first-order regression curve, with an R value of 0.98. Loadings of 50 kg/ha/day consistently produce an effluent BOD of 10 μg/L or less.

SUSPENDED SOLIDS

In wetlands, suspended solids are retained as a consequence of slow water velocities and long residence times. This contributes to sedimentation, which has been identified as the primary method of suspended solids reduction.[6] Organic colloidal and suspended solids autoflocculate and settle out in the open spaces between and around plants. Zooplankton such as *Daphnia* concentrate in these areas as they feed on these solids.[3] Wetland vegetation causes surface waters to follow a tortuous path, around and through innumerable obstacles and voids which provide sites where suspended solids may be removed by autoflocculation and/or filtration/predation.[2-4] Direct vegetative filtration also contributes to suspended solids reduction,[6] and it has been noted that suspended solids reduction improves in artificial surface flow systems as wetland plants mature.[1]

Table 1. Comparison of Pilot Project and Full-Scale Results for NPDES Parameters (Percent Change Oxidation Pond Effluent through Wetlands), 1980–1988

	First pilot project (all cells)	Second pilot project (all cells)	Full-scale operation
BOD			
Mean	11.4	13.8	12
% change	–56	–73	–55
% less 30 mg/L	100	100	100
% less 20 mg/L	84	72	81
% less 10 mg/L	37	33	18
SS			
Mean	5.3	10.8	14
% change	–85	–80	–54
% less 30 mg/L	100	100	100
% less 20 mg/L	100	93	78
% less 10 mg/L	91	43	45
Dissolved Oxygen			
Mean	1.5	1.1	5.0
% change	–73	–76	–27
pH			
Mean	6.5	6.1	7.1
% change	29	–14	–6
Theoretical retention			
Time/s (days)	1.5–30	6	8.5
1981 average	3.7		
1982 average	9.0		
Open water			
Fraction	0–10	0–25	75–90

Suspended solids were removed in the first sections of the Arcata pilot project wetlands (see Figure 1).[3] This represents a theoretical retention time of about 1 day for the Arcata system. This autoflocculation/sedimentation process builds a significant detrital bank in the influent area of the wetland. The detrital bank is about 70 to 90% of the volume in this first section after 8 years of continuous loading. The detrital bank extends in a tapered fashion 75% of the length of the cell.[3] Organic loadings of suspended solids are approximately zero-order kinetics over the range 0 to 200 kg/ha/day, reflecting this progressive accumulation through the cell length.[2]

An important concept in relation to surface flow wetlands relates to the nature of suspended solids in the influent in comparison with solids in the effluent. In wetlands, suspended solids and other material constituting waste, are transformed from an initial "effluent-derived" form to an eventual "wetland-derived" form. Thus, although the absolute suspended solids levels may be very similar, material discharging from the wetland consists of plant detritus and natural wetland products. This characteristic has significant implications in terms of the requirements in discharge permits.

The effectiveness of constructed wetlands to treat domestic effluent can best be seen by comparing the 8 years of research and monitoring in Arcata[1,3] (Table 1). Table 1 shows the removal efficiency and effluent quality of the two pilot projects and full-scale AMWS. The average BOD values for these studies were 11.4, 13.8, and 12 µg/L; the average suspended solids were 5.3, 10.8, and 14 µg/L, respectively. The variations in suspended solids can be attributed to a high fraction of open water at the AMWS compared to the pilot project's densely vegetated water volume. The effectiveness of wetland systems to remove consistently (8 years of data to date) SS (suspended solids) and BOD at a level significantly below secondary standard is noteworthy. For example, approximately 75% of the BOD and SS values were less than 20 µg/L for the three studies; approximately 45% of the suspended solids were less than 10 µg/L for the study period.

The dissolved oxygen levels in the wetlands are a function of the organic loading and the fraction of open water. In the first pilot project, the cells were totally vegetated by the end of the 2-year study. The average

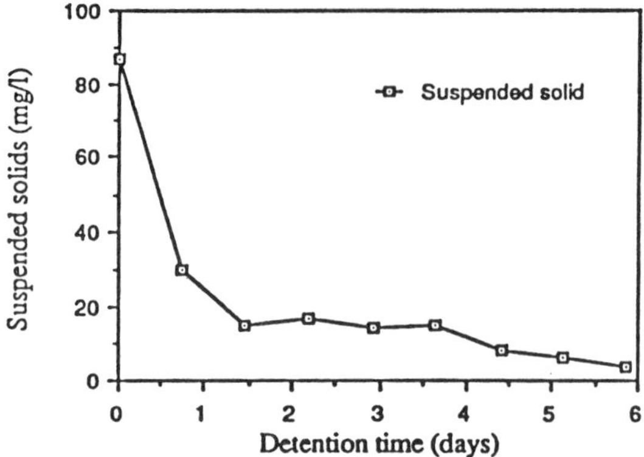

Figure 1. Suspended solids removal through Pilot Marsh 8. Samples taken at 7 weirs through the 200-ft length. The theoretical retention time of marsh cell 8 is 6 days.

dissolved oxygen level was lowest in this study, at 1.1 μg/L. Comparing this with the full-scale project where the open water fraction is 75 to 90%, the average is dissolved oxygen was 5.0 μg/L, reflecting the oxygen input from phytoplankton populations.

The pH of the wetlands cells represents the amount of anaerobic decomposition and subsequent organic acid production vs. the amount of photosynthetic activity in the cells. The more open the cell is, the greater the algal activity and the higher the pH. In Arcata's case with moderately buffered waters (alkalinity 80 to 120 μg/L, $CaCO_3$), the pH is stable at 6.5 to 6.8 (Table 1).

AQUATIC VEGETATION

Vegetation growth in the wastewater wetlands appears to be heading towards emergent and floating plant types as the dominant cover species (Table 2). Sago pondweed (*Potemogeton pectinatus*) peak biomass in Gearheart Marsh during 1985 and 1986 was 187.97 and 132.77 g/m^2, with a standard deviation of 113.08 and 88.73, respectively.[5] Peak Sago pondweed biomass for Hauser Marsh was determined only for 1986 because of the drainage of the marsh in 1985, and was 93.03 g/m^2 with a standard deviation of 47.66.

WILDLIFE BENEFITS

Some 174 species of birds were observed at the AMWS during the study.[5] Of these, 98 species were observed on the water areas and are treated in this report. The total waterbird use-days in 1984–1985 was 1.435, and 1.430 million in 1985–1976. Shorebirds represented 87.9% of all bird use-days, and waterfowl, coots, and rails accounted for 8.8%. Peak waterbird use occurred during winter, and early and late spring in both years, with 63.1 and 61.0% of the annual total bird use-days. Evidence of nesting species during this study was limited to mallards (*Anas platyrynchos*), cinnamon teal (*Anas cyanoptera*), northern shoveler (*Anas clypeata*), pied-billed grebe (*Podilymbus podiceps*), killdeer (*Charadrius vociferus*), and black-necked stilt (*Himantopus mexicanus*).

Herons and egrets used the marshes in a consistent pattern, with peak numbers occurring in early and mid-fall and lowest numbers in late spring. Herons nest in the surrounding areas; therefore, they do not leave during the summer. The highest rate of use per 5 hectares of herons occurred in Gearheart Marsh (Figure 2).

Puddle duck peak use occurred in winter both years. Puddle ducks used Gearheart Marsh at the highest rate, with Allen and Hauser Marshes trailing far behind and Klopp Lake receiving nearly no puddle duck use. The low use of Hauser Marsh in 1985–1986 coincided with the extended drainage of this marsh from early fall to the beginning of late fall, 1985 (Figure 2). The high numbers recorded in Gearheart Marsh in 1985–1986

Table 2. Precent of the Dominant Plant Species at the Arcata Marsh and Wildlife Sanctuary

	Marsh units, month, and year					
	Gearheart Marsh		Hauser Marsh		Allen Marsh	
Type of cover	April, 1985	Sept, 1985	Sept, 1987	Sept, 1987	April 1986	Sept, 1987
Open water	83.8	5.0	32.5[a]	23	70.0	36.2
Common cattail	5.5	6.0	10.5	4.3		6.3
Marsh pennywort	10.0	11.8	27.0			5.6
Sago pond weed		77.2	NV[b]	NV[b]		NV[b]
Alkali bulrush				0.8		11.9
Lesser duckweed			30.0[a]	69.6		40.0[a]
Hardstem bulrush				2.3		
Common spikerush		0.7				
Upland grass sp.					30.0	

[a] Duckweed coverage was too low because the wind had pushed it into wind rows.
[b] NV, not visible because of duckweed coverage.

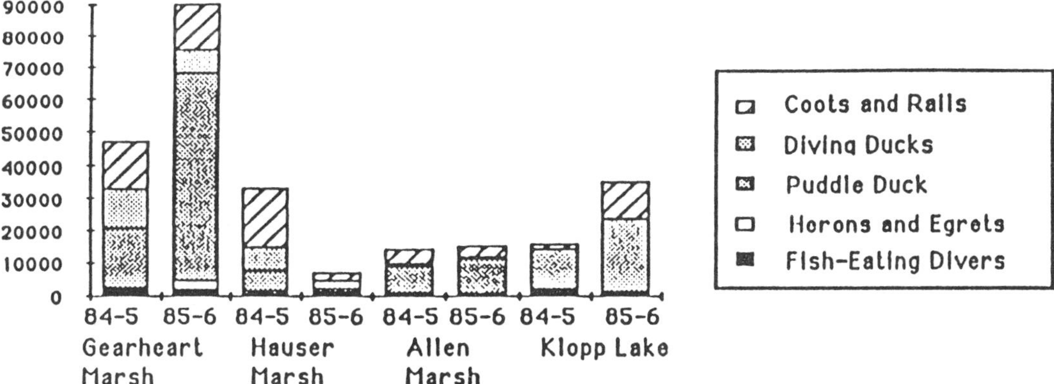

Figure 2. Total annual waterbird use-days per 5 hectares in AMWS, 1984–1985 and 1985–1986.

coincided with the drainage of this marsh for 2 days in mid-fall and then reflooding it slowly and maintaining the water level at a shallow depth allowing the birds access to the pond bottom (Figure 2).

Diving ducks are very seasonal visitors to the AMWS, arriving during late fall, peaking in winter and early spring and leaving by summer. Diving ducks used Klopp Lake at the highest rate and the frequently drained Allen Marsh the least (Figure 2). The high peak of use in Klopp Lake during winter 1985–1986 followed a growing season which produced Wigeon grass (*Ruppia spiralis*) in large amounts within the lake.

The total of bird use-days for the Marsh Project was 1,434,633 in 1984–1985 and 1,429,837 in 1985–1986 with 98 waterbird species recorded for both years (Table 3).

Based upon a normalized calculation with other wetlands in the area, the Marsh Project was used at rates more than 10 times those of either Lake Earl or South Humboldt Bay and 3 times those of the Arcata oxidation pond (Table 3). When only waterfowl use of the respective areas is considered, the Marsh Project had about 38% higher rates of use on an area basis than Lake Earl, 74% higher use than South Humboldt Bay, and 57% less use than the Arcata oxidation pond (Table 3). The total number of use days at the Marsh Project as a whole was less than the other areas only because the total area of the Marsh Project units was considerably less.

Table 3. Comparison of Bird Use Days AMWS to Other Northcoast Non-wastewater Wetlands

	Average annual (1000)			Average annual/5ha	
Study areas	Water bird use days	Waterfowl use days	Hectares	Water bird use days	Waterfowl use days
Lake Earl	3,091,305	1,467,843	923.0	17,117	8,128
So. Humboldt Bay	2,968,218	1,420,119	1634.0	12,142	4,345
Arcata Ox. Pond	884,000	137,745	22.3	50,301	30,883
Aracta Marsh Pr.	1,432,253	98,689	37.8	189,451	13,054

COMMUNITY BENEFITS

The AMWS has become a major form of low-cost recreation for bird watchers, nature lovers, fishermen, walkers, joggers, boaters, picnickers, meditators, and tourists. The area has been actively used by students from Humboldt State University and other schools as a research facility and as a field trip destination. In a 1987 user survey, the most common reasons given for going to AMWS was its value as a human sanctuary and as a place to enjoy the natural setting and ecology of a marsh. Several notes and letters from both volunteers and participants in the local Hospice program have lauded the AMWS setting and the value it has for people. For the last 10 years, the Redwood Region Audubon Society has sponsored weekly nature walks at AMWS. Approximately 900 people a year participate in these nature walks. Docents trained by the Audubon Society explain the role the wetland habitat plays in attracting birds and mammals and the role the wetlands play in the water quality management of Humboldt Bay. Interpretive programs have proved to be extremely valuable in increasing the awareness of users of the multiple benefits being derived from the multiple functions of the system. Local public schools along with Humboldt State University also use the wetlands in various educational programs ranging from general environmental awareness to research and graduate courses in wetland management.

SUMMARY

Wetland treatment effectiveness is a function of retention time and capacity of the vegetation and sediments to retain and/or cycle certain constituents.[3] In using a wetland to polish domestic secondary treated effluent, the following general guidelines are considered reliable. It has been shown that an effluent suspended solids level of 5 to 10 µg/L can be achieved with a retention period of 1 to 2 days (Table 2). A longer retention time is required for effective BOD removal. An effluent BOD value of 10 to 15 µg/L can be achieved with 4 to 8 days of retention of a secondary treated effluent. Total nitrogen levels of the order of 4 to 6 µg/L can be achieved with 10 to 12 days of retention. About 30 ac of constructed free surface wetlands can insure that a small community can consistently meet the 30/30 NPDES standard, while affording significant environmental and community benefits. Management plans must be developed and implemented to optimize the multipurposed site (Table 4).

Table 4. Gross Use Estimates, Negative and Noncompatible Activities, and Methods of Resolving Negative Impacts of Recreational Activities Observed in Aracta Marsh and Wildlife Sanctuary

Recreational use	Estimated usage	Management & mitigation interventions
Picnicing & relaxing	75–175 people a day; 15–45 min. per visit	Traffic control, parking, placement of picnic tables and refuge cans; volunteer cleanup program; integrated mosquito control programs
Birdwatching and nature studies	5–50 people a day; 2–4 h per visit	Traffic control, parking, designated areas with blinds, to enhance the experiment and to allow all-weather usage; integrated mosquito control program; docent program by Audobon Society
Jogging & Walking	20–80 people a day; 2–4 h per visit	Designated trails, all weather surface-chips and bark; integrated mosquito control program
Education	50–200 people per month; 2–4 h per day	
Fishing	Presently light and seasonal	Borrow pit next to southlake and fishing jetties designated to direct bank fishing users away from critical bird areas; integrated mosquito control program; public education to values of trophy-fishing program and values
Photo/Art	30–75 people per month; 2–4 h per visit	None

REFERENCES

1. Gearheart, R. A., S. Wilbur, J. Williams, D. Hull, et al. City of Arcata Pilot Project Second Annual Progress Report, Sept. 1981. Report to California State Water Resources Control Board, August 1982.
2. Gearheart, R. A., S. Wilbur, J. Williams, D. Hull, B. Finney, et al. Final Report City of Arcata Marsh Pilot Poroject, City of Arcata Department of Public Works, Arcata, California, April 1983.
3. Gearheart, R. A., J. Williams, H. Holbrook, and M. Ives. City of Arcata Marsh Pilot Project Wetland Bacterica Speciation and Harvesting Effects on Effluent Quality. Environmental Resources Engineering Department, Humboldt State University, Arcata, California, 1985.
4. Hammer, D. and R. H. Kadlec. Design Principles for Wetland Treatment Systems. EPA 600/S2-83-026, May 1983.
5. Higley, M. Hoopa, California.
6. Tchobanoglous, G., R. Stowell, R. Ludwig, J. Colt, and A. Knight. Aquaculture Systems for Wastewater Treatment: Seminar Proceedings and Engineering Assessment. EPA 430/9-80-006, R. K. Bastian and S. C. Reed, Eds., 1979, 35–55.

CHAPTER 63

Constructed Wastewater Wetlands: The Answer in South Dakota's Challenging Environment

J. N. Dornbush

INTRODUCTION

In South Dakota (SD), population 700,000, there are 265 community wastewater treatment facilities, with only 10 systems serving populations greater than 10,000 people. However, to meet increasingly stringent EPA discharge requirements and regulations, constructed wetlands have been installed by 40 systems since the first at Kadoka in western SD in 1987. Why? The answers are a pro-active, cooperative regulatory agency that provided design guidelines, the economic advantage for wetlands with inexpensive land costs, the widespread use of stabilization ponds, and a population that cherishes water, natural wetlands, and related wildlife in a challenging environment.

Communities and engineers solving municipal wastewater problems in SD face extreme environmental constraints — climatic, physical, economic, and regulatory in nature. Located in the northern Great Plains, SD has a continental climate with extreme summer heat of 38°C, winter cold of –30°C, and wide fluctuations in temperature and precipitation. First fall frost usually occurs in late September, while the last spring frost occurs in mid-May. Consequently, streams, lakes, ponds, and wetlands usually have ice cover from November to March.

Annual average precipitation varies from 63 cm in the southeast to 33 cm in the northwest.[1] Most precipitation occurs in the spring and early summer, with thunderstorms expected 40 to 45 times per year. Annual average precipitation may be expected to vary widely from year to year. Snowfall averages from 64 to over 115 cm, but spring may also arrive without a snow cover. Lake evaporation exceeds precipitation, varying from 81 to 112 cm with a net potential evapotranspiration water loss of 25 to 75 cm from northeast to southwest.[2]

As a result, water flow in streams and water in natural wetlands is highly variable. Floods may occur with spring snow melt or after severe thunderstorms. Unless sustained by shallow ground water aquifers, stream flows are intermittent. Drought is of major concern to agriculture, wildlife, and those few communities using a surface supply as a drinking water source. Most communities rely on highly mineralized ground water supplies from deep aquifers.

Soils vary widely, from the mountainous Black Hills in western SD, through the Missouri Plateau west of the Missouri River, to the glaciated eastern half containing the central lowlands. These soil variations plus the climate extremes and topographic features contribute to widely varying hydrologic regimes. An estimated 526,500 ha of natural wetlands in eastern SD in the Prairie Pothole Region[3] contributed to past wastewater disposal difficulties and opportunities. During extreme drought, some lakes and many wetlands dry up and most streams experience zero flow. Wastewater stabilization ponds and wetlands offer the advantage of a sustained water supply for wildlife.

Yet South Dakotans cherish the available water and classify virtually all lakes and streams for fishing and recreation, thus imposing an ammonia limit to wastewater discharges. To control phosphorus inputs, wastewater discharges to classified lakes are prohibited. Stringent ground water protection regulations have also been imposed. More recent EPA requirements of secondary treatment before discharge to natural wetlands have imparted additional regulatory hardship as 20 SD community systems discharged to natural wetlands before 1990.[4]

0-87371-550-0/93/$0.00+$.50

Economic impacts of up-grading facilities to meet the wetlands regulations in SD can be substantial. For example, a small community (population 161) recently completed a three-cell pond system at a cost of $217,000, with 55% federal participation, to replace an Imhoff tank that previously discharged to a 12-ha natural slough.[3] The economic health of many small SD communities is marginal at best. Of the 40 SD communities (population 12,448 to 97) now using constructed wetlands, about 90% declined in population in the last 10 years. Economic assistance with 75% funding as alternative treatment through the construction grants program spurred constructed wetlands to up-grade some stabilization ponds. Thus, communities in a water-short climate, experienced with stabilization ponds and natural wetlands, have embraced constructed wastewater wetlands as a treatment solution in South Dakota.

STABILIZATION PONDS

Raw wastewater lagoons or stabilization ponds were the first constructed wetlands in the Dakotas. When Fessenden, North Dakota (ND) completed community sewer facilities in 1928, the discharge was to a natural pothole in the absence of a natural stream to provide dilution disposal. Favorable aesthetic and treatment experience at this natural wetland led to the construction in 1948 of the first "modern-day lagoon built on good, sound engineering principles" to treat raw sewage at Maddock, ND.[5] Some 2 years later, Maddock, also in a pothole region, expanded an initial single cell to three cells providing adequate area for seepage and evaporation to obviate discharge.

In SD, the first engineered stabilization pond was installed at Lemmon in 1951; by 1960, 77 pond systems had been constructed serving 88,413 persons and providing 324 ha of water surface.[6] Empirical design standards for stabilization ponds had been developed by SD by 1957. Low construction and operation costs were the major factors in community selection of stabilization ponds. By 1970, the number of communities with pond systems had doubled to 144 and all 40 new collection systems in the state had selected stabilization ponds for treatment.[7] By the end of 1985, 215 SD communities (84%) used stabilization ponds, either for their complete treatment process or a part of the treatment system.[8] As of 1991, 246 of 275 SD communities employ stabilization ponds as part or all of the wastewater treatment process, with 40 constructed wetlands added to up-grade these pond systems since 1987.[9]

In the 1960s the potential benefits of raw wastewater stabilization ponds to waterfowl was evaluated.[10] With respect to duck production, these ponds and surrounding habitat with a constant water resource were found to be "at least twice as productive as some of the better natural wetlands." However, nearly 70% of these ponds were constructed in or near natural wetlands. Thus, a diverse natural environment was further enhanced by a stable wastewater supply in water-short SD. Recommended mowing of pond dikes was delayed until after August 1 by SD regulatory personnel to enhance wildlife production further.

In summary, stabilization ponds have gained wide acceptance in SD and the upper Great Plains area for the following reasons: (1) low construction costs, (2) low operation and maintenance costs, (3) ability to avoid or control effluent discharge, and (4) for wildlife enhancement. These same advantages presently apply to constructed wetlands to up-grade SD stabilization pond effluents.

CONSTRUCTED WETLANDS IN SD

Since 1987, SD communities have responded to wetland construction opportunities with 40 up-graded facilities (Table 1). Each system includes stabilization ponds for pretreatment and winter storage. At 11 facilities, mostly larger systems, wetlands were added to existing ponds to meet discharge limits. In numerous cases, complete retention of treated wastewater will occur unless excessive precipitation necessitates discharge. Other systems (18) added both ponds and wetlands to up-grade treatment.

New facilities included three that replaced systems constructed in the 1950s, incorporating natural sloughs as treatment facilities, while five replaced mechanical plants such as Imhoff tanks plus trickling filters constructed in the 1930–1950 era that discharged to natural sloughs. Newly sewered districts accounted for three new, small wetlands systems.

In summary, from 1987 to 1991, SD communities have used wastewater to create nearly 372 ha of new constructed wetlands and 65 ha of stabilization ponds. These also complement about 255 ha of previously constructed stabilization ponds. Can these constructed wetland areas be used to mitigate wetlands impacted

Table 1. Constructed Wetlands in South Dakota (1987–1991) According to Treatment Application with Added Areas of Stabilization Ponds and Constructed Wetlands

Application	Wetland systems	Population served	Original ponds (ha)	Added ponds (ha)	Added wetlands (ha)
Added wetlands to existing ponds	11	23,203	193.3	—	293.9
Added ponds plus wetlands to existing ponds	18	10,358	52.4	31.1	49.2
Previously discharged to natural wetlands	3	2209	—	14.8	10.4
Replaced mechanical plants	5	1962	—	14.7	15.8
Serve newly sewered areas	3	883	—	4.1	2.1
Totals	40	38,615	254.7	64.7	371.4

by other essential construction projects in SD in the "no net loss" of wetlands policy? Descriptions that follow may provide insight.

Huron

The largest wetlands system in SD is at Huron (population 12,448), where both municipal and meatpacking wastewater are treated. In mideastern SD on the James River, residential Huron is located on the relatively flat west bank while the pork plant and wastewater facilities are in more rolling terrain on the east bank. Ground water qualities of the area are highly mineralized, so the city relies on the river for municipal water supply supplemented by deep wells during prolonged drought.

Wastewater treatment facilities were under study to up-grade effluent quality to meet discharge requirements in the late 1980s. Facilities included a lift station with pretreatment for the municipal and pork plant wastes at the river bank site of a mechanical plant replaced in 1968. The separate waste streams are pumped to a location about 3.21 km west where anaerobic lagoons pretreat the pork plant wastes prior to joining the municipal stream in sequencing batch reactors. Six cell three-stage stabilization ponds (97 ha) provide 150-day storage for controlled gravity discharge to the river (See Figure 1).

In April 1988, pond levels were nearly overtopping the dikes, so discharge from the stage III ponds began. Due to the 1987–1988 drought, mean river flow during April, May, and June was only 0.61, 1.50, and 0.26 m^3/s — about 10% of median flows for that period.[10] Excess ammonia concentrations resulted in a widely publicized fish kill in the river, followed by discharge enforcement action by the state and EPA.

The selected plan for up-grading the Huron facilities was low-rate land application to wetlands. By spring of 1989, the site-3 wetlands had been constructed to serve as a temporary no-discharge facility. With annual domestic and pork plant flows of 6840 and 3000 m^3/da, three wetland cells of 49, 28, and 57 ha were constructed on 178 ha of land. Since the constructed wetlands affected about 5.1 ha of seasonal and temporary natural wetlands, 10.4 ha of wetland mitigation area was provided with controlled flooding to optimize waterflow and wildlife benefits.

For the wetlands design, water balance calculations used 49 and 66 cm/year for years with normal and above-normal precipitation and evaporation of 88 cm/year. Seepage losses of 0.76 cm/da were estimated for the undisturbed bottom of the wetland sites. With these conditions, a normal year would result in no discharge, with a discharge of 807,270 m^3 (a 73-da combined wastewater flow) in an above-normal year.

In early June of 1991, James River flows were in excess of 40 m^3/s, a flow that occurs only about 10% of the time based on records from 1946 to 1985.[10] Huron elected to release treated wastewater from site 1 wetlands during the period June 12 to 28. During that period, river flows decreased from 44 to 7.3 m^3/s and effluent releases from 0.57 to 0.18 m^3/s.

Although effluent from the stage-3 stabilization ponds was used to "push" wetland effluent to the outfall, influent quality was not monitored for NPDES parameters, including 5-day biochemical oxygen demand (BOD), total suspended solids (TSS), ammonia nitrogen (NH_3-N), dissolved oxygen (DO), fecal coliforms

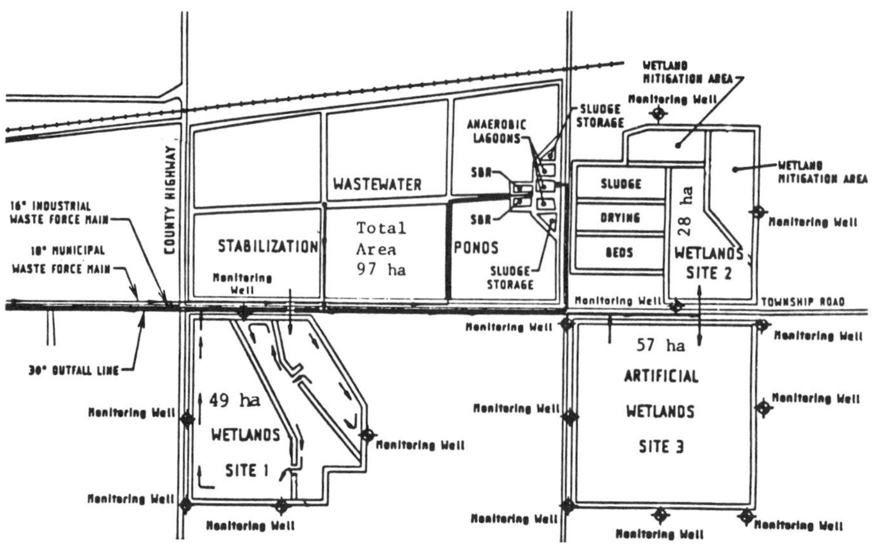

Figure 1. Huron wetlands treatment system.

(FC), pH, and temperature (Table 2). Ammonia concentrations (the critical quality parameter) were well below the levels calculated with the river flow-based NPDES limit. In total, about 17% of the annual combined wastewater design flow was discharged satisfactorily during the 16-day period.

Wetland vegetation will probably not be fully established for several years. Wildlife use of the Huron site (Figure 2) has been most encouraging, so much so that the plant superintendent has expressed concern about waterfowl waste input during the migratory season.

Belle Fourche

At Belle Fourche (population 4335) in the northern Black Hills of western SD, a four-cell, 45-ha stabilization pond system discharged to a creek classified for fishing. Treated effluent could not meet discharge limits for pH and ammonia. Also, effluent discharge was prohibited when ice covered the creek or when water was diverted to a downstream irrigation canal. Ammonia discharge limits varied from 0.34 mg/L for June to August, to 0.65 mg/L for April and October with a stream flow-based discharge rate.

Wetlands of 29 ha in 13 cells at 0.6-m depth have been fitted to the rugged topography on adjacent city-owned land with only one lift station. Wetland cell areas vary from 0.4 to 5.7 ha. Increased evapotranspiration and limited seepage are projected to provide complete retention of wastewater for a design population of 4770. Ground water will be monitored with eight wells. Completed in 1990, project costs were $1,034,000, with an EPA grant providing 55%.

Sisseton

In the northeast corner of SD, Sisseton (population 2181) was served in 1988 with a three-cell stabilization pond system (21.2 ha) discharging to the Little Minnesota River, a tributary of Big Stone Lake. Although the 1985 permit required only secondary treatment requirements, Minnesota Pollution Control Agency objected, claiming a significant phosphorus input (11.2%) to the lake. As a result, Sisseton was issued a zero discharge permit by SD.

To serve the community, an annual flow of 165,000 m^3 was calculated. A system with 12 wetland areas containing 103 ha was constructed to provide total retention. The existing pond and constructed wetland storage was calculated to be capable of handling annual design flows plus precipitation of 91 cm/year without discharge. An evaluation of 87 years of records revealed a maximum annual precipitation of 89.5 cm. Records for 1976 and 1977 had annual precipitation of 23 and 80 cm, respectively, to illustrate variability of rainfall in SD.

Table 2. Effluent Quality from Huron Constructed Wetlands — June 12–28, 1991

	Influent[a]		Effluent		
Parameter	Samples	Range	Samples	Range	NPDES limit
BOD (mg/L)	4	6–19	9	4–22	30
TSS (mg/L)	4	3–80	9	4–40	90
NH_3-N (mg/L)	2	12.9–18.9	9	0.7–2.7	5.3–13.4[b]
DO (mg/L)	—	—	9	4.8–6.8	—
FC (MPN)	—	—	6	5–160	1000
pH	4	7.6–8.5	9	7.7–8.3	6.5–9.0
Temp (°C)	4	16–24	9	21–27	—

[a] Stage-3 stabilization pond effluent samples collected May 20–June 3, 1991.
[b] Maximum NH_3-N = 0.27 + 0.17 (stream flow/effluent flow).

Figure 2. Aerial photograph of Huron wastewater treatment facility.

Two wetland cells were adjacent to the ponds (Figure 3) with wastewater pumped 3.2 to 3.9 km to the other 10 cells. Eight cells, enclosed with dikes of 1.2 m, included about 112 excavations of 920 m^2 or more to depths an extra 0.6 m to serve as open water areas for wildlife habitat enhancement with the constructed wetlands. The other four cells enclosed some natural wetland areas that were undisturbed. Additional wetland mitigation areas surrounded adjacent natural wetlands. Project costs were $1,659,000, including 60% EPA grant funds for the Sisseton wetland expansion completed in 1990.

Arlington

In 1976, a wastewater facility plan for Arlington (population 908) recommended construction of a three-cell pond system to replace an Imhoff tank, trickling filter, and final settling plant reconstructed in 1954. The

Figure 3. Aerial view of Siseton constructed wetlands: A, stabilization/storage pond plus two wetlands; B, nine wetlands cells (note excavated open-water areas).

old plant operated economically by gravity and discharged to a city-owned slough which overflowed to a creek with wetlands and a lake 18 km downstream. The city-owned slough was the economical site for the proposed ponds, with highly productive agricultural land as an alternate. The old mechanical plant could not meet discharge standards but operated with little health, aesthetic, or environmental risk. With uncertainty of future discharge requirements and wetlands policy, plus public opinion against the ag-land site, the project was delayed.

In the mid-1980s, Arlington was directed to comply with secondary treatment limits of 30/30 for BOD/SS. Innovative technology included two storage ponds of 4.3 and 2.4 ha plus a 3.4-ha constructed wetland with 30-day storage at 0.6 m depth. Islands and open water areas were provided in the constructed wetland planted with cattails. The system carved into an adjacent hill discharged to the city-owned wetlands. The system was in accord with design guidelines drafted by the SD regulatory personnel in 1986. It was constructed in 1987 and filled in 1988, during which time the natural wetland dried up.

During discharge in the late fall of 1989, wastewater analyses were performed to determine removal efficiencies during the pre-winter drawdown. Mean values of grab samples collected weekly are reported in Table 3. Wetland growth had not been established and discharge rates were high in the pre-winter drawdown. Wetland effluent was of satisfactory quality except for SS from algal growth, but removals occurred primarily in the pond system. Wildlife use has been rewarding. The innovative system cost $477,000, with a 68% federal share.

SUMMARY

Wastewater wetlands in SD started with stabilization ponds in the 1950s. To the 235 pond systems in use, constructed wetlands have economically up-graded effluents and solved difficult regulatory problems throughout the state. In water-short SD, 40 new wetland systems since 1987 have contributed 65 ha of additional ponded wastewater plus 372 ha of constructed wetlands, all of which serve as highly productive wildlife propagation areas. In all, SD is a case history of successfully applying wastewater wetland technology toward the nationwide no-net-loss-of-wetlands goal.

Table 3. Wastewater Quality at Arlington, SD Constructed Wetlands, October 11 — November 6, 1989

	Stabilization pond	Constructed wetland	
Parameter[a]	Influent	Influent	Effluent
BOD	129.4	14.4	10.8
SS	218	85	81
Nitrate-N	<0.1	<0.1	<0.1
Ammonia-N	36.2	0.13	0.08
TKN-N	42.7	3.30	2.72
Total P	8.24	0.85	0.63
Fecal Coli (organisms/100 ml)	—	520–<10	40–<10
pH range	7.47–7.73	8.61–9.07	8.63–8.93

[a] Results expressed as mg/L, except ranges of fecal coliforms and pH.

REFERENCES

1. Spuhler, W., W. F. Lytle, and D. Moe. Climate of South Dakota. Bulletin 582, Agricultural Experiment Station, South Dakota State University, Brookings, SD, November 1971.
2. Reed, S.C., E. J. Middlebrooks, and R. W. Crites. *Natural Systems for Waste Management.* McGraw Hill, New York, 1988, 211.
3. Kantrud, H. A., G. L. Krapu, and G. A. Swanson. Prairie Basin Wetlands of the Dakotas: A Community Profile. USFWS, Biological Report 85 (7.28), 71, September 1989.
4. Struck, C. W. South Dakota Department of Environment and Natural Resources, personal communication, December 1990.
5. Svore, J. H. History of raw sewage lagoons, *Proc. Symp. Waste Stabilization Lagoons.* USPHS, Kansas City, MO, August 1–5, 1960, 3–8.
6. Carl, C. E. and D. C. Kalda. Waste Stabilization Ponds in South Dakota, *Proc. Symp. Waste Stabilization Lagoons.* Kansas City, MO, August 1–5, 1960, 135–146.
7. Van Heuvelan, W. A Decade of Change in Waste Stabilization Lagoons in the Missouri River Basin, *2nd Int. Symp. Waste Treatment Lagoons.* Federal Water Quality Administration, Kansas City, MO, January 23–25, 1970, 10.
8. *South Dakota Municipal Wastewater Treatment Facilities.* Div. of Env. Qual., Dept. of Water and Natural Resources, Pierre, SD.
9. Struck, C. W. personal communication, June 1, 1991.
10. Dornbush, J. N. and J. R. Andersen. Ducks on the wastewater pond, *Water and Sewage Works.* 3(6):271, 1964.
11. Benson, R. D. Analysis of Flood-Flow Frequency, Flow Duration, and Channel-Forming Flow for the James River in South Dakota. USGS, Report 87-4208, Huron, SD, 1988.

CHAPTER 64

Application of Constructed Wetlands to Treat Wastewaters in Australia

T. H. Davies, P. D. Cottingham, and B. T. Hart

INTRODUCTION

Constructed wetlands or reed beds offer a potentially low-cost and low-maintenance biological method of wastewater treatment. The main systems currently in use around the world are predominantly designed from concepts originally reported by Kickuth who has had reed beds operating in Germany since 1972 treating both domestic and industrial wastewaters, including the wastes from a large textile processing installation.[1]

Most constructed wetlands in use today employ a horizontal flow path, although some experimental wetlands with vertical flow are operating. Horizontal flow wetlands consist of a shallow bed (ca. 0.6 m deep) containing a very porous medium placed on a waterproof membrane in an excavated trench so that the wastewaters can flow horizontally below the surface through the bed in the root and rhizome region of a suitable aquatic plant such as *Phragmites australis* (Figure 1). No harvesting of the plants is required since fallen debris composts away on the surface. For domestic wastewaters, a retention time of 3 to 5 days through the bed gives adequate treatment. This is equivalent to a bed surface area of 2.5 to 3.5 m^2 per person. Although full development of the plant's root system takes up to 3 years, significant treatment is achieved fairly quickly as the bed operates initially as a "sand filter". To avoid problems due to clogging and odor, the influent needs to have the gross solids removed before application to the bed. This can be achieved by primary settling, screening, septic tank treatment with even shorter than normal retention time, or by the use of a primary pond.

The advantage of a reed bed system over a normal sand filter (essentially the bed without the plants) is that the plants are able to transfer oxygen into the bed, creating aerobic micro-zones around the plant roots, with anaerobic zones away from them.[2] Since both aerobic and anaerobic bacteria will be present, a rapid breakdown of organic matter and removal of nitrogen by the nitrification-denitrification pathway can occur.

A wide range of materials have been used for the constructed wetland bed. Perhaps the two most important properties of the bed material are to supply a large inert surface area for the microbial population and to have sufficient hydraulic conductivity for all of the influent water to flow subsurface. Surface flow results in lower treatment efficiency because of reduced contact between the wastewater and bacteria. Suitable media include screened materials with the fines removed, such as river gravel, crushed rock (i.e., basalt, granite, and limestone), volcanic material such as pumice and scoria, and even molded plastic shapes as used in trickling filters.

Constructed wetlands have been successfully used for small communities in Europe for many years, giving excellent organic pollutant removal but variable nutrient (nitrogen and phosphorus) removal.[3] In the U.S., there are many towns (up to 5000 persons) using constructed wetlands for wastewater treatment, either using a primary settler, primary lagoon, or septic tank to remove the gross solids first.[4]

The reed bed concept has replaced absorption trenches in areas where the topography and/or soil is unsuitable; specifications for such applications have been produced.[5] A two-unit system in series is used, the first being lined (often a long and narrow bed along a side boundary), and the second, a smaller rectangular, unlined bed for final treatment and disposal. Decorative plants (lilies, cannas, ferns, irises, etc.) which thrive in wet conditions are used to produce an installation with a pleasant appearance.

Considerable care is needed in applying wetland technology to treat industrial wastewaters because of the wide variety of wastewater types that can exist. The Tennessee Valley Authority (TVA) has produced portable constructed wetlands which can be transported on trailers to industrial sites to test their suitability in handling

0-87371-550-0/93/$0.00+$.50

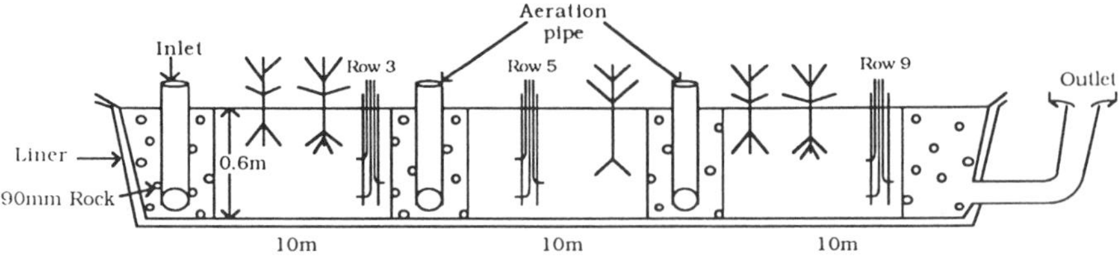

Figure 1. Longitudinal section of the bed.

the specific wastewaters for that industry.[6] The ability of such systems to break down defined "toxic organic pollutants" opens a broad field of future applications for difficult industrial wastewaters.[7] There are also several examples worldwide where constructed wetlands have been used to treat agricultural wastewaters. For example, in the U.S., a 500-pig farm unit has been set up as a demonstration facility with reed beds used to treat the pig effluent via a pond system.[8] Also, in Czechoslovakia, reed bed treatment of chicken wastes has shown removal efficiency of 80, 80, and 65% for BOD_5, total P, and total N, respectively.[9]

Large-scale constructed wetlands can be used for polishing of secondary treated effluents to produce a high-quality effluent with low levels of nitrogen and phosphorus. Orlando, Florida has such an installation of 490 ha treating up to 76 ML/day effluent from its city treatment plant with a retention time of around 30 days.[10]

In this article, we provide details of the application of constructed wetlands to treat wastewaters in Australia.

PILOT-SCALE CONSTRUCTED WETLAND

A series of constructed wetlands have been established at Frankston, and a research program funded by the Mornington Peninsula and District Water Board has been carried out. Regular monitoring on a weekly basis has been maintained from December 12, 1988, until the present, on two pairs of constructed wetlands (30 m long × 5 m wide), each pair consisting of identical beds, one planted with *Phragmites australis* and the other unplanted to act as a control. One pair of beds was constructed with 7-mm gravel as the bed material, with the influent flowing subsurface. The other pair of beds was constructed with sandy soil with relatively low hydraulic conductivity, which resulted in most of the influent flowing on the surface.

Bed Construction and Design

The beds were excavated to a depth of 0.6 m at the inlet, with the bottom having a 1% slope for the gravel beds and 2% for the sandy soil.

An impervious PVC liner was installed and sections of 90-mm coarse rock 1 m wide were placed at the inlet, outlet, and every 10 m. Each of these sections contained a 90-mm slotted pipe running across the bed at the bottom with connected vertical pipes rising out of the bed at each side (Figure 1). These were used as a method of introducing the influent to the base of the bed at the inlet, and those further down the bed as a means of introducing air for aeration experiments or chemicals for dosing experiments. The appropriate bed material was then added to give a horizontal top surface. Enough freeboard was allowed to enable flooding of the beds for weed control and to allow for build-up of debris from the plants over many years of operation, as it was not planned to harvest the plants. A submersible outlet pump connected to an adjustable electronic level controller enabled the level of the water in the bed to be controlled accurately and, for most of the time, the water was kept a few centimeters below the surface. Sampling units, having small tubes inserted into the bed at 20-cm intervals from top to bottom, were placed at the center and 1 m from the sides, every 3 m along the beds, enabling very comprehensive sampling from all parts of the bed.

General Results

Table 1 contains data relating to the performance of the beds over 11 consecutive seasons. Weekly sampling was undertaken for most of the period. The unplanted beds acted as a control and showed that although the planted beds were superior in performance, particularly as the plants became more fully developed, significant treatment was always achieved by the bed matrix alone. Although the planted sandy

Table 1. Mean Treatment Results Over 11 Seasons, December 1988–August 1991

			No. samples	Suspended solids, mg/L (% Rem)	BOD_5, mg/L (% Rem)	Ammonia-N, mg/L (% Rem)	Oxid N,[a] mg/L	Soluble P, mg/L (% Rem)
Effluent	All results		217	12 (88)	40 (78)	22 (34)	0.15	6.50 (0)
planted	Summer	1988–89	77	20 (81)	58 (67)	8 (78)	0.05	8.69 (0)
gravel	Autumn	1989	33	10 (91)	59 (67)	27 (27)	0.16	6.27 (0)
	Winter	1989	26	12 (89)	36 (76)	27 (4)	0.11	3.90 (0)
	Spring	1989	18	8 (94)	61 (71)	23 (15)	0.10	4.90 (0)
	Summer	1989–90	9	4 (96)	11 (94)	26 (0)	0.29	4.80 (0)
	Autumn	1990	12	4 (96)	19 (90)	26 (28)	0.16	4.90 (0)
	Winter	1990	12	5 (93)	27 (85)	35 (5)	0.29	5.40 (0)
	Spring	1990	12	8 (89)	27 (82)	23 (21)	0.13	8.00 (0)
	Summer	1990–91	9	4 (96)	17 (91)	25 (11)	0.25	7.30 (0)
	Autumn	1991	5	5 (93)	22 (87)	26 (28)	0.53	8.10 (0)
	Winter	1991	4	6 (93)	27 (82)	31 (0)	0.11	6.90 (0)
Effluent	All results		220	20 (81)	47 (73)	22 (33)	0.21	6.60 (0)
unplanted	Summer	1988–89	78	25 (76)	61 (66)	8 (78)	0.06	8.67 (0)
gravel	Autumn	1989	34	22 (81)	54 (70)	26 (30)	0.19	6.31 (0)
control	Winter	1989	26	24 (78)	45 (70)	28 (0)	0.13	5.52 (0)
	Spring	1989	18	11 (92)	53 (74)	21 (22)	0.27	4.65 (0)
	Summer	1989–90	10	8 (91)	20 (89)	26 (0)	0.25	5.10 (0)
	Autumn	1990	12	15 (84)	33 (83)	27 (25)	0.24	4.10 (0)
	Winter	1990	11	19 (74)	59 (72)	36 (3)	0.18	6.40 (0)
	Spring	1990	12	8 (89)	34 (77)	29 (0)	0.12	7.90 (0)
	Summer	1990–91	9	9 (90)	34 (82)	26 (7)	0.46	6.90 (0)
	Autumn	1991	5	5 (93)	23 (86)	30 (17)	0.30	8.60 (0)
	Winter	1991	4	17 (95)	57 (62)	28 (7)	0.84	6.00 (0)
Effluent	All results		221	24 (76)	23 (87)	21 (36)	0.16	5.40 (16)
planted	Summer	1988–89	78	34 (67)	32 (82)	8 (78)	0.08	5.86 (25)
sandy	Autumn	1989	36	26 (78)	27 (85)	25 (32)	0.09	4.28 (24)
soil	Winter	1989	26	17 (84)	22 (85)	24 (14)	0.10	4.45 (12)
	Spring	1989	18	17 (88)	20 (90)	20 (26)	0.14	2.68 (28)
	Summer	1989–90	10	10 (89)	12 (94)	19 (21)	0.17	—
	Autumn	1990	12	19 (79)	17 (91)	24 (33)	0.28	3.00 (38)
	Winter	1990	11	17 (77)	18 (90)	33 (11)	0.23	4.90 (17)
	Spring	1990	12	14 (80)	21 (86)	28 (4)	0.08	8.60 (0)
	Summer	1990–91	9	10 (89)	27 (86)	26 (7)	0.24	8.30 (0)
	Autumn	1991	5	20 (73)	16 (90)	32 (11)	0.14	8.50 (14)
	Winter	1991	4	14 (83)	35 (76)	25 (19)	0.79	7.50 (0)
Effluent	All results		206	40 (61)	56 (69)	21 (36)	0.12	6.20 (0)
unplanted	Summer	1988–89	70	43 (58)	42 (76)	8 (77)	0.10	6.96 (0)
sandy	Autumn	1989	34	42 (64)	48 (73)	21 (43)	0.16	5.74 (0)
soil	Winter	1989	26	35 (68)	51 (66)	26 (7)	0.09	5.10 (0)
control	Spring	1989	18	29 (79)	64 (69)	23 (15)	0.16	4.54 (0)
	Summer	1989–90	10	28 (69)	33 (82)	25 (0)	0.20	—
	Autumn	1990	12	22 (76)	56 (72)	31 (14)	0.17	4.80 (0)
	Winter	1990	11	38 (49)	79 (56)	32 (14)	0.14	5.50 (0)
	Spring	1990	11	38 (48)	80 (46)	33 (0)	0.14	7.40 (0)
	Summer	1990–91	8	56 (40)	76 (60)	26 (7)	0.10	8.70 (0)
	Autumn	1991	4	59 (20)	82 (51)	35 (3)	0.06	10.10 (0)
	Winter	1991	4	74 (10)	120 (20)	34 (0)	0.30	9.2 (0)

Note: Southern hemisphere summer, December to February.

[a] Oxid. N: nitrate + nitrite-N

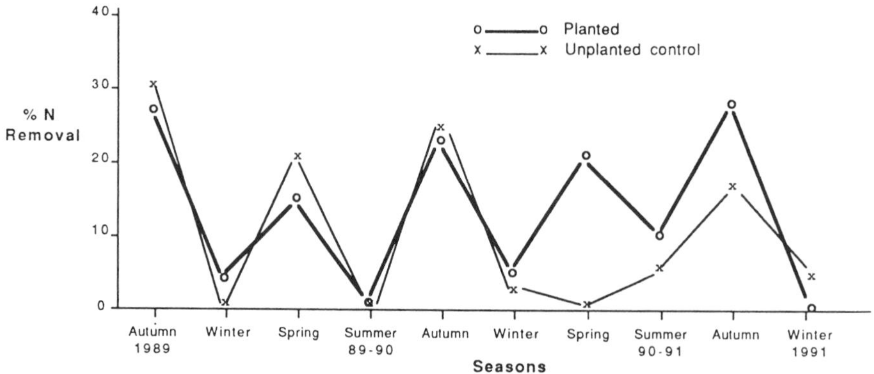

Figure 2. Ammonia-nitrogen removal for gravel bed.

soil bed had most of the water flowing on the surface because of its poor hydraulic conductivity, good treatment was still obtained as the bed acted as a "grass filtration" unit, with treatment occurring by means of the bacterial colonies on the stalks and on the surface of debris.

The detailed results for each major parameter are summarized below:

1. *Suspended solids:* The planted gravel bed, once developed, has generally achieved better than 90% removal, with the planted soil bed giving around 80% removal.
2. *BOD_5:* The planted gravel and planted soil beds have consistently given around 90% removal after a year of development.
3. *Ammonia-Nitrogen:* Although ammonia removal averaged around 35% for all beds, there was a marked seasonal influence, with autumn and spring giving peak removals and winter and summer considerably lower removals (Table 1, Figure 2). The apparent low removal in summer is probably due to the concentrating effect of evaporation and transpiration, and correcting for a modest 20% water loss of this type would give nitrogen removal slightly greater than the best autumn and spring results.
4. *Oxidized nitrogen:* These species were always low in concentration, being less than 0.8 mg/L for all beds. These low concentrations are consistent with a process where the rate of formation by nitrification is considerably slower than their rate of removal by denitrification.
5. *Soluble phosphorus:* The removal of P was insignificant in the gravel beds. This is not surprising, as the procedure of having no harvesting precluded the plants from any removal effect and the nature of the basaltic gravel similarly gave no potential for removal once the few absorption sites had been utilized.

The planted sandy soil bed was the only one which had any significant, but variable, P removal, averaging 16% removal of soluble phosphorus. This was probably due to the root system enabling access to the soil despite the low hydraulic conductivity, and adsorption onto the more absorbent soil particles.

Aeration Experiments for Improving Nitrogen Removal

Under aerobic conditions, ammonia — produced by the breakdown of proteins in domestic sewage — is oxidized by nitrifying bacteria to nitrite ions and then to nitrate ions by different microbial populations,[11] the overall process being termed *nitrification*. The nitrate ions can be converted under anaerobic conditions by different microorganisms to nitrogen gas which is lost to the atmosphere; this step is termed *denitrification*. Nitrification is generally the rate-determining process, since denitrification rates can be up to an order of magnitude greater than nitrification rates.[12] Experiments were carried out to accelerate the nitrification step using aeration, and its effectiveness was increased by a factor of three.[13]

Dye Study

The gravel beds were given a single concentrated dose of dye and the dye peak was followed through the beds, with samples of the peak being taken at 3-m intervals for BOD_5 and suspended solids (SS) analyses. Results showed that most of these parameters were removed in the first 30 to 40% of the bed (Figure 3).

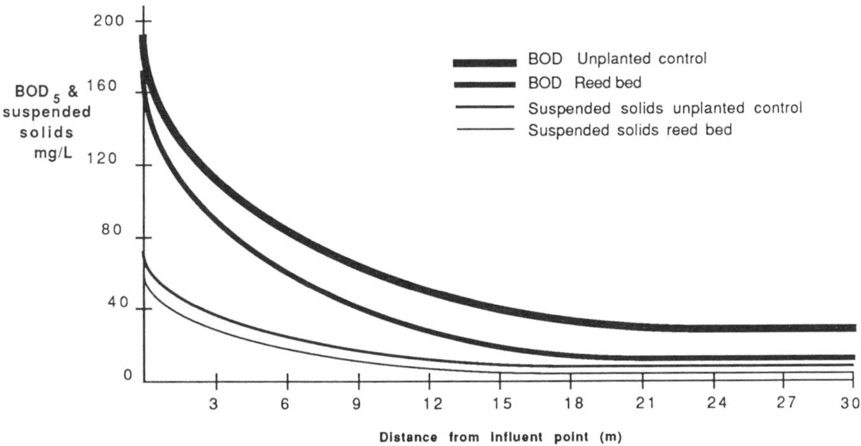

Figure 3. Progressive BOD_5 and suspended solids removal.

SINGLE HOUSEHOLD CONSTRUCTED WETLAND

In January 1990, a constructed wetland (20 m × 1.33 m) was built to treat the septic tank effluent from a three-bedroom house of a family of six persons located in Somerville in Victoria. It was made in two fully lined sections, each 10 m long, in series. Normally, the second section is unlined but to simulate the worst possible situation of poor soil absorption which in winter conditions would have zero absorbing capability, the second section was also fully lined. The plant growth was very rapid and, within 1 year, the canna lilies planted had grown into a solid mass of flowers 2.5 m tall.

Monitoring commenced in March 1990 and has continued on a monthly basis. Table 2 shows the monitoring results. BOD_5 and SS removal took more than 9 months to reach a satisfactory level, which has been maintained. Ammonia-nitrogen removal was erratic and ranged from 0 to 36%. Phosphorus removal was similarly erratic and ranged from 0 to 33%.

CONSTRUCTED WETLAND TO TREAT LANDFILL LEACHATE

A municipal waste disposal landfill has operated since the early 1900s at Lysterfield near Melbourne. After contaminated leachate is produced, it needs to be treated before disposal to a nearby creek. The raw leachate is collected and pumped into two holding dams in series before being passed through a constructed wetland system. Depending on rainfall, up to 4 ML/year of leachate will be produced.

The leachate enters a 12 m × 12.5 m bed with 25 to 30-mm diameter scoria (volcanic rock) matrix overlaid with 10 cm of sandy soil and planted with *Typha latifolia*. It then discharges to a 12 m × 5 m shallow pond and then to another bed identical to the first one before flowing down a grassy hill to a creek. Monitoring results are shown in Table 3. Good removal of suspended solids, BOD_5, ammonia, and iron have been consistently obtained.

CONCLUSIONS

From our experience with the three systems described above, we recommend a two- or three-unit wetland system, in series. Bed depth needs to be around 0.5 to 0.6 m, with horizontal top and bottom surfaces, a porous inert matrix material with a large surface area for bacterial growth, and a thin covering of sandy soil to keep the surface moist and cool: retention time of around 3 to 4 days total.

Table 2. Single Household Constructed Wetland — Somerville, Victoria

	Suspended solids			BOD_5			Ammonia			Soluble phosphorus		
Date	mg/L inlet	mg/L effluent	% removed	mg/L inlet	mg/L effluent	% removed	mg/L inlet	mg/L effluent	% removed	mg/L inlet	mg/L effluent	% removed
03/30/90	177	22	88	315	83	74	—	—	—	16	14	13
05/04/90	146	22	85	226	111	51	—	—	—	20	16	20
05/10/90	66	18	73	261	177	32	78	54	31	9	7	22
05/16/90	90	15	83	330	138	58	78	76	3	10	9	10
05/23/90	60	21	65	159	104	35	90	78	13	12	9	25
05/31/90	85	19	78	—	—	—	74	74	0	11	9	18
06/07/90	88	22	75	324	111	66	76	78	—	8	9	—
06/14/90	79	29	63	288	113	61	56	56	0	13	11	15
08/16/90	39	15	62	255	113	56	56	56	0	4	4	0
09/27/90	126	19	85	264	83	69	31	23	26	12	8	33
10/12/90	73	47	49	123	42	66	16	12	25	8	7	13
11/29/90	116	16	86	279	42	85	94	73	22	21	17	19
01/10/91	128	17	87	282	39	86	—	—	—	16	13	19
02/13/91	90	20	78	255	35	86	—	—	—	12	13	—
02/22/91	74	9	88	264	30	87	69	44	36	19	14	26
03/14/91	83	9	89	294	36	88	70	65	7	17	21	—
03/20/91	101	11	89	321	48	85	67	65	3	16	19	—
05/10/91	76	13	83	316	36	88	78	71	9	13	13	0
06/26/91	108	20	82	276	56	80	42	39	7	12	11	8
07/30/91	177	29	84	369	71	81	19	38	—	13	13	—
08/21/91	174	21	88	270	90	67	47	41	13	15	13	13

Table 3. Lysterfield Landfill Leachate Constructed Wetland

Date	Sample points	Suspended solids (mg/L)	BOD_5 (mg/L)	Ammonia-N (mg/L)	Oxidized N (mg/L)	Soluble P (mg/L)	pH	Dissolved oxygen (mg/L)	Total iron (mg/L)
05/05/90	Influent to beds	34	12	—	—	0.17	7.8	5.4	—
	Pond	9	0	—	—	0.53	7.3	5.0	—
	Effluent	11	6	—	—	0.33	8.1	6.0	—
05/23/90	Influent to beds	18	27	1.9	0.80	0.22	8.0	6.9	—
	Pond	15	0	0.2	0.26	0.33	7.4	7.4	—
	Effluent	6	3	0.9	0.23	0.29	8.3	6.4	—
06/21/90	Influent to beds	22	5	3.8	1.00	0.03	8.0	6.6	—
	Pond	7	0	0.7	0.16	0.03	8.3	9.9	—
	Effluent	4	2	0.3	0.19	0.13	7.8	5.8	—
08/16/90	Raw leachate	32	9	38.0	0.66	0.12	8.2	11.3	—
	Influent to beds	15	9	11.0	1.60	0.04	8.2	7.2	—
	Pond	16	0	1.6	0.78	0.08	8.2	12.1	—
	Effluent	9	0	0.5	0.54	0.15	7.7	7.7	—
11/22/90	Raw leachate	174	41	18.8	0.36	0.37	8.0	15.0	1.6
	Influent to beds	134	24	0.6	0.88	0.12	8.2		0.5
	Pond	38	14	5.0	0.23	0.35	8.3	13.3	0.4
	Effluent	42	47	3.0	0.05	0.58	7.8	1.8	0.3
12/13/90	Raw leachate	41	16	50.0	0.05	0.15	7.8	—	2.0
	Influent to beds	44	7	10.5	0.60	0.15	8.0	—	1.80
	Pond	20	3	9.0	0.20	0.34	8.0	—	0.5
	Effluent	9	7	10.5	0	0.74	7.7	—	0.5
01/10/91	Raw leachate	53	44	38	0.05	0.80	7.7	0.8	2.9
	Influent to beds	43	10	8	0.80	0.18	8.8	13.2	0.5
	Pond	64	11	0.1	0.04	0.25	8.6	12.9	0.4
	Effluent	10	7	6	0.03	0.60	7.8	3.9	0.4
06/13/91	Raw leachate	45	23	29.0	4.5	0.29	7.4	4.0	3.8
	Influent to beds	50	5	10.0	0.45	0.09	8.1	6.6	2.0
	Pond	55	1	4.0	0.13	0.21	8.3	9.5	1.0
	Effluent	30	1	0.1	0.03	0.43	7.5	5.7	1.0
07/17/91	Raw leachate	69	30	29.0	0.14	0.42	7.6	2.0	4.1
	Influent to beds	24	36	13.5	0.09	0.06	8.1	4.6	1.0
	Pond	20	9	5.0	1.06	0.03	8.5	9.6	0.4
	Effluent	8	3	2.5	0.18	0.08	7.7	4.5	0.4
08/21/91	Raw leachate	63	23	40.0	2.00	0.07	7.5	3.7	9.7
	Influent to beds	14	8	3.5	9.0	0.03	8.1	6.8	0.5
	Pond	25	5	3.0	7.8	0.12	7.9	7.1	0.3
	Effluent	7	1	2.0	6.5	0.05	7.7	5.6	0.10
09/11/91	Raw leachate	99	14	21.5	1.0	0.13	7.8	7.1	3.5
	Influent to beds	19	11	7.8	10.5	0.08	8.2	5.7	0.5
	Pond	21	10	2.5	5.5	0.09	8.1	7.9	0.3
	Effluent	16	4	2.0	4.5	0.09	7.9	7.2	0.1
10/09/91	Raw leachate	38	29	46.0	3.0	0.09	7.7	7.3	3.4
	Influent to beds	21	10	15.0	5.2	0.09	7.9	5.7	0.8
	Pond	22	8	12.3	6.7	0.08	7.8	5.7	0.2
	Effluent	24	10	12.3	6.0	0.11	7.7	4.4	0.1

Unit 1:

Unit 1 is fully lined, with a retention time of 1.5 to 2 days. With most of the suspended solids and BOD_5 being removed in the first section of the bed (Figure 3), where fine gravel is used, there is a tendency for the solids to block some of the void paths to give some surface flow for a few meters after 2 to 3 years. Hence, a coarse matrix 80 to 90 mm in diameter covered with a thin layer of sandy soil would be used in the first unit, with vigorous macrophytes such as *Phragmites australis* or *Typha latifolia*.

Unit 2 — Optional:

If nutrient removal is critical, a small pond about 20% of the size of the first unit should be used which can be used for aeration to accelerate nitrification and to allow chemical dosing if necessary for phosphorus removal: retention time about 0.5 day.

Unit 3:

This would be similar to the first unit but with smaller diameter matrix, 20 to 30 mm diameter, and area about two thirds of the first: retention time 1 to 1.5 days. This bed would finish the organic pollutant removal, remove nitrogen by denitrification, and act as a filter to trap any insoluble phosphate carried over from the pond.

REFERENCES

1. Kickuth, R. Das Wurelraumverfahren in der Praxis, *Landsch Stadt.* 16:145, 1984.
2. Lawson, G. J. Cultivating reeds for root zone treatment of sewage. Contract report to the Water Research Centre, Institute of Terrestrial Ecology, U.K., 1985, 20.
3. Brix, H. and H. H. Schierup. Root zone systems, operational experience of 14 Danish systems. Report to the Environmental Protection Board, Denmark, 1986.
4. Steiner, G. R., J. T. Watson, and D. A. Hammer. Constructed wetlands for municipal wastewater treatment, *Mississippi Water Resources Conf.* Jackson, MS, March 1988.
5. Watson, J. T. Constructed wetlands design guidelines for residential septic tank systems. Tennessee Valley Authority, Chattanooga, TN, 1989.
6. Davies, T. H., J. T. Watson, and D. B. Jenkins. Treatability assessment of industrial wastes, *Proc. Constructed Wetlands Conf.* IAWPRC, Cambridge, U.K., 1990, 403.
7. Wolverton, B. C. and R. C. McDonald. Biotransformation of priority pollutants using biofilms and vascular plants, *J. Mississippi Acad. Sci.* XXXI:79, 1986.
8. Hammer, D. A. and J. T. Watson. Agricultural waste treatment with constructed wetlands, *Natl. Symp. Protection of Wetlands.* Fort Collins, CO, 1988.
9. Vymazal, J. Use of reed bed systems for the treatment of concentrated wastes for agriculture, *Proc. Constructed Wetlands Conf.* IAWPRC, Cambridge, U.K., 1990, 347.
10. Swindell, C. E. and J. A. Jackson. Constructed wetlands design and operation to maximise nutrient removal capabilities, *Proc. Constructed Wetlands in Water Pollution Control Conf.* IAWPRC, Cambridge, U.K., 1990, 107.
11. Atlas, R. M. and R. Bartha. *Microbial Ecology.* Addison-Wesley, Reading, U.K., 1981.
12. Hsieh, Y. and C. L. Coultes. Nitrogen removal from freshwater wetlands, *Constructed Wetlands for Wastewater Treatment.* D. A. Hammer, Ed. Lewis Publishing, Chelsea, MI, 1989.
13. Davies, T. H. and B. T. Hart. Use of aeration to promote nitrification in reed beds treating wastewater, *Proc. Constructed Wetlands Conf.* IAWPRC, U.K., 1990, 77.

CHAPTER 65

Creating a Wetlands Wildlife Refuge from a Sewage Lagoon

R. Lofgren

INTRODUCTION

The City of San Antonio, Texas plans to clean up Mitchell Lake, an old city-owned sludge containment reservoir in South San Antonio, by constructing a wetlands and reuse reservoir. Over the years, the area has become a major stopping point along the bird migratory path known as the Central Flyway. Several volunteer groups are helping the city construct a wetlands wildlife refuge at the north end of Mitchell Lake. Treated effluent from a local sewage treatment plant will supplement rainfall as the water source.

In recognition of the environmental importance of this area and in compliance with the requirement by the Texas Water Commission to clean up the area, the city will supply the water for the project as well as pay for and supervise the construction of a pipeline to bring water to the wetlands. The city will also contour the wetlands area into a series of shallow basins so the the water will flow down over the mudflats to the lake.

Volunteers from environmentally conscious organizations will have much to contribute. They will locate and plant native vegetation in and around the wetlands. Plants will be chosen for both the aquatic and the semi-arid regions surrounding the mudflats, with emphasis on using those plants which provide good habitat for wildlife. Paths to be used by visitors will be built around the wetlands. The volunteers will design and install signs explaining the function of the wetlands and the kinds of birds to be sighted in the area. The importance of the area will be publicized through preparation and distribution of educational brochures about these wetlands. By the end of the project, an on-going volunteer society will have been organized to preserve and maintain the wetlands wildlife refuge into the future.

The Junior League of San Antonio, the League of Women Voters, the San Antonio Audubon Society, and the Bexar Audubon Society are currently taking the lead in the Wetlands Project.

BACKGROUND

The history of sewage treatment in San Antonio goes back to 1718 when the Mission San Antonio de Valero (better known as the Alamo) was founded by the Spanish. However, the first sewer system was not approved by the city council until 1894.[1] When San Antonio's first deep wells had been drilled in 1890 into the underlying artesian aquifer, providing the town with an abundance of pure water; a system for wastewater disposal was urgently needed.

The Mitchell Lake area has a long history relating to San Antonio's wastewater management. In 1897, the City of San Antonio contracted with a private firm to handle the irrigation and land disposal of its municipal sewage. When downstream irrigators became irate with the increased flow of sewage, San Antonio contracted with another private corporation to construct an open ditch from the existing sewage farm to Mitchell Lake, which was owned by the company. In 1901, "a good and serviceable dam at Mitchell Lake, sufficient to contain the surplus sewage of the City of San Antonio" was constructed.

Mitchell Lake served as San Antonio's only wastewater treatment facility until 1930. Water from the lake was used for irrigation of surrounding farms for over half a century. The Rilling Road Treatment Plant was completed in 1930, with Mitchell Lake continuing to serve as a stabilization/treatment pond for waste

0-87371-550-0/93/$0.00+$.50

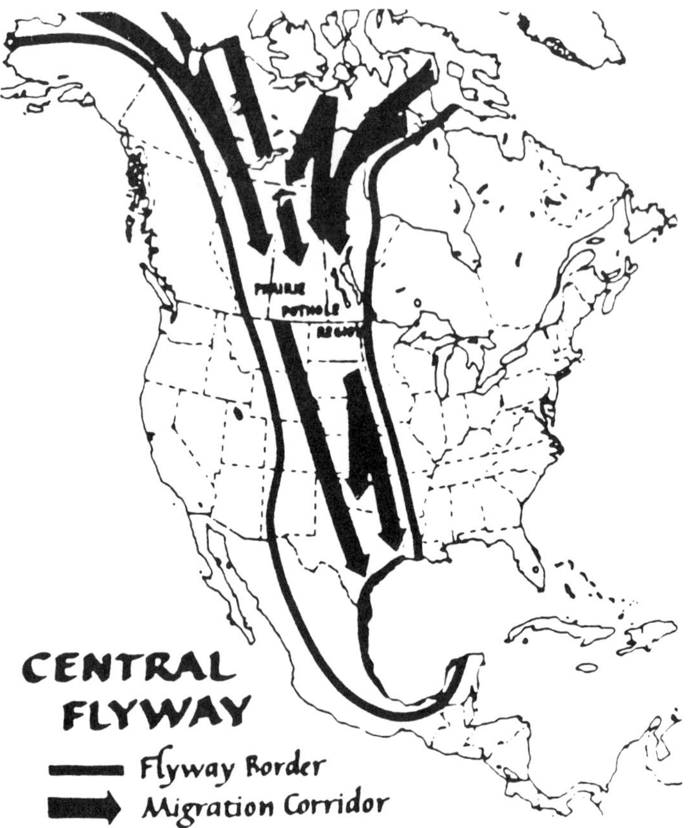

Figure 1. Central Flyway, showing Flyway border and migration corridor.

activated sludge. Due to rapid growth in San Antonio, the wastewater treatment facilities became too small for the volume of sewage. Frequently, raw sewage and sludge were discharged to the lake.

By the late 1960s, raw sewage was discharged into the lake in a continuous flow. Although the sludge material quickly stabilized and was relatively odor free, large areas of shallows were created, periodically producing odor problems.

People living in the area may have considered it a nuisance, but wildlife flourished on the rich supply of food and water available in this wetlands area. In the century since Mitchell Lake was formed, shorebirds and waterfowl migrating along the Central Flyway[2] (see Figure 1) have used the rich waters of Mitchell Lake as a resting and feeding spot in an otherwise semi-arid region. More than 270 species of birds have been sighted in this area; 30 species use the lake area as their breeding grounds. Mitchell Lake is the primary birding site recommended in regional and national guide books for bird watchers in the San Antonio area. Its reputation has brought bird watchers to San Antonio from across the U.S., Canada, Mexico, and Europe.

For over a quarter of a century, members of the San Antonio Audubon Society have recorded bird sightings in the Mitchell Lake area. They include ducks, geese, grebes, pelicans, phalaropes, herons, egrets, stilts, avocets, dowitchers, and yellow-legs.[3]

In 1973, when the Texas legislature ordered the city to "abate the nuisance", some San Antonians became concerned that changes might be made that would have an impact on wildlife in the area. Members of the San Antonio Audubon Society and others urged the city council to pass Ordinance #41789 "Designating Mitchell Lake as a refuge for shorebirds and waterfowl; and authorizing such changes and additions as are necessary to establish this refuge." It did so in 1973.

The city managed the problem first by diking off the north end of the lake to form small ponds, called polders, to receive the fresh sludge. Later, when these were filled to capacity, additional small ponds, called "decant basins", were diked off from the lake, 173 acres in all (see Figure 2).[4] After the Dos Rios Treatment Plant went on-line in 1987 and sludge no longer entered the lake, a comprehensive plan for the clean-up and management of the Mitchell Lake area was required.

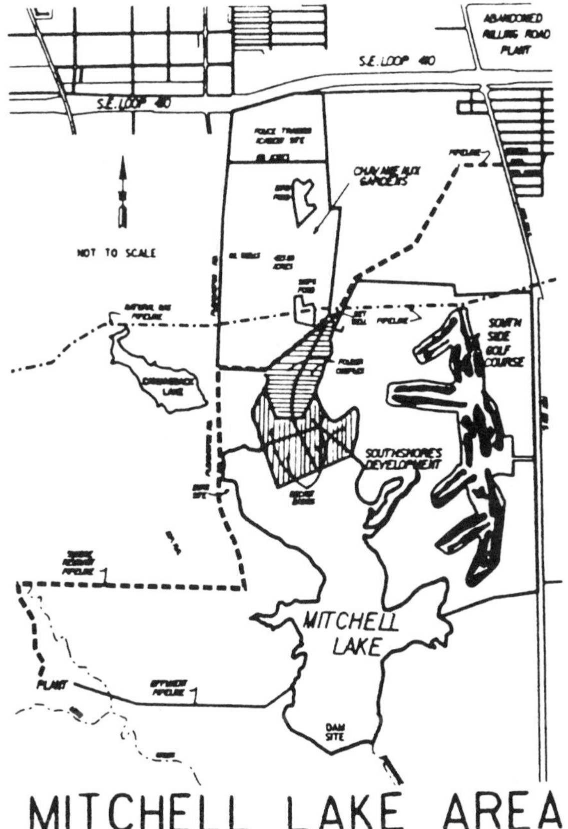

Figure 2. View of Mitchell Lake area.

The Mitchell Lake Recovery Advisory Subcommittee was created February 25, 1987, upon the recommendation of the 201 Wastewater Advisory Committee to the San Antonio City Council. The 201 Wastewater Advisory Committee charged the Subcommittee with the development of the framework for a comprehensive plan for the clean-up and on-going management of the Mitchell Lake and Chavaneaux Gardens areas, consistent with currently known goals, objectives, or mandates.[4] The Mitchell Lake area was annexed by the City of San Antonio at the end of 1987.

A comprehensive plan for the clean-up and management of the Mitchell Lake area was approved by city council in 1989. The plan called for the construction of wetlands in the polder and decant basins as the most cost-effective and desirable way to cleanup the sludge deposits there. The city's Wetlands Master Plan includes the design for the pipeline to supply water (treated effluent) from the nearby Leon Creek Treatment Plant to the wetlands area, as well as the design for the construction of the wetlands.

The Master Plan also includes the rebuilding of the century-old dam to meet current standards and future plans to develop Mitchell Lake proper into a reuse reservoir.

WETLANDS ISSUES

"Wetlands represent a very small fraction of our total land area, but they harbor an unusually large percentage of our nation's wildlife. For example, 900 species of wildlife in the U.S. require wetland habitats at some stage in their life cycle, with an even greater number using wetlands periodically. Representatives from almost all avian groups use wetlands to some extent and one third of North American bird species rely directly on wetlands for some resource." [5]

Our wetlands project directly addresses two environmental issues faced today — water conservation and preservation of wildlife habitat. Plans are being made to tackle a third issue in the future, that of water quality.

The wetlands project will use waste water from the Leon Creek Sewage Treatment Plant in South San Antonio to supplement rain water run-off. Water is a precious resource in South Texas and conservation is increasingly necessary as the human population grows. Through water reuse, this project will ensure a continued water source for the wetlands and its migratory bird population, while conserving water.

Preservation of wildlife habitat is the second issue addressed by the project. Without the influx of treated effluent, Mitchell Lake and the wetlands would soon dry up, and the birds would no longer have resting and feeding grounds in the area. Through the volunteer effort in planting the wetlands, food and nesting sites will be available in abundance. The monies spent by the city in constructing the site, the volunteers' work in educating the public about the constructed wetlands, and the creation of an organized group to protect the area will ensure that the wetlands will continue into the 21st century.[6]

In the future, we see the possibility of using this constructed wetlands area as a natural effluent treatment system, allowing organisms to utilize organic material from the effluent, resulting in cleaner water. The city of Austin, Texas, the University of Texas, and Texas A & M University are currently operating a natural effluent treatment system with very promising results. Our volunteer coalition plans to visit the Austin site and study the feasibility of such an operation at our wetlands for the benefit of the city of San Antonio.

MASTER PLAN

The two-phase Wetlands Master Plan calls for the creation of a marshland-wetland-mudflat out of the decant basins and polders with its own permanent water supply. Phase I is an Interim Plan, while the level of Mitchell Lake is low, during the construction of a dam. The lake level will be at 518 ft above sea level, while the wetlands will begin at 523 ft above sea level and drop down to 519 ft above sea level (see Figure 3).[7] Water flows into Mitchell Lake from the Leon Creek Treatment Plant. Water will be pumped from the lake into the highest decant basin. It will then trickle down through the wetlands to flow out into the lake. After the dam has been completed in late 1993 and the lake level raised to 524 ft above sea level, Phase II, the Final Plan, will be in operation (see Figure 4).[7] Water will flow by gravity from the lake into the basin with the highest elevation and follow the same path to the discharge point where it will be pumped out into the lake. The contouring of the decant basins and polders is currently being planned. The construction will be timed and designed to cause as little disruption to the habitats as possible, thus safeguarding the wildlife in the area.

CONCLUSIONS

The San Antonio Wetlands Wildlife Refuge will become an ecological treasure. As Susan McAtee Monday said, "The Mitchell Lake plan is recycling at its best. The city takes what everyone (but birders) thinks of as a swampy eyesore and makes of it a wetland treasure, a place sure to attract world class birders during spring and fall migration seasons."[8]

A truly effective program for protecting wetlands requires the cooperation of governmental agencies, developers, environmental groups, the scientific community, and active citizen support and participation. With ongoing management, this urban constructed wetlands wildlife refuge will preserve the diverse habitats that sustain so many fascinating life forms and will continue to serve the needs of migratory birds.

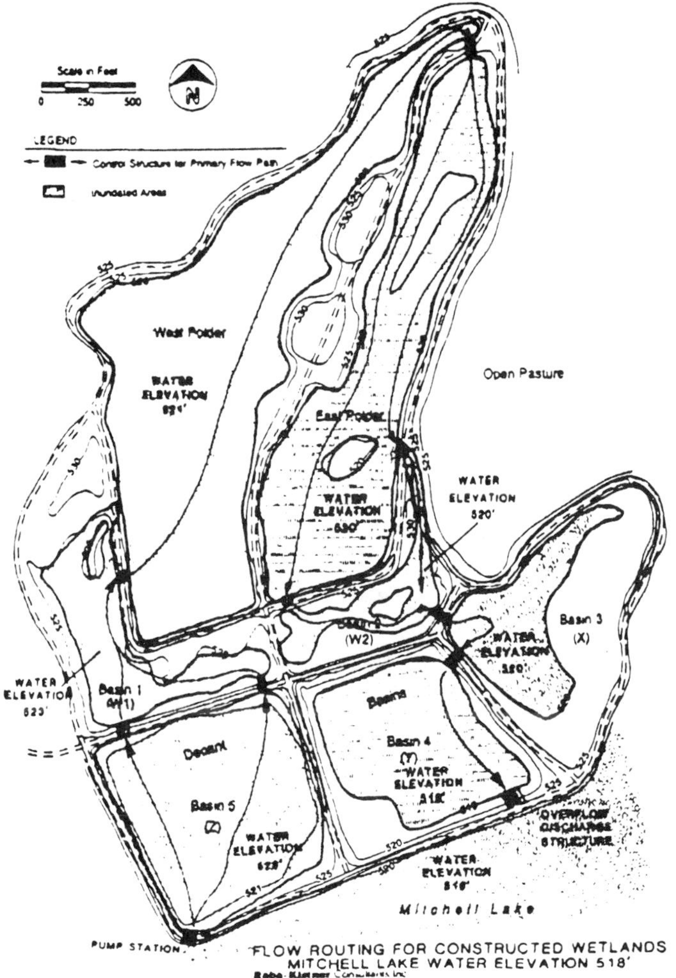

Figure 3. Phase I of the Wetlands Master Plan.

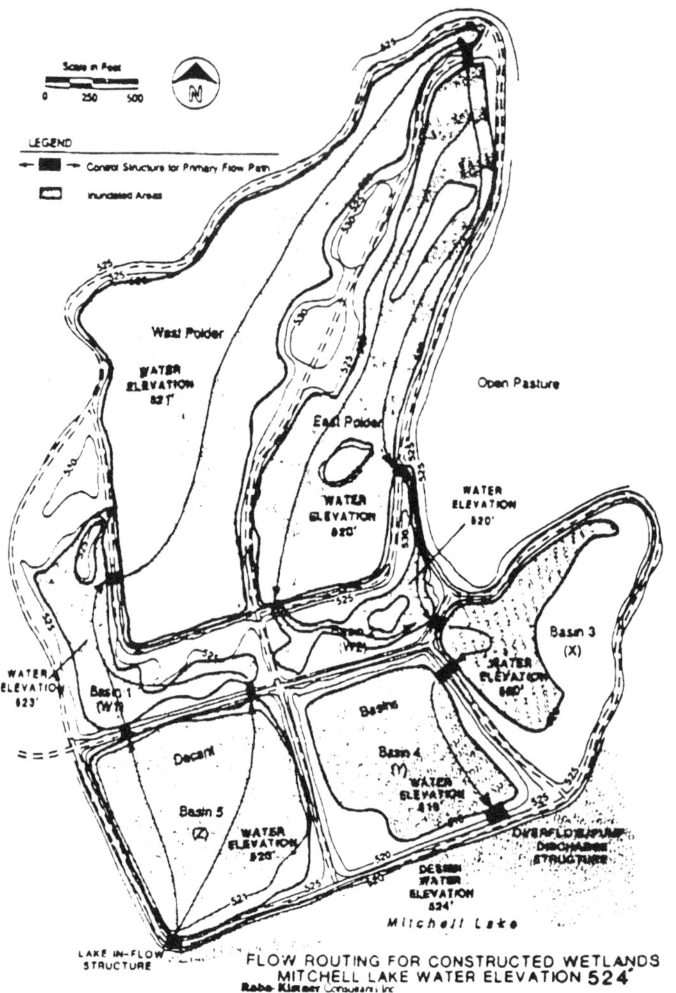

Figure 4. Phase II of the Wetlands Master Plan.

REFERENCES

1. San Antonio Department of Wastewater Management. Brief history of sewage treatment in San Antonio. A Visitor's Guide to the San Antonio Wastewater Treatment Facilities. October 1985.
2. Bellrose, F. C. and R. A. Heister. The central flyway, The *Nature Conservancy News*. 35(1):4, 1985.
3. Roney, E. Report of the Bird Records Committee of the San Antonio Audubon Society prepared for the Mitchell Lake Recovery Advisory Subcommittee, 1989.
4. Mitchell Lake Recovery Advisory Subcommittee of the 201 Wastewater Advisory Committee to the San Antonio City Council. A Comprehensive Plan for the Cleanup and Ongoing Management of Mitchell Lake and Chavaneaux Gardens, January 1989
5. Hammer, D. A. *Constructed Wetlands for Wasterwater Treatment.* Lewis Publishers, Chelsea, MI, 1989.
6. Lofgren, R. Restoring an ecological treasure, in *Bright Scrawl.* Junior League of San Antonio, 12–13, 1990.
7. Raba-Kistner Consultants to San Antonio Department of Wastewater Management. Mitchell Lake Rehabilitation Project Report, in process, 1991.
8. Monday, S. M. Mitchell Lake proposal centering on wildlife preserve, nature center. *San Antonio Light.* Section J, July 14, 1991.

CHAPTER 66

Production and Suppression of Mosquitoes in Constructed Wetlands

K. J. Tennessen

INTRODUCTION

The mosquito species of any given area occupy a vast array of habitats and, as a group, withstand extreme environmental conditions. Each species, however, is adapted to a certain type of habitat; some species exhibit a rather wide habitat range (or ecological tolerance), whereas others may be narrow in their habitat choice. Larval habitats include lake edges, ponds, ditches, bogs, swamps, marshes, springs, snowmelt pools, rock pools, slow reaches of streams, floodwater or overflow depressions, and containers (e.g., tree holes and artificial containers such as rain barrels, cans, and tires).[1] The major requirement for successful larval growth and development is water standing for some duration, which can be as little as 1 week. Food and protective cover are also critical. Owing to their short life cycles and high fecundity, mosquito populations can escalate rapidly, and in areas of human habitation, nuisance levels can be reached quickly.

Mosquitoes are common inhabitants of natural wetlands, so their invasion of constructed wetland systems should be expected. Where standing water, high nutrient levels, and vegetative cover occur in constructed wetlands, an ideal medium for larval growth exists. Gravid females quickly find places to lay eggs where such habitat might otherwise be limiting. Anaerobic, bacteria-laden water has been found to be highly attractive to ovipositing females.[2,3] The potential for explosive mosquito populations in highly enriched constructed wetlands and the annoyance and vector capabilities of some species require consideration early in the design and operation of constructed wetlands.

Constructed wetlands in the Tennessee Valley, designed and operated for studies on the treatment of domestic wastewater, were sampled to determine the species composition and population levels of invading mosquitoes. Additional objectives were to determine what conditions favor mosquito production in constructed wetlands, and to test various insecticide applications in cases where control became warranted.

METHODS

The constructed wetlands sampled for mosquito production are located in western Kentucky (KY) (Figure 1). The site at Benton (population 5,000), in Marshall County, consists of three parallel 1.4-ha cells adjacent to a 6.4-ha primary storage lagoon. The site at Pembroke (population 1,100), in Christian County, consists of two parallel 1-ha marsh-pond-meadow systems. Mean annual air temperature for the Benton area (data from Land Between the Lakes) in 1988 was 14.7°C (58.4°F); total rainfall was 59.35 in., about 12 in. above annual norm. Mean annual air temperature for Pembroke area (data from Ft. Campbell, KY) was 14.3°C (57.8°F); total rainfall was 60.38 in., 13 in. above normal.

Larvae were sampled biweekly from March 2, 1988 to February 21, 1989. At Benton, 20 standard dipper samples were taken from each of the three cells and the lagoon. At Pembroke, operation of the parallel systems was highly variable; 10 to 20 samples were taken from each segment — the marsh, the pond, and the meadow, depending on available mosquito habitats. Dips were taken from the margins, usually where there was some type of plant cover. All larvae collected were preserved, identified as to species and instar, and counted; pupae were counted but not identified.

0-87371-550-0/93/$0.00+$.50

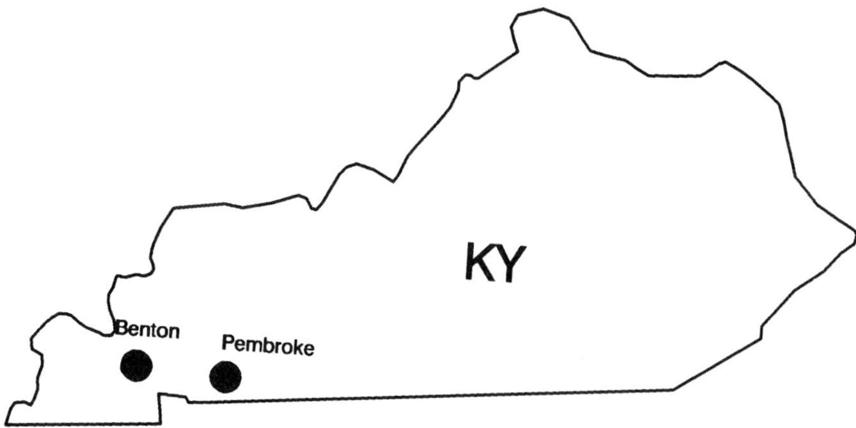

Figure 1. Outline map of Kentucky, showing locations of Benton and Pembroke constructed wetlands.

Adult landing counts were conducted at dusk, after light intensity dropped below 1 foot-candle. A biting count usually consisted of two or three 15-min sessions. Female mosquitoes were collected using a battery-operated aspirator as they landed on arms and clothing of a male assistant. Specimens were killed in chloroform and identified and counted.

Several different insecticides were tested in the wetlands to compare efficacy and to devise a treatment schedule. In August 1989, cell 1 at Benton was treated with Abate granules (5% A), an organophosphate, at the rate of 10 lbs/ac. Cell 2 was treated with *Bacillus sphaericus*, a bacterial larvicide adsorbed on corn cob grits (Abbott Labs), at the rate of 20 lbs/ac. These dry materials were dispersed into the wetlands via a blower mounted on a truck driven at a speed of 10 mph. Mosquito larvae were sampled before treatment, and 24 h, 7 days, and 14 days after treatment by taking 20 random dips per cell each time. Cell 3, which remained mostly dry, was not treated, but was sampled for larvae. In September and October 1990, these cells were slug-treated with liquid formulations of two bacterial larvicides: cell 1 with *Bacillus sphaericus* var 5A-SB (Bs) at 2.5 ppm, and cell 2 with *Bacillus thuringiensis israelensis* var. H-14 (Bti) at 5 ppm. Larvae were sampled as in the previous experiment. In June 1990, the meadow at Pembroke was also slug-treated with 2000 mL Bti.

Based on results from these tests, the wetlands at Pembroke were used to test weekly applications of Bti during the height of mosquito populations. Granular Bti was applied with a Buffalo Turbine Mity Mite backpack sprayer-duster by walking though the marsh and the meadow parts of the system. Larvae were sampled immediately prior to treatment, 24 h later, and 7 days later, beginning July 16, 1991. The first treatment was applied at a rate of 10 lbs/ac; the seven subsequent treatments were increased to 20 lbs/ac. The final treatment was made on September 2, 1991.

RESULTS AND DISCUSSION

Larval Production

Benton

Out of 1500 dipper samples taken at Benton, 806 were positive for mosquito larvae (53.7%); a total of 38,630 larvae (average 26 per sample) and 1200 pupae were collected. The number of larvae increased steadily through spring, followed by a sharp increase in early summer (Figure 2). The total population then declined in mid-July, immediately after the entire wetlands area was sprayed with a mixture of malathion and sevin, in unknown concentration, possibly under the direction of state or county health officials. Several weeks later, populations appeared to have recovered, as another peak occurred from later summer through early fall, followed by expected declines with the onset of colder weather.

There were 12 species of larval mosquitoes found in the Benton wetlands, which is nearly one-fourth of the total number of species of Culicidae known to occur in the Tennessee Valley region. *Culex pipiens* Linnaeus and *C. salinarius* Coquillett were the two most abundant species present. Successional changes in

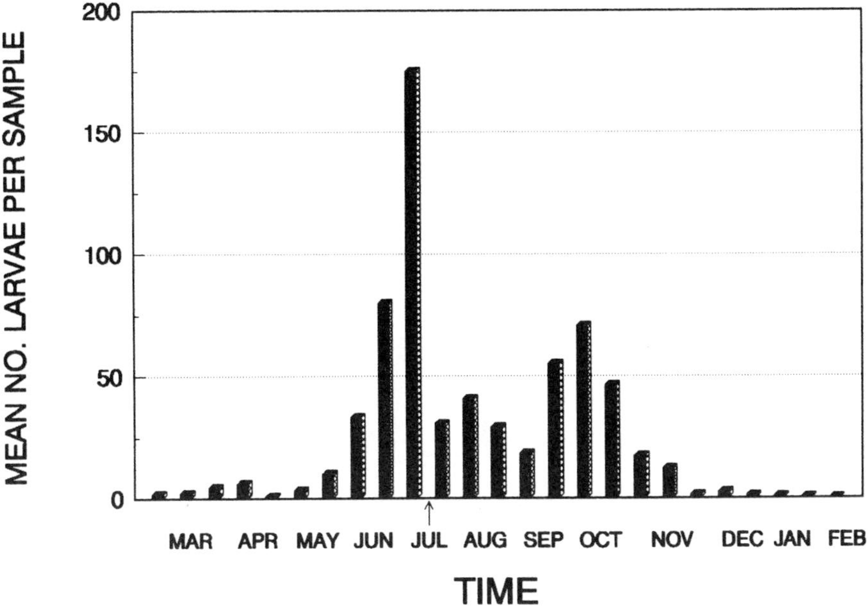

Figure 2. Mean number of mosquito larvae collected per sample in Benton constructed wetlands from March 1988 to February 1989. Arrow indicates date area was sprayed (see text).

species abundance were dramatic (Figure 3). The *Aedes* and *Psorophora* species we collected are floodwater opportunists, and appeared very early — in March. Larvae of these two species were rarely found after May. The disjunct distribution of *A. vexans* (Meigen) within the wetlands is noteworthy, as numerous larvae were found in the two cells with native clay substrate, whereas few were found in the cell with a gravel substrate. Floodwater mosquitoes should be expected to invade wetlands where water levels fluctuate due to rainfall or changes in loading. The other nine species collected, which are permanent pool inhabitants, flourished mainly during the warmer part of the breeding season. Two exceptions in this group are *Culiseta inornata* and *C. restuans*. *C. inornata* larvae were abundant from mid-March until late May, but then were not found again until late September. *C. restuans* was abundant from mid-April to mid-June, but was not collected again until mid-August. The latest to occur as larvae were *C. salinarius* and *C. inornata*, which were collected in December and January.

The lagoon was free of mosquito larvae until October 3, when large numbers (mean, 365 per sample) of *C. pipiens* and *C. salinarius* were found. The steep margins of the lagoon, lack of emergent littoral vegetation, and wave action are factors that inhibit mosquitoes. Therefore, reasons for the sudden appearance of *Culex* larvae late in the year are unknown. It is possible that because of the large build-up of *Culex* populations in the adjacent treatment cells by late summer, chance dictated that some females would oviposit in marginal, less than optimal habitats.

Pembroke

Of 611 dipper samples, 274 were positive for mosquito larvae (45%), and the average number of larvae per sample was 102. During the first 3 months of sampling, the majority of larvae were found in open influent troughs, as standing water seldom occurred in the marsh or meadow systems. The majority were *Culex pipiens*, although *C. restuans* and *C. salinarius* also were found in troughs. In June, the system was modified by replacing the open troughs with closed distributor pipes, greatly reducing the habitat for mosquito oviposition (Figure 4). Beginning in July, standing water was found in the marsh, and mosquito larvae were usually present; occasionally, the ponds also contained larvae. *Culex pipiens* again dominated, although the population was in much lower numbers than when the open troughs were in operation. No larvae were found after early December; sampling was discontinued in February 1989. No larvae of floodwater species were collected within the wetlands proper, although some were occasionally found in flooded depressions surrounding the wetlands.

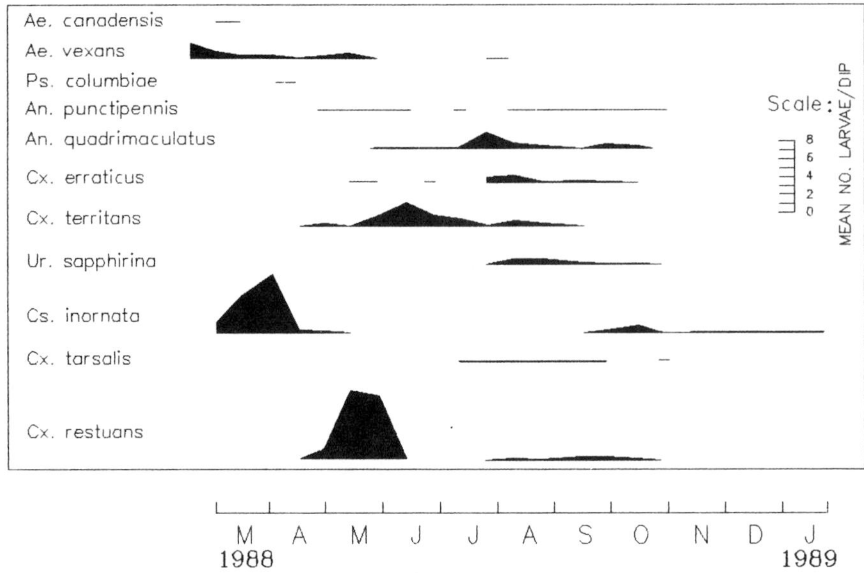

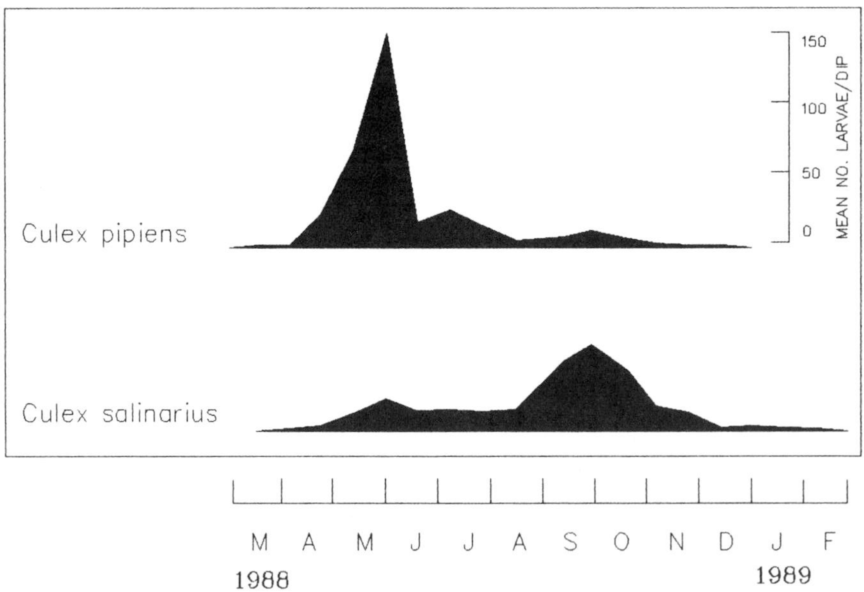

Figure 3. Seasonal change in species populations in Benton constructed wetlands from March 1988 to February 1989.

An additional factor that led to mosquito production at Pembroke was the leaking of the newly constructed ponds into an adjacent creek. During periods of low rainfall, flow in this creek was greatly diminished, resulting in stagnant pools. Wastewater from the wetland ponds seeped into these pools during much of 1988, creating enriched standing water. mosquito larvae of seven species were found here from May 11 through August 3, in greatest abundance in early July. These ponds were drained and sealed later in the year.

Mosquitoes in Constructed vs. Natural Wetlands

The mosquito composition of natural wetlands is highly variable because of extreme diversity in types of habitat that different wetlands contain. About 15 species normally can be expected to occur in natural wetlands

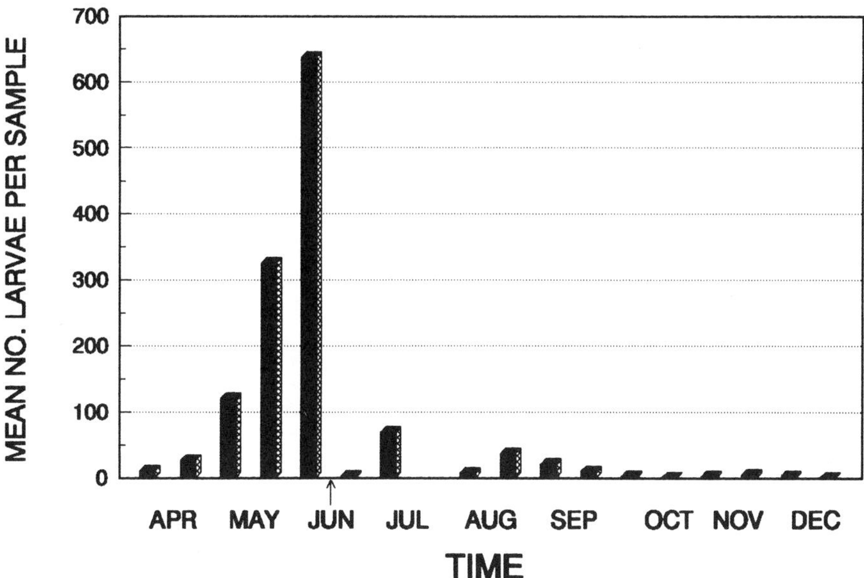

Figure 4. Mean number of mosquito larvae collected per sample in Pembroke constructed wetlands from April 1988 to December 1988. Arrow indicates date influent system was modified (see text).

in western Kentucky. The most commonly encountered species are *Anopheles punctipennis* (Say), *A. quadrimaculatus* (Say), *Culex erraticus* (Dyar and Knob), *C. restuans, C. territans* (Walker), *Coquillettidia perturbans* (Walker), *Culiseta inornata, Aedes canadensis* (Theobald), *A. vexans* (Meigen), and *Psorophora ferox* (Humboldt). As seen in the above accounts on Benton and Pembroke, most of these species were also found in the constructed wetlands. Major determining factors in species composition are permanence of water, fluctuations in water level, size of water body, stillness of the water surface, and the organic make-up of the water.

Larval density is also variable; the patchy distributions usually found depend mainly on selection of oviposition sites by gravid females, types of plant cover, food availability, and predator abundance and distribution. Limited sampling of natural wetlands in western Kentucky in 1988–1989 yielded means of 0.1 to 5.0 larvae per sample. Based on many years of sampling of mosquito larvae and recording complaints, the Vector Control Program of Tennessee Valley Authority (TVA) has determined that an average larval density of 0.25 larvae per dipper sample in areas of human habitation near reservoirs usually results in high levels of adult annoyance and concomitant adulticide application requests. Although summertime larval densities in natural wetlands exceeded this treatment threshold on occasion, they did not approach the upper mean densities recorded in the constructed wetlands (6.1 to 272 at Benton, 1.7 to 37 at Pembroke).

The great influx of organic matter and the associated production of microorganisms in the constructed wetlands appears to be the major factor influencing mosquito density differences between natural and constructed wetlands. The species found in greatest abundance (in the genus *Culex*) are known to be attracted to organic pollution,[4] and have been called "filth breeders". Factors which normally act to reduce larval populations appear to be overridden by the presence of the wastewater. For example, few if any mosquito larvae are found under dense cover of duckweed (*Lemna* spp.) in ponds and swamps. However, at the Benton and Pembroke constructed wetlands, high larval densities were often found in heavy duckweed cover. The high concentration of organic matter and microorganisms apparently negated the limiting effect of the plants. Another factor that usually exerts a negative effect on mosquito density is predator abundance. However, despite the presence of large numbers of *Gambusia affinis* (mosquitofish) at Benton and an abundance of invertebrate predators, mosquito larvae remained abundant throughout the season.

Mosquito Annoyance at Benton and Pembroke

Of the 15 species of mosquitoes collected in our dusk landing counts, *Culex salinarius* was numerically dominant at both sites (Table 1). At Benton, this species appeared as early as May 10, continued to increase

Table 1. Total Number of Adult Females Aspirated During Dusk Landing Counts in Constructed Wetlands, Based on 7 Sampling Dates at Benton and 3 at Pembroke, from May to October, 1988

Species	Benton	Pembroke
Aedes canadensis[a]	2	
Aedes triseriatus[a]	2	3
Aedes trivittatus[a]		1
Aedes vexans	6	2
Coquillettidia perturbans[b]	12	
Culex erraticus	8	23
Culex pipiens	8	
Culex salinarius	821	34
Culiseta inornata	1	
Psorophora ciliata[a]	1	
Psorophora columbiae[a]	1	1
Psorophora cyanescens[a]		7
Psorophora ferox[a]		1
Anopheles punctipennis	11	2
Anopheles quadrimaculatus	4	

[a] Larvae not collected in constructed wetlands proper.

[b] Larvae not collected in constructed wetlands proper, but probably occurred there.

throughout the summer, and reached a peak in early September (Figure 5). This late seasonal peak in adult activity coincides with the peak in the larval population (Figure 3). These mosquitoes began seeking blood meals when light intensity fell below 1 foot-candle, and were aggressive and difficult to see. Seven of the eight floodwater species taken biting, plus *Aedes triseriatus* which inhabits tree holes and artificial containers, were not found in the constructed wetland cells as larvae (see Table 1). Larvae of *Coquillettidia perturbans* require special collecting techniques[5] not used in this study, as they attach to underwater roots and stems of aquatic macrophytes. The species dominant in the larval stage, *Culex pipiens*, was rarely taken in the biting samples at either Benton or Pembroke; it is likely that birds are the favorite hosts in this area. Adult mosquitoes were rarely seen during daylight.

The dispersal capability of adult *Culex salinarius* is relatively high for the genus, as unobstructed flights of 12 km (8 mi) have been recorded, although lesser distances (up to 5 km) might be the norm.[6] Therefore, potential for annoyance at sites a few kilometers away from a large population is high, although adult density would be expected to decrease as distance from the emergence site increased. In October 1988, simultaneous dusk landing counts were conducted by three two-man crews at three locations in Benton: station 1, edge of constructed wetlands; station 2, swamp 1 km SW of the wetlands; and station 3, parking lot 1 km NW of the wetlands. The peak in biting activity occurred around 7:00 p.m. at each station, with the total number of female *C. salinarius* collected in 15 minutes as follows: station 1, 54; station 2, 62; and station 3, 39. A swamp at station 3 produced larval *C. salinarius*, and undoubtedly was a factor in the higher number of adult mosquitoes collected there. However, the data above indicate that females of this species disperse in numbers from the constructed wetlands into Benton at least within a radius of 1 km of the wetlands.

Mosquitoes and Potential Disease Transmission

The species of mosquitoes found breeding in large numbers in the constructed wetlands at Benton and Pembroke raise concern about increased potential for disease outbreak. Wherever such conditions occur, increased risk of encephalitis transmission to humans must be considered a major possibility because the vectors are being produced within flight range of human communities. A list of the different strains of encephalitis in the eastern U.S., and the mosquito species implicated in their transmission, is given in Table 2. Eight of the species collected during this study are known or suspected vectors of five strains of viral

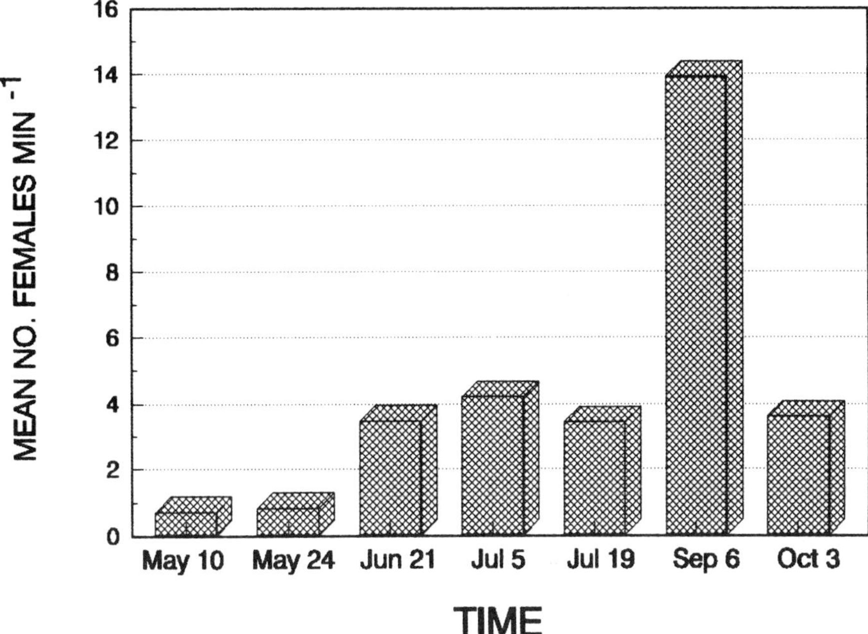

Figure 5. Seasonal change in host-seeking *Culex salinarius* aspirated from one person during dusk landing counts at Benton constructed wetlands.

encephalitis, of which the eastern equine and St. Louis strains are the most serious threats in western Kentucky. The primary hosts of many of these mosquitoes are birds, which act as reservoirs for the viruses. Human beings can become involved in the transmission cycle when infected mosquitoes seek a second blood meal from a secondary host.

Many biological and environmental factors affect encephalitis transmission rates, but a recognized major variable is vector abundance. Therefore, suppression of mosquito populations in wetlands is crucial as a disease prevention step. Several species breeding in the wetlands are also known vectors of dog heartworm disease, which is endemic in the southeastern U.S. with a high incidence of cases.

Mosquito Control in Constructed Wetlands

In most problem situations, the level of control that is considered satisfactory, or desirable, against mosquito larvae is arbitrary and difficult to establish. Decisions on the level of control needed should be based on the species of mosquitoes present, population sizes, dispersal patterns, resistance, etc. Control can be reported as a percentage of reduction or as a mean number of larvae per dip still present after treatment. For most compounds, a 95% reduction in larval numbers would be considered an effective treatment. However, 95% kill of a large initial population, where the sample mean was 30 larvae per dip, would still leave a mean of 1.5 larvae per dip. This would greatly exceed TVA's threshold of 0.25 larvae per dip (for anophelines) in large reservoirs. In the Benton wetlands, where the vast majority are *Culex* species and population sizes can be very large, a mean of 0.5 larvae (species combined) per dip after treatment was selected as a reasonable target level. Achieving this residual population density would require great efficacy of the active ingredient and optimal application techniques. Whether this magnitude of reduction would satisfactorily reduce annoyance and disease potential is unknown.

Insecticide Trials at Benton

Attempts to control mosquitoes with many kinds of insecticides in wastewater of high organic content have generally been ineffective.[13] Concentrations two or more times higher than normal usually are required to achieve adequate larval control because the toxicant molecules are adsorbed onto organic matter, and also are diluted and degraded rapidly. In spite of this knowledge, an organophosphate (Abate) and two bacterial agents

Table 2. Known or Suspected Vector Capabilities of Eight Mosquito Species Collected in the Benton and Pembroke Constructed Wetlands

Species	Virus	Other
Culex pipiens	SLE, EEE, WEE	Dog heartworm, fowl pox, bird malaria
Culex restuans	SLE, EEE	Dog heartworm
Culex salinarius	SLE, EEE	
Culex tarsalis	SLE, EEE, CE	
Culex territans	EEE	
Culiseta inornata	SLE, WEE, JBE	
Copquillettidia perturbans	EEE	
Aedes vexans	EEE, WEE, CE	Dog heartworm

Note: SLE, St. Louis encephalitis; EEE, Eastern equine encephalitis; WEE, Western equine encephalitis; CE, California group encephalitis; and JBE, Japanese B encephalitis.

From Reference 11.

(*Bacillus thuringiensis israelensis* H-14 (*Bti*) and *Bacillus sphaericus* (*Bs*)) were tested at Benton in 1989 and 1990 for comparative purposes.

Test 1 — Abate vs. *Bs,* 1989 — A slightly higher percentage of control was achieved with *Bs* (72%) than with Abate (67%) 24 h after application. The difference in efficacy of the two compounds increased through time as percentage of recovery in the Abate-treated cell rose to 80% 2 weeks after treatment; it was less than 40% in the *Bs*-treated cell. Neither compound reduced the mean number of larvae below 2.0 per dip at any time after treatment.

Test 2 — Bs vs. *Bti,* 1990 — The cell treated with *Bs* had a pretreatment population (September 18, 1990) nearly twice that in the cell treated with *Bti* (average 157 larvae per dip vs. 92 per dip). Mean number of larvae per dip 24 h after treatment dropped to 2 (97%) in the *Bti*-treated cell and 27 (85%) in the *Bs*-treated cell; however, 14 days later, the mean in the *Bti*-treated cell rose to 14, whereas it dropped and remained near 2 in the *Bs*-treated cell. The treatments were repeated after 14 days, further reducing the populations in the cells. Fourteen days after the second treatment (October 19), a mean of 1.8 larvae per dip was found in the *Bti*-treated cell vs. 0.2 larvae per dip in the *Bs*-treated cell. These results are consistent with other reports on extended control using *Bs*.[14-16]

Insecticide Treatments at Pembroke

Results of the slug injection of *Bti* into the Pembroke meadow in 1990 were extremely disappointing. Mean number of larvae per dip before treatment was 17.35. The mean, 24 h after treatment, increased to 24.8, and after 48 h remained high at 22.6. Contrary to results obtained at Benton, this slug treatment yielded absolutely no control, for reasons unknown. A possible explanation is that the flow was too rapid to permit adequate retention time of the *Bti* particles within the system for mosquito larvae to ingest a toxic dose.

Combining the data from the 8 consecutive weeks of granular *Bti* applications in 1991 revealed that the overall percentage of dips positive for larvae decreased 24 h after treatment, but then increased after 48 h (Figure 6). Mean larval density showed a similar trend (Figure 7). Although the total population was reduced by over 90%, a 24-h post-treatment average of 5 larvae per dip still remained. Mosquito population responses were remarkably similar in the marsh and meadow components (Figures 8 and 9). For the first 4 weeks of treatment, pretreatment means exceeded 25 larvae per dip, and were increasing through time despite the weekly treatments. Relatively low percentages of control were achieved during this time, and the 48-h counts showed several-fold increases. The great majority of the larvae in these latter counts was newly or recently hatched (first and second instars), evidence that gravid females continued to oviposit high numbers of egg rafts in both cells. After the 5th week, however, pretreatment larval densities decreased; 10 days after the last (8th) treatment, the counts had risen to a mean of 5 in the marsh and 43 in the meadow. These data indicate that the toxic effect of *Bti* is of short duration, and that once *Culex* mosquito populations reach a relatively high density, frequently repeated treatments with *Bti* will be necessary to reduce larval density. The population was composed of *Culex pipiens* and *C. salinarius* in nearly equal proportions.

Recommendations for a Mosquito Monitoring and Abatement Program

The best method of mosquito control is to eliminate larval habitat, either by preventing its formation or drying it up once formed (drain, fill, etc.). Although not practicable in all cases, this approach might be

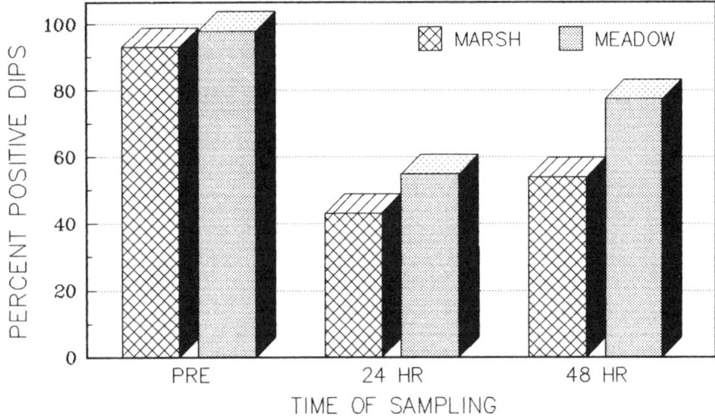

Figure 6. Overall effect of *Bti* on percentage of dips positive for larvae in Pembroke bulrush marsh; summer 1991.

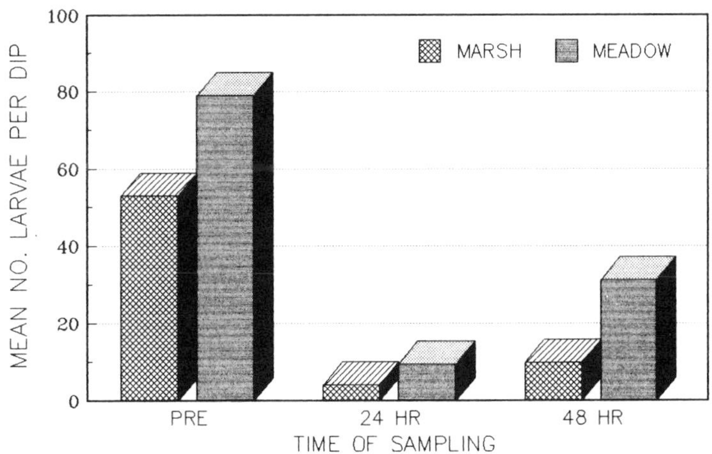

Figure 7. Overall effect of *Bti* on mean number of larvae per dip in Pembroke meadow; summer 1991.

applicable to constructed wetlands in several ways. In cells constructed with gravel substrates designed for subsurface flow, standing water would not become a problem, provided the system remained unclogged. Cells such as these could operate continuously, but should be kept under surveillance. In surface flow systems where two or more cells are constructed, alternating operation for 1-week periods could dry up the habitat before mosquito larvae and pupae completed development. Problems could occur if uneven construction of the cell bottom created depressions which would not dry up within a week. Also, this design might require twice the land area of one that operated continuously.

If surface water is present in constructed wetlands receiving domestic wastewater, a monitoring program should be set up for early detection of mosquito larvae. Weekly (or at least bimonthly) sampling trips during the mosquito season, in which dipper samples are taken ar various places, will provide information on population changes. An average count of just 0.5 *Culex* larvae per sample could be a sign of an upcoming population explosion. Training personnel on sampling techniques and recognizing mosquito larvae requires little time or funds. Actual sampling takes little time and could be performed during trips to the wetlands for other routine checks. A monitoring program might quickly pay for itself by providing data which will be used to make decisions concerning treatment.

Since most insecticides are ineffective in highly polluted waters and *Bacillus sphaericus* has not been approved by the EPA for marketing, I recommend the use of *Bti* for larval control. Based on larval monitoring, weekly treatments should begin before mean larval density reaches 0.5 *Culex* larvae per dip. In order to prevent

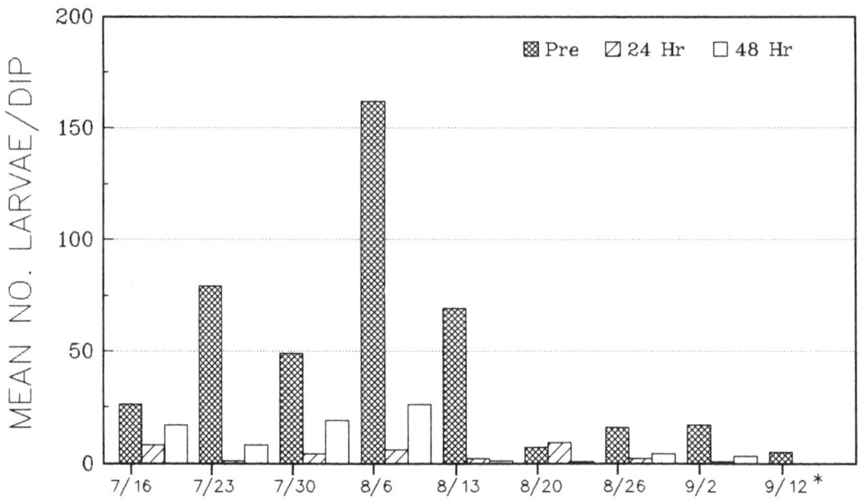

Figure 8. Effect of weekly *Bti* applications on mean number of larvae per dip in Pembroke bulrush marsh from July 16 to September 2, 1991. *, no treatment.

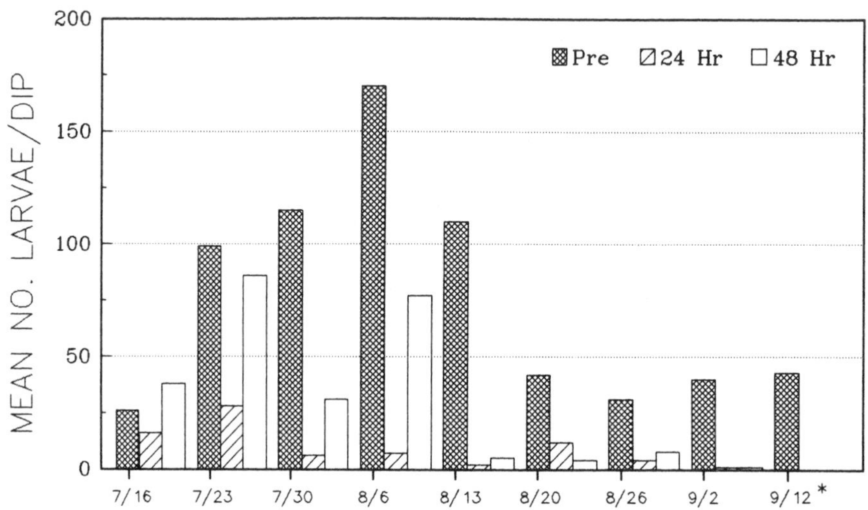

Figure 9. Effect of weekly *Bti* applications on mean number of larvae per dip in Pembroke meadow from July 16 to September 2, 1991. *, no treatment.

development to the adult stage, time between treatments should not exceed 7 days. Sampling for larvae should be continued, as treatment could be suspended if the mean drops below 0.5 per dip, saving material and money.

Several problems encountered in controlling mosquitoes in constructed wetlands can be circumvented. The dispersal limit of the backpack blower used in this study is 25 ft, and walking around the narrow dike and through the cells with tall vegetation is difficult and somewhat dangerous. Tall weeds deflected some of the granular material from intended target areas. The construction of wider dikes that could be routinely mowed would greatly enhance application. Large cells, such as at Benton, make insecticide application extremely difficult, requiring much more labor and increasing costs. At such sites, slug injection of the insecticide would save time and effort, provided there was minimal flow and adequate dispersion of the material to yield extensive, even control.

ACKNOWLEDGMENTS

I sincerely thank Martin K. Painter, who helped design and conduct the studies presented in this article. His diligence in sampling, applying treatments, recording data, and solving problems made completion of many tasks possible. I also thank Homer R. Starkey and Edward L. Snoddy for their work in the insecticide applications and associated larval sampling. Other TVA colleagues assisting in various parts of the studies were Joseph C. Cooney, Bobby R. McDuff, Diana A. Miles, Harry Porter, and Cynthia L. Russell, whose help is greatly appreciated.

REFERENCES

1. Laird, M. *The Natural History of Larval Mosquito Habitats*. Academic Press, London, 1988.
2. Gerhard, R. W. The influences of soil fermentation on oviposition site selection by mosquitoes, *Mosq. News*. 19:151, 1959.
3. Rockett, C. L. Bacteria as ovipositional attractants for *Culex pipiens* (Diptera: Culicidae), *Great Lakes Entomol*. 20:151, 1987.
4. O'Meara, G. F. and F. D. S. Evans. Seasonal patterns of abundance among three species of *Culex* mosquitoes in a south Florida wastewater lagoon, *Ann. Entomol. Soc. Am.* 76:130, 1983.
5. Morris, C. D., J. L. Callahan, and R. H. Lewis. Devices for sampling and sorting immature *Coquillettidia perturbans, J. Am. Mosq. Contr. Assoc.* 1:247, 1985.
6. Horsfall, W. R. *Mosquitoes. Their Bionomics and Relation to Disease*. Ronald Press, New York, 1955, 590.
7. Mulla, M. S. and H. A. Darwazeh. Efficacy of new insect growth regulators against mosquito larvae in dairy wastewater lagoons, *J. Am. Mosq. Cont. Assoc.* 4:322, 1988.
8. Mulla, M. S., H. A. Darwazeh, E. W. Davidson, and H. T. Dulmage. Efficacy and persistence of the microbial agent *Bacillus sphaericus* against mosquito larvae in organically enriched habitats, *Mosq. News*. 44:336, 1984.
9. Lace, L. A., J. Day, and C. M. Heitzman. Long-term effects of *Bacillus spaericus* on *Culex quinquefasciatus, J. Invertebr. Pathol.* 49:116, 1987.
10. Matanmi, B. A., B. A. Federici, and M. S. Mulla. Fate and persistence of *Bacillus spaericus* used as a mosquito larvicide in dairy wastewater lagoons, *J. Am. Mosq. Contr. Assoc.* 6:384, 1990.
11. Monath, T. P. and W. C. Reeves. *St. Louis Encephalitis*. Am. Public Health Assoc., Washington, D.C., 1980.

CHAPTER 67

The Integration of Seaweed Biofilters with Intensive Fish Ponds to Improve Water Quality and Recapture Nutrients

A. Neori, S. P. Ellner, C. E. Boyd, and M. D. Krom

INTRODUCTION

The high densities of animals in aquaculture ponds cause problems of water quality, which in turn limit the intensification.[1] Ammonia, a compound toxic to aquatic organisms, is a major metabolite produced by fish and is also a major source of N, contributing to the eutrophication of aquatic environments. Removal of dissolved N compounds, in particular ammonia but also nitrite, nitrate, and dissolved organic nitrogen (DON), improves water quality in the ponds. The removal from the mariculture effluents of dissolved N and dissolved phosphorus also reduces downstream eutrophication, which is particularly dangerous to the lush coral reefs in the highly oligotrophic Gulf of Eilat (Aqaba), Israel.

With respect to the environmental consequences of aquaculture, the following three issues are of particular importance: (1) the fraction of the nutrients in feed which is exported via the pond outflow; (2) the physical and chemical nature of these exported nutrients; and (3) the methods to minimize this export. Several studies have been made on the transformations of N and P in intensive culture of marine fish under the desert conditions of southern Israel, on the Gulf of Eilat.[1-9] These publications described various culture systems, such as high water flow tanks and ponds, and phytoplankton-rich soft/hard-bottom ponds. Neori and Krom[10] have summarized some of the data in the form of nutrient budgets of the outflow from the different culture systems. In more recent studies we have investigated seaweed biofilters as a means to recapture outflowing nutrients before their release to the environment.[11-13] In this article, we compare the exported fraction of N budgets in the different pond systems and in the newly developed recirculating fish-seaweed biofilter system.

MATERIAL AND METHODS

Details of the construction of nutrient budgets in fish tanks, an unaerated soft-bottom fishpond and an aerated hard-bottom fishpond have appeared elsewhere in the literature (see References 3, 4, and 7). Briefly, in each system, all the inputs and outputs of the discussed nutrient were sampled and quantified.

This study, with recirculating seaweed biofilters, was conducted in four 3.3-m^2, 2500-l tanks (Figure 1). The seaweed *Ulva Lactuca* (L.) was grown unattached in three of the tanks, where they were kept suspended in the water column by air diffusers. Each seaweed tank was stocked with 3kg of the thalli. Once a week, the seaweeds were harvested and the weekly growth removed. The fish, 15 kg of 15-g gilthead seabream (*Sparus aurata*) were grown in the fourth tank, which was shaded to prevent phytoplankton growth, and were fed 600 g of pellets a day. The fish tank effluents were pumped into a small sedimentation tank (400 l) and then flowed by gravity to the seaweed tanks and back into the fish tank at the rate of approximately 25,000 L/day. Clean seawater entered the system via the fish tank at a rate of 2000 L/day and the same volume (less losses) flowed out from the seaweed tanks into a 600-l, 1-m^2 polishing tank, stocked with 1 kg of the seaweeds. Water residence time in the whole system was 5 days, compared with 2.5 days in the earthen pond and 2 days in the hard-bottom pond. However, due to the water recirculation pattern, the fish tank water was exchanged nearly 11 times a day, 7% of it clean water and the rest recirculated water from the seaweed tanks.

0-87371-550-0/93/$0.00+$.50

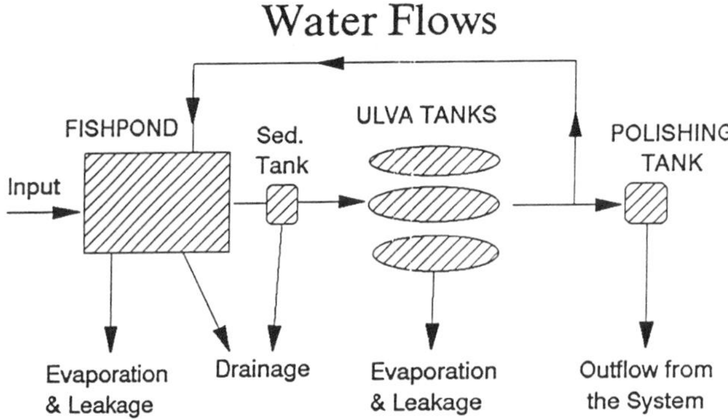

Figure 1. A schematic description of the fish-seaweed experimental system (not drawn to size).

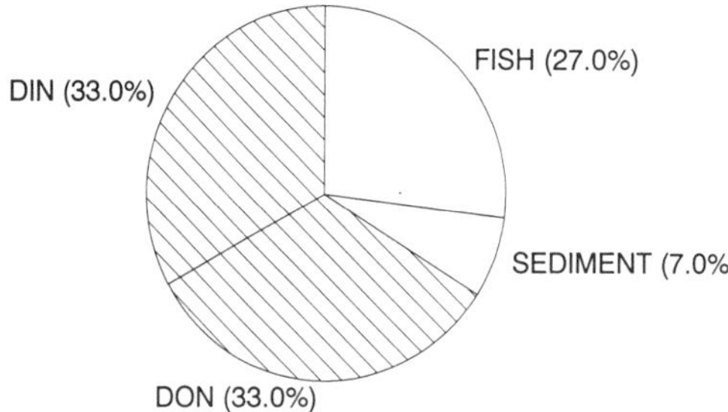

Figure 2. The average nitrogen budget determined from data collected for three sizes of gilthead seabream (*Sparus aurata*) grown in 190-l tanks under controlled conditions without any plankton. Percent values were calculated on the basis of nitrogen content in the consumed food. The striped sector is the fraction of exported nitrogen. DIN, dissolved inorganic nitrogen; DON, dissolved organic nitrogen. (Recalculated from Reference 4.)

After a stabilization period of several weeks, intensive sampling took place for 5 days. Water samples were taken every morning and afternoon from the fish tank and the seaweed tanks. The N was quantified in the water inflow to the system and in the sediment drained from the sedimentation tank. Total ammonia was measured by an electrode (Ingold). Aliquots of the samples were filtered via acid-washed GF/F (Whatmann) filters for determination of dissolved inorganic nitrogen (DIN) and dissolved organic nitrogen (DON) on a Technicon AA-2 autoanalyzer following standard procedures. The DON was measured as nitrate after an alkaline persulphate oxidation.[14] Samples for particulate N were filtered on pre-ashed GF-F filters, and other aliquots were filtered on GF/F filters for chlorophyll determination. The coefficient of variation from replicate samples was usually less than 5%.

RESULTS

Tank Experiments

Of the total N consumed by the fish in the laboratory tank (Figure 2), only 27% was assimilated by the fish. Most of the N (66%) was found dissolved in the water, half as ammonia and half as DON; 7% settled as feces. These distribution values fall within the range reported for N in other carnivorous fish.

Pond Experiments

In the earthen semi-intensive ponds, dense phytoplankton and microbial populations transformed approximately half of the dissolved nutrients to the particulate fraction (Figure 3). Only 14% of the N was assimilated

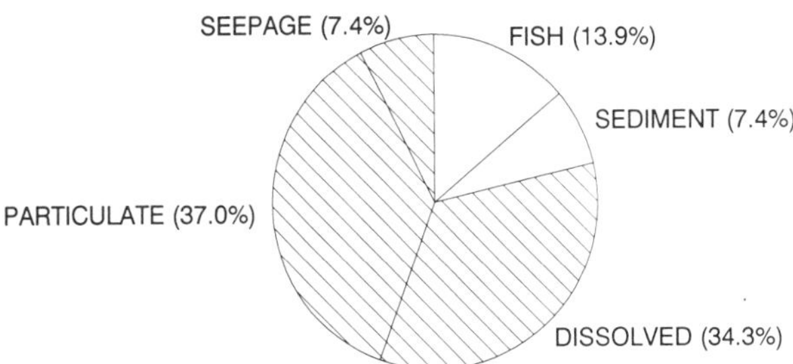

Figure 3. The average nitrogen budget determined from data collected monthly over a year for a nonaerated, semi-intensive earthen fishpond (325 m^2). Percent values were calculated on the basis of nitrogen introduced in the food and in the inflowing water. The striped sector is the fraction of exported nitrogen. (Recalculated from Reference 3.)

by the fish. This number is misleading, as in this pond much of the N input was dissolved in the inflowing water. Thus, of the food N input, the fish have assimilated nearly the same fraction as in the tank experiments in Figure 2. Over 75% of the N input to this pond flowed out with the water and approximately 7% settled on the bottom.

In the hard-bottom aerated intensive fishponds, the dense phytoplankton and microbial populations transformed most of the fish-excreted dissolved N into the particulate fraction (Figure 4). This improved water quality for the fish, yet approximately 60% of the N was still exported in the outflow. The fraction of N input assimilated by the fish (27%) equaled the value assimilated in the tank experiment. The sedimented fraction was also quite similar.

Recirculating Fish-Seaweed Biofilter System

Although water residence time was 5 days, phytoplankton populations did not develop due to the shading of the fishtank and competition by the seaweeds. Instead, the seaweeds captured and removed from the water nearly 50% of the N input (Figure 5). This value equals the value found in the whole particulate fraction in the hard-bottom pond. Only 3% of the N was found in the sediment, and less than 20% was exported in dissolved and particulate forms. The latter fraction was mostly made out of bacteria and protozoa, at approximately 4% each.

DISCUSSION

In pond systems with high water exchange rates, and in fish cages, the dissolved and sedimented fractions of the fish-excreted N are exported to the environment.[5-8] Land-based culture allows the recapture of part of this excess N.

In traditional aquaculture, it is common to rely on dense phytoplankton and bacterial populations, which develop naturally in the ponds, to take up the ammonia. This practice allows significant savings in the use of clean water per kilogram of produced fish.[5,8] The phytoplankton populations in our ponds indeed transformed much of the dissolved N to the particulate phase. While improving the water quality for the fish, this process did little to reduce the total fraction of N exported in the water outflow relative to the situation in the fish tank.

In previous studies, extending from research such as that by Ryther[15,16] and Langton,[17] it has been shown that the seaweeds *Ulva lactuca*[11,12] and *Gracilaria conferta* (unpublished results) can efficiently remove fish metabolites, in particular ammonia, from marine fishpond water. These metabolites served as the sole nutrients for the algae which grew at the year-round averaged phenomenal yield of 30 g dw/m^2/day, removing in the process 1.5 g N/m^2/day.[13] In further studies, it has been shown that the fish grew well in the treated water,[18] that is, treated water could be reused.

In our present research, we have quantified the parameters involved with the integration of *Ulva lactuca* into the intensive fishponds. The low fraction of N exported (less than 20%), is a marked improvement over

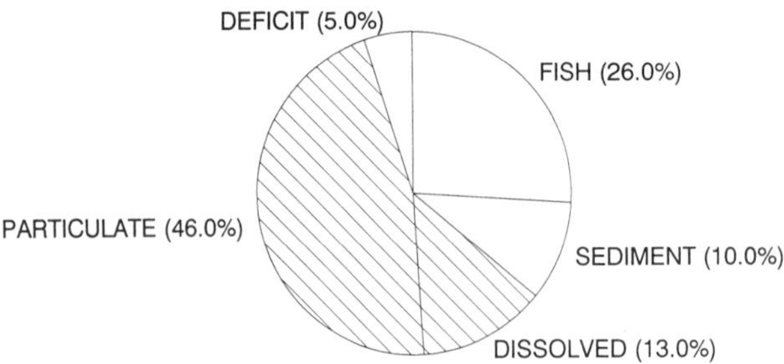

Figure 4. The average nitrogen budget determined from data collected daily over a month for an aerated, intensive hard-bottom fishpond. Percent values were calculated on the basis of nitrogen introduced in the food (nitrogen in the inflowing water was negligible). The striped sector is the fraction of exported nitrogen. (Recalculated from Reference 7.)

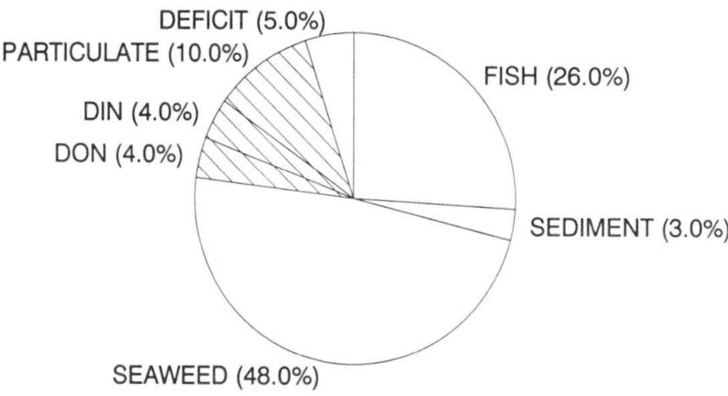

Figure 5. The average nitrogen budget determined from data collected daily over a 5-day period for an integrated fishtank-seaweed recirculating biofilter system. Percent values were calculated on the basis of nitrogen introduced in the food (nitrogen in the inflowing water was negligible). The striped sector is the fraction of exported nitrogen.

the 60 to 75% N exported from the fish tank and the phytoplankton-based fish ponds. The reduced export of N was achieved without a reduction in the area and volume intensification criteria, and with a three-fold reduction in the use of clean seawater (Table 1).

The increased residence time of the water in the system, and the large surface area provided by the seaweed fronds, enhances the activity of different bacterial processes, such as ammonification of DON and nitrification of ammonia. Compared to 33% in the form of DON excreted by the fish (Figure 2), only 4% was exported from the seaweed-fish system. The rest was converted, mostly to seaweed biomass (Figure 5).

In conclusion, the newly developed fish culture system, which incorporates recirculating seaweed biofilters, can reduce by more than two thirds the fraction of N exported from the fishpond. At the same time, it can also bring a saving of two thirds in the usage of clean seawater and produce large yields of clean and potentially valuable seaweed biomass. With such benefits, the incorporation of seaweed biofilters into mariculture systems has a promising future.

ACKNOWLEDGMENTS

We thank D. Zuber and other members of our Institute for technical help; N. Kress, L. Izraelov, M. Uko, and P. Krost for help with chemical analyses; and D. Angel for his continuous advice and help. A. Colorni critically reviewed an earlier version of the manuscript. This research was funded by BARD, Project No. I-1634-89, by the "Negev-Arava" Fund for Research and Development, and by the Israeli Ministries for Energy and Infrastructure and for Science and Technology.

Table 1. Intensification Criteria for Three Mariculture Systems on the Basis of Feeding Rate

Pond type	Feed (g/m²/day)	Feed (g/m²/day)	g feed/m³ of water exchange
Soft-bottom pond	4	7.4	18
Hard-bottom pond	45	45	94
Fish-seaweed pond[a]	42	57	300

[a] The calculations for this system include area and volume of the biofilter.

REFERENCES

1. Krom, M. D., C. B. Porter, and H. Gordin. Causes of fish mortalities in the semi-intensively operated seawater ponds in Eilat, *Isr. Aquaculture*. 49:159, 1985.
2. Motzkin, F., Y. Cohen, H. Gordin, and E. Padan. Productivity relations in seawater fish ponds: a comparison of stocked and unstocked ponds, *Mar. Ecol. Prog. Ser.* 8:203, 1982.
3. Krom, M. D., C. B. Porter, and H. Gordin. Nutrient budget of a marine fishpond in Eilat, *Isr. Aquaculture*. 51:65, 1985.
4. Porter, C. P., M. D. Krom, M. G. Robbins, L. Brickell, and A. Davidson. Ammonia excretion and total N budget for gilthead seabream (*Sparus aurata*) and its effect on water quality conditions, *Aquaculture*. 66:287, 1987.
5. Krom, M. D., J. Erez, C. B. Porter, and S. Ellner. Phytoplankton-nutrient uptake dynamics in marine fishponds under winter and summer conditions, *Aquaculture*. 76:237, 1989.
6. Krom, M. D., A. Neori, and J. van Rijn. Importance of water flow rate in controlling water quality processes in marine and freshwater fish ponds, *Isr. J. Aquacult. Bamidgeh*. 41:23, 1989.
7. Krom, M. D. and A. Neori. A total nutrient budget for an experimental intensive fishpond with circularly moving seawater, *Aquaculture*. 83:345, 1989.
8. Krom, M. D. and J. van Rijn. Water quality processes in fish culture systems: problems and possible solutions, in European Aquaculture Soc., N. DePauw, E. Jaspers, H. Ackefors, and N. Wilkins, Eds. Bredene, Belgium, 1989, 1–21.
9. Neori, A., M. D. Krom, I. Cohen, and H. Gordin. Water quality conditions and particulate chlorophyll *a* of new intensive seawater fishponds in Eilat, Israel: daily and diel variations, *Aquaculture*. 80:63, 1989.
10. Neori, A. and M. D. Krom. Nitrogen and phosphorus budgets in an intensive marine fishpond: the importance of microplankton, *Proc. Int. Symp. Feeding Fish in our Water*. Guelph, Canada, May 1990. C. B. Cowey and Y. C. Cho, Eds. 1991.
11. Vandermeulen, H. and H. Gordin. Ammonium uptake using *Ulva* (Chlorophyta) in intensive fishpond systems: mass culture and treatment of effluent, *J. Appl. Phycol.* 2:363, 1990.
12. Cohen, I. and A. Neori. *Ulva lactuca* biofilters for marine fishpond effluents. I. Ammonia uptake kinetics and nitrogen content, *Bot. Mar.* (in press).
13. Neori, A., I. Cohen, and H. Gordin. *Ulva lactuca* biofilters for marine fishpond effluents. II. Growth rate, yield and C:N ratio, *Bot. Mar.* (in press).
14. Solorzano, L. and J. Sharp. Determination of total dissolved nitrogen in natural waters, *Limnol. Oceanogr*. 24:751, 1980.
15. Ryther, J. H., J. C. Goldman, C. E. Gifford, J. E. Hugenin, A. S. Wing, J. P. Clarner, L. D. Williams, and B. E. Lapointe. Physical models of integrated waste recycling marine polyculture systems, *Aquaculture*. 5:163, 1975.
16. Ryther, J. H., T. A. DeBusk, and M. Blakeslee. Cultivation and Conversion of Marine Macroalgae. Final Subcontract Report to SERI, the U.S. Dept. of Energy, Publication SERI/STR-231-2360 DE 84004522. National Technical Information Service, U. S. Dept. of Commerce, Springfield, VA.
17. Langton, R. W., K. C. Haines, and R. E. Lyon. Ammonia-nitrogen produced by the bivalve mollusc *Tapes japonica* and its recovery by the red seaweed *Hypnea musciformis* in a tropical mariculture system, *Helgolander Wiss. Meresunters*. 30:217, 1977.
18. Porter, C. B., D. Popper, and A. Neori. Unpublished results.

CHAPTER 68

Treating Intensive Aquaculture Recycled Water with a Constructed Wetlands Filter System

W. H. Zachritz, II and R. B. Jacquez

INTRODUCTION

The U.S. aquaculture industry has expanded at an annual rate of almost 20% since 1980, with the 1990 U.S. harvest valued at nearly $1 billion.[1,2] Availability of suitable sites, treatment costs for waste discharges, and lack of high-quality water are likely to be the greatest constraints for continued aquaculture development in North America.[3] Future aquaculture systems will have to incorporate cost-effective designs for minimizing wastewater generation and maximizing water use rates while still being competitive with existing aquaculture production facilities.

State-of-the-art intensive culture systems are flow-through raceway designs that rely on high turnover rates and supplemental aeration to maintain water quality. Closed, recirculating systems for aquaculture production incorporating treatment processes for nitrogen, phosphorous, and BOD (biochemical oxygen demand) removal have not been cost effective for large-volume culture systems.[3] Development of an effective, low-cost treatment is imperative if aquaculture is to continue at the present rate of growth. Constructed wetlands filters (CWF) offer a unique alternative to meet these challenges.

Several investigators have demonstrated the ability of constructed wetlands to remove nitrogen compounds from domestic wastewater.[4-6] The mechanism for removal is generally considered to be coupled nitrification/denitrification, plant uptake and bacterial biomass incorporation, and detritus formation. In most cases, very little recirculation was used and retention times greater than 4 days were required. For treating recycled aquaculture wastewater, these retention times would result in excessively large unit process volumes with potentially high water loss rates. This is an unacceptable condition, particularly in the western U.S. where water laws restrict consumptive use and evaporation rates commonly exceed 120 cm/year.

Nitrification reactors based on fixed-film kinetics have been used for water treatment in closed aquaculture systems for finfish production.[7] In general, these systems are operated at high hydraulic loadings with continuous recycle through the filter media, and long cell residence time achieved by the attached microbial populations. There are five main factors that affect the rate of nitrification in recirculating culture systems: ammonia and nitrite concentrations, pH, temperature, alkalinity, dissolved oxygen, and salinity. But in most cases, oxygen is considered the limiting factor for effective nitrification. The CWF unit described in this article operates under conditions similar to the fixed-film reactors detailed in these studies. This article examines the development design and operational elements of a high-rate CWF system treating culture water recycle and presents recommendations for future system designs.

SYSTEM DESCRIPTION

The pilot-scale aquaculture system is located at the Southwest Technology Development Institute's geothermal greenhouse (1110 m^2) complex at New Mexico State University. The geothermal system is supplied by a 308-m deep well producing 64°C water at 262 m^3/day which is cascaded through two on-line processes (hydronic and benchtop heating systems) before being available for use at the aquaculture facility.

0-87371-550-0/93/$0.00+$.50

Typical water inlet temperatures to the aquaculture system vary from 32° to 57°C. The temperature control system has both heating and cooling components to maintain an even 12-month culture temperature of 26°C. Heating is provided by a direct supply of geothermal water applied by spray to the blending tank and contributes both culture water and heat input. Water flow and aquaculture system temperatures are monitored with an IBM XT computer and Keithley data acquisition unit. System cooling and the bulk of the culture water are supplied via the reservoir supply located outside the greenhouse.

The aquaculture system is a high-density finfish culture system using geothermal water for both heating and as a culture medium. The system is enclosed in a quonset style greenhouse ($75m^2$) with an east-west orientation to maximize winter solar gain. The culture system (Figure 1) consists of two 833-l raceways operated in an 80 to 95% variable recycle mode. The flow-through each raceway is 12.3 L/m, resulting in about 20 complete water changes per raceway per day. The present configuration for treating recycled culture water consists of sequential flow-through a down-draft clarifier to remove solids and a CWF unit to control the remaining solids, chemical oxygen demand (COD), and excess total nitrogen. A 3330-l blending tank or sump is used to equalize flows from the recycled culture water and the degassed and cooled geothermal makeup water. This blended water provides the source for the culture tank feed water supply and is delivered to the culture tanks under pressure via spray heads to provide aeration and mixing. Additional mixing is provided by a spiral paddlewheel aerator mounted transversely in the culture tanks. The combined spray and paddlewheel aerators are designed to maintain a minimum dissolved oxygen concentration in the culture tanks of 5.0 mg/L. Potential finfish production for this small system is over 227 kg/year.

At the present operating flowrate of 24.6 l.m, the 378-l clarifier has a hydraulic retention time of 0.25/h and results in total suspended solids (TSS) removals of over 95%. Settled sludge is wasted daily and allowed to surface-dry on a slow sand filter bed. Adjustable overflow weirs located in the clarifier provide controlled flows for both wasting and culture recycle water. Waste flow rates are maintained between 1.5 and 3.8 L/m, resulting in a mean flow rate to the CWF unit of 21.9 L/m. The 3.3-m long CWF units is 0.91 m deep with a surface area of 4.0 m^2 and a working volume of 2051 L. Graded 5 to 8-cm rock serves as the medium with a void volume of 39.3% and estimated surface area of 62.3 m^2/m^3. This translates into a bulk hydraulic loading rate of 7.9 m^3/m^2/day, specific hydraulic loading rate of 0.25 m^3/m^2/day, and a hydraulic retention time of 0.64/h. The plant used in this system is locally available bullrush (*Scripus californicus*), planted on about 0.5-m centers. This plant was selected because of availability, deep root penetration, and nitrogen removal effectiveness.[4] Water levels in the CWF unit are controlled by an adjustable, effluent standpipe weir.

SYSTEM OPERATION

The culture and support treatment systems have been continuously operated for the past 16 months. The systems have evolved from a strict flow-through system with no recirculation, to a 85 to 95% recycle system incorporating the present treatment train. We have successfully grown fish to the 0.5-kg size and experienced only two major system failures that were not related to the treatment train. Those failures arose from faulty temperature sensors and lack of redundant backups for the computer control.

The CWF unit has been operated under several hydraulic loading regimes and with two filter media types. Initially, a graded 2.5-cm lava rock with a 48.3% void volume was used because of its potential low-cost availability in the region, and very high surface area-to-weight ratio. Within a few weeks of operation, this medium began to clog and short-circuit the CWF unit, resulting in impaired treatment and hydraulic overloading. Examination of the medium indicated that in the aerobic zones around the influent structure, populations of attached bryozoans were clogging interstitial zones and preventing deep bed penetration. In addition, rapid plant root growth resulted in areas of the bed that were effectively clogged. The media were replaced with graded 5 to 8-cm rock and no clogging has since been observed. Plant growth on the geothermal water (total dissolved solids = 1800 mg/L) has been rapid in all cases and no indications of plant stress have been observed.

FUTURE DESIGN CONSIDERATIONS

Future designs for all aquaculture systems must incorporate several important features to ensure continued success. First and foremost, they must address controlled discharge of pollutants, particularly to protect groundwater resources, and meet permitted discharge standards. This will become a "cost of doing business" in aquaculture. Controlled discharge requirements will severely limit continued development of high volume flow-through systems and shift emphasis to closed, recirculating systems. From our experience, completely closed systems are difficult to operate, expensive to maintain, and have inherent instability associated with collective ion accumulations and loss of buffering capacity. Controlled discharge systems with water recycle rates approaching 85 to 95% appear to offer substantial advantages in system stability, water conservation and overall cost savings. Integral to this concept is the expanded use of both high- and low-rate CWF units for treatment of recycled culture water as well as discharged wastewater. The conceptual process flow, shown in Figure 2, incorporates these design considerations and presents estimates of general water quality parameters.

The high-rate nitrification CWF unit would operate in a mode similar to our existing filter; that is, remove soluble COD and convert ammonia-nitrogen to nitrate-nitrogen. However, we suggest using media with higher hydraulic conductivities. Such an arrangement could include larger rock (>8 cm) or plastic synthetic media with high specific surface areas and high hydraulic conductivities. An effective clarifier must be included to reduce the organic loading on the CWF unit and the oxygen demand for the overall culture system. The down draft design used in our system was effective, but the performance of a scaled-up system has not been tested.

The low-rate CWF unit for denitrification would remove nitrate-nitrogen and excess solids to acceptable concentrations for discharge. This unit would mix a split wastestream from the high-rate CWF unit and direct effluent from the culture units to provide conditions for denitrification. Soluble carbon coupled to low hydraulic loadings would affect removal of residual nitrate and COD in the effluent stream. Specific requirements for this system need further investigation to develop critical design refinements.

Even in arid regions, the use of CWF units for the treatment of various wastewaters has great potential as a low-cost solution for protecting scarce ground and surface water supplies while supporting expanding economic development. These systems can be designed at almost any scale of operation from home units to large, centralized treatment systems with the same degree of dependable performance. In aquaculture, these CWF units may be the viable treatment alternative for recycle culture water and effluent wastewater that continues rapid expansion of worldwide finfish production systems.

ACKNOWLEDGMENTS

A portion of this work was supported by the Southwest Center for Environmental Research and Policy. The authors wish to thank Carol Fischer and Rudi Schoenmackers of the Southwest Technology Development Institute for their assistance in reviewing this manuscript.

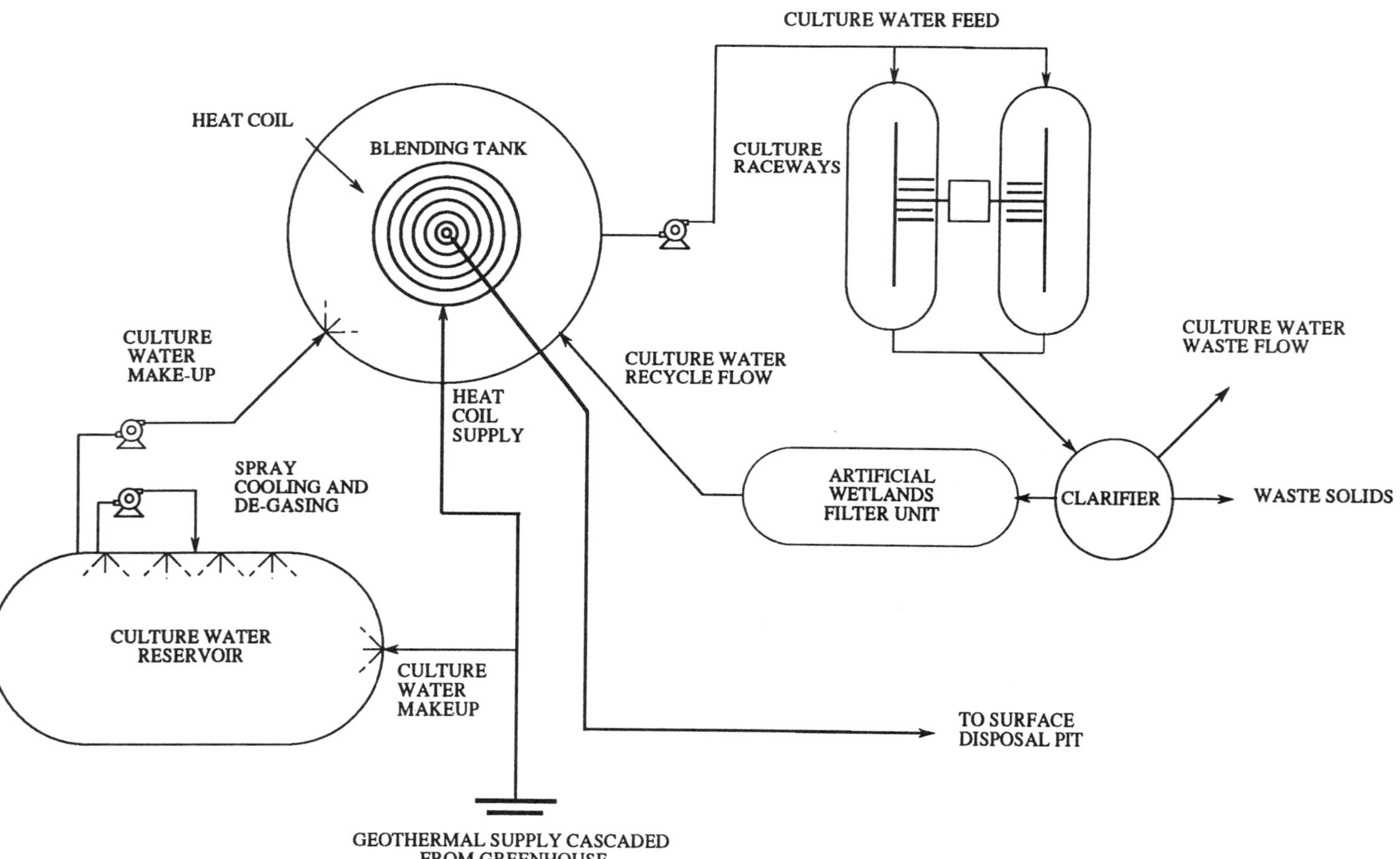

Figure 1. Geothermal-based aquaculture system process flow diagram.

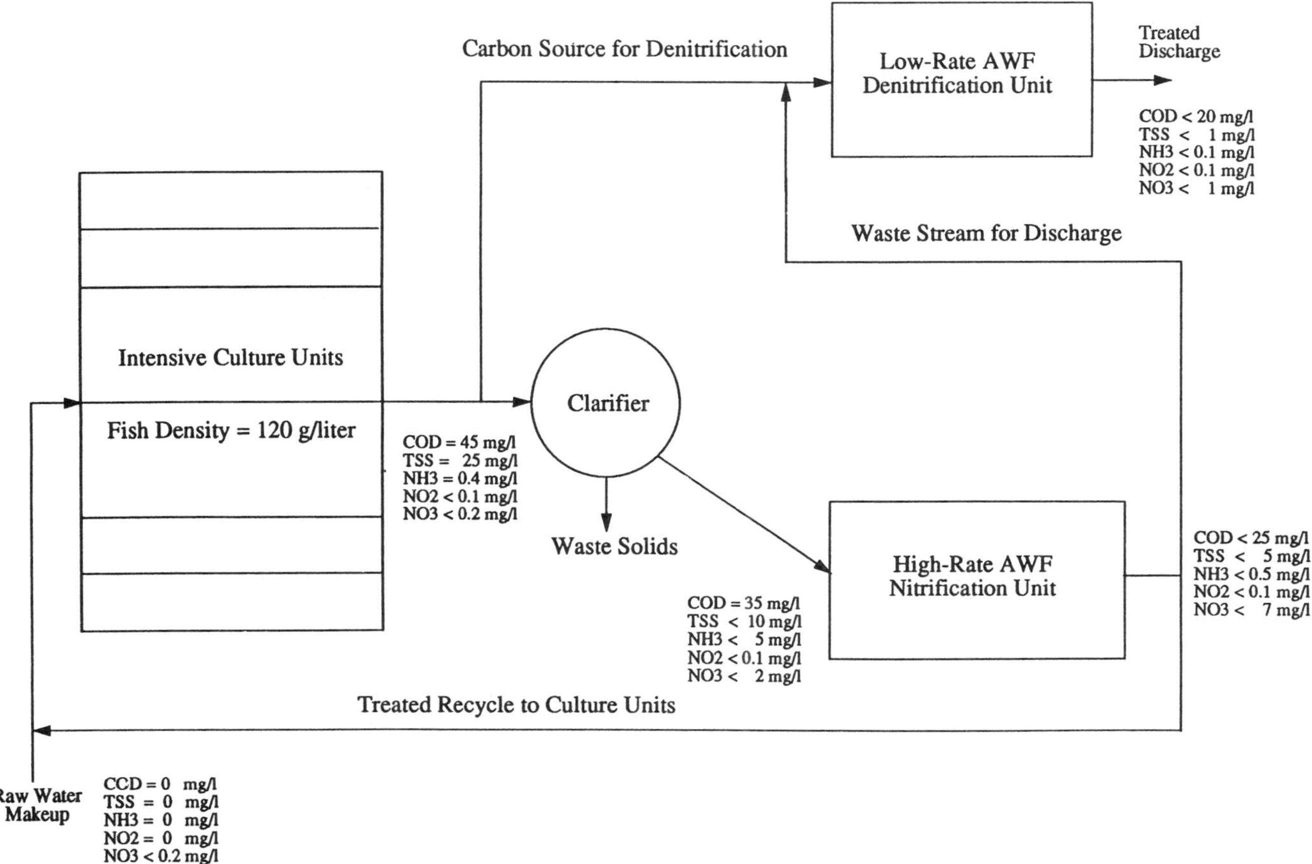

Figure 2. Process flow diagram for a constructed wetlands filter system treating intensive fish culture wastewater.

REFERENCES

1. Manning, E. and B. Kidwell. Aquaculture comes of age, *Progressive Farmer*. 26, April 1991.
2. FAO 1990 aquaculture production (1985–1988). Fisheries Cir. No. 815, Revision 2, 1990.
3. Randall, H. et al. Commercial aquaculture-AFS position statement, *Fisheries*. 16(1), 1991.
4. Gersberg, R. J., B. V. Elkins, S. R. Lyon, and C. R. Goldman. Role of aquatic plants in wastewater treatment by artificial wetlands. *Wat. Res.* 10:363, 1986.
5. Reed, S. C., E. J. Middlebrooks, and R. W. Crites. *Natural Systems for Waste Management and Treatment*. McGraw-Hill, New York, 1988.
6. Design Manual, Constructed Wetlands and Aquatic Plant Systems for Municipal Wastewater Treatment. U.S. Environmental Protection Agency. EPA-625/1-88-022 1988.
7. Bovedeur, J., A. B. Zwaga, B. G. J. Lobee, and J. H. Blom. Fixed film reactors in aquacultural water recycle systems: effect of organic matter elimination on nitrification kinetics, *Wat. Res.* 24(2):207, 1990.

Index

A

B

C

D

E

F

G

H

I

M

N

O

P

Q

R

T

U

V

W

X

Z